Norbert Sewald and Hans-Dieter Jakubke

Peptides: Chemistry and Biology

Norbert Sewald and Hans-Dieter Jakubke

Peptides: Chemistry and Biology

WILEY-VCH

Prof. Dr. Norbert Sewald
Universität Bielefeld
Fakultät für Chemie
Organ. und Bioorgan. Chemie
PO Box 100131
33501 Bielefeld, Germany

Prof. em. Dr. Hans-Dieter Jakubke
Universität Leipzig
Institut für Biochemie
Private address:
Albert-Richter-Straße 12
01465 Dresden-Langebrück, Germany

Cover
The cover picture shows the TPR1 domain of Hop in complex with -Gly-Pro-Thr-Ile-Glu-Glu-Val-Asp-OH (GPTIEEVD). TPR domains participate in the ordered assembly of Hsp70-Hsp90 multichaperone complexes.
The TPR1 domain of the adaptor protein Hop specifically recognizes the C-terminal heptapeptide -Pro-Thr-Ile-Glu-Glu-Val-Asp-OH (PTIEEVD) of the chaperone Hsp70 while the TPR2A domain of Hop binds the C-terminal pentapeptide -Met-Glu-Glu-Val-Asp-OH (MEEVD) of the chaperone Hsp90. The EEVD motif is conserved in all soluble forms of eukaryotic Hsp70 and Hsp90 proteins.
Peptide binding is mediated with the EEVD motif. Both carboxy groups of the C-terminal aspartate anchor the peptide by electrostatic interactions. The hydrophobic residues located N-terminally within the peptide are critical for specificity.
[C. Scheufler, A. Brinker, G. Bourenkov, S. Pegoraro, L. Moroder, H. Bartunik, F. U. Hartl, I. Moarefi, Structure of TPR domain-peptide complexes: critical elements in the assembly of the Hsp70-Hsp90 multichaperone machine, Cell 2000, 101, 199; PDB entry 1ELW (http://www.rcsb.org/pdb/)]

Library of Congress Card No.: applied for

British Library Cataloguing-in-Publication Data
A catalogue record for this book is available from the British Library.

Die Deutsche Bibliothek – CIP-Cataloguing-in-Publication Data
A catalogue record for this publication is available from Die Deutsche Bibliothek.

Printed in the Federal Republic of Germany
Printed on acid-free paper

Typesetting K+V Fotosatz GmbH, Beerfelden
Printing betz-druck gmbH, Darmstadt
Bookbinding J. Schäffer GmbH & Co.KG, Grünstadt

ISBN 3-527-30405-3

Contents

Preface

The past decades have witnessed an enormous development in peptide chemistry with regard not only to the isolation, synthesis, structure identification, and elucidation of the mode of action of peptides, but also to their application as tools within the life sciences. Peptides have proved to be of interest not only in biochemistry, but also in chemistry, biology, pharmacology, medicinal chemistry, biotechnology, and gene technology.

These important natural products span a broad range with respect to their complexity. As the different amino acids are connected via peptide bonds to produce a peptide or a protein, then many different sequences are possible – depending on the number of different building blocks and on the length of the peptide. As all peptides display a high degree of conformational diversity, it follows that many diverse and highly specific structures can be observed.

Whilst many previously published monographs have dealt exclusively with the synthetic aspects of peptide chemistry, this new book also covers its biological aspects, as well as related areas of peptidomimetics and combinatorial chemistry. The book is based on a monograph which was produced in the German language by Hans-Dieter Jakubke: *Peptide, Chemie und Biologie* (Spektrum Akademischer Verlag, Heidelberg, Berlin, Oxford), and first published in 1996. In this new publication, much of the material has been completely reorganized and many very recently investigated aspects and topics have been added. We have made every effort to produce a practically new book, in a modern format, in order to provide the reader with profound and detailed knowledge of this field of research. The glossary, which takes the form of a concise encyclopedia, contains data on more than 500 physiologically active peptides and proteins, and comprises about 20% of the book's content.

Our book covers many different issues of peptide chemistry and biology, and is devoted to those students and scientists from many different disciplines who might seek quick reference to an essential point. In this way it provides the reader with concise, up-to-date information, as well as including many new references for those who wish to obtain a deeper insight into any particular issue. In this book, the "virtual barrier" between peptides and proteins has been eliminated because, from the viewpoint of the synthesis or biological function of these compounds, such a barrier does not exist.

This monograph represents a personal view of the authors on peptide chemistry and biology. We are aware however that, despite all our efforts, it is impossible to include all aspects of peptide research in one volume. We are not under the illusion that the text, although carefully prepared, is completely free of errors. Indeed, some colleagues and readers might feel that the choice of priorities, the treatment of different aspects of peptide research, or the depth of presentation may not always be as expected. In any case, comments, criticisms and suggestions are appreciated and highly welcome for further editions.

Several people have contributed considerably to the manuscript. All the graphical material was prepared by Dr. Katherina Stembera, who also typed large sections of the manuscript, provided valuable comments, and carried out all the formatting. We appreciate the kindness of Professor Robert Bruce Merrifield, Dr. Bernhard Streb and Dr. Rainer Obermeier for providing photographic material for our book. Margot Müller and Helga Niermann typed parts of the text. Dr. Frank Schumann and Dr. Jörg Schröder contributed Figures 2.19 and 2.25, respectively. We also thank Dirk Bächle, Kai Jenssen, Micha Jost, Dr. Jörg Schröder and Ulf Strijowski for comments and proofreading parts of the manuscript.

Dr. Gudrun Walter, Maike Petersen, Dr. Bill Down, and Hans-Jörg Maier took care that the manuscript was converted into this book in a rather short period of time, without complications.

Bielefeld
and
Dresden-Langebrück
April 2002

Norbert Sewald

Hans-Dieter Jakubke

Abbreviations

aa	amino acid
AA	antamanide (anti-amanita peptide)
Aad	2-aminoadipic acid
βAad	3-aminoadipic acid
AAP	antimicrobial animal peptides
aatRS	amino-acyl tRNA synthetase
Ab	antibody
Abu	aminobutyric acid
A_2bu	2,4-diaminobutyric acid
Abz	aminobenzoic acid
Ac	acetyl
ACE	angiotensin-converting enzyme
AChR	acetylcholine receptor
Acm	acetamidomethyl
ACP	acyl carrier protein
ACTH	corticotropin
AD	Alzheimer's disease
Ada	adamantyl
Adoc	adamantyloxycarbonyl
ADP	adenosine diphosphate
Aet	aminoethyl
Ag	antigen
AGaloc	tetra-O-acetyl-β-D-galactopyranosyloxycarbonyl
AGE	advanced glycation end products
AGloc	tetra-O-acetyl-D-glucopyranosyloxycarbonyl
AGRP	agouti-related protein
Ahx	2-aminohexanoic acid (norleucine)
εAhx	6-aminohexanoic acid
AHZ	alanyl-histidinato zinc (β)
Aib	α-aminoisobutyric acid
AIDS	acquired immune deficiency syndrome
AIle (aIle)	alloisoleucine
AKH	adipokinetic hormone

Al	allyl (used only in 3-letter code names)
Ala	alanine
βAla	β-alanine
Aloc	allyloxycarbonyl
Alom	allyloxymethyl
AMP	adenosine monophosphate
AM-PS	aminomethyl polystyrene
ANF	atrial natriuretic factor
ANP	atrial natriuretic peptide
Ans	anthracene-9-sulfonyl
Aoc	1-azabicyclo[3.3.0]octane-2-carboxylic acid
AOE	(S)-2-amino-8-oxo-(S)-9,10-epoxidecanoic acid
AOP	7-azabenzotriazol-1-yloxytris(dimethylamino)phosphonium hexafluorophosphate [1)]
Apa	6-aminopenicillanic acid
APM	aspartame
Apm	2-aminopimelic acid
AQP	aquaporin
Ar	aryl
Arg	arginine
Asn	asparagine
Asp	aspartic acid
Aβ	amyloid-β
Asu	aminosuberic acid
At	azabenzotriazolyl
AT	angiotensin
ATP	adenosine triphosphate
AVP	8-arginine-vasopressin
Bac5	bactenecin
BAL	backbone amide linker
BBB	blood brain barrier
Bet	α-betainyl
BGloc	tetrabenzylglucosyloxycarbonyl
BGP	bone Gla protein
BHA	benzhydrylamine
Bip	biphenyl-4-sulfonyl
BK	bradykinin
BLP	bombinin-like peptide
BMP	brain morphogenetic protein
BNP	brain natriuretic peptide
Boc	*tert*-butoxycarbonyl
BOI	2-[(1*H*-benzotriazol-1-yl)oxy]-1,3-dimethylimidazolidinium hexafluorophosphate
Bom	benzyloxymethyl

BOP	benzotriazol-1-yloxytris(dimethylamino)phosphonium hexafluorophosphate
Bpoc	2-(biphenyl-4-yl)prop-2-yloxycarbonyl
BPTI	basic pancreatic trypsin inhibitor
BroP	bromotris(dimethylamino)phosphonium hexafluorophosphate
BSA	bovine serum albumin
Bsmoc	1,1-dioxobenzo[b]thiophen-2-ylmethoxycarbonyl
Bspoc	2-(*tert*-butylsulfonyl)-2-propenyloxycarbonyl
Bt	(benzylsulfanylmethyl)
Btb	1-*tert*-butoxycarbonyl-2,3,4,5-tetrachlorobenzoyl
Btm	benzylsulfanylmethyl (benzylthiomethyl)
Bu	butyl
tBu	*tert*-butyl
Bum	*tert*-butoxymethyl
Bz	benzoyl
Bzl	benzyl (Bn in contemporary organic synthesis)
Bzl(4-Me)	4-methylbenzyl
CADD	computer-aided drug design
CaM	calmodulin
Cam	carboxamidomethyl (carbamoylmethyl, aminocarbonylmethyl)
CAMD	computer-aided molecular design
CAMM	computer-assisted molecular modeling
cAMP	cyclic AMP
CBD	chitin binding domain
CCAP	crustacean cardioactive peptide
CCK	cholecystokinin
CD	circular dichroism
CD4	cell surface glycoprotein 4
CD8	cell surface glycoprotein 8
cDNA	complementary DNA
CDR	complementary determining region
CE	capillary electrophoresis
CF_3-BOP	6-(trifluoromethyl)benzotriazol-1-yloxytris(dimethylamino)-phosphonium hexafluorophosphate
CF_3-HBTU	2-[6-(trifluoromethyl)benzotriazol-1-yl]-1,1,3,3-tetramethyluronium hexafluorophosphate [2)]
CF_3-PyBOP	6-(trifluoromethyl)benzotriazol-1-yloxytripyrrolidinophosphonium hexafluorophosphate
CG	chorionic gonadotropin
Cg	chromogranin
cGMP	cyclic guanosine monophosphate
CGRP	calcitonin gene related peptide
Cha	cyclohexylalanine
Cho	choline

cHp	cycloheptyl
CID	chemically ionized desorption
CIEF	capillary isoelectric focusing
Cit	citrulline
CJD	Creutzfeldt Jakob disease
2,6-Cl_2Bzl	2,6-dichlorobenzyl
CLIP	corticotropin-like intermediate lobe peptide
Clt-Cl	2-chlorotritylchloride
CLTR	2-chlorotrityl resin
Cm	carboxymethyl
CM	casomorphin or chorionic mammotropin
CN	calcineurin
CNBr	cyanogen bromide
Cne	2-cyanoethyl
CNP	C-type natriuretic peptide
CNS	central nervous system
cOc	cyclooctyl
COSY	correlated NMR spectroscopy
Cpa	4-chlorophenylalanine
CP	carboxypeptidase
cPe	cyclopentyl
CPP	cell-penetrating peptide
CPS	convergent peptide synthesis
CRF	corticotropin releasing factor
CRIF	corticotropin release-inhibiting factor
CRL	cerulein
CsA	cyclosporin A
CSF	colony stimulating factor
CSPPS	convergent solid-phase peptide synthesis
CST	cortistatin
CT	calcitonin or charge-transfer
Cy	cyclohexyl
Cya	cysteic acid
Cyp	cyclophilin
Cys	cysteine
$(Cys)_2$	cystine
CZE	capillary zone electrophoresis
Dab	α,γ-diaminobutyric acid
DABCO	1,4-diazabicyclo[2.2.2]octane
DABSCL	4-(dimethylamino)azobenzene-4′-sulfonyl chloride
DABITC	4-(dimethylamino)azobenzene-4′-isothiocyanate
DAG	diacylglycerol
DAST	N,N-diethylaminosulfur trifluoride
DBF	dibenzofulvene

DBIP	diazepam-binding inhibitor peptide
DBU	1,8-diazabicyclo[5.4.0]undec-7-ene
Dcb	dichlorobenzyl
DCC	N,N′-dicyclohexylcarbodiimide
DCHA	dicyclohexylamine
DCM	dichloromethane
DCME	dichloromethyl methyl ether
Dcp	α,α-dicyclopropylglycine
DCU	N,N′-dicyclohexylurea
DDAVP	[1-deamino,D-Arg^8]vasopressin
Dde	1-(4,4-dimethyl-2,6-dioxocyclohexylidene)ethyl
Ddz	α,α-dimethyl-3,5-dimethoxybenzyloxycarbonyl
DEA	diethylamine
DEAE	diethylaminoethanol
Deg	$C^{\alpha,\alpha}$-diethylglycine
DEPBT	3-(diethoxyphosphoryloxy)-1,2,3-benzotriazine-4(3*H*)-one
DEPC	diethyl pyrocarbonate
DFIH	2-fluoro-4,5-dihydro-1,3-dimethyl-1*H*-imidazolium hexafluorophosphate
Dha	α,β-didehydroalanine (more commonly, α,β-dehydroalanine)
Dhbt	3,4-dihydro-4-oxobenzotriazin-3-yl
DIC	N,N′-diisopropylcarbodiimide
DIPEA	diisopropylethylamine
DKP	diketopiperazine
DMA	dimethylacetamide
Dmab	4-{[1-(4,4-dimethyl-2,6-dioxocyclohexylidene)-3-methylbutyl]amino}benzyl
DMAE	2-(dimethylamino)ethanol
DMAP	4-(dimethylamino)pyridine
2,4-Dmb	2,4-dimethoxybenzyl
3,4-Dmb	3,4-dimethoxybenzyl
DMBHA	dimethoxybenzhydrylamine
DMF	N,N-dimethylformamide
Dmh	2,6-dimethylhept-4-yl
Dmp	2,4-dimethoxyphenyl
DMS	dimethyl sulfide
DMSO	dimethyl sulfoxide
Dmt	2′,6′-dimethyltyrosine
DNA	deoxyribonucleic acid
Dnp	dinitrophenyl
Dns	5-(dimethylamino)naphthalene-1-sulfonyl (dansyl)
Doc	2,4-dimethylpent-3-yloxycarbonyl
DOPA	3,4-dihydroxyphenylalanine
Dpa	3,3-diphenylalanine
Dpg	$C^{\alpha,\alpha}$-diphenylglycine

Dpm	diphenylmethyl (benzhydryl)
DPPA	diphenyl phosphorazidate
Dpr	2,3-diaminopropionic acid
DPTU	N,N′-diphenylthiourea
DSC	di(N-succinimidyl)carbonate
DSIP	delta sleep-inducing peptide
DSK	drosulfakinin
Dsu	(2*S*,7*S*)-2,7-diaminosuberic acid
Dts	dithiasuccinyl
DTT	dithiothreitol
DVB	divinylbenzene
Dyn	dynorphin
E^+	electrophile (or E-X)
<E	pyroglutamic acid
EC	ethylcarbamoyl
ECD	extracellular domain
ECE	endothelin converting enzyme
ECEPP	Empirical Conformational Energy Program for Peptides
ECGF	endothelial cell growth factor
ECM	extracellular matrix
ECP	ecdysteroid carrier protein
ED_{50}	median effective dosis
EDC	N-ethyl-N′-(3-dimethylaminopropyl)carbodiimide hydrochloride
EDF	epidermal growth factor or erythrocyte differentiation factor
EDFR	epidermal growth factor receptor
EDT	ethanedithiol
EDTA	ethylenediaminetetraacetic acid
EEDQ	ethyl 2-ethoxy-1,2-dihydroquinoline-1-carboxylate
EF	elongation factor
EGF	epidermal growth factor
EH	eclosion hormone
ELAM	endothelial leukocyte adhesion molecule
ELH	egg laying hormone
ELISA	enzyme linked immunosorbent assay
EM	electron microscopy
EMIT	enzyme multiplied immunoassay technology
EMSA	electrophoretic mobility shift assay
ENK	enkephalin
EPL	expressed protein ligation
EPO	erythropoietin
EPR	electron paramagnetic resonance
ER	endoplasmatic reticulum
ES	electrospray
ESI-MS	electrospray ionisation mass spectrometry

ES-MS	electrospray mass spectrometry
ESR	electron spin resonance
ET	endothelin
Et	ethyl
ETH	ecdysis-triggering hormone
Etm	ethoxymethyl
Fa	3-(2-furyl)acryloyl
Fab	antigen binding Ig fragment
FAB-MS	fast atom bombardment mass spectrometry
FACS	fluorescence-activated cell sorter
FAD	flavin adenine dinucleotide, oxidized form
$FADH_2$	flavin adenine dinucleotide, reduced form
Farn	farnesyl
FaRP	FMRFamide-related peptide
Fbg	fibrinogen
Fc	ferrocenyl
Fd	ferredoxin
FGF	fibroblast growth factor
FITC	fluoresceinyl isothiocyanate
FKBP	FK506 binding protein
Fm	9-fluorenylmethyl
FMDV	foot-and-mouth disease virus
Fmoc	9-fluorenylmethoxycarbonyl
FN	fibronectin
For	formyl
FP	fluorescence polarisation
Fpa	4-fluorophenylalanine
FPLC	fast protein liquid chromatography
FPP	farnesyl pyrophosphate
FRET	fluorescence resonance energy transfer
FRL	formin-related peptide
FS	follistatin
FSF	fibrin-stabilizing factor
FSH	follicle-stimulating hormone
FTase	farnesyltransferase
FTIR	Fourier-transform infrared
GABA	γ-aminobutyric acid
Gal	galanin, galactose
GalT	galactosyl transferase
GC	gas chromatography
gCSF	granulocyte colony stimulating factor
GFC	gel filtration chromatography
GFP	green fluorescent protein

GGTase	geranylgeranyltransferase
GH	growth hormone
GHRH	growth hormone releasing hormone
GHRP	growth hormone releasing peptide
GHS	growth hormone secretagogue
GHS-R	growth hormone secretagogue receptor
GIP	gastric inhibitory polypeptide
Gla	4-carboxyglutamic acid
GLC	gas liquid chromatography
Glc	glucosyl, glucose
GlcNAc	N-acetyl-D-glucosamine
Gln	glutamine
GLP	glucagon-like peptide
Glu	glutamic acid
Gly	glycine
GMP	guanosine monophosphate
GnRH	gonadotropin releasing hormone
GPC	gel permeation chromatography
GPCR	G-protein coupled receptor
GPI	glycosylphosphatidylinositol or guinea pig ileum
GRF	growth hormone releasing factor
GRP	gastrin-releasing peptide
GRPP	glicentin-related pancreatic peptide
GSF	glutathion-S-transferase
GSH	glutathione reduced
GSSG	glutathione oxidized
GT	gastrin
GTP	guanosine triphosphate
Gva	δ-guanidinovaleric acid
hXaa	homoamino acid
h	human
HA	head activator
HAL	5-(4-hydroxymethyl-3,5-dimethoxy)-valeric acid (derived hypersensitive acid-labile linker)
HAMDU	O-(7-azabenzotriazol-1-yl)-1,3-dimethylimidazolidinium hexafluorophosphate [1)]
HAMTU	O-(7-azabenzotriazol-1-yl)-1,3-dimethyl-1,3-trimethyleneuronium hexafluorophosphate [1, 2)]
HAPipU	O-(7-azabenzotriazol-1-yl)-1,1,3,3-bis(pentamethylene)uronium hexafluorophosphate [1, 2)]
HAPyTU	S-(7-azabenzotriazol-1-yl)-1,1,3,3-bis(tetramethylene)thiouronium hexafluorophosphate [1, 2)]
HAPyU	O-(7-azabenzotriazol-1-yl)-1,1,3,3-bis(tetramethylene)uronium hexafluorophosphate [1, 2)]

hArg	homoarginine
HATTU	S-(7-azabenzotriazol-1-yl)-1,1,3,3-tetramethylthiouronium hexafluorophosphate
HATU	O-(7-azabenzotriazol-1-yl)-1,1,3,3-tetramethyluronium hexafluorophosphate; correct IUPAC name: 1-[bis-(dimethylamino)methyliumyl]-1*H*-1,2,3-triazolo[4,5-b]pyridin-3-oxide hexafluorophosphate or 1-[(dimethylamino)-(dimethyliminium) methyl]-1*H*-1,2,3-triazolo[4,5-*b*]pyridin-3-oxide hexafluorophosphate
HAV	hepatitis A virus
Hb	hemoglobin
HBMDU	O-(benzotriazol-1-yl)-1,3-dimethyl-1,3-dimethyleneuronium hexafluorophosphate [2)]
HBPyU	O-(benzotriazol-1-yl)-1,1,3,3-bis(tetramethylene)uronium hexafluorophosphate [2)]
HBsAg	hepatitis B virus surface antigen
HBTU	2-(1*H*-benzotriazol-1-yl)-1,1,3,3-tetramethyluronium hexafluorophosphate [2)]; correct IUPAC name: 3-[Bis(dimethylamino)methyliumyl]-3H-benzotriazol-1-oxide hexafluorophosphate
HBV	hepatitis B virus
Hbz	2-hydroxybenzyl
Hci	homocitrulline
HCRT	hypocretin
HCV	hepatitis C virus
hCys	homocysteine
HDCOTU	O-(dicyanomethylenamino)-1,1,3,3-tetramethyluronium hexafluorophosphate
HDL	high density lipoprotein
HDTU	O-(3,4-dihydro-4-oxo-1,2,3-benzotriazin-3-yl)-1,1,3,3-tetramethyluronium hexafluorophosphate
Hep	heptyl
Hepes	N-(2-hydroxyethyl)piperazine-N′-2-ethanesulfonic acid
HEPP	human IgE pentapeptide
HF	hydrogen fluoride
HFBA	heptafluorobutyric acid
HFIP	hexafluoroisopropanol
HG	human little-gastrin
hGH	human growth hormone
HIC	hydrophobic interaction chromatography
Hip	hippuryl, hippuric acid
His	histidine
HIV	human immunodeficiency virus
HMB	hydroxymethylbenzoic acid
Hmb	2-hydroxy-4-methoxybenzyl

HMFS	N-[9-(hydroxymethyl)-2-fluorenyl]succinamic acid
HMK	high molecular weight kininogen
HMP	4-(hydroxymethyl)phenoxy
HMPA	hexamethylphosphoric triamide
HMPAA	4-(hydroxymethyl)phenoxyacetic acid
HMPB	4-(4-hydroxymethyl-3-methoxyphenoxy)butyric acid
HMPPA	3-[4-(hydroxymethyl)phenoxy]propanoic acid
HMPT	hexamethylphosphorous triamide
HMQC	heteronuclear multiple quantum coherence spectroscopy
Hnb	2-hydroxy-6-nitrobenzyl
HOAt	1-hydroxy-7-aza-1*H*-benzotriazole
HOBt	1-hydroxy-1*H*-benzotriazole
Hoc	cyclohexyloxycarbonyl
HOCt	ethyl 1-hydroxy-1*H*-1,2,3-triazole-4-carboxylate
HODhbt	3-hydroxy-1,2,3-benzotriazin-4(3*H*)-one
HOPip	N-hydroxypiperidine
HOPPipU	2-[2-oxo-1(2*H*)pyridyl]-1,1,3,3-bis(pentamethylene)uronium hexafluorophosphate
HOSu	N-hydroxysuccinimide
HOTU	O-[(ethoxycarbonyl)cyanomethyleneamino]-1,1,3,3-tetramethyluronium hexafluorophosphate
HPCE	high performance capillary electrophoresis
HPLC	high performance liquid chromatography
Hpp	5-hydroxy-1-(4′-nitrophenyl)pyrazole
HppTU	2-[1-(4′-nitrophenyl)-1*H*-pyrazol-5-yl]-1,1,3,3-tetramethyluronium tetrafluoroborate
HPSEC	high performance size exclusion chromatography
hPTH	human parathyroid hormone
HpyClU	chloro-1,1,3,3-bis(tetramethylene)-uronium hexafluorophosphate
HRMS	high resolution mass spectrometry
HS	hymenstatin
HSA	human serum albumin
HSAB	hard and soft acids and bases
Hse	homoserine
Hsp	heat shock protein
HSPS	high speed peptide synthesis
HSV	herpes simplex virus
HTS	high-throughput screening
HVE	high voltage electrophoresis
Hyp	hydroxyproline
Hyv	α-hydroxyisovaleric acid
iBu	isobutyl
IC	inhibitory concentration
iC	isocapronic

ICAM	intracellular adhesion molecule
IEC	ion-exchange chromatography
IF	initiation factor
IFN	interferon
Ig	immunoglobulin
IGF	insulin-like growth factor
IHB	inhibin
IIDQ	1-isobutoxycarbonyl-2-isobutoxy-1,2-dihydroquinoline
IL	interleukin
Ile	isoleucine
im	imidazole
IMAC	immobilized metal ion affinity chromatography
iMds	2-methoxy-4,6-dimethylbenzenesulfonyl
in	indole
iNoc/iNOC	isonicotinyloxycarbonyl (4-pyridylmethoxycarbonyl)
Ioa	isooctanoic acid
IP	inositol phosphate
Ipc	isopinocamphenyl
IPL	intein-mediated protein ligation
IPNS	isopenicillin N synthase
iPr	isopropyl
IRaa	internal reference amino acid
IRMA	immunoradiometric assay
IS-MS	ion spray mass spectrometry
IU	international unit
Iva	isovaline
IvDde	1-(4,4-dimethyl-2,6-dioxocyclohexyliden)-3-methylbutyl
kB	kilo base pair
kDa	kilodalton
KGF	keratinocyte growth factor
K_M	Michaelis constant
Lac	lactic acid
Lan	lanthionine
LAP	leucine aminopeptidase
Lau	lauroyl
LD_{50}	lethal dose 50%
LDL	low density lipoprotein
LD-MS	laser desorption mass spectrometry
LDToF	laser desorption time-of-flight
LEC	ligand-exchange chromatography
Leu	leucine
LFA-1	leukocyte function-associated antigen-1
LH	luteinizing hormone

LHRH	luteinizing hormone releasing hormone
LIF	leukemia inhibitory factor
LPH	lipotropic hormone
LPS	lipopolysaccharide
LSF	lung surfactant factor
LSI-MS	liquid secondary ion mass spectrometry
LVP	8-lysine vasopressin
Lys	lysine
MA	mixed anhydride
MAb	monoclonal antibody
Mal	maleoyl
MALDI-MS	matrix-assisted laser desorption/ionization mass spectrometry
MALDI-ToF	matrix-assisted laser desorption/ionization time-of-flight
MAP	membrane-anchored protein or multiple antigen peptides
MARS	multiple automatic robotic synthesizer
Mba	2-mercaptobenzoic acid
Mbc	4′-methyl-2,2′-bipyridine-4-carboxylic acid
Mbh	dimethoxybenzhydryl
MBHA	methoxybenzhydrylamine
MBHAR	methoxybenzhydrylamine resin
Mbom	4-methoxybenzyloxymethyl
MBP	maltose binding protein or myelin basic protein
Mbs	4-methoxybenzenesulfonyl
Mca	(7-methoxycoumarin-4-yl)acetyl
MCDP	mast cell degranulating peptide
MCH	melanin-concentrating hormone
MCPBA	3-chloroperoxybenzoic acid
MCPS	multiple constrained peptide synthesis
MD	molecular dynamics
Mdc	4-methoxy-2,6-dimethylbenzenesulfonyl
MDP	maduropeptin
Me	methyl
β-ME	β-mercaptoethanol
MeBHA	methylbenzhydrylamine
Mee	methoxyethoxyethyl
Mem	(2-methoxyethoxy)methyl
Men	menthyl
Menpoc	1-(6-nitro-1,3-benzodioxol-5-yl)ethoxycarbonyl
Menvoc	1-(4,5-dimethoxy-2-nitrophenyl)ethoxycarbonyl
MeO	methoxy
MeOSu	methoxysuccinyl
Mes	mesityl
Met	methionine
MGP	matrix Gla protein

MHC	major histocompatibility complex
MIC	minimum inhibitory concentration
MIF	melanotropin release-inhibiting factor
MIH	melanotropin release-inhibiting hormone
MK	midkine
MMA	metamorphisin A
mMIF	macrophage migration inhibitory factor
Mmt	4-methoxytrityl
Mnp	2-methyl-2-(2′-nitrophenoxy)propionyl
Moa	6-methyloctanoic acid
Mob	4-methoxybenzyl
Moc	methoxycarbonyl
MoEt	2-(N-morpholino)ethyl
Mom	methoxymethyl
Mot	motilin
Mp	4-methoxyphenyl
MP	myelopeptides
Mpa	3-sulfanylpropanoyl (3-mercaptopropionyl)
Mpe	3-methylpent-3-yl
Mpg	3-methoxypropylglycine
MPGF	major proglucagon fragment
MPLC	medium pressure liquid chromatography
Mpr	3-mercaptopropionic acid
MPS	multiple peptide synthesis
MPTA	dimethylphosphinothioyl azide
MRF	melanotropin-releasing factor
MRH	melanotropin-releasing hormone
MRIH	melanotropin release-inhibiting hormone
mRNA	messenger ribonucleic acid
MS	mass spectrometry
Ms	mesyl(methanesulfonyl)
Msbs	4-(methanesulfonyl)benzenesulfonyl
Msc	2-(methylsulfonyl)ethoxycarbonyl
MSH	melanocyte stimulating hormone (melanotropin)
MSNT	1-(mesitylenesulfonyl)-3-nitro-1*H*-1,2,4-triazole
Msob	4-(methylsulfinyl)benzyl
MsOH	methanesulfonic acid
Mspoc	2-(methylsulfonyl)-3-phenyl-2-propenyloxycarbonyl
Msz	4-(methylsulfinyl)benzyloxycarbonyl
Mtb	2,4,6-trimethoxybenzenesulfonyl
MTBE	methyl *tert*-butyl ether
Mtbs	3,5-di-*tert*-butyl-4-methoxybenzenesulfonyl
Mte	4-methoxy-2,3,5,6-tetramethylbenzenesulfonyl
Mthp	4-methoxytetrahydropyran-4-yl-methylthiomethyl
MTLRP	motilin-related peptide

Mtr	4-methoxy-2,3,6-trimethylbenzenesulfonyl
Mts	2,4,6-trimethylbenzenesulfonyl (mesitylsulfonyl)
Mtt	4-methyltrityl
Mtz	4-(methylthio)benzyl
Mur	muramic acid
MVD	mouse vas deferens
Mwt	molecular weight
Mz	4-(methoxyphenylazo)benzyloxycarbonyl
βNA	β-naphthylamide
pNA	4-nitroanilide
NADH	nicotinamide adenine dinucleotide (reduced)
NADPH	nicotinamide adenine dinucleotide phosphate (reduced)
Nal	2-naphthylalanine
Nbb	nitrobenzamidobenzyl
Nboc	2-nitrobenzyloxycarbonyl
NBS	N-bromosuccinimide
Nbs	nitrobenzenesulfonyl
Nbz	4-nitrobenzyl
2Nbz	2-nitrobenzyl
NC	nociceptin
NCA	α-amino acid N-carboxyanhydride
NCL	native chemical ligation
Nde	1-(4-nitro-1,3-dioxoindan-2-ylidene)ethyl
NEM	N-ethylmaleimide
Neu	neuraminic acid
NeuNAc	N-acetylneuraminic acid
^{15}N-HSQC	^{15}N heteronuclear single quantum correlation
NGF	nerve growth factor
Nic	nicotinoyl
NIDDM	noninsulin-dependent (type II) diabetes mellitus
NK	neurokinin
Nle	norleucine
NM	neuromedin
NMA	N-methylaniline
NMDA	N-methyl-D-aspartate
NMM	N-methylmorpholin
NMP	N-methyl-2-pyrrolidinone
NMR	nuclear magnetic resonance
NN	neuromedin N
Noc	4-nitrocinnamyloxycarbonyl
NOE	nuclear Overhauser effect
NOESY	nuclear Overhauser enhanced spectroscopy
NOP	6-nitrobenzotriazol-1-yloxytris(dimethylamino)phosphonium hexafluorophosphate

Np	4-nitrophenyl
NP	neurophysin
Npa	nitrophenoxyacetyl
Npe	2-(4-nitrophenyl)ethyl
NPF	neuropeptide F
NPFF	neuropeptide FF
NPK	neuropeptide K
Npoc	(6-nitro-1,3-benzodioxol-5-yl)-methyloxycarbonyl
Nps	2-nitrophenylsulfenyl
NPY	neuropeptide Y
Npys	3-nitro-2-pyridylsulfanyl
NP_{γ}	neuropeptide γ
Nsc	2-[(4-nitrophenyl)sulfonyl]ethoxycarbonyl
NT	neurotensin
3-NT	3-nitrotyrosine
NTA	N-thiocarboxy anhydride or nitrilotriacetic acid
Nu/Nuc^{-}	nucleophile
Nva	norvaline
Nvoc	4,5-dimethoxy-2-nitrobenzyloxycarbonyl
OAI	allyloxy
OBt	1*H*-1,2,3-benzotriazol-1-yloxy
OBzl	benzyloxy
OCM	oncostatin M
Oct	octanoyl
OEt	ethoxy
OGP	osteogenic growth peptide
OGp	4-guanidinophenyloxy
OMe	methoxy
ONbz	4-nitrobenzyloxy
O2Nbz	2-nitrobenzyloxy
ONC	onconase
ONdc	5-norbornene-2,3-dicarboximidooxy
ONp	4-nitrophenyloxy
O2Np	2-nitrophenyloxy
ONs	2-nitro-4-sulfophenoxy
OPA	2-phthaldialdehyde
OPcp	pentachlorophenyloxy
OPfp	pentafluorophenyloxy
OPip	1-piperidinooxy
O2Py	2-pyridyloxy
Orn	ornithine
OSu	succinimidooxy
OtBu	*tert*-butoxy
OTcp	2,4,5-trichlorophenyloxy

Ox	1,3-oxazolidine
OXT/OT	oxytocin
Pa	palmitic acid
Pab	4-alkoxybenzyl
Pac	phenacyl
PACAP	pituitary adenylate cyclase activating polypeptide
PAGE	polyacrylamide gel electrophoresis
Pal	3-pyridylalanine
PAL	4-aminomethyl-3,5-(dimethoxy)phenoxy
PAM	4-(hydroxymethyl)phenylacetic acid (resin linker) or peptidylglycine α-amidating monooxygenase
PAOB	4-phenylacetoxybenzyl
PAPS	3′-phosphoadenosine-5′-phosphosulfate
Pbf	2,2,4,6,7-pentamethyldihydrobenzofuran-5-yl-sulfonyl
PD-ECGF	platelet-derived endothelial cell growth factor
PDGF	platelet-derived growth factor
PDH	pigment dispersing hormone
PDI	protein disulfide isomerase
PD-MS	plasma desorption mass spectrometry
Pdpm	pyridyldiphenylmethyl
PEG	poly(ethylene glycol)
PEGA	poly(ethylene glycol)-dimethylacrylamide copolymer
Pen	penicillamine (3-mercaptovaline)
PEt	polyethylene
Pfp	pentafluorophenyl
PfPyU	O-pentafluorophenyl-1,1,3,3-bis(tetramethylene)uronium hexafluorophosphate
PfTU	O-pentafluorophenyl-1,1,3,3-tetramethyluronium hexafluorophosphate
PG	protecting group
PGLa	peptidyl-glycyl-leucine carboxamide
pGlu	pyroglutamic acid
Ph	phenyl
Pha	phenoxyacetyl
Phac	phenylacetyl
Phacm	phenylacetamidomethyl
Phe	phenylalanine
Phen	phenanthrene-3-sulfonyl
PhFl	9-phenylfluoren-9-yl
Phg	phenylglycine
PHI	peptide histidine isoleucine amide
Phlac	phenyllactic acid
Phpa	3-phenylpropanoic acid
Phth	phthaloyl

pI	isoelectric point
PI3K	phosphatidylinositol-3-kinase
Pic	4-picolyl
PicCO	picolinoyl
Picoc	4-picolyloxycarbonyl
PIH	prolactin-release-inhibiting hormone
Pip	pipecolic acid (homoproline)
PITC	phenyl isothiocyanate
Piv	pivaloyl
PKA	proteinkinase A
PKS	phytosulfokinin
PL	placenta lactogen
PLC	phospholipase C
Plm	palmitoyl
PLR	peptide leucine arginine amide
Pmc	2,2,5,7,8-pentamethylchroman-6-yl-sulfonyl
Pme	2,3,4,5,6-pentamethylbenzenesulfonyl
3-Pn	pent-3-yl
pNA	4-nitroaniline
PNA	peptide nucleic acid
Poc	cyclopentyloxycarbonyl
Pom	pivaloyloxymethyl
POMC	proopiomelanocortin
PP	pancreatic polypeptide
PPIase	peptidyl prolyl *cis/trans* isomerase
PPOA	(4-propionylphenoxy)acetic acid
PPST	tyrosyl protein sulfotransferase
Ppt	diphenylphosphorothionyl
Pr	propyl
Pra	propargylglycine
PRH	prolactin-releasing hormone
PRL	prolactin
Pro	proline
PrRP	prolactin-releasing peptide
PS	polystyrene
PSA	preformed symmetrical anhydride
Psec	2-(phenylsulfonyl)ethoxycarbonyl
PS-POE	polystyrenepolyoxyethylene
PS-SCL	positional scanning synthetic combinatorial libraries
PST	pancreastatin
PTB	phosphotyrosine binding domain
PTC	phenylthiocarbamyl
PTH	phenylthiohydantoin or parathyroid hormone
PTHrP	parathyroid hormone-related hormone
PTK	protein-tyrosine kinase

PTP	protein-tyrosine phosphatase
PVA	polyvinyl alcohol
Py	pyridine
PyAOP	7-azabenzotriazol-1-yloxytripyrrolidinophosphonium hexafluorophosphate
PyBOP	benzotriazol-1-yloxytripyrrolidinophosphonium hexafluorophosphate
PyBroP	bromotripyrrolidinophosphonium hexafluorophosphate
PyCloP	chlorotripyrrolidinophosphonium hexafluorophosphate
PyFOP	6-fluorobenzotriazol-1-yloxytripyrrolidinophosphonium hexafluorophosphate
PyNOP	6-nitrobenzotriazol-1-yloxytripyrrolidinophosphonium hexafluorophosphate
Pyoc	pyridylethoxycarbonyl
Pyr	pyrrolidide
PYY	peptide tyrosine tyrosine
Pz	4-(phenyldiazenyl)benzyloxycarbonyl
QCl	5-chloro-8-quinolyl
RAFT	regioselectively addressable functionalized template
RAMP	receptor activity modifying protein or (*R*)-1-amino-2-(methoxymethyl)-pyrrolidine
rDNA	recombinant DNA
RER	rough endoplasmatic reticulum
RET	resonance energy transfer
R_f	retention factor (TLC)
RGD	fibrinogen binding sequence (-Arg-Gly-Asp-)
RIA	radioimmunoassay
RLX	relaxin
RNA	ribonucleic acid
RNase	ribonuclease
ROE	rotating frame nuclear Overhauser effect
ROESY	rotating frame nuclear Overhauser enhanced spectroscopy
RP	reversed phase
RPCH	red pigment concentrating hormone
RP-HPLC	reversed phase high performance liquid chromatography
SA	symmetrical anhydride
Saa	sugar amino acid
SABR	Structure Activity Bioavailability Relationships
Sar	sarcosine
SAR	Structure Activity Relationship
SASRIN	super acid sensitive resin
sBu	*sec*-butyl

SBzl	thiobenzyl
SCAL	safety catch acid-labile linker or safety catch amide linker
Scg-MT	schistomyotropin
SCL	synthetic combinatorial libraries
SDB	styrene divinylbenzene
SDS	sodium dodecylsulfate
Sec	selenocysteine (one-letter symbol U)
SEC	size-exclusion chromatography
SEM	2-(trimethylsilyl)ethoxymethyl
Ser	serine
SES	2-(trimethylsilyl)ethanesulfonyl
SH2	Src homology 2 domain
Shk	stichodactyla toxin
SIH	somatotropin release-inhibiting hormone
SLC	sublethal concentration
Sle^x	sialyl-$Lewis^x$
SM	somatomedin
Smc	S-methylcysteine
SMPS	simultaneous multiple peptide synthesis
SN	secretoneurin
SP	substance P
SP5	splenopentin
SPCL	synthetic peptide combinatorial library
SPOCC	solid phase organic combinatorial chemistry
SPPS	solid-phase peptide synthesis
SR	sarcoplasmic reticulum
SRH	somatotropin releasing hormone
SRTX	sarafotoxin
ssDNA	single stranded DNA
SST	somatostatin
Sta	statine, (3S,4S)-4-amino-3-hydroxy-6-methylhexanoic acid
StBu	*tert*-butylsulfanyl
STH	somatotropin
STI	soybean trypsin inhibitor
Su	succinimide
Suc	succinoyl
Sulfmoc	2-sulfo-9-fluorenylmethoxycarbonyl
SVG	sauvagine
Tac	2-toluidinocarbonyl
Tacm	trimethylacetamidomethyl
TASP	template-assembled synthetic protein
TbFmoc	tetrabenzo-9-fluorenylmethoxycarbonyl
TBAF	tetrabutylammonium fluoride

TBPipU	2-(benzotriazol-1-yl)-1,1,3,3-bis(pentamethylene)uronium tetrafluoroborate
TBPyU	O-(benzotriazol-1-yl)-1,1,3,3-tetramethyluronium tetrafluoroborate
Tbs	3-*tert*-butyl-4-methoxybenzenesulfonyl
Tbtr	4,4′,4″-tris(benzoyloxy)trityl
TBTA	*tert.*-butyl trichloroacetimidate
TBTU	2-(1H-benzotriazol-1-yl)-1,1,3,3-tetramethyluronium tetrafluoroborate
tBu	*tert*-butyl
t-Bumeoc	1-(3,5-di-*tert*-butylphenyl)-1-methylethoxycarbonyl
TCE	2,2,2-trichloroethyl
TCL	thin layer chromatography
Tcp	trichlorophenyl
TDBTU	O-(3,4-dihydro-4-oxo-1,2,3-benzotriazin-3-yl)-1,1,3,3-tetramethyluronium tetrafluoroborate
TEA	triethylamine
TEMPO	2,2,6,6-tetramethylpiperidin-1-yloxy
Teoc	2-(trimethylsilyl)ethoxycarbonyl
TES	triethylsilyl
Tf	trifluoromethanesulfonyl
TFA	trifluoroacetic acid
TFAA	trifluoroacetic anhydride
TFE	trifluoroethanol
TFFH	tetramethylfluoroformamidinium hexafluorophosphate
Tfm	trifluoromethyl
TFMSA	trifluoromethane sulfonic acid
TfOH	trifluoromethanesulfonic acid
TGF	transforming growth factor
TH	thymus hormone
thexyl	1,1,2-trimethylpropyl
THF	tetrahydrofuran
Thi	thienylalanine
Thr	threonine
Thx	thyroxine
Thz	thiazolidine-4-carboxylic acid
Tic/TIC	1,2,3,4-tetrahydroisoquinoline-3-carboxylic acid
Tip	2,4,6-triisopropylbenzenesulfonyl
TIS	Triisopropylsilane
TLE	thin layer electrophoresis
TM	thrombomodulin
Tmb	trimethoxybenzyl
TMBHA	4-(benzyloxy)-2′,4′-dimethoxybenzhydrylamine
TMS	trimethylsilyl
TMSBr	trimethylsilyl bromide

TMSE	2-(trimethylsilyl)ethyl
TMSOTf	trimethylsilyl trifluoromethanesulfonate
Tmtr	tris(4-methoxyphenyl)methyl
Tn	troponin
TNBS	2,4,6-trinitrobenzenesulfonic acid
TNF	tumor necrosis factor
TNTU	2-(5-norbornene-2,3-dicarboximido)-1,1,3,3-tetramethyluronium tetrafluoroborate
TOAC	4-amino-4-carboxy-2,2,6,6-tetramethylpiperidinyl-1-oxyl
TOCSY	total correlation spectroscopy
Tol	tolyl
TOPPipU	2-[2-oxo-1,2-dihydropyridyl]-1,1,3,3-bis-(pentamethylene)uronium tetrafluoroborate
Tos	tosyl
TOTU	O-[(ethoxycarbonyl)cyanomethyleneamino]-1,1,3,3-tetramethyl-uronium tetrafluoroborate
TP	thymopoietin
TPST	tyrosine protein sulfotransferase
TP5	thymopentin
tPA	tissue plasminogen activator
TPTU	O-(1,2-dihydro-2-oxo-1-pyridyl)-N,N,N′,N′-tetramethyluronium tetrafluoroborate
t_R	retention time
TRCOSY	transferred rotational correlated NMR spectroscopy
TRF	time-resolved fluorescence
TRH	thyrotropin releasing hormone
tRNA	transfer RNA
Tris	tris(hydroxymethyl)aminomethane
TRNOE	transferred nuclear Overhauser effect
Troc	2,2,2-trichloroethoxycarbonyl
Trp	tryptophan
Trt	triphenylmethyl (trityl)
TS	thrombospondin
Tse	2-(4-toluenesulfonyl)ethyl
TSH	thyroid stimulating hormone (thyrotropin)
Tsoc	2-(4-toluenesulfonyl)ethoxycarbonyl
TSTU	2-(succinimido)-1,1,3,3-tetramethyluronium tetrafluoroborate
Tyr	tyrosine
UCN	urocortin
UF	ultrafiltration
uHTS	ultra-high throughput screening
UK	urokinase
UNCA	urethane protected α-amino acid N-carboxy anhydride
uPA	urokinase-type plasminogen activator

UT	urotensin
UV	ultraviolet
Val	valine
VIC	vasoactive intestinal contractor
VIP	vasoactive intestinal peptide
VLDL	very low density lipoprotein
Vn	vitronectin
VP	vasopressin
WSCI	water-soluble carbodiimide, e.g. [1-ethyl-3-(3′-dimethylaminopropyl)carbodiimide hydrochloride]
Xaa	any amino acid
Xan	9*H*-xanthen-9-yl
XRP	xenopsin-related peptide
Z	benzyloxycarbonyl
Z(2-Br)	2-bromobenzyloxycarbonyl
Z(2-Cl)	2-chlorobenzyloxycarbonyl
Z(Br)	4-bromobenzyloxycarbonyl
Z(NO_2)	4-nitrobenzyloxycarbonyl
Z(OAc)	acetoxybenzyloxycarbonyl
Z(OAcPh)	phenylacetoxybenzyloxycarbonyl
Z(OMe)	4-methoxybenzyloxycarbonyl
Z(SMe)	4-(methylsulfonyl)benzyloxycarbonyl
Zte	1-benzyloxycarbonylamino-2,2,2-trichloroethyl

1) The fragment name 7-azabenzotriazole is used for simplicity, despite the fact that the correct IUPAC nomenclature requires it to be named as triazolopyridine (cf. HATU)

2) Many benzotriazole and 7-azabenzotriazole-based uronium salts have been shown to exist as guanidium salts in solution. For simplicity, they still are named as uronium salts (cf. HBTU)

1
Introduction and Background

Peptide research has experienced considerable development during the past few decades. The progress in this important discipline of natural product chemistry is reflected in a flood of scientific data. The number of scientific publications per year increased from about 10000 in the year 1980 to presently more than 20000 papers. The introduction of new international scientific journals in this research area reflects this remarkable development.

A very useful bibliography on peptide research was published by John H. Jones [1]. The Houben-Weyl sampler volume E 22 "Synthesis of Peptides and Peptidomimetics" edited by Murray Goodman (Editor-in-Chief), Arthur Felix, Luis Moroder and Claudio Toniolo [2] represents the most actual and exhaustive general treatise in this field. This work is a tribute to the 100th aniversary of Emil Fischer's first synthesis of peptides and is the successor of two Houben-Weyl volumes in German language edited by Erich Wünsch in 1974 [3].

A number of very important physiological and biochemical functions of life are influenced by peptides. Peptides are involved as neurotransmitters, neuromodulators, and hormones in receptor-mediated signal transduction. More than 100 peptides with functions in the central and peripheral nervous systems, in immunological processes, in the cardiovascular system, and in the intestine are known. Peptides influence cell-cell communication upon interaction with receptors, and are involved in a number of biochemical processes, for example metabolism, pain, reproduction, and immune response.

The increasing knowledge of the manifold modes of action of bioactive peptides led to an increased interest of pharmacology and medical sciences in this class of compounds. The isolation and targeted application of these endogenous substances as potential intrinsic drugs is gaining importance for the treatment of pathologic processes. New therapeutic methods based on peptides for a series of diseases give rise to the hope that diseases, where peptides play a functional role, can be amenable to therapy.

Peptide chemistry considerable contributes to research in the life science area. Synthetic peptides serve as antigens to raise antibodies, as enzyme substrates to map the active site requirements of an enzyme under investigation, or as enzyme inhibitors to influence signaling pathways in biochemical research or pathologic processes in medical research. Peptide ligands immobilized to a solid matrix may facilitate specific protein purification. Protein-protein interaction can be manipulated by small synthetic peptides. The "peptide dissection approach" uses relatively short peptide fragments that are part of a protein sequence. The synthetic peptides are investigated for their ability to fold independently, with the aim to improve the knowledge on protein folding.

The isolation of peptides from natural sources often is problematic, however. In many cases, the concentration of peptide mediators ranges from 10^{-15} to 10^{-12} mol per mg fresh weight of tissue. Therefore, only highly sensitive assay methods such as immunohistochemical techniques render cellular localization possible. Although not all relevant bioactive peptides occur in such low concentrations, isolation methods generally suffer from disadvantages, such as the limited availability of human tissue sources. Complicated logistics during collection or storage of the corresponding organs, e.g., porcine or bovine pancreas for insulin production, additionally imposes difficulties on the utilization of natural sources. Possible contamination of tissue used for the isolation of therapeutic peptides and proteins with pathogenic viruses is an enormous health hazard. Factor VIII preparations for treatment of hemophilia patients isolated from natural sources have been contaminated with human immunodeficiency virus (HIV), while impure growth hormone preparations isolated from human hypophyses after autopsy have led to the transmission of central nerve system diseases (Creutzfeld-Jacob disease). Nowadays, many therapeutic peptides and proteins are produced by recombinant techniques. Immunological incompatibilities of peptide drugs obtained from animal sources have also been observed. Consequently, the development of processes for the synthesis of peptide drugs must be pursued with high priority.

Chemical peptide synthesis is the classical method which has been mainly developed during the past four decades, although the foundations were laid in the early 20th century by Theodor Curtius and Emil Fischer. Synthesis has often been the final structural proof of many peptides isolated only in minute amounts from natural sources.

The production of polypeptides and proteins by recombinant techniques has also contributed important progress in terms of methodology. Genetically engineered pharmaproteins verify the concept of therapy with endogenous protein drugs (endopharmaceuticals). Cardiovascular diseases, tumors, auto-immune diseases and infectious diseases are the most important indications. Classical peptide synthesis has, however, not been questioned by the emergence of these techniques. Small peptides, like the artificial sweetener aspartame (which has an annual production of more than 5000 tons) and peptides of medium size remain the objectives of classical synthesis, not to mention derivatives with non-proteinogenic amino acids or selectively labeled (^{13}C, ^{15}N) amino acid residues for structural investigations using nuclear magnetic resonance (NMR).

The demand for synthetic peptides in biological applications is steadily increasing. The new targets do not allow for an isolated position of peptide chemistry exclusively oriented toward synthesis. Modern interdisciplinary science and research require synthesis, analysis, isolation, structure determination, conformational analysis and molecular modeling as integrated components of a cooperation between biologists, biochemists, pharmacologists, medical scientists, biophysicists, and bioinformaticians. Studies on structure-activity relationships involve a large number of synthetic peptide analogues with sequence variation and the introduction of nonproteinogenic buildings blocks. The ingenious concept of solid-phase peptide synthesis has exerted considerable impact on the life sciences, whilst methods of combinatorial peptide synthesis allow for the simultaneous creation of peptide libraries which contain at least several hundreds of different peptides. The high yields and purities enable both *in-vitro* and *in-vivo* screening of biological activity to be carried out. Special techniques enable the creation of peptide libraries that contain several hundred thousands of peptides; these techniques offer an interesting approach in the screening of new lead structures in pharmaceutical developments.

Peptide drugs, however, can be applied therapeutically only to a limited extent because of their chemical and enzymatic labilities. Many peptides are inactive when applied orally, and even parenteral application (intravenous or subcutaneous injection) is often not efficient because proteolytic degradation occurs on the locus of the application. Application via mucous membranes (e.g., nasal) is promising. Despite the utilization of special depot formulations and new applications systems (computer-programmed minipump implants, iontophoretic methods, etc.) a major strategy in peptide chemistry is directed towards chemical modification in order to increase its chemical and enzymatic stability, to prolong the time of action, and to increase activity and selectivity towards the receptor.

The synthesis of analogues of bioactive peptides with unusual amino acid building blocks, linker or spacer molecules and modified peptide bonds is directed towards the development of potent agonists and antagonists of endogenous peptides. Once the amino acids of a protein that are essential for the specific biological mode of action have been revealed, these pharmacophoric groups may be incorporated into a small peptide. The development of orally active drugs is an important target. Rational drug design has contributed extensively in the development of protease-resistant structural variants of endogenous peptides, and in this context the incorporation of D-amino acids, the modification of covalent bonds, and the formation of ring structures (cyclopeptides) must be mentioned.

Peptidomimetics imitate bioactive peptides. The original peptide structure can hardly be recognized in these molecules, which induce a physiological effect by specific interaction with the corresponding receptor. Hence, a peptide structure may be transformed into a nonpeptide drug. This task is another timely challenge for peptide chemists, because only sufficient knowledge of the biologically active conformation of a peptide drug and of the interaction with the specific receptor enable the rational design of such peptide mimetics.

The variety of the tasks described herein renders peptide research an important and attractive discipline of modern life sciences. Despite the development of gene technology, peptide chemistry will have excellent future prospects because gene technology and peptide chemistry are complementary approaches.

References

1 J.H. Jones, *J. Pept. Sci.* **2000**, 6, 201.

2 M. Goodman, A. Felix, L. Moroder, C. Toniolo, *Synthesis of Peptides and Peptidomimetics* in Houben-Weyl-Methoden der organischen Chemie, Vol. E 22, K.H. Büchel (Ed.), Thieme, Stuttgart, **2002**.

3 E. Wünsch, *Synthese von Peptiden,* in Houben-Weyl-Methoden der organischen Chemie, Vol. 15, 1/2, E. Müller (Ed.), Thieme, Stuttgart, **1974**.

2
Fundamental Chemical and Structural Principles

2.1
Definitions and Main Conformational Features of the Peptide Bond

Peptides **1** formally are polymers of amino acids, connected by amide bonds (peptide bonds) between the carboxy group of one building block and the amino group of the following block.

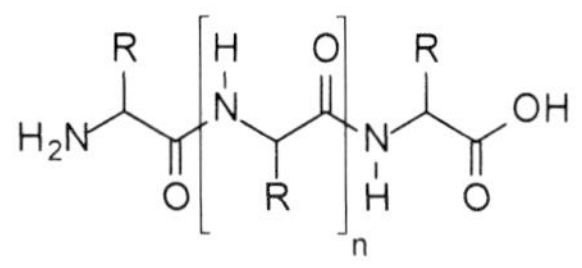

1

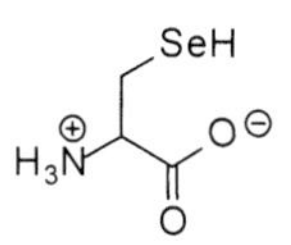

2

Natural peptides and proteins encoded by DNA usually contain 21 different α-amino acids (including the imino acid proline and the rare amino acid selenocysteine, **2**). The different side chains R of amino acids fundamentally contribute to their biochemical mode of action. A collection of the names, structures, three-letter code, and one-letter code abbreviations of these proteinogenic amino acids is given on the inside front cover of this book. Selenocysteine **2**, which is found both in prokaryotes and eukaryotes, is encoded by a special tRNA with the anticodon UCA recognizing UGA triplets on mRNA, and is incorporated into proteins by ribosomal synthesis. The UGA codon usually serves as a stop codon.

Besides the great variety of linear peptides there are cyclic peptides, macrocycles composed of amino acids, which occur in different ring sizes. Formally, cyclic peptides **3** are formed upon formation of a peptide bond between the amino and carboxy termini of a linear peptide.

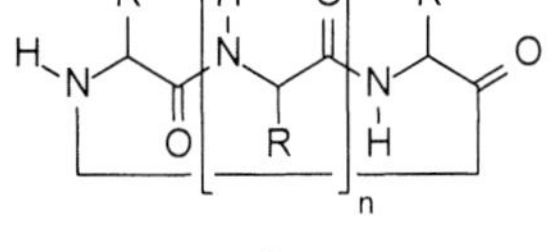

3

In 1951, Pauling and Corey proved by X-ray crystallography of amino acids, amino acid amides, and simple linear peptides that the C–N bond length in a peptide bond is shorter than a regular single bond. The resonance delocalization confers partial double bond character onto the C–N bond. The conformation of the peptide backbone is characterized by the three torsion angles φ [C(=O)–N–C$^{\alpha}$–C(=O)], ψ [N–C$^{\alpha}$–C(=O)–N], and ω [C$^{\alpha}$–C(=O)–N–C⁻], as depicted in Fig. 2.1.

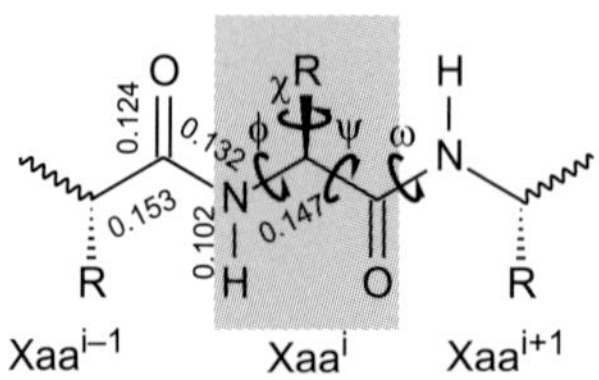

Fig. 2.1 Torsion angles φ, ψ, ω, and χ^1 and bond lengths of the amino acid Xaa^1 in a peptide.

The free rotation around the C–N amide bond is drastically restricted because of the partial double bond character with a rotational barrier of ~105 kJ mol^{-1}. Consequently, two rotamers of the peptide bond exist (Fig. 2.2): the *trans*-configured peptide bond (ω=180°) and the *cis*-configured peptide bond (ω=0°). The former is energetically favored by 8 kJ mol^{-1} and is found in most peptides that do not contain proline. In cases where the amide group of the imino acid proline is involved in a peptide bond, the energy of the *trans*-configured Xaa-Pro bond is increased. Consequently, the energy difference between the *cis* and *trans* isomers decreases.

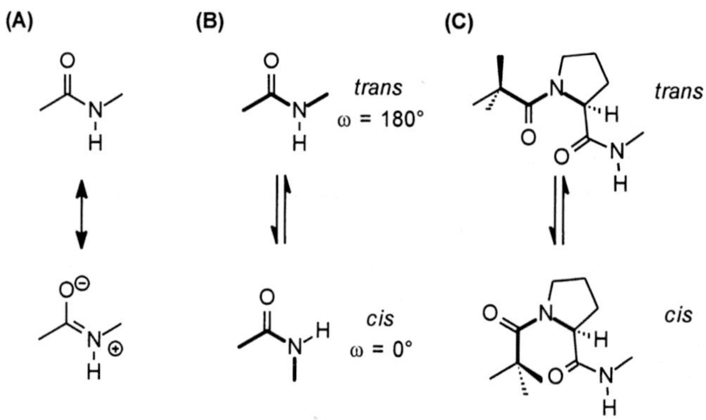

Fig. 2.2 (A) Resonance stabilization and (B) *cis/trans* isomerization of the peptide bond (C) *cis/trans* Isomers of a Xaa-Pro bond.

The percentage of *cis*-configured Xaa-Pro bonds (6.5%) is approximately two orders of magnitude higher compared to *cis* peptide bonds between all other amino acids (0.05% *cis*). However, several examples are known where a peptide bond configuration in proteins has been assigned erroneously to be *trans* in X-ray crystallographic studies. The *cis/trans* isomerization of peptide bonds involving the imino group of proline usually takes place in many proteins, and has a half-life between 10 and 1000 s. Peptidyl prolyl-*cis/trans* isomerases (PPIases) have been shown to accelerate significantly this conformational transition in cellular systems. These enzymes catalyze rotation around a C–N bond of the peptide moiety situated N-terminally to proline (Xaa-Pro). Hence, they catalyze a new type of enzymatic reaction which is of enormous importance for cellular functions [1].

Cis peptide bonds are present in the diketopiperazines **4**, which can be considered as cyclic dipeptides. Cyclic tripeptides with three *cis* peptide bonds are stable. As proline does not stabilize *trans*-configured peptide bonds, cyclo-$(Pro)_3$ and cyclo-(-Pro-Pro-Sar-) **5** can be synthesized.

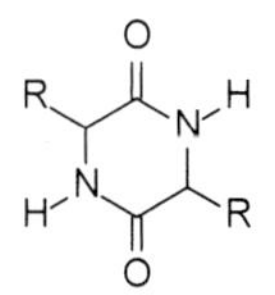

4

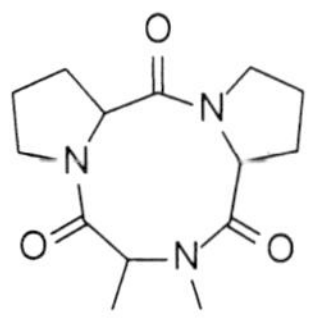

5

2.2
Building Blocks, Classification, and Nomenclature

Peptides are classified with Greek prefixes as di-, tri-, tetra-, penta-, … octa-, nona-, decapeptides, etc., according to the number of amino acid residues incorporated. In longer peptides, the Greek prefix may be replaced by Arabic figures; for example, a decapeptide may be called 10-peptide, while a dodecapeptide is called 12-peptide.

Formerly, peptides containing fewer than 10 amino acid residues were classified as oligopeptides (Greek oligos = few). Peptides with 10–100 amino acids residues were called polypeptides.

From a chemical point of view a differentiation between polypeptides and proteins is ambiguous. According to the currently accepted nomenclature rules, "oligopeptides" are composed of fewer than 15 amino acids, "polypeptides" contain approximately 15–50 amino acids residues, and the expression "protein" is used for derivatives containing more than 50 amino acids.

The nomenclature formally considers peptides as N-acyl amino acids. Only the amino acid residue at the carboxy terminus of the peptide chain keeps the original name without suffix, all others are used with the original name and the suffix -yl (Fig. 2.3). Consequently, peptide **6** is called alanyl-lysyl-glutamyl-tyrosyl-leucine.

A further simplification of a peptide formula is achieved by the three-letter code for amino acids (see inside cover). Linear peptide sequences usually are written horizontally, starting with the amino terminus on the left side and the carboxy terminus on the right side. When nothing is shown attached to either side of the three-letter symbol it should be understood that the amino group (always on the left) and carboxy group, respectively, are unmodified. This can be emphasized, e.g., Ala-Ala = H-Ala-Ala-OH. Indicating free termini by presenting the terminal group is wrong. H_2N-Ala-Ala-COOH implies a hydrazino group at one end and an α-keto acid derivative at the other. Representation of a free terminal carboxy group by writing H on the right is also wrong, because that implies a C-terminal aldehyde. Side chains are understood to be unsubstituted if nothing is shown, but a substituent can be indicated by use of brackets or attachment by a vertical bond up or down.

In peptide **6** alanine is the N-terminal amino acid and leucine the C-terminal amino acid. The three-letter code Ala-Lys-Glu-Tyr-Leu represents the pentapeptide **6** independent of the ionization state. If discrete ionic states of peptides should be emphasized, formulae **7**, **8**, or **9** may be used for the anion, the zwitterion, and the cation, respectively.

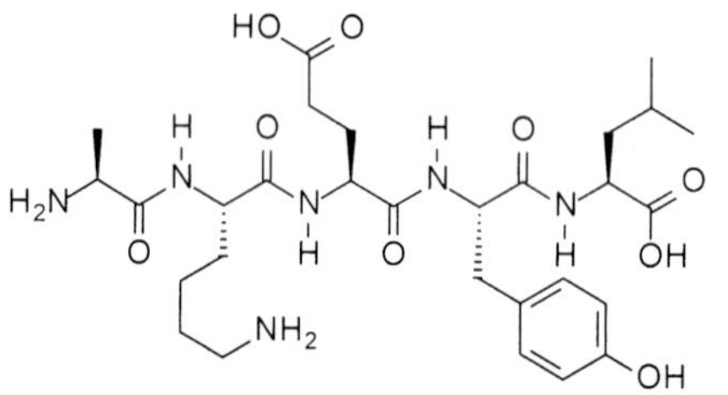

6

Alanyl—lysyl—glutamyl—tyrosyl—leucine

H–Ala–Lys–Glu–Tyr–Leu–OH **6** $\oplus H_2$–Ala–Lys–Glu–Tyr–Leu–$O^{\ominus}$ **8**

H–Ala–Lys–Glu–Tyr–Leu–$O^{\ominus}$ **7** $\oplus H_2$–Ala–Lys–Glu–Tyr–Leu–OH **9**

Fig. 2.3 Structural formula, nomenclature, and three-letter code of the hypothetical pentapeptide iupaciubicin **6** in different ionic states (**7–9**).

The three-letter code usually precludes that trifunctional amino acids with additional amino or carboxy functions located in the side chains (Lys, Glu, Asp) are connected by α-peptide bonds. This means, that the peptide bond is regularly formed between the C-1 (CO) of one amino acid and N-2 (N^{α}) of another amino acid. Special attention must be paid to abbreviations of isopeptide bonds in the three-letter code. The biochemically important peptide, glutathione **10** (Fig. 2.4) comprises, besides an α-peptide bond, also a γ-peptide bond. α-Glutamyl-lysine, N^{ε}-α-glutamyl-lysine, and N^{ε}-γ-glutamyl-lysine are constitution isomers of glutathione.

The side-chain substituents are displayed, if necessary, in the three-letter code by an abbreviation of the corresponding substituent which is displayed above or below the amino acid symbol, or in brackets immediately after the three-letter abbreviation. The pentapeptide derivative **11** serves as an example of this system of abbreviations, including protecting groups.

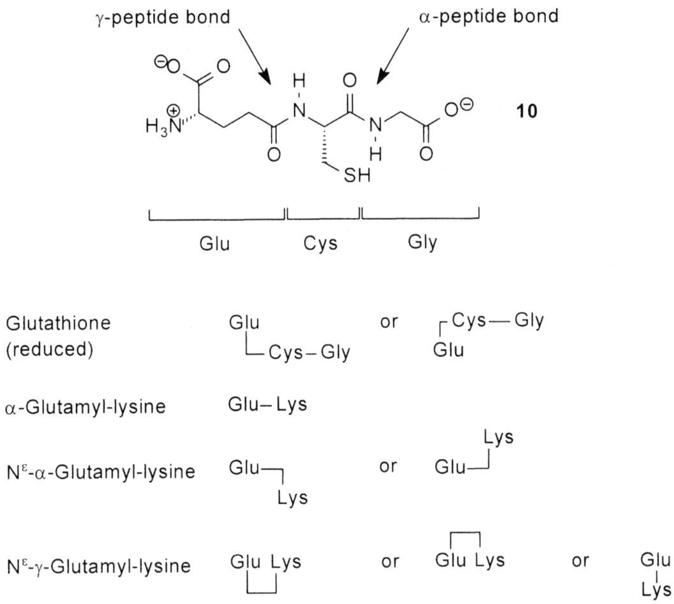

Fig. 2.4 Different connectivities of amino acids with additional side-chain functional groups lead to constitutional isomers, as exemplified for glutathione **10.**

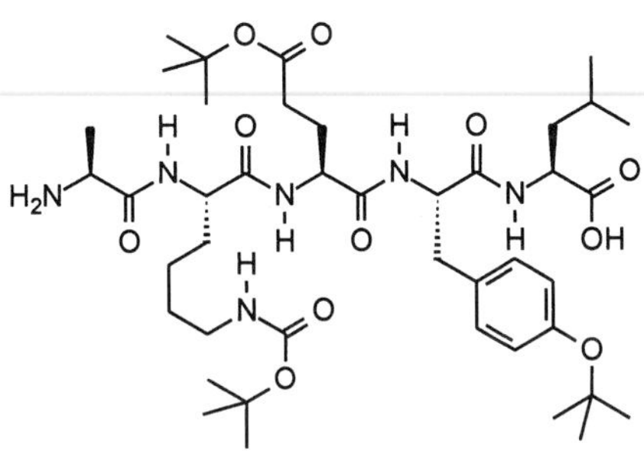

11

H–Ala—Lys(Boc)—Glu(OtBu)—Tyr(tBu)—Leu–OH

Note that atoms which belong integrally to an amino acid are not shown in the abbreviation. For example, the amino group in lysine (**11**) is not displayed in the three-letter code presentation. An abbreviation Lys(NHBoc) would imply a Boc-protected ε hydrazino group, and the abbreviation Tyr(OtBu) would imply a *tert*-butylperoxo substituent. The number and sequence of amino acids that are connected to a peptide or a protein are called the "primary structure". If the sequence of a peptide is completely known, the three-letter code symbols are listed sequentially, divided by a hyphen "–", which symbolizes the peptide bond. Notably, if the full peptide sequence is given, for instance Ala-Lys-Glu-Tyr-Leu, no dashes are written at the termini. However, if a partial sequence within a peptide chain, e.g., -Ala-Lys-Glu-Tyr-Leu- is written, additional dashes are added at both termini. If partial sequences of a peptide have not yet been elucidated, or if the presence of one amino acid in a special position has not been identified unambiguously, the tentative amino acid present in this position is given in brackets, separated by commas, as shown in **12**.

H–Ala–Lys–Glu–Tyr–Leu–(Tyr,Leu,Lys,Ala,Glu)–Leu–
Tyr–Glu–Lys–Ala–OH

12

Covalent bonds between side-chain functional groups are also possible for the amino acid cysteine. A disulfide bond between two thiol groups of cysteine is formed upon oxidation to give a cystine residue. Different types of disulfide bridges can be distinguished (cf. Section 6.2). Intramolecular (intrachain) disulfide bridges are formed between two cysteine residues in one peptide chain, while intermolecular (interchain) disulfide bonds are formed between two different peptide chains, as shown in **13**.

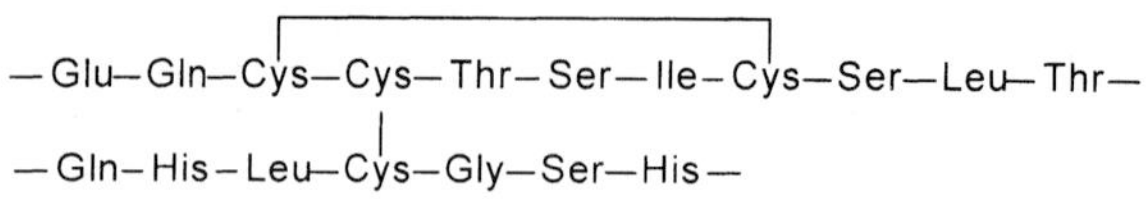

13

As nonproteinogenic building blocks such as hydroxycarboxylic acids, D-amino acids, and N-alkyl-amino acids may occur in peptides, an extension of the original definition for a peptide is required. Primarily, homomeric peptides and heteromeric peptides must be distinguished. While the former are composed exclusively of proteinogenic amino acids, the latter also contain nonproteinogenic building blocks.

Analogues of peptides, in which the –CO–NH-group that joins residues is replaced by another grouping, may be named by placing Greek psi (ψ), followed by the replacing group in parentheses, between the residue symbols where the change occurs. Examples:

Ala–[ψ](NH–CO)–Ala for H_3N^+–CHMe–NH–CO–CHMe–COO^-

Ala–[ψ](CH=CH, *trans*)–Ala for H_3N^+–CHMe–CH=CH–CHMe–COO^-

Further differentiation is made according to the nature of the chemical bond into homodetic peptides that contain exclusively peptide bonds (N$^\alpha$ amide bonds) and heterodetic peptides that may also contain ester, disulfide, or thioester bonds. The sequence of a cyclic homodetic homomeric peptide can be written in two different variants: following the prefix "cyclo-", the sequence is annotated in the three-letter code, set in brackets, and backbone cyclization is symbolized by a hyphen "–" before the first and after the last sequence position given. Gramicidin S can be written as shown in **14**.

cyclo–(–Val–Orn–Leu–D-Phe–Pro–Val–Orn–Leu–D-Phe–Pro–) **14**

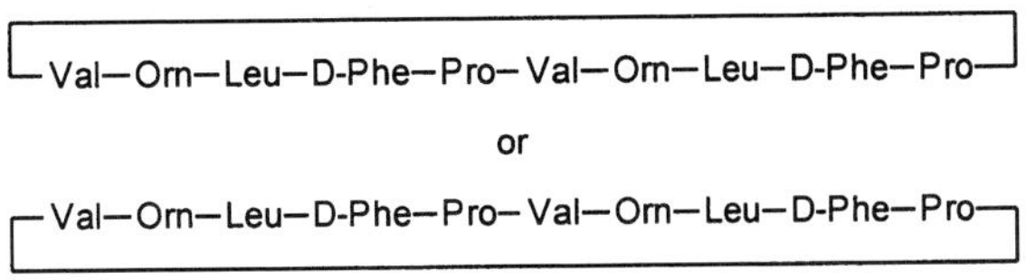

15

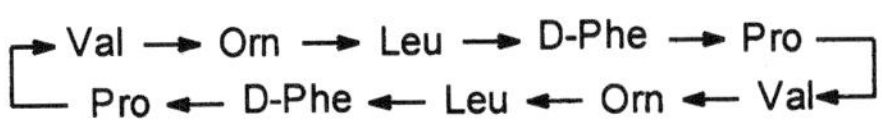

16

Alternatively, the sequence of the same peptide may be given as in **15**, where cyclization is symbolized by a line above or below the sequence. In the case of a peptide cyclization that occurs via the backbone, this should be symbolized by horizontal lines starting from the N- and C-termini. A third possibility to display a cyclic peptide is shown in **16**, where the direction of a peptide chain (N → C) has to be symbolized by an arrow (→) that points towards the amino acid located in the direction of the C-terminus. Cyclic heterodetic homomeric peptides are treated similarly in the three-letter annotation.

As analogues of biologically active peptides are very often synthesized in order to study the relationship between structure and activity, a brief introduction into the rules for the nomenclature of synthetic analogues of natural peptides should be given here in accordance with the suggestions by IUPAC-IUB Joint Commission on Biochemical Nomenclature [2]. The most important rules are detailed in Tab. 2.1 for the example of a hypothetical peptide with the name iupaciubin with the sequence Ala-Lys-Glu-Tyr-Leu.

2.3
Analysis of the Covalent Structure of Peptides and Proteins

According to Linderstrom-Lang, the structural description of proteins can be considered at four levels of organization (Fig. 2.5).

Primary structure, which is the subject of this chapter, comprises the number and sequence of amino acids connected consecutively by peptide bonds within the peptide chain. Secondary structure describes the local three-dimensional arrangement of the peptide backbone. Tertiary structure describes the three-dimensional structure or overall shape of a single peptide chain resulting from the intramolecular interactions between secondary structure elements. The term quaternary structure (not shown in Fig. 2.5) refers to the spatial arrangement of two or more polypeptide chains associated by noncovalent interactions, or in special cases linked by disulfide bonds, forming definite oligomer complexes. The term "domain" is applied to describe globular clusters within a protein molecule with more than ~200 amino acid residues.

Tab. 2.1 Important nomenclature rules for peptide analogues.

Description	*Name/Three-letter code*
1. The *exchange of amino acid residues* in a peptide is symbolized by the trivial name of the corresponding peptide preceded by the full name of the amino acid replacement and its position given in square brackets. Alternatively, the three-letter code may be used instead of the full name of the amino acid. Multiple exchange is treated analogously.	Ala^1-Lys^2-Glu^3-Phe^4-Leu^5 [4-phenylalanine]iupaciubin [Phe^4]iupaciubin
2. The *extension of a peptide* may occur N-terminally (A) as well as C-terminally (B). The modified name is generated according to the previously discussed rules.	(A) Arg-Ala^1-Lys-Glu-Tyr-Leu^5 Arginyl-iupaciubin (B) Ala^1-Lys-Glu-Tyr-Leu^5-Met Iupaciubyl-methionin
3. An *insertion of an additional amino acid residue* is indicated by the prefix "endo" in combination with the number of the sequence position.	Ala-Lys^2-Thr^{2a}-Glu^3-Tyr-Leu Endo-2a-threonine-iupaciubin Endo-Thr^{2a}-iupaciubin
4. The *omission of an amino acid residue* is symbolized by the prefix "des" and the position.	Ala-Lys^2-Tyr^4-Leu De-3-glutamic acid-iupaciubin De-Glu^3-iupaciubin
5. *Substitutions* on side-chain amino groups (A) or side-chain carboxy groups (B) are symbolized considering the general nomenclature rules.	(A) Val—┐ Ala-Lys-Glu-Tyr-Leu Nε^2-Valyl-iupaciubin Nε^2-Val-iupaciubin (B) ┌—Val Ala-Lys-Glu-Tyr-Leu N-(Iupaciubin-Cδ^3-yl)-valine Iupaciubin-Cδ^3-yl-Val
6. The nomenclature for *partial sequences* that are derived from peptides with a trivial name uses the trivial name, followed by the numbers of sequence positions of the first and last amino acid within the partial sequence in brackets.	Lys^2-Glu^3-Tyr^4 Iupaciubin-(2-4)-peptide

A pure, homogeneous compound is the precondition for structural or biochemical studies involving peptides. Pharmacologically active peptides for therapeutic application have to meet even more strict requirements. Before embarking on a discussion on structural analysis, it is useful to describe a few general methods that are specific for the separation and purification of peptides and proteins.

2.3.1 Separation and Purification

2.3.1.1 Separation Principles

Analysis and purification of naturally occurring peptides and of synthetic peptides relies on a series of separation techniques. In general, separation procedures are

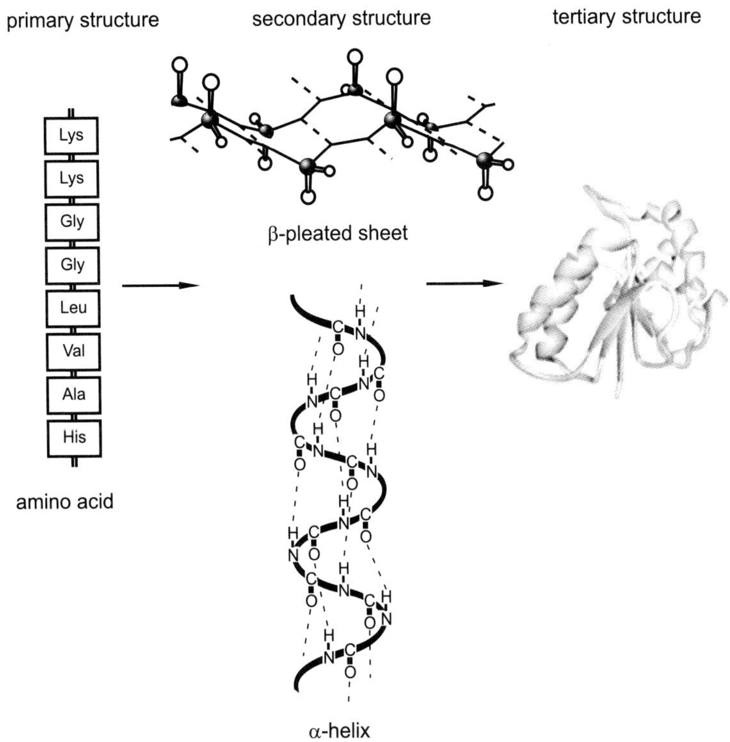

Fig. 2.5 Important peptide structures: Primary, secondary, and tertiary structures.

either directed to the preparative level, in order to isolate one or more individual components from a mixture for further investigations, or to the analytical level, with the goal of identifying and determining the relative amounts of some or all of the mixture components. In carrying out any preparative separation, studies on the analytical level are the initial steps to optimize separation conditions. Separation methods for peptide mixtures should provide samples of high purity, whereas analytical methods include not only the evaluation of the final product of peptide synthesis, but also the monitoring of intermediates with respect to chemical and stereochemical purity. Nearly all practical procedures in the peptide or protein field are based on separation of solute components. The partition of components between a solution and a solid surface is the most often exploited principle, but due to the existence of charged forms in aqueous solution ion-exchange and electrophoresis also belong to the dominant separation procedures, together with separation on the basis of molecular size. Additionally, adsorption chromatography (especially salt-promoted adsorption chromatography) has not lost its importance.

A recent comprehensive review by C. K. Larive et al. [3] focuses on selected applications of the separation and analysis of peptides and proteins published during 1997–1998. The review highlights the state of the art in this field. Selected

Tab. 2.2 Selected methods for separation and purification of peptides and proteins.

Method	***Remarks***
Reversed-phase HPLC (RP-HPLC)	Most popular HPLC variant used for separation of peptides and proteins; suitable for the assessment of the level of heterogeneity.
Ion-exchange chromatography (IEC)	Most commonly practised method for protein purification.
Size-exclusion chromatography (SEC)	Separation is solely based on molecular size; larger molecules elute more rapidly than smaller ones.
Affinity chromatography (AC)	A bioselective ligand chemically bound to an inert matrix retains the target component with selective affinity to the ligand.
Capillary electrophoresis (CE)	Separation of peptides and proteins is based on their differential migration in an electric field.
Multidimensional separations [4–6]	2D gel electrophoresis and multidimensional chromatography approaches are capable of more accurate quantification of the analyte, and are more compatible with mass spectrometry.
Ultrafiltration (UF) [7, 8]	Method for rapid concentration of protein solutions; lack of selectivity has severely restricted the use of UF for protein fractionation.
Two-phase systems for protein separation and purification [9, 10]	Hydrophobic partitioning of proteins in aqueous two-phase systems containing poly(ethylene glycol) and hydrophobically modified dextrans.

separation and purification methods are compiled in Tab. 2.2, and methods of structural analysis will be described in Section 2.5.

Thin-layer chromatography, the simplest technique for peptide analysis, is performed in various solvent systems and uses different detection systems, often followed by electrophoresis. In particular, free peptides can be examined by paper electrophoresis or by thin-layer electrophoresis, either in acidic solvent (dilute acetic acid) or under basic conditions. In the case of more than trace amounts of impurities, the product must be further purified prior to final evaluation. Chromatographic methods [11] and countercurrent distribution according to Craig [12] are the most commonly used methods for the purification of free peptides and of blocked intermediates. The classical crystallization procedure remains the simplest and most effective approach, but suffers from the low tendency of peptides to crystallize.

Normal-phase liquid chromatography [13] is used preferentially for the separation of hydrophilic peptides as they are very often not retained sufficiently on standard reverse-phase-high performance liquid chromatography (RP-HPLC) packing material. The introduction of HPLC both on an analytical and a preparative scale opened a new area in separation and analysis of peptides and proteins [14–18], and continues to be the method of choice [19–21]. Insoluble, hydrophilic support materials are used as the stationary phase in normal-phase HPLC. The major difference between "high-performance" or "high-pressure" chromatography and "low-pressure" bench-top chromatography is that HPLC employs columns and pumps that allow the application of very high pressures, with the advantage that particles with 3–10 μm mean diameter can be used as packing material for the columns. These conditions allow for superior resolution within relatively short times (minutes) for HPLC, compared to hours or even days for low-pressure chromatography systems. In RP-HPLC, the solid stationary phase is derivatized with nonpolar hydrophobic groups so that the elution conditions are the reverse of normal-phase liquid chromatography. Alkylsilanes with between four and 18 carbon atoms (referred to as C-4 to C-18 columns) are used for derivatization of the stationary phases. Retention of the solute occurs via hydrophobic interactions with the column support, and elution is accomplished by decreasing the ionic nature, or increasing the hydrophobicity of the eluant. Commercially available reversed-phase columns allow for rapid separation and detection of the components present in a mixture.

Ion-exchange chromatography (IEC) is easy to use for protein purification due to its high scale-up potential [22, 23]. The separation principle of IEC is based on interaction of the protein's net charge with the charged groups on the surface of the packing materials. Polystyrene and cellulose, as well as acrylamide and dextran polymers, serve as the preferred support materials for the ion exchanger. They are functionalized by quaternary amines, diethylaminoethyl (DEAE) or polyethylenimine for anion exchange, and sulfonate or carboxylate groups for cation exchange.

Hydrophobic interaction chromatography (HIC) is, besides thiophilic adsorption chromatography and electron donor-acceptor chromatography, the dominant method of salt-promoted adsorption chromatography [24].

Size-exclusion chromatography (SEC) [25–28] continues to be an efficient separation method for proteins, though in peptide purification its resolving power is somewhat limited. Although low-molecular weight impurities can be separated without problems from a crude mixture of peptides, the separation of a target peptide from a closely related peptide mixture is practicably not possible. In aqueous separation systems, SEC is also named gel filtration chromatography (GFC), whereas the alternative term gel permeation chromatography (GPC) is related to the application in nonaqueous separation systems.

In affinity chromatography, pioneered by Cuatrecasas [29] a low-molecular weight biospecific ligand is linked via a spacer to an inert, porous matrix, such as agarose gel, glass beads, polyacrylamide, or cross-linked dextrans. Monospecific ligands (analyte given in brackets) include hormones (receptors), antibodies (antigens), enzyme inhibitors (enzymes), or proteins (recombinant fusion proteins). Group-specific ligands (binding partner given in brackets) include, for example, lectins (glycoproteins), protein A and protein G (immunoglobulins G), calmodulin (Ca^{2+}-binding proteins), or dyes (enzymes). In general, affinity chromatography is useful if a high degree of specificity is required [30, 31], for example in the isolation of a target protein present in a low concentration in a biological fluid or a cellular extract. Affinity methods have become popular for biological recognition using peptide combinatorial libraries (see Chapter 8) to optimize affinity-based separations [32]. Antibodies are widely used to isolate and purify the corresponding antigen. In contrast, it is also possible to isolate specific antibodies using an immobilized antigen on a column.

Capillary electrophoresis (CE) or capillary zone electrophoresis (CZE), pioneered by Jorgenson and Lukacs [33], has been widely developed for the separation of peptides and proteins, including recombinant proteins [34, 35]. CE separates peptides and proteins based on their differential migration in solution in an electric field [36]. In accordance with ion-exchange HPLC, the separation is a function of the charge properties of the compound to be separated. However, the physical basis of separation is different in that it is performed in a typical capillary column with a dimension of 50 μm i.d. × 30–100 cm length and a volume of 0.6 2 μL, and the injection volume is limited to <20 nL. A common set-up for peptide separation uses a buffer at pH 2.0. All charged peptides are cationic and migrate to the cathode, and under these conditions the effect of electroendoosmosis is minimized. The mentioned limited sample injection volume in CE limits the detection of minor components in complex mixtures, and suitable preconcentration methods are often needed. Capillary isoelectric focusing (CIEF) can provide high-efficiency separations, and an improved detection sensitivity for complex peptide mixtures [37].

2.3.1.2 **Purification Techniques**

The application of a peptide or protein either for scientific investigations or commercial purposes usually requires isolation in a homogeneous state. Purification is defined as a process where the target molecule is separated from other compounds. Separation and purification formally are different categories, but in practice they overlap to a great extent. Therefore, methods described in the preceding section are included in purification schemes. The purity is an essential prerequisite both for structural analysis and for reproducible results in biological investigations. Homogeneity and correct covalent structure must be the main goals as peptides and proteins have become an important class of potent therapeutic drugs (see Chapter 9).

Purification procedures can take considerable amounts of time and effort. Unfortunately, evidence is provided in the literature that insufficient attention is often paid to the correct evaluation of the final product. Insufficiently pure compounds may lead to totally erroneous conclusions. Residues such as Trp, Met, N-terminal Gln, and Cys are known to cause potential problems from unwanted reactions during synthesis, work-up, handling, or storage. Cys, Met, and Trp are susceptible to oxidation. Only the oxidation of Cys and Met usually is reversible, which is generally not true for Trp. Gln in the N-terminal position of a peptide sequence is capable of forming a lactam structure, often termed pyroglutamate. In several publications, peptides have been claimed to be either crude or pure, but often different categories of purity are used, such as homogeneous, very pure, and chromatographically pure. Doubtlessly, the different evaluation criteria are applied to each of these categories. In practice, the extent of evaluation needed depends on the peptide's use. Application as an antigen in the production of polyclonal antisera requires a smaller degree of peptide purity, compared to a product used for conformational studies or special investigations on biological activity.

An accurate and specific assay for the component of interest is an essential prerequisite for peptide and protein purification. The quantitative result obtained from an assay is defined as the "activity". This term is normally used for enzyme assays, but it can also be applied to the results of other assays. Enzyme assays generally are based on a specific reaction catalyzed by the enzyme, whilst assays for other proteins rely on their physical properties. For example, the purification of proteins bearing highly chromogenic cofactors can be easily monitored using spectrophotometric methods. A highly specific assay uses antibodies against the peptide or protein. The specific activity is a quantitative measure for purity, and is defined as the ratio of the amount of activity in a protein solution to the amount of total protein in the solution. The specific activity increases during purification by removing contaminating proteins that lack assay activity. A protein is enriched during purification, and will be completely pure after attaining the highest specific activity for that protein.

Most protein-purification protocols involve multiple steps employing various techniques [38]. To obtain a cell-free solution of an intracellular protein, the cell must first be broken in order to release its contents. Insoluble particles, including membranes, can then be removed by centrifugation or other bulk methods. Soluble proteins are separated and purified from the resulting crude solution. The release of membrane-bound proteins from the membrane can sometimes be performed with the use of detergents.

The purification of bacterial proteins starts with the physical separation of the protein from other bacterial nonprotein cell constituents. Generally, proteins secreted into the extracellular medium can be separated and purified more easily. In the case of a protein solution, most purification protocols start with a precipitation step by means of alcohols, heat, or salt. Ammonium sulfate precipitations are often used, since this salt causes precipitation of many proteins without affecting their biological activity. Most purification methods subsequently involve various forms of chromatography.

Immobilized metal affinity chromatography (IMAC), pioneered by Porath et al. [39], continues to be very popular for the large-scale purification of proteins. Iminodiacetic acid or nitrilotriacetic acid (NTA), to which Ni^{2+} is typically chelated, are the most widely used support materials. IMAC is especially appropriate for recombinant proteins where the codons of (usually) six histidine residues (His-tag) are appended to the cDNA [40, 41]. Small-scale purification of proteins can easily performed by the attachment of NTA to magnetic beads [42].

Numerous applications of preparative chromatography are cited in reviews for the purification of both recombinant proteins [43] and unstable proteins [44]. Countercurrent chromatography has also been used for protein purification [45, 46]. Further information on protein purification has been summarized in several excellent monographs [38, 47–50].

A variety of methods has found application in the purification of potential peptide drugs, such as crystallization, countercurrent distribution, partition chromatography, GPC, low-pressure HIC, IEC and RP-HPLC.

As mentioned above, crystallization should be one of the most powerful purification methods, but its use is mainly limited to small oligopeptides up to pentapeptides. Countercurrent distribution and partition chromatography have now been largely substituted by more powerful purification procedures, such as preparative RP-HPLC. At present, most peptides with 50 residues or less can be purified using this technique. In practice, purification schemes are very often used which consist of two complementary techniques. For example, a crude synthetic peptide should first be liberated from by-products resulting from final deprotection steps which are mostly uncharged, low-molecular weight compounds. For this purpose, IEC, GPC, or HIC are the methods of choice. For a necessary final purification step, RP-HPLC should be applied as a complementary technique. Such a two-step purification scheme seems to be superior to the application of various RP-HPLC steps using different mobile phases.

2.3.1.3 **Stability Problems**

Peptides and proteins differ from most other chemical compounds. Their chemical [51], physical [52], and enzymatic [53] instability presents a challenge, for example in the development of more stable peptide and protein drugs. In most cases the pure compounds are obtained as amorphous solids, preferentially by lyophilization (freeze-drying). Generally, peptides and proteins tend to retain variable amounts of water and acids (as counterions or residual acid) resulting from the final stage of purification or isolation. In addition, peptides are preferably stored as their corresponding acetates or hydrochlorides. Special stability programs for bulk peptide drugs are performed by testing their storage behavior at various temperatures for different periods of time [54]. Decomposition or deterioration may occur as a result of oxidation, absorption or release of moisture, exposure to heat or light, or interactions with surfaces.

Solid-state formulations are generally more stable compared to the corresponding aqueous formulations. In the latter case, the nature of the solvent, concentration, pH, and temperature have a great influence on stability. Adsorption to the container, inactivation, racemization, oxidation, deamidation, chain cleavage, diketopiperazine formation, and rearrangements are the most well-known effects causing instability of peptides in solution. Surprisingly, several cases are known where protein stability in the solid state is less than or comparable to that observed in solution [55, 56]. In the solid state, chemical instability is caused similarly to that in solution (bond cleavage, bond formation, rearrangement or substitution). Typical reactions are deamidation of Asn and Gln [57], oxidation at sulfur atoms of Cys and Met, disulfide exchange at Cys, peptide bond cleavage, β-elimination, and dimerization/aggregation [58]. Peptides and proteins in food, as well as in pharmaceutical formulations, may undergo reactions with nonprotein compounds. Loss of Lys in food proteins is primarily attributed to the Maillard reaction

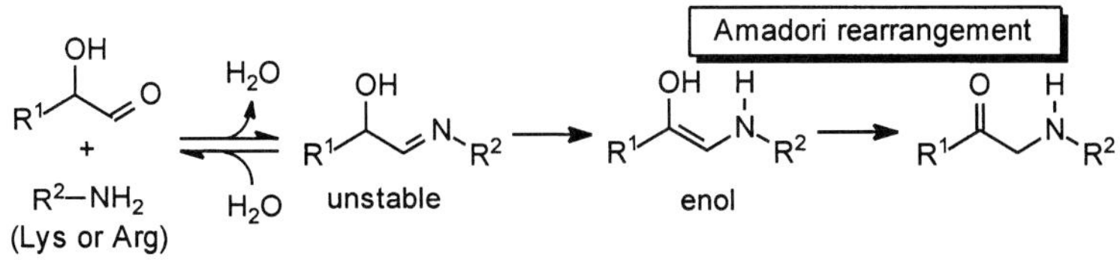

Fig. 2.6 The Amadori rearrangement as the initial step of the Maillard reaction.

(Fig. 2.6), but other amino acid residues such as Arg, Asn, and Gln are also capable of reacting with reducing sugars.

In the first step, a condensation reaction between the carbonyl group of a reducing sugar and an amino group forms an N-substituted glycosylamine; this is followed by conversion to a Schiff base. This may also be a problem in drug formulations. Subsequent cyclization and Amadori rearrangement yields the typical derivatives which cause discoloration of the formulation [59]. The reasons that these and other chemical degradation reactions can still occur in "dry" formulations is unclear. Moisture, temperature, and formulation excipients (e.g., polymers) are the main factors influencing peptide and protein chemical instability in the solid state. It has been suggested that proton activity in the solid state may have a similar influence on stability as does pH in the solution state. Details on the solid-state stability of proteins and peptides have been recently reviewed by Lai and Topp [60].

2.3.1.4 Evaluation of Homogeneity

After final purification, proof of homogeneity and structural characterization (see Section 2.5) are necessary to ascertain that the desired product, and not structural modifications of it, has been isolated. Numerous analytical methods are available to evaluate homogeneity and the covalent structure.

Fig. 2.7 shows schematically the relative contribution of analytical methods regarding the goals to assess homogeneity and structural integrity. Amino acid analysis (see Section 2.3.2.3) is necessary for the determination of the covalent struc-

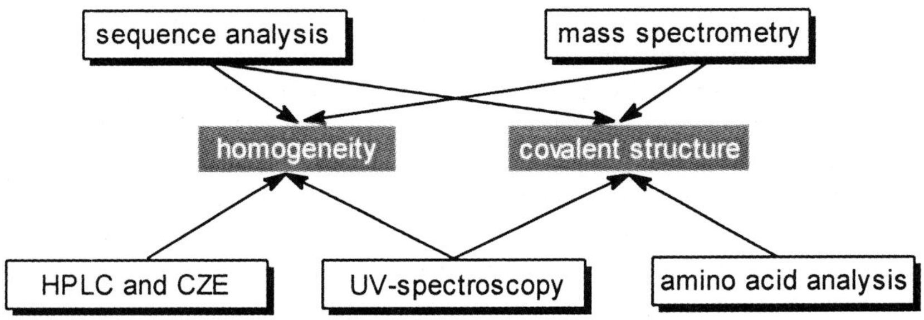

Fig. 2.7 Analytical techniques used for the determination of homogeneity and covalent structure of peptides.

ture, whereas sequence analysis (see Section 2.3.2.5) contributes to the evaluation of both homogeneity and covalent structure. Methods of structural analysis will be discussed in Section 2.5. Mass spectrometry with the dominant ionization methods matrix-assisted laser desorption/ionization (MALDI) and electrospray ionization (ESI), and the coupling of ESI to separation techniques, is an important tool in the analysis of peptides and proteins [61]. Nuclear magnetic resonance (NMR) spectroscopy (Section 2.5.3) is one of the premier analytical tools for structure elucidation. In particular, the use of multidimensional NMR methods for the structure determination of peptides and small proteins in solution has become routine. Ultraviolet spectroscopy (UV) is used to monitor the integrity of aromatic amino acids (particularly Trp), and is a routine tool for monitoring peptide purification. However, the information on the covalent structure of a peptide is very limited.

2.3.2 Primary Structure Determination

Amino acid sequence analysis is an essential component of protein structure determination [62–64]. The elucidation of the primary structure of insulin by Frederick Sanger in the early 1950s was a breakthrough of enormous biochemical significance and earned him the first of two Nobel prizes. Since that period, the amino acid sequences of several thousand proteins have been elucidated. Refinement and automation of the procedures nowadays allow for sequence analysis of proteins on the 10 pmol level, whereas the experimental work involved in determining the insulin structure by Sanger and his coworkers took more than a decade and required about 100 g insulin.

In 1980, Sanger earned his second Nobel prize for the development of the technology for rapid DNA sequencing. Nucleic acid sequencing initially lagged far behind the corresponding techniques for peptides and proteins. Sanger's chain termination method (or dideoxy method) [65] and the chemical cleavage method according to Maxam and Gilbert [66] proved to be efficient methods for DNA sequencing. Nowadays, DNA sequencing is easier and faster than protein sequencing; furthermore, this new technique allows even the sequence determination of proteins that have never been isolated. Despite this progress, direct peptide and protein sequencing remains an indispensable tool for several reasons. Doubtlessly, the determination of amino acid sequences by degradation methods is a conclusive necessity to confirm the structure of isolated and synthetic peptides. Furthermore, determination of the primary structure from the DNA sequence does not provide information on post-translational modifications of proteins and the location of disulfide bonds. It must also be mentioned that a common error in DNA sequencing is the inadvertent insertion or deletion of a single nucleotide which requires the need to identify the open reading frame in the DNA structure. In addition, it is often not easy to identify and isolate the nucleic acid that encodes the protein of interest. Determining the amino acid sequence of a portion of the protein allows the chemical synthesis of the corresponding DNA fragment that can be used to identify and isolate the whole gene.

2.3.2.1 End Group Analysis

Before starting sequence analysis it is necessary to determine the number of different peptide chains in a protein by the analysis of N-terminal and/or C-terminal residues. Of course, this problem may be complicated if the amino group at the N-terminus and the carboxy group at the C-terminus are chemically blocked, or the peptide to be analyzed is cyclic.

N-terminal analysis can be performed using several chemical and enzymatic methods. Chemical methods are mostly based on the transformation or blocking of the N-terminal α-amino function, followed by hydrolysis, separation and characterization of the terminal amino acid derivative (Fig. 2.8).

The reagent 2,4-dinitrofluorobenzene was first used by Sanger [67], and this dinitrophenyl (DNP) method (Fig. 2.8 A) has found widespread application. After la-

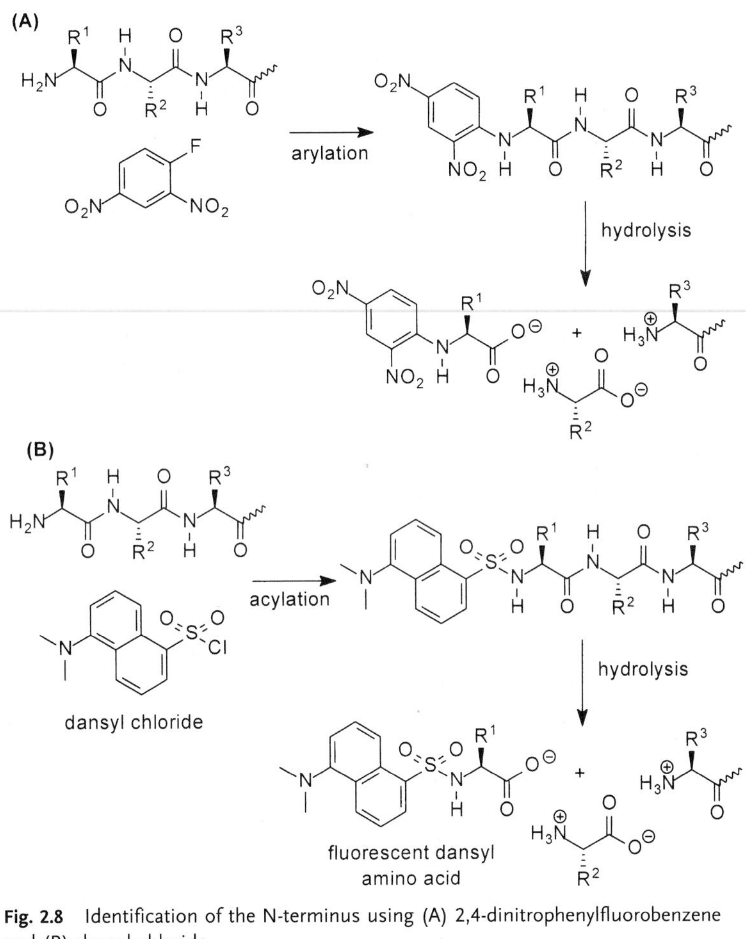

Fig. 2.8 Identification of the N-terminus using (A) 2,4-dinitrophenylfluorobenzene and (B) dansyl chloride.

beling the N-terminus of the peptide chains, complete hydrolysis liberates the DNP amino acids that can be identified by chromatography.

The dansyl (1-dimethylaminonaphthalene-5-sulfonyl) method [68] (Fig. 2.8 B) is more sensitive; dansyl chloride provides dansyl amino acids which are strongly fluorescent and can be identified chromatographically at the 100 pmol level. In addition, a number of less important chemical methods for blocking N-terminal amino acids, such as arylsulfonation, carbamoylation or carboxymethylation, and aminopeptidases as an enzymatic alternative can be used for N-terminal analysis. Enzymatic release of the N-terminal amino acid requires the presence of an unblocked α-amino function. Although porcine kidney leucyl aminopeptidase (LAP) hydrolyzes N-terminal L-amino acids (including glycine), peptide chains that bear L-leucyl residues are the preferred substrates [69]. Unfortunately, this enzyme does not release N-terminal arginine and lysine or any amino acid that is followed by proline. Furthermore, three aminopeptidases from *Bacillus stearothermophilus* [70] are suitable for the cleavage of N-terminal amino acid residues. Aminopeptidase I (API) shows a broad specificity; it releases not only neutral (preferentially aliphatic and aromatic), but also acidic and basic amino acids, including proline, from the N-terminus. Aminopeptidase N, which in 1980 was renamed to aminopeptidase M, shows a preference for neutral amino acids [71].

Without doubt, Edman degradation (see below) is one of the most powerful procedures for N-terminal residue identification.

C-terminal analysis can be performed both chemically and enzymatically, but has not gained as much importance as N-terminal procedures. Using the hydrazine method [72] according to Akabori, all amino acids (apart from the C-terminal ones) are converted into hydrazides on treatment with anhydrous hydrazine for 20–100 h at 90 °C in the presence of an acidic ion-exchange resin. Only the C-terminal amino acid residue is released as the free amino acid, and this can be identified chromatographically (Fig. 2.9).

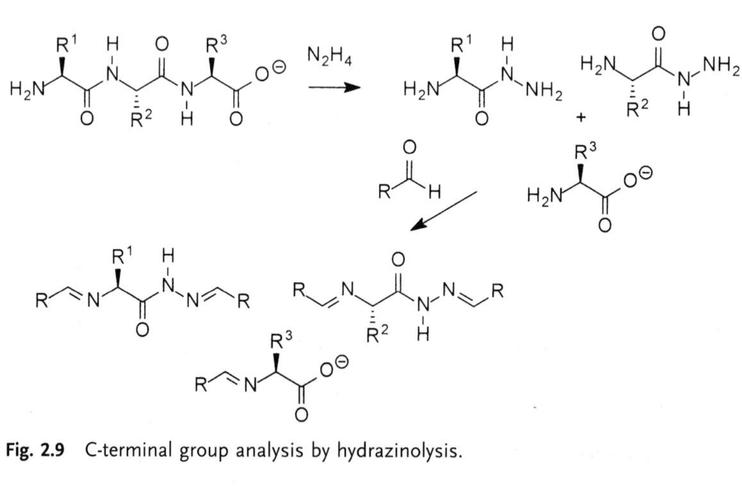

Fig. 2.9 C-terminal group analysis by hydrazinolysis.

C-terminal Cys, Gln, Asn, and Trp cannot be identified by this method, and Arg will be partly converted to Orn. Another procedure is based on the treatment of the peptide with ammonium thiocyanate and acetic anhydride to give 1-acyl-2-thiohydantoins by cyclization (cf. Section 2.3.2.6). The 2-thiohydantoin derived from the C-terminal amino acid is formed after mild alkaline hydrolysis in addition to the N-acylpeptide containing one residue less.

The carboxypeptidase (CP) approach is more efficient (Fig. 2.10), with CP A and CP B from bovine pancreas showing different, but complementary, specificities.

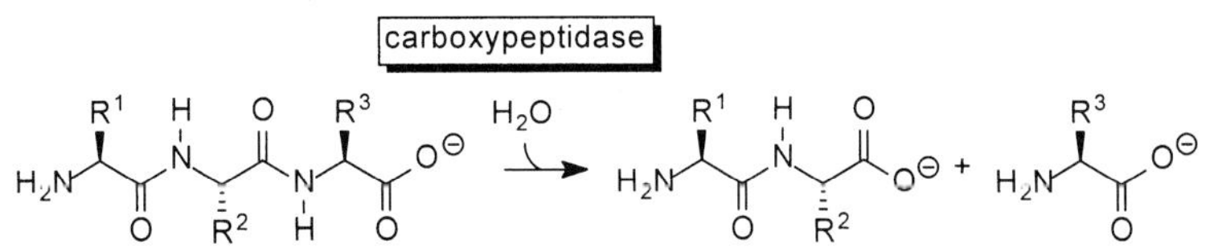

Fig. 2.10 C-terminal group analysis by carboxypeptidase.

CP A preferentially liberates C-terminal residues with aromatic side chains, but not Pro, Arg, Lys, and His, whereas CP B releases only C-terminal Arg, Lys, and His. The serine peptidase CP Y is much less specific and removes all residues, but Gly and Pro only very slowly.

2.3.2.2 **Cleavage of Disulfide Bonds**

Cleavage of disulfide bonds is necessary to separate disulfide-linked peptide chains, and also to destroy the native peptide or protein conformation in the case that it is stabilized by those bonds. The cleavage procedure involves hydrolysis of the protein under conditions that minimize the risk of disulfide bond exchange. Intra- or inter-chain disulfide bonds can be cleaved by reduction or oxidation (Fig. 2.11).

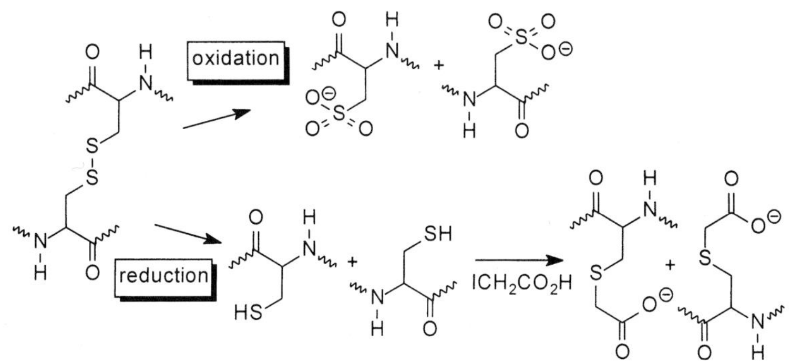

Fig. 2.11 Oxidative and reductive cleavage of disulfide bonds.

Oxidation using performic acid, also pioneered by Sanger, converts all Cys residues to cysteic acid residues. Since cysteic acid is stable both under acidic and basic conditions, it is possible to determine the total Cys content as cysteic acid. Unwanted oxidation of Met residues to methionine sulfoxide/sulfone, as well as partial degradation of the Trp side chain, are serious disadvantages. The reductive cleavage of disulfide bonds is mostly achieved by treatment with a large excess of a suitable thiol, e.g., 2-mercaptoethanol 1,4-dithiothreitol/1,4-dithioerythritol (Cleland's reagent). The resulting free thiol groups should be blocked by alkylation with iodoacetic acid in order to prevent reoxidation in air. The location of the disulfide bonds is carried out in the final step of sequence analysis.

2.3.2.3 **Analysis of Amino Acid Composition**

Before starting the sequence analysis of a peptide chain it is necessary to know its amino acid composition. This analysis [73, 74] can be performed by complete hydrolysis, followed by quantitative analysis of the liberated amino acids. For this purpose numerous chemical and enzymatic protocols are known, but none of these procedures alone is fully satisfactory. Besides hydrolysis with 6 M hydrochloric acid at 120 °C for 12 h, or with dilute alkali (2–4 M NaOH) at 100 °C for 4–8 h, mixtures of peptidases can also be used for complete peptide hydrolysis. Under acidic hydrolysis conditions, the polypeptide is dissolved in 6 M HCl and sealed in an evacuated tube to minimize the destruction of particular amino acids. Tryptophan and cysteine/cystine are especially sensitive towards oxygen. For the complete release of aliphatic amino acids, long hydrolysis times up to 100 h are sometimes required. Unfortunately, under such harsh conditions hydroxy-substituted amino acids (Ser, Thr, Tyr) are partially degraded, and Trp will be largely destroyed. Furthermore, Gln and Asn are converted into Glu and Asp plus NH_4^+, with the consequence that only the amounts of Asx (=Asn+Asp), Glx (=Glu+Gln), and NH_4^+ (=Asn+Gln) can be independently determined. Although Trp largely survives alkaline hydrolysis and, therefore, allows the content of this amino acid to be determined, this procedure causes decomposition particularly of Ser and Trp, and Arg and Cys might also be damaged. Pronase consisting of a mixture of relatively unspecific peptidases from *Streptomyces griseus* is often used for enzymatic hydrolysis. However, the amount of peptidase used should not be more than ~1% by weight of the polypeptide to be hydrolyzed. Otherwise, self-degrading by-products might contaminate the final digest. The enzymatic procedure is mostly used for the determination of Trp, Asn and Gln because of the reasons mentioned above.

Based on the pioneering work of Stein and Moore, amino acid analysis has been automated. Such automated equipment separates amino acids by IEC, and the Stein and Moore protocol employs post-column derivatization with ninhydrin to give the blue color of Ruhemann's violet (Fig. 2.12).

The intensity of the color can be measured spectrophotometrically, thus providing quantitative detection of each of the separated components. Nowadays, instruments are in use for quantitative amino acid analysis which are based on partition chromatography, such as HPLC [75, 76] and gas-liquid chromatography (GLC). The deriva-

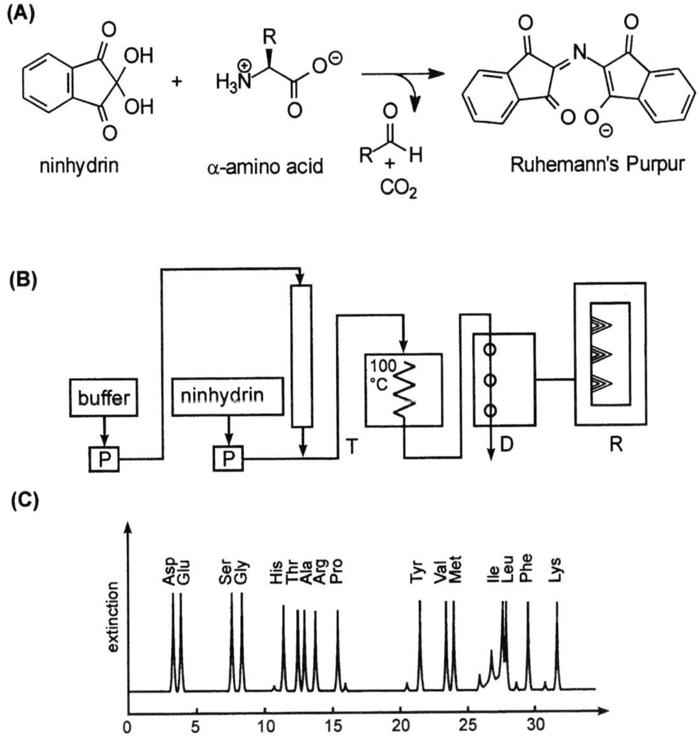

Fig. 2.12 Principle of automatic amino acid analysis. (A) Ninhydrin reaction. (B) Analyzer scheme. P=pump, T=Teflon tubing, D=detector, R=recorder. (C) Cation exchange HPLC amino acid analysis.

tization of amino acids is necessary to introduce fluorophoric groups which conveniently permit quantitation of UV/visible absorption or fluorescence characteristics for HPLC, or to convert amino acids into volatile derivatives for GLC. Various derivatization protocols for HPLC utilize fluorescent and UV-absorbing tags, e.g., dansyl, Fmoc, fluoresceamine, iso-indolyl [*o*-phthaldialdehyde (OPA) + 2-mercaptoethanol] condensation products, and Edman's reagent (see Section 2.3.2.5). The complete analysis of an amino acid mixture can be performed using a modern automated amino acid analyzer in < 60 min, with a sensitivity as low as 1 pmol of each amino acid.

2.3.2.4 **Selective Methods of Cleaving Peptide Bonds**

Efficient sequence analysis requires fragments of the protein to be analyzed, since stepwise degradation is limited to 40–80 residues (see Section 2.3.2.5). Therefore, longer polypeptide chains must be cleaved into fragments suitable for stepwise analysis by either enzymatic or chemical methods.

Several highly specific endopeptidases which allow for complete and specific fragmentation are used for enzymatic cleavages. Trypsin cleaves at the carboxy side of lysyl and arginyl peptide bonds, thereby providing fragments with C-terminal Lys or Arg residues. In principle, it is possible to restrict cleavage to arginyl bonds by blocking the Lys ε-amino group. In principle, selective blocking of the ε-amino function can be performed by any acylating agent, as trypsin recognizes its specific substrate by a positively charged side-chain function. Using citraconic anhydride or ethyl trifluoroacetate, the appropriate protecting group can be deblocked by morpholine or very mild acid treatment, and this allows exposure of the Lys side-chain function for a second tryptic hydrolysis. Interestingly, Cys may be aminoalkylated by a β-haloamine, e.g., 2-bromoethylamine, yielding a positively charged residue that enables tryptic cleavage. In contrast to trypsin, thrombin is more specific and only capable of cleaving a limited number of Arg bonds. Unfortunately, hydrolysis by this enzyme sometimes proceeds slowly and results in incomplete degradation of the substrate. Clostripain from the anaerobic bacterium *Clostridium histolyticum* is well known for its selective hydrolysis of arginyl bonds, but lysyl bonds are cleaved at a lower rate. The V8 protease from *Staphylococcus aureus* is highly specific for the cleavage of glutamyl bonds and is, therefore, widely used in sequence analysis. The less-specific chymotrypsin provides an alternative set of peptide fragments bearing aromatic or hydrophobic aliphatic amino acids at the C-terminus. In general, too many small peptides will be formed, and this does not facilitate determination of the primary structure of a protein. The same is also true for other less-specific endopeptidases, e.g., papain, subtilisin or pepsin, but these enzymes gained importance for isolating small peptide fragments containing disulfide bonds, phosphoserine or glycosyl moieties in their side chains.

For selective chemical cleavages, cyanogen bromide (BrCN) and *N*-bromosuccinimide are the favorite reagents in common use (Fig. 2.13). Cleavage with BrCN [77] under acidic conditions (0.1 M HCl or 70% formic acid) denatures the protein and causes the formation of a peptidyl homoserine lactone from Met residues under release of the aminoacyl peptide. It should be taken into consideration that the C-terminal homoserine lactone as a reactive species may be used to anchor the peptide to an insoluble support for solid-phase sequencing (Section 2.3.2.5).

Trp is, like Met, another amino acid that occurs only rarely in proteins. Cleavage of tryptophyl bonds will, therefore, also form larger fragments. *N*-Bromosuccinimide cleaves not only tryptophyl bonds but also tyrosyl bonds [78]. The use of either 2-(2-nitrophenylsulphenyl)-3-methyl-3-bromoindolenine, which is generated *in situ* from 2-(nitrophenylsulphenyl)-3-methyl-indole, *N*-bromosuccinimide [79], or 2-iososobenzoic acid [80] allows for a more selective cleavage of tryptophyl bonds.

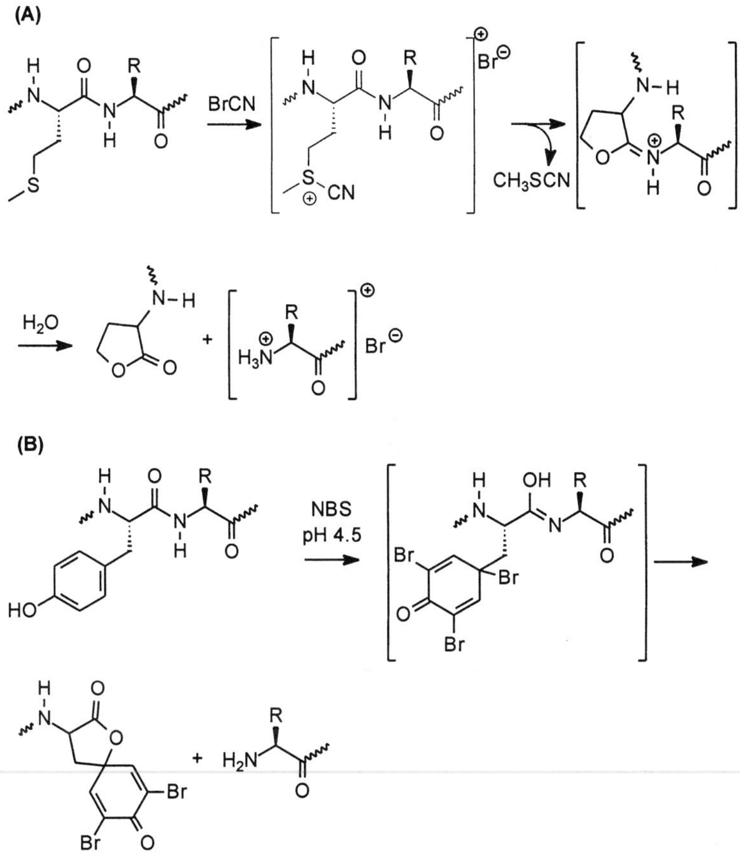

Fig. 2.13 Chemical sequence-specific cleavage of peptide bonds with (A) bromocyan (BrCN) and (B) *N*-bromosuccinimide (NBS).

2.3.2.5 N-Terminal Sequence Analysis (Edman Degradation)

The most efficient method for stepwise degradation of a peptide chain starting from the N-terminus was developed by Pehr Edman in 1949 [81]. This cyclic process consists of three steps: coupling, degradation, and conversion. Each cycle involves the reaction of the α-amino group at the N-terminus of a peptide chain with phenyl isothiocyanate (PITC) under slightly basic conditions, thereby forming the phenylthiocarbamoyl (PTC) adduct (Fig. 2.14).

Extractive removal of excess PITC is followed by treatment with an anhydrous strong acid such as trifluoroacetic acid to cleave the N-terminal amino acid to give its 2-anilino-5(4H)-thiazolone derivative, without attacking the rest of the peptide bonds within the chain. The 5(4H)-thiazolone derivative is extracted and undergoes rearrangement in hot trifluoroacetic acid to give the more stable 3-phenyl-2-thiohydantoin (PTH). The PTH-amino acids can be identified by TLC on silica gel, by RP-HPLC on a column with UV detection (2-anilino-5(4H)-thiazolone can

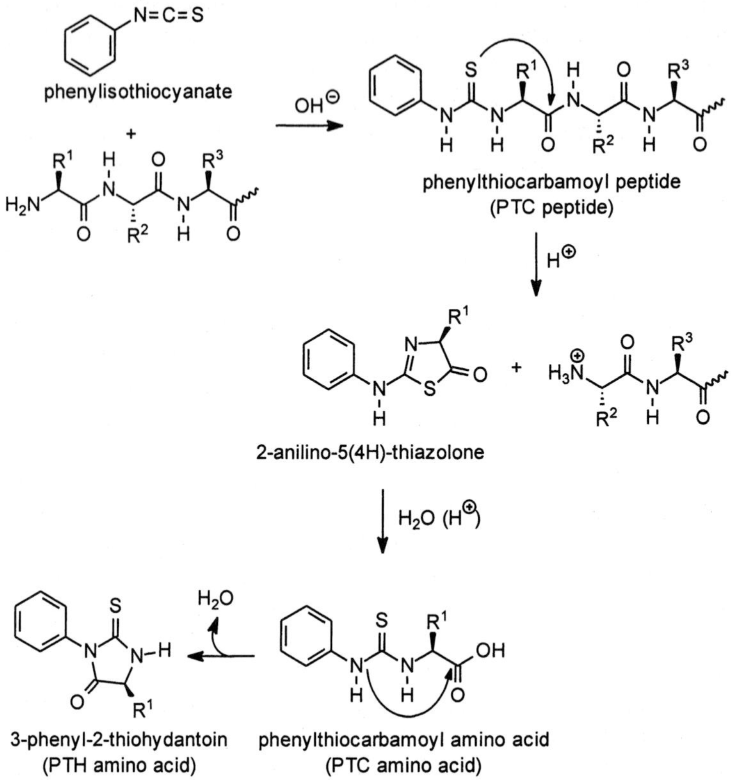

Fig. 2.14 The Edman method for N-terminal stepwise peptide degradation.

be identified and quantified similarly), or by GLC also in combination with MS. In order to obtain good results, precisely standardized conditions are required. The repetitive yield is defined as the total yield of one cycle. Unfortunately, the yield of PTH cannot achieve 100% at every step because of incomplete reactions and losses during the manipulations. Consequently, small amounts of PTHs corresponding to preceding sequence positions will be formed. At a stage when the yield of PTH from the newly exposed N-terminus is very low and the number of PTHs resulting from incompletely degraded chains is very high, interpretation of the results is invalid.

The Edman degradation has been improved by the development of the spinning-cup or liquid-phase sequencer [82]. Using this equipment, the analysis material is spread by centrifugal forces as a thin film on the inner wall of a spinning cup. The reagents and solvents are fed automatically to the bottom of the glass cup through a valve assembly and a special feeding line. The extracting solvents reach a groove in the cup, from where they are removed and leave the cup through an effluent line. All operations such as dissolution, concentration, drying under vacuum and extraction are carried out in the same system. The truncated peptide remains in the reaction vessel. Automated Edman degradation achieves repetitive yields of >90%. Nowadays "normal" repetitive yields of approximately 95% allow 30–40 cycles to be performed, with reliable results.

The amount of material required for analysis can be significantly reduced by using a polymeric quaternary ammonium salt, called polybrene. This adheres strongly both to the analyte and to glass, and results in a form of immobilization of the peptide material.

Another breakthrough in sequence analysis was the development of the gas phase sequencer by Hewick et al. [83] in 1981. Using this simple principle, the amount of material used for analysis could be reduced even further. A chemically inert disc made of glass-fiber, sometimes coated with polybrene, is used for the application of the analysis sample. Exact quantities of basic and acidic reagents, respectively, are delivered as a vapor in a stream of argon or nitrogen, and then added to the reaction cell at programmed times. Under these conditions the peptide loss can be significantly minimized, and such an instrument is capable of processing up to one residue per hour. The thiazolinone derivative is automatically removed and converted to the PTH derivative. The pulsed-liquid sequencer is a variant of the gas-phase sequencer. The acid is delivered as a liquid for very rapid degradation, which requires an accurately measured quantity sufficient to moisten the protein sample and to prevent it from being washed out. Such a procedure shortens the cycle by up to 30 min. Modern sequencers need, on average, as little as 10 pmol peptide or protein, though this must be free of salts and detergents.

The solid-phase sequencer, described by Laursen et al. [84], is based on a type of "reversed Merrifield technique" (see Section 4.5). The peptide is covalently immobilized on a suitable support, and this allows for simple separation of excess phenyl isothiocyanate and 2-anilino-5(4H)-thiazolone. Initially, aminopolystyrene was applied as the matrix, but neither this material nor polyacrylamide could fulfil the requirements for optimum solid supports. Controlled-pore glass treated with 3-aminopropyltriethoxysilane displays improved properties for covalent attachment of peptides to the free amino groups of the support by different coupling methods. Despite new attempts to use improved coupling methods for covalent fixation of the peptide to membranes, the solid-phase sequencing principle has not yet achieved the value of the gas-phase sequencer.

2.3.2.6 **C-Terminal Sequence Analysis**

The identification of C-terminal sequences is a useful complementation to the Edman degradation procedure, especially in the investigation of N-terminally blocked peptides and proteins. Furthermore, C-terminal sequence information is of general interest for the control of recombinant protein products, for identification of post-translational truncations, for the confirmation of DNA sequence data, and for assisting the design of oligonucleotide probes for molecular cloning studies.

According to Schlack and Kumpf [85], the treatment of a N-acyl peptide with ammonium thiocyanate and acetic anhydride leads to cyclization at the C-termi-

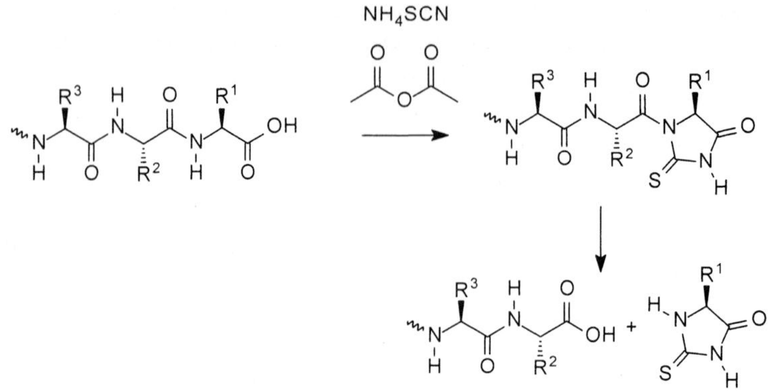

Fig. 2.15 The Schlack-Kumpf method for C-terminal stepwise peptide degradation.

nus, forming 1-acyl-2-thiohydantoins. Mild alkaline hydrolysis yields the 2-thiohydantoin corresponding to the C-terminal amino acid residue and the N-acyl peptide containing one amino acid residue less (Fig. 2.15).

Further developed instruments [86] with improved chemistry for the activation and cleavage steps [87, 88] are now in use for C-terminal sequencing. Compared to Edman chemistry, C-terminal degradation techniques are less efficient and give rise to various side reactions. In particular, the yields are significantly lower than those obtained by Edman degradation. Proline normally stops C-terminal degradation and requires special derivatization. Using a C-terminal sequencer in which most parameters have been modified has resulted in improved sensitivity, length of degradation, Pro passage, and combined application with Edman degradation [89]. C-Terminal sequence analysis is now also possible even on the 10 pmol level. Efficient combination of N- and C-terminal sequencing using the same analysis sample and degradation of more than 10 residues is feasible. The present standard should allow routine analysis of most proteins, and hence C-terminal sequence analysis has become a useful complement to N-terminal degradation and MS.

Enzymatic degradation by carboxypeptidase is an attractive method for C-terminal sequencing, both by determination of the released amino acids and by identifying the truncated peptides with mass spectrometry.

2.3.2.7 **Mass Spectrometry**

Mass spectrometry has proved to be a very useful tool in the analysis of peptides and proteins [90], as well as in the rapidly developing area of proteome analysis [91]. MALDI [92] and ESI [93] currently are the dominant methods for ionization of biomacromolecules. The latter technique may be also be coupled to separation techniques such as HPLC and capillary electrophoresis (CE). The different techniques of biomolecular MS enable an exact determination to be made of the molecular mass of a peptide or protein, and with high sensitivity and resolution.

Nanospray techniques have subsequently pushed the detection limits for peptide sequencing into the attomolar range [94]. The advances of the genome-sequencing projects and bioinformatics methods have also changed the position of MS, which is now becoming a deeply integrated tool in proteomics. Proteins separated on two-dimensional polyacrylamide gels can be identified at the subpicomolar level [95]. In this approach, the protein present in the gel is digested by treatment with a protease (e.g., trypsin) and the resulting peptide fragments (peptide fingerprint, peptide mass map) are identified by searching against databases. MALDI peptide mapping operates efficiently when a statistically significant number of peptide peaks is detected and assigned unambiguously in the digest. Alternatively, nanoelectrospray MS can be applied if MALDI peptide mapping fails to identify the protein.

In particular, MALDI-MS and ESI-MS offer an alternative to the classical Edman peptide sequencing. The soft desorption of these techniques does not induce fragmentation processes, and allows the transfer of high-molecular mass polypeptide ions into the gas phase [96, 97]. MALDI often is combined with a very sensitive time-of flight (ToF) detector. As a rule, short peptides can be directly sequenced, for example by MALDI-ToF techniques via post-source decay (PSD), while preceding cleavage to give suitable fragments is an imperative prerequisite for proteins and longer polypeptides. As indicated above (Section 2.3.2.6), either the released amino acid or the truncated peptides obtained from C-terminal sequencing using carboxypeptidase can be identified by MS coupled with various instrumental variations such as fast-atom bombardment [98], plasma desorption [99], ESI [100], or MALDI-MS.

Tandem mass spectrometry (MS/MS, or MS^n) was originally developed for the analysis of the molecular structure of single ions. These ions are selected by mass and further fragmented by employing collision-induced dissociation (CID) and subsequent analysis of the resulting fragments. MS/MS is routinely employed to determine the site and nature of a modification (post-translational or chemical). Targeted chemical modification of proteins may be detected by MS, and has been utilized as a probe of protein secondary structure or for the characterization of active sites. Coupling to high-performance separation techniques allows quantitation. MS/MS is a suitable method for the sequence analysis of a peptide, and is proving indispensable for proteome analysis. Peptides are prone to fragment at amide bonds after low-energy collisions, and this results in a predictable fragmentation pattern. Consequently, sequence information can be obtained by this technique because most amino acids have unique masses. Only the pairs of leucine/isoleucine and lysine/glutamine cannot be distinguished. Chemically modified amino acids have in most cases a molecular mass that is not identical with one of the naturally occurring amino acids and hence can be readily identified in the fragmentation pattern. Furthermore, disulfide bond assignment in proteins is possible using MALDI/MS.

The application of an ion-trap instrument for MS^3 experiments and a computer algorithm for automated data analysis have led to a novel concept of two-dimensional fragment correlation MS and its application in peptide sequencing [101].

Even though the daughter ion (MS^2) spectrum of a peptide contains the sequence information of the peptide, it is very difficult to decipher the MS^2 spectrum due to the difficulty in distinguishing the N-terminal fragments from the C-terminal fragments. This problem can be solved by taking a grand-daughter ion (MS^3) spectrum of a particular daughter ion, since all fragment ions of the opposite terminus are eliminated in this spectrum. The sequence of the peptide can be derived from a two-dimensional plot of the MS^2 spectrum versus the intersection spectra (2-D fragment correlation mass spectrum). Using this technique, about 78% of the sequence of a tryptic digest of cytochrome c could be determined. Interestingly, this *de-novo* sequencing approach works with complex mixtures, does not require any additional wet chemistry step, and should be fully automated in the near future.

2.3.2.8 **Peptide Ladder Sequencing**

Peptide ladder sequencing combines ladder-generating chemistry and MS. The principle of peptide ladder sequencing based on Edman degradation with MALDI-MS was first developed by Chait et al. [102] in 1993.

N-terminal ladder sequencing requires a modified Edman procedure in which in every step the peptide is incompletely degraded to yield continuously a mixture of one amino acid-shortened peptides. Such a peptide ladder is formed when Edman degradation with PITC (cf. Fig. 2.14) is performed in the presence of 5% phenyl isocyanate (PIC) as terminating reagent.

The phenylthiocarbamoyl peptide is cleaved in the presence of a strong acid to give the 5(4H)-thiazolone, while the phenylcarbamoyl peptide, formed from PIC, is stable under these conditions. During the next cycles the content of phenylcarbamoyl peptides (PC peptides) increases to a statistical mixture, thereby forming the peptide ladder. Analysis of the mixture using MALDI-MS allows for direct sequence determination from the successive mass differences of the peptide ladder (Fig. 2.16).

Since, contrary to the classical Edman procedure, quantitative derivatization is not necessary, one degradation cycle can be performed within approximately 5 min. Furthermore, application of the volatile trifluoroethyl isothiocyanate resulted in a significant optimization of this procedure and allowing for peptide sequencing at the femtomole level [103]. The equal masses of Leu and Ile, and the small mass differences between Glu/Gln and Asp/Asn, represent serious problems. The latter requires sufficient mass resolution, provided for example by modern MALDI-ToF spectrometers in a range up to 5000 daltons.

C-terminal ladder sequencing is based on the same principle, and initially was combined with C-terminal sequence analysis after carboxypeptidase treatment (Section 2.3.2.6). This procedure was applied both by time-dependent [104,105] and concentration-dependent [106] CP digestion. Another C-terminal ladder sequencing approach uses Schlack-Kumpf chemistry (Section 2.3.2.6) coupled with MALDI-MS analysis of the truncated peptides [107]. This chemical one-pot degradation procedure is applied without purification steps, and no repetitions must be

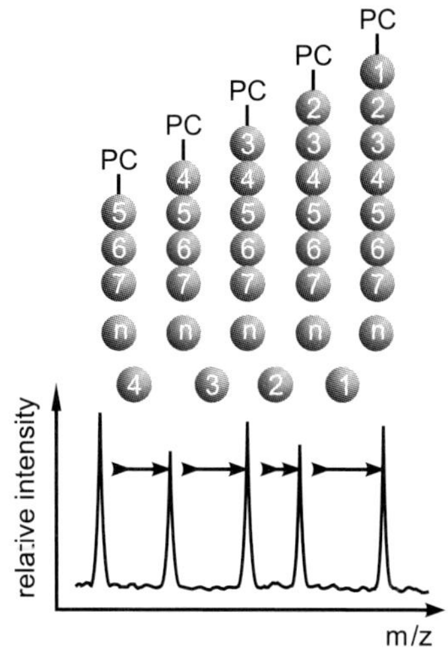

Fig. 2.16 Derivation of the amino acid sequence from the peptide ladder from mass differences analyzed by MALDI-MS. PC = phenylcarbamoyl.

performed to obtain ladders of C-terminal fragments – so-called ragged ends. It could be demonstrated that the 20 common proteinogenic amino acid residues are compatible with this technique, but only up to eight residues of C-terminal sequences can be determined. MALDI instruments with delayed extraction [108, 109] allow for the discrimination of all amino acids except Leu and Ile. Lys and Gln, having the same mass, can be distinguished after chemical acetylation with the formation of N^{ε}-acetyllysine.

2.3.2.9 **Assignment of Disulfide Bonds and Peptide Fragment Ordering**

In order to determine disulfide bond positions present in a peptide or protein, the latter is hydrolyzed under conditions such that the risk of disulfide exchange is minimized. The classical approach involves enzymatic cleavage using peptidases of low specificity, e.g., thermolysin or pepsin. The pairs of proteolytic fragments linked by disulfide bonds are then separated by diagonal electrophoresis. This technique includes electrophoretic separation in two dimensions using identical conditions. After electrophoresis in the first dimension, the electropherogram is exposed to performic acid vapor in order to oxidize cystine residues to cysteic acid (see Section 2.3.2.2). Electrophoresis in the second dimension is then carried out under the same conditions. Those peptide fragments not modified by performic acid are located on the diagonal of the electropherogram, because their migration is similar in both directions. In contrast, cystine-containing peptide fragments will be oxidatively cleaved to give two new peptides that occur off-diagonally. Normally, fragments with an intrachain disulfide give only one new product, whereas fragments connected by interchain disulfide bonds are transformed to two new peptide products. After isolation of the parent disulfide-linked peptide fragment, the disulfide bond is cleaved, followed by alkylation and fragment sequence analysis. Alternative procedures are based on RP-HPLC and MS.

The strategy for sequence determination of a peptide or protein subsequent to enzymatic cleavage requires the correct ordering of the peptide fragments to be sequenced individually. This can be performed by comparing the sequences of two sets of overlapping peptide fragments that result from polypeptide cleavage with different sequence specificities (Section 2.3.2.4). The principle is shown in Tab. 2.3.

Amino acid analysis in the case of Tab. 2.3 gives the result that the polypeptide consists of 30 amino acids. The peptide has Phe at its N-terminus, as determined by Sanger's end-group analysis. Since the absence of Met excludes selective cleavage with BrCN (Section 2.3.2.4), enzymatic cleavage can be performed with chymotrypsin and trypsin, respectively. The pattern of five overlapping fragments leads to the conclusion that the original sequence corresponds to the B chain of human insulin.

Tab. 2.3 Determination of the primary structure of a polypeptide by comparing the sequences of two sets of overlapping peptide fragments.

Analytical steps	***Conclusion***
Amino acid analysis	
Ala 1 Gln 1 Leu 4 Ser 1 Arg 1 Glu 2 Lys 1 Thr 2 Asn 1 Gly 3 Phe 3 Tyr 2 Cys 2 His 2 Pro 1 Val 3	The polypeptide consists of 30 amino acids.
Determination of terminal groups by Sanger	
Phe	Phe is the N-terminal amino acid residue.
Enzymatic cleavage	
Cleavage with chymotrypsin A Val-Asn-Gln-His-Leu-Cys-Gly-Ser-His-Leu-Val-Glu-Ala-Leu-Tyr B Leu-Val-Cys-Gly-Glu-Arg-Gly-Phe C Thr-Pro-Lys-Thr D 2 Phe, 1 Tyr Cleavage with trypsin E Phe-Val-Asn-Gln-His-Leu-Cys-Gly-Ser-His-Leu-Val-Glu-Ala- Leu-Tyr-Leu-Val-Cys-Gly-Glu-Arg F Gly-Phe-Phe-Tyr-Thr-Pro-Lys G Thr	Fragment C forms the C-terminus of the original peptide, because it does not contain either an aromatic amino acid (chymotrypsin cleavage) or a Lys/Arg residue (trypsin cleavage) at the C-terminus.
Sequence analysis and elucidation	
A B C FVNQHLCGSHLVEALYLVCGERGFFYTPKT E F	The polypeptide has the sequence of human insulin B chain.

2.3.2.10 **Location of Post-Translational Modifications and Bound Cofactors**

Post-translationally modified proteins (see Section 3.2.2) and those with permanently associated prosthetic groups fulfil important functions in biochemical pathways. In order to determine the location of modified amino acids, the protein must be degraded under conditions that are similar to those described for assignment of disulfide bonds (Section 2.3.2.9). Especially mild conditions are required to detect, for example, γ-carboxyglutamic acid (Gla) in various proteins as it is decarboxylated very easily under acidic conditions to give glutamic acid. Many proteins (and especially enzymes) contain covalently bound nonpeptide cofactors, e.g., N^{ε}-lipoyllysine in dihydrolipoyl transacetylase, 8α-histidylflavin in succinate dehydrogenase, and biotin linked to the enzyme via the ε-amino group of a lysine residue. Subsequent to specific peptide chain cleavage reactions (as mentioned above), the resulting fragments containing the modified building block may be efficiently analyzed using various MS-based techniques (Section 2.3.2.7).

Nonenzymatic post-translational modifications cause protein degradation both *in vivo* as well as *in vitro*. The formation of 3-nitrotyrosine (3-NT), protein carbonyls, advanced glycation end-products (AGE), oxidation of methionine to methionine sulfoxide as well as tyrosine to dityrosine are selected examples that require exact quantification and characterization. The presence of 3-NT-containing proteins in tissue results from the reaction with reactive nitrogen species that are formed *in vivo* [110]. Besides reduction of 3-NT to 3-aminotyrosine, protein tyrosine nitration can be characterized directly by MS of a purified full-length or proteolytically digested peptide or protein [111, 112]. Protein carbonyls may be formed under oxidative stress by direct conversion of an amino acid side chain into a carbonyl, or by covalent attachment of carbonyl-carrying molecules like 4-hydroxynonenal. The detection of a protein carbonyl can be performed after derivatization with 2,4-DNP and either spectrophotometric analysis or staining with an anti-DNP antibody [113]. AGE formation is associated, for example, with aging and diabetes. Post-translational modifications of proteins (glycosylation, oxidation, phosphorylation) may be detected by MS. Electrospray and MALDI-ToF mass spectrometry have been applied to the direct analysis of, for example, glycosylated proteins. Methylglyoxal-dependent modifications to N^{ε}-carboxymethyllysine could be detected in human lens proteins as a result of age-dependent reactions [114]. Higher contents of N^{ε}-carboxymethyllysine-protein adducts have also been found in the peripheral nerves of human diabetics [115]. General strategies for the separation and identification of glycated proteins have been reviewed recently [116]. Further oxidative modifications result from the oxidation of methionine to methionine sulfoxide, or tyrosine to dityrosine [117]. Hydroxyl radical oxidation is characterized by the formation of 3-hydroxyvaline and 5-hydroxyvaline, as well as 3,4-dihydroxyphenylalanine (DOPA), *o*- and *m*-tyrosine, and dityrosine [118]. The oxidation of Met to methionine disulfoxide and of Tyr to dityrosine, as well as the hydroxylation of aromatic amino acid side chains in proteins, are often correlated with pathologic phenomena and can be detected by MS/MS [119]. Deamidation [120] and diketopiperazine formation [121] complete the nonenzymatic post-translational modifications.

2.4
Three-Dimensional Structure

The structural features of a polypeptide and protein chain are described in terms of primary, secondary, and tertiary structure. This description is, to a limited extent, also valid for peptides. While the primary structure classifies the number and sequence of the amino acid residues in a peptide or protein chain, the secondary structure describes ordered conformations of periodic (helix, sheet) or non-periodic (turn) nature. Often, these regular structures are stabilized by hydrogen bonds between hydrogen bond donors (NH) and acceptors (C=O) of peptide bonds. The three-dimensional arrangement of secondary structure elements is called the tertiary structure.

2.4.1
Secondary Structure

The peptide chain conformation preferred under physiological conditions is dominated by the energetically favored torsion angles φ, ψ, and ω, together with additional stabilizing factors such as hydrogen bonds and hydrophobic contacts. In comparison to an alkyl chain, the number of accessible torsion angles for each backbone bond – and hence the number of possible conformations of a peptide chain – is restricted as a consequence of the partial double bond character of the amide bond with a significant rotational barrier (see Section 2.1). The accessible torsion angle regions of φ and ψ are displayed in Ramachandran plots. Substituents at the amide bond can be positioned to give *cis* (ω=0°) or *trans* (ω=180°) configuration, respectively.

A hydrogen bond **17** basically is formed between the NH group (hydrogen bond donor) and the carbonyl oxygen atom (hydrogen bond acceptor) of peptide bonds.

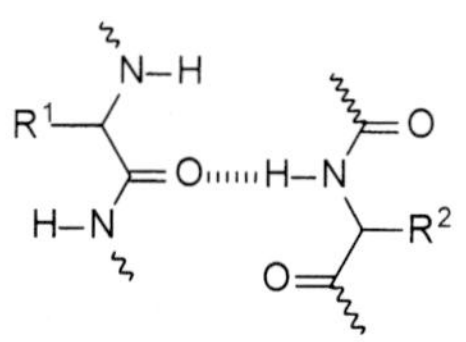

17

The amide bond resonance confers a negative partial charge onto the carbonyl oxygen, and a neighboring NH proton comes into contact with the orbitals of the carbonyl oxygen. The distance between the oxygen and the nitrogen atom in a hydrogen bond is ~280 pm. The energy of a single hydrogen bond is quite low (20 kJ mol^{-1}), compared to a covalent bond (200–400 kJ mol^{-1}). However, in most secondary structure elements stabilized by hydrogen bonding it is multiple rather than single hydrogen bonds that are formed, and it is these multiple interactions of such a cooperative system that result in considerable stabilization.

2.4.1.1 **Helix**

The helix is a widely occurring secondary structure element comprising a screw-like arrangement of the peptide backbone that is stabilized by intramolecular hydrogen bonds aligned in parallel to the helix axis. A helix is characterized by a well-defined number of amino acid residues per turn (n), the helix pitch (h, repeat distance), and the number of skeleton atoms incorporated into the "ring" formed by the intramolecular hydrogen bond. Helices are chiral objects, the direction of the helical turn being given by the letters "P" (plus) for clockwise and "M" (minus) for anti-clockwise helices.

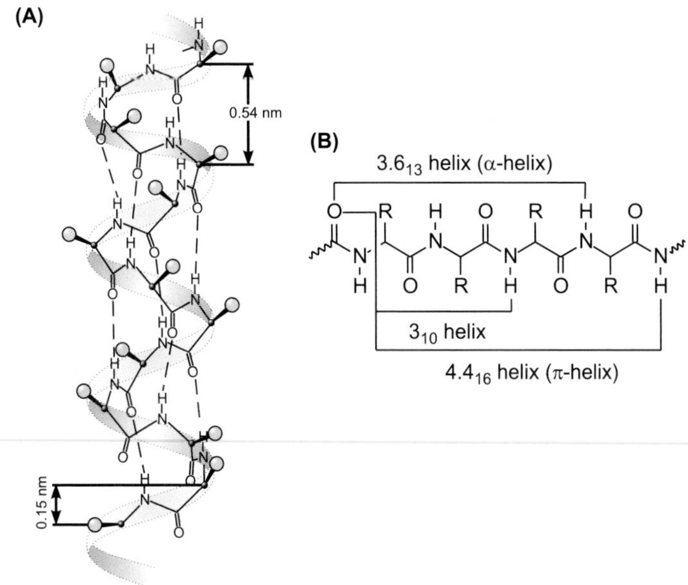

Fig. 2.17 Hydrogen bond pattern in α-helical peptides (A) and schematic view of the hydrogen bond pattern in different helices (B).

The most common form is the α-helix (Fig. 2.17), which was originally proposed by Pauling and Corey based on theoretical investigations regarding the X-ray diffraction patterns of α-keratins. The α-helix comprises a spiral arrangement of the peptide backbone with 3.6 amino acid residues per turn (n=3.6), a helix pitch (h) of 540 pm, and the torsion angles $\varphi=57°$, $\psi=47°$. It is stabilized by hydrogen bonds directed backwards from a C-terminal NH to a N-terminal CO ($NH^{i+4} \rightarrow CO^{i}$) forming a 13-membered "ring" (Fig. 2.17). Consequently, the full nomenclature of an α-helix composed of L-amino acid residues is 3.6_{13}-P-helix. The amino acid side chains are oriented perpendicularly to the helix axis in order to reduce steric strain. Other, less prominent helix types are 3_{10}-helix (3_{10}-P-helix, $\varphi=-60°$, $\psi=-30°$), π-helix (4.4_{16}-P-helix, $\varphi=-57°$, $\psi=-70°$), and γ-helix (5.1_{17}-helix).

The nature of the amino acid side chain is of crucial importance for helix stability. Helix compatibility of a series of amino acids has been examined [122]. The amino acids proline and hydroxyproline as well as other (nonproteinogenic) N-alkylated amino acids are not able to act as hydrogen bond donors, and display high helix-breaking properties. However, they do occur in the collagen triple helix. Glycine also does not have any conformational bias towards helix formation, whereas many other amino acids (Ala, Val, Leu, Phe, Trp, Met, His, Gln) are highly compatible with helical structures. The general criteria for helix stabilization by amino acid residues are: (i) the steric requirements of the amino acid side chain; (ii) electrostatic interactions between charged amino acid side-chain functionalities; (iii) interactions between distant amino acid side chains ($i \leftrightarrow i+3$ or $i \leftrightarrow i+4$); (iv) the presence of proline residues; and (v) interactions between the amino acid residues at the helix termini and the electrostatic dipole moment of the helix.

α-Helices can only be formed by peptide chains of homochiral building blocks. They contain exclusively D- or exclusively L-amino acids. The right-handed α-helix (from L-amino acids) is the preferred conformation, for energetic and stereochemical reasons. A minimum number of amino acids is required for helix formation.

2.4.1.2 β-Sheet

The hydrogen bond pattern of β-sheets differs fundamentally from that of helical structures, with hydrogen bonds being formed between two neighboring polypeptide chains. Two major variants of β-sheet structures may be distinguished.

- The parallel β-sheet, where the chains are aligned in a parallel manner (Fig. 2.18 A).
- The antiparallel β-sheet, where two neighboring peptide chains connected by hydrogen bonds are aligned in an antiparallel manner (Fig. 2.18 B).

An ideal β-sheet structure of a peptide chain is characterized by ϕ and ψ angles of $\pm 180°$.

A hypothetical fully extended oligoglycine chain is characterized by the angles $\phi = -180°$ and $\psi = 180°$. This structure, however, cannot be accommodated without distortion when side chains are present. In this case, an antiparallel β-pleated sheet displays torsion angles $\phi = -139°$ and $\psi = 135°$ (Fig. 2.18). β-Pleated sheets are found in silk fibroin and other β-keratins, as well as in several domains of globular proteins. The side chains of amino acids involved in β-sheet formation are aligned in an alternating manner towards both sides of the β-sheet. β-Sheet structure is much more complex than a simple ribbon diagram might imply. β-Sheets may occur in a twisted, curled, or backfolded form [123]. β-Strands may alter the direction of the main chain dramatically by 180° in a β-turn, or more subtly in a β-bulge. β-Sheets usually exhibit a right-handed twist, favored by nonbonded intrastrand interactions and interstrand geometric constraints. With respect to tertiary structure, layers of β-sheets usually are oriented relative to each other either at a small angle (–30°) in aligned β-sheet packing, or close to 90° in orthogonal β-sheet packing [124]. Statistical studies of proteins of known structure

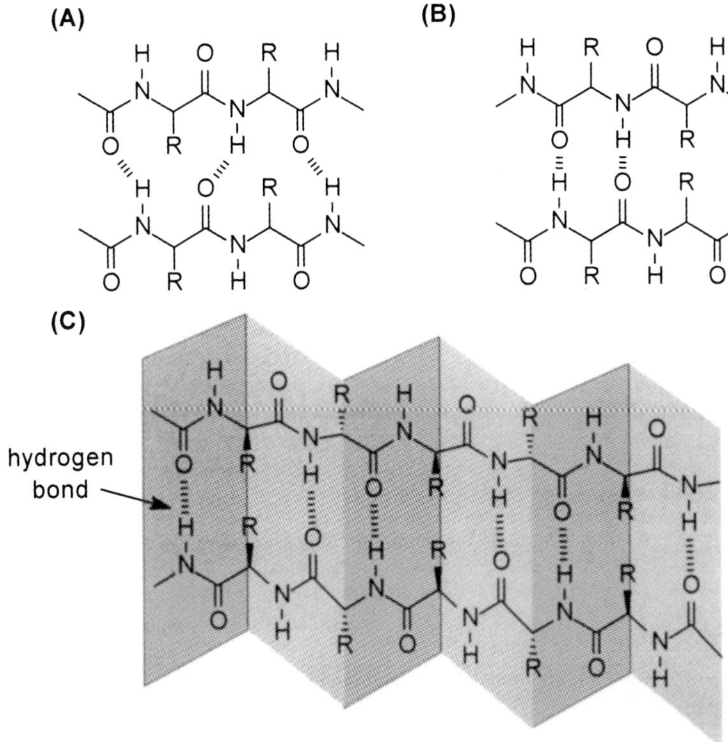

Fig. 2.18 Hydrogen bond pattern in parallel (A) and antiparallel (B, C) β-pleated sheets.

revealed that β-branched and aromatic amino acids most frequently occur in β-sheets, while Gly and Pro tend to be poor β-sheet-forming residues. β-Sheets often occur in the hydrophobic core of proteins. Consequently, the β-sheet-forming propensities of amino acids may reflect the hydrophobic requirement more than a real β-sheet-forming propensity. The β-pleated sheet has been postulated as a structure into which any amino acid could substitute.

2.4.1.3 **Turns**

Loops of a polypeptide chain are characterized by an inversion of the chain direction. The characteristic secondary structure element of a loop is, therefore, the reverse turn. A polypeptide chain cannot fold into compact globular structure without involving tight turns that usually occur on the exposed surface of proteins. Hence, turns provide useful information to defined template structures for the design of new molecules such as drugs, pesticides, and antigens. Turns are also found in small peptides. Often, but not necessarily, they are stabilized by a hydrogen bond between an amino group located C-terminally and a carboxy group located N-terminally. Turns are classified according to the number of amino acid residues involved as γ-turns (three amino acids), β-turns (four amino acids), α-turns (five amino acids), or π-turns (six amino acids) (Fig. 2.19). The characterization of α-turns involves dihedral angles of three amino acids (i+1, i+2, i+3). Consequently, nine α-turn types have been classified by Pavone et al. [125] (Tab. 2.4).

A general criterion for the existence of a β-turn is that the distance of the atoms C^{α} (i) and Cα (i+3) is smaller than 7 Å.

β-Turns can be further classified according to the characteristic dihedral angles ϕ and ψ (Fig. 2.20; Tab. 2.4) of the second (i+1) and the third (i+2 amino acid). Originally β-turns were classified into types I, II, and III and the pseudo-mirror images (I′, II′, III′) [126]. Subsequently, the definition was broadened and the number of β-turns types was increased from six to 10 (I, I′, II, II′, III, III′, IV, V, VI, VII). However, as type III β-turns are the basic structural elements of the 3_{10}-helix, type III β-turns have been eliminated from this classification. These loop structures differ, therefore, in the spatial orientation not only of the NH and CO functions of the amino acids in positions i+1 and i+2, but also of the side chains. The three-dimensional side-chain orientation is given by the vectors

Tab. 2.4 Characteristic torsion angles in the most important secondary structures.

Type	φ_{i+1}	ψ_{i+1}	φ_{i+2}	ψ_{i+2}	φ_{i+3}	ψ_{i+3}
γ-turn	75°	–64°				
γ^i-turn	–79°	69°				
βI-turn	–60°	–30°	–90°	0°		
βI′-turn	60°	30°	90°	0°		
βII-turn	–60°	120°	80°	0°		
βII′-turn	60°	–120°	–80°	0°		
βIV-turn	61°	10°	–53°	17°		
βVIa-turn (1)	–60°	120°	–90°	0°		
βVIa-turn (2)	–120°	120°	60°	0°		
βVIb-turn	–135°	135°	–75°	160°		
βVIII-turn	–60°	–30°	–120°	120°		
I-α_{RS}-turn	–60°	–29°	–72°	–29°	–96°	–20°
I-α_{LS}-turn	48°	42°	67°	33°	70°	32°
II-α_{RS}-turn	–59°	129°	88°	–16°	–91°	–32°
II-α_{LS}-turn	53°	–137°	–95°	81°	57°	38°
I-α_{RU}-turn	59°	–157°	–67°	–29°	–68°	–39°
I-α_{LU}-turn	–61°	158°	64°	37°	62°	39°
II-α_{RU}-turn	54°	39°	67°	–5°	–125°	–34°
II-α_{LU}-turn	–65°	–20°	–90°	16°	86°	37°
I-α_C-turn	–103°	143°	–85°	2°	–54°	–39°
3_{10}-helix	–60°	–30°				
α-Helix	–57°	–47°				
π-Helix	–57°	–70°				
Polyproline II helix	–75°	145°				
Antiparallel β-pleated sheet	–139°	135°				
Parallel β-pleated sheet	–119°	113°				

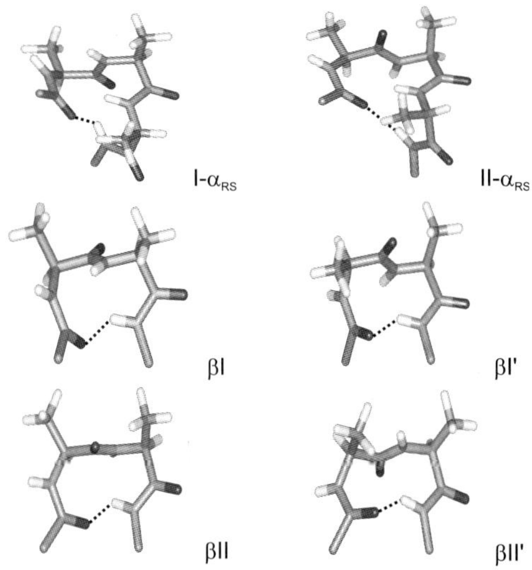

Fig. 2.19. Selected α- and β-turns.

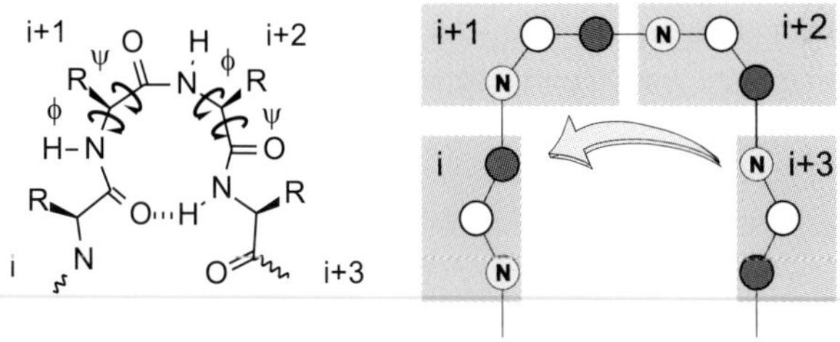

Fig. 2.20 Characteristic torsion angles in β-turns.

$C^{\alpha} \rightarrow C^{\beta}$. If functional groups are present in the side chain, they may be crucial for peptide-receptor interactions.

Some types of β-turns are stabilized intrinsically by certain amino acids. Proline has the highest tendency to occur in a reverse turn. The pyrrolidine ring in L-proline restricts the dihedral angle φ to –60°: Therefore, proline with a *trans*-configured peptide bond is found preferentially in position i+1 of βI- or βII-turns. Proline with a *cis*-peptide bond occurs in position i+2 of a βVIa-turn, which is also named proline-turn. Mainly D-proline, but also D-amino acids, in general have a high preference to occur in position i+1 of a βII′-turn. Glycine often is considered as "proteinogenic D-amino acid", because it is, like D-amino acids, often found in positions i+1 or i+2, respectively, of a turn.

2.4.1.4 **Amphiphilic Structures**

Amphiphilic (amphipathic) compounds are at the same time both hydrophilic and hydrophobic (Fig. 2.21). In amphiphilic helices, one side of the helix mainly presents hydrophobic residues, while the other side mainly contains hydrophilic residues. Interactions between hydrophobic residues in an aqueous environment mediate molecular self-assembly of amphiphilic peptides and stabilize the aggregate.

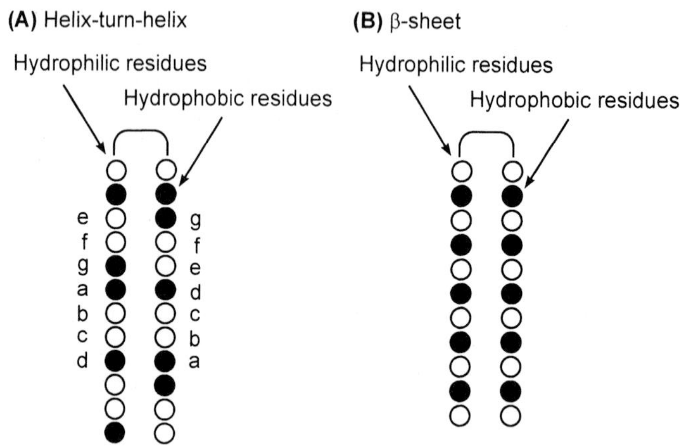

Fig. 2.21 Positioning of hydrophobic and hydrophilic residues within an amino acid sequence that fold as helix-turn-helix (α-helical hairpin, A) or β-sheet (β-hairpin, B) motif.

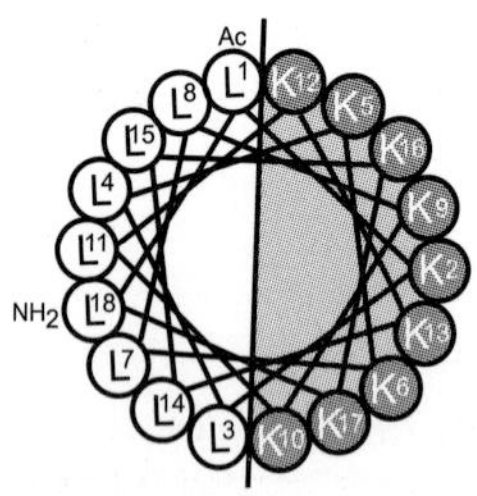

Fig. 2.22 Sequence and Schiffer-Edmundson's helical wheel representation of the amphipathic α-helical 18-peptide LKLLKKLLKKLKKLLKKL [127].

Correct placement of hydrophobic and hydrophilic residues within an amino acid sequence promotes the formation of amphiphilic helices (Fig. 2.22). Helices may associate as α-helical hairpins (helix-turn-helix), coiled coils [128], or helix bundles. The association may be either inter- or intramolecular. In a hydrophilic surrounding, the hydrophobic residues assist the formation of a secondary structure (α-helix, β-sheet; Fig. 2.21) and form a stabilizing core between the single secondary structure elements. The hydrophobic surface areas are not exposed to the aqueous (hydrophilic) environment because of the association of two or more helices via a hydrophobic interhelix interface.

Amphiphilic β-sheets usually are composed of alternating polar and nonpolar residues within the β-strand sequence. Consequently, association of the β-strands gives an amphiphilic β-hairpin or β-sheet where the hydrophobic faces may associate in a sandwich-like fashion. Amphiphilic helices also interact with lipid membranes. The 26-peptide melittin is present in a monomeric form at low concentrations in aqueous media of low ionic strength, where it is largely in a random coil conformation according to circular dichroism (CD) spectroscopy. However, at the lipid bilayer interface, melittin adopts a highly helical conformation. Peptides such as melittin belong to the defense system of a variety of species, and enhance the permeability of the lipid membrane by directly disturbing the lipid matrix. They usually disrupt the transmembrane electrochemical gradient. Magainins and cecropins have been shown to adopt α-helical secondary structure in membrane environments [129]. Many amphiphilic peptides are thought to form transmembrane helical bundles, though whether amphiphilic peptides align along the lipid bilayer plane or whether they form transmembrane helical structures remains a controversial issue. A lipid bilayer is ~30 Å thick, which corresponds to the length of a 20-peptide α-helical structure. Hydrophobic or amphiphilic peptide segments of about 20 amino acids length are often found in natural ion channel proteins. These peptides are regarded as being able to penetrate membranes perpendicularly [130]. The detergent-like characteristics of amphiphilic helical peptides might provide an alternative explanation for their cytotoxic activity [131].

2.4.2 Tertiary Structure

In helices or β-sheets, the conformation of a polypeptide chain is determined not only by hydrogen bonds but also by additional interactions and bonds that stabilize the chain's three-dimensional structure (Fig. 2.23). Metal chelation by different groups within a protein fold (e.g., zinc finger) is a further stabilizing factor.

The disulfide bond is the second type of covalent bond besides the amide bond in a polypeptide chain. It contributes sequence-specifically to formation of the so-called tertiary structure, the conformation of the full peptide chain. A disulfide bond is formed by oxidation of the SH groups of two cysteine residues. Intramolecular disulfide bonds are formed within a single polypeptide chain, while intermolecular disulfide bonds occur between different peptide chains. A torsion angle of

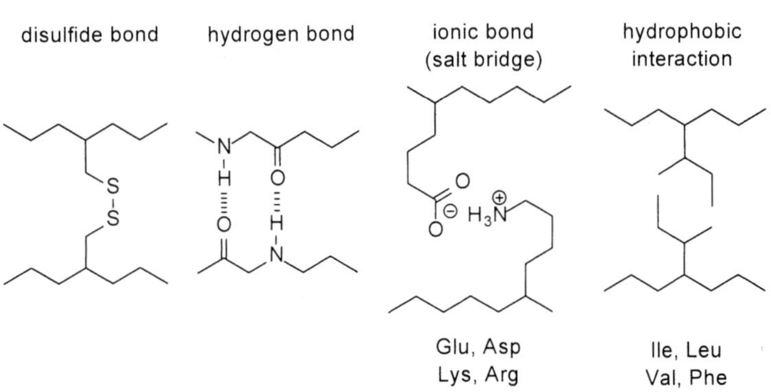

Fig. 2.23 Stabilizing interchain interactions between amino acid side chains.

~110° is observed at the disulfide bridge. The disulfide bonds are not cleaved thermally because of the high bond energy of 200 kJ mol^{-1} (cf. C–C: 330 kJ mol^{-1}, C–N: 260 kJ mol^{-1}, C–H: 410 kJ mol^{-1}). However, they can be cleaved reductively by an excess of reducing agents such as thiols (e.g., dithiothreitol, DTT).

Hydrogen bonds may be formed not only between structural elements of the peptide bonds (as shown for secondary structures), but also between suitable side-chain functionalities of trifunctional amino acids.

The functional groups in the side chains of acidic or basic amino acids are completely or partially ionized at physiological pH. Therefore, electrostatic interactions are observed between acidic and basic groups. These ionic bonds (salt bridges) between carboxylate groups (aspartyl, glutamyl or the C-terminus) and N-protonated residues (arginine, lysine, histidine or the N-terminus) with a bond energy of ~40–85 kJ mol^{-1} influence peptide conformation. Electrostatic interactions are also extended to the hydrate shell. Moreover, ion-dipole and dipole-dipole interactions are observed; these occur because of electrostatic interactions between polarizable groups (SH groups, OH groups) with relatively small binding contributions. Generally, electrostatic interactions are highly dependent on the pH, salt concentration, and dielectric constant of the medium.

Hydrophobic interactions are eminently important in the stabilization of peptide chain conformations. Amino acids with aliphatic or aromatic side chains are characterized by apolar regions with uneven electron distribution. Between these residues, temporary asymmetric electron distribution (temporary dipole formation) results in van der Waals bonds which contribute slightly to the stabilization of the conformation. The formation of these hydrophobic bonds can only occur in the presence of water molecules. Hydrophobic regions are covered in aqueous solution by a molecular layer of highly ordered water molecules. State 1 in Fig. 2.24 is thermodynamically disfavored because of the higher degree of order.

Although both hydrophobic residues in State 2 are ordered to a higher degree, the degree of order of the water molecules has decreased dramatically. A transition from State 1 to State 2 depends on a decrease in free enthalpy, ΔG, of the system, and is connected with a clear increase of the entropy of the system ($\Delta S > 0$), as the equation $\Delta G = \Delta H - T\Delta S$ applies. ΔH in this case is quite small compared to $T\Delta S$, and ΔG consequently becomes negative. The free enthalpy is, therefore, influenced crucially by the term $T\Delta S$. The increase in entropy consequently is the driving force for hydrophobic interactions.

In general, the three-dimensional arrangement of a peptide chain in a globular polypeptide or protein is characterized by a relatively small content of periodical structural elements (α-helix, β-sheet) and shows unsymmetrical and irregular structure. The cooperativity between hydrogen bonds and hydrophobic interactions and other noncovalent interactions basically is the reason for the formation of stable three-dimensional structures. Under physiological conditions, thermodynamically stable conformations of a biologically active peptide with a minimum of free enthalpy occur.

Tertiary structure formation is based on supersecondary structure elements, these being formed by the association of secondary structures. The helix-turn-he-

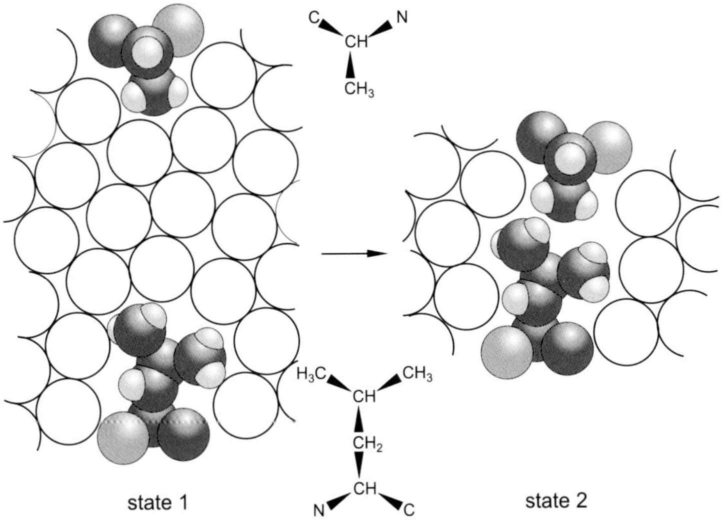

Fig. 2.24 Hydrophobic interactions between Ala and Leu side chains in an aqueous medium.

lix motif (αα, Figure 2.25 A), the β-hairpin motif (ββ, Figure 2.25 B), the Greek key motif (β_4), and the βαβ motif belong to this type of element. The increasing amount of protein structural data led to the classification of tertiary protein folds. To date, more than 500 distinct protein tertiary folds have been characterized, and these represent one-third of all existing globular folds. The rigid framework of secondary structure elements is the best defined part of a protein structure. The special organization of secondary structure elements (topology) may be characterized into several classes. Tertiary structure is formed by packing secondary structure elements into one or several compact globular units (domains). Tertiary fold family classification [132] is used in different databases (SCOP [133] and CATH [134]). The overall agreement between these databases suggests the existence of a natural logic in structural classification. In the CATH database, structures are grouped into fold families depending on both overall shape and connectivity of the secondary structure. In general, three classes of domains (tertiary structure elements) can be distinguished (Fig. 2.25):

1. Structures containing only α-helices (e.g., 7-helix bundles, Fig. 2.25 C; α-superhelix, Fig. 2.25 F)
2. Structures containing exclusively antiparallel β-pleated sheets (e.g., β-propellers, Fig. 2.25 D)
3. Structures containing α-helices and β-sheets (e.g., TIM barrel, Fig. 2.25E; α,β-superhelix, Fig. 2.25 G)

Many efforts have been made to predict protein structural classes. In contrast to α-helices and β-sheets, very few methods have been reported for predicting tight turns [135].

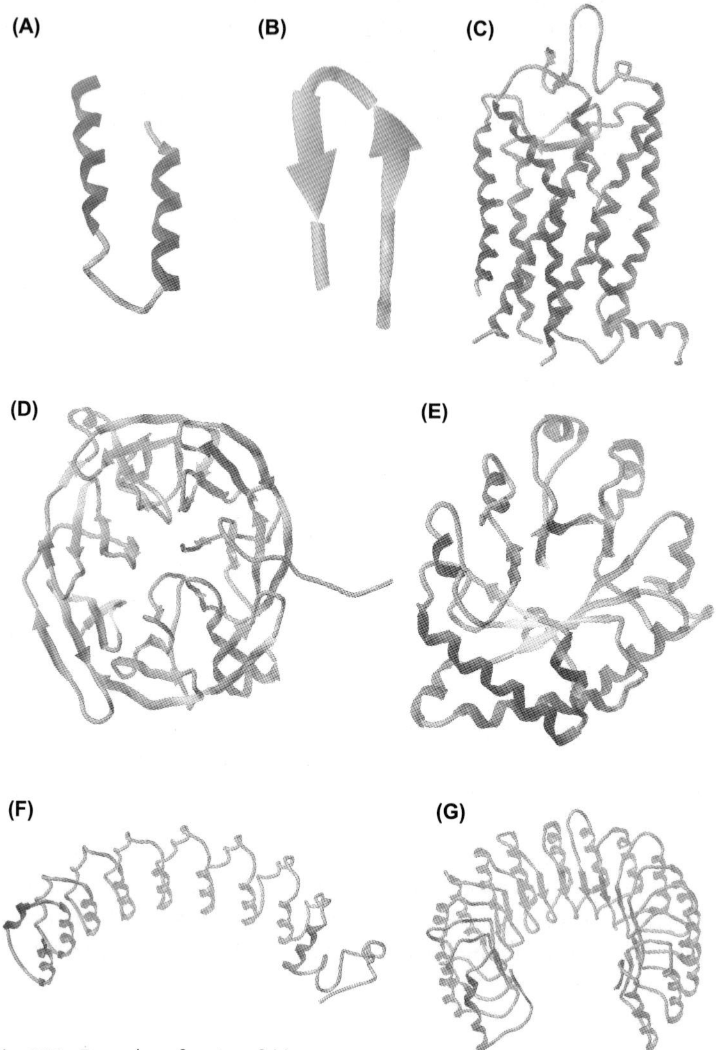

Fig. 2.25 Examples of tertiary folds.

2.4.2.1 **Structure Prediction**

Approaches to protein structure prediction are based on the thermodynamic hypothesis which postulates the native state of a protein as the state of lowest free energy under physiological conditions [136]. Studies of protein folding (structure prediction, fold recognition, homology modeling, and homology design) generally make use of some form of effective energy function.

Considering the fact that a protein consisting of 100 amino acids may adopt three possible conformations per amino acid residue, and that the time for a single conformational transition is 10^{-13} s, the average time a protein would need to find the optimum conformation for all 100 residues by searching the full conformational space is 10^{27} years. In fact, protein folding takes place on a time scale of seconds to minutes. This contradiction has been called the Levinthal paradox, which states that there is insufficient time to search randomly the entire conformational space available to a polypeptide chain in the unfolded form [137]. Levinthal concluded that protein folding follows multiply branched pathways where local secondary structure is formed on the first stage, which is governed by the interaction of neighboring amino acid side chains. Hence, secondary structure formation is considered to be an early event in protein folding. Alternatively, it has been proposed that initially a hydrophobic collapse takes place to form a partially folded structure from which secondary structure subsequently forms. Protein structure prediction remains the "holy grail" of protein chemistry. Until recently, with few exceptions, the prediction of protein structure has been more of a conceptual than a practical importance. Predictions were rarely accurate enough to deduce biological function or to facilitate the structure-based design of new pharmaceuticals.

Protein structure prediction basically relies on two strategies: (i) *ab initio* prediction [138]; and (ii) homology modeling [139, 140]. Homology models are based primarily on the database information. Threading, or fold-recognition methods lie between these two extremes, and involve the identification of a structural template that most closely resembles the structure in question. In cases where homologous (>30% homology) or weakly homologous sequences of known structure are not available, the most successful methods for structure prediction rely on the prediction of secondary structure and local structure motifs. Secondary structure prediction is gaining increasing importance for the prediction of protein structure and function [141]. Secondary structures of peptides and proteins are partly predictable from local sequence information based on knowledge of the intrinsic propensities of amino acids to form a helix or a β-sheet. The prediction of ambivalent propensities can assist in the definition of regions that are prone to undergo conformational transitions. Although primary protein sequence information may provide an educated guess about function (especially when well-characterized homologous sequences exist), incomplete annotation, false inheritance and multiple structural and functional domains may disturb the interpretation of database searches based on primary sequence alone [142].

2.5 Methods of Structural Analysis

The three-dimensional structure of a peptide or a protein is the crucial determinant of its biological activity. As the various genome projects are steadily approaching their final goal, attention now returns to the functions of the proteins encoded by the genes. This will increasingly transform structural biology into structural genomics [143]. Especially in drug design and molecular medicine, the mere information of a gene sequence is not sufficient to obtain information about a corresponding protein on the molecular level.

Besides the structural analysis methodology discussed in this chapter, techniques for biomolecular interaction analysis such as surface plasmon resonance [144], fluorescence correlation spectroscopy [145], and microcalorimetry are steadily gaining importance.

2.5.1 Circular Dichroism

Linear polarized light consists of two circular polarized components of opposite helicity but identical frequency, speed, and intensity. When linear polarized light passes through an optically active medium, for instance a solution containing one enantiomer of an optically active compound, the speed of light in matter is different for the left and right circular polarized components (different refractive indices). In such a case a net rotation of the plane of polarization is observed for the linear polarized light. Consequently, enantiomerically pure or enriched optically active compounds can be characterized by the optical rotation index and optical rotatory dispersion.

However, it is not only the speed of the two circular polarized components but also the extinction by chiral chromophores that may be different. If this is the case, elliptically polarized light is observed. CD spectroscopy detects the wavelength dependence of this ellipticity, and positive or negative CD is observed when either the right- or the left-circular polarized component is absorbed more strongly.

CD is a method of choice for the quick determination of protein and peptide secondary structure. Proteins are often composed of the two classical secondary structure elements, α-helix and β-sheet, in complex combinations. Besides these ordered regions, other parts of the protein or peptide may exist in a random coil state. CD spectroscopy is a highly sensitive method that is able to distinguish between α-helical, β-sheet and random coil conformations. Although the information that can be obtained by CD spectroscopy is somewhat limited compared to NMR or X-ray diffraction, CD data are valuable as a preliminary guide to peptide and protein conformation and conformational transitions under a wide range of conditions [146].

Peptides and proteins that lack non-amino acid chromophores (e.g., prosthetic groups) do not exhibit absorption or CD bands at wavelengths above 300 nm. The amide group is the most prominent chromophore of peptides and proteins to be observed by CD spectroscopy. Two electronic transitions of the amide chromophore have been characterized. The n-π^* transition is usually quite weak and occurs as a negative band around 220 nm. The energy (wavelength) of the amide n-π^* transition is sensitive to hydrogen bond formation. The π-π^* transition usually is stronger, and is registered as a positive band around 192 nm and a negative band around 210 nm.

The proportions of α-helical secondary structure, β-sheet conformation and random coil can be determined by CD spectroscopy. An α-helical conformation usually is characterized by a negative band at 222 nm (n-π^*), a negative band at

208 nm, and a positive band at 192 nm. Short peptides usually do not form stable helices in solution; however, it has been shown that the addition of 2,2,2-trifluoroethanol (TFE) leads to an increase in the helix content of most peptides [147]. As discussed in Section 2.4.1.2, β-sheets in proteins are less well-defined, compared to the α-helix, and can be formed either in a parallel or antiparallel manner. β-Sheets display a characteristic negative band at 216 nm and a positive band of comparable size close to 195 nm. Random coil conformations (unordered conformations) are usually characterized by a strong negative CD band just below 200 nm. Several algorithms are available that allow computational secondary structure analysis of peptides and proteins by fitting the observed spectrum with the combination of the characteristic absorption of the three secondary structures mentioned above. One interesting approach is to deconstruct a protein into a series of synthetic peptides that are then analyzed by CD [148].

In special cases, aromatic side chains of amino acids and disulfide bridges may also serve as chromophores that can give rise to CD bands.

2.5.2
Infrared Spectroscopy

Fourier transform infrared spectroscopy (FT-IR) is nowadays commonly applied to peptides and proteins, and is mainly used to estimate the content of secondary structure elements. A common approach towards these studies has been to prepare peptides that correspond to protein epitopes, and to examine the structure under various conditions (solvent systems, biomembrane models). Additionally, turn-forming model peptides [149–151] and helix models [152, 153] have been investigated. Furthermore, protein hydration and the structural integrity of lyophilized proteins are often examined by FT-IR. The samples may be investigated in solution, in membrane-like environments, and in the solid state, for example after adsorption onto a solid surface. In the latter cases the technique of attenuated total reflectance (ATR) is often involved [154]; this is especially valuable when the analyte displays low solubility or tends to associate in higher concentrations. The sample is applied in the solid state, but may be hydratized as in its natural environment.

ATR–FT-IR is one most powerful methods for recording IR spectra of biological materials in general, and for biological membranes in particular [155]. It can also be applied to proteins that cannot be studied by X-ray crystallography or NMR. ATR–FT-IR requires only very small amounts of material (1–100 μg) and provides spectra within minutes. It is especially suited for studies concerning structure, orientation, and conformational transitions in peptides and membrane proteins. Furthermore, temperature, pressure and pH may be varied in this type of investigation and, additionally, specific ligands may be added. No external chromophores are necessary.

The amide N–H stretching band (νNH), the amide I band around 1615 cm^{-1} (νC=O), and the amide II band around 1550 cm^{-1} (δNH) are the characteristic features that usually are examined by FT-IR spectroscopy of peptides and proteins.

Recently, side-chain carboxy groups of proteins (Asp, Glu) and the OH group of the tyrosine side chain have become targets of FT-IR investigations. Signal deconvolution techniques are usually applied in order to identify the distinct absorption frequencies. The formation of hydrogen bonds characteristically influences the free amide vibrations mentioned above, and allows distinction to be made between α-helical, β-sheet, and random coil conformation.

IR spectroscopy allows monitoring of the exchange rate of amide protons, and hence provides collective data for all amino acid residues present in a protein or peptide. Polarized IR spectroscopy provides information on the orientation of parts of a protein molecule present in an ordered environment.

The replacement of an amide hydrogen by deuterium is highly sensitive to changes in the environment. Moreover, the exchange kinetics may be used to detect conformational changes in the protein structure. Protons exposed on the protein surface will undergo hydrogen/deuterium exchange much more rapidly than those present in the protein core. Amide protons present in flexible regions buried in the protein or involved in secondary structure formation are characterized by medium exchange rates. It has been shown that careful analysis of the amide I band during the deuteration process provides information that helps to assign the exchanging protons to a secondary structure type [155].

2.5.3
NMR Spectroscopy [156–158]

NMR spectroscopy is one of the most widely used analytical techniques for structure elucidation [159]. Nowadays, more-dimensional NMR methods are used routinely for the resonance assignment and structure determination of peptides and small proteins. While ^{1}H is the nucleus to be detected in unlabeled peptides and proteins, proteins uniformly labeled with ^{13}C and ^{15}N provide further information and allow for the application of heteronuclear NMR techniques. NMR studies directed towards the elucidation of the three-dimensional protein structure rely, especially for proteins, on isotope labeling with ^{13}C and ^{15}N in connection with three- and more-dimensional NMR methods [160]. These labeled proteins usually are provided by the application of overexpression systems utilizing isotope-enriched culture media.

For some time, the limit for the elucidation of the three-dimensional protein structure by NMR with respect to the molecular mass was considered to be 30 kDa. However, recent developments and the construction of ultra-high-field superconducting magnets have extended the molecular mass range of molecules to be investigated with NMR well beyond 100 kDa. NMR spectra of such large proteins usually suffer from considerable line broadening, but this has been overcome by the development of transversal relaxation-optimized spectroscopy (TROSY) developed by Wüthrich et al. [161]. At molecular masses above $\sim$20 kDa, spin diffusion becomes a limiting problem because of the longer correlation time of the protein. Consequently, the transversal relaxation time becomes short, which leads to line broadening. These limitations may be overcome by random partial deuteration of proteins [162] and by using the TROSY technique. The molecular mass is not the only determinant for successful NMR structure determination. Before NMR studies may be conducted, investigations must be carried out in order to establish whether the peptide or protein forms aggregates, and whether the folded state of the protein is stable under the experimental conditions.

NMR studies can be applied either in solution phase or in solid phase. Solution-phase NMR uses noncrystalline samples. The peptide or protein is dissolved in aqueous or nonaqueous solvents that may also contain detergents (for the analysis of proteins in a membrane-like environment). In recent years, solid-state NMR has been increasingly used in the area of membrane protein structure elucidation [3]. Solid-state NMR spectroscopy is an attractive method to investigate peptides that are immobilized on solid surfaces, or peptide aggregates such as amyloid fibrils [131].

Crystallization of the proteins, which remains the major obstacle in X-ray structural analysis, is not necessary in NMR studies. The solution conditions (pH, temperature, buffers) can be varied over a wide range. Nowadays, even NMR investigations on the folding pathway of a protein are amenable. In such cases, partially folded proteins are often considered as models for transient species formed during kinetic refolding [163]. Moreover, NMR studies provide a dynamic picture of the protein under investigation.

The chemical shift value is one of the classical NMR parameters used in structural analysis. Insufficient signal dispersion of larger molecules, however, requires the application of additional parameters. Typically, scalar couplings (through-bond connection) and dipolar couplings (through-space connection, nuclear Overhauser effect, NOE) are used for the assignment of the nuclei. In particular, NOE information provides valuable data on the spatial relationship between two nuclei. These internuclear distances (e.g., interproton) usually are indispensable for elucidation of the three-dimensional structure and are used together with other geometric constraints (covalent bond distances and angles) for the computation of three-dimensional protein or peptide structure.

To determine the three-dimensional structure of a peptide, initially all signals in the NMR spectra are assigned to the amino acid residues present. If the peptide sequence is known – which is usually the case for compounds obtained by synthesis or overexpression – the primary structure can be verified by inter-residue NOE signals. If the sequence is unknown, it may be established by analysis of NOESY spectra.

The crosspeak volume integrals of NOESY spectra also provide valuable distance information for the corresponding nuclei. Calibration of these crosspeak volumes with those of known internuclear distance (e.g., geminal protons) leads to a direct conversion of the crosspeak volume values to interproton distances. The sequential distances $d(H^{\alpha}, H^{N})$, $d(H^{N}, H^{N})$, and $d(H^{\beta}, H^{N})$ depend on the torsion angles of the bonds involved. Consequently, the coupling constants between two protons provide information on the torsion angle between these two protons according to the Karplus equation. Regular secondary structures are also characterized by a variety of medium-range or long-range backbone interproton distances that are sufficiently short to be observed by NOE. Helices and turns usually display short sequential and medium-range ^1H–^1H distances, while β-sheets usually display short sequential and long-range backbone ^1H–^1H distances. The characteristic internuclear distances in helices, β-sheets and β-turns are compiled in the classical monograph by Wüthrich [156].

NMR studies are especially helpful in determining the three-dimensional peptide structure when applied to conformationally constrained peptides. This is the case either for cyclic peptides or for peptides containing sterically hindered amino acids. Linear unconstrained peptides usually are too flexible to adopt a preferred conformation in solution.

The amide N–H exchange rates in D_2O solutions and the temperature coefficients of the NH proton chemical shift are further valuable data for conformational analysis of peptides and proteins. An NH proton that is exposed to the solvent will undergo much faster exchange by deuterons in D_2O. Plots of the amide chemical shift versus temperature usually are linear, and their slope is referred to as the temperature coefficient. In general, the temperature coefficients of amide proton resonances for extended peptide chain structures are in the order of –6 to –10 ppb K^{-1}, while greater values of the temperature coefficient (>–4 ppb K^{-1}) are correlated with solvent-inaccessible environments or hydrogen bonds. Consequently, higher temperature coefficients should correlate with slowly exchanging amide protons, and a combination of these data can be used in order to identify regular secondary structures, especially in the case of small linear or cyclic peptides.

Automation programs and tools for the recognition of spin systems have been designed on the bases of patterns recognition techniques [164] in order to assist sequence-specific assignments. Once an ensemble of three-dimensional structures has been calculated from the NMR data, it has to be refined with respect to geometry and constraint violations. In addition to the determination of three-dimensional protein structures in solution, NMR provides valuable information on local structure, conformational dynamics and on the interaction of the protein with small molecules. Consequently, NMR is nowadays a highly versatile tool, for example in industrial drug research, because it can detect very weak ligand-protein interactions that are characterized by only millimolar binding constants [165, 166]. In this context, transferred NOE experiments should be mentioned especially, as they may reveal the protein-bound conformation of a small-molecular weight ligand, provided that a rapid equilibrium between the bound and unbound state exists [167]. Furthermore, the SAR by NMR technique (cf. Section 7.1) as described by Fesik et al. has proved to be a valuable tool in drug discovery.

2.5.4
X-Ray Crystallography

X-ray crystallography permits determination of the three-dimensional structure of molecules at very high resolution. Some proteins have been characterized by this technique at a resolution <1 Å, while the standard resolution of newly determined protein structures is in the range of 2 Å. Meanwhile, more than 15 000 structures of biomolecules have been deposited in the protein structure database (PDB) [168, 169]. X-ray crystallography relies on the diffraction of an X-ray beam at the single crystal lattice. The wavelength range of X-rays corresponds to the size of the diffracting structures (atomic radii and lattice constants), and the observed diffraction pattern results from a superposition of the diffracted beams. The rotating-crystal method uses monochromatic X-rays at a constant wavelength with variation of the angle of incidence, while the Laue methods apply polychromatic X-rays of different wavelengths at a constant angle of incidence. The latter method, when used in combination with a cyclotron beam as the radiation source, provides sufficient data for structural analysis within a short period of time. Ultimately, this will enable time-resolved X-ray analysis of proteins in order to study catalytic mechanisms, for example. Nowadays, intense, highly focused X-rays from integrated cyclotron radiation beamlines facilitate the traditionally tedious and time-consuming process of structural analysis [170].

The protein to be characterized must first be purified and crystallized. Sequence analysis prior to X-ray diffraction experiments is helpful to support the assignment. Protein crystallization can be seen as a multi-parameter optimization process where the optimum concentration and purity of the protein, the concentration of the precipitation agent, ionic strength, pH value, buffer, and temperature may be varied. The presence of detergents is necessary for the crystallization of some proteins, and consequently protein crystallization is the crucial bottleneck in structural analysis by X-ray diffraction. Additionally, novel methods for automatic crystallization and automatic crystallographic data analysis [171] facilitate this method and will eventually provide methodology for high-throughput determination of three-dimensional protein structures.

The reflex pattern obtained upon X-ray diffraction provides information on the crystal lattice constants, the symmetry of the crystal, and its space group. The number of observed reflexes should be as high as possible. After data scaling, the phase problem must be solved. The phase determination uses, for example, heavy metal ions incorporated into the crystal (soaking), but in other cases a known protein structure with sequence homology >35% to the protein under investigation may be used as a starting structure for the refinement. Once a suitable molecular model has been obtained, it is further refined, and then finally validated to provide a three-dimensional structure that is related as closely as possible to the electron density map obtained from X-ray diffraction.

In contrast to proteins that are highly hydrated even in the crystalline state, peptide crystals usually display limited hydration. Smaller peptide molecules usually provide very good crystals, and a resolution of <1 Å is observed in X-ray analysis. On occasion, multiple conformations may be encountered in peptide crystals due to the flexibility of peptides and to similar conformer energies. These conformers may be present in the same crystal, or in separate crystals [172].

2.5.5 UV Fluorescence Spectroscopy

Fluorescence spectroscopy [173] is widely used in peptide and protein chemistry, either observing intrinsic fluorophors (Trp, Tyr), fluorescent cofactors, or extrinsic fluorophors that are used to label the protein. As fluorescence spectroscopy involves electronic transitions, it can be applied to study very fast processes. Interactions of proteins with other proteins [173], nucleic acids [174], small ligands, and membranes can be monitored, as well as protein folding [175] and conformational transitions [176]. The association of peptides can be monitored by fluorescence quenching [177]. Usually, fluorescence intensity, anisotropy, and emission wavelength may be observed. Intracellular processes may be observed by fluorescence spectroscopy when the protein to be examined is expressed by genetic engineering as a fusion protein with green fluorescent protein (GFP) [178].

Fluorescence resonance energy transfer (FRET) is one technique involving fluorescence spectroscopy that is applied increasingly [179]. The fluorescence of one chromophore present in a molecule can be quenched by interaction with other chromophores. This is not a process that involves collision or complex formation that is usually required for electronic coupling of the fluorophor and the quencher molecule. In FRET, energy is transferred across even bigger distances of up to 10 nm. Efficient energy transfer according to the FRET mechanism is only possible when the absorption range of the acceptor chromophore corresponds to the fluorescence range of donor chromophore. The donor must be a fluorescent moiety with a sufficiently long fluorescence lifetime. Moreover, the donor and acceptor must be oriented correctly with respect to each other. It is noteworthy that FRET is a radiationless process and proportional to R^{-6}, where R is the distance between the donor and the acceptor. Therefore, the distance R can be determined in FRET experiments. FRET can be used to monitor interactions between biomolecules. In addition, the cleavage of synthetic substrates by proteases may be easily monitored, even in a highly parallel manner by labeling the substrate with the fluorescence donor on one end and the fluorescence acceptor on the other end [180].

Fluorescence correlation spectroscopy (FCS) is a method used to monitor biomolecular interactions. Single molecules can be detected because a confocal volume element in the femtoliter range is observed. Concentrations between 10^{-9}–10^{-15} M can be detected, and the diffusion times of fluorescently labeled molecules will change upon binding to a potential binding partner. The diffusion time observed is subsequently correlated with the molecular size. This method allows the determination of kinetic parameters [145].

2.6 References

1 G. Fischer, *Angew. Chem.* **1994**, *106*, 1479; *Angew. Chem. Int. Ed. Engl.* **1994**, *33*, 1415

2 IUPAC-IUB Joint Commission on Biochemical Nomenclature (JCBN), Nomenclature and Symbolism for Amino Acids and Peptides, Recommendations 1983, *Eur. J. Biochem.* **1994**, *138*, 9

3 C.K. Larive, S.M. Lunte, M. Zhong, M.D. Perkins, G.S. Wilson, G. Gokulrangan, T. Williams, F. Afroz, C. Schöneich, T.S. Derrick et al., *Anal. Chem.* **1999**, *71*, 389

4 N.G. Anderson, N.L. Anderson, *Anal. Biochem.* **1978**, *85*, 331

5 T. Okazaki, Y. Yanagisawa, T. Nagai, *Clin. Chim. Acta* **1997**, *258*, 137

6 G.J. Opiteck, J.W. Jorgenson, M.A. Mosely, R.J. Anderegg, *J. Microcolumn Sep.* **1998**, *10*, 365

7 M.R. Torres, A.J. Ramos, E. Soriano, *Bioprocess Eng.* **1998**, *19*, 213

8 U. Mathias, R. Marco, M. Uwe, *J. Membr. Sci.* **1996**, *120*, 239

9 F. Hachem, B.A. Andrews, J.A. Asenjo, *Enzyme Microb. Technol.* **1996**, *19*, 507

10 M. Lu, F. Tjerneld, *J. Chromatogr. A* **1997**, *766*, 99

11 G. Zweig, J. Sherma, *Handbook of Chromatography*, CRC Press, Boca Raton, **1972**

12 L.C. Craig, T.P. King, *Fed. Proc.* **1958**, *17*, 1126

13 T. Yoshida, *J. Chromatogr.* **1998**, *88*, 105

14 W.S. Handcock, *Handbook of HPLC for the Separation of Amino Acids, Peptides, and Proteins, Vols. 1 and 2*, CRC Press, Boca Raton, **1984**

15 A. Henschen, K.P. Hupe, F. Lottspeich, W. Voelter, *High Performance Liquid Chromatography in Biochemistry*, VCH, Weinheim, **1985**

16 M.T.W. Hearn, HPLC of Proteins, *Peptides, and Polynucleotides*, VCH, New York, **1991**

17 K.K. Unger, E. Weber, *Handbuch der HPLC*, GIT, Darmstadt, **1995**

18 C.T. Mant, R.S. Hodges, *High Performance Liquid Chromatography of Peptides and Proteins: Separation, Analysis, and Conformation*, CRC Press, Boca Raton, **1991**

19 W.R. LaCourse, C.O. Dasenbrock, *Anal. Chem.* **1998**, *70*, 37

20 M. Hanson, K.K. Unger, C.T. Mant, R.S. Hodges, *Trends Anal. Chem.* **1996**, *15*, 102

21 T. Sereda, C.T. Mant, R.S. Hodges, *J. Chromatogr. A* **1997**, *776*, 153

22 E. Kalsson, L. Rydon, J. Brewer in: *Protein Purification*, J.-C. Janson, L. Rydon (Eds.), Wiley, New York, **1998**

23 P.R. Levison, C. Mumford, M. Streater, A.B. Nielsen, N.D. Pathirana, S.E. Badger, *J. Chromatogr. A* **1997**, *760*, 151

24 A. Vailaya, C. Horvath, *Ind. Eng. Chem. Res.* **1996**, *35*, 2964

25 B. Lindquist, T. Storgards, *Nature* **1955**, *175*, 511

26 J. Porath, P. Flodin, *Nature* **1959**, *183*, 1657

27 J. Porath, J. Protein Chem. **1997**, *16*, 463

28 H.G. Barth, B.E. Boyers, C. Jackson, *Anal. Chem.* **1998**, *70*, 251

29 P. Cuatrecasas, *Adv. Enzymol.* **1972**, *36*, 29

30 K.J. Miller, A.C. Herman, *Anal. Chem.* **1996**, *68*, 3077

31 D. Zhou, H. Zou, J. Ni, L. Yang, L. Jia, Q. Zhang, Y. Zhang, *Anal. Chem.* **1999**, *71*, 115

32 J.A. Buettner, C.A. Dadd, B. L. Masecar, D. J. Hammond, *Int. J. Peptide Protein Res.* **1996**, *47*, 70

33 J.W. Jorgenson, K.D. Lukacs, *Anal. Chem.* **1981**, *53*, 1298

34 K.M. Hutterer, J.W. Jorgenson, *Anal. Chem.* **1999**, *71*, 1293

35 G. Teshima, S.L. Wu, *Methods Enzymol.* **1996**, *271*

36 M. Hadley, M. Gilges, J. Senior, A. Shab, P. Camilleri, *J. Chromatogr. B* **2000**, *745*, 177

37 Y. Shen, J. Berger, G.A. Anderson, R. D. Smith, *Anal. Chem.* **2000**, *72*, 2154

38 D.M. Bollag, M.D. Rozycki, S.J. Edelstein, *Protein Methods*, Wiley, New York, **1996**

39 J. Porath, J. Carlsson, I. Olsson, G. Belfrage, *Nature* **1975**, *258*, 598
40 N. K. Shin, D.-Y. Kim, C.-S. Shin, M.-S. Hong, J. Lee, H.-C. Shin, *J. Biotechnol.* **1998**, *62*, 143
41 H. Rogl, K. Kosemund, W. Kühlbrandt, I. Collinson, *FEBS Lett.* **1998**, *432*, 21
42 Z. Ji, D. I. Pinon, L. J. Miller, *Anal. Biochem.* **1996**, *240*, 197
43 J. E. Dyr, J. Suttnar, *J. Chromatogr. B* **1997**, *699*, 383
44 M. Kaufmann, *J. Chromatogr. B* **1997**, *699*, 347
45 Y. Ito, Y. Ma, *J. Chromatogr. A* **1996**, *753*, 1
46 Y. Ma, Y. Ito, *Anal. Chim. Acta* **1997**, *352*, 411
47 J. C. Janson, L. Rydon, *Protein Purification*, VCH, Weinheim, **1989**
48 M. P. Deutscher, *Protein Purification*, Academic Press, New York, **1990**
49 R. K. Scopes, *Protein Purification*, Springer, Heidelberg, **1994**
50 S. Doonan, *Protein Purification Protocols*, Humana Press, **1996**
51 J. W. van Nispen in: *Topics in Pharmaceutical Sciences*, D. D. Breimer, P. Speiser (Eds.), Elsevier, Amsterdam, **1987**
52 H. R. Costantino, R. Langer, A. M. Klibanov, *J. Pharm. Sci.* **1994**, *83*, 1662
53 J. W. van Nispen, R. Pinder, *Annu. Rep. Med. Chem.* **1986**, *21*, 51
54 L. Chafetz, *Pharm. Technol.* **1984**, 24
55 R. Pearlman, T. J. Nguyen, *J. Pharm. Pharmacol.* **1992**, *44*, 178
56 R. G. Strickley, G. C. Visor, L. Lin, L. Gu, *Pharm. Res.* **1989**, *6*, 971
57 N. Bhatt, K. Patel, R. T. Borchardt, *Pharm. Res.* **1990**, *7*, 593
58 G. M. Jordan, S. Yoshioka, T. Terao, *J. Pharm. Pharmacol.* **1994**, *46*, 182
59 M. J. Hageman, *Drug Dev. Ind. Pharm.* **1988**, *14*, 2047
60 M. C. Lai, E. M. Topp, *J. Pharm. Sci.* **1999**, *88*, 489
61 A. L. Burlingame, R. K., Boyd, S. J. Gaskell, *Anal. Chem.* **1996**, *68*, 559
62 G. Allan, Sequencing of Proteins and Peptides in: *Laboratory Techniques in Biochemistry and Molecular Biology, Vol. 9*, R. M. Burdon, P. H. van Knippenberg (Eds.), Elsevier, Amsterdam, **1989**
63 J. B. C. Findlay, M. J. Geisow (Eds.), *Protein Sequencing, A Practical Approach*, IRL Press, Oxford, **1989**
64 F. Sanger, *Annu. Rev. Biochem.* **1988**, *57*, 1
65 F. Sanger, S. Nicklen, A. R. Coulson, *Proc. Natl. Acad. Sci. USA* **1977**, *74*, 5463
66 A. M. Maxam, W. Gilbert, *Proc. Natl. Acad. Sci. USA* **1977**, *74*, 560
67 F. Sanger, *Biochem. J.* **1945**, *39*, 507
68 W. R. Gray, B. S. Hartley, *Biochem. J.* **1963**, *89*, 59
69 H. Hanson, D. Glasser, M. Ludewig, H. G. Mannsfeldt, M. John, *Hoppe-Seyler's Z. Physiol. Chem.* **1967**, *348*, 689
70 H. Zuber, G. Roncari, *Angew. Chem.* 1967, 6, 906; *Angew. Chem. Int. Ed. Engl.* **1967**, *6*, 880
71 H. Feracci, S. Maroux, *Biochim. Biophys. Acta* **1980**, *599*, 448
72 S. Akabori, K. Ohno, T. Ikenaka, Y. Okada, H. Hanafusa, I. Haruna, A. Tsugita, K. Sugae, T. Matsushima, *Bull. Chem. Soc. Japan* **1956**, *29*, 507
73 S. Blackburn, *Amino Acid Determination*, M. Dekker, New York, **1978**
74 G. C. Barrett, *Chemistry and Biochemistry of the Amino Acids*, Chapman & Hall, London, **1985**
75 W. S. Hancock, *Handbook of HPLC for the Separation of Amino Acids, Peptides, and Proteins*, CRC Press, Boca Raton, **1984**
76 K. Ashman, A. Bosserhoff in: *Modern Methods in Protein Chemistry*, H. Tschesche (Ed.), deGruyter, Berlin, **1985**
77 E. Gross, B. Witkop, *J. Am. Chem. Soc.* **1961**, *83*, 1510
78 A. Patchornik, W. B. Lawson, B. Witkop, *J. Am. Chem. Soc.* **1958**, *80*, 4747
79 A. Fontana, W. E. Savie, M. Zombonin in: *Methods in Peptide and Protein Analysis*, C. Birr (Ed.), Elsevier/North Holland Biomedical Press, Amsterdam, **1980**
80 W. C. Mahoney, P. K. Smith, M. A. Hermodson, *Biochemistry* **1981**, *20*, 443
81 P. Edman, *Arch. Biochem.* **1949**, *22*, 475
82 P. Edman, G. Begg, *Eur. J. Biochem.* **1967**, *1*, 80
83 R. W. Hewick, M. W. Hundapiller, L. E. Hood, W. J. Dreyer, *J. Biol. Chem.* **1981**, *256*, 7990

84 R. A. Laursen, A. G. Bonner, *Chem. Eng. News* **1970**, *48*, 52

85 P. Schlack, W. Kumpf, *Hoppe-Seyler's Z. Physiol. Chem.* **1926**, *154*, 125

86 G. R. Stark, *Biochemistry* **1968**, *7*, 1796

87 V. L. Boyd, M. Bozzini, G. Zon, R. L. Noble, R. J. Mattaliano, *Anal. Biochem.* **1992**, *206*, 344

88 J. M. Bailey, N. R. Shenoy, M. Ronk, J. E. Shively, *Protein Sci.* **1992**, *1*, 68

89 T. Bergmann, E. Cederlund, H. Jörnvall, *Anal. Biochem.* **2001**, *290*, 74

90 G. Siuzdak, *Mass Spectrometry for Biotechnology*, Academic Press, San Diego, **1996**

91 T. Rabilloud, *Proteome Research: Two-Dimensional Gel Electrophoresis and Identification Methods*, Springer, Berlin, **2000**

92 M. Karas, F. Hillenkamp, *Anal. Chem.* **1988**, *60*, 2299

93 J. Fenn, M. Mann, C. K. Meng, S. F. Wong, C. M. Whitehouse, *Science* **1985**, *246*, 64

94 M. Wilm, A. Shevchenko, T. Houthaeve, S. Breit, L. Schweigerer, T. Fotsis, M. Mann, *Nature* **1996**, *379*, 466

95 A. Shevchenko, A. Loboda, A. Shevchenko, W. Ens, K. G. Standing, *Anal. Chem.* **2000**, *72*, 2132

96 B. T. Chait, S. B. H. Kent, *Science* **1992**, *257*, 78

97 R. W. Nelson, M. A. McLean, T. W. Hutchens, *Anal. Chem.* **1994**, *66*, 1408

98 C. V. Bradley, D. H. Williams, M. R. Hanley, *Biochem. Biophys. Res. Commun.* **1982**, *104*, 1223

99 K. Klarskov, K. Breddam, P. Roepstorff, *Anal. Biochem.* **1989**, *180*, 28

100 C. E. Smith, K. L. Duffin in: *Techniques in Protein Chemistry*, R. H. Angeletti (Ed.), Academic Press, San Diego, **1993**

101 Z. Zhang, J. S. McElvain, *Anal. Chem.* **2000**, *72*, 2337

102 B. T. Chait, R. Wang, R. C. Beavis, S. B. H. Kent, *Science* **1993**, *262*, 89

103 M. Bartlet-Jones, W. A. Jeffery, H. F. Hansen, D. J. Pappin, *Rapid. Commun. Mass Spect.* **1994**, *8*, 737

104 B. Thiede, B. Wittmann-Liebold, M. Bienert, E. Krause, *FEBS Lett.* **1995**, *357*, 65

105 H. A. Woods, A. Y. C. Huang, R. J. Cotter, G. R. Pasternack, D. M. Pardoli, E. M. Jaffee, *Anal. Biochem.* **1995**, *226*, 15

106 D. H. Patterson, G. E. Tarr, F. E. Regnier, S. A. Martin, *Anal. Chem.* **1995**, *67*, 3971

107 B. Thiede, J. Salnikov, B. Wittmann-Liebold, *Eur. J. Biochem.* **1997**, *244*, 750

108 R. S. Brown, J. J. Lennon, *Anal. Chem.* **1995**, *67*, 1998

109 R. M. Whittal, L. Li, *Anal. Chem.* **1995**, *67*, 1950

110 B. Halliwell, *FEBS Lett.* **1997**, *411*, 157

111 D. Yi, G. A. Smythe, B. C. Blount, M. W. Duncan, *Arch. Biochem. Biophys.* **1997**, *344*, 253

112 J. P. Crow, Y. Z. Ye, M. Strong, M. Kirk, S. Barnes, J. S. Beckmann, *J. Neurochem.* **1997**, *69*, 1945

113 R. L. Levine, J. A. Williams, E. R. Stadtman, E. Shacter, *Methods Enzymol.* **1994**, *233*, 346

114 M. Ahmed, E. Brinkmann Frye, T. P. Degenhardt, S. R. Thorpe, J. W. Baynes, *Biochem. J.* **1997**, *324*, 565

115 K. Sugimoto, Y. Nishizawa, S. Horiuchi, S. Yagihashi, *Diabetologia* **1997**, *40*, 1380

116 I. Miksik, Z. Deyl, *J. Chromatogr. B*, **1997**, *699*, 311

117 C.-C. Chao, Y.-S. Ma, E. R. Stadtman, *Proc. Natl. Acad. Sci. USA* **1997**, *94*, 2969

118 S. Fu, R. T. Dean, *Biochem. J.* **1997**, *324*, 41

119 K. L. Schey, L. Finley, *Acc. Chem. Res.* **2000**, *33*, 299

120 R. Zand, M. X. Li, X. Jin, D. Lubman, *Biochemistry* **1998**, *37*, 2441

121 S. Capasso, L. Mozzarella, *Peptides* **1998**, *19*, 389

122 D. Obrecht, M. Altorfer, U. Bohdal, J. Daly, W. Huber, A. Labhardt, C. Lehmann, K. Müller, R. Ruffieux, P. Schönholzer, C. Spiegler, C. Zumbrunn, *Biopolymers* **1997**, *42*, 575

123 G. D. Rose, L. M. Gierasch, J. A. Smith, *Adv. Protein Chem.* **1985**, *37*, 1

124 C. K. Smith, L. Regan, *Acc. Chem. Res.* **1997**, *30*, 153

125 V. Pavone, G. Gaeta, A. Lombardi, F. Nastri, O. Maglio, *Biopolymers* **1996**, *38*, 705

126 C. M. Venkatachalam, *Biopolymers* **1968**, 6, 1425

127 S. Castano, I. Cornut, K. Büttner, J.L. Dasseux, J. Dufourcq, *Biochim. Biophys. Acta* **1999**, *1416*, 161
128 P. Burkhard, J. Stetefeld, S.V. Strelkov, *Trends Cell Biol.* **2001**, *11*, 82.
129 B. Bechinger, *J. Membr. Biol.* **1997**, *156*, 197
130 S. Futaki, *Biopolymers* **1998**, *47*, 75
131 B. Bechinger, R. Kinder, M. Helmle, T.C.B. Vogt, U. Harzer, S. Schinzel, *Biopolymers* **1999**, *51*, 174
132 C. Zhang, C. DeLisi, *Cell. Mol. Life Sci.* **2001**, *58*, 72
133 http://scop.mrc-lmb.cam.ac.uk/scop
134 http://www.biochem.ucl.ac.uk/bsm/cath_new/
135 K.-C. Chou, *Anal. Biochem.* **2000**, *286*, 1
136 T. Lazaridis, M. Karplus, *Curr. Opin. Struct. Biol.* **2000**, *10*, 139
137 C. Levinthal, *J. Chim. Phys.* **1968**, *65*, 44
138 R. Bonneau, D. Baker, *Annu. Rev. Biophys. Biomol. Struct.* **2001**, *30*, 173
139 B. Al-Lazikani, J. Jung, Z. Xiang, B. Honig, *Curr. Opin. Chem. Biol.* **2001**, *5*, 51
140 P. Koehl, *Curr. Opin. Struct. Biol.* **2001**, *11*, 348
141 B. Rost, *J. Struct. Biol.* **2001**, *134*, 204
142 T.F. Smith, X. Zhang, *Nature Biotechnol.* **1997**, *15*, 1222
143 S. Sunyaev, W. Lathe III, P. Bork, *Curr. Opin. Struct. Biol.* **2001**, *11*, 125
144 R.L. Rich, D.G. Myszka, *Curr. Opin. Biotechnol.* **2000**, *11*, 54
145 T. Winkler, U. Kettling, A. Koltermann, M. Eigen, *Proc. Natl. Acad. Sci. USA* **1999**, *96*, 1375
146 R.W. Woody in: *Circular Dichroism, Principles and Applications*, K. Nakanishi, N. Berova, R.W. Woody (Eds.), VCH, New York, **1994**
147 J.W. Nelson, N.R. Kallenbach, *Proteins Struct. Funct. Genet.* **1986**, *1*, 211
148 P. Wittung-Stafsheded, *Biochim. Biophys. Acta* **1998**, *1382*, 324
149 E. Vass, M. Kurz, R.K. Konat, M. Hollosi, *Spectrochim. Acta A* **1998**, *54*, 773
150 E. Vass, E. Lang, J. Samu, Z. Majer, M. Kajtar-Peredy, M. Mak, L. Radics, M. Hollosi, *J. Mol. Struct.* **1998**, *440*, 59
151 E. Vass, J. Samu, Z. Majer, M. Hollosi, *J. Mol. Struct.* **1998**, *408*, 207
152 G. Yoder, P. Pancoska, T.A. Keiderling, *Biochemistry* **1997**, *36*, 15123
153 D.K. Graff, B. Pastrana-Rios, S.Y. Venyaminov, F.G. Prendegast, *J. Am. Chem. Soc.* **1997**, *119*, 11282
154 K. Oberg, A.L. Fink, *Anal. Biochem.* **1998**, *256*, 92
155 C. Vigano, L. Manciu, F. Buyse, E. Goormaghtigh, J.-M. Ruysschaert, *Biopolymers* **2000**, *55*, 373
156 K. Wüthrich, *NMR of Proteins and Nucleic Acids*, Wiley, New York, **1986**
157 G.C.K. Roberts, *NMR of Macromolecules*, IRL Press, Oxford, **1993**
158 J.N.S. Evans, *Biomolecular NMR Spectroscopy*, Oxford University Press, Oxford, **1995**
159 G.T. Montelione, D. Zheng, Y.J. Huang, K.C. Gunsalus, T. Szyperski, *Nature Struct. Biol., Struct. Genomics Suppl.* **2000**, 982
160 K.H. Gardner, L.E. Kay, *Annu. Rev. Biophys. Biomol. Struct.* **1998**, *27*, 357
161 K. Pervushin, R. Riek, G. Wider, K. Wüthrich, *Proc. Natl. Acad. Sci. USA* **1997**, *94*, 12386
162 S.J. Archer, D.J. Domaille, E.D. Laue, *Annu. Rep. Med. Chem.* **1996**, *31*, 299
163 J. Balbach, V. Forge, N.A. van Nuland, S.L. Winder, P.J. Hore, C.M. Dobson, *Nature Struct. Biol.* **1995**, *2*, 865
164 D. Croft, J. Kemmink, K.-P. Nedig, H. Oschkinat, *J. Biomol. NMR* **1997**, *10*, 207
165 T. Diercks, M. Coles, H. Kessler, *Curr. Opin. Chem. Biol.* **2001**, *5*, 285
166 J.M. Moore, *Biopolymers* **1999**, *51*, 221
167 F. Ni, *Prog. NMR Spectrosc.* **1994**, *26*, 517
168 H.M. Berman, J. Westbrook, Z. Feng, G. Gilliland, T.N. Bhat, H. Weissig, I.N. Shindyalov, P.E. Bourne, *Nucleic Acids Res.* **2000**, *28*, 235
169 http://www.rcsb.org/pdb/
170 E. Abola, P. Kuhn, T. Earnest, R.C. Stevens, *Nature Struct. Biol., Struct. Genomics Suppl.* **2000**, 973
171 V.S. Lamzin, A. Perrakis, *Nature Struct. Biol., Struct. Genomics Suppl.* **2000**, 978
172 I.L. Karle, *Acc. Chem. Res.* **1999**, *32*, 693
173 M.P. Brown, C. Royer, *Curr. Opin. Biotechnol.* **1997**, *8*, 45
174 J.J. Hill, C.A. Royer, *Methods Enzymol.* **1997**, *278*, 390

175 M. R. Eftink, M. C. R. Shastry, *Methods Enzymol.* **1997**, *278*, 258
176 M. R. Eftink, *Biochemistry* **1998**, *63*, 276
177 A. Karlström, A. Undén, *Biopolymers* **1997**, *41*, 1
178 S.-H. Park, R. T. Raines, *Protein Sci.* **1997**, *6*, 2344
179 D. L. Andrews, A. A. Demidov, *Resonance Energy Transfer*, Wiley, Chichester, **1999**
180 H. R. Stennicke, M. Renatus, M. Meldal, G. S. Salvesen, *Biochem. J.* **2000**, *350*, 563

3
Biologically Active Peptides

3.1
Occurrence and Biological Roles [1–4]

Although Emil Fischer had already predicted at the beginning of the last century that organic chemistry and biology would, in the near future, turn to peptide and protein research, interest in this field was rather poor until the early 1940s. At that time, the number of known biologically attractive peptides was very limited.

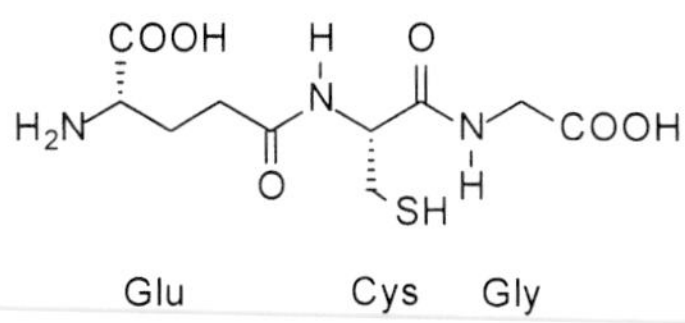

1

In addition to glutathione **1** (a fascinating compound for people interested in bioorganic chemistry) and carnosine (β-Ala-His), only gramicidin S and a few other mold-derived compounds (e.g., enniatins) had been discovered. As early as 1888, glutathione was observed as a reducing agent in yeast by Rey-Pailhade. It was isolated from liver, yeast and muscle by Hopkins in 1921 and synthesized by Harington and Mead in 1935 [4]. This remarkable event, and also the synthesis of carnosine by Sifferd and du Vigneaud [6] in the same year, had profited from the invention some years earlier of the benzyloxycarbonyl group as an amino-protecting group. The amino acid sequence of gramicidin S **2** was elucidated as early as 1947 as being a cyclic decapeptide with a repeated sequence of two pentapeptides [7]. An understanding of its biosynthesis was achieved at a much later date, however [8].

cyclo[Val^1–Orn–Leu–D-Phe–Pro^5–Val–Orn–Leu–D-Phe–Pro^{10}]

2

The importance and broad functional role of peptides in life processes became apparent only in the 1950s and early 1960s, when the continuous development of increasingly sensitive analytical methods and techniques for isolation and purification

Tab. 3.1 Important peptide hormones.

Peptide hormone	*Natural source*	*Number of aa*	*Reference(s)*
Insulin	Pancreas	51	12
Glucagon	Pancreas	29	13
Corticotropin	Adenohypophysis	39	14
α-Melanotropin	Adenohypophysis	14	15
β-Melanotropin	Adenohypophysis	14	16
Oxytocin	Neurohypophysis	9	17, 18
Vasopressin	Neurohypophysis	9	19, 20
Angiotensin	Blood plasma	10/8	21, 22

signaled the start of a new era in this research field. Size-exclusion chromatography [9], chromatography on cellulose-based ion-exchangers [10], countercurrent distribution [11] and other methods developed in various areas of biochemistry complemented the techniques for peptide isolation that had been developed previously.

The details of several important peptide hormones, the amino acid sequences of which were elucidated during the 1950s, are listed in Tab. 3.1. The isolation of peptides from natural sources has been – and will remain in the future – a necessity of peptide research, as was stated by the unforgotten Josef Rudinger at the Third American Peptide Symposium in 1972:

"In the work on the smaller biologically active peptides, isolation is to my mind still the most difficult and critical phase and its practitioners have my sincere respect and admiration."

The deduction of amino acid sequences from the experimentally determined cDNA sequences of peptides that are synthesized ribosomally has not changed the importance of peptide isolation. Unexpected post-translational modifications and uncertainties concerning the cleavage sites of the precursors are, besides other problems, important points to maintain and to improve the methods of peptide isolation from natural sources.

Until now, innumerable peptides with a variety of biological and physiological effects have been detected, isolated, characterized and mostly synthesized. Doubtlessly, the sequence determination [17, 18] and, especially, the chemical synthesis of oxytocin **3** performed by the Nobel prize winner V. du Vigneaud [23], was a historical milestone in peptide research, since the synthetic hormone had a biological activity indistinguishable from that of the natural hormone isolated from the neurohypophysis.

$$\text{H–}\overset{\ulcorner}{\text{Cys}}{}^{1}\text{–Tyr–Ile–Gln–Asn}^{5}\text{–}\overset{\urcorner}{\text{Cys}}\text{–Pro–Leu–Gly–NH}_2$$

3

Peptides and their higher correlates, the proteins, fulfil crucial functions in almost all processes of the living cell. The distinction between what constitutes a peptide and what constitutes a protein should be increasingly only of academic interest, and becomes increasingly unclear as peptides increase in length, both from the synthetic and structural points of view. It holds true to state that proteins are large peptides. With increasing length, peptides have a greater tendency to exhibit elements of secondary structure forming preferred solution conformations. Furthermore, the problems connected with the chemical synthesis of proteins in the range of 100 to 150 amino acid residues are basically the same as in the synthesis of peptides or, in a general sense, of peptide chemistry.

Many of these naturally occurring bioactive molecules act as enzymes [24–27]. These biocatalysts mediate very precisely and effectively all chemical reactions within and outside the cells. Enzymes differ from ordinary chemical catalysts by having higher reaction rates, greater reaction specificity, milder reaction conditions, and last – but not least – the capacity of regulation. The capability of binding their substrates specifically through geometrically and physically complementary interactions permit enzymes to act absolutely stereospecifically in substrate binding and catalysis.

Complex chemical signaling systems by hormones [28, 29] and neurotransmitters forms the basis of the biochemical communication [30] used by all living organisms to coordinate their activities at any level of their organization. In the early days, it was essential when working with peptide hormones to avoid degradation by proteolytic enzymes that might be co-extracted from the appropriate tissues. Without doubt, the extraction technique using ethanol containing hydrochloric acid that was first employed in the extraction of insulin from pancreatic tissue was a major breakthrough. Alternatively, in the case of relatively thermostable peptides the proteolytic tissue enzymes could be inactivated by brief boiling and subsequent extraction with cold 0.2 M HCl. The simultaneous isolation of gastrin by Gregory and Tracy in Liverpool and secretin by Jorpes and Mutt in Stockholm was described in the early 1960s. Human secretin was sequenced after its isolation from 181 g of intestinal tissue obtained after surgery. The isolation of a species variant of a known peptide hormone is much easier to perform when guided by radioimmunoassay – a highly sensitive technique introduced by Yalow and Berson [31].

Peptides and proteins are responsible for the regulation of biochemical processes in complex organisms. Fig. 3.1 outlines such regulation by peptides. Autocrine hormones (A) act on the same cell that released them as demonstrated, for example, by the T-cell proliferation stimulating interleukin 2. Paracrine hormones (B), which are also named local mediators, are directed to surrounding cells by diffusion. Endocrine hormones (C) are synthesized and transported to distant cells via the bloodstream; examples include insulin and glucagon. Furthermore, neurotransmitters (D) – many of which are also peptides (e.g., β-endorphin, enkephalins, somatostatin) – elicit chemically transmitted nerve impulses across most synapses. Neurohormones (E) act as mediators on nerve cells. The brain contains many neuroactive peptides with complicated relationships, though to date only a fraction of these have been discovered [32]. Various peptides originally considered to be brain peptides exerting multiple effects in the central nervous system (CNS) have also been found in the gut, and in other locations. In other words, peptides

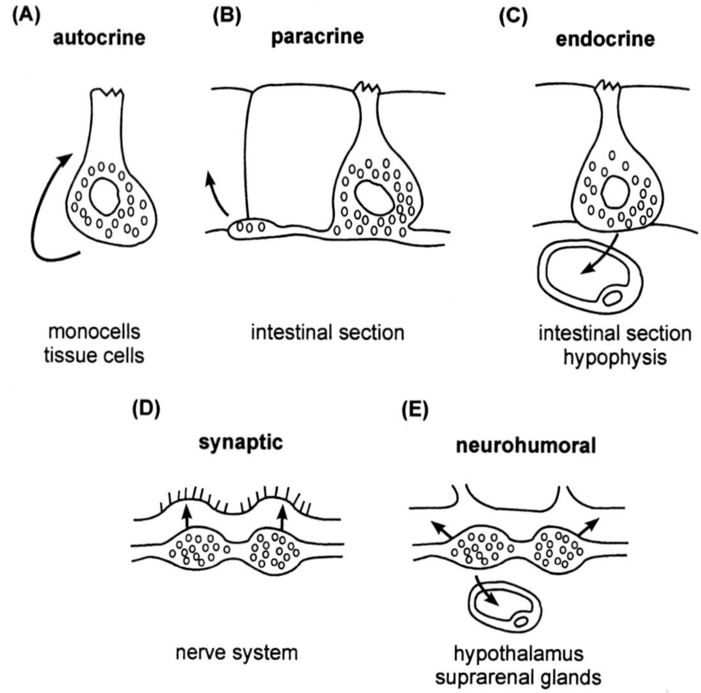

Fig. 3.1 Biochemical communications exerted by peptides and proteins.

occurring in one place exert effects in other places. An interactive blood-brain barrier (BBB) helps to regulate the passage of numerous peptides from the periphery to the CNS and *vice versa* [33]. Based on their lipophilicity and other physicochemical properties, many peptides are capable to cross the BBB by simple diffusion, whereas other peptides cross by saturable transport systems. Without doubt, the BBB provides a regulatory system involved in controlling the communication between the CNS and the rest of the body.

Biological membranes [34–36] contain a large number of proteins, such as integral proteins with nonpolar surface regions which allow hydrophobic association with the bilayer core, and peripheral proteins which bind to integral proteins on the membrane surface by polar interactions. For example, membrane proteins are involved in controlling the permeability of the membrane and supporting the transport of solutes across it against thermodynamic gradients. Due to the presence of these nonpolar cores, biological membranes are highly impermeable to most ionic and polar molecules; hence, the transport of these substances requires the action of specific transport proteins. For the transport of various ions (e.g., Na^+, K^+, Ca^{2+}, Cl^-, and metabolites, for example amino acids, pyruvate, nucleotides, and sugars), specific transmembrane transport proteins are required. Channel-forming ionophores, for example gramicidin A, H-CO-Val-Gly-Ala-D-Leu-Ala-D-Val-Val-D-Val-Trp-D-Leu^{10}-Trp-D-Leu-Trp-D-Leu-Trp-NH-CH_2-CH_2-OH, permits the

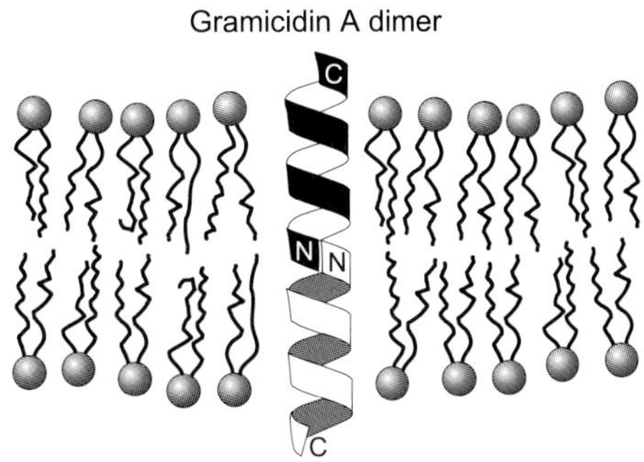

Fig. 3.2 Schematic representation of a transmembrane channel formed by a gramicidin A dimer.

passage of protons and alkali cations, but this channel is blocked by Ca^{2+}. The transmembrane channel is formed by dimerization of gramicidin A in a head-to-head fashion (Fig. 3.2).

The bacterial porins are trimeric transmembrane proteins each consisting mainly of a 16-stranded antiparallel β-barrel forming a solvent-accessible channel along its barrel axis. Special carrier proteins transport smaller metabolic intermediates from one location to another. Electrons generated (e.g., by oxidation of $NADH+H^+$ and $FADH_2$ in the citric acid cycle) pass through four protein complexes, termed the electron-transport chain. The free energy released by the electron transport is conserved by the generation of an electrochemical proton gradient across the inner mitochondrial membrane. The energy stored in the proton gradient is utilized by the proton-pumping F_1F_0-ATPase, an oligomeric protein, in the synthesis of ATP [37]. The tetrameric protein hemoglobin is a very important oxygen delivery system which transports the required amount of oxygen from the lungs, gills, or skin to the tissue for use in respiration [38].

Very specialized proteins are involved in the immune system used by vertebrates to defend themselves against other organisms [39–42]. Cellular immunity which targets parasites, fungi, virally infected cells, and foreign tissue is mediated by T lymphocytes or T cells, whereas humoral immunity directed against bacterial infections, and the extracellular phases of viral infections is mediated by a huge and diverse collection of antibodies or immunoglobulins. The latter are produced by B lymphocytes or B cells. Antibodies are glycoproteins consisting of two identical light (L) chains and two identical heavy (H) chains. There are two types of light chain ($L=\kappa$ or λ), each of which can be associated with any of five types of heavy chains (α, δ, ε, γ, μ). Secreted human immunoglobulins (Ig) consist of the following classes, in which the Greek letter designates the type of heavy chain corresponding to the class of Ig [IgA: $(L_2\alpha_2)_{n=1-3}$, IgD: $L_2\delta_2$, IgE: $L_2\varepsilon_2$, IgM: $(L_2\mu_2)_5$, IgG: $L_2\gamma_2$]. Four IgG subclasses (IgG_{1-4}) exist, differing in their γ chains. Each heavy chain contains a N-linked oligosaccharide. The four subunits of the Ig associate by disulfide bonds as well as by noncovalent interactions, thereby forming a Y-shaped symmetric dimer as confirmed by electron microscopy and X-ray diffraction.

As shown schematically in Fig. 3.3, each light (L) chain and heavy (H) chain consists of a variable (V) region (V_L and V_H) and the remaining constant region.

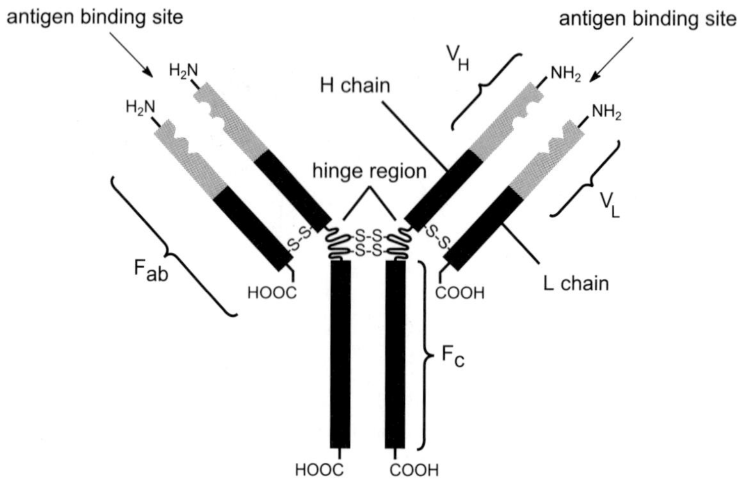

Fig. 3.3 Schematic diagram of a IgA antibody molecule.

The two identical F_{ab} segments form with their terminal variable regions the two antigen-binding sites of the antibody. Papain cleaves the IgG molecule in its hinge region, yielding two F_{ab} fragments and one F_c fragment. A virtually unlimited variety of antigen-binding sites are generated from the immune system by somatic recombination as well as somatic mutation. This generation of antibody diversity provides a shield against almost any antigen attacking the organism.

An antigenic determinant that is commonly (but not very precisely) called an epitope characterizes the portion of an antigen that makes contact with a particular antibody or T-cell receptor. There is a fundamental difference between epitopes recognized by T cells and those recognized by antibodies or B cells. T cells recognize protein antigens in a fragmented form on the surface of antigen-presenting cells. The cellular immune response begins when a macrophage engulfs and partially digests a foreign antigen and then displays the resulting antigenic fragments on its surface.

Two types of cell-surface proteins, known as major histocompatibility complex (MHC) proteins, are engaged in the immune response [43]. The MHC proteins function as markers of individuality; they are the antigen-presenting markers that allow the immune system to distinguish body cells from invading antigens and cells of the immune system from other cells, respectively. MHC proteins are the products of the major histocompatibility gene complex; they are integral membrane glycoproteins with the biological function to bind peptides and present them as T-cell epitopes to T-cell receptors. In order to indicate their function in the discrimination between own and foreign, the products of the MHC gene cluster are also termed antigens. Class I MHC molecules occur on most cells, whereas class MHC II molecules are found on macrophages and B lymphocytes (Fig. 3.4). The peptide fragments are presented to the T-cell receptor either via a class I or class II MHC molecule. As a rule, class I MHC molecules present pep-

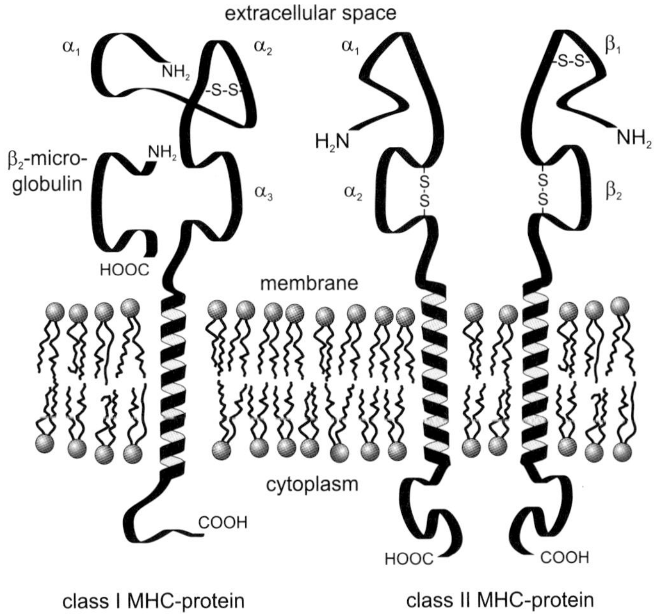

Fig. 3.4 Major histocompatibility complex (MHC) proteins.

tides to CD8 (predominantly cytotoxic) T cells, and class II MHC molecules to CD4 (mainly helper) T cells. Since MHC proteins have domains which structurally resemble immunoglobulins and T-cell receptors, the genes encoding these proteins form a gene superfamily.

The complement system serves as an essential biological defense system. It is directed against foreign invaders by eliminating foreign cells via complement fixation – that means killing foreign cells by binding and lyzing their cell membranes, by inducing the phagocytosis of foreign particles (opsonization), and triggering local acute inflammatory processes. The complement system consists of approximately 20 plasma proteins and is characterized by two related activation pathways: the antibody-dependent classical pathway; and the antibody-independent alternative pathway. Similar to the blood clotting system, both pathways mainly comprise the sequential activation of a series of serine proteases. Each activated protease acts only on the next component of the cascade. Most of the complement proteins names bear the upper-case letter "C" followed by a component number, and active proteases are indicated by a bar over the component number [44]. The classical pathway consists of the recognition unit (C1q, C1r, C1s) assembling on cell surface-bound antibody-antigen complexes, the activation unit (C2, C3, C4) amplifying the recognition event via a proteolytic cascade, and the membrane attack unit (C5–C9) that punctures the antibody-marked plasma membrane of the cell; this results in cell lysis and death. The alternative pathway is thought to provide the initial response to the invasion of microorganisms since it is activated in the absence of antibodies. It uses a couple of the same components as the classical pathway. The alternative pathway leads to the assembly of the membrane attack unit, but it proceeds via a series of reactions which are activated by whole bacteria, polymers of bacterial origin, and virus-infected host cells.

Blood clotting (blood coagulation) involves a double cascade of proteolytic reactions resulting in the formation of a gel (blood clot, thrombus) which is sufficiently dense to prevent bleeding from a wound [44]. In this bifurcated cascade, nearly 20 different blood coagulation factors are involved. Most of these substances are plasma glycoproteins which are synthesized in the liver. The human blood coagulation factors are historically designated by Roman numerals, and the activated form is indicated by the subscript "a", e.g., XII_a for the activated form of factor XII. With exception of fibrin, the active clotting factors are serine proteases. The human blood coagulation factors are listed in Tab. 3.2. The primary stages of the human blood clotting cascade is formally divided into the so-called intrinsic and extrinsic pathways. In the intrinsic pathway, the protein components are contained in the blood, whereas in the extrinsic pathway one of its important factors (VII) occurs in the tissues. The Stuart factor (X) may be activated by factor IX_a (a product of the intrinsic pathway) or by factor VII_a (a product of the intrinsic pathway). The initiation of blood clotting in the intrinsic pathway is performed by the contact system in which XII (in the presence of high-molecular weight kininogen, HMK) is activated by adsorption to a negatively charged surface such as glass or kaolin *in vitro*. Collagen and platelet membranes cause the same effect *in vivo*. In the last two steps, which require a phospholipid membrane surface and Ca^{2+}, factor XI_a activates factor IX, and the latter activates factor X in the presence of factor VIII_a. The start of the extrinsic pathway is characterized by proconvertin (VII) activation either by factor XII_a or thrombin. The VII_a formed promotes the activation of X in the presence of Ca^{2+}, phospholipid membrane, and tissue factor III.

Fibrous proteins are responsible for the mechanical properties of bone, hair, skin, horn, tendon, muscle, feather, tooth, and nails. Keratin which forms protofibrils consists of two pairs of α-helices [45]. The members of each pair are twisted together into a left-handed coil. Silk fibroin [46] exists as a semicrystalline array of antiparallel β-sheets. Collagen – the major protein component of connective tissue – forms a triple helical structure, and the molecules aggregate in a staggered array forming fibrils which are additionally cross-linked [45]. Last, but not least, elastin [47] – a protein with elastic properties – forms a three-dimensional network of fiber cross-linked by allysine aldol, lysinonorleucine, desmosine, and isodesmosine.

Toxic peptides and proteins are used by various species for defense purposes, or are employed in struggles for limited resources. Several venoms (e.g., of bees, wasps, and snakes) contain well-characterized peptides and proteins. Both in Europe and in the United States, the notorious toadstool *Amanita phalloides* is responsible for 95% of the casualties occurring after ingestion of poisonous fungus. After World War II, Theodor Wieland and coworkers were engaged in the elucidation of the constituents of the poisonous *Amanita* mushrooms. At the very end of these research activities, the formulae and three-dimensional structures of the toxic constituents of the green death cap *Amanita phalloides* and of the destroying

Tab. 3.2 Blood coagulation factors.

No.	Common name	M_r (kDa)	Properties and functions
I	Fibrinogen	~ 340	$(A\alpha)_2(B\beta)_2$; A and B represent the N-terminal 16- and 14-residue fibrinopeptides which are cleaved by thrombin, forming the fibrin monomer $\alpha_2\beta_2\gamma_2$
II	Prothrombin II_a is Thrombin	~ 72	Zymogen of thrombin; glycoprotein with 579 aa, several disulfide bonds; three-domain structure (N-terminal 40-residue Gla-domain with 10 Gla residues, and two ~ 115-residue kringle domains); Factor X_a cleaves between Arg^{271}-Thr^{272} and Arg^{320}-Ile^{321} releasing the N-terminal propeptide 1–271 and thrombin connected with the A and B peptides linked by one of the disulfide bonds
III	Tissue factor or thromboplasmin	~ 30	Membrane glycoprotein occurring in many tissues; involved in the intrinsic pathway
IV	Calcium ions		Promote the binding of the factors IX, X, VII and II to acidic phospholipids of cell membranes where activation is performed. Stabilizing functions for I, V, and other factors during activation, and for the subunit dissociation of XIII
V	Proaccelerin V_a is accelerin	~ 350	V_a promotes the binding of X and II to platelets, where the activation of X is performed, II is converted to II_a
VII	Proconvertin	~ 50	Involved in the extrinsic pathway; VII_a mediates the activation of X significantly in presence of Ca^{2+}, phospholipid membrane and III; VII_a activates also IX
VIII	Antihemophilic factor	~ 265	Activated VIII acts as an accessory factor during the activation of X by activated Christmas factor (IX_a); VIII forms a complex with the von Willebrandt factor during circulation in blood
IX	Christmas factor	~ 56	Single-chain Gla-containing glycoprotein; IX_a activated by XI_a and VII_a, respectively, activates factor X proteolytically, and X_a cleaves II in the preceding step of the clotting cascade
X	Stuart factor	~ 55	X may be activated via either the intrinsic pathway (IX_a + VIII_a + Ca^{2+}) or the extrinsic pathway (VII_a + III + Ca^{2+})
XI	Plasma thromboplastin antecedent (PTA)	~ 136	Two-chain glycoprotein joined by disulfide bond(s); XI belongs to the four glycoproteins of the so-called contact system

Tab. 3.2 (continued)

No.	Common name	M_r (kDa)	Properties and functions
XII	Hageman factor	~ 75	Single-chain glycoprotein acting as the first factor in the intrinsic pathway; XII is activated by plasmin, kallikrein, and the high-molecular weight kininogen (HMK); XII belongs to proteins of the contact system
XIII	Fibrin-stabilizing factor (FSF)	~ 300/ ~ 160	FSF is a transamidase present both in platelets and plasma; platelet FSF consists of two a chains (~ 75 kDa), whereas plasma FSF has two additional b chains (~ 88 kDa); XIII_a cross-links the rather fragile soft clots to the more stable hard clots
	Prekallikrein	~ 69	Zymogen of the serine protease kallikrein which activates the Hageman factor (XII)
	High-molecular weight kininogen (HMK)	~ 70	Activation yields a kinin that is involved in the activation of XII; HMK belongs to the proteins of the contact system

angels, *Amanita virosa* and their relatives, could be characterized as peptides and were named phallotoxins and amatoxins [48]. The basic formulae of the most important peptides of poisonous *Amanita* mushrooms are shown in Fig. 3.5. More details are given in Section 3.3.4 and in the Glossary.

Interestingly, antamanide **4**, which is a cyclic decapeptide of *A. phalloides*, protects experimental animals from intoxication by *Amanita*-derived phalloidin.

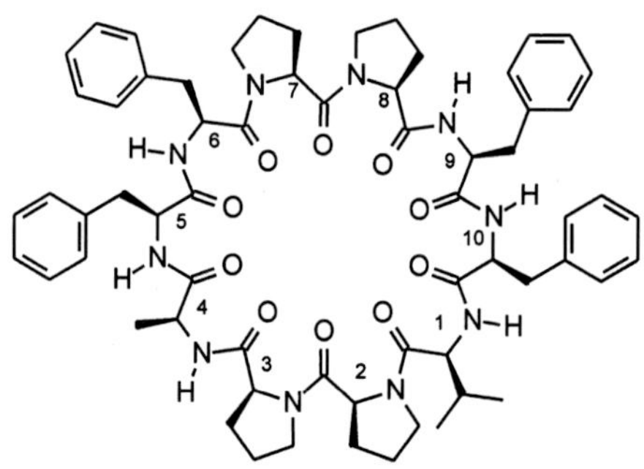

4

Frog skin has been used for medical purposes for centuries, and indeed is still used today in some South American countries. Amphibian skin is a rich source of various bioactive compounds, including various peptides which are produced by holocrine-type serous glands in the integument. The peptides are stored as granules in the lumen of the cells and are released upon specific stimulation. They are involved in the regulation of physiological functions of the skin, or in de-

(A)

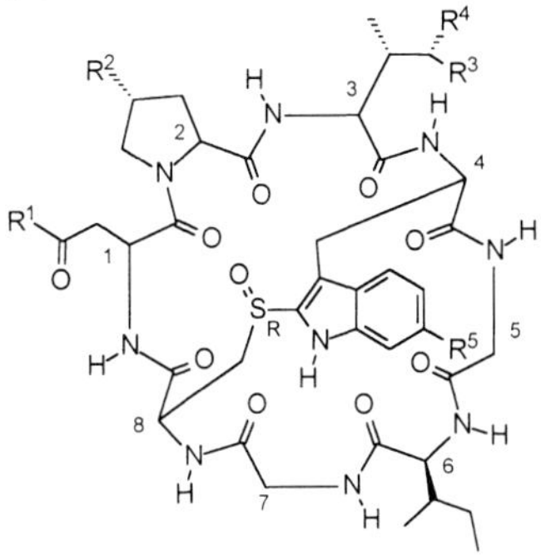

	R^1	R^2	R^3	R^4	R^5
α-Amanitin	NH_2	OH	CH_2OH	OH	OH
β-Amanitin	OH	OH	CH_2OH	OH	OH
γ-Amanitin	NH_2	OH	CH_3	OH	OH
ε-Amanitin	OH	OH	CH_3	OH	OH
Amanin	OH	OH	CH_2OH	OH	H

(B)

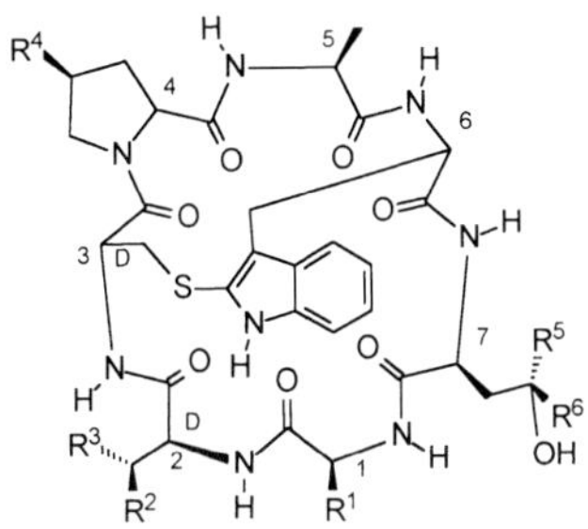

	R^1	R^2	R^3	R^4	R^5	R^6
Phalloin	CH_3	CH_3	OH	OH	CH_3	CH_3
Phalloidin	CH_3	CH_3	OH	OH	CH_2OH	CH_3
Phallisin	CH_3	CH_3	OH	OH	CH_2OH	CH_2OH
Prophalloin	CH_3	CH_3	OH	H	CH_3	CH_3
Phallacin	$CH(CH_3)_2$	OH	CO_2H	OH	CH_3	CH_3

Fig. 3.5 Selected naturally occurring toxic peptides of *Amanita* mushrooms. (A) amatoxins; (B) phallotoxins.

fense against predators or microorganisms. A systematic study on the occurrence of bioactive peptides in more than 100 amphibian species was performed in the early 1970s by Vittorio Erspamer [49], and led to the discovery of a huge number of peptides with a range of pharmacological activities. Normally, peptides originating from ribosome-mediated biosynthesis contain only L-amino acids, whereas peptides occurring D-amino acids are very rare. The first example of a D-amino acid-containing peptide discovered in an animal was dermorphin **5**, which was isolated from the skin of the South American frog *Phyllomedusa sauvagei* by Erspamer and his colleagues in 1981.

H–Tyr^1–D-Ala–Phe–Gly–Tyr^5–Pro–Ser–NH_2

5

Further examples are the neuroexcitatory peptide achatin-I (H-Gly-D-Phe-Ala-Asp-OH) from the ganglia of the African giant snail *Achatina fulica*; fulicin (H-Phe-D-Asn-Glu-Phe-Val-NH_2) was also found in *A. fulica*.

During the past decades, over 500 peptides have been detected in microorganisms, with *Bacillus* sp. and *Actinomyces* sp. the main sources. In contrast, only a few plants and animals are yet known as sources of these types of peptides.

Peptide antibiotics [50–52] have found applications in various fields ranging from the classical β-lactam antibiotics to immunosuppressors such as cyclosporin to teichoplanin or daptomycin, for example. Peptide antibiotics can be classified into nonribosomally synthesized peptides (e.g., bacitracins, polymyxins, gramicidins) and ribosomally synthesized peptides, such as the lantibiotics and related peptides [53]. The first totally chemical synthesis of a natural lanthionine peptide – nisin – was described by Tetsuo Shiba and coworkers in 1987 [54]. Nonribosomally synthesized peptides are very often drastically modified, and are largely produced by bacteria. Ribosomally synthesized peptide antibiotics act as a major component of the natural host defense molecules of the producing species. In 1962, Kiss and Michl [55] noted for the first time the occurrence of antimicrobial and hemolytic peptides in the skin secretions of *Bombina variegata*, and this led to the discovery of the antimicrobial peptide bombinin [56]. A database of antimicrobial peptides from amphibian skin [57] is available at: http://www.bbcm.univ.trieste.it/~tossi/search.html.

During the late 1980s, the magainins were isolated from the African clawed frog *Xenopus laevus* by Michael Zasloff [58]. These antimicrobial peptides, which are obtained from various amphibian species, represent the best-studied class of peptide antibiotics [59]. They are considered as effector molecules of innate immunity acting as the first line of defense against bacterial infections [60, 61]. The most studied mammalian peptides are the defensins [62]. The α-defensins (also known as "classical defensins") and β-defensins are predominantly β-sheet structures stabilized by three disulfide bonds, and contain a high arginine content [63]. The human α-defensins, HNP-1 (30 aa), HNP-2 (29 aa), and HNP-3 (30 aa) are constituents of the microbicidal granulae of neutrophils [64]. Thionins were the first antimicrobial peptides to be isolated from plants [65] and were found to be toxic against both Gram-positive and Gram-negative bacteria, yeast, fungi and various mammalian cell types [66]. Plant defensins consisting of 27 to 84 amino acid residues have eight disulfide-linked cysteines comprising a triple-stranded antiparallel β-sheet structure with only one α-helix. They have a high antifungal activity [67]. Cecropins [68] are insect-derived linear peptides found in the hemolymph of the giant silk moth (*Hyalopora cecropia*), and form α-helices in solution. These positively charged peptides form voltage-dependent ion channels in planar lipid membranes, but are not lethal to mammalian cells at microbicidal levels. Surprisingly, a porcine cecropin has been discovered in the upper intestinal tract.

Last – but not least – it should be mentioned that during the course of chemical evolution, peptides were produced ahead of all other oligomer precursors of biomolecules [69]. The formation of peptides under primordial Earth conditions have been simulated experimentally to investigate the real chances for the formation of precursor molecules under such conditions. Based on the results obtained, it might be interesting to seek the connection between self-organizing and self-reproducing processes in terms of biological evolution. It is well known that the formation of peptides from amino acids in aqueous solution is a thermodynamically and kinetically unfavorable process. Various condensation reactions have been proposed which can be categorized into melt processes, heterogeneous systems involving mineral catalysis, and condensation reactions induced by various reagents in homogeneous phase, respectively. In particular, "salt-induced peptide formation" (SIPF) seems to be the simplest way of obtaining peptides in aqueous solution, and the possible combination of SIPF with clay mineral catalysis might provide insight into the chemical evolution of peptides on the primitive Earth. Although at present highly speculative, when combined with the finding that peptides can self-replicate [70], further research in this area might lead to the vision of a "protein world" as the first steps leading to life on Earth.

3.2 Biosynthesis

3.2.1 Ribosomal Synthesis

Although the formation of a peptide bond is known to be a relatively simple chemical reaction, the biosynthesis of polypeptides is a very complex process. Ribosomal peptide synthesis involves deoxyribonucleic acid (DNA) encoding genetic information, and two different types of ribonucleic acids which convey (as messenger RNA, mRNA) the genctic information from the nucleus to the ribosome, and carry (as transfer RNA, tRNA) the specific amino acids appended enzymatically to the site of peptide bond formation of the ribosome [71]. The sequence of amino acid residues for each polypeptide is like a blueprint stored in the encoding DNA of the genes in the chromosomes [72]. Each of the 21 proteinogenic amino acids in a polypeptide is encoded by a triad of codon bases in the gene [73, 74].

In the first step of ribosomal synthesis, named transcription, mRNA is synthesized enzymatically under the direction of the DNA template by copying the encoded sequence of deoxyribonucleotides onto a molecule of mRNA with the complementary sequence of ribonucleotides. RNA polymerase initiates transcription on the antisense strand of a gene at a position designed by its promoter. Although prokaryotic mRNA transcripts do not require additional processing, eukaryotic mRNAs bear an enzymatically appended 5′ cap and very often an enzymatically generated poly(A) tail. In addition, the introns of primary eukaryotic mRNA transcripts are removed by splicing mechanisms.

Amino acid activation is the first step in the aminoacylation process of tRNA, catalyzed by aminoacyl-tRNA synthetases (aatRS). An amino acid reacts with adenosine triphosphate (ATP) under elimination of pyrophosphate to yield a mixed anhydride, the aminoacyl adenylate, which normally remains tightly bound to the enzyme (Fig. 3.6 A).

In the second step, the highly activated aminoacyl moiety is transferred to the appropriate tRNA, thereby forming the aminoacyl-tRNA and liberating adenosine

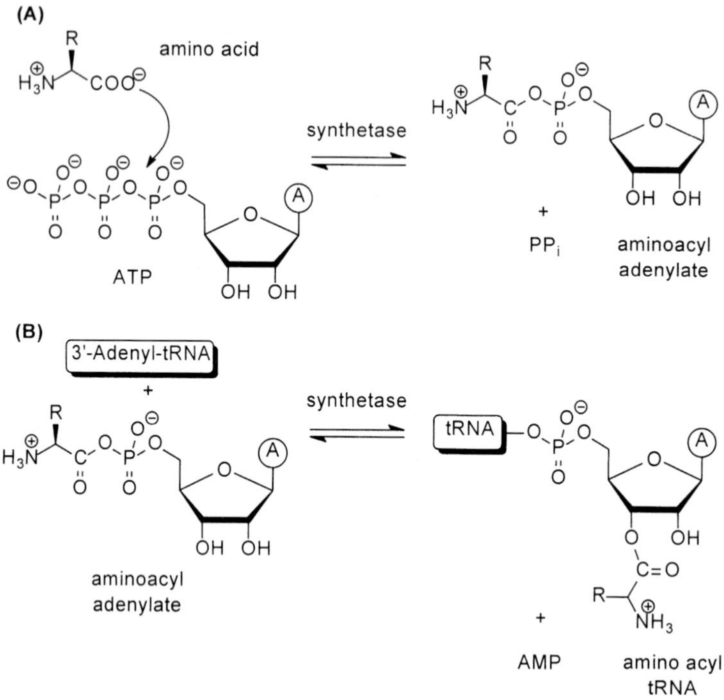

Fig. 3.6 Two-step process of tRNA aminoacylation catalyzed by aminoacyl-tRNA synthetases (aatRS).

monophosphate (AMP) (Fig. 3.6 B). The reaction is driven to completion by the pyrophosphatase-catalyzed hydrolysis of inorganic pyrophosphate generated in the first step. At least one specific tRNA and one aminoacyl-tRNA synthetase exist for every amino acid. Two classes of these enzymes are known, each containing ten members. The accurate translation process requires two exact recognition steps. The first step is the choice of the correct amino acid for the covalent attachment to a tRNA, this being catalyzed by the appropriate aatRS. The second step is recognition of the amino acid-loaded tRNA, as specified by the mRNA sequence. A proofreading or editing step by the appropriate aatRS enhances the fidelity of tRNA loading with its cognate amino acid at the expense of ATP hydrolysis. The tRNA vary in length from 60 to 95 ribonucleotides ($M_r \sim$ 18–28 KDa) and contain up to 25% of modified bases. In all tRNA the acceptor group for the amino acid is terminated by the sequence CCA with a free 3′-OH group. In aminoacyl-tRNA the amino acid is esterified to the 2′- or 3′-OH group of their 3′-terminal ribose moiety.

In the translation process, the appropriate tRNA is selected only through codon-anticodon interactions, without any participation of the aminoacyl group. The tRNA can recognize several degenerate codons through wobble base pairing, according to the wobble hypothesis suggested by Crick. He proposed that the first two codon-anticodon pairs show normal Watson–Crick geometry, whereas the third codon-anticodon pairing allows limited conformation adjustments in its pairing geometry based on a small amount of flexibility or "wobble" in the third codon position.

Ribosomal peptide synthesis proceeds stepwise from N- to C-terminus by reading the mRNAs in the 5′ → 3′ direction. The ribosome consists of a small and a large subunit, and contains (e.g., in the simple *E. coli* version) three rRNA molecules and 52 protein molecules. Eukaryotic ribosomes are larger and more complex than prokaryotic ribosomes, however. The ribosomes that are active in protein synthesis are tandemly arranged on the mRNA like beads on a string, named polyribosomes or polysomes.

In prokaryotic peptide synthesis, N^{α}-formylmethionine (fMet) is the N-terminal residue in the chain initiation process. The appropriate $tRNA_f^{Met}$ differs from $tRNA_m^{Met}$ which is bearing internal methionine residues. Initiation is a complex three-stage process requiring the two ribosomal subunits, fMet-$tRNA_f^{Met}$, and the initiation factors IF-1, IF-2, and IF-3. The ribosome has various tRNA binding sites. The P site, which normally binds the *p*eptidyl-tRNA, is in the initiation step occupied with fMet-$tRNA_f^{Met}$ whereas the A site binds the incoming *a*minoacyl-tRNA. Furthermore, the ribosome contains a third t-RNA-binding site, the E site (derived from *e*xit), which temporarily binds the outgoing tRNA.

Chain elongation is a three-stage process adding amino acid residues step by step, starting from the carboxy group of fMet at a rate of up to 40 residues per second. Three elongation factors (EF-Tu, EF-Ts, and EF-G) and GTP are involved in the elongation cycle. Peptide bond formation occurs in the second stage of this cycle by nucleophilic attack on the ester moiety of the peptidyl-tRNA in the P site by the amino group of the aminoacyl-tRNA in the A site (Fig. 3.7).

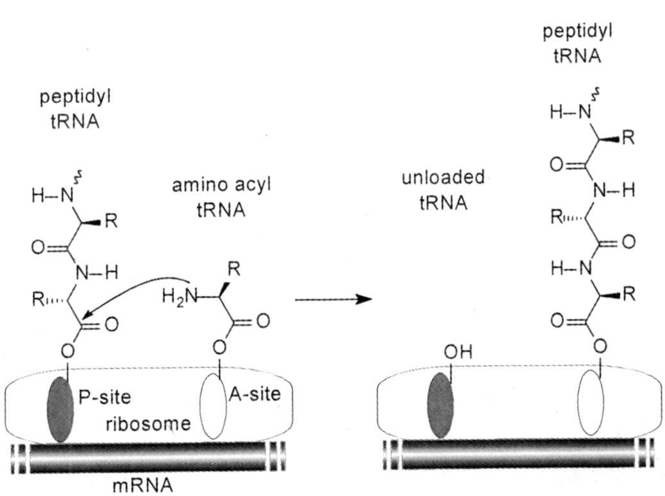

Fig. 3.7 Schematic view of the ribosomal peptidyl transferase reaction forming new peptide bond in translation process.

This process does not require ATP, and is catalyzed by the peptidyl transferase. Since the catalytic activity appears to be resident on the 23S rRNA, this process is probably catalyzed rather by a ribozyme than by an enzyme. The resulting unloaded P site tRNA is then transferred to the E site (not shown in Fig. 3.7) and, simultaneously, in the so-called translocation process the peptidyl-tRNA from the A site is transferred together with its bound mRNA to the P site, thus preparing the ribosome for the next elongation cycle.

Chain termination in *E. coli* starts when the termination codons UAA, UGA, or UAG are recognized by the release factors RF-1 (UAA and UAG) and RF-2 (UAA and UGA). In addition, RF-3, which is a GTP-binding protein, stimulates together with GTP the binding process of the other two releasing factors. As a result of the binding process the peptidyl transferase hydrolyzes the ester bond of the peptidyl-tRNA, and the synthesized polypeptide is released. The resulting unloaded tRNA and the release factors dissociate from the ribosome with concomitant hydrolysis of the RF-3-bound GTP to GDP and P_i. The translation process in eukaryotes resembles that in prokaryotes, but differs from it in certain details.

3.2.2
Post-Translational Modification

Post-translational covalent modifications occurring in nature include acetylation, hydroxylation, methylation, glycosylation, sulfatation, iodination, carboxylation, phosphorylation, nucleotidylation, ADP-ribosylation, and numerous other types.

This post-translational processing is an essential prerequisite for mature polypeptides or proteins. Normally, the translation product cannot be considered as a functional protein. After the assembly of the complete sequence of a protein, some of the amino acid building blocks may be involved in post-translational modifications. Most modifications are performed after release of the polypeptide from the ribosome, but modifications such as disulfide bridge formation or N-terminal acetylation very often occur in the nascent polypeptide chain. Enzymes catalyzing processing reactions are mainly located in the endoplasmic reticulum (ER) and Golgi apparatus; among these are enzymes which catalyze disulfide bridge formation, iodination, and glycosylation. Peptide chain folding is stabilized mainly by noncovalent interactions, and in many cases the association of subunits forming oligomeric proteins is an important event after translation. In general, many proteins are modified by limited proteolysis and/or by derivatization of specific amino acid residues.

3.2.2.1 Enzymatic Cleavage of Peptide Bonds

Limited proteolytic cleavage by specific peptidases is a common process in post-translational modification. Although it seems to be a waste of cellular resources, proteolytic reactions belong to the most important types in the maturation of proteins.

Proteolytic removal of the N-terminal fMet or Met (in eukaryotic proteins) residues most likely occurs in all synthesized polypeptides shortly after their release from the ribosome. Only few mature proteins contain the formyl or even the N-terminal methionine residue.

- Eukaryotic membrane-bound ribosomes synthesize transmembrane proteins and proteins that are destined for secretion. The question arises how these proteins are differentiated from soluble and mitochondrial proteins assembled by free ribosomes. According to the signal hypothesis, established by the Nobel laureate Günter Blobel, all secreted proteins are synthesized with N-terminal signal peptide sequences of 13 to 36 predominantly hydrophobic amino acid residues. Immediately after the signal peptide sequence of the preprotein enters the lumen of the rough endoplasmic reticulum (RER), it is specifically cleaved by a membrane-bound signal peptidase, thereby releasing the signal peptide (pre-sequence). Likewise, many proteins are synthesized as inactive precursors – termed proproteins – that must be activated by limited proteolysis that cleaves the so-called propeptide sequence. Normally, the initial biosynthesis product is termed a prepro-protein. The activation of proenzymes (zymogens) of the complement system and proteolysis in the cascade system of blood coagulation should be mentioned in this context. Some polypeptides are synthesized as segments of so-called polyproteins containing the sequences of two or more polypeptides. This applies to polypeptide hormones, ubiquitin, and virus proteins (such as those causing AIDS or poliomyelitis):
- Ubiquitin is synthesized as several tandem repeats, termed polyubiquitin.
- Proopimelanocortin (POMC) is a prototype of a polyprotein with a cleavage pattern that varies among different tissues, yielding a different set of peptide hormones.
- The hormone insulin is derived from an inactive single-chain, 84-residue precursor termed proinsulin. Correct formation of the disulfide bonds is performed efficiently *in vivo* by assembly of the longer chain of proinsulin containing the A and B chains, together with an internal segment known as the C chain. The active hormone is formed after proteolytic excision of the C chain from proinsulin.

Another example is the biosynthesis of collagen, a fibrous triple-helix protein that forms the major extracellular component of connective tissue. N- and C-terminal sequences each containing about 100 residues are constituents of the polypeptide chains of procollagen. These propeptide sequences direct the formation of the collagen triple helix, and they are removed by amino- and carboxyprocollagen peptidases after folding of the triple helix. In addition, collagen assembly requires chemical modification of Pro and Lys residues, and this will be discussed below.

3.2.2.2 **Hydroxylation**

The assembly of collagen in the course of collagen biosynthesis is a very good example of protein maturation by post-translational modification. When the nascent polypeptides of procollagen move into the RER of the fibroblasts, the sequence-specific enzymes prolyl hydroxylase and lysyl hydroxylase catalyze the regioselective hydroxylation of proline and lysine residues to give 4-hydroxy- or 3-hydroxyproline (Hyp) and 5-hydroxylysine (Hyl), respectively. Furthermore, glycosyl transferases catalyze the attachment of sugar residues to Hyl residues. These modifications occur prior to triple helix formation, because the hydroxylases and glycosyl transferases involved do not act on the helical structure.

Besides the examples mentioned above, polypeptides and proteins of biological interest often include structural elements exceeding the 21 genetically encoded amino acid residues. Both the side-chain functionalities and the terminal amino and carboxy groups of proteins are covalently modified. More than 150 different types are known. Thirteen of the encoded amino acids contain functional groups at the side chain as possible modification sites.

3.2.2.3 **Carboxylation**

Carboxylation of glutamate building blocks in certain proteins to give γ-carboxyglutamate (Gla) residues is catalyzed by vitamin K-dependent carboxylases. Some proenzymes in the blood-clotting cascade undergo this type of modification. Human prothrombin contains in its Gla domain sequence 6–32 no fewer than 10 Gla residues. The synthesis of Gla from Glu, known as the liver reaction cycle [75], requires vitamin K hydroquinone as an essential cofactor that is transformed into vitamin K-2,3-epoxide during Gla synthesis. This epoxide is converted to vitamin K in two sequential reactions. Proteins containing Gla residues involved in Ca^{2+} binding have been detected also in other tissues following the discovery of Gla residues in clotting factors.

3.2.2.4 **Glycosylation**

Glycosylation of proteins is more abundant than all other types of post-translational modifications taken together. Glycoproteins contain covalently linked carbohydrate moieties in proportions between <1 and >70 mass %, and can be found both in soluble and membrane-bound forms in all cells, but also in the extracellular matrix and in extracellular liquids. They are characterized by microheterogeneity with respect to the carbohydrate portion. Glycoproteins occur naturally in a variety of forms (glycoforms [76]) that share the same sequence, but differ in both the nature and site of glycosylation. These microheterogeneous mixtures complicate the determination of the exact function in structure-activity relationships.

Especially in eukaryotic cells, many secreted and membrane-associated proteins, as well as those located inside membranous organelles, are glycosylated. Protein glycosylation plays an important role in biological processes [77–81], for example in protein folding, cellular differentiation, cell-cell communication, and the sliminess of mucosa. The hydrophilic carbohydrate moieties exert considerable influence on the structure, polarity, and solubility of the proteins. The addition of large glycan structures to the protein backbone may dramatically alter its conformation and consequently influence the protein functions. Furthermore, it may substantially affect the thermal stability and targeting/clearance properties of the glycosylated protein. Co- or post-translational modification of proteins with oligosaccharide moieties may also change the folding pathway of the polypeptide chain by interactions between the protein and the oligosaccharide part.

The higher steric requirements, shielding of potential cleavage sites, and charged oligosaccharide groups confer higher stability towards proteolytic degradation of the protein. While native coeruloplasmin has a half-life of 54 h, the enzymatic degradation of the terminal *N*-acetylneuraminic acid with neuraminidase reduces the half-life to 5 min [82]. A galactose-recognizing receptor system (asialoglycoprotein receptor) is responsible for rapid degradation in the liver. This receptor system recognizes the terminal galactose residue exposed by cleavage of *N*-acetylneuraminic acid and initiates lysosomal degradation. By contrast, ribosomal enzymes in the Golgi complex are addressed for transport into the lysosomes by the attachment of mannose-6-phosphate units to the terminus of the oligosaccharide chains [83].

The carbohydrate proportion of glycoproteins in tumor cells is often changed significantly in comparison with regular cells. The investigation of so-called tumor-associated antigens provides promising starting points for diagnostics and immune therapy [84].

Carbohydrate moieties on cell surfaces can be considered as molecular codes. Although often only seven or eight monosaccharide units occur in a single glycosyl residue of a mammalian cell [85], the multifunctionality of these monomers results in an unusually high number of possible complex structures. Considering the possibility of branched derivatives, the stereochemistry of the glycosidic bond, together with modifications of hydroxylic and amino groups of the protein, millions of different structures are possible even for tetrasaccharides. These many different ways of encoding information may all lead to different biological functions.

On the cell surface many carbohydrate groups of glycoconjugates are involved in different types of biochemical recognition processes responsible for growth, development, infection, immune response, cell adhesion, formation of metastases, and different events of signal transduction. Viral invasion also may involve membrane-bound glycoproteins. Erroneous glycosylation is thought to be associated with autoimmune diseases, infectious diseases, or cancer. Such compounds often occur in minute concentrations and are very difficult to isolate, to characterize, and to synthezise. The carbohydrate portions of glycoproteins are of major importance in the realm of cell-cell and cell-matrix recognition. Cells utilize glycoproteins for encoding information on the protein-folding pathway, the localization (intracellular organels, cell surface, or protein export), or on the recognition of other proteins. Carbohydrate-mediated cell adhesion is also of major biological importance and may be induced either by tissue damage or infections [86]. The interaction of the glycoprotein E-selectin (endothelial leukocyte adhesion molecule, ELAM-1) and an oligosaccharide on the surface of neutrophilic cells represents such an adhesion process. E-selectin is expressed on the surface of endothelial cells and recognizes the tetrasaccharide sialyl-Lewis^x (Sle^x), that forms the terminus of a glycolipid on the surface of neutrophilic cells [87, 88]. Cell adhesion is initiated by cytokines or other inflammatory compounds (e.g., leukotrienes, toxins, lipopolysaccharides) that induce the formation of E-selectin. The interaction between E-selectin on the endothelial cell and the complementary receptor on leukocytes present in the bloodstream leads to retardation of the leukocyte in the so-called rolling process on the surface of the endothelium. In addition, adhesion between the leukocytes and the endothelial cells is mediated by interaction of integrins present on the leukocyte and proteins such as intercellular adhesion molecule 1 (ICAM-1) containing an RGD sequence (-Arg-Gly-Asp-), or the protein VCAM-1 containing the recognition sequences LDV (-Leu-Asp-Val-) or IDSP (-Ile-Asp-Ser-Pro-). As a consequence of these interactions, the leukocyte is attached to the endothelium, undergoes extravasation, and resolves the inflammatory process.

The tetrasaccharide sialyl-Lewis^x is also found on the surface of several tumor cells. Cancer cells clearly utilize this type of adhesion for metastasizing processes, and spread with the bloodstream throughout the body. Synthetic sialyl-Lewis^x derivatives compete in solution with leukocytes for binding of E-selectin. As a consequence, these derivatives may act as antiadhesive agents and are regarded as potential antitumor or anti-inflammatory agents.

In eukaryotes, the protein portion of glycoproteins is synthesized ribosomally, and the attachment of the carbohydrate moieties occurs either co- or post-translationally. The saccharide residues in glycoproteins are covalently linked to the protein backbone either across N- (via asparagine **6**) or O-glycosidic bonds (via serine, or threonine **7** and **8**).

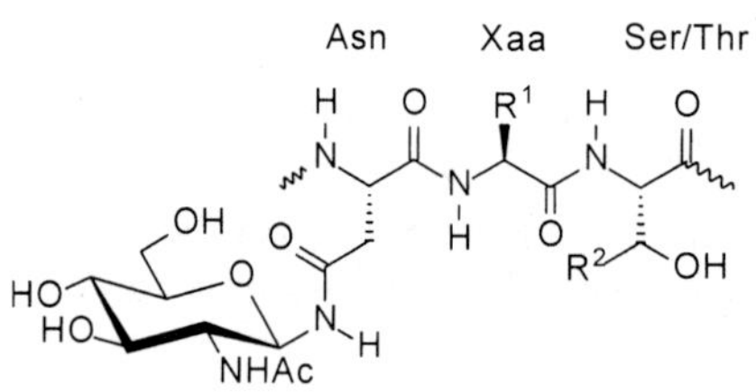

6

R^2 = H (Ser) or CH_3 (Thr)

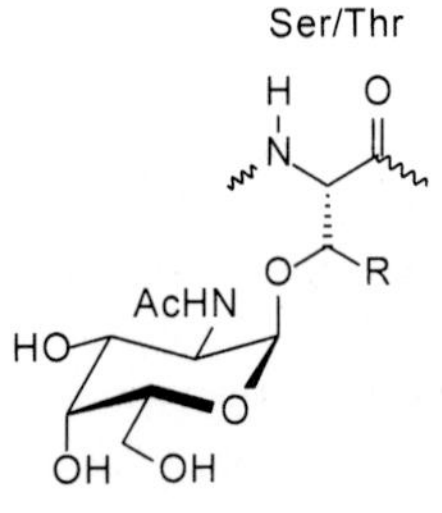

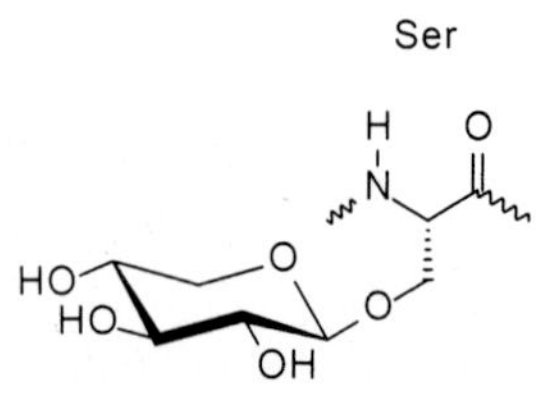

7 **8**

R = H (Ser) or CH_3 (Thr)

The carbohydrate moiety of N-glycosides is attached to the amide side chain of asparagine via a β-glycosidic bond. The first monosaccharide unit usually is an *N*-acetylglucosamine residue **6**. At least three different types of N-glycosides can be distinguished: complex, high-mannose, and hybrid (Fig. 3.8).

These types are all characterized by a $Man_3GlcNAc_2$ core sequence attached to the protein via an N-glycosidic bond to an Asn residue within the sequence Asn-Xaa-Ser/Thr, where Xaa is any amino acid residue except Pro (and sometimes Asp) [89]. On occasion, C^6 of the first GlcNAc residue is glycosylated with α-L-fucose.

The biosynthesis [90] of N-linked glycoproteins begins in the ER with the multistep formation of a lipid-linked precursor that consists of dolichol pyrophosphate linked to a common 14-residue core oligosaccharide of the composition (*N*-acetylglucosamine)$_2$(mannose)$_9$(glucose)$_3$. Each of the 14 monosaccharide units is added stepwise to the dolichol pyrophosphate carrier, catalyzed by a unique glycosyltransferase. The saccharide moiety is then transferred to an Asn residue of the

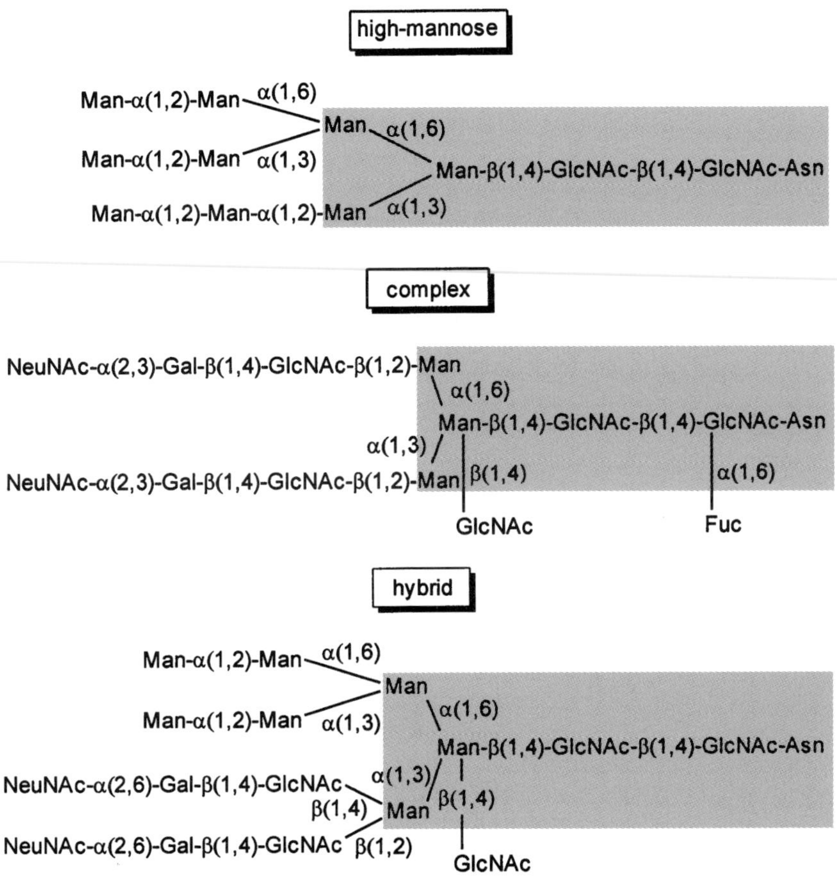

Fig. 3.8 Structural motifs of N-glycopeptides.

growing peptide chain. Processing begins in the ER by enzymatic removal ("trimming") of three glucose residues and one mannose residue. The glycoprotein is then transported in membranous vesicles to the Golgi apparatus in order for processing to be continued. Trimming of further mannose residues followed by attachment of various other monosaccharide units finally provides the typical primary structures of N-linked oligosaccharides as shown in Fig. 3.8.

The saccharide moiety of the different core types in O-linked glycoproteins (Fig. 3.9) is attached via an α-O-glycosidic bond between *N*-acetylgalactosamine and Ser or Thr (**7**), or between xylose and Ser (**8**). Tyrosine O-glycosylation has been reported; 5-hydroxylysine residues in collagen may also be glycosylated. Other less-common linkage types are known, including surprising C-linkages to the C2 position of the tryptophan indole group. Some noncoded amino acids (e.g., 4-hydroxyproline, 5-hydroxylysine, and τ-histidinoalanine) have also been found at the linkage site in glycopeptides. The carbohydrate moiety can be a single monosaccharide or a more complex, sometimes branched, oligosaccharide consisting of up to about 20 monosaccharide units.

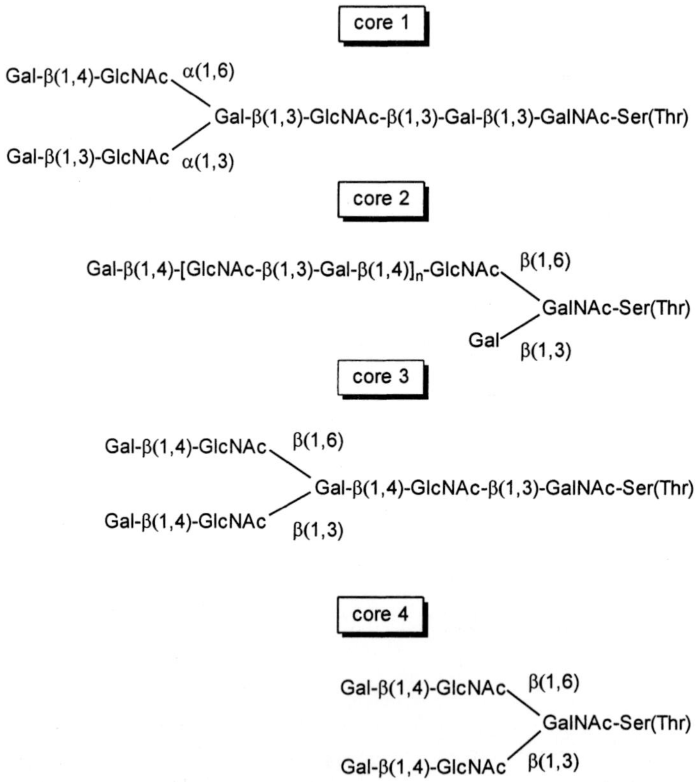

Fig. 3.9 Structural motifs of O-glycopeptides.

Synthesis of O-linked oligosaccharides is carried out in the Golgi apparatus by sequential coupling of specific monosaccharide units to certain Ser or Thr residues [91].

The urgent need for alternative sources of homogeneous glycoproteins is a challenge for chemical glycoprotein and glycopeptide synthesis (see Section 6.3).

3.2.2.5 **Amidation**

Amidation of bioactive peptides is performed using the bifunctional enzyme peptidylglycine α-amidating monooxygenase (PAM, EC 1.14.17.3) by N-oxidative cleavage of a glycine-extended precursor [92]. The N-terminal domain of PAM is the ascorbate-dependent peptidylglycine α-hydroxylating monooxygenase (PHM) that catalyzes the stereospecific hydroxylation of the glycine C^{α} in peptidylglycine substrates, whereas the peptidyl-α-hydroxyglycine α-amidating lyase (PAL) is responsible for generation of the α-amidated peptide product and glyoxylate (Fig. 3.10). The isolation of peptide amides with all 20 amino acid amides underline the broad substrate specificity of PAM.

3.2.2.6 **Phosphorylation**

Nature widely uses the phosphorylation and dephosphorylation of serine, threonine, and tyrosine residues in polypeptides and proteins as a universal mecha-

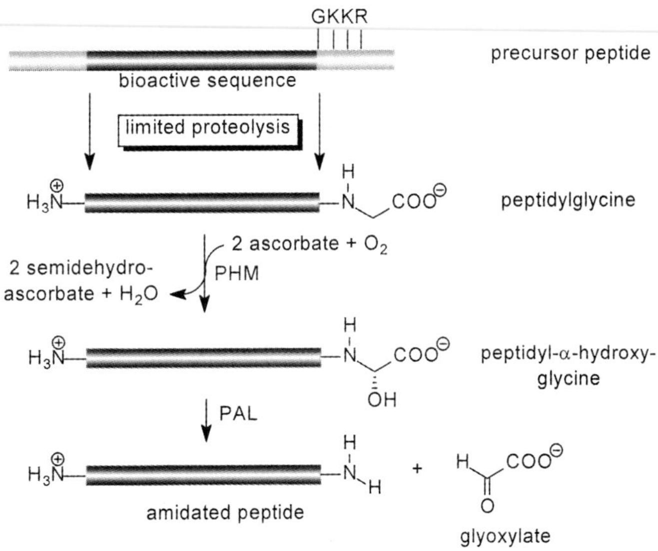

Fig. 3.10 Amidation reaction catalyzed by the bifunctional enzyme peptidylglycine α-hydroxylating monooxygenase (PAM). PAL = peptidyl-α-hydroxyglycine α-amidating lyase; PHM = peptidylglycine α-hydroxylating monooxygenase.

nism in the regulation of many cellular processes, for example in the activation or deactivation of enzymes in signal transduction and protein trafficking. Protein phosphorylation in prokaryotic systems mediates chemosensing of extracellular signals which eventually leads to chemotaxis.

Protein kinases and protein phosphatases respectively effect the phosphorylation and dephosphorylation of serine, threonine, and tyrosine residues present in proteins. All organisms utilize this phenomenon for the regulation of a variety of intracellular processes. Protein-tyrosine kinases (PTK) convert an extracellular signal (binding of a ligand to a membrane-bound receptor) into an intracellular signal (phosphorylation of tyrosine residues situated on specific proteins) [93–95]. Covalent protein phosphorylation is important for many purposes, including control of the cell cycle, regulation of cell growth and proliferation, regulation of transcription, and signal transduction. Furthermore, the biological functions of many receptors for hormones, neurotransmitters, peptide-derived drugs, and growth factors are known to be regulated by reversible phosphorylation. The phosphorylation of a single Ser residue is sometimes sufficient to regulate the enzyme activity, a classical example being the control of glycogen synthase and glycogen phosphorylase. These enzymes are regulated by phosphorylation/dephosphorylation integrated in amplifying cascades controlled by the hormones insulin, glucagon, and epinephrine, and also Ca^{2+}. Other phosphoproteins include casein, ovalbumin, phosphovitin, and vitellin.

Phosphorylated tyrosine residues (pTyr) serve for example as recognition sites for binding by other signaling proteins. These are often assembled in a modular fashion like the Src homology 2 (SH2) domain or phosphotyrosine-binding domains (PTB). Both of these display high affinity towards ligands containing pTyr. SH2 domains are clusters of approximately 100 amino acids binding to peptides, phosphorylated on tyrosine residues, with an affinity in the range of 10–100 nM [96]. In principle, two categories of SH2 domains can be distinguished according to the recognition motif:

1. Class 1 SH2 domains recognize sequences of the type pTyr-Xaa-Xaa-Ile/Pro (Xaa: hydrophilic amino acids).
2. Class 2 SH2 domains bind with high affinity to the motif pTyr-Yaa-Zaa-Yaa (Yaa: hydrophobic amino acid, Zaa: any amino acid).

pTyr phosphate groups are removed by protein tyrosine phosphatases (PTPase). Consequently, physiological or pathological events in this type of signaling cascade may involve either phosphorylation (protein tyrosine kinases), recognition and binding of phosphorylated proteins (SH2 and PTB domains), and dephosphorylation (protein tyrosine phosphatases). All three phenomena may be addressed in order to influence possible pathological events [97]. While tyrosine phosphorylation only represents a minor subset of protein phosphorylations of a cell, it profoundly influences signal transduction and, hence, cellular responses such as differentiation, mitogenesis, migration, and survival [98]. Inappropriate tyrosine phosphorylation is closely connected with pathological phenomena, including oncogenesis. While only 0.03% of protein-bound phosphate is found on tyrosine residues in normal cells, this percentage is significantly (10- to 20-fold) increased in some malignant tumors.

3.2.2.7 Lipidation

Lipidation of proteins is performed by the enzyme-catalyzed covalent attachment of different types of lipid groups (prenylation, acylation with fatty acids, and glycosylphosphatidylinositol groups). Lipid-modified proteins are very often attached to cell membranes, with the lipid moiety anchoring the protein to membranes and mediating protein-protein interactions. In many cases these lipid-linked proteins are involved in the transduction of extracellular signals across the plasma membrane and into the nucleus.

Prenylation of polypeptides and proteins with polyisoprenoids belongs to the functionally most important post-translational modification [99–101]. S-Farnesylation and S-geranylgeranylation – the two different types of prenylation – are principally similar in their behavior, but they differ in specificity with regard to the C-terminal sequence of proteins. Prenylation can be performed via a covalent attachment of either a C_{15} farnesyl or C_{20} geranylgeranyl moiety to the cysteine residue of CAAX motif present in the protein to be modified (A: aliphatic amino acid, X: variable residue). Farnesylation, catalyzed by the farnesyltransferase (FTase), is preferred for proteins with the CAAX motif where X is either Met, Ser, Gln or Ala, whereas proteins with X=Leu are geranylgeranylated under the catalysis of the geranylgeranyltransferase (GGTase-I). FTase catalyzes transfer of the C_{15} farnesyl moiety from farnesyl pyrophosphate (FPP) to a protein containing the C-terminal CVLS motif yielding the farnesylated protein **9**.

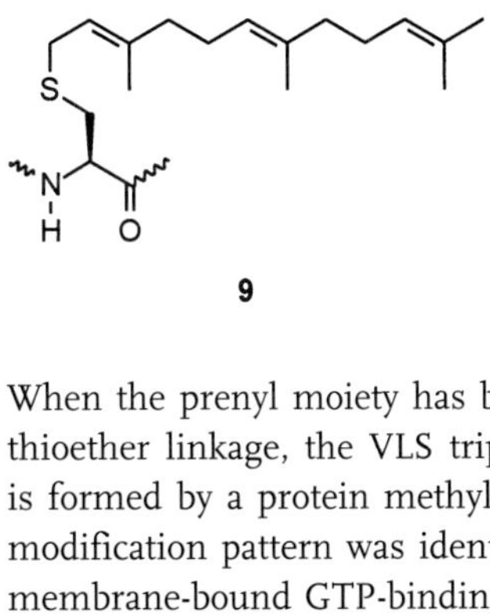

9

When the prenyl moiety has been attached to the Cys residue of the protein via a thioether linkage, the VLS tripeptide is proteolytically cleaved and a methyl ester is formed by a protein methyltransferase at the new C-terminal Cys residue. This modification pattern was identified in Ras proteins [102]. Ras proteins are plasma membrane-bound GTP-binding proteins that influence numerous signal transduction processes and play a crucial role as a molecular switch involved in tumor formation [103].

Acylation with fatty acids is performed either co- or post-translationally in various protein families that cover important functional properties [104]. The acyl moieties are linked either via thioesters or via amide bonds. N-Acylation has been found either in the form of myristoyl (C_{14}) groups linked to a N-terminal glycine of a protein **10** [105], or on occasion as palmitoyl moieties attached to lysine ε-amino groups.

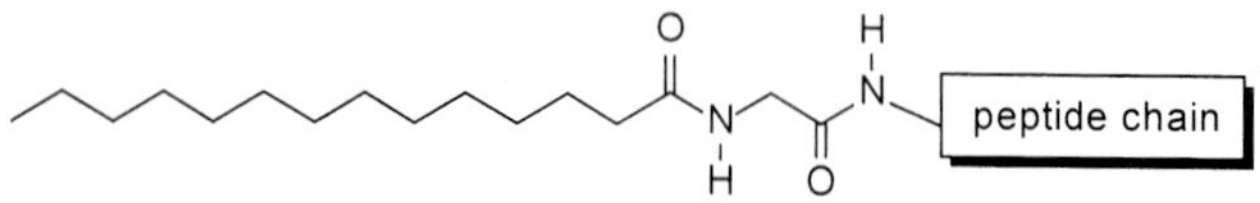

10

N-Terminal acetylation via common biochemical acetylating agents leads to N-acetylated peptides such as α-melanotropin (α-MSH), or some proteins. In histone H4, the N-terminal Ser is N-acetylated and may be O-phosphorylated. Some other Lys residues in positions 5, 8 and 12 of this protein are N^{ε}-acetylated. Interestingly, the side-chain amino function Lys^{20} is either mono- or dimethylated. N^{ε}-Palmitoyl-lysine has been found in adenylate cyclase from *Bordatella pertussis,* where it is formed enzymatically by a specific enzyme [106]. Palmitoyl (C_{16}) groups are most commonly linked to specific cysteine residues in protein S-acylation [107], this modification occurring post-translationally in the cytosol. Palmitoylated proteins are occasionally also prenylated. Thioester-linked palmitoic acid may undergo acylation/deacylation cycles due to the labile nature of the thioester bond. In addition, only a few O-acylated peptides or proteins have been elucidated [108].

In GPI-linked proteins **11** the glycosylphosphatidylinositol moieties anchor the protein to the exterior surface of the eukaryotic plasma membrane.

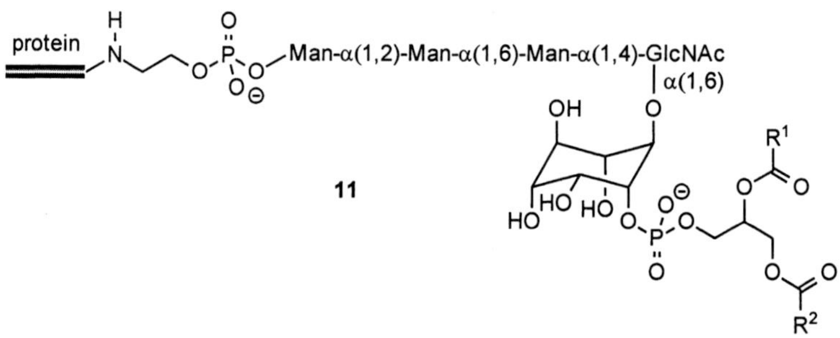

11

GPI-linked proteins belong are associated with, for example, enzymes, recognition antigens, receptors, and immune system proteins. The glycan is a core tetrasaccharide that varies with the identity of the protein. In addition, diversity in the fatty acid residues (R^1 and R^2) must be stated.

3.2.2.8 **Pyroglutamyl Formation**

Pyroglutamic acid (<Glu, pGlu, Pyr, or Glp) formation via intramolecular cyclization of N-terminal Gln residues is another example of post-translational modification. The aminolytic cyclization of a N-terminal glutamine ($R=NH_2$) in peptides irreversibly gives terminated pyroglutamyl peptides **12**.

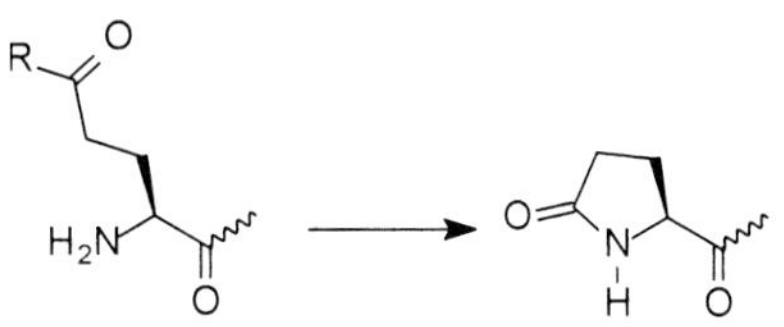

12

In addition, a N-terminal γ-alkyl ester of glutamic acid (R = OAlkyl) may be converted into the same derivative. Naturally occurring pyroglutamyl peptides include eisenin (<Glu-Gln-Ala-OH), pelvetin (<Glu-Gln-Gln) from algae, or the liberins thyrotropin-releasing hormone (TRH) (<Glu-His-Pro-NH_2) and gonadoliberin. In contrast, pyroglutamyl formation is a serious side reaction in peptide synthesis.

3.2.2.9 **Sulfatation**

Sulfatation of tyrosine residues **13** is performed in cholecystokinin, gastrin, and related peptide hormones.

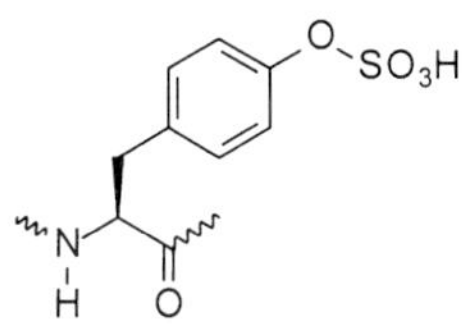

13

Protein tyrosine sulfatation is emerging as a widespread post-translational modification in multicellular eukaryotes. Today, proteins of different types and different modes of action are known to be sulfated at tyrosine residues, among them α-choriogonadotropin, heparin cofactor II, and gastrin II. Secretory proteins are the most abundant, but not exclusive, in-vivo substrates for tyrosine sulfatation [109].

Proteins that contain tyrosine residues sulfated at the phenolic hydroxy function participate in protein-protein interactions, mediated by recognition of the sulfate group. Tyrosine sulfatation is clearly one of the key modulators mediating inflammatory leukocyte adhesion. Furthermore, chemokine receptors may undergo tyrosine sulfatation, which seems to be crucial for the recognition by other proteins.

The modification is generated by tyrosyl protein sulfotransferases (TPST) in the Golgi apparatus using 3′-phosphoadenosine-5′-phosphosulfate (PAPS) as the sulfate donor (Fig. 3.11).

This does not seem to be a comparable system of dynamic regulation, as in tyrosine phosphorylation. However, the functional significance of tyrosine sulfatation remains the subject of many ongoing investigations.

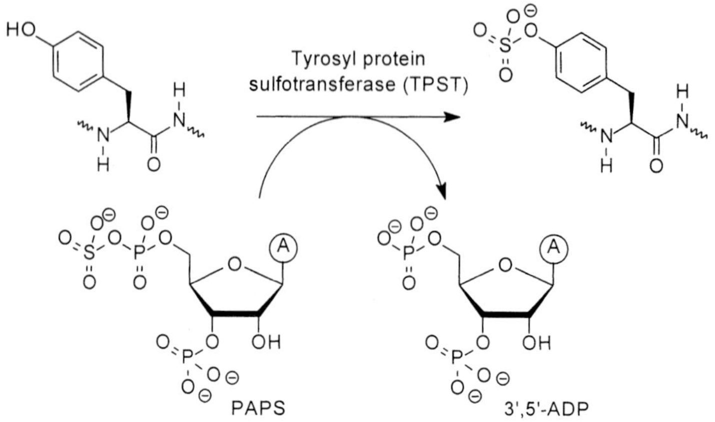

Fig. 3.11 Tyrosyl protein sulfotransferase (TPST) catalyzed sulfation.

3.2.3
Nonribosomal Synthesis

A vast number of structurally diverse linear, cyclic and branched cyclic peptides, lipopeptides, depsipeptides and peptidolactones are produced by microbial organisms via the nonribosomal pathway. Nonribosomal peptide synthesis occurs without predetermination of the amino acid sequence by nucleic acids, and is mechanistically performed according to the template-directed peptide synthesis. The existence of a poly- or multienzymatic pathway to peptides was first suggested by Fritz Lipmann in the mid-1950s [110], and this idea was later defined as being template-directed peptide synthesis catalyzed by large multifunctional enzymes (peptide synthetases). The peptide synthetases show a unique modular arrangement of functional units composed of individual domains catalyzing the successive condensation of their substrates by adenylation, thiolation, modification, and transpeptidation (for selected reviews, see [111–113]). The nonribosomal multienzyme thiotemplate mechanism is shown in simplified form in Fig. 3.12.

According to this mechanism, peptide bond formation is performed on multienzymes in a two-step process that includes ATP-dependent aminoacyl adenylation with similarities to the amino acid activation procedure catalyzed by aminoacyl-tRNA synthetases, and aminoacyl thioesterification at specifically reactive thiol groups of the multienzyme. The initially formed unstable aminoacyl adenylate is subsequently transferred to a thiotemplate site, where it is bound as a thioester to the cysteamine moiety of an enzyme-bound 4′-phosphopantetheine (4′-PP), which is covalently bound to each thiolation domain. It has been demonstrated that peptide synthetases require post-translational modifications in order to become catalytically active. The active holoforms are formed from the inactive apoproteins by post-translational transfer of the 4′-PP residue of coenzyme A to the hydroxy group of a highly conserved serine residue. This is located at the C-terminal region of each substrate activating unit, and is defined as the acylation or thiolation

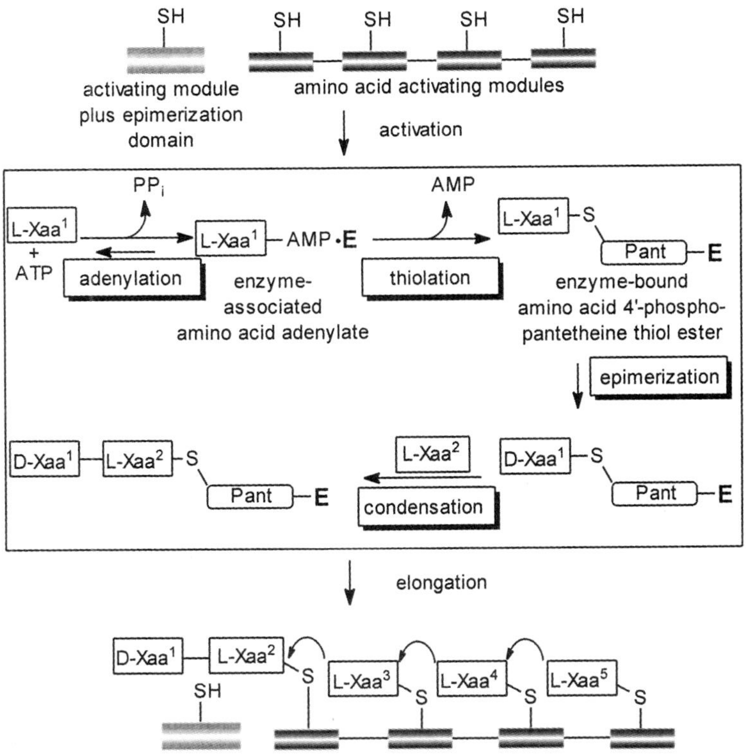

Fig. 3.12 Simple scheme of nonribosomal peptide synthesis.

domain. In this binding state, the thiol-activated amino acid derivative can undergo modifications such as epimerization (as shown in Fig. 3.12) or N-methylation. Transfer of the thioester-activated carboxy group of one residue to the adjacent amino group of the next amino acid results in a stepwise elongation in a series of transpeptidation reactions. This biosynthesis system can be described as a template-driven assembly. A series of very large multifunctional peptide synthetases characterized by a modular arrangement perform the synthesis in an ordered fashion. Only the protein template is genetically encoded. A single peptide synthetase gene contains four to six modules, allowing the addition of four to six building blocks. Each module is capable of recognizing a residue, followed by its activation (if necessary), its modification, and its addition to the growing peptide chain. Such a minimal module can activate both amino acid or hydroxy acid residues. This model, which involves multiple carriers at multifunctional templates, was derived from investigations of the gramicidin S synthetase multienzyme complex. Based on this mechanism, various peptide products containing either encoded amino acids, or hydroxyamino acids, D-configured amino acids, or other unusual derivatives, or amino acids with further modifications such as acylation, glycosylation, N-methylation, heterocyclic ring formation, can be formed. There is evidence that more than 300 different building blocks have been found in nonribosomally synthesized peptide structures. Nonribosomally synthesized peptides are produced exclusively by microorganisms, and the structural diversity is immense. Gram-positive bacteria of the species *Bacillus, Actinomyces* and *Streptomyces* belong to the most important producers.

Gramicidin S and the tyrocidin peptides produced by *Bacillus brevis* strains represent best-investigated examples related to the biochemistry of nonribosomal peptide synthesis. In gramicidin S **2**, two identical pentapeptides are linked via head-to-tail linkage. In contrast, the tyrocidins contain the sequence Val-Orn-Leu-Phe-Pro only as one copy, while the second pentapeptide of tyrocidins A, B, C, and D is assembled from a different set of amino acid building blocks with variation in specific positions. Gramicidin S and tyrocidin underline the relationship between peptide structure and multienzyme assembly of cyclic peptides in *Bacillus brevis* strains. Both peptides share a pentapeptide sequence that is encoded by two and three multienzymes, respectively.

The cyclosporin synthetase from the fungus *Beauveria nivea* is one of the largest known integrated enzyme structure, with a molecular weight of ~ 1400 KDa. This synthetase catalyzes at least about 40 reaction steps during the final assembly of the undecapeptide chain of cyclosporin A and its cyclization [114].

3.3 Selected Bioactive Peptide Families

3.3.1 Peptide and Protein Hormones

Living organisms require a highly complex chemical signaling system in order to coordinate activities at any level of their organization. Intercellular biochemical communication is mainly based on the action of hormones and neurotransmitters. The relationship between the hypothalamus, hypophysis, and target glands or target tissues in humans is shown schematically in Fig. 3.13.

The chemically heterogeneous group of hormones contains a vast number of peptides, proteins, and amino acids.

Peptide and protein hormones exert their action via binding to specific receptors. As a rule, the receptor (R) is a membrane-bound protein that is capable of binding its ligand (L) according to the law of mass action ($R + L \rightarrow [RL]$). The binding parameter of a radiolabeled ligand to a receptor can be determined from a plot of bound ligand (B)/free ligand (F) versus bound ligand ($B/F = (B_{max} - B)/K_L$). This is known as a Scatchard plot, named after its originator, George Scatchard. K_L is operationally defined as the ligand concentration at which the receptor is half-maximally occupied by ligand, and can be determined from tangential slope $-1/K_L$.

The hormone–receptor interactions stimulate the synthesis and activation of specific enzymes via signaling cascades involving second messengers such as cAMP, cGMP, diacylglycerol (DAG), inositol triphosphate (IP_3), and Ca^{2+}. Many

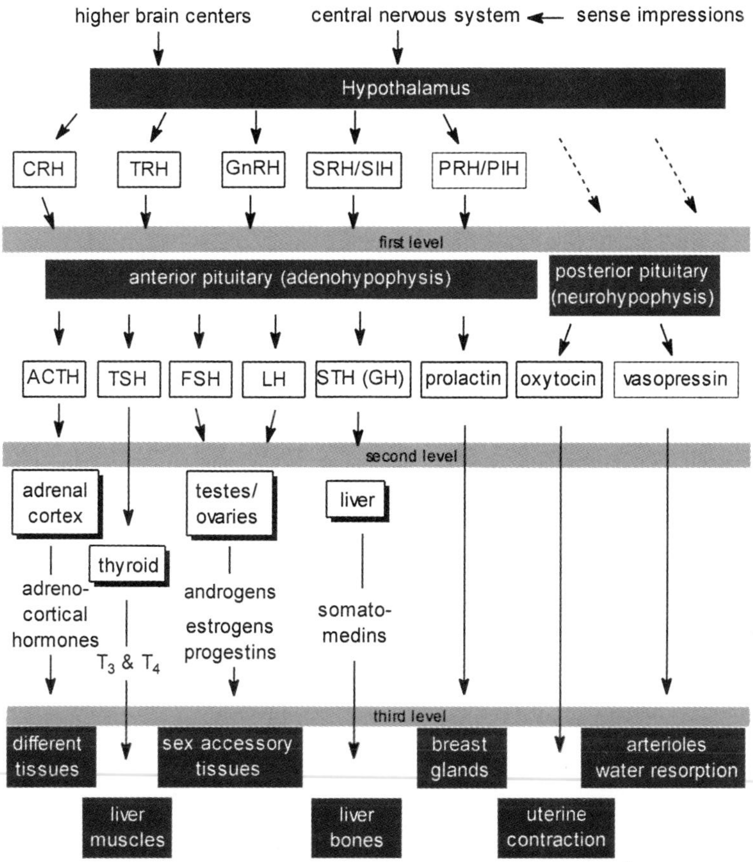

Fig. 3.13 Relationship between hypothalamus, hypophysis and target tissues including the action of the hormones of the second and third level.

peptide and protein hormone receptors mediate adenylate cyclase activation and inhibition, respectively, via membrane-bound G-proteins. The release of prostaglandins, steroid hormones, thyroid hormones, peptides, and glycoproteins can also be stimulated by peptide and protein hormones. Sometimes, internalization of a receptor is induced by an excess of a peptide hormone which leads to disappearance of the receptor – described as receptor down-regulation. This results in temporary stimulation followed by inhibition of the target cells. The term peptide agonist is given for analogues of native peptide hormones that trigger the hormone signal in the same manner. Peptide antagonists are analogues that act as competitive inhibitors, occupy the appropriate receptor, displace the agonist from the receptor, but do not transmit the hormone signal.

3.3.1.1 Liberins and Statins

A group of peptide hormones synthesized in the neurons of various distinct nuclei of the hypothalamus were initially named as releasing factors and release-inhibiting factors. Nowadays, these molecules are called releasing hormones or, according to an IUPAC-IUB recommendation, liberins. Moreover, the release-inhibiting factors are better designated as release-inhibiting hormones or statins (Tab. 3.3). The hypothalamus is the lowest part of the midbrain, and in certain nuclear areas nervous excitation is transformed into hormonal signals, for example releasing hormones and neurohypophyseal hormones. The releasing hormones, which are stored in the eminence of the hypothalamus in nanogram quantities, have a half-life in the order of a few minutes. The hormones are delivered via a

Tab. 3.3 Hypothalamic-releasing hormones (liberins) [a] and release-inhibiting hormones (statins) [b].

Name [b]	*Synonym*	*Abbreviation*	*Major effects*
Thyroliberin	Thyrotropin-releasing hormone	TRH	Stimulates TSH release
Gonadoliberin	Gonadotropin-releasing hormone	GnRH	Stimulates both lutropin and follitropin release
Luliberin	Luteinizing hormone-releasing hormone	LH-RH	
Folliliberin	Follicle stimulating hormone-releasing hormone	FSH-RH	
Corticoliberin	Corticotropin-releasing hormone	CRH	Stimulates corticotropin (ACTH) release
Prolactoliberin	Prolactin-releasing hormone	PRH	Stimulates prolactin release
Prolactostatin	Prolactin-release inhibiting hormone	PIH	Inhibits prolactin release
Somatoliberin	Somatotropin-releasing hormone (Growth hormone-releasing hormone, GRH)	SRH	Stimulates somatotropin (growth hormone) release
Somatostatin	Somatotropin-release inhibiting hormone	SIH	Inhibits somatotropin (growth hormone) release
Melanotropin	Melanotropin-releasing hormone	MRH	Stimulates melanotropin release
Melanostatin	Melanotropin-release inhibiting hormone	MIH	Inhibits melanotropin release

a) Instead of releasing hormone and release-inhibiting hormone, the terms releasing factor (RF) and release-inhibiting factor (RIF) are also in use.
b) According to IUPAC-IUB recommendation.

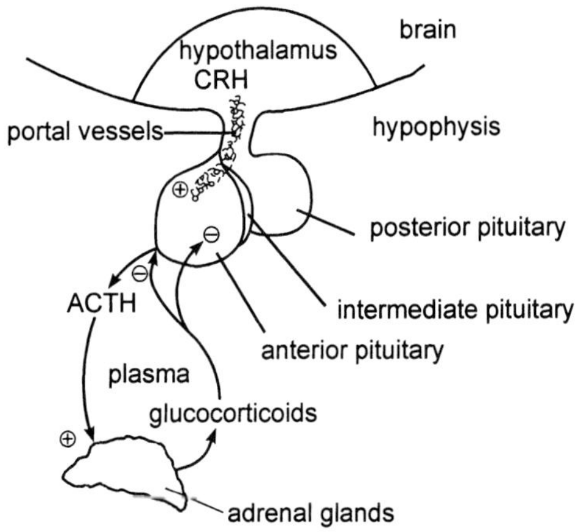

Fig. 3.14 Representative example for the regulatory relationships between the hypothalamus, the anterior pituitary, and target tissue (adrenal cortex) demonstrated for the axis corticotropin-releasing hormone (CRH), adrenocorticotropin (ACTH), and adrenocortical hormones. + = stimulation, − = inhibition.

direct circulatory connection to the anterior pituitary (adenohypophysis), where they stimulate or inhibit the release of the appropriate trophic hormone into the bloodstream. In most cases the trophic hormones stimulate the appropriate endocrine glands to secrete the corresponding endocrine hormones.

Fig. 3.14 shows as a representative example the hormone control circuit for the synthesis of adrenocorticotropin (ACTH), including the feedback regulation. The corticotropin-releasing hormone (CRH) formed in the hypothalamus is released into the capillaries of the portal vessels and transported to the anterior pituitary where it stimulates the production of POMC, a polyprotein containing the sequences of ACTH and several other hormones. The post-translationally released ACTH is secreted into the plasma, interacts with receptors at the surface of the adrenal lobe, and stimulates especially the synthesis of glucocorticoids. ACTH inhibits the release of CRH, and the glucocorticoids inhibit the release of both ACTH and CRH. Furthermore, dopaminergic neurons originating in the hypothalamus have major regulatory function on the synthesis and release of POMC.

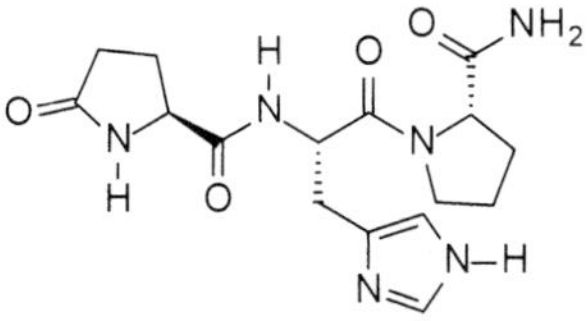

<Glu–His–Pro–NH_2

14

Thyroliberin **14** was initially discovered in the nervous system (brain), pancreas, intestine, and the prostate gland. It is formed as prepro-TRH which, in the rat, consists of 225 residues and contains five copies of the TRH sequence. After release of the TRH sequences by limited proteolysis, the N-terminal glutamine is cyclized to pyroglutamic acid, whilst the C-terminal glycine is converted to the amide function under catalysis of peptidylglycine-amidating monooxygenase. TRH stimulates the anterior pituitary to produce and secrete the trophic hormone thyrotropin, which in turn stimulates the thyroid gland to synthesize and secrete thyroxine (T_4) and triiodothyronine (T_3). The latter two hormones cause a negative feedback on the secretion of TRH and thyrotropin. Further biological effects of TRH are the secretion of prolactin, the inhibition of glucagon in the pancreas, contraction of the urinary bladder, vasodilatation in the cardiovascular system, and action as a neurotransmitter in the CNS, with various specific effects. TRH has found application as a diagnostic for thyroid and hypophyseal functions. Some TRH preparations for oral, parenteral, and intranasal application are available commercially to treat spinal injury, schizophrenia, epilepsy, motor neuron disease, depression, and circulatory shock. Some analogues are more effective than native TRH; for example, [2-(*N*-Me-His)]TRH and the corresponding *N*-amylamide possess eight- and ten-fold the potency of the native hormone, respectively. In contrast, a higher number of analogues which have central activity but which do not release TSH have been described. Pyroglutamyl-peptidase II (thyro-liberiase), which cleaves specifically the pGlu-His bond, is an important enzyme in the breakdown of TRH. Metabolites of TRH, such as cyclo-(His-Pro) and pGlu-His-Gly-OH show a central appetite-inhibiting effect. The latter compound, named anorexigenic peptide, was found in the urine of women suffering from anorexia nervosa. Furthermore, a correlation between increased cyclo-(His-Pro) plasma levels with weight loss in bulimia and anorexia nervosa has been reported. TRH was the first releasing hormone to be isolated, and its structure was elucidated independently in 1969 by the groups of Andrew Schally [115] and Roger Guillemin [116] using extracts of hypothalami from approximately 3 million animals (sheep and pigs).

Gonadoliberin (gonadotropin-releasing hormone, GnRH) **15** stimulates the adenohypophysis in a Ca^{2+}-dependent process to release the gonadotropins, for example lutropin (luteinizing hormone, LH) and follitropin (follicle-stimulating hormone, FSH).

<Glu^1—His—Trp—Ser—Tyr^5—Gly—Leu—Arg—Pro—Gly^{10}—NH_2

15

The sequence of gonadoliberin is formed initially as a part of the precursor protein (prepro-GnRH) consisting of 92 amino acid residues. The release of GnRH from the precursor requires C-terminal amidation by peptidylglycine α-amidating monooxygenase. GnRH occurs not only in the hypothalamus but also in the brain, liver, heart, pancreas, kidneys and adrenal glands, gonads, and small intestine. Many thousands of analogues have been synthesized and tested biologically. Several superagonists (e.g., leuprolide, nafarelin, buserelin) have found clinical application for disorders such as prostate cancer or endometriosis, and many antagonists are undergoing evaluation as both male and female contraceptive agents.

H–Ser^{1}–Glu–Glu–Pro–Pro^{5}–Ile–Ser–Leu–Asp–Leu^{10}–Thr–Phe–His–Leu–
Leu^{15}–Arg–Glu–Val–Leu–Glu^{20}–Met–Ala–Arg–Ala–Glu^{25}–Gln–Leu–
Ala–Gln–Gln^{30}–Ala–His–Ser–Asn–Arg^{35}–Lys–Leu–Met–Glu–Ile–Ile–NH_2

16

Corticoliberin (CRH) **16** occurs in nerve fibers throughout the brain, but it has been also found in the adrenal medulla, pancreas, placenta, and testes. It stimulates the synthesis and release of POMC in the adenohypophysis and placenta, which is subsequently degraded to form, for example, corticotropin (ACTH), β-endorphin, and melanotropins. The 41-polypeptide amide shows structural similarities to sauvagine from the skin of *Phyllomedusa sauvagii*, and urotensin I from the urophysis of the white sucker (*Catostomus commersoni*), respectively.

Three of the seven adenohypophysis hormones (somatotropin, prolactin, and melanotropin) do not act on endocrine glands, but rather influence tissue mechanisms directly. Due to the lack of feedback mechanisms, control is performed by the action of appropriate release-inhibiting hormones.

Somatoliberin (somatotropin-releasing hormone, SRH; also known as growth hormone-releasing hormone, GRH) **17** is a 44-polypeptide amide which is released proteolytically from prepro-SRH, followed by post-translational amidation.

H–Tyr^{1}–Ala–Asp–Ala–Ile^{5}–Phe–Thr–Asn–Ser–Tyr^{10}–Arg–Lys–Val–Leu–
Gly^{15}–Gln–Leu–Ser–Ala–Arg^{20}–Lys–Leu–Leu–Gln–Asp^{25}–Ile–Met–Ser–
Arg–Gln^{30}–Gln–Gly–Glu–Ser–Asn^{35}–Gln–Glu–Arg–Gly–Ala^{40}–Arg–Ala–
Arg–Leu–NH_2

17

SRH stimulates the release of somatotropin (growth hormone) from the anterior pituitary. Somatostatin is the corresponding release-inhibiting hormone (somatotropin release-inhibiting hormone, SIH) **18** which is formed primarily as prepro-SRIH and processed to a heterodetic cyclic 14 peptide.

H–Ala^{1}–Gly–Cys–Lys–Asn^{5}–Phe–Phe–
Trp–Lys–Thr^{10}–Phe–Thr–Ser–Cys–OH

18

SIH inhibits the secretion of somatotropin, but also shows inhibitory activity, for example on the release and action of thyrotropin, prolactin, insulin, glucagon, gastrin, cholecystokinin, and motilin.

The secretion of prolactin from the hypophysis is regulated by the hypothalamic peptide hormones prolactoliberin and prolactostatin, which are of unknown structure. Melanoliberin most likely corresponds to the enzymatic degradation product of oxytocin (H-Cys-Tyr-Ile-Gln-Asn-Cys-OH), whereas melanostatin should be identical with the C-terminal tripeptide of oxytocin (H-Pro-Leu-Gly-NH_2).

3.3.1.2 **Pituitary Hormones**

The hypophysis or pituitary gland is a vertebrate endocrine gland located at the base of the brain, and connected to the midbrain by the hypophyseal stalk. The hypophysis consists of the anterior pituitary (adenohypophysis), the middle part, and the posterior pituitary (neurohypophysis). As shown above (cf. Fig. 3.13), the secretion of pituitary hormones (also termed trophic hormones) is controlled by the hypothalamus. The pituitary hormones (Tab. 3.4) are released into the bloodstream and stimulate their corresponding glands or tissues (adrenal cortex, thyroid, testes/ovaries, liver, and other special tissues) to secrete the appropriate endocrine hormones. Alternatively, they may exert their biological effect directly, as shown by somatotropin (growth hormone).

The binding of human growth hormone (hGH), a 191-residue protein, activates its receptor (620 aa) to stimulate growth and metabolism in muscle, bone, and cartilage cells. The receptor consists of a N-terminal extracellular ligand binding domain (hGHbp, 238 aa), a single (probably helical) transmembrane segment, and a C-terminal cytoplasmic domain. The X-ray structure analysis of a complex of hGH with the extracellular domain of its receptor revealed that this complex consists of two molecules of hGHbp with one molecule of bound hGH. The ligand-induced receptor dimerization is important for the specific mechanism of signal transduction, which might also be verified for other growth factors. GH promotes the longitudinal growth and regeneration of bone, and a mainly tumor-promoted overproduction of GH causes excessive growth (gigantism). In adults GH influences only the growth of soft tissues, and this results in enlarged hands and feet, as well as thickened facial features (acromegaly). Insufficient growth (dwarfism) based on GH deficiency in the period before skeletal maturity can only be treated by application of human GH, as animal GH is without effect in humans. Nowadays, Met-h-GH and h-GH are produced by recombinant DNA techniques and used in the therapy of hypophyseal dwarfism. Last, but not least, it should be mentioned that GH stimulates the liver and kidneys to synthesize somatomedins, these being polypeptide growth factors ($M_r \sim$ 7–10 KDa) with growth-promoting activity. In particular, recombinant somatomedins are important in the treatment of hypophyseal dwarfism.

POMC synthesized both in the anterior and intermediate lobe contains seven different peptide hormones, among them corticotropin (ACTH) and melanotropin. The tissue-specific post-translational processing of POMC in the anterior and intermediate lobes of the pituitary provides the signal sequence, the N-terminal fragment, corticotropin, and β-lipoprotein (β-LPH), whereas in the intermediate lobe only further proteolytic cleavage occurs, yielding γ-MSH (from the N-terminal fragment), α-MSH and corticotropin-like intermediate lobe peptide (CLIP from corticotropin), as well as β-endorphin together with γ-lipotropin (γ-MSH from β-LPH). Further information regarding hypophyseal hormones is available in Tab. 3.4; general information on hormones is also available in the Glossary.

Tab. 3.4 Hypophyseal hormones [a].

Name [b]	*Synonyms*	*Abbrev.*	*Major effects*	*Structure*
Thyrotropin	Thyroid-stimulating hormone	TSH	Synthesis and secretion of the thyroid hormones	Glycoprotein [c] α chain: 96 aa β chain: 112 aa
Lutropin	Luteinizing hormone	LH	Acts together with FSH in stimulation growth of the gonads and the synthesis of the sex hormones	Glycoprotein [c] α chain: 96 aa β chain: 121 aa
Follitropin	Follicle-stimulating hormone	FSH	Responsible for development and function of male and female gonads; promotes spermatogenesis in male testes, and controls maturation of female follicle	Glycoprotein [c] α chain: 96 aa β chain: 111 aa
Prolactin	Lactotropin Lactogenic hormone Luteotropic hormone Luteotropin	PRL LTH	Increases in female mammals the milk production, and initiates materal behavior	Single-chain protein [c] (198 aa)
Somatotropin	Somatotropic hormone Growth hormone	STH GH	Promotes species-specific longitudinal growth and bone regeneration; stimulates synthesis of somatomedins	Single-chain protein [c] (191 aa); two disulfide bonds
Lipotropin	Lipotropic hormone	LPH	Stimulates lipid metabolism, especially, fatty acid mobilization from lipid depots	Single-chain protein β-LPH: 91 aa
Corticotropin	Adreno-corticotropic hormone	ACTH	Stimulates in the adrenal cortex the synthesis of glucocorticoids and mineralocorticoids	Single peptide chain: 39 aa
Melanotropin	Melanocyte-stimulating hormone	MSH	Causes dispersion of melanin in the melanophores of the skin of cold-blooded vertebrates; α-MSH is present extensively in the CNS of humans and other animals	Mammalian α-MSH [d]: 13 aa β-MSH: 18-22 aa

a) Not included: oxytocin and vasopressin which are formed in hypothalamus and stored in the posterior pituitary.
b) According to IUPAC-IUB recommendation.
c) Human proteins.
d) Occurs primarily in the hypophyseal pars intermedia and in the hypothalamus.

3.3.1.3 Neurohypophyseal Hormones

The neurohypophysis is the posterior lobe of the pituitary, and is anatomically distinct from the adenohypophysis. It secretes oxytocin (OT) **3** and vasopressin (VP) **19**, two homologous heterodetic cyclic peptides that are primarily synthesized in the hypothalamus.

H–Cys^{1}–Tyr–Phe–Gln–Asn^{5}–Cys–Pro–Arg–Gly–NH_2

19

Oxytocin is synthesized in the nucleus paraventricularis, and vasopressin in the nucleus supraopticus, of the hypothalamus in the form of appropriate neurophysin precursor proteins (neurophysin I and neurophysin II). They are transported in neurosecretory vesicles via the tractus paraventriculo-hypophyseus to the posterior pituitary. The latter acts as a storage and release facility for both peptide hormones. Stimulation causes proteolytic release of the prohormone form and secretion into the blood. Oxytocin mediates contraction of the smooth muscle of the uterus and milk ejection by contraction of the breast glands. Vasopressin, also known as antidiuretic hormone, stimulates the kidneys to retain water and increases blood pressure. Vasopressin release is mainly controlled by osmoreceptors that monitor the osmotic pressure of the blood. Bovine and other mammalian pituitaries contain [Arg^{8}]vasopressin (AVP), whereas porcine pituitary has [Lys^{8}]vasopressin (LVP). From phylogenetic investigations could be concluded that vasotocin (Cys^{1}-Tyr-Ile-Gln-Asn^{5}-Cys-Pro-Arg-Gly-NH_2 (disulfide bond: Cys^{1}-Cys^{6}) is the evolutionary precursor of the neurohypophyseal hormones. The latter is responsible for the regulation of water and mineral metabolism of lower vertebrates. During the transition from the cyclostomata to fish, a gene duplication occurred which resulted in the formation of oxytocin and vasopressin. Only the replacement of Ile^{3} in vasotocin by Phe led to vasopressin as the mammalian hormone. The number of evolutionary precursors for oxytocin is much more demanding with the naturally occurring vertebrate hormones aspartocin, valitocin, glumitocin, isotocin, and mesotocin.

Many analogues of the neurohypophyseal hormones have been synthesized in the course of extensive investigation into their structure and activity [117]. Some selected examples are given below. The commercially produced 1-deamino-[D-Arg^{8}]vasopressin (DDAVP) shows 400 times the effect of VP on the kidneys, whereas its effect on blood pressure is practicably negligible. The most widely used AVP antagonists of vasopressor responses are [(1-mercaptocyclohex-1-yl)-acetic acid, 2-*O*-methyltyrosine]AVP, and [1-deamino-penicillamine, 2-*O*-methyltyrosine]AVP. One of the most potent oxytocin antagonists in the uterine receptor is the bicyclo analogue, cyclo-(1-6,4-8)[Mpa^{1},Glu^{4},Lys^{8}]oxytocin. [Pen^{1},D-Pen^{2},Thr^{4},Orn^{8}]OT is a potent, long-lasting antagonist both *in vitro* and *in vivo*.

3.3.1.4 Gastrointestinal Hormones

At present, more than 30 peptide hormone genes are known to be expressed throughout the digestive tract. Consequently, the gut is the largest endocrine hormone-producing organ in the body, both in terms of the number of endocrine cells and the number of hormones [118, 119]. According to the classical concept of gastrointestinal hormones which prevailed until the 1970s with only three known hormones of secretin, gastrin, and cholecystokinin (CCK), a gut hormone was defined as a substance produced by one type of endocrine cell dispersed in a relatively well-characterized region of the gastrointestinal tract. From here, it is released to the bloodstream upon an appropriate stimulus, and reaches its target organ in order to elicit an acute response. Modern developments in cellular and molecular biology demand a new definition of gastrointestinal hormones, however [120]. In fact, the well-known widespread expression of gastrointestinal hormone genes outside the gastrointestinal tract makes these peptide hormones multifunctional regulators of general physiological and biochemical interest, whilst acting simultaneously as acute metabolic hormones, local growth factors, neurotransmitters, and fertility factors. The resulting complexity has been further increased by the fact that individual genes for gut peptides encode various peptides, which release in a tissue- and cell-specific manner a very large number of different, biologically active peptides. Originating from one common ancestral gene, the gastrointestinal hormones form structural homology groups.

Since several peptide genes are expressed both in the gut and in the pancreas, it has been assumed that in onto- and phylogenetic terms the pancreas is of intestinal origin. The major gastroenteropancreatic peptide families according to a proposal of Rehfeld [120] are listed in Tab. 3.5. The degree of homology varies from family to family. In the gastrin family, the decisive homology is concentrated in and around the common C-terminal tetrapeptide amide (Trp-Met-Asp-Phe-NH_2). O-Sulfated tyrosyl residues adjacent to this conserved active site characterize this family. In the PP-fold family, the three members PP, PYY, and NPY display sequence similarities between 45 and 70%. The characteristic three-dimensional PP fold consists of a polyproline-like helix (residues 1–8) and an amphiphilic α-helix (residues 15–30). Both helices are joined together by a type I β-turn (residues 9–12), supported by hydrophobic interactions.

Secretin was long believed to exist only as a 27mer peptide amide, until three additional secretins – which have 28, 30, and 71 residues and are released from prepro-secretin (132 residues) and show full biological activity – were identified.

The CCK gene encodes a prepro-peptide (115 residues) which is processed to six CCK peptides with 8, 22, 33, 39, 58, and 83 amino acid residues, respectively. All of these have the same C-terminal bioactive 8-peptide sequence.

The endoproteolytic cleavage of prepro-glucagon occurs cell specifically. Pancreatic A cells form glucagon, and intestinal L cells the glucagon-like peptides GLP-I and GLP-II.

To summarize briefly, in vertebrates gastrointestinal hormones should not be regarded as local hormones; rather, they participate in the coordination and regulation of many functions. More than 100 bioactive peptides are involved in intercel-

Tab. 3.5 Gastroenteropancreatic peptide families according to Rehfeld [120].

Secretin family	***Gastrin family***
Secretin	Gastrin
Glucagon [a]	Cholecystokinin (CCK)
Glucagon-like peptides (GLPs) [a]	Cerulein [b]
Growth hormone-releasing hormone	Cionin [b]
Pituitary adenylyl cyclase-activating hormone	PP-fold family
Gastric inhibitory polypeptide (GIP)	Pancreatic polypeptide (PP)
Vasoactive intestinal polypeptide (VIP) [a]	Peptide tyrosine tyrosine (PYY)
Peptide histidine isoleucine (PHI) [a]	Neuropeptide Y (NPY)
Insulin family	***Tachykinin family***
Insulin	Substance P
Insulin-like growth factor I	Neurokinin A
Insulin-like growth factor II	Neurokinin B
Relaxin	
Somatostatin family	***EGF family***
Somatostatin	Epidermal growth factor (EGF)
Corticostatin	Transforming growth factor α (TGF α)
	Amphiregulin

a) Encoded by one gene, respectively.
b) Nonmammalian peptide.

lular regulation, ranging from local control of growth and differentiation to acute systemic effects. Indeed, as cell biology techniques advance, it is likely that applications in clinical oncology, neurobiology, and psychiatry will be developed from studies of gastrointestinal endocrinology. Further information related to gastrointestinal hormones is provided in the Glossary.

3.3.1.5 **Pancreatic Islet Hormones**

Pancreatic islet hormones are responsible for regulating not only the storage of glucose and fatty acids, but also their release. Three different types of cells are found in the pancreatic islets of Langerhans, and each secretes a specific polypeptide hormone. The α-cells secrete glucagon, the β cells insulin, and the δ cells somatostatin.

Insulin acts together with its counterpart glucagon (which has largely opposing effects) to maintain the correct concentration of blood glucose. An increased blood glucose level (hyperglycemia) is the physiological signal for the synthesis and secretion of this hormone. Insulin increases cell permeability for glucose and other monosaccharides, as well as fatty acids and amino acids. Furthermore, it stimulates liver, muscle, and adipose cells to store these metabolites for further use in the form of glycogen, protein, and fat. In diabetes mellitus, insulin is either not secreted in sufficient concentrations or does not efficiently stimulate its target cells, and an abnormally high concentration of ketone bodies (ketosis) is one of the most dangerous effects. Two major forms of diabetes mellitus are known:

- Insulin-dependent (type I) diabetes mellitus (IDM) is caused by a deficiency of pancreatic β cells. IDM (also termed juvenile-onset diabetes mellitus) often strikes suddenly in childhood. In genetically susceptible individuals an autoimmune response selectively destroys the pancreatic β cells. Daily postprandial insulin injections are required in order for these individuals to survive. It is known that certain genetic variants of the class II major histocompatability complex (MHC) proteins are particularly common in IDM. It has been suggested that autoimmunity against β cells may be induced by a foreign antigen with immunological resemblance to the β-cell component. The class II MHC protein binds the antigen and initiates a vigorous and prolonged immune response, attacking also the β cells of the pancreas.
- Noninsulin-dependent (type II) diabetes mellitus (NIDDM) is strongly correlated with obesity, and may be associated with loss of fully active insulin receptors on normally insulin-responsive cells. Individuals with NIDDM have a normal or even greatly elevated insulin level resulting from malnutrition which may suppress the biosynthesis of the insulin receptor.

Insulin was discovered in 1921 by Banting (Nobel Prize 1923) and Best, and was purified and crystallized by Abel five years later. The elucidation of the primary structure was described in 1955 by Frederic Sanger, who was awarded the Nobel Prize in 1958 for this pioneering work. He obtained this award a second time for the development of the chain-terminating method for DNA sequence analysis in 1980. Insulin consists of the A chain (*a*cidic chain) containing 21 residues, and the B chain (*b*asic chain) with 30 residues. The two chains are connected by two interchain disulfide bridges ($Cys^{7}{}_{A}$-$Cys^{7}{}_{B}$, and $Cys^{20}{}_{A}$-$Cys^{19}{}_{B}$). One more intrachain disulfide bond is located between Cys^{6}-Cys^{11} within the A chain (Fig. 3.15). Although most species have only one type of insulin, three rodents (laboratory rat, mouse, spiny mouse) and two fish (toadfish, tuna) have two distinct insulin types.

In biosynthesis, the mature mRNA transcript encodes prepro-insulin bearing a N-terminal hydrophobic, 16-residue signal sequence. The latter is removed by a signal peptidase when prepro-insulin traverses the membrane into the lumen of the ER. The resulting proinsulin containing the disulfide bonds is transported to the Golgi complex, where insulin is formed after enzymatic excision of the connecting peptide (C-peptide) between the C-terminal A chain and the N-terminal B chain. Insulin secretion as a response to elevated blood glucose levels is triggered by an increase in the concentration of free cytosolic Ca^{2+}. Inositol triphosphate as the messenger releases Ca^{2+} from internal stores. As a rule, the biosynthesis of insulin is stimulated at glucose concentrations >2–4 mM, whereas glucose concentrations 4–6 mM are required for the stimulation of insulin secretion. The normal concentration of circulating insulin in human blood, which can be determined radioimmunologically, is 1 ng mL^{-1}. The liver is the major site of insulin degradation, but most peripheral tissues also contain insulin-degrading enzymes.

Glucagon

H^{1}SQGT^{5}FTSDY^{10}SKYLD^{15}SRRAQ^{20}DFVQW^{25}LMNT

Insulin

G^{1}IVEQ^{5}CCASV^{10}CSLYQ^{15}LENYC^{20}N

F^{1}VNQH^{5}LCGSH^{10}LVEAL^{15}YLVCG^{20}ERGFF^{25}YTPKA^{30}

metabolic influence of

Glucagon	Insulin
⊕ glycogen breakdown (liver)	⊕ absorption of glucose (muscles)
⊖ glycogen synthesis (liver)	⊕ absorption of glucose (liver)
⊖ glycolysis (liver)	⊕ glycogen synthesis
⊕ gluconeogenesis (liver)	⊖ degradation of glycogen (liver, muscles)
⊕ mobilization of fatty acids (fatty tissue)	⊕ glycolysis and production of acetyl-CoA (liver, muscles)
	⊕ fatty acid synthesis (liver)
	⊕ triacylglycerine synthesis (fatty tissue)

Fig. 3.15 Primary structures of the peptide hormones insulin (bovine) and glucagon, and their metabolic influence.

The insulin receptor is an integral membrane glycoprotein ($M_r \sim 300$ KDa) comprising two α and two β subunits joined together by disulfide bridges: [β-subunit-S-S-α-subunit]-S-S-[α-subunit-S-S-β-subunit]. The α subunit consists of 719 residues ($M_r \sim 82.5$ KDa), and the β-subunit contains 620 residues ($M_r \sim 69.7$ KDa). The extracellular insulin-binding domain is located in the cysteine-rich domain between Asn^{230} and Ile^{285}. The β subunits span through the membrane, whereas each α subunit binds one molecule of insulin externally on the membrane surface. Insulin binding activates a tyrosine kinase in the intracellular domain of the β subunit, which catalyzes tyrosine phosphorylation in the β subunit itself, and also in other proteins. The structure of the tetrameric insulin receptor is derived from a single biosynthesis protein precursor. Proteolytic processing, formation of disulfide linkages, and glycosylation occur both in the ER and Golgi apparatus, followed by transport to the plasma membrane. The insulin receptor, and especially the receptor tyrosine kinase of the β subunit, shows similarities to receptors of other protein growth factors, such as IGF-1 receptor, EGF-receptor, PDGF-receptor, and NGF-receptor.

Glucagon is a single-chain 29-peptide (Fig. 3.15) that acts as a hormone antagonist of insulin. It controls glycogen metabolism in the liver, whereas in muscles and various tissues this is controlled by insulin and the adrenal hormones epinephrine and norepinephrine. A blood glucose level <5 mM causes the pancreatic α cells to secrete glucagon into the bloodstream. Glucagon receptors located on liver cell surfaces bind glucagon, thereby activating adenylate cyclase. The resulting increase in cAMP concentration inside the cells triggers glycogen breakdown.

The increasing intracellular concentration of glucose-6-phosphate is adjusted by glucose-6-phosphatase-catalyzed hydrolysis to give glucose, which enters the bloodstream and maintains the blood glucose level. The glucagon gene encodes preproglucagon which is processed by cell-specific proteolytic cleavages in pancreatic α cells either to the genuine pancreatic glucagon or to glucagon-like peptides (GLPI and II) in intestinal L cells.

Somatostatin secreted by the δ cells of the pancreatic islets is synthesized by the neurons of the hypothalamus as the growth hormone release-inhibiting hormone (see Section 3.3.1.1). Inhibition of the release of both insulin and glucagon from the β cells and α cells, respectively, underlines a paracrine function of somatostatin in the pancreas.

3.3.1.6 Further Physiologically Relevant Peptide Hormones

Three hormones are involved in maintaining Ca^{2+} homeostasis, the normal extracellular calcium concentration being ~ 1.2 mM. Ca^{2+} is an essential ion that is involved in many biological processes, for example as a second messenger mediating hormone signals, as a necessary cofactor for a number of extracellular enzymes, as a structure-stabilizing factor for proteins and lipids in cell membranes, cytoplasm, organelles and chromosomes, and as a constituent of bone as hydroxyapatite, $Ca_5(PO_4)_3OH$. Indeed, bones are the main Ca^{2+} reservoir of the body. Besides vitamin D, two peptide hormones are implicated in controlling Ca^{2+} metabolism.

Parathyroid hormone (PTH) is a 84-aa polypeptide which is secreted by the parathyroid gland. PTH increases the Ca^{2+} serum level by promoting its resorption from bone and kidney, as well as increasing its dietary absorption from the intestine. PTH stimulates bone resorption by osteoclasts, and also inhibits collagen synthesis by osteoblasts. Furthermore, it promotes the production of the active form of vitamin D in the kidney that, in turn, increases the export of intestinal Ca^{2+} to the blood. PTH displays essentially the opposite effect of calcitonin.

Calcitonin (CT) **20** is a 32-peptide amide formed in the parafollicular gland of the mammalian thyroid, and in the ultimobranchial gland of nonmammalian species.

H–Cys^{1}–Gly–Asn–Leu–Ser^{5}–Thr–Cys–Met–Leu–Gly^{10}–Thr–Tyr–Thr–
Gln–Asp^{15}–Phe–Asn–Lys–Phe–His^{20}–Thr–Phe–Pro–Gln–Thr^{25}–Ala–Ile–
Gly–Val–Gly^{30}–Ala–Pro–NH_2

20

CT acts as an antagonist of PTH and is released in response to a rising Ca^{2+} level. CT causes the serum Ca^{2+} concentration to be reduced by inhibiting Ca^{2+} resorption from bone and kidney and promoting the incorporation of Ca^{2+} into bones.

Chorionic gonadotropin (CG) is an important proteohormone formed in the placenta during pregnancy. Human CG (hCG) is a glycoprotein (M_r ~ 30 KDa) consisting of two subunits (α/β), and stimulates the ovaries to produce the progesterone necessary to maintain pregnancy. As hCG may be detected using monoclonal antibodies both in maternal urine or plasma, an immunoassay is feasible as a pregnancy test.

Chorionic mammotropin (CM) – also known as placenta lactogen (PL) or human lactogen – is formed in increasing concentrations during pregnancy. CM is a single-chain proteohormone ($M_r \sim 22$ KDa; 191 aa) which acts in a similar manner (albeit more weakly) to human somatotropin (growth hormone); it also appears to play an important role in fetal development. CM also stimulates the development of breast tissue, without any effect on milk secretion. The secretion of CM decreases after childbirth, at which time human prolactin stimulates milk secretion.

Relaxin is a peptide hormone formed primarily in the serum and tissues of pregnant mammals. Two different genes (H1 and H2) have been identified for human relaxin. Relaxin-2 (Fig. 3.16) is synthesized in the ovaries during pregnancy, and is the main source of plasma relaxin. The two relaxins are heterodetic cyclic peptides consisting of an A chain (24 aa) and a B chain (32 aa) that are connected by disulfide bridges. Relaxin-1 and -2 differ in sequence from each other and both show structural similarity to insulin. The biosynthetic precursor is the single-chain protein, prorelaxin. In the advanced stages of pregnancy relaxin causes contraction of the uterus, leading to childbirth.

The insulin-like growth factors (IGF) are structurally related to both insulin and relaxin. These earlier termed somatomedins display insulin-like activities and stimulate cartilage growth. Their synthesis in the liver is induced by somatotropin (growth hormone). IGF-1 like proinsulin is a single-chain, 70-peptide molecule ($M_r \sim 7.6$ KDa) with three intrachain disulfide bridges. In contrast to insulin and relaxin, the C-peptide (connecting peptide) is not cleaved. IGF-2 is also a single-chain polypeptide ($M_r \sim 7.5$ KDa) containing 67 residues and three intrachain disulfide bridges. The action of the somatomedins is mediated by the IGF-1 receptor ($M_r \sim 400$ KDa) and the IGF-2 receptor ($M_r \sim 250$ KDa), respectively. The IGF-1 receptor (IGF receptor type I) structurally resembles the insulin receptor ($\alpha_2\beta_2$). The extracellular α subunit contains a cysteine-rich region and 11 glycosylation sites. The β subunit is glycosylated in its extracellular region, and contains a transmembrane domain with 24 amino acid residues together with an intracellular domain exhibiting tyrosine kinase activity. In contrast, the IGF-2 receptor (IGF receptor type II), like the mannose-6-phosphate receptor, is a single-chain protein with a larger extracellular domain and 19 glycosylation sites. The much shorter intracellular, hydrophilic domain can be phosphorylated at various sites.

Blood pressure-regulating peptides possess a wide variety of physiological actions. Various peptide hormones, including endothelins, angiotensin II, and

<E^{1}LYSA^{5}LANKC^{10}CHVGC^{15}TKRSL^{20}ARFC

S^{1}WMEE^{5}VIKLC^{10}GRELV^{15}RAQIA^{20}ICGMS^{25}TWSKR^{30}SL

Fig. 3.16 Primary structure of human relaxin-2.

[Arg]vasopressin stimulate vasoconstriction and thus increase blood pressure. They are involved in homeostasis together with vasodilators such as plasma kinins, substance P, and atrial natriuretic peptide.

Endothelins (ET) released from endothelial cells are peptides with vasoconstrictor potency. They are found in the kidneys, hypophysis, hypothalamus, and also in human milk. Three different human ETs have been found.

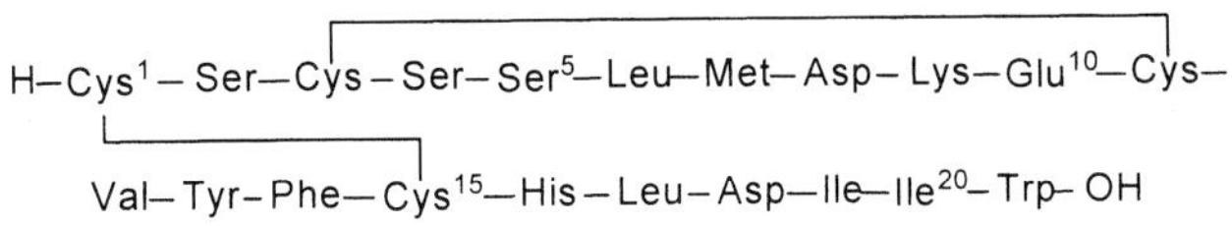

21

Endothelin-1 **21** is a 21-peptide which was first isolated in 1988 from cultivated endothelial cells [121] and which is ten-fold more potent than angiotensin II. Endothelin-2, ([Trp^6,Leu^7]ET-1) and endothelin-3 ([$Thr^{2,5}$,Phe^4,$Tyr^{6,14}$,Lys^7]ET-1) were discovered one year later in other tissues (brain, suprarenal glands, kidneys, intestine). ET-1 is synthesized as prepro-endothelin (212 residues) and released via proendothelin (38 residues). The latter bears the sequence of ET-1 in the N-terminal region, and the release of the active peptide involves scission of a Trp-Val bond by the endothelin-converting enzyme (ECE). The nature of this phosphoramidon-sensitive protease has not yet been precisely elucidated, but it should be similar to enkephalinase (EC 3.4.24.11) or related to neprilysin. Big endothelin II and III contain 37 and 41 amino acids, respectively. From the two classes of known receptors, ET_A receptors occur primarily in peripheral tissues (e.g., lung, heart, aorta) which mediate vasoconstriction. On the other hand, ET_B receptor subtypes are found in the CNS and endothelial cells, and appear to be linked to vasodilatation. Endothelin analogues (selective agonists or antagonists) with selected affinity to ET_A and ET_B receptors as demonstrated by structure-activity relationships studies, should be valuable in the development of antihypertensive drugs.

The angiotensin-kinin system plays an important role in the regulation of blood pressure. The tissue hormones belonging to the classes of the angiotensins and kinins are released into biological fluids from precursor proteins by the action of proteases. Kinins – which lower blood pressure – are formed from α-globulin fractions of the plasma (kininogens) by proteases called kallikreins. These proteases have various substrate specificities, occur in the plasma (and also in the pancreas, kidneys, and other organs), and are released from inactive precursors. The plasma kinins comprise bradykinin (kinin 9) **22**, kallidin (kinin 10) **23** and methionyl-lysyl-bradykinin (kinin 11) **24**. There are only quantitative differences in the pharmacological effects of the three plasma kinins.

H–Arg^{1}–Pro–Pro–Gly–Phe^{5}–Ser–Pro–Phe–Arg–OH **22**

H–Lys^{1}–Arg–Pro–Pro–Gly^{5}–Phe–Ser–Pro–Phe–Arg^{10}–OH **23**

H–Met^{1}–Lys–Arg–Pro–Pro^{5}–Gly–Phe–Ser–Pro–Phe^{10}–Arg–OH **24**

Kinins are also found in lower animals, for example in the skin of various amphibia (bradykinin, [Thr^{6}]bradykinin, C-terminally shortened or extended bradykinins).

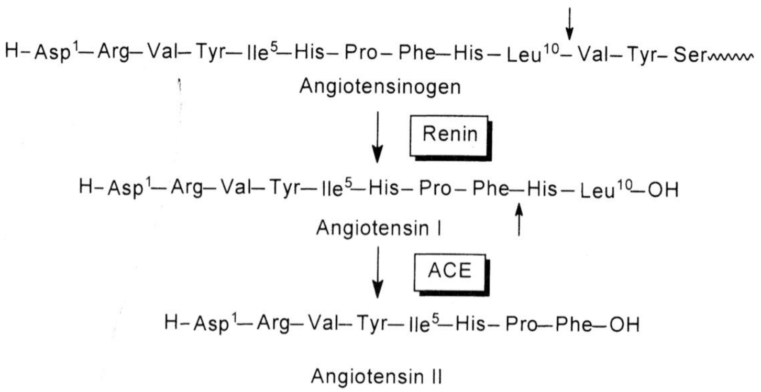

Fig. 3.17 Two-step formation of the active angiotensin II. ACE = angiotensin-converting enzyme.

Angiotensins (AT) occur both in the periphery and in the brain. Angiotensin I is proteolytically released by renin from angiotensinogen, a plasma protein of the α_2-globulin fraction (Fig. 3.17). The inactive angiotensin I (AT I) in turn is transformed into the highly active octapeptide angiotensin II by angiotensin converting enzyme (ACE). AT II causes elevation of blood pressure in mammals. AT II is primarily formed in the lungs, but also in the kidneys, hypophysis, hypothalamus, adrenal cortex, ovarian follicle, and testes. AT II exerts its action by binding to the AT1 and AT2 receptors, respectively, in a variety of target tissues. Structure-activity relationships studies have led to a vast number of agonists and antagonists being identified, though only two are worthy of mention at this point. The agonist [Asn^{1},Val^{5}]AT II (Hypertensin®, CIBA) is administered in collapse or shock situations in order to restore normal blood pressure as quickly as possible. In contrast, the antagonist Saralasin® (Sarenin), H-Sar-Arg-Val-Tyr-Val-His-Pro-Ala-OH, has found application in the diagnosis of AT II-dependent forms of hypertonia; it is also used to treat donor kidneys before transplantation in order to minimize loss of function of the transplanted organ as a result of ischemia.

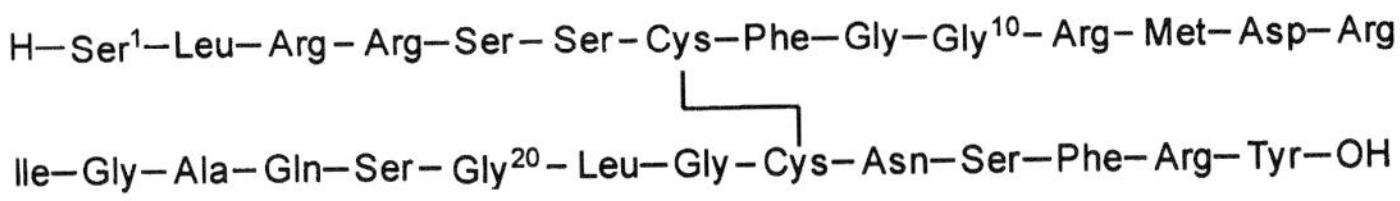

25

The atrial natriuretic peptide (ANP) **25** (also named cardionatrin I, atrial natriuretic factor (ANF), or atriopeptin) is a 28-peptide produced by the atrial heart muscle. It belongs to the family of natriuretic peptides that exerts natriuretic, diuretic, and vasodilating effects. Besides ANP, the family of natriuretic peptides comprises mainly the brain natriuretic peptide (BNP) and the C-type natriuretic peptide (CNP). Human ANP is primarily synthesized as prepro-ANP (151 aa) and as pro-ANP (126 aa) stored within specific granules of the atrial cardiocytes. It is assumed that circulating human ANP is secreted from the atria and then cleaved proteolytically to the 28 peptide. Specific receptors are located in blood vessels, kidneys, and adrenal cortex. The binding of ANP at the extracellular domain of a transmembrane protein guanylate cyclase stimulates intracellular synthesis of cGMP that in turn activates protein kinase G. ANP is a potent hypotensive and diuretic (natriuretic) agent.

3.3.2
Neuropeptides

As shown above, peptides are ubiquitous bioactive compounds that occur throughout the body, and in almost every type of organism. Several peptides previously considered to be brain peptides have also been found in the gut. Other peptides, which in the past were considered to be gut peptides are now known to be located also in the brain, and hence it is difficult to make a clear-cut definition of neuropeptides. CCK has long been known as a gastrointestinal hormone that mediates digestive functions and feeding behavior, but in 1975 it was also identified in rat brain. In particular, CCK-8 is the most abundant form in the brain acting as a neuromodulator and/or neurotransmitter. The definition of neuropeptides based on putative physiological effects is normally based on historical observations rather than on a detailed understanding of the general significance of the biological effect. The appearance of separate pools of a peptide in two (or more) discrete biological compartments is of evolutionary significance.

Neuropeptides are synthesized in the form of large protein precursors (Fig. 3.18) that undergo proteolytic processing to yield the bioactive peptides. The signal peptide sequence is required for vectorial transport across the membranes of the ER. The conformation of the signal peptide, typically a β-turn, is important for the recognition of the exact cleavage site. The eukaryotic signal peptidase is an integral membrane protein with strong sequence similarity to the related enzyme isolated from bacteria. Precursor proteins of opioid peptides, tachykinins, and some releasing hormones have been characterized. Prior to the release of the

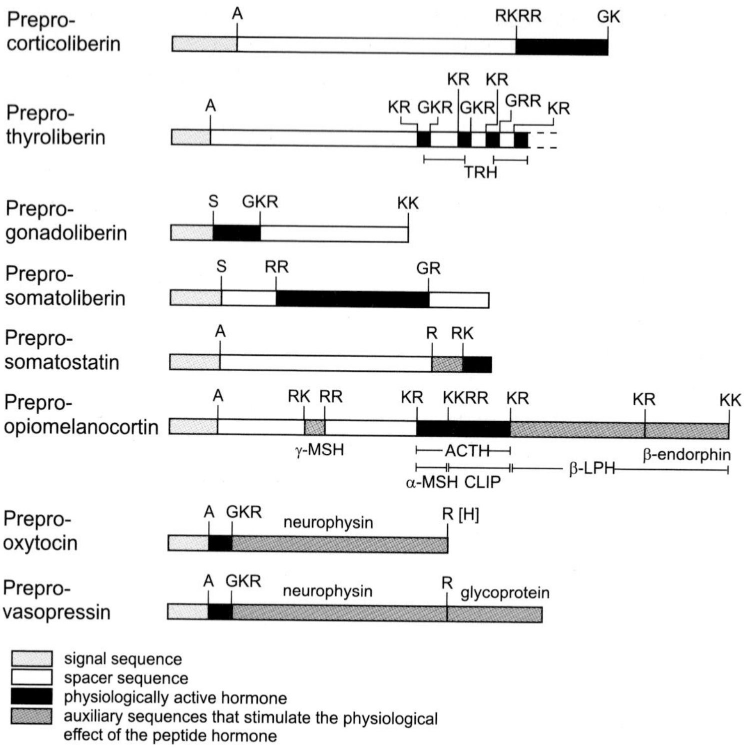

Fig. 3.18 Polyproteins as precursors of neuropeptides.

bioactive peptide, propeptides may be modified by amidation, acylation, or glycosylation. The cleavage of propeptides at pairs of basic amino acids or single basic residues is catalyzed by trypsin-like endopeptidases (prohormone converting enzymes). Post-translational processing can generate a number of active peptides from a single precursor protein [122] which has for example been demonstrated for POMC [123], this being the common precursor for ACTH, β-endorphin, MSH, corticotropin-like intermediate lobe peptide (CLIP), and related peptides.

Neuropeptides are stored in intracellular vesicles and act as neurohormones upon release into the bloodstream. Neuropeptides released into the synaptic cleft are neuromodulators that inhibit the action of excitatory neurotransmitters, or neuromediators which prolong the action of neurotransmitters. In the extracellular space, neuropeptides are involved both in pharmacodynamic and pharmacokinetic cascades. Neuropeptides interact with selective receptors on the target cells. The homeostasis of neuropeptides is characterized by the equilibrium between active regulation of cellular function via receptor interactions and their own catabolism. Opioid δ-, μ-, and κ-receptors, as well as receptors for bradykinin, tachykinins, angiotensin, neuropeptide Y, vasointestinal peptide, and somatostatin have been cloned and characterized. These G-protein-coupled receptors (GPCR) share as a common feature seven transmembrane helices. An extracellular signal is transmitted into the cell and amplified by coupling to G proteins [124].

Peptides are capable of exerting multiple effects in the CNS and elsewhere. The important effects of peptides in the CNS have been underlined by increasing numbers of publications that have been reviewed for the years 1980–1985 [125], 1986–1993 [126], and 1994–1999 [127, 128]. Examples of CNS effects caused by peptides (selected members in brackets) in humans and experimental animals are involved, for example, in:

- feeding (β-endorphin, motilin, corticotropin-releasing hormone, NPY, CCK, galanin, GLP-1, neuropeptide FF, bombesin, amylin, somatostatin, galanin, β-casomorphin, angiotensin II, enterostatin, oxytocin, gastrin-releasing peptide, pituitary adenylate cyclase-activating peptide);
- behavior (α-MSH, calcitonin gene-related peptide, NPY, melanin-concentrating hormone, vasopressin, neurotensin, CCK-8, oxytocin, ACTH);
- stress (atrial natriuretic factor, opioid peptides, NPY, substance P);
- thermogenesis (NPY);
- thermoregulation (dermorphin, NPY, delta sleep-inducing peptide, vasopressin, α-MSH, neurotensin, cyclo-(His-Pro);
- sleep (delta sleep-inducing peptide, vasopressin, corticotropin-like intermediate lobe peptide);
- learning (substance P);
- memory (vasopressin, galanin, CCK);
- aggression (α-MSH);
- alcohol uptake (tachykinin);
- anorectic effects (amylin, calcitonin gene-related peptide);
- anxiety (α-MSH, motilin, melanin-concentrating hormone, NPY);
- drinking (angiotensin II, tachykinins, opioid peptides);
- pain (neurotensin, vasopressin, CCK);
- sexual behavior (α-MSH, melanin-concentrating hormone); and
- locomotion (neurotensin, NPY, amylin, vasopressin, melanin-concentrating hormone).

3.3.2.1 **Opioid Peptides**

Opiates such as morphine **26** have been widely used by clinicians both for the blockade of severe pain and for anesthesia.

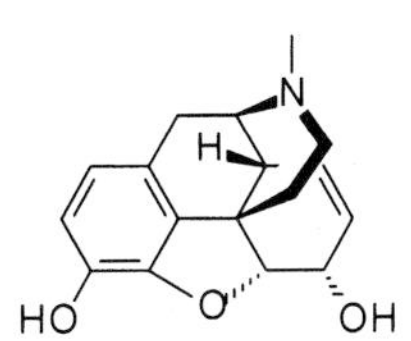

26

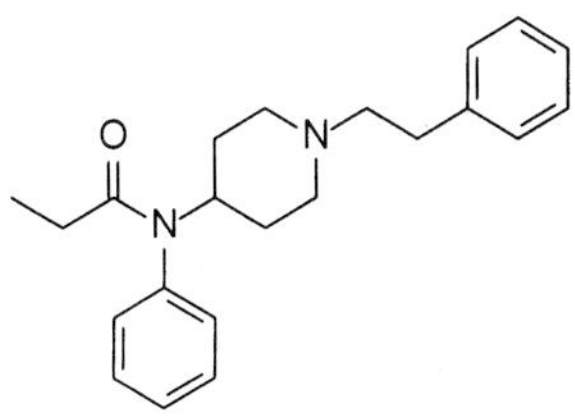

27

Although at present synthetic or semisynthetic morphine analogues such as the phenylpiperidine derivative fentanyl **27** are used [129], opium extracts have been the natural source for morphine for several thousand years. Indeed, it was surprising to find that nature provided morphine not only as a plant product but also (as structurally related opiates) in mammals, albeit in very small amounts and (presumably) during certain metabolic periods [130].

Collier first postulated the existence of "endogenous morphine" in 1972 at the International Congress of Pharmacology in San Francisco:

"The receptor for a foreign drug is really the receptor for a humoral substance with which the foreign molecule also interacts.... We do not yet know the natural function of the macromolecule(s) with which morphine interacts."

The binding of endogenous opioid receptor ligands was independently demonstrated one year later by three groups [131–133], and in 1975 the first endogenous peptide ligands were discovered by Hughes et al. [134], followed by the characterization of four major types (μ, δ, κ, and σ) of opioid receptors some years later [135, 136].

The opioid peptides first isolated from porcine brain were named methionine enkephalin (Met-enkephalin) **28** and leucine enkephalin (Leu-enkephalin) **29**. These compounds were found to occur naturally and bind as physiological agonists to the opioid receptors in the brain.

H–Tyr^1–Gly–Gly–Phe–Met^5–OH

28

H–Tyr^1–Gly–Gly–Phe–Leu^5–OH

29

Besides the enkephalins (enkephalos = brain), which were the first opioid peptides to be discovered, the endorphins and dynorphins are the most important of these compounds. They have also been termed "typical" opioid peptides that originate from three large precursor proteins, as shown in Fig. 3.19. These typical opioid peptides share the N-terminal sequence H-Tyr-Gly-Gly-Phe-, and most of the members bind to more than one type of opioid receptors, though certain selectivities are clear (see below). The so-called "atypical" opioid peptides originate from a variety of precursor proteins with different N-terminal amino acid sequences and a conserved N-terminal tyrosine residue [137].

The proenkephalin sequence contains four copies of Met-enkephalin, one of Leu-enkephalin, and two extended forms of Met-enkephalin. Prodynorphin (proenkephalin B) is the precursor for α-neoendorphin and the dynorphins, and contains three copies of Leu-enkephalin. Prepro-opimelanocortin, which is transformed into POMC by cleavage of the signal sequence, is the source of ACTH and β-lipotropin. These hormones are cleaved to give smaller bioactive peptides in a tissue-specific manner (Fig. 3.20).

POMC is proteolytically cleaved in the anterior lobe, yielding the N-terminal fragment, ACTH, and β-lipotropin (β-LPH). However, the latter polypeptide hormones are further cleaved to yield γ-MSH, α-MSH, CLIP, γ-LPH, and β-endorphin in the intermediate lobe only. The resulting range of hormones (Tab. 3.6) shows different activities, underlining the fact that the peptides formed both in

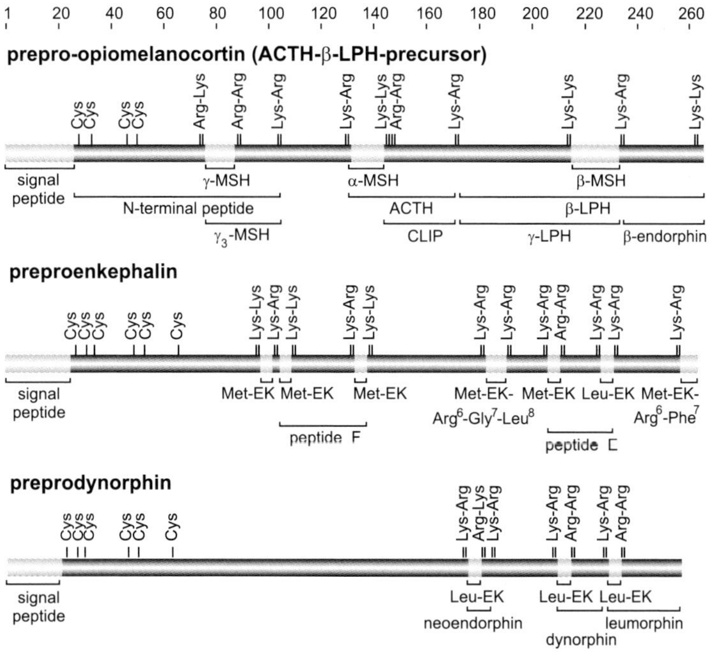

Fig. 3.19 Schematic representation of the sequences of opioid peptide precursors. EK=enkephalin.

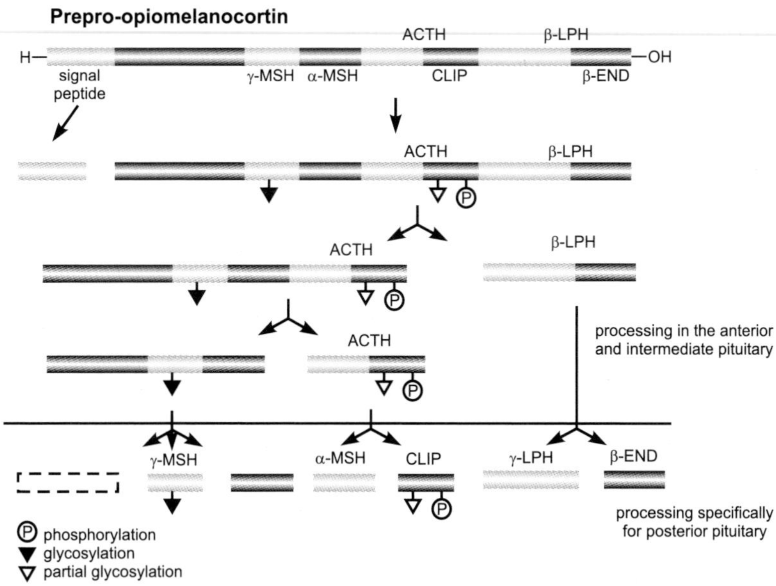

Fig. 3.20 Tissue-specific processing of proopiomelanocortin (POMC) in the anterior pituitary and intermediate pituitary.

Tab. 3.6 Different sets of opioid peptides released from polyprotein precursors.

Proopiomelanocortin	*Proenkephalin*	*Prodynorphin*
ACTH	Met-enkephalin	Dynorphin A (1–17)
β-Lipotropin [a)]	Leu-enkephalin	Dynorphin A (1–8)
γ-MSH	Met-enkephalin-Arg^6-Phe^7	α-Neoendorphin [b)]
α-MSH	Met-enkephalin-Arg^6-Gly^7-Leu^8	Dynorphin B (Rimorphin)
Corticotropin-like intermediate lobe peptide (CLIP)		Dynorphin B29 (Leumorphin)
		Leu-enkephalin

a) β-Lipotropin contains β-MSH and the endorphins (α, β, γ, δ).
b) α-Neoendorphin contains β-neoendorphin.

the anterior and intermediate lobes and the corresponding cleavage products of the proenkephalins are physiologically distinct.

The N-terminal tyrosine of the opioid peptides is indispensable for binding to opioid receptors. However, the activation of the peptide receptor complex requires the N-terminal dipeptide sequence which is directly related to the tyramine moiety of morphine **26** or other opiate alkaloids. Besides this dipeptide "message part", the opioid peptides contain – according to the proposal of Schwyzer – the "address" sequence which is composed of Gly-Phe or Phe that is responsible for their high affinity to opioid receptors. The C-terminal sequence part differentiates the affinity of opioid peptides toward the receptor types.

Morphine **26** shows relatively high affinity toward μ-receptors, where two subtypes (μ_1, μ_2) have been proposed. Fentanyl **27** similarly binds with high selectivity to the μ-receptor. Naloxone **30**, naltrexone **31**, and nalorphine **32** are the most commonly used opioid antagonists with selectivity toward the μ-receptor.

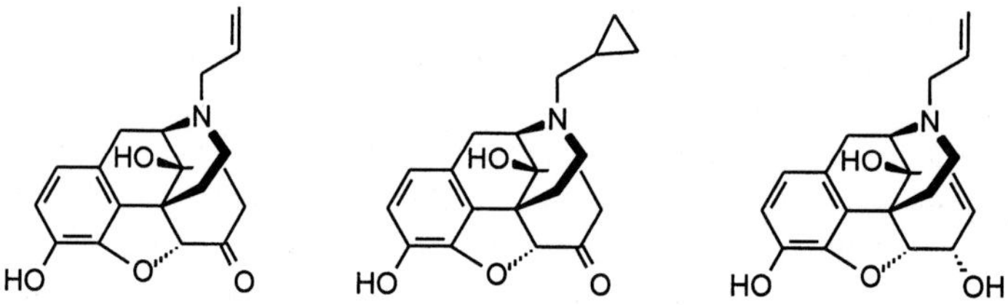

30 **31** **32**

However, they also show affinity toward δ- and κ-receptors, reversing the effects of agonists on these receptors. δ-Receptor subtypes have been classified into δ_1 and δ_2 or into δ_{cx} and δ_{ncx}, whereas for the κ-receptor the subtypes κ_1, κ_2, κ_3, κ_{1a}, κ_{1b}, κ_{2a}, and κ_{2b} have been described [138]. Until now, only one single gene has been identified for each of the three opioid receptor types. Alternative splicing and post-translational modifications are responsible for receptor subtype expression, but this remains to be verified with suitable pharmacological data. Although β-endorphin (Fig. 3.21) has been suggested as the endogenous agonist ligand for

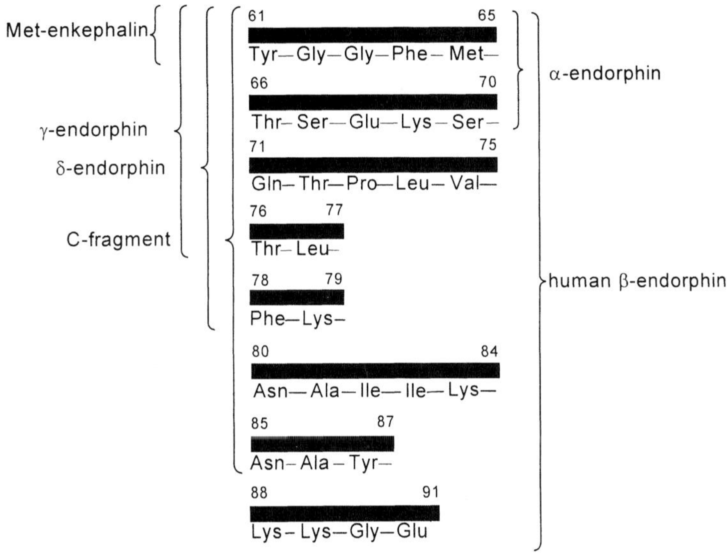

Fig. 3.21 Primary structure of β-endorphin (sequence 61–91 of β-lipotropin) and related sequences of the other endorphins. The N-terminal sequence 61–65 of β-lipotropin corresponds to methionine enkephalin.

the μ-receptor, its affinities for μ- and δ-receptor are almost equal. Dermorphin **5** appears to be the most μ-selective endogenous opioid peptide, but it also shows some affinity for the δ-receptor.

Since the first discovery of endogenous opioid peptides, the search for new opioid ligands with improved selectivity has remained a major target for many pharmaceutical companies. The main problem is to identify an opioid ligand that does not have the typical, well-known adverse side effects of addiction, intolerance, respiratory depression, and constipation. The opioid drugs currently in use exert their analgesic effect primarily via interaction with μ opioid receptors. Although these μ-type drugs are highly effective in alleviating severe pain, the adverse side effects outlined above represent serious disadvantages. Furthermore, whilst κ opioid agonists are potent analgesics, they are known to cause psychotomimetic and dysphoric effects. Based on the fact that κ- and μ-selective opioids produce many of their side effects via interactions with central opioid receptors, the development of peripheral- rather than central-acting analgesics might seem useful. Furthermore, it could be shown that δ agonists produce analgesic effects with fewer side effects, and it has been suggested that opioids with a mixed μ-agonist/δ-antagonist profile might be analgesic compounds with a low propensity for addiction and intolerance. New opioid peptide analogues with interesting activity profiles could be obtained based on the concept of conformational restriction [139, 140]. For example, H-Tyr-D-Phe-Phe-NH_2 (DALDA) and H-Tyr-D-Phe-Phe-Phe-NH_2 show both quite high μ-agonist potency and μ-receptor selectivity. Following the discovery of the prototype δ-opioid antagonists TIPP (H-Tyr-Tic-Phe-Phe-OH) and TIP (H-Tyr-Tic-Phe-OH) containing 1,2,3,4-tetrahydroisoquinoline-2-carboxylic acid (Tic) in position 2, extensive investigations of structure-activity relationships have led to the development of further members of the TIPP opioid peptide family comprising δ antagonists, δ agonists, and mixed μ agonist/δ antagonists [141]. In particular, the μ agonist/δ antagonist H-Dmt-Ticψ[CH_2NH]Phe-Phe-NH_2 (DIPP-NH_2[ψ]) with N-terminal 2′,6′-dimethyltyrosine (Dmt) was expected to be a potent analgesic with a low propensity for intolerance and physical dependence.

Opioid peptides and the corresponding receptors are also present in the heart and nerves, where they can easily be involved in the modulation of cardiac function. Indeed, the heart may be considered as a complex neuroendocrine (opioid peptides, NPY, VIP) or mechanoendocrine (atrial natriuretic peptide) organ [142, 143]. It is assumed that cardiac opioid peptides may have autocrine, paracrine, and endocrine functions.

Exorphins are opioid peptides produced from food upon digestion. The first example of a peptide with opioid activity derived from food proteins was β-casomorphin-7 (β-CM-7) **33**, which was isolated from commercial casein peptone [144].

H–Tyr^1–Pro–Phe–Pro–Gly^5–Pro–Ile–OH

33

β-CM-7 corresponds to the partial sequence 60–66 of bovine-β-casein A_2. Morphiceptin (β-CM-4 amide), H-Tyr-Pro-Phe-Pro-NH_2, which had already been synthesized prior to isolation from commercial enzymatic digest of casein, is a much more potent μ opioid agonist than β-CM-7 [145]. The β-CM show various effects on the CNS and cardiovascular, endocrine, and gastrointestinal systems. Further milk protein-derived opioid peptides and other exorphins, such as hemorphins and gluten exorphins, have been isolated and appropriate analogues synthesized and tested.

Interestingly, hemoglobin might serve as a source not only of the hemorphins but also of endogenous bioactive peptides, according to the so-called concept of the tissue-specific peptide pool which has been proposed to describe a novel system of peptidergic regulation [146]. Besides hemoglobin, a huge variety of other functional proteins should generate proteolytically derived fragments that contribute to tissue homeostasis.

3.3.2.2 **Tachykinins**

The term tachykinins describes a family of peptides that shares a common C-terminal sequence (-Phe-Xaa-Gly-Leu-Met-NH_2). They exist both in mammalian and nonmammalian species. Substance P (SP) has for a long time been the only peptide of the tachykinin family to be detected in mammals [147]. In 1931, von Euler and Gaddum discovered a hypotensive substance in brain extracts that stimulated smooth muscle. The name substance P was derived from the dried *p*owder obtained from the alcoholic extract, and the sequence was elucidated in 1970 by Leeman et al. In the early 1980s, several groups described the existence of novel mammalian members of the tachykinin family [148]. However, the different names used by different authors to label the new members of the family and the appropriate receptors initiated a troublesome story of "nomenclature mismatch" which creates problems to this day [149].

The most important structurally related mammalian tachykinins SP, neurokinin A (NKA), and neurokinin B (NKB) that are widely distributed within the central and peripheral nervous systems are listed in Tab. 3.7. The prepro-tachykinin gene gives rise via alternative splicing of its primary transcript to the mRNA for either α-prepro-tachykinin containing substance P as the only tachykinin, or else for β-prepro-tachykinin or the shortened γ-prepro-tachykinin. The latter contain two SP as well as NKA separated by intervening peptides of different length. In addition to the main members of the mammalian tachykinins listed in Tab. 3.7, neuropeptide K (NPK), a N-terminally extended form of NKA (extension: DADSSIEKQV^{10} ALLKALYGHG^{20}QISHKR) and neuropeptide γ (DAGHGQISHK^{10}RHKTDSFV GL^{20}Ma), which corresponds to the amino acid sequence of γ-prepro-tachykinin, also belong to this family.

The biological response of the tachykinins is mediated by specific G protein-coupled receptors (GPCR) with distinct pharmacological features connected with several chronic diseases such as asthma, rheumatoid arthritis, inflammatory bowel disorders, pain, and psychiatric disorders. Three tachykinin receptors (NK_1, NK_2, and NK_3) have been identified based on marked differences of agonist potencies in different tissues. SP is the preferred agonist for the NK_1 receptor, NKA for the NK_2 receptor, and NKB for the NK_3 receptor. Today, it is accepted that NKA is also a high-affinity endogenous ligand for the NK_1 receptor at various synapses and/or neuroeffector junctions [150]. A novel member of the tachykinin receptor family, termed NK_4, might be a subtype of the NK_3 receptor. Since the mammalian tachykinins have been implicated in a wide variety of biological functions, nonpeptide antagonists of the receptor types may provide opportunities for the treatment of diseases mentioned above [151].

Nonmammalian tachykinins [152, 153] that occur in cold-blooded animals also show a wide spectrum of pharmacological activities in certain mammalian tissue preparations. Selected members are listed in Tab. 3.8. In 1952, Erspamer isolated a substance subsequently named eledoisin from extracts of salivary glands of the cuttlefish (*Eledone moschata*). Ten years later its structure was elucidated, starting

Tab. 3.7 Primary structures of the most important mammalian tachykinins.

Name	***Sequence***
Substance P (SP)	H-Arg-Pro-Lys-Pro-Gln-Gln-Phe-Phe-Gly-Leu-Met-NH_2
Neurokinin A (NKA) [a)]	H-His-Lys-Thr-Asp-Ser-Phe-Val-Gly-Leu-Met-NH_2
Neurokinin B (NKB) [b)]	H-Asp-Met-His-Asp-Phe-Phe-Val-Gly-Leu-Met-NH_2

a) Also known as substance K, neurokinin α, and neuromedin L.
b) Also known as neurokinin β, and neuromedin K.

Tab. 3.8 Primary structures of selected nonmammalian tachykinins.

Name	***Orign***	***Sequence***
Eledoisin	*Eledone moschata*	<EPSKDAFIGLMa
Uperolin	*Uperoleia rugosa*	<EPDPNAFYGLMa
Physalaemin	*Physalaemus fuscumaculatus*	<EADPNKFYGLMa
Phyllomedusin	*Phyllomedusa bicolor*	<ENPNRFIGLMa
Kassinin	*Kassina senegalesis*	DVPKSDQFVGLMa
Scyliorhinin I	*Scyliorhinus caniculus*	AKFDKFYGLMa
Scyliorhinin II	*Scyliorhinus caniculus*	SPSNSKCPDGPDCFVGLMa [a)]

a) Disulfide bond between C^7–C^{13}.

from 10 000 pairs of posterior salivary glands of 1450 kg of the eight-armed cephalopod [154]. Eledoisin causes contraction of smooth muscle preparations and marked peripheral vasodilatation, resulting in a blood pressure decrease. Physalaemin acts similarly, but is three to four times more potent. Five tachykinins have been isolated from the skin of the Australian frog *Pseudophyrne güntheri*. Whereas three of these are similar to kassinin, the two other peptides are similar to SP. Locustatachykinin I (GPSGFYGVRa) and locustatachykinin II (APLSGFYGVRa) from the insect *Locusta migratoria* are characterized by significant changes in the C-terminal sequence part compared with the other tachykinins.

3.3.2.3 **Further Selected Neuroactive Peptides**

The dipeptide kyotorphin **34** has been isolated from bovine brain.

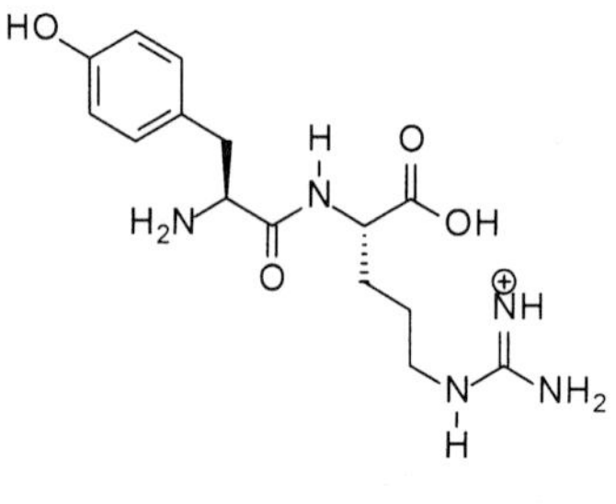

H–Tyr–Arg–OH

34

The sequence is part of neokyotorphin, H-Thr-Ser-Lys-Tyr-Arg-OH, which has also been found in bovine brain and seems to be the propeptide of kyotorphin. It has been shown that angiotensin-converting enzyme (ACE) is capable of cleaving neokyotorphin, yielding kyotorphin. Kyotorphin shows analgesic activity and promotes the release of Met-enkephalin.

H–Arg^1–Tyr–Leu–Pro–Thr^5–OH

35

In 1975, 180 µg of a pentapeptide called proctolin **35** were isolated from the proctodeal muscles of 125 000 cockroaches (~ 125 kg) *Periplaneta americana* [155]. Proctolin acts as an excitatory neurotransmitter and initiates strong contractions at the proctodaeum at a concentration of only 10^{-9} mol L^{-1}.

In 1972, a crustacean color-change hormone, the so-called red pigment-concentrating hormone (RPCH) **36**, was described as the first neuropeptide from invertebrates [156], having been discovered in the eye stalks of the prawn *Pandalus borealis*.

<Glu^1–Leu–Asn–Phe–Ser^5–Pro–Gly–Trp–NH_2

36

<Glu^1–Leu–Asn–Phe–Thr^5–Pro–Asn–Trp–Glu–Thr–NH_2

37

Some years later, the isolation of adipokinetic hormone (AKH) **37** from the corpora cardiaca of the locust was described [157]. AKH has a sequence similar to the prawn RPCH, but it is important for insect flight. Interestingly, RPCH/AKH form a group of peptides with widespread occurrence in arthropod species [158].

Coelenterates contain a multitude of neuropeptides. Two different neuropeptides were isolated from the simple sea anemone species *Anthopleura elegantissima*; these partly contained N-terminal amino groups acylated by a phenyllactyl moiety, e.g., antho-Kamide **38**, antho-RIamide I (L-3-phenyllactyl-Tyr-Arg-Ile-NH_2), and antho-RNamide I (L-3-phenyllactyl-Leu-Arg-Asn-NH_2) [159].

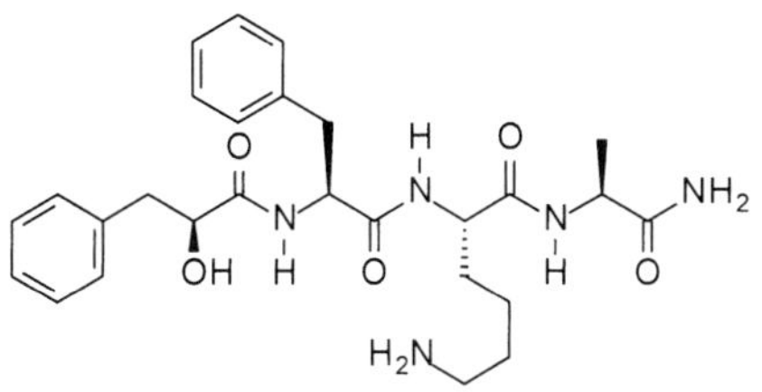

38

In 1975, the delta-sleep-inducing peptide (DSIP) **39** was isolated from the blood of sleeping rabbits. Intraventricular infusion of DSIP into the brain of rats causes δ-slow-wave sleep.

H–Trp^1–Ala–Gly–Gly–Asn^5–Ala–Ser–Gly–Glu–OH **39**

H–Gln^1–Ala–Thr–Val–Gly^5–Asn–Val–Asn–Thr–Asp^{10}–Arg–Pro–Gly–
Leu–Leu^{15}–Asp–Leu–Lys–OH **40**

In 1984, the diazepam-binding inhibitor peptide (DBIP) **40** was isolated from brain extracts. In contrast to the 1H-benzo-1,4-diazepines (e.g., diazepam, chlordiazepoxide), interaction of DBIP with the benzodiazepine receptors exerts anxiety; hence this "anxiety peptide" acts as an endogenous antagonist.

<Glu^{1}–Leu–Tyr–Glu–Asn^{5}–Lys–Pro–Arg–Arg–Pro^{10}–Tyr–Ile–Leu–OH

41

Neurotensin (NT) **41**, which occurs in the brain and gastrointestinal tract, was originally isolated from calf hypothalamus in 1971 [160]. NT acts as a neuromodulator of dopamine transmission and on anterior pituitary hormone secretion. NT exerts potent hypothermic and analgesic effects in the brain. Furthermore, it is a paracrine and endocrine modulator of the digestive tract and the cardiovascular system of mammals. NT is synthesized together with neuromedin N and the neurotensin-like peptide (H-Lys-Leu-Pro-Leu-Val-Leu-OH) as a larger precursor. Neuromedin N (NN) **42** exhibits a biological activity profile similar to that of NT, mediated by the same NT receptor subtypes [161, 162].

H–Lys^{1}–Ile–Pro–Tyr–Ile^{5}–Leu–OH

42

Processing of pro-NT/NN in the brain gives rise to NN and NT, whereas in the gut processing mainly leads to the formation of NT and large NN, a large peptide containing the NN sequence at its C-terminus [163]. Neuropeptide FF (NPFF) **43** has been initially detected in bovine brain using antisera directed against the molluscan peptide FMRFamide.

H–Phe^{1}–Leu–Phe–Gln–Pro^{5}–Gln–Arg–Phe–NH_2

43

NPFF is involved in opiate-induced analgesia, morphine tolerance, and abstinence. For these reasons NPFF has been referred to as a morphine-modulating peptide [164].

H–Tyr^{1}–Pro–Ser–Lys–Pro^{5}–Asp–Asn–Pro–Gly–Glu^{10}–Asp–Ala–Pro–
Ala–Glu^{15}–Asp–Leu–Ala–Arg–Tyr^{20}–Tyr–Ser–Ala–Leu–Arg^{25}–His–Tyr–
Ile–Asn–Leu^{30}–Ile–Tyr–Arg–Gln–Arg^{35}–Tyr–NH_2

44

Neuropeptide Y (NPY) **44** belongs to the PP fold family, the name being given because of the two tyrosine residues at the C- and N-termini. NPY, which is widely distributed throughout the central and peripheral nervous systems, is processed from a 97 aa prepro-polypeptide. Four human NPY receptor subtypes (Y_1, Y_2, Y_4/PP1, and Y_5) are known, and an additional Y_6 receptor has been characterized in mice. The main actions of NPY are stimulation of food uptake via the Y_5 receptor and increase of intracellular Ca^{2+} concentration in vascular smooth muscle cells, though many other biological effects of NPY have been described [165, 166].

3.3.3 Peptide Antibiotics

Antibiotics, derived from the Greek *antibios* which means "against life", are a chemically heterogeneous group of substances produced by microorganisms (bacteria and fungi) which kill or inhibit the growth of other microorganisms. Initially, peptide antibiotics [50–52] can be classified according to their chemical structure into linear and cyclic compounds, which can further subdivided into homomeric peptide antibiotics exclusively composed of amino acids, and heteromeric ones additionally containing nonamino acid-derived building blocks. Furthermore, homodetic and heterodetic peptide antibiotics can be distinguished based on the character of the covalent bonds. For example, tyrocidins, gramicidin S, bacitracins, cyclosporins, viomycin, and capreomycin belong to the homomeric homodetic cyclic peptide antibiotics. Linear peptides, such as gramicidins A–C, bleomycin, and peplomycin are found in the family of heteromeric peptide antibiotics. The representatives of cyclic peptides can be either homodetic (polymyxins, colistines, etc.) or heterodetic (vancomycin, dactinomycin, ristocetin A, etc.). The hundreds of known peptide antibiotics can additionally be classified according to the mode of their biosynthesis (cf. Sections 3.2.1 and 3.2.2, respectively) into nonribosomally synthesized peptides and ribosomally synthesized peptides [52, 167].

3.3.3.1 Nonribosomally Synthesized Peptide Antibiotics

Nonribosomally synthesized peptide antibiotics are largely produced by bacteria, though a few representatives are produced by *Streptomyces* or lower fungi. They display some unusual structural features which may in part account for their antibiotic properties. Peptide antibiotics often contain nonproteinogenic amino acids. In general, building blocks with D-configuration, N-methylation or nonstandard structural features are incompatible with ribosomal synthesis. Many peptide antibiotics are cyclic and branched-cyclic peptidolactones and depsipeptides. Such peptide structures are not found in animal cells. Because of the cyclic structures and the high content of nonproteinogenic constituents, these antibiotics are resistant to proteolysis.

The antibiotic action targets various metabolic areas, including nucleic acid and protein biosynthesis, energy metabolism, cell-wall biosynthesis, and nutrient uptake. The streptogranins, for example, act as protein synthesis inhibitors, while the Gram-positive bacteria-specific bacitracin inhibits the transfer of peptidoglycan (cell wall) precursors to bactoprenol pyrophosphate. Although traditionally gramicidin S has been considered to be selective against Gram-positive bacteria, it has been shown also to have high activity against Gram-negative bacteria and the fungus *Candida albicans*. Most peptide antibiotics are relatively toxic, and so few are used clinically:

- Bacitracin and polymyxin B have been used in antibacterial chemotherapy.
- Thiopeptin, thiostrepton, and enduracidin are used as stock feed additives and veterinary drugs.

- The cationic lipopeptide colistin (polymyxin E) was modified in order to reduce its systemic toxicity; the resultant product, the methosulfate derivative colimycin, has been used as an aerosol formulation against *Pseudomonas aeruginosa* lung infections.

The *modification* of existing peptide antibiotics represents one approach to increase the potential of therapeutic applications. For example, the semisynthetically modified water-soluble compounds dalfopristin and quinupristin are derived from the potent, but rather insoluble, streptogranins – which were first discovered in the 1950s. They have been clinically tested in combination as a parenteral agent against resistant Gram-positive bacteria.

A second interesting approach comprises a combination of peptide synthesis modules to yield novel structures. These can, for example, be used as templates for chemical synthesis and diversity, and analogues of gramicidin S with variations in amino acid sequence, ring size, charge, etc., have been designed and synthesized in this way. Such compounds display enhanced selectivity towards bacteria than towards mammalian cells.

Although clearly not a peptide antibiotic, penicillin should be mentioned in this context, as the biosynthesis of penicillin N **45** and cephalosporin C **46** in *Cephalosporium acremonium* starts from cysteine and valine, together with α-aminoadipic acid, catalyzed by the ACV synthetase.

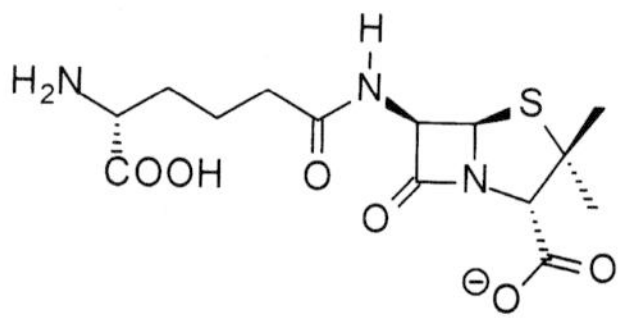

45

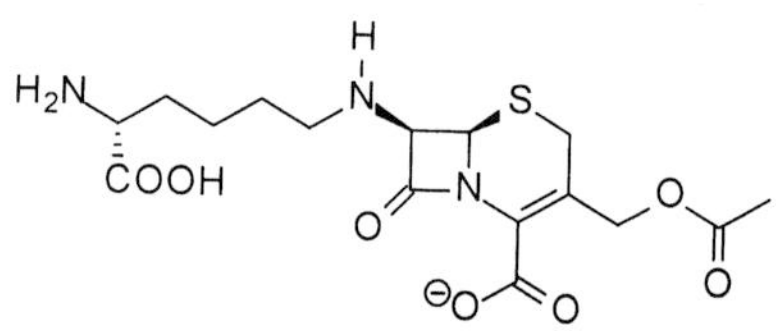

46

The enzyme involved not only catalyzes the formation of the two peptide bonds but also epimerizes the valine residue, yielding the tripeptide δ-(L-α-aminoadipoyl)-L-Cys-D-Val (ACV) **47** as an intermediate.

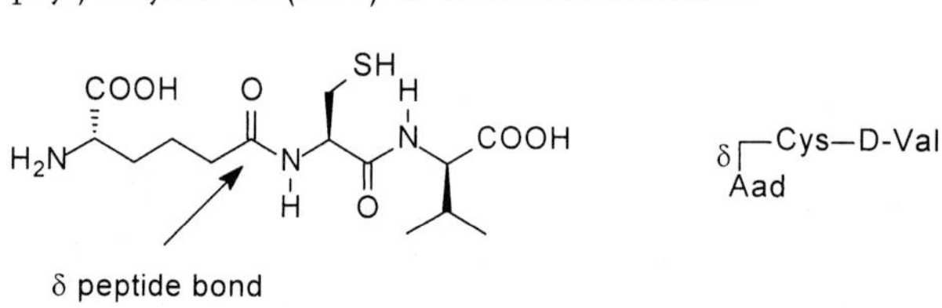

47

The next step in the biosynthesis of **45** involves oxidative cyclization of **47** by the enzyme isopenicillin N synthase (IPNS) in the presence of Fe^{2+} ions, ascorbate, and oxygen [168].

The growth-inhibitory properties of penicillins were first observed in 1928 by Alexander Fleming, in a *Staphylococcus* culture, and the application of penicillin against bacterial infections was first tested in humans in 1941.

All penicillins are characterized by the bicyclic β-lactam-thiazolidine structure, while the acyl moiety of the 6-aminopenicillanic acid (6-APA) **48** is variable. The latter has no antibiotic activity, and can be isolated as a fermentation product of *Penicillium chrysogenum*, though it is also available via enzymatic hydrolysis of benzylpenicillin (penicillin G) **49** using a bacterial acylase.

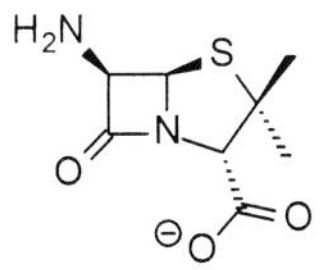

48

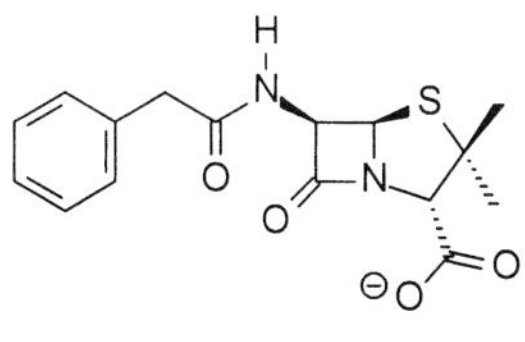

49

Thousands of new penicillin derivatives have been prepared in a semisynthetic approach using the enzymatic conversion of **49** into 6-aminopenicillanic acid **48** followed by acylation of the free amino group.

The structural diversity within the nonribosomally synthesized peptide antibiotics is extremely high, and does not permit strict definition of the families of related peptide antibiotics. These compounds vary in size, ranging from modified dipeptides (e.g., bacilysin or the phosphopeptide alaphosphin **50**) to peptides containing 16 to 17 amino acid residues (e.g., thiostrepton and enduracidin).

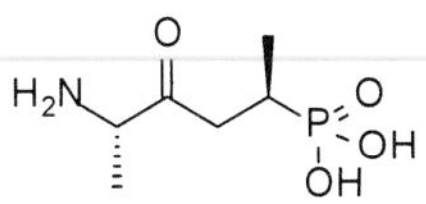

50

For this reason, only some well-investigated examples of peptide antibiotics have been selected for description in this chapter and in Chapter 4, or are compiled in the Glossary.

- Gramicidins A–C and alamethicin are linear membrane active peptide antibiotics. They form well-defined transmembrane channels and, therefore, are interesting model systems for the study of both peptide membrane interactions and mechanisms of voltage-gated transmembrane channel formation. The valine-rich gramicidin A **51** consists of alternating D- and L-amino acids, and is blocked at both termini. Monovalent cations (alkali ions, NH_4^+, and H^+) permeate through the channel, formed by a dimer of **51** (left-handed antiparallel double-stranded helix, cf. Fig. 3.2).

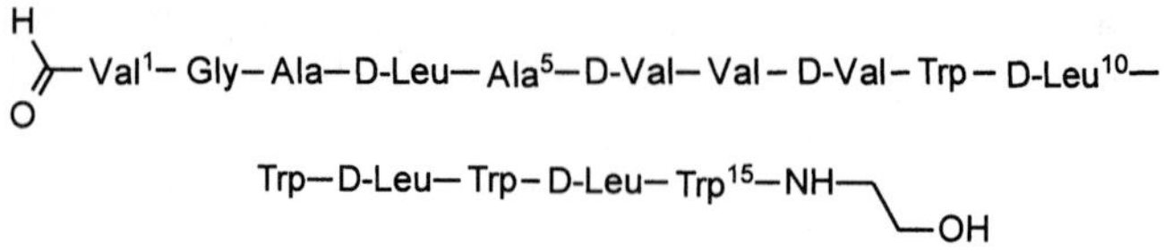

51

Ac-Aib^{1}–Pro–Aib–Ala–Aib^{5}–Ala–Gln–Aib–Val–Aib^{10}–

Gly–Leu–Aib–Pro–Val^{15}–Aib–Aib–Glu–Gln–Pheol^{20}

52

- Alamethicin **52** contains several aminoisobutyric acid (Aib) residues and L- phenylalaniol (Pheol) at the C-terminus. It is produced by the fungus *Trichoderma viride*. Alamethicin shows a α-helical structure with a bend in the helix axis at the proline residue, and spontaneously inserts into lipid bilayers.
- Valinomycin, a cyclic dodecadepsipeptide, has a high affinity for the K^+ ion which is coordinated by six valine carbonyl oxygen atoms (Fig. 3.22). Since the amino acid side chains provide an overall exterior, valinomycin is capable of crossing the hydrophobic core of the membrane.

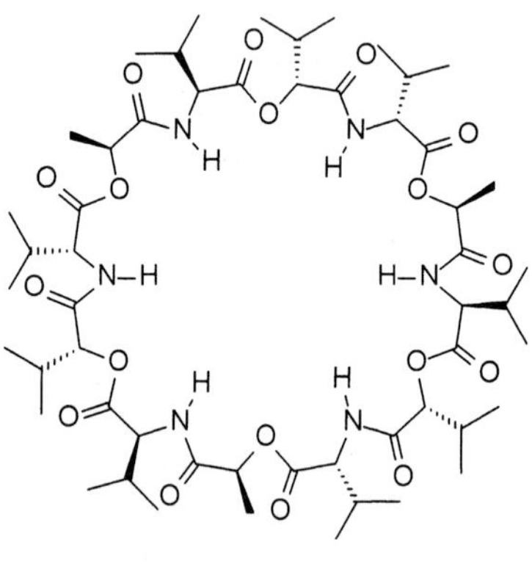

Fig. 3.22 Valinomycin.

- Bacitracin A **53** is the main component of a mixture of branched-cyclic dodecapeptides produced by *Bacillus licheniformis*. Bacitracin has been used as a component of various antiseptic combinations, and is also included in diagnostic agar formulations used for growth-inhibition assays of Gram-positive bacteria.

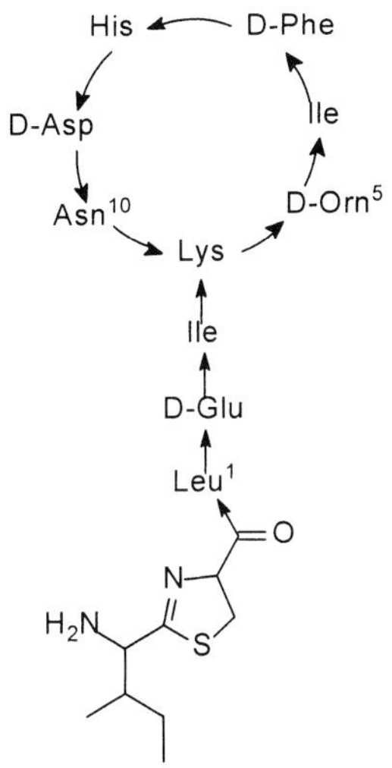

53

- Polymyxins consist of ten amino acids, six of them being L-,α,γ-diaminobutyric acid (Dab). The N-terminus is always acylated with methyloctanoic acid (Moa) or isooctanoic acid (Ioa). *Bacillus polymyxa* is the producer of these branched-cyclic peptide antibiotics. Polymyxin B exists in the modifications B_1 **54** and B_2 (Ioa instead of Moa). Further polymyxins contain serine in position 3, leucine in position 6, and isoleucine or threonine in position 7. The polymyxins act preferentially against Gram-negative bacteria. Polymyxin B is characterized by a strong synergistic effect on the activity of many other antibiotics as it promotes their access to bacterial cells.

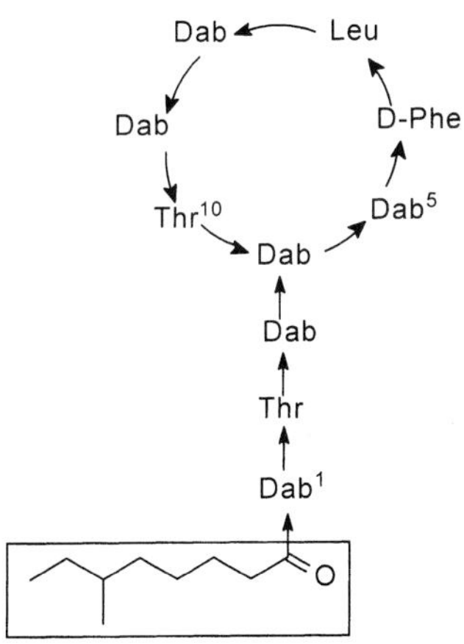

6-Methyloctanoic acid (MOA)

54

- Surfactin **55** is a biodetergent produced from *Bacillus subtilis.* It shows some structural similarities to the previously mentioned branched-cyclic lipopeptides.

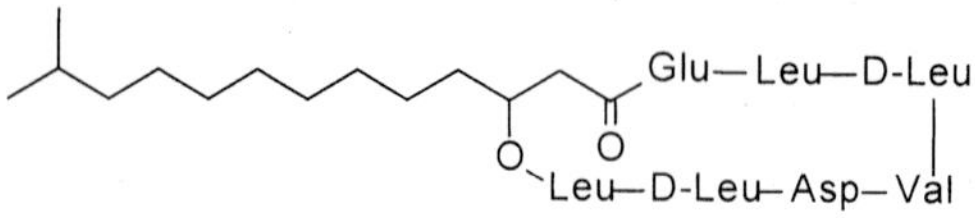

55

- Cyclosporin A (CsA), the main component of the 25 naturally occurring cyclosporins, is shown in Fig. 3.23 [169–171]. CsA is a cyclic undecapeptide which exerts antifungal, antiparasitic, anti-inflammatory, and immunosuppressive activities. CsA has been used therapeutically after organ transplantation, and in the treatment of autoimmune diseases. The cyclosporins are produced by the fungus *Beauveria nivea* (previously known as *Tolypocladium inflatum*). The cellular cyclosporin-binding protein, designated as cyclophilin, the activity of which is inhibited by CsA, is identical to the peptidyl-prolyl *cis/trans* isomerase (PPIase). Based on the first total chemical synthesis by Wenger [172], several hundred CsA analogues have been synthesized since 1984. The biosynthesis of CsA is carried out by a nonribosomal mechanism catalyzed by the multienzyme cyclosporin synthetase [114].

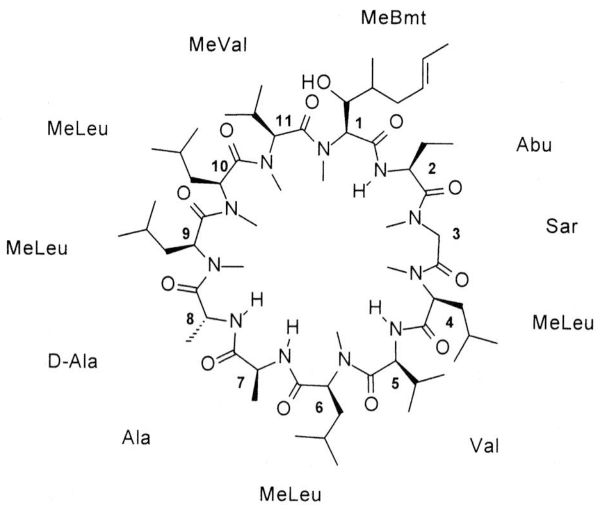

Fig. 3.23 Primary structure of cyclosporin A. MeBmt = (4*R*)-4[(E)-2-butenyl]-4-*N*-dimethyl-L-threonine; Abu = L-α-aminobutyric acid.

3.3.3.2 **Ribosomally Synthesized Peptide Antibiotics**

This class of peptide antibiotics plays a major role in natural host defenses. In general, these peptides do not exceed 50 amino acid residues, they are rich in positively charged amino acids, and are distinctly amphiphilic. β-Structures stabilized by disulfide bonds, amphipathic α-helices, extended structures and loops are characteristic structural features. A short survey of important families and representatives of defense peptides of eukaryotes was provided in Section 3.1. Hence, only some well-studied examples have been selected for a further short description. More detailed information is available in the Glossary.

Boman suggested the classification of these structurally heterogeneous peptides into five different groups [173]. The cecropins [68] and the magainins [59, 60] are prototypes of linear, preferentially helical peptides lacking cysteine. Linear proline- and arginine-rich peptides form an extremely heterogeneous group which are found for example in bovine neutrophils, pig intestine, honeybee, *Drosophila*, and other insects. Representatives are *Bac5*, RFRPPIRRPP^{10}IRPPFYPPFR^{20}PPIRPPIF PP^{30}IRPPFRPPLGPFPa, indolicidin, ILPWKWPWWP^{10}WRRa, and apidaecin Ia, GNNRPVYIPQ^{10}PRPPHPRI. The peptides of this group are equally active against Gram-negative and Gram-positive bacteria. The magainin-related brevenins and bactenectin, RLCRIVVIRV^{10}CR (disulfide bond: C^3–C^{11}), from bovine neutrophils belong to the group of peptides containing one disulfide bond. These peptides show highly specific activity against bacteria, but are also toxic towards eukaryotic cells. Peptides with more than one disulfide bond and β-sheet structures comprise a large group, including the best-studied families of animal and insect defensins [62–64]. Doubtlessly, S–S bond formation is very important in restricting the conformational freedom of smaller peptides and stabilizing certain conformations essential for biological activity. However, the thioether bridges in lantibiotics (see below) may fulfil the same function and should represent another version of sulfide-derived bridges influencing peptide conformations. Finally, peptides originating from nonantibiotic proteins form the fifth group in the Boman classification that includes GIP(7-42) and DBI(32-86), these being proteolytic fragments of gastric inhibitory polypeptide (GIP) and the diazepam-binding inhibitor (DBI), respectively.

In general, eukaryotic defense peptides that are produced constitutively are targeted to either secretory glands or to storage in granulae. This mainly applies to frog skin peptides, tachyplesins, and mammalian defensins. In contrast, the biosynthesis of many insect peptides (e.g., cecropins) occurs in response to an acute challenge by living bacteria or macromolecules (peptidoglycan or lipopolysaccharides) derived from bacteria. In both cases, the defense peptides are synthesized as inactive prepro-peptides like the gene encoded bacterial peptides (see below). There is generally only one processing step in bacteria to remove the propeptide part or leader peptide sequence, whereas the usual two-step maturation process is observed for eukaryotic defense peptides.

Bacteriocins are antimicrobial peptides produced by bacteria either without or with post-translational modification. The bacteriocins of Gram-positive bacteria [174] are divided into five groups according to structural features. Lantibiotics form one of these groups, the name being derived from "lanthionine-containing antibiotic peptides" [175–177]. Although the existence of the lantibiotic nisin has been known since 1928, and subtilin (another group member) was discovered in 1944, it was not until the early 1970s that the complete structure of these two lan-

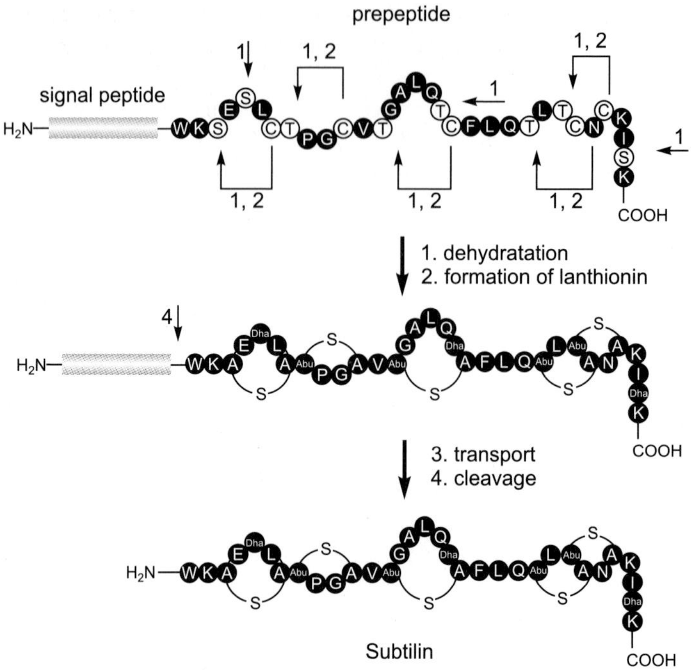

Fig. 3.24 Schematic representation of lantibiotic biosynthesis demonstrated for subtilin.

tibiotics was elucidated in the pioneering studies of Erhard Gross. The lantibiotics consist of a minimum 19 and a maximum 38 amino acids. In biosynthesis, the precursor peptides are first modified and then proteolytically activated, as shown schematically for subtilin in Fig. 3.24.

The gene organization for the biosynthetic machinery is performed in clusters, including information for the antibiotic prepeptide, the modifying enzymes, and accessory functions (e.g., special proteases and ABC transporter, immunity factors and regulatory peptides). According to Jung, the lantibiotics are subdivided into type A lantibiotics (e.g., nisin and subtilin), and type B lantibiotics (e.g., cinnamycin, ancovenin, and duramycins). Type A lantibiotics are elongated, helical amphiphiles, the action of which is directed to pore formation in the cytoplasmic membrane of susceptible bacteria. In contrast, the type B peptides are more compact, form globular structures, and generally interfere with a variety of membrane-bound enzymes.

3.3.4
Peptide Toxins

During evolution, toxic peptides and proteins have been developed by many species, either for defense against predators or for attack in aggressive competition for limited nutrient resources. Peptide and protein toxins are mostly low-molecu-

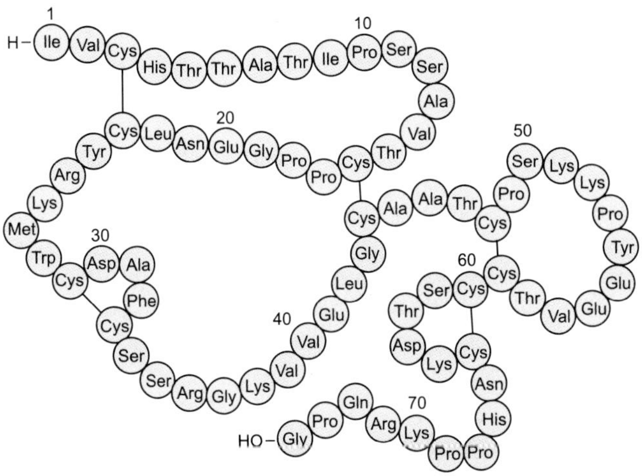

Fig. 3.25 Primary structure of α-bungarotoxin.

lar weight, single-chain compounds produced by snakes and invertebrates, as well as by virulent strains of bacteria and some plants.

The isolation of toxic peptides from venoms of various snakes was first attempted in the late 1960s. After elucidation of the amino acid sequence of a neurotoxin from the cobra *Naja nigricollis* [178], the primary structure of α-bungarotoxin (Fig. 3.25) was described in 1971 [179]. The latter toxin was isolated from the venom of the Chinese krait *Bungarus multicinctus*. α-Bungarotoxin strongly binds to the nicotinic acetylcholine receptor, and has been a useful tool in the characterization of this receptor [180]. The neurotoxic 71-peptide α-cobratoxin from the venom of the Thai cobra *Naja naja siamensis* [181] shows similar activity, and binds specifically to the acetylcholine receptor, thereby inhibiting its opening.

Conotoxins comprise the peptide neurotoxins of the venom of the fish-hunting sea snail (genus *Conus*) [182]. The conotoxins contain on average 9–29 amino acid residues, with a cysteine content between 22 and 50%, as shown for α-conotoxin GI **56**. During biting, the toxin is injected into and paralyzes the prey fish very quickly.

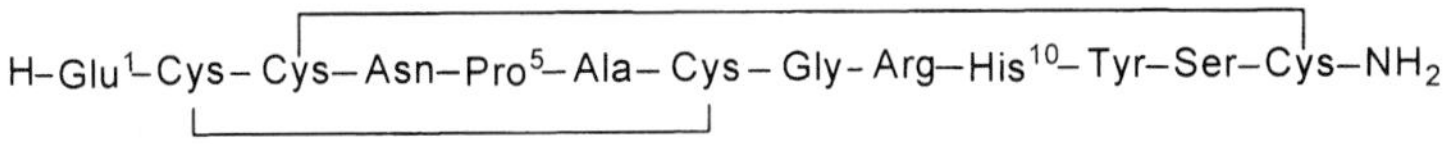

56

These toxic peptides have been classified into α-, μ-, ω-conotoxins, sleep conotoxins, convulsant conotoxins, conotoxins K, and the conopressins. The conopressins are basic 9-peptide amides with structural similarity to vasopressin. Furthermore, the sea snail contains the neurotoxic conantokins, also known as sleeper peptides, based on their sleep-inducing action after injection into mice. Conantokin G **57**, which contains several γ-carboxyglutamic acid (Gla) residues, and other members of this group are antagonists of the NMDA receptor in the brain, this being a subtype of the glutamate receptor.

H–Gly¹– Glu– Gla– Gla—Leu⁵—Gln– Gla – Asn—Gln – Gla¹⁰ – Leu—Ile—Arg–
Gla—Lys¹⁵– Ser– Asn—NH₂

57

More than 700 scorpion species produce venoms from which about 40% are capable to damage mammals, including human beings. The primary neurotoxic components of scorpion venoms are basic peptides ($M_r \sim 8$ KDa). Independent of chain length, all the toxins contain four intrachain disulfide bridges. For example, toxin V from *Leiurus quinquestriatus quinquestriatus* (LppV, Middle East), and toxin II from *Androctonus australi Hector* (AaH, North Africa) are both 64 peptides that cause membrane potential-dependent slackening of Na^+ channel activation. This effect results from a voltage-dependent high-affinity binding of the scorpion toxin on the Na^+ channel. In contrast to the α-toxins, β-toxins bind voltage independently on the Na^+ channel.

Spider peptide toxins paralyze insects by blocking the neuromuscular transmission mediated via glutamate receptors. Argiotoxin from *Argiope lobata* was the first spider venom to be structurally elucidated. The curtatocins from the venom of *Hololena curta* are cysteine-rich peptide amides with 36–38 amino acid residues. The toxic effect of these peptides results from an irreversible presynaptic neuromuscular blockade.

Sea anemone toxins are divided into type I and type II toxins that cause neurotoxicity by binding on neural and muscular Na^+ channels, respectively. The 49-peptide anthopleurin A (Ax-I) is a type I toxin, whereas the 48-peptide toxin Sh-I from *Stichodactyla helianthus* is a type II toxin.

Bee venom contains various neurotoxic and cytolytic peptides in addition to the enzymes hyaluronidase and phospholipase A2 [183]. The 26-peptide amide mellitin corresponds to the sequence 44–69 of prepro-mellitin of the queen-bee. Mellitin occurs in the venom in 50-fold molar excess over other constituents. The hemolytic and surface tension-decreasing activity of mellitin is based on the distribution of hydrophobic amino acid residues in the N-terminal sequence part, and hydrophilic building blocks in the C-terminal part. Despite lacking similarity in the primary structure, the bombolitins from bumblebee venom and the 14-peptide amide mastoparan from wasp venom (*Vespula lewisii*) show mellitin-like effects on cell membranes. Mastoparan induces the release of catecholamines and serotonin from specific tissues, and has also mast cell-degranulating activity. The mast cell-degranulating peptide (MCDP) is a 22-peptide amide containing two disulfide bridges. According to the name, it causes degranulation of mast cells and the release of high histamine concentrations into the tissue. Additionally, MCDP shows anti-inflammatory activity that is 100-fold that of hydrocortisone. The 18-peptide amide apamin is the third peptide component of bee venom (besides mellitin and MCDP) and, together with MCDP, is responsible for the venom's neurotoxicity. Since apamin specifically blocks both one type of Ca^{2+}-dependent K^+ channel and one type of Ca^{2+} channel, it serves as a valuable tool in neurophysiological studies.

In contrast to the peptide toxins described above, the poisonous constituents of the notorious toadstool *Amanita phalloides* are complex cyclic peptides. Doubtless, most fatal intoxications by mushrooms occur after ingestion of the *Amanita* species. The structural elucidation, synthesis, structure-activity relationship studies and biochemical studies were mainly carried out by Theodor Wieland and coworkers [48]. This team was able to explore the molecular events by which the amatoxins and phallotoxins exert their biological and toxic activities.

Bicyclic 8-peptides, called amatoxins – from which up to nine individuals have been isolated so far – are solely responsible for such effects. The naturally occurring amatoxins (α-amanitin, β-amanitin, γ-amanitin, ε-amanitin, amanin, amanin amide, amanullin, amanullinic acid, proamanullin), are derived from one parent molecule (cf. Fig. 3.5), and differ only by the number of hydroxy groups and by an amide carboxy exchange. As an example, α-amanitin is termed systematically cyclo-(L-asparaginyl-trans-4-hydroxy-L-prolyl-(*R*)-4,5-dihydroxy-L-isoleucyl-6-hydroxy-2-mercapto-L-tryptophyl-glycyl-L-cysteinyl-)-cyclo-(4-8)-sulfide-(*R*)-*S*-oxide. The lethal dose of amatoxins in humans could be estimated from accidents to be about 0.1 mg kg^{-1} body weight, or even lower. Wieland et al. [184] have developed a simple test for the discrimination between poisonous *Amanita* (except *A. visosa* specimens) and edible fungi as follows:

- Press a piece of the mushroom (cap or stalk) onto crude paper (news-print).
- Mark the spot with a pencil, and after drying moisten the spot with a drop of concentrated hydrochloric acid.
- A greenish-blue color developing within 5–10 min indicates the presence of almost all amatoxins.

A second group of less-toxic peptides, termed phallotoxins, occurs beside the amatoxins in all toxic *Amanita* species. The lethal doses are generally higher compared with those of the amatoxins. The basic structural formula of the phallotoxins is also shown in Fig. 3.5. All seven naturally occurring members (phalloidin, phalloin, phallisin, prophalloin, phallacin, phallacidin, phallisacin) of this group are derived from the same cyclcopeptide backbone consisting of seven amino acids and cross-linked from residue 3 to residue 6 by tryptathionine. For example, the systematic name of phalloidin (R^1, R^2, R^6=CH_3; R^2, R^3=OH; R^5=CH_2OH) is cyclo(L-alanyl-D-threonyl-L-cysteinyl-*cis*-4-hydroxy-L-prolyl-L-alanyl-2-mercapto-L-tryptophyl-4,5-dihydroxyleucyl)cyclo(3-6)-sulfide. As mentioned above, the cyclic 10-peptide antamanide **4**, which is also found in *Amanita phalloides*, protects experimental animals from intoxication by phalloidin. Based on their strongly specific potential of inhibiting eukaryotic RNA polymerase II (B), the amatoxins have importance as diagnostic tools for recognizing whether a cellular process is dependent on the function of those enzymes. The binding of phallotoxins to F-actin stabilizes filamentous structures of the cytoskeleton; therefore, by conjugation with fluorescent molecules the phallotoxins can be used to visualize such cellular structures.

3.4 References

1 G. Schmidt, *Top. Curr. Chem.* **1986**, *136*, 109
2 D.T. Krieger, *Science* **1983**, *222*, 975
3 J.H. Schwarz in: *Principles of Neural Science*, E.R. Kandel, J.H. Schwarz, T.M. Jessel (Eds.), Elsevier, Amsterdam, **1991**
4 V. Mutt in: *Peptides 1992*, C.H. Schneider, A.N. Eberle (Eds.), Escom, Leiden, **1993**
5 C.R. Harington, T.H. Mead, *Biochem. J.* **1935**, *29*, 1602
6 R.H. Siffert, V. du Vigneaud, *J. Biol. Chem.* **1935**, 108, 753
7 R. Consden, A.H. Gordon, A.J.P. Martin, R.L.M. Synge, *Biochem. J.* **1947**, 41, 596
8 W. Gevers, H. Kleinkauf, F. Lipmann, *Proc. Natl. Acad. Sci. USA* **1969**, *63*, 1335
9 J. Porath, P. Flodin, *Nature* **1959**, *183*, 1657
10 E.A. Peterson, H.A. Sober, *J. Am. Chem. Soc.* **1956**, *78*, 751
11 L.C. Craig, T.P. King, *Fed. Proc.* **1958**, *17*, 1126
12 A.P. Ryle, F. Sanger, L.F. Smith, R. Kitai, *Biochem. J.* **1955**, *60*, 541
13 W.W. Bromer, L.G. Sinn, A. Staub, O.K. Behrens, *J. Am. Chem. Soc.* **1956**, *78*, 3858
14 C.H. Li, I.I. Geschwind, R.D. Cole, I.D. Raacke, J.I. Harris, J.S. Dixon, *Nature* **1955**, *176*, 687
15 J.I. Harris, P. Roos, *Nature* **1956**, *178*, 90
16 J.I. Harris, A.B. Lerner, *Nature* **1957**, *179*, 1346
17 V. du Vigneaud, C. Ressler, S. Trippett, *J. Biol. Chem.* **1953**, *205*, 949
18 H. Tuppy, *Biochim. Biophys. Acta* **1953**, *11*, 449
19 V. du Vigneaud, H.C. Lawler, E.A. Popenoe, *J. Am. Chem. Soc.* **1953**, *75*, 4880
20 R. Acher, J. Chauvet, *Biochim. Biophys. Acta* **1953**, *12*, 487
21 D.F. Elliot, W.S. Peart, *Nature* **1956**, *177*, 527
22 L.T. Skeggs, K.E. Lentz, J.R. Kahn, N.P. Shumway, *J. Exp. Med.* **1956**, *104*, 193
23 V. du Vigneaud, C. Ressler, J.M. Swan, C.W. Roberts, P.G. Katsoyannis, S. Gordon, *J. Am. Chem. Soc.* **1953**, *75*, 4879
24 W.P. Jencks, *Catalysis in Chemistry and Enzymology*, Dover Publications, New York, **1987**
25 C. Walsh, *Enzymatic Reaction Mechanism*, Freeman, New York, **1979**
26 M.L. Bender, R.J. Bergeron, M. Komiyama, *The Bioorganic Chemistry of Enzymatic Catalysis*, Wiley, New York, **1984**
27 T.C. Bruice, S.J. Benkovic, *Biochemistry* **2000**, *39*, 6267
28 L. Carpo, *Hormones*, Freeman, New York, **1985**
29 W. König, *Peptide and Protein Hormones: Structure, Regulation, Activity*, VCH, Weinheim, **1993**
30 H.F. Bradford, *Chemical Neurobiology*, Freeman, New York, **1986**
31 R.S. Yalow, S.A. Berson, *In vitro Procedures with Radioisotopes in Medicine*, IAEA, Wien, **1970**
32 W. Pan, A.J. Kastin, W.A. Banks, J.E. Zadina, *Peptides* **1999**, *20*, 1127
33 A.J. Kastin, W. Pan, L.M. Maness, W.A. Banks, *Brain Res.* **1999**, *848*, 96
34 J.B. Finean, R. Coleman, R.H. Mitchell, *Membranes and Their Cellular Functions*, Blackwell Scientific Publications, Oxford, **1984**
35 M. Bretscher, *Sci. Am.* **1985**, *253*(4), 100
36 M.K. Jain, *Introduction to Biological Membranes*, Wiley, New York, **1988**
37 Y. Hatefi, *Annu. Rev. Biochem.* **1985**, *54*, 1015
38 M.F. Perutz, *Sci. Am.* **1978**, *239*, 92
39 F.W. Alt, T.K. Blackwell, G.D. Yancopoulos, *Science* **1987**, *238*, 1079
40 G. Kroemer, A.C. Martinez, *Immunol. Today* **1992**, *13*, 401
41 J. Kuby, *Immunology*, Freeman, New York, **1994**
42 T. Hunkapiller, L. Hood, *Adv. Immunol.* **1989**, *44*, 1
43 L.D. Barber, P. Parham, *Annu. Rev. Cell Biol.* **1993**, *9*, 163
44 T. Halkier, *Mechanisms in Blood Coagulation, Fibrinolysis, and the Complement Sys-*

tem, Cambridge University Press, Cambridge, **1991**

45 M. E. Nimni, *Collagen*, CRC Press, Boca Raton, **1988**

46 D. Kaplan, W. W. Adams, B. Farmer, C. Viney, *Silk Proteins*, American Chemical Society, New York, **1994**

47 L. Roberts, W. Hornebeck (Eds.), *Elastin and Elastases*, CRC Press, Boca Raton, **1989**

48 T. Wieland, *Peptides of Poisonous Amanita Mushrooms*, Springer, New York, **1986**

49 V. Erspamer in: *Amphibian Biology*, Vol. 1, H. Heatwole (Ed.), Surray Beatty & Sons, Chipping Northon, **1994**

50 R. E. W. Hancock, *Lancet* **1997**, *349*, 418

51 H. Kleinkauf, H. von Döhren, *Bio chemistry of Peptide Antibiotics*, de Gruyter, Berlin, **1990**

52 R. E. W. Hancock, D. S. Chapple, *Antimicrob. Agents Chemother.* **1999**, *43*, 1317

53 R. W. Jack, G. Bierbaum, H.-G. Sahl, *Lantibiotics and Related Peptides*, Springer, New York, **1998**

54 K. Fukase, M. Kitazawa, A. Sano, K. Shimbo, H. Fujita, S. Horimoto, T. Wakamiya, T. Shiba, *Tetrahedron Lett.* **1988**, *29*, 795

55 G. Kiss, H. Michl, *Toxicon* **1962**, **1**, 33

56 A. Csordas, H. Michl, *Monatsh. Chem.* **1970**, *101*, 182

57 M. Simmaco, G. Mignogna, D. Barra, *Biopolymers* **1998**, *47*, 435

58 M. Zasloff, *Proc. Natl. Acad. Sci. USA* **1987**, *84*, 5449

59 D. Barra, M. Simmaco, *Trends Biotechnol.* **1995**, *13*, 205

60 M. Zasloff, *Curr. Opin. Immunol.* **1992**, *4*, 3

61 D. Barra, M. Simmaco, H. Boman, *FEBS Lett.* **1998**, *430*, 130

62 T. Ganz, M. E. Selsted, R. I. Lehrer, *Eur. J. Haematol.* **1990**, *44*, 1

63 S. H. White, W. C. Wimley, M. E. Selsted, *Curr. Opin. Struct. Biol.* **1995**, *5*, 521

64 R. I. Lehrer, A. K. Lichtenstein, T. Ganz, *Annu. Rev. Immunol.* **1993**, *11*, 105

65 R. Fernandez de Caleya, B. Gonzalez-Pascual, F. Garcia-Olmedo, P. Carbonero, *Appl. Microbiol.* **1972**, *23*, 998

66 W. F. Broekaert, B. P. A. Cammue, M. F. C. Debolle, K. Thevissen, G. W. Desamblanx, R. W. Osborne, *Crit. Rev. Plant Sci.* **1997**, *16*, 297

67 H. G. Boman, D. Hultmark, *Annu. Rev. Microbiol.* **1987**, *41*, 103

68 H. G. Boman, *Annu. Rev. Immunol.* **1995**, *13*, 61

69 B. M. Rode, *Peptides* **1999**, *20*, 773

70 D. E. Lee, J. R. Granja, J. A. Martinez, K. Severin, M. R. Ghadiri, *Nature* **1996**, *382*, 525

71 R. L. P. Adams, J. P. Knowler, D. P. Leader, *The Biochemistry of the Nucleic Acids*, Chapman & Hall, London, **1992**

72 S. Benzer, *Sci. Am.* **1962**, *206*, 70

73 M. W. Nierenberg, *Sci. Am.* **1963**, *208*, 80

74 F. H. C. Crick, *Sci. Am.* **1962**, *207*, 66

75 J. W. Suttie, *Annu. Rev. Biochem.* **1985**, *54*, 472

76 T. W. Rademacher, R. B. Parekh, R. A. Dwek, *Annu. Rev. Biochem.* **1988**, *57*, 785

77 A. Varki, *Glycobiology* **1993**, *3*, 97

78 B. Imperiali, *Acc. Chem. Res.* **1997**, *30*, 452

79 B. Imperiali, S. E. O'Connor, *Curr. Opin. Chem. Biol.* **1999**, *3*, 643

80 P. Sears, C.-H. Wong, *Cell. Mol. Life. Sci.* **1998**, *54*, 223

81 C. M. Taylor, *Tetrahedron* **1998**, *54*, 11317

82 G. Ashwell, A. G. Morell, *Adv. Enzymol. Relat. Areas Mol.* **1974**, *41*, 99

83 K. V. Figura, A. Hasilik, *Trends Biochem. Sci.* **1984**, *9*, 29

84 G. F. Springer, *Science* **1984**, *224*, 1198

85 A. T. Beyer, J. E. Sadler, J. I. Rearick, J. C. Paulson, R. L. Hill, *Adv. Enzymol.* **1981**, *52*, 24

86 L. A. Lasky, *Science* **1992**, *258*, 964

87 M. L. Phillips, E. Nudelman, R. C. A. Gaeta, M. Perez, A. K. Singhal, S. Hakomori, J. C. Paulson, *Science* **1990**, *250*, 1130

88 G. Waltz, A. Arufto, W. Kalanus, M. Bevilacqua, B. Seed, *Science* **1990**, *250*, 1132

89 G. W. Hart, K. Brew, G. A. Grant, R. A. Bradshaw, W. Lennarz, *J. Biol. Chem.* **1979**, *254*, 9747

90 R. T. Schwartz, R. Datema, *Adv. Carbohydr. Chem. Biochem.* **1982**, *40*, 287

91 F. Kornfeld, S. Kornfeld, *Annu. Rev. Biochem.* **1985**, *53*, 631

92 S.T. Prigge, A.S. Kolhekar, B.A. Eiper, R.E. Mains, L.M. Amzel, *Science* **1997**, *278*, 1300

93 F. Marks, *Protein Phosphorylation*, VCH, Weinheim, **1996**

94 M.J. Hubbard, P. Cohen, *Trends Biochem. Sci.* **1993**, *18*, 172

95 P. Cohen, *Trends Biochem. Sci.* **1992**, *17*, 408

96 B. Schaffhausen, *Biochim. Biophys. Acta* **1995**, *1242*, 61

97 T.R. Bruke, Z.-J. Yao, Jr., D.-G. Liu, J. Voigt, Y. Gao, *Biopolymers* **2001**, *60*, 32

98 D.J. Dunican, P. Doherty, *Biopolymers* **2001**, *60*, 45

99 L.A. Beck, T.J. Hosick, M. Sinensky, *J. Cell. Biol.* **1988**, *107*, 1307

100 S.L. Wolda, J.A. Glomset, *J. Biol. Chem.* **1988**, *263*, 5997

101 D.M. Leonard, *J. Med. Chem.* **1997**, *40*, 2971

102 D.R. Lowy, B.M. Willumsen, *Annu. Rev. Biochem.* **1993**, *62*, 851

103 A. Wittinghofer, H. Waldmann, *Angew. Chem.* **2000**, *112*, 4360; *Angew. Chem. Int. Ed. Engl.* **2000**, *39*, 4192

104 P.J. Casey, X.Y. Buss, *Methods Enzymol.* **1995**, *250*, 435

105 M.D. Resh, *Biochim. Biophys. Acta* **1999**, *1451*, 1

106 M. Hackett, L. Guo, J. Shabanowitz, D. Hunt, E.L. Hewlett, *Science* **1994**, *266*, 433

107 J.T. Dunphy, M.E. Linder, *Biochim. Biophys. Acta* **1998**, *1436*, 245

108 W.D. Branton, C.G. Fields, V.L. VanDrisse, G.B. Fields, *Tetrahedron Lett.* **1993**, *34*, 4885

109 A. Hille, T. Braulke, K. von Figura, W.B. Huttner, *Eur. J. Biochem.* **1990**, *188*, 577

110 F. Lipmann in: *The Mechanism of Enzyme Action*, W.D. McElroy, B. Glass (Eds.), Hopkins, Baltimore, **1954**

111 H. Kleinkauf, H. von Döhren, *Eur. J. Biochem.* **1990**, *192*, 1

112 M.A. Marahiel, T. Stachelhaus, H.D. Mootz, *Chem. Rev.* **1997**, *97*, 2651

113 M.A. Marahiel, *Chemistry & Biology* **1997**, *4*, 561

114 A. Lawen, J. Dittmann, B. Schmidt, D. Riesner, H. Kleinkauf, *Biochimie* **1992**, *74*, 511

115 R.M.G. Nair, J.F. Barrett, C.Y. Bowers, A.V. Schally, *Biochemistry* **1970**, *9*, 1103

116 R. Burgus, T.F. Dunn, D. Desiderio, D.N. Ward, W. Vale, R. Guillemin, *Nature* **1970**, *226*, 321

117 V.J. Hruby, D. Patel in: *Peptides: Synthesis, Structures, and Applications*, B. Gutte (Ed.), Academic Press, New York, **1995**, 247

118 S.G. Schultz, G.M. Makblout, *Handbook of Physiology: The Gastrointestinal System II*, Bethesda, MD: American Physiological Society, **1989**, 87

119 J.H. Walsh, G.J. Dockray, *Gut Peptides*, Raven, New York, **1994**

120 J.F. Rehfeld, *Physiol. Rev.* **1998**, *78*, 1087

121 M. Yanagisawa, H. Kuribara, S. Kimura, Y. Tomobe, M. Kobayasshi, Y. Mitsui, Y. Yazaki, K. Goto, T. Masaki, *Nature* **1988**, *332*, 411

122 H. Gainer, J.T. Russel, Y.P. Loh, *Neuroendocrinology* **1985**, *40*, 171

123 R.E. Mains, B.A. Eipper, *J. Biol. Chem.* **1979**, *254*, 7885

124 C. Humblet, T. Mirzadegan, *Annu. Rep. Med. Chem.* **1992**, *27*, 291

125 J.E. Zadina, W.A. Banks, A.J. Kastin, *Peptides* **1986**, *7*, 497

126 B. Ahmed, A.J. Kastin, W.A. Banks, J.E. Zadina, *Peptides* **1994**, *15*, 1105

127 W. Pan, A.J. Kastin, W.A. Banks, J.E. Zadina, *Peptides* **1999**, *20*, 1127

128 A.L. Vaccarino, G.A. Olson, R.D. Olson, A.J. Kastin, *Peptides* **1999**, *20*, 1527

129 P.A.J. Janssen, *Br. J. Anaesth.* **1962**, *34*, 260

130 S. Spector, J. Donnerer in: *Handbook of Experimental Pharmacology*, A. Herz, H. Akil, E.J. Simon (Eds.), Springer, Berlin, **1993**

131 C.B. Pert, S.H. Snyder, *Proc. Natl. Acad. Sci. USA* **1973**, *70*, 2243

132 E.J. Simon, J.M. Hiller, I. Edelman, *Proc. Natl. Acad. Sci. USA* **1973**, *70*, 1947

133 L. Terenius, *Acta Pharmacol. Toxicol.* **1973**, *33*, 377

134 J. Hughes, T.W. Smith, H.W. Kosterlitz, L.A. Morgan, H.R. Morris, *Nature* **1975**, *258*, 577

135 G.W. Pasternak, *The Opioate Receptors*, Humana Press, Totowa, New York, **1988**

136 R.J. Knapp, K.N. Hawkins, G.K. Lui, J.E. Shook, J.S. Heyman, F. Porreca, V.J. Hruby, T.F. Burks, H.I. Yamamura, *Adv. Pain Res. Ther.* **1990**, *14*, 45

137 H. Teschemacher in: *Handbook of Experimental Pharmacology*, A. Herz, H. Akil, E.J. Simon (Eds.), Springer, Berlin, **1993**

138 C.J. Fowler, G.L. Fraser, *Neurochem. Int.* **1994**, *24*, 401

139 P.W. Schiller in: *Medicinal Chemistry*, Vol. 28, G.P. Ellis, G.B. West (Eds.), Elsevier, Amsterdam, **1991**

140 V.J. Hruby, C.A. Gehrig, *Med. Res. Rev.* **1989**, *9*, 343

141 P.W. Schiller, G. Weltrowska, I. Berezowska, T.M.-D. Nguyen, B.C. Wilkes, C. Lemieux, N.N. Chung, *Biopolymers* **1999**, *51*, 411

142 W.R. Millington, D.W. Rosenthal, C.V. Unal, C. Nyquist-Battie, *Cardiovasc. Res.* **1999**, *43*, 107

143 B.A. Barron, *Cardiovasc. Res.* **1999**, *43*, 13

144 V. Brantl, H. Teschemacher, A. Henschen, F. Lottspeich, *Hoppe-Seyler's Z. Physiol. Chem.* **1979**, *360*, 1211

145 K.J. Chang, Y.F. Su, D.A. Brent, J.K. Chang, *J. Biol. Chem.* **1985**, *125*, 309

146 V.T. Ivanov, A.A. Karelin, M.M. Philippova, I.V. Nazimov, V.Z. Pletnov, *Inc. Biopoly.* **1997**, *43*, 171

147 B. Pernow, *Pharmacol. Rev.* **1983**, *35*, 85

148 J.E. Maggio, *Annu. Rev. Neurosci.* **1988**, *11*, 13

149 C.A. Maggi, *Trends Pharmacol. Sci.* **2000**, *21*, 173

150 C.A. Maggi, T.W. Schwarz, *Trends Pharmacol. Sci.* **1997**, *18*, 351

151 Z. Gao, N.P. Peet, *Curr. Med. Chem.* **1999**, *6*, 374

152 V. Erspamer, *Annu. Rev. Pharmacol.* 1971, *11*, 327

153 V. Erspamer, P. Melchiorri, *Pure Appl. Chem.* **1973**, *35*, 463

154 V. Erspamer, A. Anastasi, *Experientia* **1962**, *18*, 58

155 A.N. Starratt, B.E. Brown, *Life Sci.* **1975**, *17*, 1253

156 P. Fernlund, L. Josefsson, *Science* **1972**, *177*, 173

157 J.V. Stone, W. Mordue, K.E. Batley, H.R. Morris, *Nature* **1976**, *263*, 207

158 M.H. Schaffer, B.E. Noyes, *BioEssays* **1987**, *7*, 67

159 C.J.P. Grimmelikhuijzen, K.L. Rinehart, E. Jacob, D. Graff, R.K. Reinscheid, H.-P. Nothacker, A.L. Staley, *Proc. Natl. Acad. Sci. USA* **1990**, *87*, 5410

160 R.E. Carraway, S.E. Leeman, *Biochem. J.* **1973**, *248*, 6854

161 N. Minamino, K. Kangawa, H. Matsuo, *Biochem. Biophys. Res. Commun.* **1984**, *122*, 542

162 J.-P. Vincent, J. Mazella, P. Kitabgi, *Trends Pharmacol. Sci.* **1999**, *20*, 302

163 P.R. Dobner, D.L. Barber, L. Villa-Komaroff, C. McKierman, *Proc. Natl. Acad. Sci. USA* **1987**, *84*, 3516

164 M. Allard, R. Rousselot, M.C. Lombard, D.T. Theodosis, *Peptides* **1999**, *20*, 327

165 T.S. Gray, J.E. Morley, *Life Sci.* **1986**, *38*, 389

166 L. Grundemar, S. Bloon, *Neuropeptide Y and Drug Developments*, Academic Press, New York, **1997**

167 W. van't Hof, E.C.I. Veerman, E.J. Helmerhorst, A.V.N. Amerongen, *Biol. Chem. Hoppe-Seyler* **2001**, *382*, 597

168 P.L. Roach, I.J. Clifton, C.M.H. Hensgens, N. Shibata, C.J. Schofield, J. Hajdu, J.E. Baldwin, *Nature* **1997**, *387*, 827

169 J.F. Borel, *Prog. Allergy* **1986**, *38*, 9

170 B.D. Kahan, *Cyclosporin: Biological Activity and Clinical Applications*, Grune and Stratton, Orlando, **1984**

171 R. Schindler, *Cyclosporin in Autoimmune Diseases*, Springer, Berlin, **1985**

172 R.M. Wenger, *Helv. Chim. Acta* **1984**, *67*, 502

173 H.G. Boman, *Annu. Rev. Immunol.* **1995**, *13*, 61

174 J.R. Tagg, A.S. Dajani, L.W. Wannamaker, *Bacteriol. Rev.* **1976**, *40*, 722

175 R. James, C. Lazdunski, F. Pattus, *Bacteriocins, Microcins and Lantibiotics*, Springer, Berlin, **1992**

176 R.N.H. Konings, C.W. Hilbers, Lantibiotics: an unique group of antibiotic peptides, *A. Leeuwenhoek, Int. J. Gen. Microbiol.* **1996**, *69*, 87

177 G. Jung, H.-G. Sahl, *Nisin and Novel Lantibiotics*, Escom, Leiden, **1991**

178 D. L. Eaker, J. Porath, *Jpn. J. Microbiol.* **1967**, *11*, 353

179 D. Mebs, K. Narita, C. Y. Lee, *Biochem. Biophys. Res. Commun.* **1971**, *44*, 711

180 J.-P. Changeux, A. Devillers-Thiery, P. Chemoulli, *Science* **1984**, *225*, 1335

181 E. Karlson, H. Arnberg, D. Eaker, *Eur. J. Biochem.* **1971**, *21*, 1

182 W. R. Gray, B. M. Olivera, *Annu. Rev. Biochem.* **1988**, *57*, 665

183 R. C. Hider, *Endeavour, New Series* **1988**, *12*, 60

184 T. Wieland, L. Wirth, E. Fischer, *Liebigs Ann. Chem.* **1949**, *564*, 152

4
Peptide Synthesis

4.1
Principles and Objectives

As the biosynthesis even of complicated proteins *in vivo* occurs within seconds or minutes, the relatively tedious classical chemical synthesis of peptides and proteins in the laboratory appears to be a somewhat hopeless enterprise, especially with regard to speed. The first chemical synthesis of insulin during the early 1960s took over two years to complete. Today, peptide and protein synthesis have become much more efficient, not only by the use of recombinant DNA techniques but also with the advent of automated solid-phase peptide synthesis and multiple peptide synthesis.

Peptides and proteins can be considered as encoded biomolecules. The translation products are determined genetically and synthesized for highly specific functions. The above-mentioned seemingly hopeless competition between a peptide chemist and nature takes on a completely new aspect when peptides and proteins can be synthesized which are not available by using biosynthetic mechanisms. Hence, the initial quest is to identify the main targets of peptide synthesis.

4.1.1
Main Targets of Peptide Synthesis

4.1.1.1 Confirmation of Suggested Primary Structures

There is a general chemical principle that total synthesis is the most convincing structural proof. Despite application of the latest techniques, misinterpretations in the sequence determination of peptides have in the past led to incorrect suggestions of primary structures, most notably for corticotropin, human growth hormone, and motilin. Erroneous sequences have been identified upon comparison of products obtained by total syntheses and by isolation from natural sources. Moreover, in some cases the final structural proof has been obtained only by chemical syntheses.

4.1.1.2 Design of Bioactive Peptide Drugs

Numerous analogues of peptides have been synthesized in order to identify the structural parameters for biological activity. In the case of oxytocin **1**, early efforts succeeded in obtaining an analogue by the exchange of glutamine in position 4 with threonine, the analogue subsequently displaying much higher biological activity compared with the native hormone. [Thr^4]oxytocin is thought to undergo a much more efficient interaction with the receptor compared with the native hormone, and hence this phenomenon has been considered a "chemical mutation" which would have succeeded in evolution, but probably only in the distant future.

$$\text{H–}\overline{\text{Cys}^1\text{–Tyr–Ile–Gln–Asn}^5\text{–Cys}}\text{–Pro–Leu–Gly–NH}_2$$

1

A systematic amino acid exchange is hampered by the variety of possible combinations of proteinogenic amino acids in bioactive peptides; however, combinatorial chemistry has today opened a series of completely new perspectives. In addition to the exchange of amino acids by other proteinogenic buildings blocks, the modification by isofunctional or isosteric amino acids is also a viable approach. An isofunctional amino acid exchange, for example of lysine by ornithine or diaminobutyric acid, of arginine by homoarginine, or the replacement of hydroxy acids (serine by threonine) and amino dicarboxylic acids (aspartic acid by glutamic acid) and *vice versa* has revealed information concerning the influence of side-chain length on biological activity. The role of amino acid side-chain structure, or of its chemical functionality, can be examined by using directed exchange with isosteric amino acids, for instance, Val/Thr, Leu/Asp, Met/Nle, Cys/Ser, Arg/Cit, Asn/Asp, Gln/Glu or His/β-(pyrazolyl)alanine. Extensive investigations on the influence of the side chain on biological activity have been performed, and it has been observed that peptide drugs with a biologically active N-terminal (e.g., corticotropin, parathyrin) or C-terminal partial sequence (e.g., eledoisin, angiotensin, secretin) occur in nature. Nonetheless, some peptides exist in which the complete sequence is required for biological activity (e.g., calcitonin, bradykinin, oxytocin, vasopressin). Modifications of the C- and N-terminal amino and carboxy groups, of functional groups in side-chain positions, and of the configuration of the peptide backbone are of central interest in structure-activity relationships. All such chemical modifications of a bioactive peptide drug have a major influence on the conformation, so that conformational studies of native drugs in comparison with their modified analogues are central targets of timely research.

In general, peptides are much more soluble and flexible than proteins. Usually, small acyclic peptides do not prefer certain conformations in solution. Longer peptides that are able to form stable secondary structures often occur in equilibrium with partially folded structures. Structural concepts and theoretical methods have considerably influenced peptide research during the past years, and only the broad methodological spectrum of peptide synthesis has made such concepts evident. The conformational analysis of linear peptides and analogues with modified structure is not only very difficult but also is often ambiguous, because linear peptides are rather flexible unless stable secondary structures such as helices, β-sheets, turns, and other folds are present. As a result, conformational studies are usually performed with cyclic oligopeptides (often using NMR spectroscopy) because these materials display a greater degree of conformational homogeneity. Most chemical modifications of bioactive compounds are directed towards the design of new peptide drugs. In principle, peptides would be ideal drugs as they often interact in a highly specific manner with the corresponding receptor. However, peptides are rapidly degraded *in vivo* by the action of luminal, pancreatic, cytosolic, or lysosomal proteases, and are then processed and excreted by the liver. In addition, many body barriers demonstrate a high diffusional resistance towards the uptake of larger molecules, and suitable carrier systems rarely exist. Consequently, the main disadvantages of using a peptide drug are the low bioavailability and short half-life, especially after oral administration. Possibilities exist by which peptide degradation by proteases can be reduced, and uptake enhanced. For example, chemical modification (e.g., cyclization, D-amino acid substitution, or the incorporation of peptide mimetics) leads to the production of compounds with higher stability *in vivo*. Other strategies aimed at stabilizing peptides *in vivo* in order to increase their half-life include the coadministration of protease inhibitors, or the use of specific formulation such as emulsions, liposomes, or nanoparticles. Another alternative is to use a pro-drug to alter the peptide's physico-chemical properties.

Peptides may be used as lead structures in order to create nonpeptidic bioactive compounds, the process often being supported by conformational studies and molecular modeling. In the course of this structure-based design, an amino acid sequence of a bioactive peptide is transformed into a nonpeptide lead compound. Here, functional and structural information is combined and the peptide backbone topography with respect to the pharmacophoric groups is reproduced in the small nonpeptide lead compound. These substances must display pharmacological activity, especially after oral administration, and must imitate the peptide in terms of its receptor interaction (agonist activity); alternately, it may block this interaction (antagonist activity). Peptide mimetics may also act as enzyme inhibitors.

4.1.1.3 Preparation of Pharmacologically Active Peptides and Proteins

Based on economic considerations, pharmacologically active peptides, such as oxytocin **1**, corticotropin **2**, and secretin **3**, have been synthesized chemically. Synthetic secretin costs about 10% less than the natural product isolated from porcine intestine, and a similar situation applies to many other peptide drugs. In addition to economic reasons, the better accessibility via a synthetic approach is important because many peptide drugs occur naturally only in nanogram quantities. Moreover, chemical synthesis is often preferred in the case of peptides and proteins that occur specifically in humans, as isolation following autopsy is generally not a viable route. In the case of corticotropin **2**, secretin **3**, and glucagon **4**, the products synthesized chemically are characterized by a higher purity than those isolated from natural sources. Often, the separation of peptides with similar sequences and antagonist or other activity is not possible using standard isolation and purification

H–Ser^{1}–Tyr–Ser–Met–Glu^{5}–His–Phe–Arg–Trp–Gly^{10}–Lys–Pro–Val–Gly–
Lys^{15}–Lys–Arg–Arg–Pro–Val^{20}–Lys–Val–Tyr–Pro–Asn^{25}–Gly–Ala–
Glu–Asp–Glu^{30}–Ser–Ala–Glu–Ala–Phe^{35}–Pro–Leu–Glu–Phe–OH

2

H–His^{1}–Ser–Asp–Gly–Thr^{5}–Phe–Thr–Ser–Glu–Leu^{10}–Ser–Arg–Leu–Arg–
Asp^{15}–Ser–Ala–Arg–Leu–Gln^{20}–Arg–Leu–Leu–Gln–Gly^{25}–Leu–Val–NH_2

3

H–His^{1}–Ser–Gln–Gly–Thr^{5}–Phe–Thr–Ser–Asp–Tyr^{10}–Ser–Lys–Tyr–Leu–
Asp^{15}–Ser–Arg–Arg–Ala–Gln^{20}–Asp–Phe–Val–Gln–Trp^{25}–Leu–Met–
Asn–Thr–OH

4

methods. With the development of recombinant techniques, the industrial production of pharmacologically active proteins using biotechnological methods has become a viable route on a preparative scale. In particular, immune modulators and tissue proteins represent the focal point of promising new concepts in tumor therapy. In addition, diagnostics are being directed increasingly towards enzymes and monoclonal antibodies, while vaccines based on specialized proteins and peptides are also under development. Today, pharmacological research and medicine stand on the brink of a series of completely new developments, with the most important therapeutic areas likely to be those of cardiovascular conditions, tumors, and autoimmune or infectious diseases.

4.1.1.4 **Synthesis of Model Peptides**

Tailor-made model peptides serve to study conformational behavior, using a variety of physico-chemical methods. Peptide substrates may also be used in studies of enzymology, while investigations into the antigenicity of larger proteins can be made using specialized multiple peptide synthesis techniques (see Chapter 8). Peptide epitope mapping especially is directed towards diagnostic targets and vaccine development. Protein-protein interactions often rely on the specific recognition of surface-exposed amino acid residues, which may be presented in either of two ways (Fig. 4.1):

- as a continuous epitope, where the amino acids involved occur consecutively in the primary structure;
- as a discontinuous epitope, where the crucial amino acids are located in different positions within the protein sequence.

An overlapping series of short peptide sequences (6–15 amino acids) is synthesized on the basis of the protein primary structure. Binding studies then reveal sequences necessary for protein-protein recognition.

In future, model peptides will serve as valuable tools in investigations of the tertiary structure of polypeptide chains. The question of whether the general rules of protein folding derived from the primary structure (amino acid sequence) can be

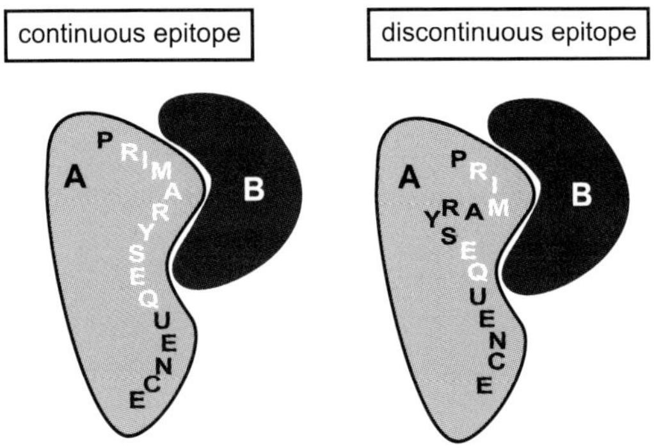

Fig. 4.1 Schematic view of the interaction between two proteins A and B via continuous or discontinuous epitopes.

maintained remains a controversial issue. Synthetic model peptides imitating protein surfaces can also be synthesized, and the flexibility of the peptide backbone can be reduced by conformational restrictions.

4.1.2
Basic Principles of Peptide Bond Formation

The formation of a peptide bond – resulting in a dipeptide – is seemingly a very simple chemical process. The two component amino acids are connected by a peptide (amide) bond with the elimination of water (Fig. 4.2).

The synthesis of a peptide bond under mild reaction conditions can only be achieved after activation of the carboxy component (A) of one amino acid. The second amino acid (as the amino component, B) attacks the activated carboxy component in a nucleophilic attack, with formation of the dipeptide (A-B, Fig. 4.3). If the amino function of the carboxy component (A) is unprotected, then formation of the peptide bond occurs in an uncontrolled manner (Fig. 4.3, lower part). Linear and cyclic peptides are formed as undesired by-products, together with the target compound A-B. Consequently, during the course of peptide synthesis all functional groups not involved in peptide bond formation must be blocked, both temporarily and reversibly.

Peptide synthesis – the formation of each peptide bond – is therefore a three-step procedure (Fig. 4.4).

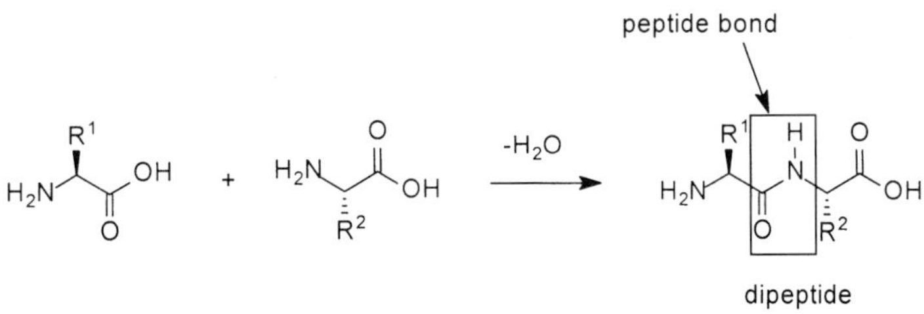

Fig. 4.2 Simplified scheme of peptide bond formation.

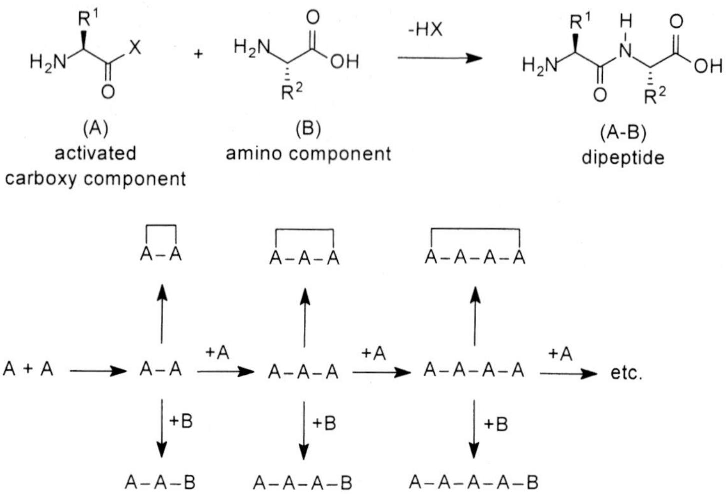

Fig. 4.3 Schematic view of possible side reactions of a peptide synthesis involving an activated N^{α}-unprotected carboxy component (A). Linear and cyclic oligomers are formed besides the desired dipeptide A-B. X=activating group; R^1, R^2=amino acid side chains.

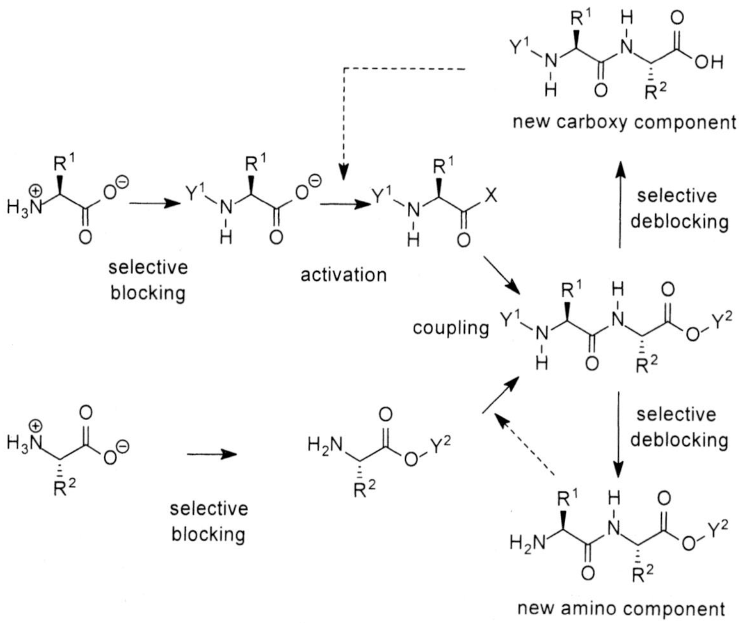

Fig. 4.4 The multi-step process of peptide synthesis. Y^1=amino-protecting group; Y^2=carboxy protecting group; R^1, R^2=amino acid side chains.

1. In the first step, the preparation of a partially protected amino acid is required. After this protection, the zwitterionic structure of the amino acids is no longer present.
2. The second step, the formation of the peptide bond, occurs in two partial steps. The N-protected amino acid must be activated at the carboxy function in order to be converted into a reactive intermediate. Subsequently, the formation of the peptide bond occurs. This coupling reaction can be performed either as a one-pot reaction or in two separate, consecutive steps.
3. In the third reaction step, selective or total cleavage of the protecting groups is carried out. Although total deprotection is required only when the peptide chain has been fully assembled, selective cleavage of the protecting groups is usually necessary in order to continue the peptide synthesis.

Peptide synthesis becomes further complicated by the fact that ten of the proteinogenic amino acids (Ser, Thr, Tyr, Asp, Glu, Lys, Arg, His, Sec, and Cys) have functional groups in the side chain which need to be selectively protected. Because of the different requirements with respect to selectivity, a distinction must be made between the intermediary (temporary, or transient) and semipermanent protecting groups. The intermediary groups are used for temporary protection of the amino or carboxy function involved in subsequent bond formations. These groups must be cleaved selectively under conditions that do not interfere with the stability of peptide bonds already present, or that of semipermanent protecting groups at amino acid side chains. The semipermanent protecting groups are usually cleaved only at the end of the peptide synthesis or, occasionally, also at an intermediate stage. The different types of protecting groups will be described in Section 4.2.

Activation of the carboxy component and subsequent bond formation (the coupling reaction) should, under ideal conditions, occur at a high reaction rate and without racemization or formation of by-products. High yields are required upon application of equimolar amounts of both components, but unfortunately no such chemical coupling method yet exists which meets these requirements. The number of appropriate methods applicable to practical synthesis is rather small compared to the high number of coupling methods described. Further details will be provided in Section 4.3.

During the course of peptide synthesis, a number of reactions occur which involve functional groups that are usually connected to a chiral center (glycine is the only exception), and consequently there is a potential risk of racemization (see Section 4.4).

Cleavage of the protecting groups is the last step in the peptide synthesis cycle. Except for the synthesis of dipeptides, where total deblocking is required, the selective cleavage of protecting groups in order to extend the peptide chain is of major importance. The synthetic strategy requires thoughtful planning, which is described as the "strategy and tactics of peptide synthesis" (see Chapter 5). Depending on the strategy chosen, either the protecting group of the N^{α}-amino function is cleaved selectively, or the protecting group of the carboxy function is cleaved. The term "strategy" refers to the sequence of condensation reactions between single amino acids to produce a peptide. Usually, a distinction is made between stepwise synthesis and segment condensation (fragment condensation). When peptide synthesis is performed in solution (also referred to as "conventional synthesis"), for example with difficult sequences, the stepwise peptide chain elongation is limited in most cases to the assembly of shorter fragments. When the synthesis of a longer peptide is planned, the target molecule must be divided into appropriate segments. Suitable fragments have to be identified, perhaps where only minimal C-terminal epimerization occurs during fragment condensation. Following stepwise assembly of the single segments, the latter are connected to produce the target compound. The tactics of peptide synthesis also include selection of the most suitable combination of protecting groups, together with application of the best coupling methods for each segment condensation.

Solid-phase peptide synthesis (SPPS) is, in its original version, a variant of the stepwise synthesis of peptides and proteins. It is a synthetic concept where the growing peptide chain is attached to an insoluble polymeric support, and was first described in 1963 by Robert Bruce Merrifield. Today, this method is referred to as the Merrifield synthesis in honor of the 1984 Nobel laureate (for details, see Section 4.5). Nowadays, segment condensations may also be performed on polymeric support (see Chapter 5, Section 3).

4.2
Protection of Functional Groups

Temporary and semipermanent protecting groups are distinguished with respect to cleavage selectivity. A temporary protecting group must fulfil the following requirements:

1. Introduction of the protecting group leads to an amino acid derivative that is no longer present in a zwitterionic structure.
2. Cleavage must occur without hampering the stability of semipermanent protecting groups or peptide bonds.
3. Racemization must not occur during all necessary operations.
4. Stability and characterization of protected intermediates must be favorable.
5. Solubility of the protected amino acid components must be favorable.

According to the tactical requirements, semipermanent protecting groups are usually cleaved at the end of the peptide synthesis. In this reaction, only those deblocking reagents that do not damage the final product can be used. The requirements with respect to cleavage selectivity are less stringent because temporary protecting groups may also be cleaved in the final deblocking step. Criteria (3), (4), and (5) must also be applied to the semipermanent protecting groups.

The selection of protecting groups must be carried out in accordance with the strategy envisaged for the synthesis of a particular peptide (see Chapter 5). Protecting group orthogonality is a criterion that is especially important for the success of a peptide synthesis.

Protection schemes rely on either quantitative rate differences of deprotection reactions, or on mechanistically different deprotection reactions in the case of orthogonal protecting groups [1]. Orthogonality means that subsets of protecting groups present in a molecule can be cleaved selectively under certain reaction conditions, while all other protecting groups remain intact. Often, orthogonality refers to a mechanism-based distinction, for example between semipermanent and temporary protecting groups. It generally provides the opportunity of substantially milder overall reaction conditions.

In the following section, the most important protecting groups are discussed in the context of the functional group to be blocked.

4.2.1
N^{α}-amino Protection

Protecting groups of this type are used for the N^{α}-amino group of all proteinogenic amino acids (including the imino group of proline), for N^{ω}-amino groups (e.g., lysine, ornithine) and also, if necessary, for the temporary blocking of N-acyl amino acid hydrazides and other functional groups. In principle, an amino group can be blocked reversibly by acylation, alkylation, and alkyl-acylation. Protecting groups based on sulfur and phosphorus derivatives have also been developed. Salt formation at the amino group does not offer a genuine protection during the operations of peptide synthesis. Several hundred different amino-protecting groups have been developed during the past decades – which confirms the non-existence of a universal, ideal amino-protecting group. A classification can be made based on the structure of the protecting group, or on the cleavage conditions: acidolysis, base cleavage, reduction/oxidation, nucleophilic substitution, and photolysis. Because most peptides are sufficiently stable under moderately acidic conditions, amino-protecting groups with different labilities towards acidic deblocking agents are preferred. Alternately, the base-labile compound 9-fluorenylmethoxycarbonyl group (Fmoc) has found widespread application. It has been estimated that, in practical terms, over 80% of all syntheses use these two deblocking methods. Some protecting groups may be cleaved by hydrogenolysis. Protection of the amino group can also be achieved easily by acylation and by alkylation. The disadvantage of structurally and chemically simple acyl residues (e.g., acetyl-, benzoyl-, monochloroacetyl-) is that the carboxamide group of the protecting function is chemically very similar to the peptide bond, which renders selective cleavage very difficult or even impossible, and this eventually led to the development of urethane-type amino-protecting groups.

4.2.1.1 Alkoxycarbonyl-Type (Urethane-Type) Protecting Groups

A variety of urethane-type amino-protecting groups, which are cleavable under different reaction conditions, is available. Selected representatives are listed in Tab. 4.1; the most prominent groups are discussed in the following section.

Tab. 4.1 Selected amino-protecting groups of the urethane type Y-O-(C=O)-NH-R^1.

Group	*Symbol*	*Y-O*	*Cleavage conditions*	*Reference(s)*
Benzyloxycarbonyl	Z	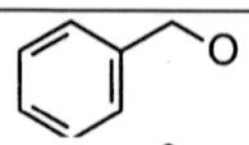	H_2/Pd, HBr/AcOH, Na/liq. NH_3	2
4-Methoxybenzyloxycarbonyl	Z(OMe)	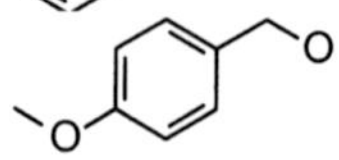	TFA; H_2/Pd; Na/liq. NH_3	3, 4
2-Nitrobenzyloxycarbonyl	Z(2-NO_2)	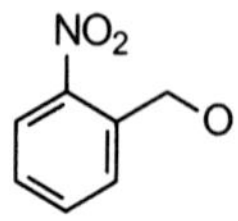	H_2/Pd (more labile than Z) HBr/AcOH (more stable than Z) photolysis	5
4-Nitrobenzyloxycarbonyl	Z(NO_2)	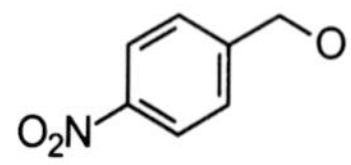	Like Z, but more stable to HBr/HOAc	6, 7
Chlorobenzyloxycarbonyl	Z(Cl) Z(3-Cl) Z(2-Cl) Z(2,4-Cl)	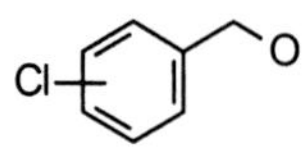	Like Z, but more stable to H_2/Pd or HBr/AcOH TFA/CH_2Cl_2	8–12
3,5-Dimethoxybenzyloxycarbonyl	Z(3,5-OMe)	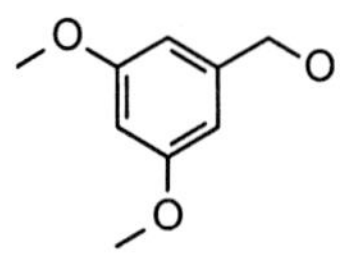	Photolysis	13
α,α-Dimethyl-3,5-dimethoxybenzyloxycarbonyl	Ddz	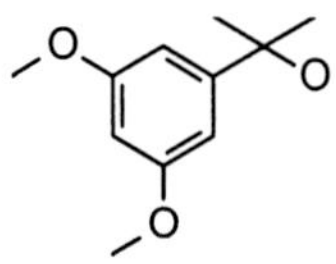	Photolysis, 5% TFA in CH_2Cl_2	14
6-Nitroveratryloxycarbonyl (4,5-Dimethoxy-2-nitrobenzyloxycarbonyl)	Nvoc	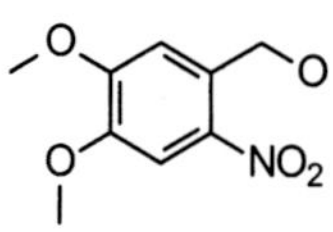	Like Z Photolysis	5
4-(Phenyldiazenyl)-benzyloxycarbonyl	Pz	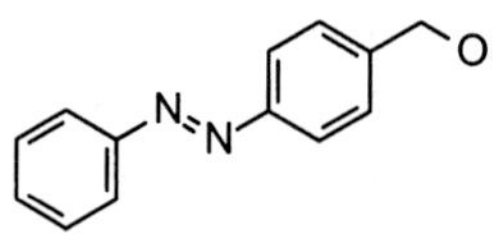	H_2/Pd, HBr/AcOH, Na/liq. NH_3	15
α-Methyl-2,4,5-trimethylbenzyloxycarbonyl	Tmz	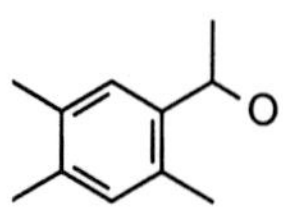	3% TFA in $CHCl_3$	16
Benzisoxazol-5-yloxycarbonyl	Bic	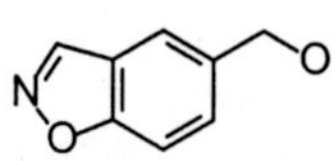	Isomerization with triethylamine in DMF and solvolysis in aq. buffer (pH 7)	17

Tab. 4.1 (continued)

Group	***Symbol***	***Y-O***	***Cleavage conditions***	***Refer-ence(s)***
2-(Biphenyl-4-yl)-2-propoxycarbonyl	Bpoc	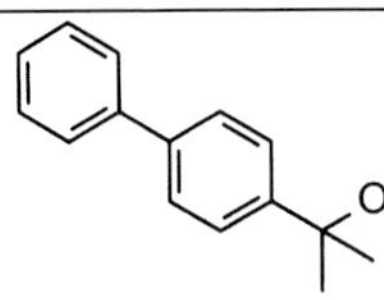	80% AcOH $MgClO_4$	18, 19
(4-Phenylazophenyl)-isopropoxycarbonyl	Azoc	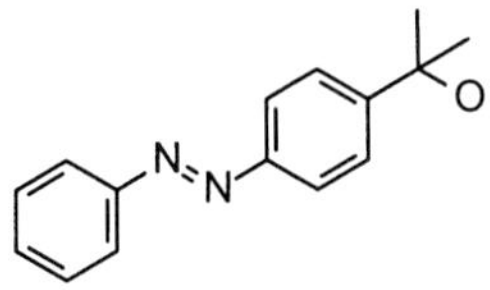	Like Bpoc	20
Isonicotinyl-oxycarbonyl	iNoc	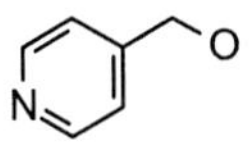	Zn/AcOH, H_2/Pd, acid stable	21
tert-Butoxycarbonyl	Boc		TFA, TFA/CH_2Cl_2, HCl/organic solvent	3, 22
2-Cyano-*tert*-butoxy-carbonyl	Cyoc	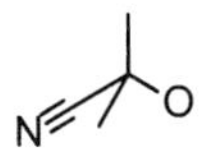	Weak base (aq. K_2CO_3, triethyl-amine)	23
2,2,2-Trichloro-*tert*-butoxycarbonyl	Tcboc	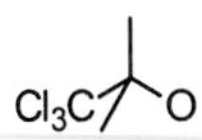	Co(I)-phthalocyanine Zn/AcOH	24
Adamantyl-1-oxy-carbonyl	Adoc	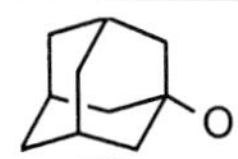	TFA	25
1-(1-Adamantyl)-1-methylethoxy-carbonyl	Adpoc	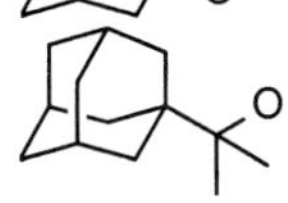	3% TFA in CH_2Cl_2, labile to H_2/Pd	26
Isobornyloxy-carbonyl	Iboc	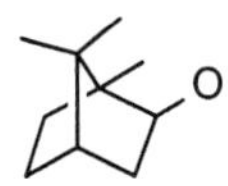	TFA, stable to H_2/Pd and bases	27
Fluorenyl-9-methoxy-carbonyl	Fmoc	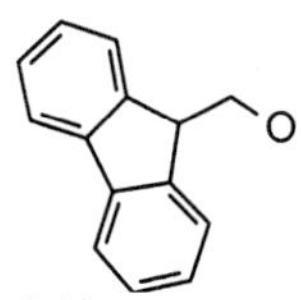	Piperidine, DBU, 2-aminoethanol, Morpholine, Liq. NH_3	28, 29
(2-Nitrofluoren-9-yl)methoxycarbonyl	Fmoc(NO_2)	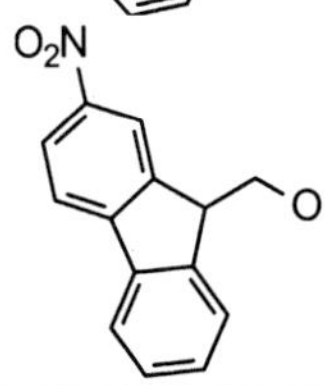	Photolysis	30

Tab. 4.1 (continued)

Group	*Symbol*	*Y-O*	*Cleavage conditions*	*Reference(s)*
2-(4-Toluenesulfonyl)-ethoxycarbonyl	Tsoc	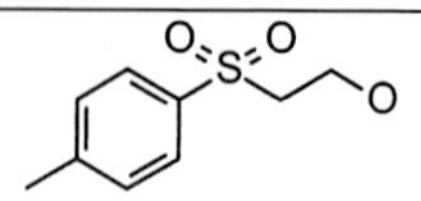	$NaOC_2H_5$/ethanol	31
Methylsulfonyl-ethoxycarbonyl	Msc	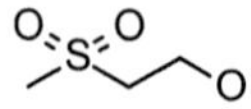	Base-catalyzed β-elimination	32
2-(4-Nitrophenylsulfonyl)ethoxycarbonyl	Nsc	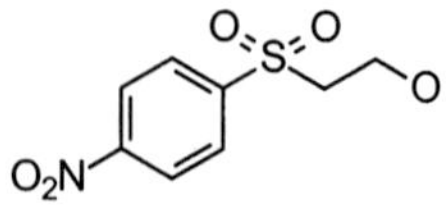	Like Fmoc	33–35
2-(*tert*-Butylsulfonyl)-2-propenyloxy-carbonyl	Bspoc	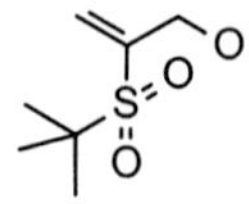	Nucleophile	36
1,1-Dioxobenzo[b]-thien-2-ylmeth-oxycarbonyl	Bsmoc	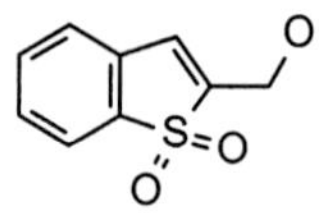	Nucleophile, e.g., 2% tris(2-amino-ethyl)amine	37, 38
2-(Methylsulfonyl)-3-phenyl-2-propenyl-oxycarbonyl	Mspoc	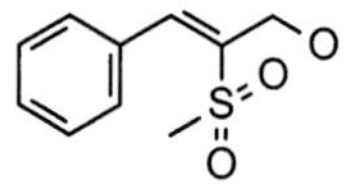	Nucleophile	39
Allyloxycarbonyl	Aloc	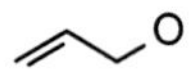	Pd^0/nucleophile	40–42
2-(Trimethylsilyl)-ethoxycarbonyl	Teoc	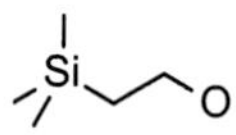	F^-, e.g. TBAF, TFA	43
Triisopropylsilyl-ethoxycarbonyl	Tipseoc	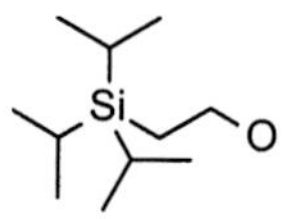	F^-, e.g. TBAF	44
Piperidinyloxy-carbonyl	Pipoc	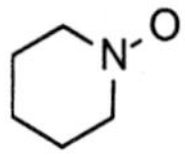	Electrolysis H_2/Pd	45
Cyclopentyloxy-carbonyl	Poc	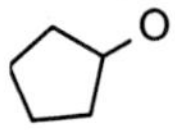	HBr/AcOH, Na/liq. NH_3, stable to H_2/Pd	3
3-Nitro-1,5-dioxa-spiro[5.5]undec-3-ylmethoxycarbonyl	PTnm	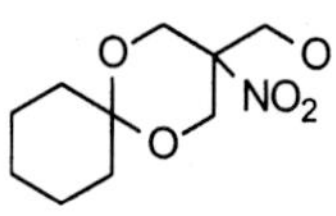	"Relay deprotection": H_3O^+, then pH 8.5	46
2-Ethynyl-2-propyl-oxycarbonyl	Epoc	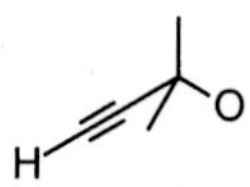	H_2, Lindlar-catalyst	47

Benzyloxycarbonyl group The introduction of the benzyloxycarbonyl group into peptide synthesis provided a major breakthrough in the development of modern peptide chemistry. Today, this amino-protecting group is among those used most frequently, and is abbreviated as Cbz (or "Z", in honor of its inventor, Leonidas Zervas). In older references, Cbz is also referred to as the carbobenzoxy (Cbo) group. Introduction of the Z group in amino acids is usually performed by reaction with benzylchloroformate **5** (the correct names benzyloxycarbonyl chloride or benzylchlorocarbonate are seldom used) under Schotten-Baumann conditions in the presence of NaOH, $NaHCO_3$ or MgO, mostly in aqueous organic solvents under vigorous stirring. Amino acid esters can be converted into the Z-protected derivatives in chloroform. Selective protection in the case of amino acids with functional groups in the side chain often requires special reaction conditions. Arginine may be selectively N^{α}-monoprotected using benzyl chloroformate in a buffered $NaHCO_3$-NaOH solution at pH 9–11 [48]. Usually, Z-amino acids are obtained in high yields in crystalline form, but products obtained as an oil can be crystallized as salts with dicyclohexylamine.

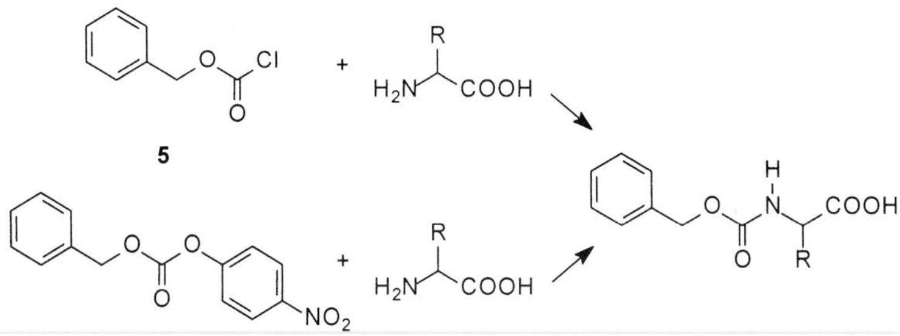

Z–Xaa–OH

Benzyl-4-nitrophenylcarbonate **6** and similarly activated benzyl derivatives can also be used to introduce this group. Cleavage of the Z group is achieved by acidolysis using hydrogen bromide in acetic acid, liquid HBr, boiling trifluoroacetic acid (TFA), HBr/TFA, liquid hydrogen fluoride (HF), or by catalytic hydrogenation using different catalysts under a broad scope of reaction conditions (Fig. 4.5). The catalytic hydrogenation proceeds smoothly in organic solvents (e.g., acetic acid, alcohols, dimethylformamide (DMF)) or in aqueous organic solution with palladium black, palladium on charcoal, or palladium on barium sulfate. Only toluene and carbon dioxide are formed in addition to the desired free peptide. Cessation of CO_2 evolution or H_2 consumption are used to monitor the cleavage reaction.

Although hydrogenolytic cleavage of Z groups fails when cysteine/cystine is present in the peptide, the reaction is successful in the presence of $BF_3 \cdot OEt_2$. Acidolysis is performed preferably using 2 N hydrogen bromide in acetic acid, although several other variants (e.g., hydrogen chloride, hydrogen iodide) and solvents (e.g., dioxan, nitromethane, TFA, tetrachloromethane) have been suggested. Side reactions can also occur for this standard method; for example, O-acetylations have been observed at threonine or serine, while S-transetherifications may occur in the presence of methionine. Tryptophan or nitroarginine building blocks

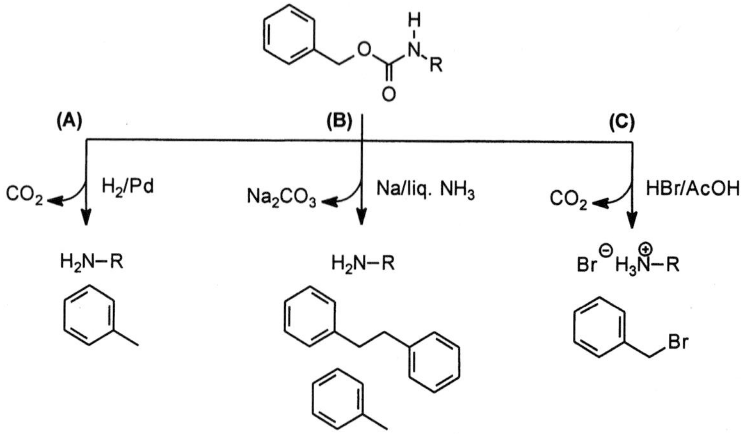

Fig. 4.5 Cleavage conditions of the benzyloxycarbonyl group. H_2N-R=amino acid.

may be destroyed, and benzyl esters or carboxamide groups are sometimes cleaved under these conditions. However, these side reactions may be minimized under certain reaction conditions. The Z group can also be cleaved with anhydrous liquid hydrogen fluoride (HF method).

More than 20 different benzyloxycarbonyl type groups with a variety of substituents have been described, mainly with regard to tuned sensitivity towards acidolytic cleavage reagents. The 4-methoxybenzyloxycarbonyl residue provides a methodological advantage because the protected amino acids are crystalline. It can be cleaved in the presence of the Z group with anhydrous TFA at temperatures below 0 °C. Z groups with halogen substituents in the 3- or 4-position of the phenyl ring are also of interest because they lead to higher stability of the modified Z group towards acidic reagents. This can be used to better differentiate between protecting groups of the Z type and Boc type (see below). The (3,5-dimethoxybenzyl)oxycarbonyl, (2-nitrobenzyl)oxycarbonyl, and (6-nitroveratryl)oxycarbonyl groups can be cleaved photochemically.

***tert*-Butoxycarbonyl group** The *tert*-butoxycarbonyl group (Boc) is, like the Z and the Fmoc group, one of the most important amino-protecting groups. It should also be mentioned that amino-protecting groups of the urethane type safeguard racemization-free peptide bond formations during the stepwise assembly of a peptide chain starting from the C-terminus.

The *tert*-butoxycarbonyl azide **7** is an excellent acylation reagent. It reacts with the salts of amino acids in water/dioxan mixtures in the presence of tertiary amines or magnesium oxides or under pH control with 2 N to 4 N NaOH to give the Boc-protected amino acids Boc-Xaa-OH, but also with amino acid esters in pyridine.

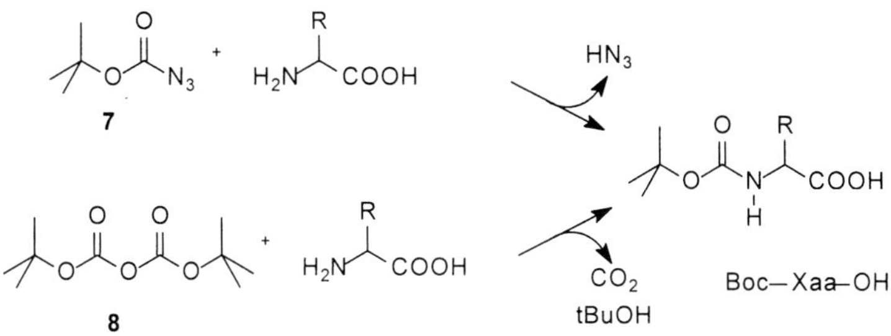

pH-Stat-controlled reaction conditions are advantageous. Although this acylation protocol has been proved to be very efficient, further methods for the introduction of the Boc group have been described, mainly because of the explosive nature of azides. The *tert*-butoxycarbonyl fluoride obtained from chlorocarbonyl fluoride and *tert*-butanol at –25 °C has attracted special attention because it gives Boc-protected amino acid in high yields under pH-stat conditions. *tert*-Butyl-*S*-4,6-dimethylpyrimidyl-2-thiocarbonate, 2-*tert*-butoxycarbonyloximino-2-phenylacetonitrile and, preferably, di-*tert*-butyldicarbonate $(Boc)_2O$ **8** are also used as reagents. Amino acid salts can be used in aqueous solution with solubilizing additives such as dioxan or tetrahydrofuran.

The Boc protecting group is compatible with most coupling methods for peptide synthesis. It can be cleaved under mild acidolytic conditions, and is resistant towards catalytic hydrogenation, alkaline hydrolysis and reduction with Na/liquid ammonia. Hydrogen chloride in acetic acid, dioxan, ether, and ethyl acetate are important cleavage reagents. A cleavage method frequently used is treatment with non-aqueous TFA at temperatures around 0 °C, which smoothly cleaves Boc groups (Fig. 4.6).

Undesired *tert*-butylations may occur during acidolysis at nucleophilic sites of the peptide chain. For instance, the indole system of tryptophan or the thioether group of methionine can be *tert*-butylated. The intermediate carbenium ions responsible for these side reactions are usually trapped using scavengers such as anisole or thioanisole. Further deblocking reagents are $BF_3 \cdot OEt_2$, 2-mercaptoethanesulfonic acid, aqueous TFA, or 98% formic acid. However, a smooth reaction has not been observed in all cases. This variety of reagents can be used when other protected functional groups and the Z group are present.

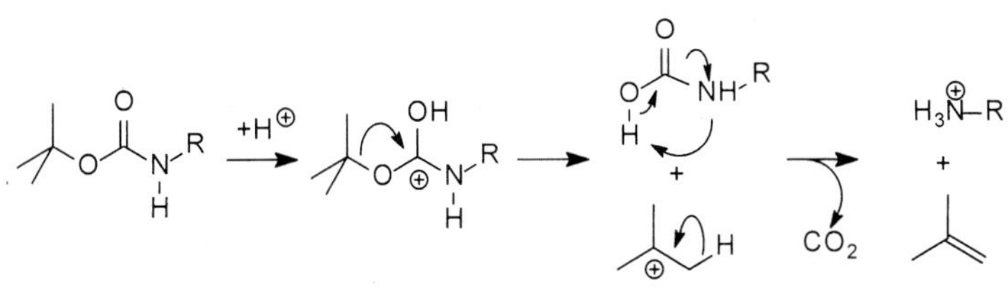

Fig. 4.6 Mechanism of acidolytic cleavage of the *tert*-butoxycarbonyl group. H_2N-R = amino acid or peptide.

Modified analogues of the *tert*-butoxycarbonyl group have been described on many occasions, and some representatives are discussed in the following section. The (α,α-dimethyl-3,5-dimethoxybenzyl)oxycarbonyl group (Ddz) can be cleaved both acidolytically with 5% TFA in methylene chloride, and also photolytically. Therefore, it provides an interesting variant for special applications. The 2-(4-biphenylyl)-2-propoxycarbonyl residue (Bpoc) can be cleaved with 80% acetic acid, and is therefore often combined with protecting groups for ω-amino-, hydroxy-, and carboxy functionalities based on the *tert*-butyl type. The 2-ethynyl-2-propyloxycarbonyl group is cleaved by catalytic hydrogenation (with Lindlar catalyst), even in the presence of methionine-containing peptides. Alternately, it can be cleaved by acidic hydrolysis under the formation of a cyclic carbonate. A different variant is the 2-cyano-*tert*-butoxycarbonyl residue (Cyoc) which can be cleaved by β-elimination under weakly basic conditions.

9-Fluorenylmethoxycarbonyl group The 9-fluorenylmethoxycarbonyl group (Fmoc) is the only amino-protecting group of the urethane type of widespread practical importance that can be cleaved under mildly basic conditions. The sensitivity of Fmoc-protected amino acids (Fig. 4.7; **9**), especially towards secondary amines, enables deblocking of the amino group with dilute solutions of piperidine or diethylamine in DMF under mild conditions. Usually, 20% piperidine in DMF is applied

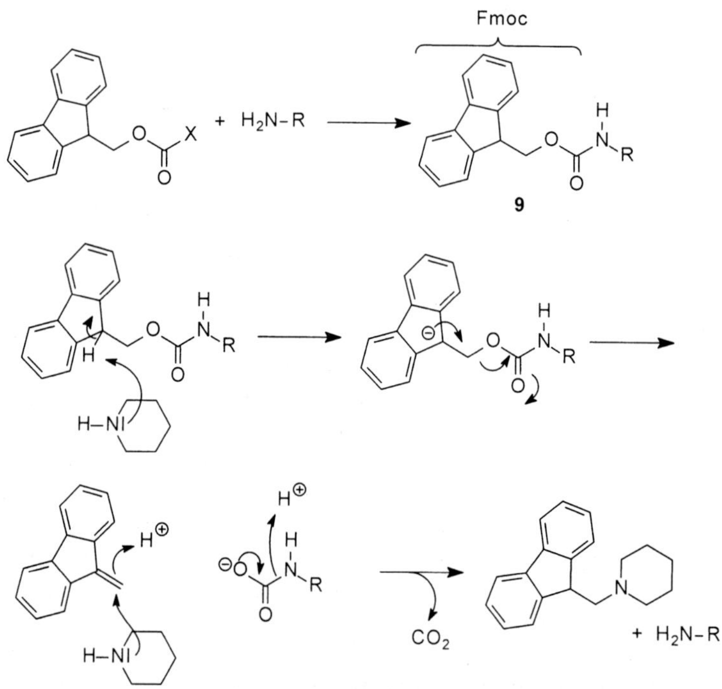

Fig. 4.7 Mechanism of Fmoc cleavage. H_2N-R = amino acid or peptide.

routinely where the cleavage of the Fmoc group occurs within seconds at room temperature. Other reagents, such as 1,8-diazabicyclo[5.4.0]undec-7-ene (DBU) or fluoride ion may be used as an alternative. The reaction proceeds according to a E1cB mechanism, with initial proton abstraction to give the stabilized dibenzocyclopentadienyl anion. The dibenzofulvene formed reacts with piperidine to give a stable adduct as a by-product of the cleavage reaction (Fig. 4.7). In cases where no primary or secondary amine is used, the addition of 2% piperidine is recommended to trap the dibenzofulvene.

Under these conditions, neither Z nor Boc groups or other modern amino-protecting groups usually applied are attacked. In order to prevent premature cleavage under peptide-coupling conditions, addition of the weakly acidic additive 1-hydroxybenzotriazole has been recommended. $CaCl_2$ addition has been reported to increase Fmoc lifetime, and hence may be used for the synthesis of Fmoc-protected peptide segments by saponification of C-terminal esters or linkers [49]. Interestingly, the Fmoc group is not totally inert towards standard hydrogenolytic cleavage conditions.

The reagents Fmoc-Cl and, alternately, Fmoc-OSu (9-fluorenylmethyl-*N*-succinimidyl carbonate) have been recommended for the formation of Fmoc-protected amino acids. Although, the Fmoc group was introduced into peptide chemistry in 1970, its application in the Merrifield solid-phase peptide synthesis was reported much later. Nowadays, the Fmoc tactics is one of the most preferred synthetic concepts in solid-phase peptide synthesis. The acid stability, combined with the mildly basic cleavage conditions, allows for the application of acid-labile semipermanent protecting groups of the *tert*-butyl type for side-chain functionalities in solid-phase peptide synthesis.

The purification of synthetic peptides and proteins may be facilitated when a polyaromatic analogue of the Fmoc group is attached to the N-terminus [50].

New base-labile N^{α}-protecting groups The 2-(4-nitrophenylsulfonyl)ethoxycarbonyl group Nsc (see Tab. 4.1) has recently been considered as a valuable alternative to the Fmoc group, because it is more base-stable and, therefore, less sensitive towards weakly basic media [51–53]. Nsc is more hydrophilic than Fmoc, and less prone to aspartimide formation [51]. Like Fmoc, Nsc forms in the deprotection reaction an olefin that subsequently is trapped by the secondary base piperidine. In the Fmoc case, the formation of the dibenzofulvene piperidine adduct is reversible, while the corresponding adduct in the Nsc case is formed irreversibly.

Carpino et al. developed the new base-labile amino-protecting groups Bspoc, Bsmoc, and Mspoc (see Tab. 4.1) [36–39]. Cleavage relies on the conjugate addition of a nucleophile (e.g., secondary amine) with concomitant liberation of the carbamate. Bsmoc is stable towards acids and tertiary amines, but is cleaved readily with secondary amines. As the deprotection step simultaneously is the scavenging step, fewer side reactions should occur.

Protecting groups that are orthogonal to Boc and Fmoc The N^{α}-allyloxycarbonyl (Aloc) group [40–42] may be used for the synthesis of labile derivatives (see Sec-

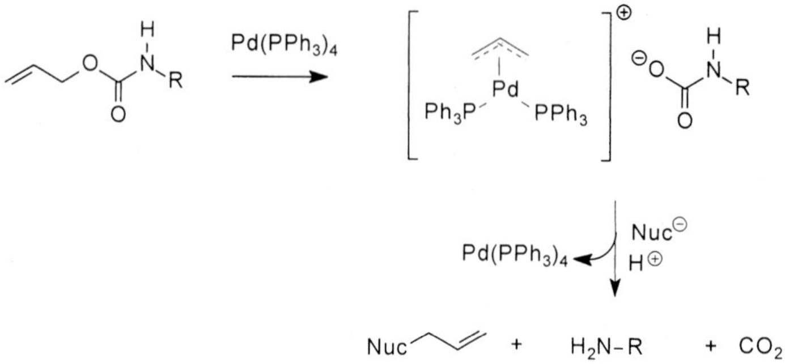

Fig. 4.8 Mechanism of Aloc cleavage. H_2N-R=amino acid or peptide.

tions 6.3–6.5). It is completely orthogonal to Boc and Fmoc and, therefore, finds application in the synthesis of cyclic peptides (see Chapter 6.1). The deprotection step involves palladium(0)-catalyzed allyl transfer (Fig. 4.8) to various nucleophiles (e.g., amines, amine-borane complexes, carboxylates, organosilanes, organostannanes) [54]. Amine-borane complexes are especially suited as allyl scavengers under neutral conditions without undesired N-allylation of the deprotected product [54]. Tandem deprotection-acylation of N^{α}-Aloc derivatives has also been reported [55].

The dithiasuccinoyl group Dts **10** [56–58] may formally be regarded as a urethane thio analogue that can be cleaved by treatment with thiols. It is stable to acids and bases and, consequently, orthogonal to Boc or Fmoc.

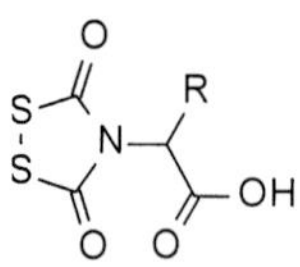

10
Dts

4.2.1.2 **Carboxamide-Type Protecting Groups**

Representatives of the carboxamide-type protecting groups are the formyl group for **11** (R=H) which can be cleaved by solvolysis, oxidation, or hydrazinolysis, and the trifluoroacetyl group Tfa **11** (R=CF_3), which is labile towards alkaline conditions.

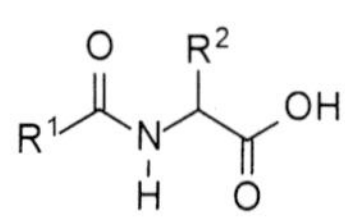

11 For: R^1 = H
Tfa: R^1 = CF_3

4.2.1.3 **Sulfonamide and Sulfenamide-Type Protecting Groups**

Arenesulfonyl groups feature some properties that render them potential amino-protecting groups. In the past, mainly the 4-toluolsulfonyl (tosyl, Tos) group **12** has been used, though this suffers from the disadvantage that it is only cleavable by reduction with Na/NH_3. In principle, N^{α}-sulfonamide protecting groups allow for N-alkylation of amino acids [59, 60] and do not suffer from racemization via the oxazolone mechanism (cf. Section 4.4.2) [61]. N^{α}-Arenesulfonyl amino acid chlorides are highly reactive species that are well suited for peptide synthesis [62]. Consequently, N^{α}-arenesulfonamide-type protecting groups experienced a renaissance with the introduction of the *o*- and *p*-nitrobenzenesulfonyl groups (*o*Nbs **13**, *p*Nbs **14**) [59, 63] and of dNbs **15** [64] and Bts **16** [60]. These protecting groups can be removed by treatment with thiophenol or alkanethiols, and hence are orthogonal to *tert*-butyl and 9-fluorenylmethyl-type protecting groups.

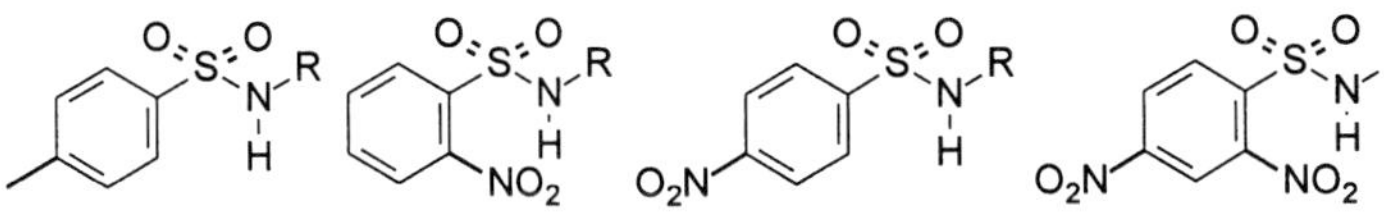

12 Tos | **13** oNbs | **14** pNbs | **15** dNbs

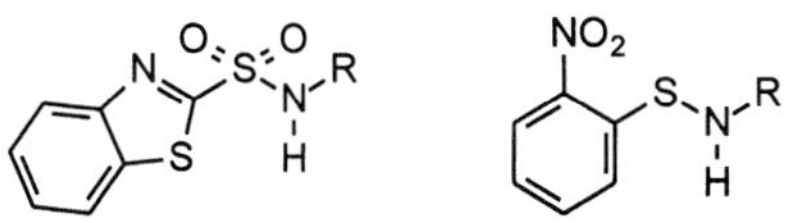

16 Bts | **17** Nps

The 2-nitrophenylsulfanyl (Nps) group **17**, which has been used in solution-phase peptide synthesis, may be removed by HCl in inert solvents or with the use of thiol reagents [65–67].

4.2.1.4 Alkyl-Type Protecting Groups

The protecting groups 1-(4,4-dimethyl-2,6-dioxocyclohexylidene)ethyl Dde **18** (R^2=CH_3) [68, 69], 1-(4,4-dimethyl-2,6-dioxocyclohexylidene)-3-methylbutyl ivDde **18** (R^2=iBu) [70], and 1-(4-nitro-1,3-dioxoindan-2-ylidene)ethyl Nde [71] are alkyl-type protecting groups that are orthogonal to Boc and Fmoc, because they are cleaved by hydrazinolysis.

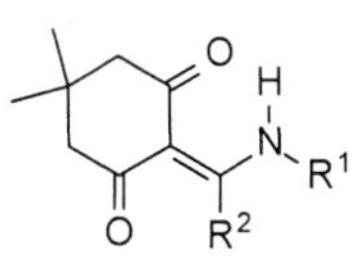

R^1NH_2: amino acid

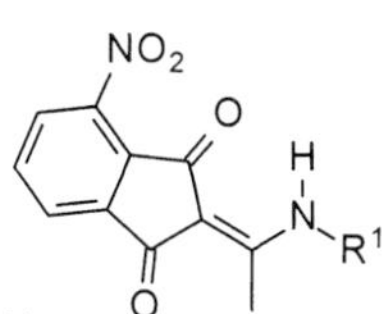

18 R^2 = CH_3 Dde
R^2 = $CH_2CH(CH_3)_2$ ivDde

19 Nde

The most popular protecting group of the alkyl type is the triphenylmethyl (Trt) group **20** that can be cleaved acidolytically under mild conditions. The activation of Trt-protected amino acids is very complicated however because of the steric hindrance. Mono- or disubstituted N-benzyl protecting groups **21, 22** have received very little attention in peptide synthesis, but they can be cleaved by catalytic hydrogenation in the presence of palladium black at elevated temperatures. This fact probably led to the finding that the exchange of an ethyl by a benzyl residue in the ethoxycarbonyl group, as described by Fischer, could result in an amino-protecting group being cleavable by hydrogenolysis. This paved the way for the protecting groups of the urethane type (as the protecting groups based on carbamates are referred to).

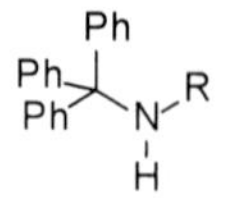

Trt **20**

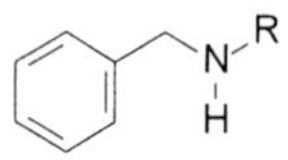

Bzl **21**

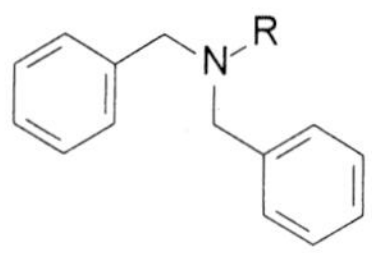

$(Bzl)_2$ **22**

4.2.2
C^{α}-Carboxy Protection

There are few exceptions (e.g., [72–74], cf. Section 5.1.1) from the usual synthetic strategy where a peptide chain is assembled from the C-terminus to the N-terminus (C → N direction). Consequently, C-terminal protection has not been investigated as extensively as N-terminal protection. C^{α}-carboxy-protecting groups in solution-phase synthesis, and also as linker moieties in solid-phase synthesis, must be orthogonal to the temporary N^{α}-protecting group.

In principle, it is possible to perform peptide synthesis without complete protection of the carboxy group of amino components. Removal of the zwitterionic structure of the corresponding amino acid or the peptide is achieved by salt formation. Here, alkali or earth alkali metal salts of amino acids can usually be reacted in water or aqueous dioxan with the corresponding activated carboxy components. The corresponding carboxylate salts of tertiary bases (*N*-methyl morpholine, *N*-ethyl piperidine, triethylamine, 1,1,3,3-tetramethylguanidine, etc.) have been recognized as important for peptide synthesis in organic solvents, especially in DMF. The mixed anhydride and azide methods are, besides active esters, best suited for the so-called salt coupling. A major advantage is that, after the coupling reaction, the carboxy group is liberated by acidification. This protocol is not universally applicable, as side reactions or solubility problems may occur without genuine protection of the carboxy group of the amino component used in the reaction.

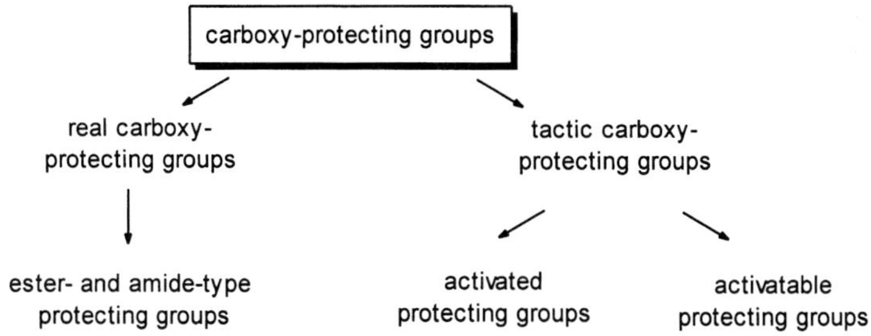

Fig. 4.9 Classification of carboxy-protecting groups.

Esters of the methyl, ethyl, benzyl, *tert*-butyl, or aryl type are appropriate for the reversible blocking of the carboxy group.

Carboxy-protecting groups may be divided into two categories on the basis of strategic-tactic criteria (Fig. 4.9):

- The first group contains protecting groups which can be deblocked normally after the end of the synthesis by regeneration of the free carboxy group.
- The second group contains derivatives that are activated towards aminolysis at the end of a synthesis, either immediately or after special chemical transformation.

Although these two different types have been classified as genuine and nongenuine carboxy-protecting groups, a better discrimination between them may be to refer to "real" and "tactic" carboxy-protecting groups. The expression "tactic" protecting groups emphasizes the application of these functionalities in special synthetic strategies where the mere role of a protecting group is exceeded.

4.2.2.1 **Esters**

Activated carboxy-protecting groups can be used only under special conditions for peptide synthesis. In order to safeguard the desired reaction pathway, the activation method used for the coupling must have a higher acylation potential compared with the activated carboxy-protecting group of the amino component. This so-called "backing-off" strategy preferentially makes use of active esters as carboxy-protecting groups, and the mixed anhydride method for coupling. In this case, the N-protected carboxy component is transformed into an active ester. The N^{α}-protecting group is then cleaved and subsequently acylated with another N-protected amino acid to form the protected dipeptide with the active ester moiety located at the C-terminus. The mixed anhydride method is mainly applied in these cases. Successive repetition of these steps allows the formation of segments in a stepwise manner, starting from the C-terminus and without practical risk of racemization. These segments can then be connected with other segments via the activated C-terminus to give larger peptides. Carboxy-protecting groups that potentially can be activated behave like real protecting groups during the coupling reaction, and the risk of undesired side reactions by uncontrolled self-condensation can be neglected in practical cases. One example of an ester that is transformed into the reactive form immediately before the envisaged coupling step is the 4-(methylthio)phenyl ester, which can be transformed oxidatively into the 4-(methylsulfonyl)phenyl ester.

Most of the currently known carboxylic protecting groups are esters derived from either primary, secondary, or tertiary alcohols (Tab. 4.2).

Methyl and ethyl esters have been used previously for peptide synthesis by Curtius and Fischer, and can be obtained easily on reaction of the amino acid with the corresponding alcohol and thionyl chloride at low temperature. Presumably, the alkyl chlorosulfinate (AlkO-SO-Cl) is formed as a reactive intermediate from $SOCl_2$ and the alcohol with cleavage of HCl. This intermediate reacts with the amino acid, liberating SO_2 and HCl to give the amino acid ester hydrochloride. Benzyl esters can be prepared in a similar manner. Using a different approach, cesium or tertiary ammonium salts of N^{α}-protected amino acids can be esterified with an alkyl halide. Methyl esters are also easily obtained with diazomethane.

Cleavage of these protecting groups on completion of a peptide synthesis proceeds by mild alkaline hydrolysis in mixtures of water and an organic solvent such as dioxan, methanol, ethanol, acetone, or DMF to solubilize the starting materials. The alkyl esters are usually applied only for the synthesis of short-chain peptides, as hydrolytic cleavage becomes very difficult with increasing chain length, and extreme saponification conditions (excess of alkali and long reaction times) are necessary. Consequently, racemization and other side reactions are observed. Enzyme-catalyzed deprotections often provide ingenious alternatives. Both methyl and ethyl ester resist hydrogenolysis and mild acidolysis. Transformation of these esters into the corresponding hydrazides (the starting materials for azide synthesis) using hydrazine is possible, as is the transformation into amides by aminolysis.

Benzyl esters of free amino acids are obtained by direct esterification with benzyl alcohol in the presence of acidic catalysts (e.g., 4-toluenesulfonic acid, hydrogen chloride, benzene sulfonic acid, polyphosphoric acid). The water formed during the course of the esterification is usually removed by azeotropic distillation (with benzene, toluene, or tetrachloromethane). Benzyl esters of N-protected amino acids are obtained after activation of the carboxylic group with *N,N'*-dicyclohexyl carbodiimide (DCC) or thionyl chloride and reaction with benzyl alcohol, but also by transesterification reactions. Benzyl ester derivatives can be cleaved by alkaline hydrolysis, catalytic hydrogenation, treatment with saturated HBr/acetic acid (12 h at room temperature or 1–2 h at 50–60 °C), by treatment with sodium in liquid ammonia, or with liquid hydrogen fluoride.

Free amino acids are transformed into the *tert*-butyl esters by addition of isobutene under acidic condition or by transesterification with *tert*-butyl acetate. The hydroxyl groups of serine or threonine are transformed under these conditions into the *tert*-butyl ethers, likewise. N-protected amino acids such as Z-amino acids can be reacted with isobutene in the presence of catalytic amounts of sulfuric acid or with *tert*-butyl acetate in the presence of perchloric acid to give the corresponding *tert*-butyl ester. The *tert*-butyl ester group, which is easily cleavable by acidolysis

Tab. 4.2 Selected carboxy-protecting groups of the ester type Y-O-(C=O)-R.

Group	*Symbol*	*Y-O*	*Cleavage conditions*	*Reference(s)* [a]
Methyl	Me	—O	Alkaline or enzymatic hydrolysis	
Ethyl	Et	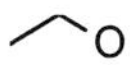	Alkaline or enzymatic hydrolysis	
Benzyl	Bzl	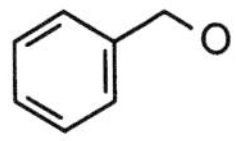	H_2/Pd, HBr/AcOH, Liq. HF, Alkaline hydrolysis	
4-Nitrobenzyl	Nbz	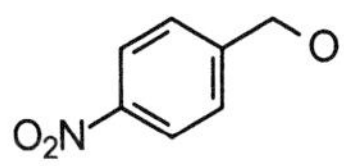	H_2/Pd, alkaline hydrolysis, stable to HBr/AcOH	
4-Methoxybenzyl	Mob	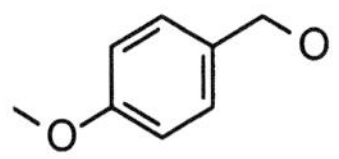	TFA/anisole, HCl/nitromethane, H_2/Pd, liq. HF, alkaline hydrolysis	
2,4-Dimethoxybenzyl	2,4-Dmb	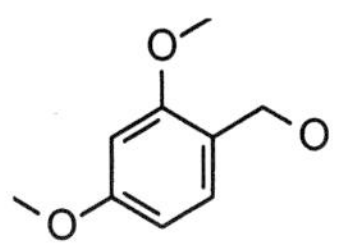	1% TFA/CH_2Cl_2	
o-Chlorotrityl	Trt(2-Cl)	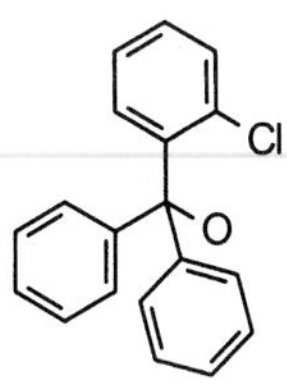	1% TFA/CH_2Cl_2	
Pyridyl-4-methyl-(4-picolyl)	Pic	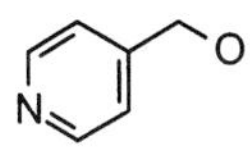	H_2/Pd, alkaline hydrolysis, electrolytic reduction	
2-(Toluene-4-sulfonyl)-ethyl	Tse	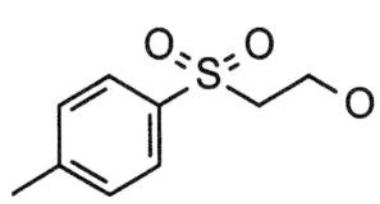	β-elimination in H_2O/dioxan and Na_2CO_3	
Phenacyl	Pac	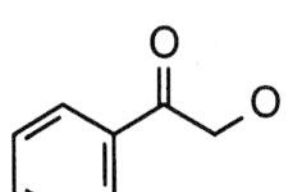	Sodium thiophenolate, H_2/Pd, Zn/AcOH	
4-Methoxyphenacyl	Pac(OMe)	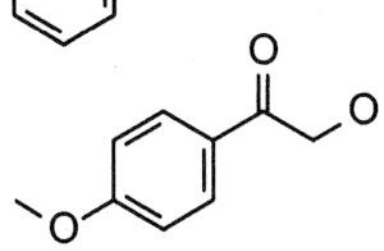	UV-photolysis at 20 °C	

Tab. 4.2 (continued)

Group	*Symbol*	*Y-O*	*Cleavage conditions*	*Reference(s)* [a]
Diphenylmethyl (Benzhydryl)	Dpm	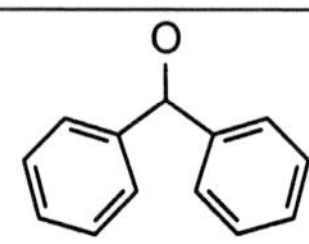	H_2/Pd, TFA at 0 °C, HCl/AcOH, $BF_3 \cdot OEt_2$/AcOH at 25 °C	
tert-Butyl	tBu	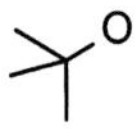	TFA, HCl/AcOH, HBr/AcOH, $BF_3 \cdot OEt_2$/AcOH	
Cyclohexyl	Cy		Strong acids	
1-Adamantyl	1-Ada		TFA	
2-Adamantyl	2-Ada	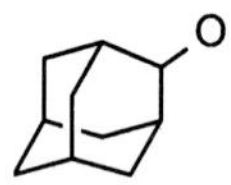	Strong acids	
Dicyclopropylmethyl	Dcpm	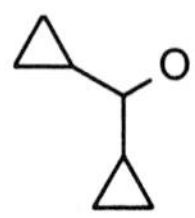	1% TFA/CH_2Cl_2	82
9-Phenylfluoren-9-yl	Pf	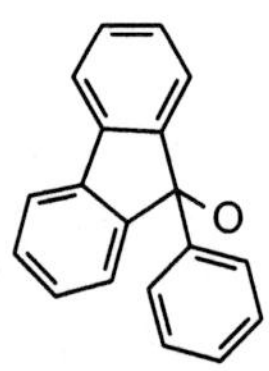	>1% TFA/scavenger/CH_2Cl_2	75
9-Fluorenylmethyl	Fm	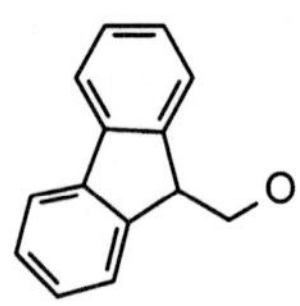	Piperidine, TBAF	76, 77
2-Trimethylsilylethyl	TMSE	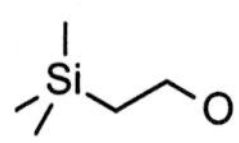	TBAF	78, 79
2-Phenyl-2-trimethylsilylethyl	PTMSE	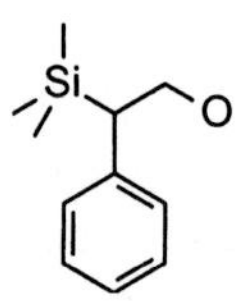	TBAF	80

Tab. 4.2 (continued)

Group	*Symbol*	*Y-O*	*Cleavage conditions*	*Reference(s)* [a]
Allyl	Al	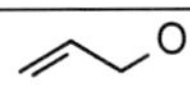	Pd(0), nucleophile	41
4-{N-[1-(4,4-dimethyl-2,6-dioxocyclohexylidene)-3-methylbutyl]-amino}benzyl	Dmab	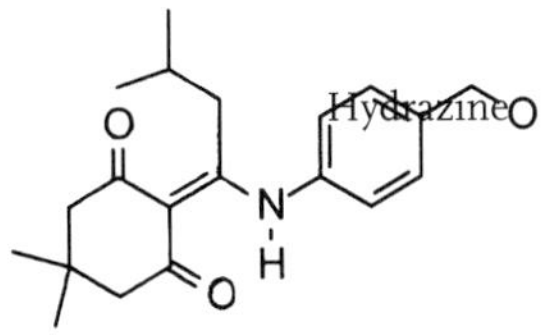	Hydrazine	81

a) For additional information see [83–86].

(TFA, HCl/AcOH, $BF_3 \cdot OEt_2$/AcOH), is resistant towards hydrogenolysis and quite stable towards alkaline hydrolysis, hydrazinolysis, and ammonolysis.

A new carboxy-protecting group, which was introduced by Carpino [82], was the dicyclopropylmethyl group (Dcpm). This can be cleaved with 1% TFA in dichloromethane, even in the presence of side chain-protecting groups of the *tert*-butyl type.

The above-mentioned representatives have been used in many practical applications. Polymeric esters in either soluble or insoluble form represent the prototype of anchoring groups in peptide synthesis on solid supports. Consequently, this type of carboxy-protecting group is of crucial importance from a methodological point of view. The carboxy-protecting groups mentioned can be used both as α- and as ω-carboxy-protecting groups in the case of amino dicarboxylic acids. Peptides with a C-terminal amino dicarboxylic acid can be assembled very easily, because in this case both carboxy groups can be blocked for solution synthesis with the same protecting group. Peptides with an internal or N-terminal amino dicarboxylic acid cannot be assembled without taking differential blocking measures [83–86].

In certain specialized cases, for example the head-to-tail cyclization of peptides (cf. Section 6.1), the use is required of special carboxy-protecting groups with a third dimension in orthogonality. This may rely on differences in cleavage kinetics, as are present in the combination of super acid-sensitive esters such as 2,4-dimethoxybenzyl ester (2,4-Dmb), 2-chlorotrityl [Trt(2-Cl)] or the corresponding linkers with Fmoc/tBu chemistry. Different cleavage mechanisms lead to complete orthogonality, as for example in the application of allyl esters or the Dmab group [81] in Fmoc/tBu chemistry.

4.2.2.2 **Amides and Hydrazides**

Although hydrazides may also be used for C^{α} and side-chain carboxy group protection, further azide couplings are usually obligatory under such circumstances. N′-substituted hydrazides are also of great importance in this context. The prepa-

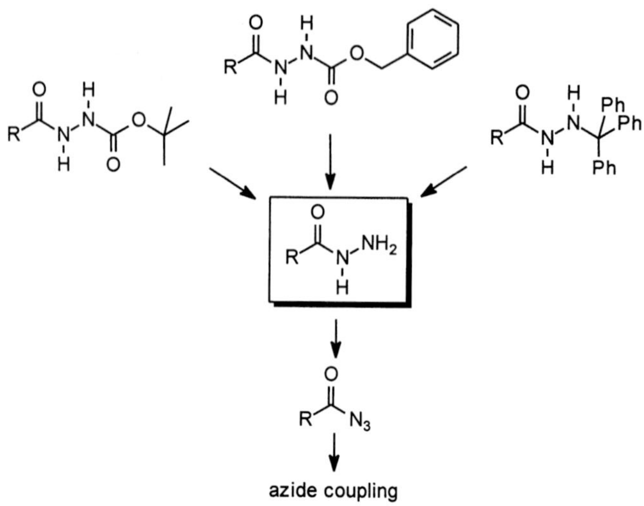

Fig. 4.10 Differently protected hydrazide groups as potentially activatable carboxy-protecting groups. R-CO = peptidyl residue.

ration is achieved by reaction of the N-protected amino acid with a partially blocked hydrazine derivative using the carbodiimide or mixed anhydride method. After cleavage of the amino-protecting group, the corresponding derivative is used for a segment synthesis and selective deblocking of the hydrazide protecting group is performed on completion of the synthesis. The resulting N-terminally protected, C-terminal hydrazide can then be transformed into the azide and used as an activated carboxy component (Fig. 4.10).

Phenylhydrazides of protected amino acids or peptides are cleaved oxidatively by a tyrosinase isolated from mushrooms to provide the free acid [87].

Carboxamides principally are suited as carboxy-protected amino acid or peptide derivatives. Selective chemical methods for cleavage have been lacking before highly specific peptide amidases became available. The carboxamide group is more easily hydrolyzed than a peptide bond, although because of the rather small differences selective cleavage is seldom successful.

4.2.3
C-Terminal and Backbone N^{α}-carboxamide Protection

Additional protection of backbone amide groups may be necessary to avoid prominent side reactions such as aspartimide formation (cf. Sections 4.2.4.3 and 4.2.4.9) or interchain aggregation of the resin-bound peptides containing "difficult sequences" (cf. Section 4.5.4.3). This modification appears to prevent or disrupt unwanted hydrogen-bonding networks, both by removal of the native amide hydrogen and by alteration of the backbone conformation based on the tertiary amide bond that is formed. Weygand et al. [88] used the 2,4-dimethoxybenzyl group for

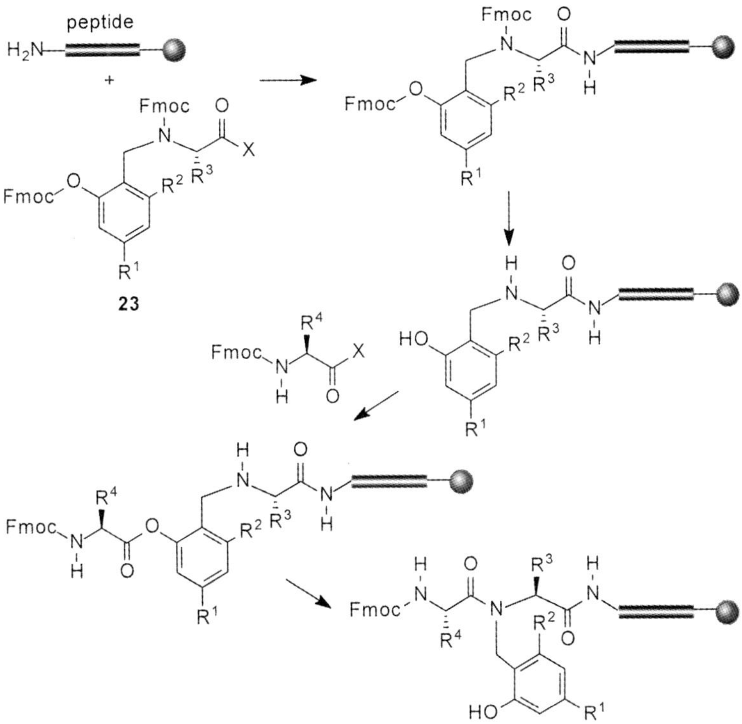

Fig. 4.11 Acyl capture and O → N acyl transfer involving Hmb (R^1=OMe; R^2=H) or Hnb (R^1=H; R^2=NO_2). R^3, R^4=amino acid side chains; X=activating group.

this purpose. Based on the reversible N^{α}-substitution effect studied by Weygand et al. [88], Narita et al. [89], and Bartl et al. [90], an elegant solution to this problem of difficult sequences was developed by Sheppard et al. These authors introduced the 2-hydroxy-4-methoxybenzyl group (Hmb) that has proven very useful for this purpose [91, 92]. It comprises an acid-labile N-benzyl-type protecting group with an additional 2-hydroxy substituent. Therefore, acyl capture of an activated amino acid by this substituent is followed by peptide bond-forming intramolecular O → N acyl transfer (Fig. 4.11). Improved acyl transfer kinetics is observed in the case of the photolabile 2-hydroxy-6-nitrobenzyl group (Hnb) [93]. The incorporation of either Hmb or Hnb or 3-methylsulfinyl-4-methoxy-6-hydroxybenzyl (SiMb)-protected derivatives [94] at every fifth to seventh sequence position during SPPS using preformed *N*, *O*-bis-(Fmoc)-*N*-(Hmb/Hnb) amino acid derivatives **23** seems to be sufficient in order to suppress aggregation.

Alternatively, pseudoproline dipeptide building blocks Fmoc-Xaa-Ser(Ox)-OH or Fmoc-Xaa-Thr(Ox)-OH **24**, as developed by Mutter et al. [95–97], may be employed. This approach has been reported to be superior to Hmb protection, provided that the sequence is compatible with the incorporation of the pseudoproline dipeptides analogues [98].

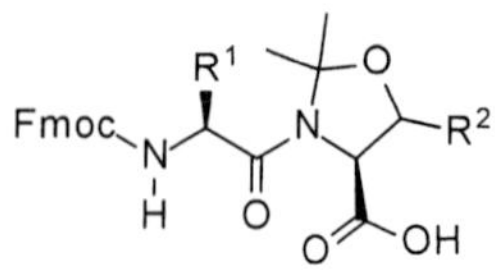

R^1 = amino acid side chain
R^2 = H Ser(Ox)
Me Thr(Ox)

24

4.2.4
Side-chain Protection

As the side chain functionality of an amino acid exerts a crucial influence on the reactivity of the α-amino or α-carboxy group, it is especially important that those side-chain functionalities which give rise to undesired side reactions be protected in any situation. In particular, ω-amino or ω-carboxy groups of diamino carboxylic acids or amino dicarboxylic acids must be mentioned in this context. Selective blocking represents a crucial problem in peptide chemistry, and selection of the protecting group combination is a question that must be addressed by synthesis tactics. The ω-protecting groups of trifunctional amino acids are termed "semi-permanent protecting groups" (see Section 4.1.2) because they are usually only cleaved on completion of peptide synthesis. The thiol group of cysteine, and usually also the guanidino group of arginine, also require semipermanent masking. In other cases, side reactions arising from side-chain functional groups can be suppressed or minimized to a certain extent by using special reaction conditions. Despite this possibility of minimal protection, maximum protection is preferred in a practical sense, and this applies especially to SPPS.

4.2.4.1 Guanidino Protection

Although the guanidino group of arginine is usually protected by protonation under normal reaction conditions due to its strongly basic character, the low solubility of the corresponding derivatives in organic solvents hampers peptide synthesis. Partial acylation of the guanidino group may occur under these conditions, and represents a crucial disadvantage.

Derivatives of arginine hydrobromide can be used in principle as amino components in peptide synthesis. If required, N^{α}-benzyloxycarbonyl-protected arginine hydrobromide can be obtained very easily by treatment of Z-Arg-OH with 1.4 N HBr in methanol and subsequent precipitation with absolute ether.

Ideally, all three side-chain nitrogen atoms should be protected, but most strategies rely on N^{ω}- or N^{ω}-, N^{δ}-protection, respectively. Despite the existence of a variety of blocking possibilities described for the strongly basic guanidino group, no ideal protecting group exists until now. The different protecting groups can be divided into four classes: nitro; urethane (acyl); arenesulfonyl; and trityl types. However, the latter derivatives are poorly soluble in organic solvents and hence are not widely used.

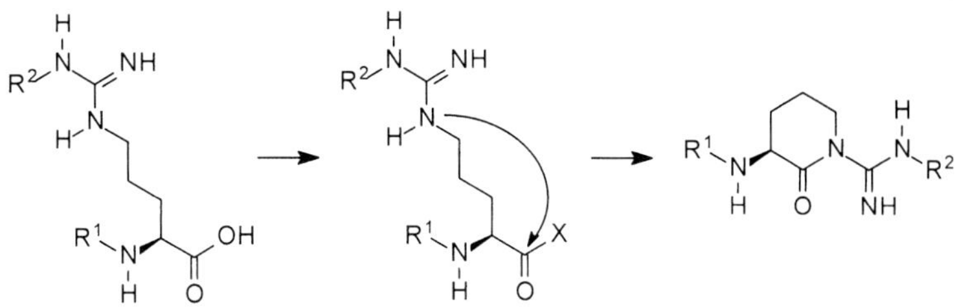

25 **26**

Fig. 4.12 δ-Lactam formation in carboxy-activated arginine derivatives.

The ω-nitro group is stable toward TFA, HBr/AcOH, and alkali. Treatment with liquid HF and several reductive methods (Zn/AcOH, electrolysis) or catalytic hydrogenation with Pd or Raney-nickel are suitable methods for cleavage. ω-Nitroarginine is prone to other side reactions during acylation and cleavage. Lactam formation (Fig. 4.12; **25** → **26**) and consecutive aminolysis have been observed upon activation of protected ω-nitro arginine derivatives ($R^2 = NO_2$).

Protecting groups of the urethane type have been successfully applied to achieve N^{ω}-mono-acylation. In this case, the synthesis of disubstituted N^{α},N^{ω}-diacyl compounds can be performed most easily.

Z-Arg(Z)-OH can only be coupled with hydrochlorides of the amino component (additional protonation of the guanidino group) to give peptide derivatives because in other cases lactam formation by attack of the N^{δ} to the activated carboxy group has been observed.

Differently substituted N^{α},N^{ω}-diacyl derivatives such as Z-Arg[Z(4-NO_2)]-OH or Z-Arg(Boc)-OH [99] show very little tendency towards lactam formation. The differences in deblocking selectivity – the 4-nitrobenzyloxycarbonyl group is stable towards HBr/AcOH – permit a variety of peptide synthetic applications. Derivatives on the basis of Z-Arg(ω,δ-Z_2)-OH, and especially the corresponding 4-nitrophenyl- or N-hydroxysuccinimidyl esters, are excellent building blocks for arginine-containing peptides [83]. The N^{ω},N^{δ}-di(1-adamantyloxycarbonyl) derivative Adoc is obtained from Z-Arg-OH and 1-adamantyl chloroformate [100]. Subsequently, different N^{α}-protecting groups may be introduced after hydrogenolytic cleavage of the Z group. However, the guanidino group is in this protection scheme and in the case of the N^{ω}-Boc derivative still sufficiently nucleophilic to be acylated. Consequently, substantial conversion of arginine residues to ornithine residues via acylation of the unprotected ω-nitrogen during coupling and subsequent intramolecular decomposition during deprotection has been observed [101]. N^{ω},N^{δ}-Bis-Boc protection is not in all cases a viable alternative because of the rather poor coupling efficiency.

A single arene sulfonyl moiety offers complete protection of the guanidino group. Consequently, members of this class belong to the most commonly used guanidino protecting groups. Albeit undesired lactam formation upon activation of the carboxy group cannot be excluded completely by N^{ω}-monoprotection using arene sulfonyl groups, it may be suppressed sufficiently by the application of additives such as 1-hydroxybenzotriazole. The 4-toluenesulfonyl group **27** can only be removed by acidolysis with liquid HF or by reduction with Na/liquid NH_3. Arene sulfonyl-type protecting groups for the guanidino function have been modified with respect to acid lability by the introduction of electron-donating substituents on the aryl residue. Groups such as 4-methoxybenzenesulfonyl **28** (Mbs) [102], 2,4,6-trimethylbenzenesulfonyl **29** (Mts) [103], 4-methoxy-2,3,6-trimethylbenzenesulfonyl (Mtr) **30** [104], 2,2,5,7,8-pentamethylchroman-6-sulfonyl (Pmc) **31** [105], and 2,2,4,6,7-pentamethyldihydrobenzofuran-5-sulfonyl (Pbf) **32** [106] have become very popular in SPPS.

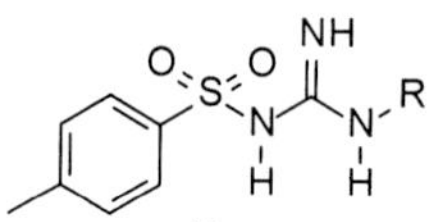

Ts
27

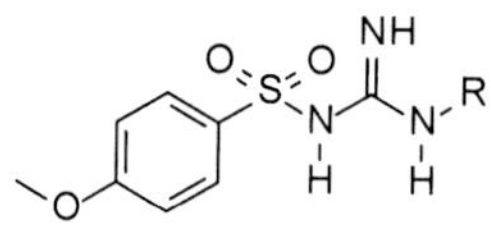

Mbs
28

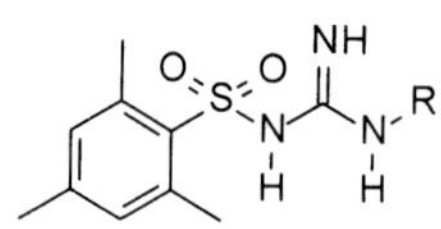

Mts
29

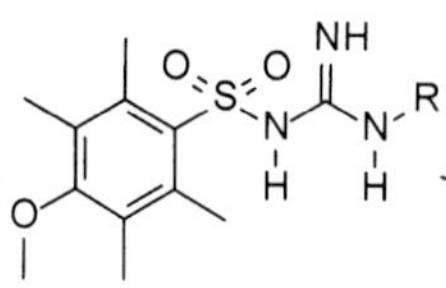

Mtr
30

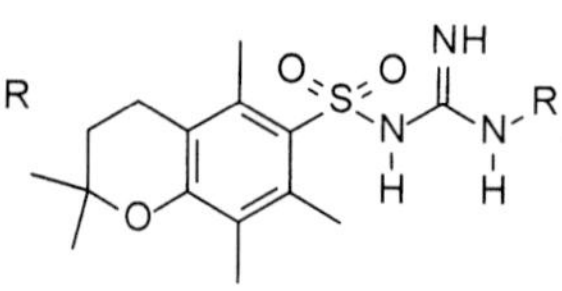

Pmc
31

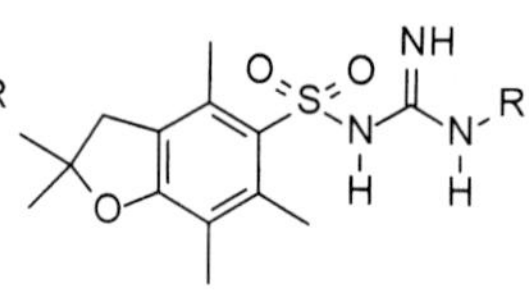

Pbf
32

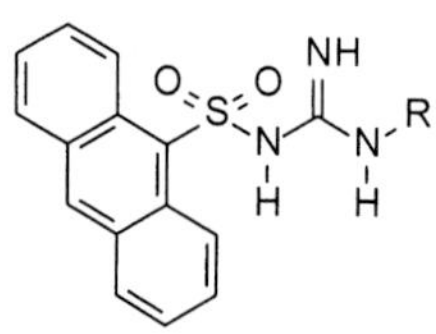

Ans
33

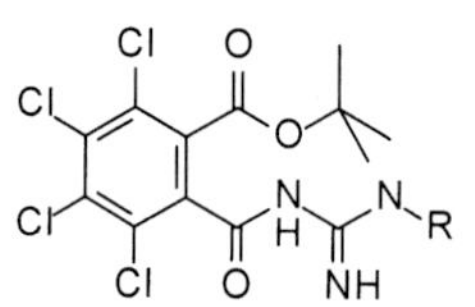

Btb
34

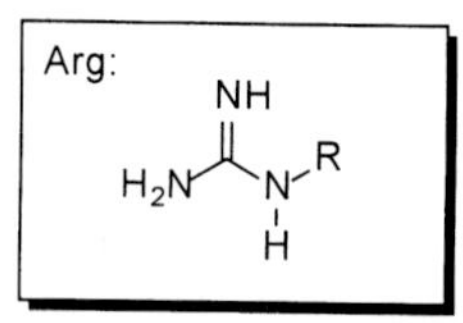

Acid lability increases in the series Tos < Mbs < Mts < Mtr < Pmc < Pbf, with Pbf displaying the best deprotection kinetics.

The Mts group is highly compatible with Boc as the temporary protecting group and can (unlike Tos) be removed with trifluoromethane sulfonic acid (TFMSA)/TFA/thioanisole. Mtr has often been used in the Fmoc tactics, but requires prolonged cleavage time with TFA/thioanisole. The acid lability of the Pmc group is similar to that of the Boc group; therefore, Boc (e.g., for Lys) and Pmc or Pbf (for Arg) may be advantageously combined as side chain-protecting groups. The risk of tryptophan, serine, or threonine sulfonation has been shown to be minimized by the addition of suitable scavengers [107].

The 9-anthracenesulfonyl (Ans) group **33**, which is cleaved by mild reducing agents [108], and the 1-*tert*-butoxycarbonyl-2,3,4,5-tetrachlorobenzoyl (Btb) group **34** [109] have also been successfully applied in both liquid-phase and solid-phase peptide synthesis.

Alternatively, the guanidino group can be introduced by selective guanylation of ornithine-containing peptides on completion of peptide synthesis. Besides other guanylation reagents, 1-amidino-3,5-dimethylpyrazole **35** is well suited for this purpose.

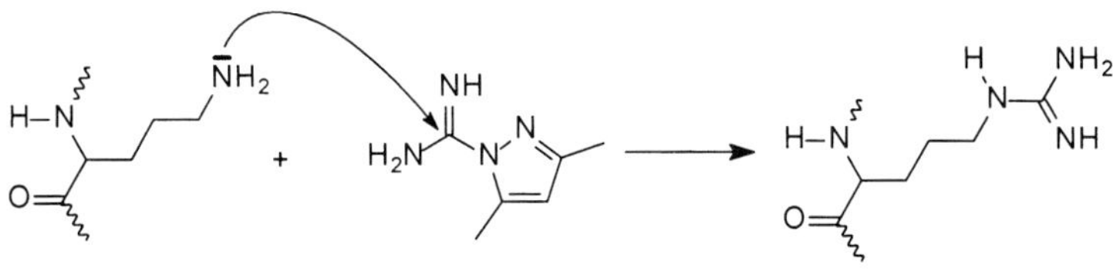

Orn-containing peptide | **35** | Arg-containing peptide

4.2.4.2 ω-Amino Protection

Diaminocarboxylic acids such as lysine and ornithine often require orthogonal protection schemes for the α- and ω-amino groups, as the protection of the ε-amino group is mandatory. Appropriate protecting groups are presented in Tab. 4.1. The ε-amino group of lysine is more basic and more nucleophilic than the α-amino group, and can be converted into the ε-imine **36** with one equivalent of benzaldehyde in the presence of one equivalent of LiOH. Under basic reaction conditions the free α-amino group can then be selectively acylated. Subsequent cleavage of the imine to form, e.g., Z-Lys(H)-OH **37** allows for orthogonal protection of the ε-amino group, for example with Boc_2O to give Z-Lys(Boc)-OH. Alternatively, the α-amino and α-carboxy groups may be simultaneously deactivated by the formation of a chelate complex **38** with copper(II) ions **38**. Subsequently, the ε-amino group can then be selectively acylated to give, for example, H-Lys(Boc)-OH **39**.

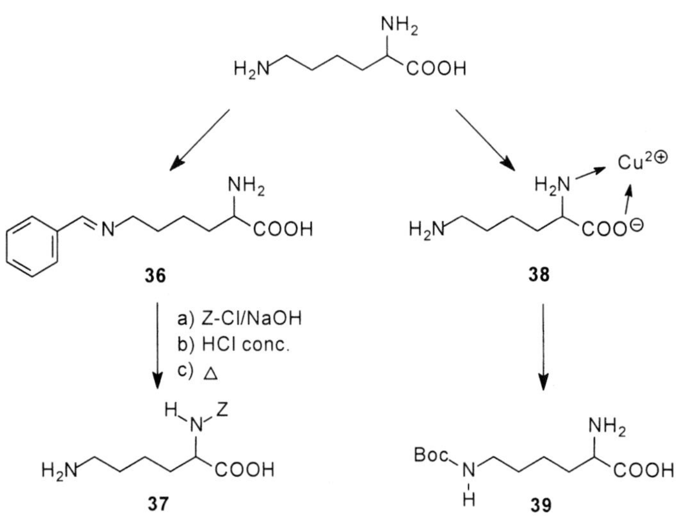

In Boc tactics, the N^{ε}-amino group of lysine usually is protected with the 2-chlorobenzyloxycarbonyl group [Z(2-Cl)], which displays 60-fold less acid lability compared to the Z group. This prevents premature deprotection during acidolytic Boc cleavage in the course of peptide chain assembly.

In the Fmoc approach, lysine side-chain protection as the Boc derivative can be regarded as the optimum combination.

On occasion, completely orthogonal protection of the N^{ε}-amino group is necessary, and in these special cases the trifluoroacetyl group (Tfa), the 3-nitro-2-pyridinesulfenyl group (Npys), the 1-(4,4-dimethyl-2,6-dioxocyclohexylidene)ethyl (Dde) **18** [68, 69], or the allyloxycarbonyl (Aloc) [40, 42] groups are available. The Tfa and Npys groups are very compatible with Boc chemistry, as cleavage occurs in the first case with piperidine and in the latter case with triphenylphosphine or 2-pyridinethiol-1-oxide. Dde **18** ($R^2=CH_3$) and the related ivDde **18** ($R^2=iBu$) [70] groups are orthogonal to Boc and Fmoc, because they are removed with dilute hydrazine solutions. Cleavage results in the formation of a pyrazole derivative, and this can be monitored spectroscopically at 300 nm.

Aloc is an orthogonal protecting group to be applied both in Boc and Fmoc chemistry, as deprotection occurs upon treatment with a suitable nucleophile in the presence of a palladium catalyst.

A new relay deprotection strategy using the PTnm group (Tab. 4.1) for N^{ε}-amino protection of lysine has been developed by Ramage et al. [46].

4.2.4.3 ω-Carboxy Protection

Orthogonal protection is required in most cases for the aminodicarboxylic acids aspartic acid and glutamic acid. Formation of isoaspartyl peptides via the corresponding succinimide intermediate is the most prominent side reaction of Asp side-chain esters. Here, the nitrogen atom of the amino acid located C-terminally with respect to the Asp residue attacks the Asp side-chain ester moiety. Subsequent ring opening with a nucleophile (e.g., water or amines such as piperidine) yields a mixture of α-aspartyl **40** and β-aspartyl peptides **41** [110], a phenomenon that is also observed upon aging of Asp-containing proteins in biological systems.

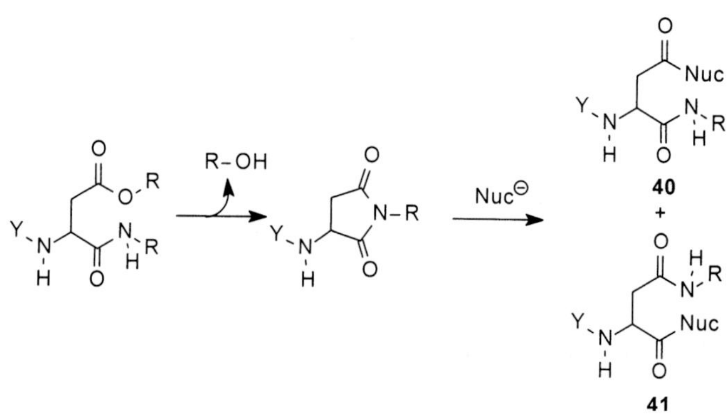

This problem may in most cases be circumvented in Fmoc chemistry by application of the β-*tert*-butyl ester of Asp. However, several publications have reported that, under certain circumstances, aspartimide formation and subsequent side reactions occur in Fmoc protocols, even with *tert*-butyl-protected derivatives. Aspartimide formation appears to be sequence-dependent [111, 112]. The main factors influencing this side reaction are the base employed, the β-carboxy protecting group, and the residue located C-terminally next to the Asp residue. Sequences containing Asp(OtBu)-Gly, Asp(OtBu)-Thr(tBu), Asp(OtBu)-Cys(Acm), Asp(OtBu)-Asn(Trt), and Asp(OtBu)-Asp(OtBu), respectively, seem to be most problematic [113]. The base deprotonates the Asp-Xaa amide group, which then attacks the side-chain ester. Aspartimides are prone to base-catalyzed epimerization and to nucleophilic ring opening (e.g., by piperidine), giving mixtures of α- and β-piperidines, aspartyl- and isoaspartyl peptides together with the corresponding diastereomers formed by epimerization. Addition of HOBt has been shown to minimize these side reactions, but if they cannot be completely suppressed then additional protection (e.g., with Hmb, Hnb) of the amide nitrogen becomes obligatory (cf. Section 4.2.3).

In Boc chemistry the β-cyclohexyl ester Asp(OCy) that is nowadays favored over the benzyl ester efficiently prevents aspartimide formation, even in sensitive sequences. Prolonged treatment with strong acids during protecting group cleavage may, however, be accompanied by cyclization to the aspartimide.

Glutamic acid usually is protected at the side chain with the same groups as used for aspartic acid. Glu derivatives are less prone to undergo this type of cyclization; however, during HF cleavage in Boc chemistry dehydration to the acylium ion and further side reactions are sometimes observed.

Regioselective acid-catalyzed esterification at the ω-carboxy group (e.g., with benzyl alcohol) can be achieved using H_2SO_4 as the acid, and yields approximately 75% of the desired side-chain ester derivative **43** of glutamic acid **42**. This mono-

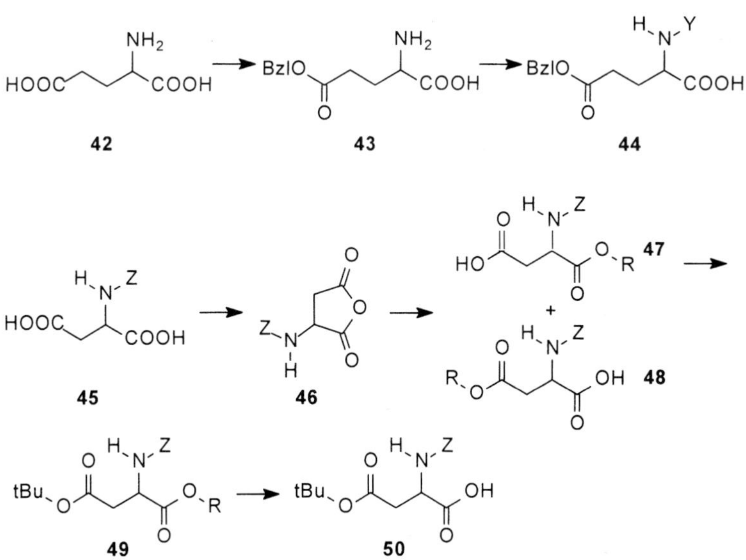

ester can then be converted into the N^{α}-urethane-protected derivative **44**. The protocol may also be applied to derivatives of aspartic acid.

Alternatively, Z-protected aspartic acid is converted into the cyclic anhydride **46** upon treatment with acetic acid. After alcoholysis and separation of the two regioisomers **47/48**, the α ester Z-Asp(OH)-OR **47** is converted to the *tert*-butyl derivative Z-Asp(OtBu)-OR **49** and subsequently saponified by alkali to give Z-Asp(OtBu)-OH **50**. The Z group may then be exchanged by other amino-protecting groups such as Fmoc [114].

The β-2,4-dimethyl-3-pentyl ester has been suggested as novel side chain-protecting group for aspartic acid, and is supposed to prevent undesired side reactions – especially during solid-phase synthesis [115]. If orthogonal protection of the side chains of Asp and Glu is required, viable alternatives are available with the base-labile 9-fluorenylmethyl ester (OFm) [77], the allyl ester (OAl) [40], which is cleaved by palladium-catalyzed transfer to a suitable nucleophile, and Dmab, which is cleaved by hydrazinolysis [81].

4.2.4.4 **Thiol Protection** [116]

The structural requirements of cysteine-containing peptides further complicate peptide synthesis. In some cases, the Cys residue is needed in its free thiol form, whilst other peptide products require regioselective intra- or intermolecular disulfide bridge formation. Solutions to this problem will be addressed in Section 6.2.

The high nucleophilicity, the ease of oxidation, and the acidic character of the cysteine thiol group require selective, semipermanent blocking of this functional group during all synthetic operations (Tab. 4.3).

The S-benzyl group was introduced in 1930 by du Vigneaud in peptide chemistry for the synthesis of oxytocin, and proved to be a major breakthrough for further synthetic projects involving peptides containing either cysteine or cystine. As this protecting group can be cleaved only by reduction with sodium in liquid ammonia, significant damage to longer peptides may occur because of the rather drastic reaction conditions, as was observed in the first total synthesis of insulin.

S-benzyl-protected cysteine residues are especially prone to racemization, however. Epimerization has been observed for resin attachment of protected (Acm, tBu, Trt, StBu, Tacm, Bzl(4-Me)) Cys residues and even for internal Cys residues during repetitive cleavage of the N^{α}-protecting group in Fmoc tactics [138–140].

Several substituted benzyl derivatives have been introduced into peptide chemistry in order to facilitate cleavage of S-benzyl-type protecting groups. The 4-methylbenzyl residue [Bzl(4-Me)] [118] is cleaved by strong acids and thallium trifluoroacetate, respectively. Higher acid lability is displayed by the 4-methoxybenzyl derivative Mob, which is stable toward TFA and iodine, respectively, but is cleaved by boiling TFA, HF, Na/liq. NH_3, mercury(II) salts, TFMSA, etc. The 4-methylbenzyl and the 4-methoxybenzyl derivatives require treatment with liquid HF for cleavage and, therefore, often are employed in Boc chemistry. The 2,4,6-trimethylbenzyl (Tmb) group and the 2,4,6-trimethoxybenzyl group are cleaved by acidolysis with TFA and consequently are compatible with Fmoc chemistry [120]. Further poten-

Tab. 4.3 Thiol-protecting groups Y-SR.

Group	***Symbol***	***Y-S***	***Cleavage conditions***	***Reference(s)***
Benzyl	Bzl	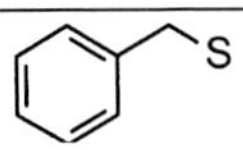	Na/liq. NH_3	117
4-Methylbenzyl	Bzl(4-Me)	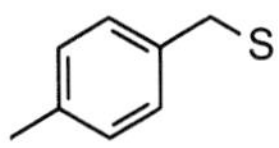	Strong acids TI(TFA)$_3$	118
4-Methoxybenzyl	Mob	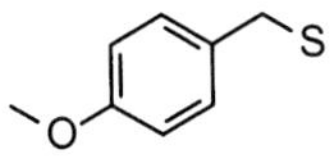	HF TFA, TFMSA Hg^{II}	119
2,4,6-Trimethoxybenzyl	Tmb	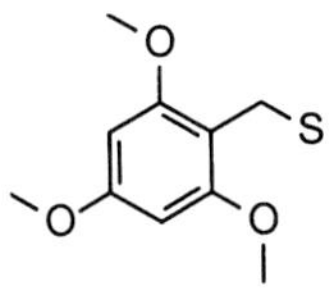	TFA	120
Diphenylmethyl	Dpm	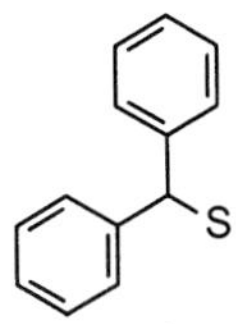	TFA HBr/AcOH HF Nps-Cl followed by reduction (B/C)[a]	121, 122
Trityl	Trt	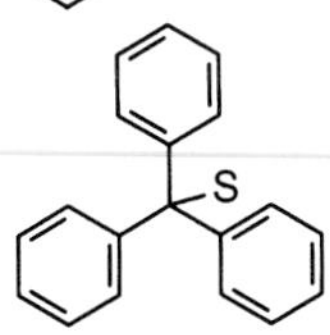	HF, TFA, TFA/AcOH I_2, Hg^{II} $AgNO_3$ Nps-Cl, then reduction (A/B/C)[a]	123–126
tert-Butyl	tBu		Nps-Cl	127
Acetamidomethyl	Acm	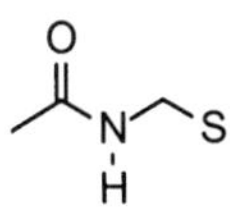	Hg^{II} I_2, Tl^{III} Nps-Cl followed by reduction (A/B/C)[a)]	128, 129
Trimethylacetamidomethyl	Tacm	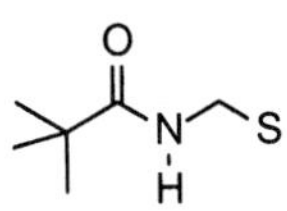	Hg^{II} I_2, $AgBF_4$	130
9-Fluorenylmethyl	Fm	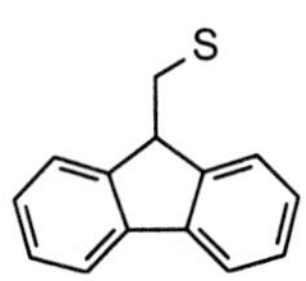	Piperidine	77

Tab. 4.3 (continuted)

Group	*Symbol*	*Y-S*	*Cleavage conditions*	*Reference(s)*
tert-Butylsulfanyl	StBu	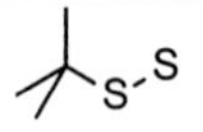	Thiolysis Phosphines	131–134
3-Nitro-2-pyridyl-sulfanyl	Npys	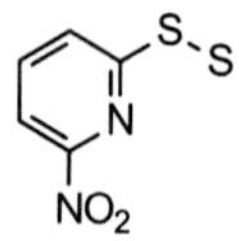	Thiolysis "low" HF	135
Allyloxycarbonyl-aminomethyl	Alocam	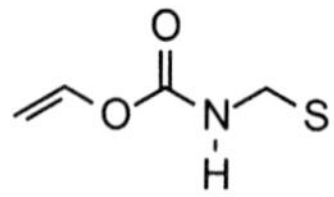	Pd^0/Bu_3SnH	136
9*H*-Xanthen-9-yl	Xan	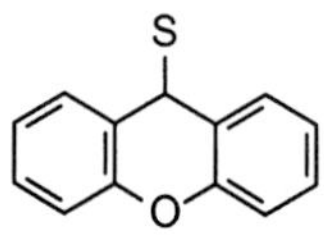	$TFA/CH_2Cl_2/Et_3SiH$ I_2	137

a) A = iodolysis; B = rhodanolysis; C = Kamber method.

tial representatives of this class of compounds have been developed as acid-labile thiol-protecting groups. However, many of them were found to be too resistant toward acids and so were not used in any practical application.

Thiol-protecting groups of the S-acyl type do not have any practical relevance, because acyl migrations from sulfur to nitrogen are energetically favored and represent prominent side reactions.

The triphenylmethyl group (trityl group, Trt) has received much attention, as it may be cleaved by silver salts, mercury(II) salts, iodine, or TFA in the presence of ethanethiol, alkyl sulfenyl chlorides, etc. Deprotection with TFA provides the free thiol, which is in particular useful in the synthesis of peptide antigens for conjugation to carrier proteins. Considerable base-catalyzed racemization has been observed upon in-situ activation of Fmoc-Cys(Trt)-OH with uronium reagents such as TBTU [141, 142].

The *tert*-butyl group has also been used for thiol protection [127, 143] as it is labile towards both TFMSA and mercury(II) salts.

A different type of thiol-protecting group is found in the N-acyl N,S-acetals, most of which are compatible with both Boc and Fmoc chemistry. The acetamidomethyl (Acm) group is completely stable towards acidolysis, and is removed by treatment with mercury(II) salts at pH 4, thallium(III) trifluoroacetate, or iodine. Oxidizing agents such as iodine simultaneously induce disulfide formation. Structural analogues of the Acm group are the chloroacetamidomethyl group, the isobutyrylamidomethyl group, and the 2,2-dimethylpropionamidomethyl group (trimethylacetamidomethyl group, Tacm) [130]. The combination of thiol-protecting groups with incongruous cleavage conditions enables differential disulfide formation in peptides containing more than two cysteine residues.

In particular, Trt, Acm, and Tacm represent thiol-protecting groups of eminent importance, as they allow for a simultaneous oxidative coupling to give cystine derivatives. Iodolysis (treatment with iodine, method A, Tab. 4.3 [144]), rhodanolysis (treatment with rhodanides, method B, Tab. 4.3 [145]), and the Kamber method, method C, Tab. 4.3 [146] are suitable for the simultaneous conversion of a thiol-protected peptide to the corresponding cystine (disulfide) derivative without separate deprotection. Oxidative cleavage by iodine, however, may be accompanied by methionine S-oxidation, iodination of Tyr residues [147], and the formation of tryptophan thioesters [148]. Additionally, iodine or thallium(III) oxidation may cause Acm migration from Cys to Ser, Thr, or Gln side chains [149].

Disulfide derivatives also serve as protecting groups for thiols. The ethylsulfanyl (R=EtS), the *tert*-butylsulfanyl (R=tBuS), and the 3-nitro-2-pyridylsulfanyl groups belong to this class of unsymmetrical disulfides, and can be selectively reduced by thiols such as 2-mercaptoethanol or by phosphines. In the former case of a thiol–disulfide exchange reaction, the excess of a free thiol acting as a deblocking reagent shifts the equilibrium in the desired direction. Removal of the S-alkylthio group occurs selectively, as many other protecting groups (including several thiol-protecting groups) are stable toward thiolysis.

The general protection scheme of thiol groups by an S-sulfo moiety ($S\text{-}SO_3^-$) should also be mentioned in this context [150], because the corresponding cysteine S-sulfo derivatives obtained from oxidative sulfolysis from S-alkylthio or S-arylthio cysteine derivatives possess sufficient stability for some peptide synthetic procedures. This temporary blocking is removed by reduction with thioglycolic acid at pH 5.

Protected cysteine residues are prone to several side reactions: β-elimination leading to the formation of dehydroalanine derivatives is favored under basic reaction conditions where subsequent conjugate addition of nucleophiles such as piperidine takes place. Oxidation and alkylation may also occur.

More than 70 variants for the protection of the cysteine thiol group have been described, and further details may be found in several monographs [83, 116]. An interesting thiol-protecting scheme for the formation of intrachain disulfide bridges has been demonstrated in an endothelin synthesis [151].

4.2.4.5 Imidazole Protection

Histidine is one of the most problematic building blocks in peptide synthesis [152]. Side reactions may occur when protection of the imidazole ring of histidine is omitted, because these derivatives are easily N-acylated. Some histidine derivatives are also relatively insoluble in the solvents normally used in peptide synthesis. Undesired side reactions such as racemization, formation of cyclic imidazolides, and N-guanylations of the imidazole ring have been observed, especially upon activation of C-terminal histidine residues. These problems may be avoided by reversible masking of the imidazole function (Tab. 4.4).

Tab. 4.4 Imidazole protecting groups Y-N^{im}.

Group	*Symbol*	*Y-N^im*	*Cleavage conditions*	*Refer-ence(s)*
Benzyl	Bzl	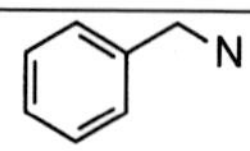	Na/liq. NH_3	153
2,4-Dinitrophenyl	Dnp	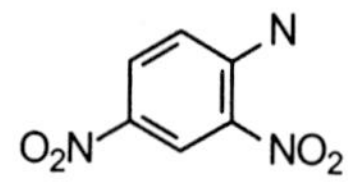	2-mercaptoethanol, thiophenol stable to acids	154
Benzyloxymethyl	Bom	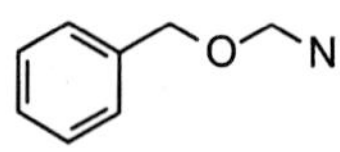	strong acids	155
tert-Butoxycarbonyl	Boc	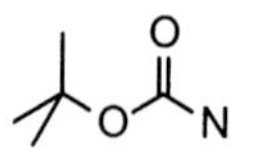	TFA, HBr/AcOH HF stable to HCl/dioxan	156
Adamantyl-1-oxycar-bonyl	Adoc	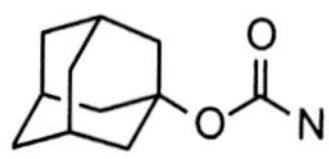	TFA	25, 157
Triphenylmethyl	Trt	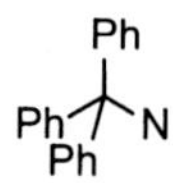	TFA	158, 159
Diphenylmethyl	Dpm	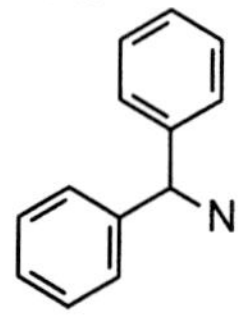	6 N HBr/AcOH (3 h) HCOOH (10 min) TFA (1 h)	160
Pyridyldiphenyl-methyl	Pdpm	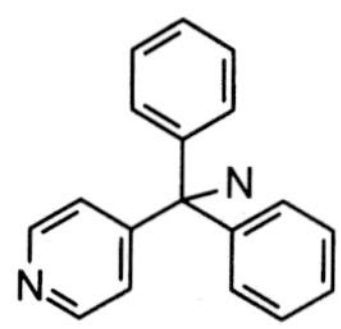	catalytic or electroly-tic reduction Zn/AcOH	77
4-Toluenesulfonyl (Tosyl)	Tos	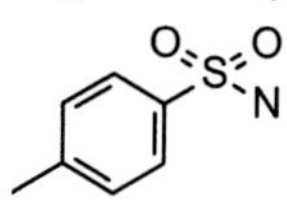	strong acids	161
4-Methoxybenzene-sulfonyl	Mbs	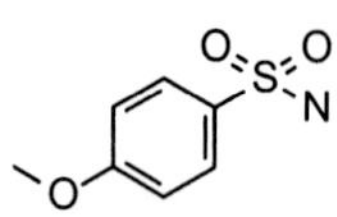	TFA/$(CH_3)_2S$	162
tert-Butoxymethyl	Bum	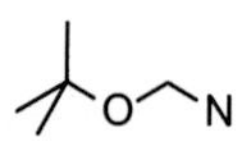	TFA stable to piperidine	163
Allyl	Al	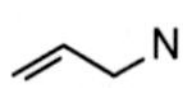	Pd^0, nucleophile	164
Allyloxymethyl	Alom	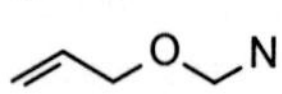	Pd^0, nucleophile	165

In the past, the imidazole group was protected mainly as the N-benzyl derivative – a method that suffers from a very limited scope of cleavage conditions. This problem was overcome by the development of other protecting groups such as the 2,4-dinitrophenyl group, which is acid stable and can be cleaved by thiolysis (2-mercaptoethanol at pH 8 for 1 h; thiophenol) or by hydrazinolysis.

Imidazole-protecting groups of the urethane type (Z, Boc, Adoc) are stable towards HCl in organic solvents, but as N-acyl imidazolides they may give rise to undesired acyl transfer reactions. The pronounced lability towards aminolysis and hydrolysis leads to problems with these permanent protecting groups during other deblocking reactions.

The N^{im} Boc group is stable toward piperidine and is, therefore, suited for the application in SPPS according to Fmoc tactics. It is cleaved by HBr/HOAc, TFA, or liquid HF, respectively. The structurally related Adoc group is removed by treatment with TFA.

Arylsulfonyl groups have increasingly received attention as imidazole-protecting groups. The N^{im}-(4-toluenesulfonyl) group often is used in combination with Boc as the temporary protecting group. It is cleaved under the general deprotection conditions with liquid HF or TFMSA, but it has been reported to be susceptible to cleavage by HOBt. Modification of the aryl residue with electron-donating groups increases the acid lability of the arylsulfonyl protecting group, as has already been discussed in connection with the arylsulfonyl-based guanidino protecting groups. The 4-methoxybenzenesulfonyl (Mbs) group and the 2,4,5-trimethylbenzenesulfonyl group, respectively, are labile toward TFA in the presence of dimethylsulfide.

Some confusion exists in the literature concerning protection of the two heterocyclic nitrogen atoms (N^{π}/N^{τ}, **51**). While the two regioisomeric derivatives are formed during alkylation in a ratio of 3:1, steric hindrance of the alkylation reagent favors N^{τ}-alkylation. Trityl chloride reacts regioselectively at the N^{τ} position, and N^{τ}-trityl-protected histidine **52** belongs to the most suitable acid-labile protecting groups that are very compatible with Fmoc tactics.

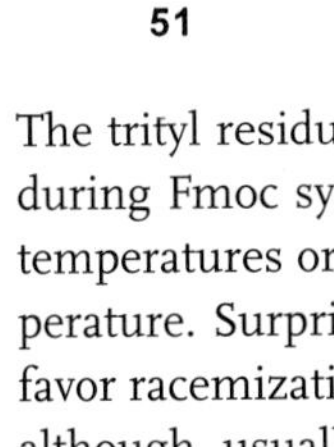

51

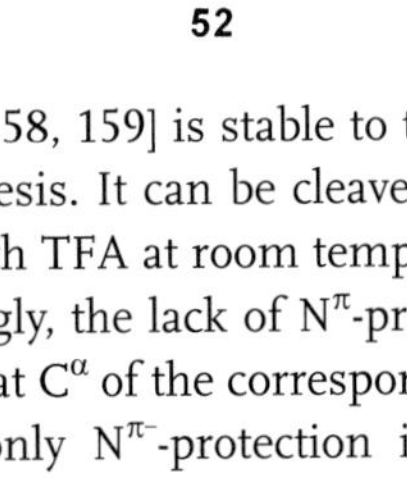

52

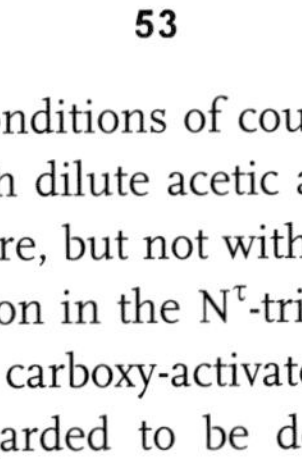

53

The trityl residue [158, 159] is stable to the conditions of coupling and deprotection during Fmoc synthesis. It can be cleaved with dilute acetic acid at slightly elevated temperatures or with TFA at room temperature, but not with 1 N HCl at room temperature. Surprisingly, the lack of N^{π}-protection in the N^{τ}-trityl derivatives does not favor racemization at C^{α} of the corresponding carboxy-activated histidine derivatives, although usually only N^{π}-protection is regarded to be devoid of racemization.

Although the N^{τ}-trityl group prevents racemization by its steric bulk to a certain extent [166], epimerization has nevertheless been observed, for example upon condensation of Fmoc-His(Trt)-OH to a 4-benzyloxybenzylalcohol linker [167].

In this respect, the N^{π}-benzyloxymethyl (Bom) group **53** (R = benzyl) is often used in combination with N^{α}-Boc protection [168]. This N^{π} protection of the imidazole significantly decreases the risk of racemization [169]. The Bom group is resistant toward TFA, but it can be easily cleaved with HF, TFMSA, or HBr/HOAc – which renders it especially suited for Boc tactics. Formaldehyde is liberated during HF deprotection, and may damage residues such as N-terminally located cysteine.

The N^{π}-*tert*-butoxymethyl (Bum) group **53** (R = *tert*-butyl) that may be applied in N^{α}-Fmoc tactics is characterized by higher acid lability, though the preparation of the protected derivative is less straightforward. Application of the N^{π}-allyl [164] and N^{π}-allyloxymethyl [165] groups has also been suggested.

4.2.4.6 **Hydroxy Protection**

In principle, protection of the primary and secondary hydroxy function of serine and threonine (and of the aromatic hydroxy group of tyrosine) is not necessary, especially when these residues are located in the amino component. However, a substantial excess of the acyl component, as is generally used for example in SPPS, may lead to partial O-acylation (especially of tyrosine residues) as the basic reaction conditions generate the highly nucleophilic phenolate anion. The azide method was regarded as the most useful variant for the activation of hydroxy-substituted amino acids and of peptides with C-terminal hydroxy-substituted amino acids. However, aromatic nitrosations in the case of tyrosine or side reactions based on Curtius rearrangements have been observed during acyl azide synthesis. Tyrosine is also prone to electrophilic aromatic substitution during deprotection procedures. An increased risk of racemization has been reported for carboxy components containing C-terminal tyrosine.

The application of active esters originally failed with these derivatives, as N-protected active esters with unprotected hydroxyl side-chain functionalities could not be obtained in sufficient purity. However, nowadays the corresponding 2,4-dinitrophenylesters and other active ester derivatives (e.g., Z-Ser-OPcp, Boc-Thr-OSu) can be obtained satisfactorily. On occasion, the syntheses applying components with free hydroxy groups suffer from further undesired side reactions, such as N → O aminoacyl migrations, O-acetylations upon deprotection (e.g., with HBr/HOAc), and racemization under the basic conditions of hydrazinolysis. In any case, the advantages of a global protection scheme with the imminent risk of incomplete deprotection reactions should be compared individually to the above-mentioned strategy with partial protection.

The blocking of hydroxy groups with benzyl or *tert*-butyl type groups is of eminent practical importance (Tab. 4.5). O-Benzyl threonine is obtained from threonine by reaction with benzyl alcohol in the presence of 4-toluenesulfonic acid and subsequent alkaline saponification of the concomitantly formed benzyl ester [180]. O-Benzyl tyrosine is formed by direct alkylation of the tyrosine-copper complex with benzyl

Tab. 4.5 Hydroxy-protecting groups Y-OR.

Group	***Symbol***	***Y-O***	***Cleavage conditions***	***Reference(s)***
Benzyl	Bzl	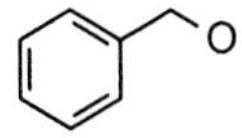	H_2/Pd HF Na/liq. NH_3 HBr/dioxan	170
2,6-Dichlorobenzyl	Dcb	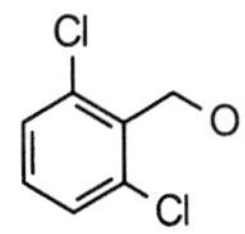	Strong acids, stable to TFA	171
Diphenylmethyl (benzhydryl)	Dpm	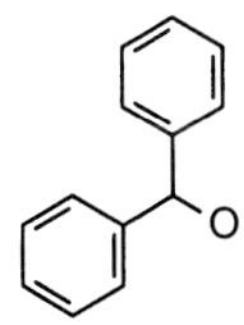	H_2/Pd TFA (reflux)	172
Cyclohexyl	Cy		TFMSA/thioanisole	173
2-Bromobenzyloxycarbonyl	Z(2-Br)	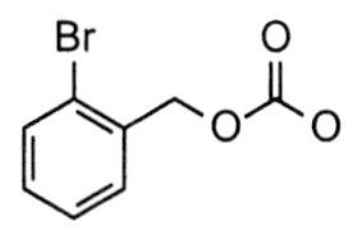	Strong acids, stable to TFA	174, 175
tert-Butyl	tBu		TFA HCl/TFA conc. HCl (0 °C, 10 min)	176
1-Benzyloxycarbonylamino-2,2,2-trifluoroethyl	Zte	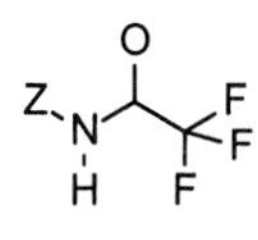	H_2/Pd Acidolysis	177
Methylthiomethyl	Mtm	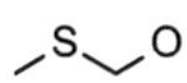	CH_3I in wet acetone ($NaHCO_3$)	178
Allyl	Al	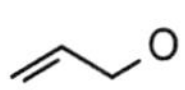	Pd^0, nucleophile	41
Allyloxycarbonyl	Aloc	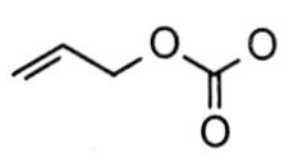	Pd^0, nucleophile	41, 179

bromide [181]. Interestingly, cleavage of the O-benzyl group in Z-Tyr(Bzl)-MeO during hydrogenolysis can be efficiently suppressed by the addition of 2,2′-dipyridyl [182].

tert-Butyl ethers of hydroxy amino acids are very easily accessible by acid-catalyzed *tert*-butylation of amino acid benzyl esters with isobutene and subsequent hydrogenolysis [183].

O-Benzyl groups can be cleaved by hydrogenolysis or with liquid HF, and hence are appropriate for Boc chemistry. During acidolysis, highly reactive benzyl cations are formed that often give rise to side reactions, but this can be suppressed by the addition of cation scavengers such as anisole or resorcine. Furthermore, acidolytic deprotection reactions sometimes may be incomplete. Tyrosine residues to be used for Boc chemistry are normally protected using either the 2,6-dichlorobenzyl (Dcb, Tab. 4.5) or the 2-bromobenzyloxycarbonyl (Z(2-Br), Tab. 4.5) groups. The 2-bromobenzyloxycarbonyl group [174, 175] is stable towards nucleophiles and TFA, and can be removed together with other side chain-protecting groups under strongly acidolytic conditions on completion of the synthesis.

The *tert*-butyl ether of Ser, Thr, or Tyr residues is very compatible with Fmoc chemistry. It is removed with TFA (1–2 h at room temperature), with HCl/TFA, and with concentrated HCl at 0 °C under an inert gas atmosphere within 10 min. Fmoc-Ser(tBu)-OH has been reported to undergo racemization under continuous-flow solid-phase synthesis involving preactivation. The correct choice of coupling reagent and the base employed (collidine) reduces racemization to an acceptable extent, however [184].

4.2.4.7 **Thioether Protection**

The methionine thioether group usually does not cause severe complications during peptide synthesis, and the introduction of amino- or carboxy-protecting groups generally succeeds without difficulties. Side reactions that occur during deprotection reactions with methionine-containing peptides are much more problematic. Partial S-demethylation producing homocysteine derivatives has been reported during deblocking reactions with sodium in liquid ammonia, though this method is rarely used nowadays. S-Alkylations may occur on acidolytic cleavage of benzyl- or *tert*-butyl-type protecting groups. The reaction of the benzyl or *tert*-butyl cations, respectively, with the thioether group gives rise to the formation of sulfonium salts, and this sometimes cannot be avoided completely, even by the addition of scavengers such as anisole or ethyl methylsulfide. An exchange of S-alkyl groups has also been reported in this context. The sulfonium salt **54** may react to give S-benzyl homocysteine. The sulfonium salt **55** has been unambiguously identified. S-Methylation of methionine to give the sulfonium salt has been recommended as a protecting measure [185].

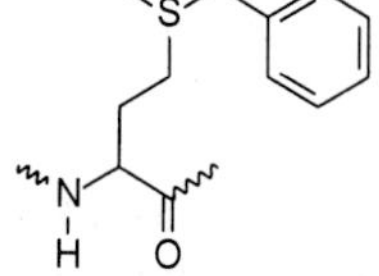

54

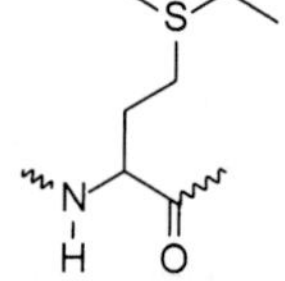

55

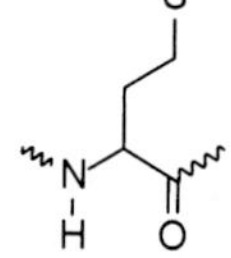

56

The oxidation sensitivity of the methionine thioether group represents another problem. Oxidizing agents such as peroxides or oxygen (or even dimethylsulfoxide (DMSO), a solvent that is frequently used in peptide chemistry), may lead to methionine *S*-oxide (**56**) formation. There are reports that biologically active peptides become devoid of their activity upon oxidation of a methionine residue. The sulfoxide formation may be reduced or even completely avoided when oxidants are strictly excluded or special cleavage cocktails are used [186].

On the other hand, methionine sulfoxide formation has been suggested as a protecting measure for the thioether moiety. S-Oxidation leads to the creation of a new stereogenic center, and diastereomeric peptides containing methionine S-oxides have been separated by crystallization of the picrates. The sulfoxide group can be reduced, for example with thioglycolic acid, *N*-methylmercaptoacetamide, or ammonium fluoride/2-mercaptoethanol (the latter method does not affect disulfide bonds).

Hydrogenolytic cleavage of protecting groups is possible for methionine-containing peptides despite the presence of the methionine sulfur, when a base or $BF_3 \cdot OEt_2$ is added or by hydrogen transfer reaction using formic acid as the hydrogen donor.

Methionine is normally used both in Boc and in Fmoc chemistry without further side-chain protection. Oxidation is avoided in Fmoc chemistry by adding ethyl methylsulfide or thioanisole during cleavage and deprotection.

4.2.4.8 **Indole Protection**

The indole ring of tryptophan usually does not require any protecting measures. A broad variety of synthetic conditions permit the incorporation of N^{in}-unprotected tryptophan both as the amino or as the carboxy component. However, when the azide method is used for peptide coupling, N^{in}-nitrosation may occur during the synthesis of the acyl azide from the hydrazide, especially in the case of incorrect stoichiometry. The indole system is prone to oxidation by either oxygen or peroxides, and therefore an inert gas atmosphere and absolutely peroxide-free solvents are required. Side reactions at the indole ring (e.g., electrophilic aromatic substitutions) have been observed during some deprotection reactions: *tert*-butylation may occur both at N^{in} and at different positions of the indole ring upon acidolysis of *tert*-butyl-type protecting groups. This also applies to other reactions involving carbenium ions [187]. In most cases, these side reactions are suppressed by the addition of scavengers.

There are few possibilities available for indole ring protection. N^{in}-Formylation increases the stability against alkylation and oxidation [188, 189] and can be removed with dilute aqueous piperidine solution, TFMSA, hydrazine, or hydroxylamine. Consequently, the formyl derivatives are applicable in Boc chemistry. Under basic reaction conditions the formyl group may migrate to free α-amino groups.

In principle, acylation of the indole nitrogen with the Z-group is also a viable alternative. This derivative is stable towards some acids, but is cleaved slowly by TFA and more rapidly by HF. Deprotection by hydrogenolysis and hydrolysis, respectively, is possible.

Although tryptophan may be used without side-chain protection in Fmoc protocols, problems may arise during final deprotection. While indole *tert*-butylation usually is suppressed by the addition of scavengers, sulfonation by the cleavage products of arenesulfonyl-type guanidino-protecting groups may be deleterious. In these cases, Fmoc-Trp(Boc)-OH provides a viable alternative of preventing Trp sulfonation, even in the case of peptides containing multiple arginine residues [190]. Indole Boc protection also diminishes the propensity of trialkylsilane scavengers to reduce the indole ring to an indoline ring.

Further possibilities for indole protection are available [191–193].

4.2.4.9 ω-Amide Protection

From a chemical point of view, the ω-carboxamide groups of glutamine and asparagine, as well as the C-terminal amides of amino acids and peptides, are rather unreactive functionalities that do not require further protection. While this is usually true for the C-terminal carboxamides, side reactions sometimes occur at the side-chain carboxamides of Asn and Gln during peptide synthesis. Especially under the conditions of carbodiimide couplings, undesired nitrile formation from the carboxamide is very prominent (Fig. 4.13). Activating reagents such as BOP, PyBOP, and HBTU are also reported to favor this dehydratation, which can be suppressed by the addition of HOBt or HOAt to the coupling reaction. As this side reaction occurs only during activation of Asn or Gln residues because of mechanistic considerations, it should not cause any problem once these residues have been incorporated into the peptide chain.

Hydrazinolytic conditions may convert the primary amides into the corresponding hydrazides, and the conversion of carboxamides into esters during solvolytic deprotection reactions has also been observed.

Cyclization reactions of asparagine in a peptide chain involving the β-carboxamide group with release of ammonia have also been reported. The tendency towards deamidation depends both on the primary and secondary structure, as well as on the solvent [194–196]. The aspartimide derivative **57** may subsequently undergo hydrolysis to give a mixture of α-aspartyl and β-aspartyl peptides (**58**, **59**).

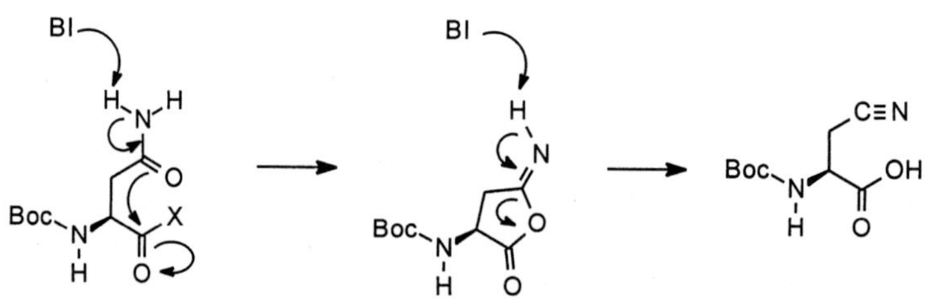

Fig. 4.13 Mechanism for the dehydration of carboxy-activated asparagine. Glutamine dehydration follows an analogous mechanism. B = base; X = electron-withdrawing activating group.

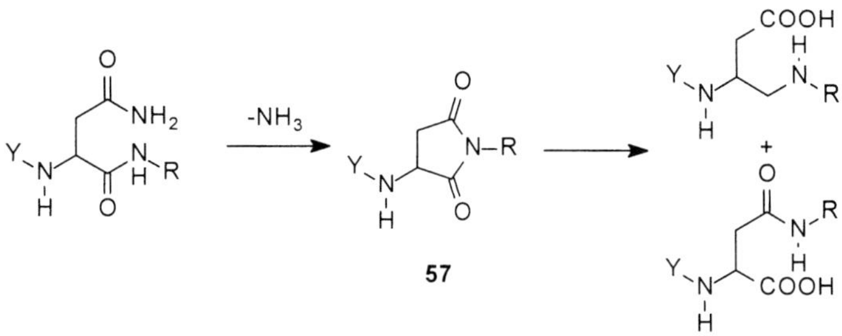

Peptides containing glutamine in the N-terminal position are easily converted into pyroglutamyl peptides **60**. This side reaction is catalyzed by acids and is, therefore, especially problematic in Boc tactics. It may be avoided by the application of high TFA concentrations for Boc cleavage. Base-catalyzed pyroglutamate formation in Fmoc chemistry is less prominent.

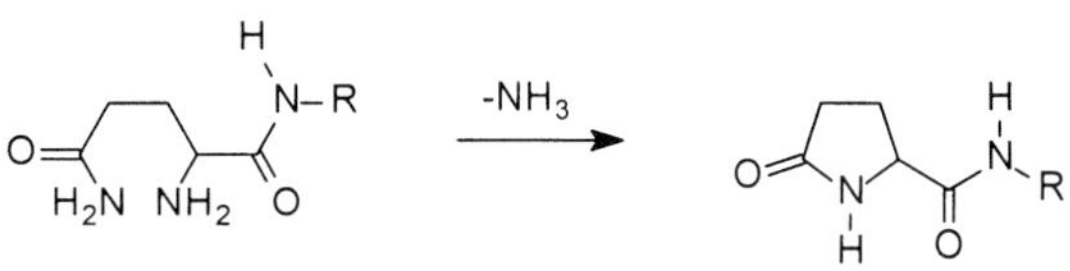

60

Most of the side reactions described above can be suppressed by reversible blocking of the side-chain carboxamide group, which also confers better solubility in organic solvents on the peptides – which is indispensable for normal peptide synthetic operations. Unprotected amide groups tend to form intra- or intermolecular hydrogen bonds, and this results in lower solubility in organic solvents and higher solubility in aqueous solvents. Substituted N-benzyl derivatives that can be cleaved under acidolytic conditions are well suited for this purpose. N-Methoxybenzyl residues or the N-diphenylmethyl protecting group can be introduced directly by the reaction of N^{α}-protected asparagine or glutamine, respectively, with the benzyl or diphenylmethyl alcohol (Fig. 4.14A). Alternately, the corresponding N^{α}/C^{α}-protected aminodicarboxylic acids (aspartic acid or glutamic acid) are transformed into the side chain-protected carboxamides with benzyl amine or diphenylamine derivatives and DCC as coupling reagent (Fig. 4.14B).

All amide-protecting groups are cleaved with liquid HF. Cleavage with TFA at ambient temperature is also possible for some derivatives, except for the 4-methoxybenzyl and the diphenylmethyl protecting groups. The addition of anisole as a scavenger favors clean cleavage reactions, as the intermediate benzyl-type cations are trapped.

The N-trimethoxybenzyl (Tmb) and the N-triphenylmethyl (Trt) groups are stable towards piperidine and can be removed by treatment with TFA [197], though Trp alkylation may be detrimental with Tmb [198]. The trityl cation usually does not alkylate Trp or other sensitive side chains. Hence, Trt and Tmb are very appropriate carboxamide-protecting groups for solid-phase synthesis in combination with Fmoc as the temporary protecting group.

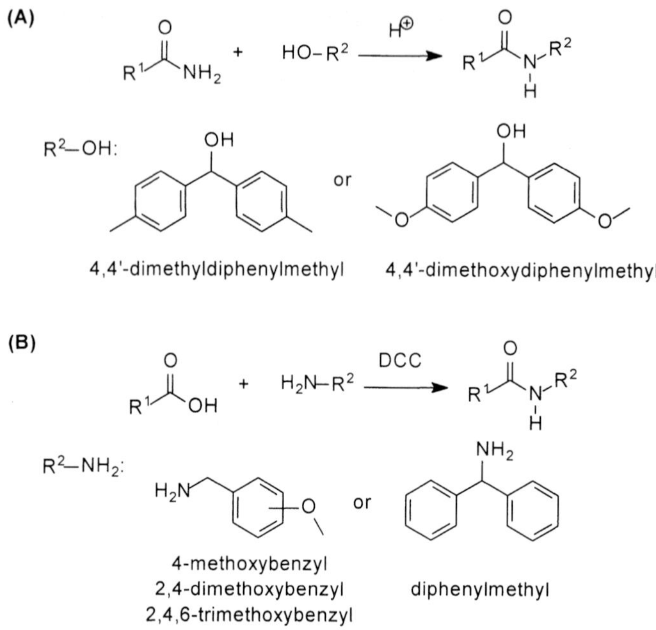

Fig. 4.14 Introduction of ω-amide-protecting groups into amino acid derivatives via N-alkylation of a carboxamide (A) or amide formation (B).

The base-stable 9-xanthenyl (Xan) or the 4,4′-dimethoxydiphenylmethyl (Mbh) groups have found some application for amide protection in Boc chemistry, as they can be removed with strong acids. However, they display somewhat limited acid stability, and this results in progressive cleavage during the consecutive operations of peptide synthesis according to the Boc protocol.

4.2.5
Enzyme-labile Protecting Groups

The synthesis of acid- and base-labile peptide derivatives and peptide conjugates requires the application of protecting groups that can be cleaved under very mild or even neutral reaction conditions. In particular, glyco-, phospho-, and lipopeptides as well as sulfated peptides (cf. Sections 6.3–6.6) belong to this class of sensitive compounds. The treatment of such derivatives with bases often leads to β-elimination reactions in the case of serine and threonine derivatives, or to other undesired side reactions. Treatment with acids in several cases causes degradation of the nonpeptide moiety of the conjugate.

Therefore, these targets may require the application of enzyme-labile protecting groups.

The first peptide synthesis using the Bz-Phe residue as an enzyme-labile protecting group to be cleaved with chymotrypsin was described by Holley in 1955 [199]. Subsequently, the development of enzyme-labile blocking groups has experienced a logical progression from the early specialized application to most sophisticated synthetic strategies [200–202]. Enzymatic manipulation of protecting groups comprises more than simple deprotection, as regiospecificity and stereospecificity are favorable features of enzyme reactions. In addition, because of the mild reaction conditions enzymes can compete successfully with chemical reagents in several cases of protecting group manipulation. The specificity of the biocatalyst safeguards that neither the peptide bonds nor other protecting groups present are attacked. Consequently, enzymatic protecting group techniques represent a particularly efficient methodology for this type of sensitive compound [203–207].

4.2.5.1 Enzyme-labile N^{α}-amino Protection

The first enzyme-labile urethane type protecting groups were developed by Waldmann et al. [205]. The overall strategy utilized in many of these derivatives employs an enzyme-labile bond, where cleavage may additionally result in a fragmentation reaction of the urethane, thus liberating the free amino group (Fig. 4.15).

The tetrabenzylglucosyloxycarbonyl group (BGloc) [208] comprises O-benzyl-protected glucose as the carbohydrate moiety which is attached to the urethane oxygen. After removal of the benzyl groups, a mixture of α-glucosidase from baker's yeast and β-glucosidase from almonds results in cleavage of the glycosidic bond, thus liberating the carbamic acid (Fig. 4.16). Based on this concept, a whole set of enzyme-labile carbohydrate-based protecting groups can be generated that may be selectively removed using a glycosidase. Furthermore, tetra-*O*-acetyl-β-D-glucopyranosyl-oxycarbonyl (AGloc) and tetra-*O*-acetyl-β-D-galactopyranosyl-oxycarbonyl (AGaloc) have been developed as urethane-type protecting groups. These are regarded as second-generation enzyme-labile urethane-type protecting groups which are readily attached to the desired amino acid and, after further synthetic steps, may be cleaved in two separate or sequential enzymatic steps [209].

The acetoxybenzyloxycarbonyl group [Z(OAc)] [210, 211] and the chemically related phenylacetyloxybenzyloxycarbonyl group [Z(OAcPh)] [212] are cleaved by an esterase or lipase [Z(OAc), Fig. 4.17] or penicillin G acylase [Z(OAcPh)]. These enzymes recognize the enzyme-labile ester bond and selectively cleave it. Conse-

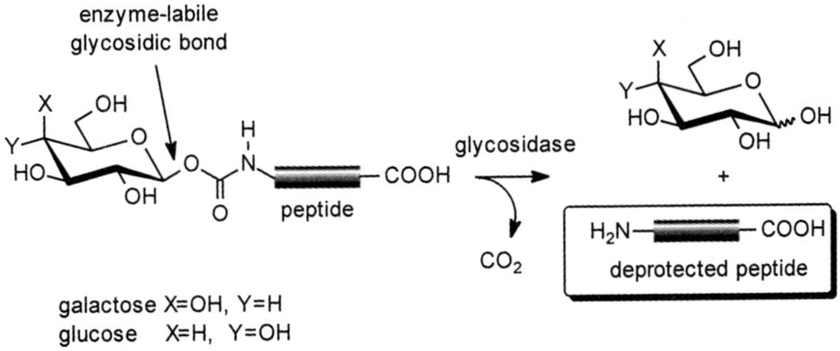

Fig. 4.15 Enzyme-labile carbohydrate-derived protecting groups.

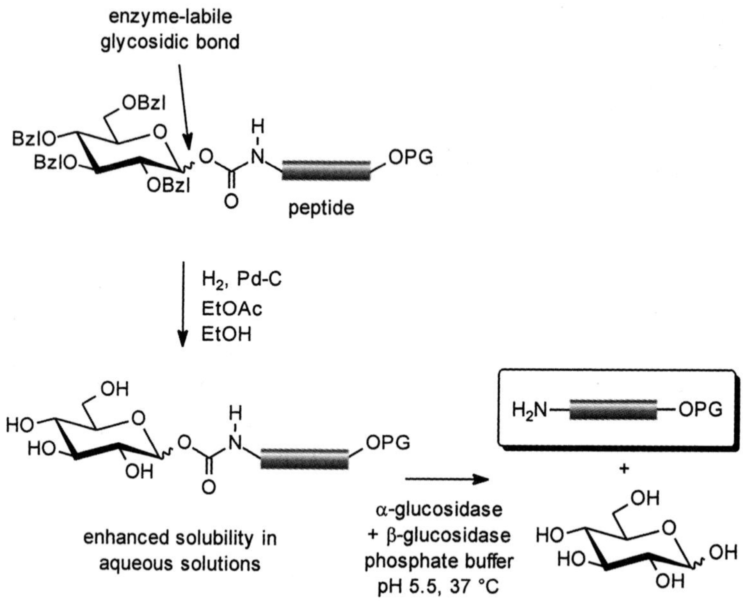

Fig. 4.16 Enzymatic cleavage of the BGloc protecting group (PG: protecting group).

quently, a *p*-hydroxybenzylurethane is formed that spontaneously undergoes fragmentation with release of the desired amino group. Depending on the acyl residue selected (acetate or phenylacetate), enzymes with different selectivity may be used.

4.2.5.2 Enzyme-labile C$^{\alpha}$-carboxy Protection and Enzyme-labile Linker Moieties

Esters form one class of carboxy-protecting groups that are labile towards cleavage by enzymes, and can be removed by lipases. The lipase from *Mucor javanicus* selectively deblocks heptyl esters of the C-terminus in sensitive amino acid derivatives, without any undesired side reactions [213, 214]. However, some problems are associated with the use of the very hydrophobic heptyl ester, these being mainly due to the low solubility of the substrate and, hence, the low turnover rate, especially in the presence of further hydrophobic amino acids. The more hydrophilic 2-(*N*-morpholino)-ethyl ester (MoEt) and the methoxyethoxyethyl ester (Mee) circumvent this problem. The Mee ester may be cleaved by lipase-catalyzed hydrolysis or even papain-catalyzed hydrolysis [215, 216].

Waldmann's group was able to transfer the principle of deprotection by enzyme-induced fragmentation of enzyme-labile urethane-type protecting groups also to an enzyme-labile ester group [217]. Cleavage of the 4-phenylacetoxybenzyl (PAOB) ester is mechanistically related to Z(OAcPh) cleavage (Fig. 4.17), and is initiated by the action of penicillin G acylase at pH 7 and room temperature, with liberation of phenylacetate. The phenolate formed concomitantly undergoes fragmenta-

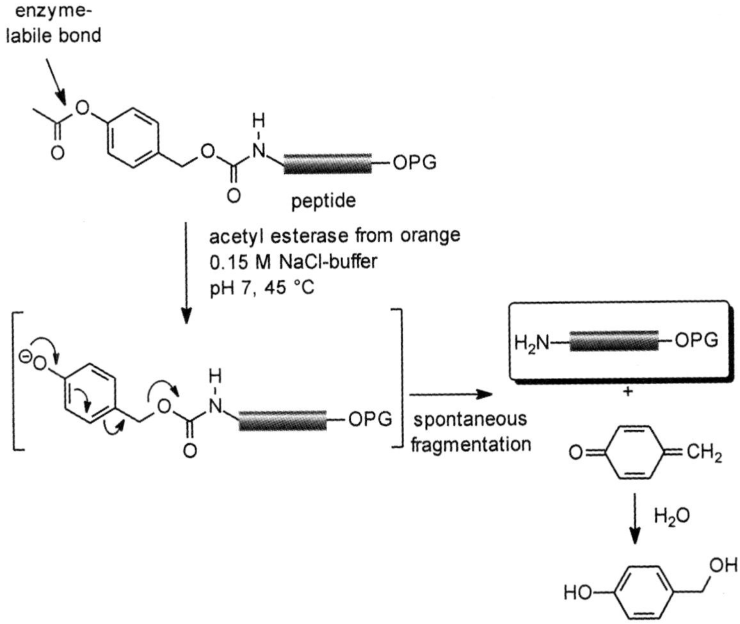

Fig. 4.17 Mechanism of enzymatic Z(OAc) protecting group cleavage (PG: protecting group).

tion to give the desired C-terminally deblocked dipeptide and the quinone methide, which is efficiently trapped by water, forming 4-hydroxybenzyl alcohol.

Another strategy uses amino acid choline esters (Fig. 4.18) that may be cleaved by the enzyme butyryl choline esterase. This commercially available biocatalyst recognizes positively charged choline esters and cleaves them under very mild conditions [218, 219]. Ester groups of sensitive peptide conjugates, such as phosphorylated and glycosylated peptides [220], lipidated peptides [221, 222] and nucleopeptides [223] can be hydrolyzed under very mild conditions with choline esterases, such as acetylcholine esterase (AChE) and butyrylcholine esterase (BChE).

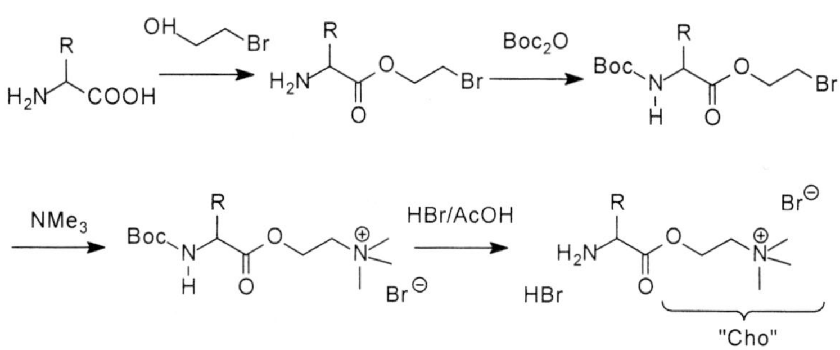

Fig. 4.18 Synthesis of amino acid choline esters.

4.2.6
Protecting Group Compatibility

Protecting group orthogonality [1] is required for the different types of protecting groups that are used for a particular synthesis. Orthogonality means that each class of protecting group (temporary, semipermanent) is cleaved independently from the other. Therefore, the chemical cleavage mechanisms should be different for an optimum degree of orthogonality. However, in most cases strict orthogonality cannot be achieved easily and the protecting schemes rely on differences in cleavage kinetics.

The protecting group schemes most frequently used in solid-phase synthesis, and which often are also suitable for solution-phase syntheses, are detailed briefly in Section 4.5.3.

For example, the application of Boc as the temporary protecting group excludes the use of semipermanent protecting groups at the side chains and at the C-terminus with comparable acid-lability. Consequently, semipermanent protection in Boc tactics must employ functional groups that resist the conditions necessary for the iterative Boc cleavages, such as benzyl-type groups. This tactics is, therefore, called the Boc/Bzl tactics or protection scheme.

Fmoc as the temporary protecting group is cleaved after each coupling step by treatment with bases. In such cases, side-chain and C-terminal protection may employ *tert*-butyl-type groups. Hence, this is referred to as the Fmoc/tBu tactics or protection scheme.

Special protection schemes may also be required, for example in the synthesis of fully side chain-protected peptides, or for subsequent cyclization reactions.

4.3
Peptide Bond Formation

Peptide bond formation is a nucleophilic substitution reaction of an amino group (nucleophile) at a carboxy group involving a tetrahedral intermediate. However, carboxylic acids react at room temperature with ammonia or amines to give an ammonium salt instead of a carboxamide. Therefore, the carboxy component **61** ($X\text{-}R^2 = OH$) must be activated prior to peptide bond formation. Furthermore, the peptide coupling reaction must be performed under mild conditions, and at room temperature.

Activation of the carboxy component (an increase of the electrophilicity) is achieved by the introduction of electron-accepting moieties. Groups which exert either an inductive (-I) effect or mesomeric (-M) effect (or both) decrease the electron density at the C=O group, thereby favoring the nucleophilic attack of the amino component (**62**). The latter attacks with its nitrogen lonepair the electrophilic position of the carboxy group to give the tetrahedral zwitterionic intermediate **63**. Peptide bond formation is then completed by dissociation of the leaving group (nucleofuge R^2X^-) from **63**. The leaving group capacity (nucleofugicity) is another factor which influences the reaction rate.

The variation of the leaving group XR^2 provides a broad spectrum of methods for peptide bond formation, and this methodological arsenal has been summarized in several monographs [83–86].

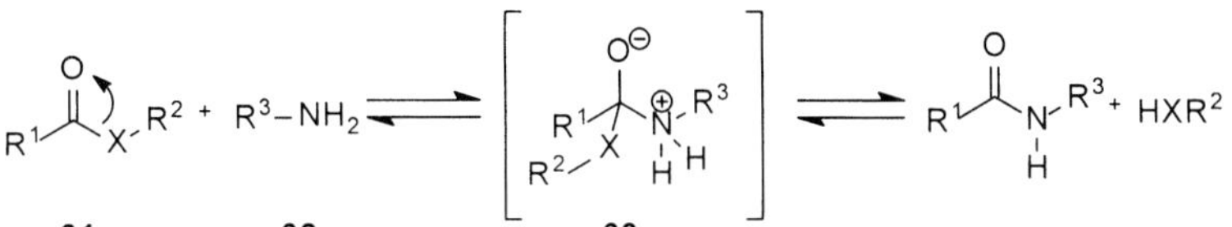

61 **62** **63**

It is beyond the scope of this book to discuss all the available methods in detail. Rather, attention will be focused on the reagents that are generally accepted in practical applications. Among these are the acyl azides, anhydrides, carbodiimides, active esters, and acyl halides, together with phosphonium and uronium reagents.

4.3.1
Acyl Azides

The acyl azide method [224] was introduced into peptide chemistry as early as 1902 by Theodor Curtius, and hence is one of the oldest coupling methods. Besides free amino acids and peptides, which are coupled with an acyl azide in alkaline aqueous solution, amino acid esters can be used in organic solvents. In contrast to many other coupling methods, the addition of an auxiliary base or a second equivalent of the amino component in order to trap the hydrazoic acid is not necessary.

For a long time the azide method was considered to be the only racemization-free coupling method, and it experienced a tremendous renaissance following the introduction of selectively cleavable amino-protecting groups. The starting materials for this method are the crystalline amino acid or peptide hydrazides **64**, respectively, that are easily accessible from the corresponding esters by hydrazinolysis. These hydrazides are transformed into the azides by N-nitrosation with one equivalent of sodium nitrite in hydrochloric acid at –10 °C and simultaneous water elimination. Alternately, mixtures of acetic acid, tetrahydrofuran, or DMF

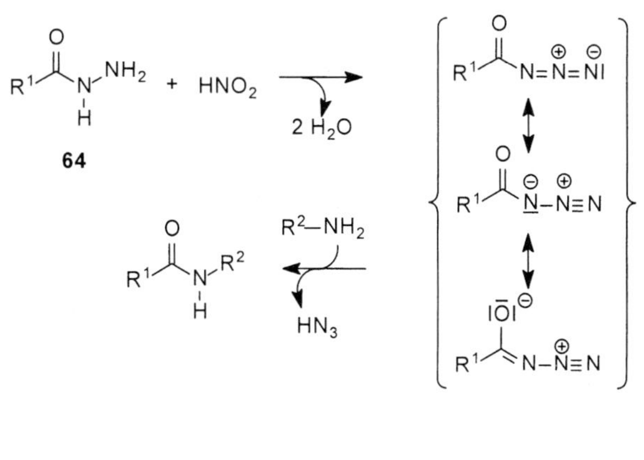

64 **65**

with hydrochloric acid have been used. The azide **65** is extracted from the aqueous layer with ethyl acetate, washed, dried, and reacted with the corresponding amino component. Some azides can be precipitated upon dilution with ice water.

Diphenylphosphoryl azide (DPPA) may also be used for the synthesis of acyl azides [225]. The Honzl-Rudinger method uses *tert*-butyl nitrite for the N-nitrosation and allows for the application of organic solvents in azide couplings [226]. The coupling reaction should be performed at low temperature because of the inherent thermal instability of acyl azides. At higher temperature, the Curtius rearrangement, i.e., conversion of the acyl azide to an isocyanate, becomes a prominent side reaction and eventually leads to the undesired formation of ureas. As a consequence of the low reaction temperature (e.g., 4 °C), the reaction rate is rather low and the peptide coupling reactions often take several days to reach completion.

Hydrazinolysis of ester groups in longer N-terminally protected peptides is often difficult, and consequently the application of orthogonally N′-protected hydrazine derivatives is the method of choice. The peptide segments assembled according to this backing-off strategy can be used for azide couplings after selective deblocking of the hydrazide group.

Although, as mentioned previously, the azide method has been considered as the method with the lowest tendency towards racemization, an excess of base in the reaction is accompanied by considerable racemization. Hence, any contact with bases during the coupling reaction must be avoided; for example, ammonium salts of the amino component should be neutralized with *N,N*-diisopropylamine or *N*-alkylmorpholine instead of triethylamine.

Despite all the limitations mentioned, the azide method is still important, especially for segment condensations, because of its low tendency towards racemization, the applicability with O-unprotected serine or threonine components, and the highly variable use of N′-protected hydrazides.

4.3.2
Anhydrides

Basic considerations towards the application of anhydrides in peptide synthesis date back to the early investigations of Theodor Curtius in 1881 in the synthesis of hippuric acid. Curtius obtained Bz-Gly_n-OH ($n = 2–6$) besides hippuric acid upon reaction of silver glycinate with benzoyl chloride. In those early days it was argued that N-benzoyl amino acids or N-benzoyl peptides formed unsymmetric anhydrides with benzoic acid as reactive intermediates when treated with benzoyl chloride.

About 70 years later, Theodor Wieland took advantage of these findings when his group made the mixed anhydride method available for modern peptide synthesis. Nowadays, besides this method, symmetrical anhydrides and N-carboxy anhydrides (NCA, Leuchs anhydrides) formed intramolecularly from the carboxy groups of the amino acid and a carbamic acid are used in peptide couplings. Last, but not least, it should be mentioned that unsymmetrical anhydrides are often involved in biochemical acylation reactions.

4.3.2.1 **Mixed Anhydrides**

Carboxylic acids and inorganic acids, respectively, are used for the formation of mixed anhydrides [227]. However, only a few representatives of the variety of synthetic variants and acids used have found widespread practical application, and in most cases alkyl chloroformates (or correctly, alkyl chlorocarbonates) are used. Ethyl chloroformate (ethyl chlorocarbonate), which in the past was used very frequently, has nowadays mainly been replaced by isobutyl chloroformate (isobutyl chlorocarbonate). The mechanism of the mixed anhydride method is shown in Fig. 4.19.

The regiochemistry of the aminolysis reaction of the mixed anhydride, which is initially formed from the carboxy component and the chloroformate, depends on the electrophilicity and/or the steric situation of the two competing carboxy groups. In the case of mixed anhydrides formed from a N-protected amino acid carboxylate (carboxy component) and an alkyl chlorocarbonate (activating component, e.g., from alkyl chloroformate), the amine nucleophile predominantly at-

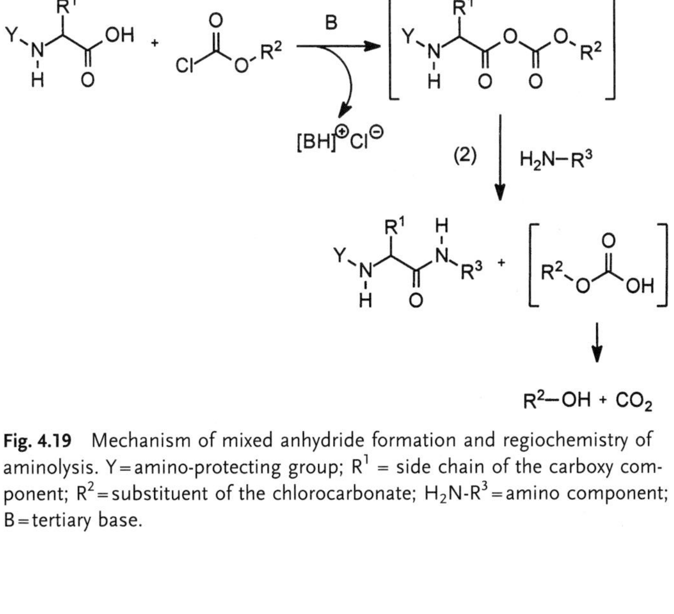

Fig. 4.19 Mechanism of mixed anhydride formation and regiochemistry of aminolysis. Y = amino-protecting group; R^1 = side chain of the carboxy component; R^2 = substituent of the chlorocarbonate; $H_2N\text{-}R^3$ = amino component; B = tertiary base.

tacks the carboxy group of the amino acid component with formation of the desired peptide derivative and release of the activating component as free acid (Fig. 4.19, reaction 2). On application of an alkyl chloroformate (R^1=isobutyl, ethyl, etc.), the free monoalkyl carbonic acid formed is unstable and immediately decomposes into carbon dioxide and the corresponding alcohol. However, there are some reports of the opposite regiochemistry (Fig. 4.19, reaction 1) of the nucleophilic attack, with formation of a urethane and regeneration of the N-protected amino acid component.

In order to form a mixed anhydride, N-protected amino acids or peptides are dissolved in dichloromethane, tetrahydrofuran, dioxan, acetonitrile, ethyl acetate, or DMF, respectively, and treated with one equivalent of a tertiary base (*N*-methylpiperidine, *N*-methylmorpholine, *N*-ethylmorpholine, etc.). The required alkyl chloroformate is then added at temperatures between –15 and –5 °C under vigorous stirring to form the unsymmetrical anhydride (activation). After a short activation period, the nucleophilic amino acid component is added. If it is used as an ammonium salt (which requires more base), then overstoichiometric application of base must be avoided. The mixed anhydride method can be performed very easily and is one of the most powerful coupling methods, provided that the reaction protocol described above is followed strictly.

Profound studies on the stability of mixed anhydrides, on the minimization of the undesired urethane formation and racemization which were carried out by Benoiton and coworkers eventually led to an improved mechanistic understanding and to an increase in efficiency of this coupling method, which now finds widespread application. An investigation into the reasons for excess urethane formation (Fig. 4.19, reaction 1), especially after the activation of isoleucyl or valinyl residues, led to the conclusion that this prominent side reaction can be prevented by using dichloromethane as the solvent and *N*-methylpiperidine as the tertiary base. The relatively high stability of mixed anhydrides towards hydrolysis allows for their purification by washing the organic phase containing the mixed anhydride with water [228–230]. The stability of mixed anhydrides obtained with alkyl chloroformates also depends on the alkyl group present. Mixed anhydrides from Boc-, Z-, and Fmoc-protected amino acids and isopropyl chloroformate can be purified and isolated. These are much more stable compared to the derivatives obtained with ethyl or isobutyl chloroformate [231, 232]. In the absence of a suitable nucleophile, the decomposition of a mixed anhydride in organic solvents starts with cyclization to produce 2-alkoxy-5(4*H*)-oxazolones, with liberation of carbon dioxide and the alcohol R^2-OH [233, 234]. Symmetrical anhydrides and esters are formed as by-products.

Several interesting aspects with respect to the practical realization of a mixed anhydride coupling should be noted: Although aqueous DMF is a good solvent both for the formation of the mixed anhydride and for the subsequent coupling reaction, it promotes racemization to a much greater extent than either tetrahydrofuran or halogenated solvents, as has been shown in the reaction of Z-Gly-Xaa-OH (Xaa=Ala, Leu, Val, Phe) with H-Val-OEt [231]. Isopropyl chloroformates are superior to ethyl or isobutyl chloroformates. Interestingly, activation with ethyl chloroformate in DMF or *N*-methylpyrrolidone leads to less racemization than activation with isobutyl chloroformate. Nevertheless, mixed anhydride formation with ethyl chloroformate, and also the application of triethylamine as a tertiary base, currently are of very little practical importance. Side reactions during mixed anhydride couplings have been observed initially upon activation of N^{α}-tosyl-, N^{α}-trityl, and N^{α}-trifluoroacetyl-protected amino acids, respectively.

Mixed anhydride formation with pivalic acid (2,2-dimethylpropionic acid) as the activation reagent is sometimes recommended, especially for N^{α}-acyl-protected asparagine. These unsymmetrical anhydrides are prepared analogously from N^{α}-acyl amino acids and pivaloyl chloride, and react in high yields with amino nucleophiles. The strong +I effect of the *tert*-butyl group in pivalic acid reduces the electrophilicity at this carbonyl group and leads, together with the steric hindrance, to the desired regioselective attack of the nucleophile on the activated amino acid.

For mechanistic reasons, the application of acylphosphonium salts as reactive intermediates in peptide couplings must be mentioned in this context (cf. Section 4.3.6) [235, 236].

4.3.2.2 **Symmetrical Anhydrides**

Symmetrical anhydrides of N^{α}-acyl amino acids **66** are highly reactive intermediates for peptide bond formation. Ambiguous regioselectivity upon reaction with amine nucleophiles is not possible, in contrast to the mixed anhydride method. However, a maximum peptide yield of 50% (with reference to the carboxy component) can be obtained.

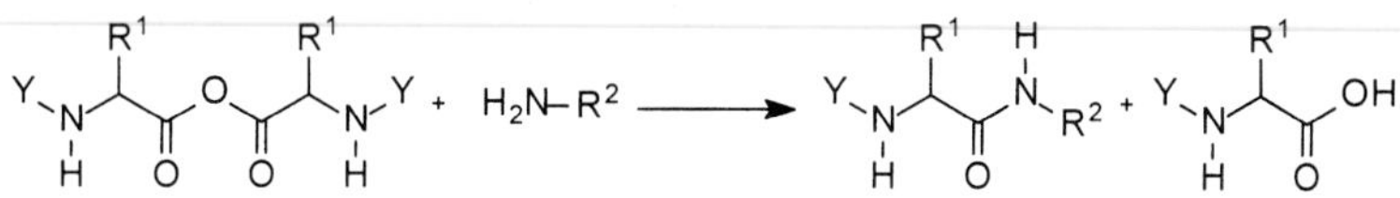

66

Although the free N^{α}-acyl amino acid formed on aminolysis of a symmetrical anhydride together with the desired peptide can be recovered by extraction with saturated sodium bicarbonate solution, the practical importance of this method initially was very small. Symmetrical anhydrides can be obtained from N^{α}-protected amino acids with phosgene or, more conveniently, with carbodiimides. The reaction of two equivalents of N^{α}-acyl amino acid with one equivalent of carbodiimide favors formation of the corresponding symmetrical anhydride that may be isolated, but can also be used in the subsequent coupling reaction without further purification. Symmetrical anhydrides based on N^{α}-alkoxycarbonyl amino acids are stable towards hydrolysis. This permits purification analogously, as described above for the mixed anhydrides.

The general interest in symmetrical anhydrides for stepwise peptide chain elongation increased as Boc-protected amino acid derivatives became commercially available at reasonable prices. Although crystalline symmetrical anhydrides can be obtained commercially, in-situ preparation remains the variant of choice.

4.3.2.3 **N-Carboxy Anhydrides**

Simultaneous activation of the carboxy group and acyl protection of an amino acid is found in the N-carboxy anhydrides (NCA), which were discovered in 1906 by Hermann Leuchs. In the German literature they are, therefore, also often called Leuchs-anhydrides. In principle, these derivatives should possess ideal preconditions for application in peptide synthesis.

The first N-carboxy anhydrides (1,3-oxazolidine-2,5-diones) **67** were obtained by thermal elimination of chloroethane from *N*-(ethoxycarbonyl) amino acid chloride.

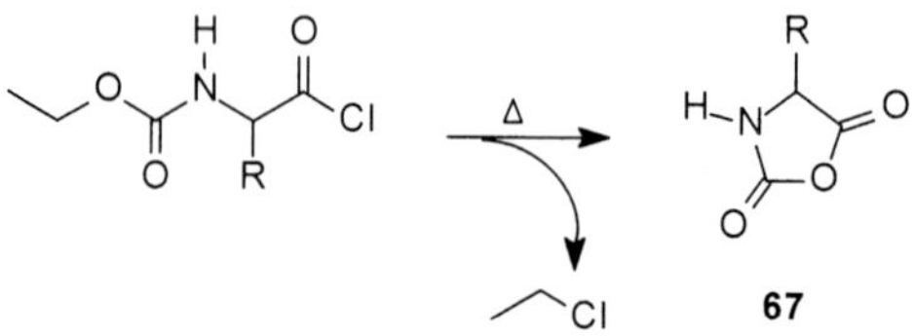

67

An elegant procedure for the preparation of these derivatives is the reaction of free amino acids with phosgene via the corresponding carbamoyl chloride as an intermediate. However, even traces of water polymerize N-carboxy anhydrides, because the carbamic acid initially formed spontaneously decarboxylates to give a free amine that acts as a nucleophile in further ring-opening reactions. Therefore, the practical application of the NCA method in peptide synthesis was seriously limited until 1966, when correct reaction conditions were published which allowed for controlled peptide synthesis with N-carboxy anhydrides in an aqueous medium [237]. As shown in Fig. 4.20, amino acids and peptides are rapidly acylated by N-carboxy anhydrides at low temperature and at pH 10.2. The intermediate peptide carbamates are decarboxylated by a decrease in the pH to values between 3 and 5. The next coupling cycle starts with an increase of the pH to 10.2, and proceeds by addition of the next N-carboxy anhydride. Vigorous stirring of the reaction mixture is necessary in order to minimize the risk of undesired carboxylate exchange between the intermediary peptide carbamate and the amino component. Exact control of the pH value (pH 10.2–10.5 for amino acids, and pH 10.2 for peptides) is another precondition, as hydantoins are formed as by-products at $pH > 10.5$.

The sulfur analogues of the N-carboxy anhydrides, N-thiocarboxy anhydrides (NTA), can be favorably applied in peptide syntheses, because of the higher stability of the thiocarbamate salts. The acylations can be performed at pH values as low as 9–9.5, which prevents the possible hydrolytic conversion to the hydantoins. The NCA/NTA method is especially suited for segment syntheses without isolation of reaction intermediates. Trifunctional amino acids (except lysine and cysteine) do not require side-chain protection. This method has been used to assemble several segments of the ribonuclease S-protein that were then combined, using the azide method to produce the full-length S-protein [238].

Recently, the NCA methodology has again attracted significant interest, as urethane-protected N-carboxy anhydrides (UNCA) have been synthesized and applied to peptide syntheses [239]. NCA can be acylated at the ring nitrogen with suitable

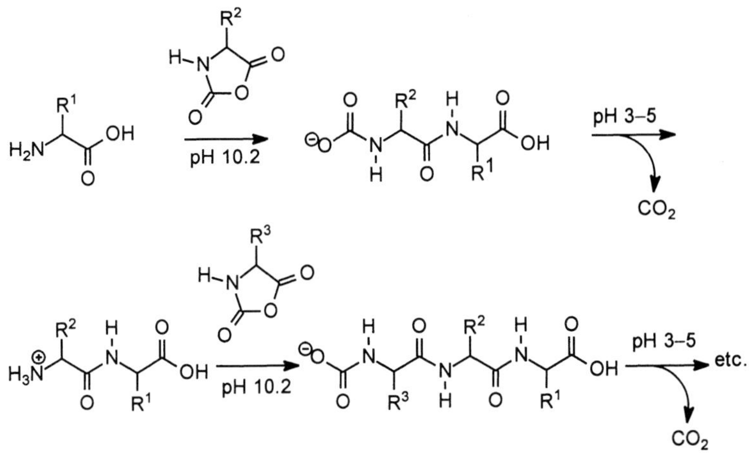

Fig. 4.20 Controlled peptide synthesis with N-carboxy anhydrides in aqueous medium.

reagents for the introduction of a Boc, Z, or Fmoc group in an aprotic solvent in the presence of a tertiary base to give the corresponding UNCA **68** (Y=Boc, Z, Fmoc). UNCA can be obtained for most amino acids as crystalline compounds, and are stable in the absence of water. They display high reactivity towards nucleophiles, and form the required peptide bonds at high rates in most anhydrous solvents (with the exception of alcohols) normally used in peptide synthesis, both in solution or on a polymeric support. Carbon dioxide is the only by-product, and there is no risk of either oligo- or polymerization, because the amino terminus of the growing peptide chain is still urethane-protected after the coupling reaction.

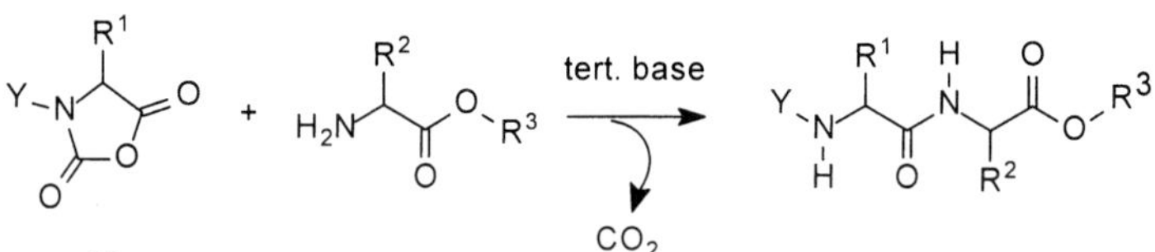

68

Likewise, N-trityl- and N-phenylfluorenyl-protected NCA have been recently reported [240].

4.3.3
Carbodiimides

Carbodiimides can be used for the condensation of a carboxy group and an amino group [241, 242]. Initially *N,N'*-dicyclohexyl carbodiimide (DCC) was used and is now well established, because it is relatively cheap and soluble in the solvents used in peptides synthesis. During peptide bond formation the carbodiimide is converted into the corresponding urea derivative, which in the case of *N,N'*-dicyclohexyl urea precipitates from the reaction solution. Clearly, the different reaction rates of aminolysis and hydrolysis of the reactive intermediate after carbodiimide activation enable peptide couplings in aqueous media. The reaction mechanism for carbodiimide-mediated peptide couplings has been established in extensive studies performed by several groups, and is shown in Fig. 4.21. The carboxylate anion adds to the protonated carbodiimide with formation of a highly reactive *O*-acylisourea **69**; this has not yet been isolated, but its existence is postulated by close analogy to stable compounds of this class. The *O*-acylisourea reacts with the amino component to produce the protected peptide and the urea derivative (Fig. 4.21, reaction A). Alternatively, the *O*-acylisourea **69** – in equilibrium with the N-protonated form **70** – is attacked by a second carboxylate as the nucleophile forming the symmetrical amino acid anhydride **72** together with N,N′-disubstituted urea (Fig. 4.21, reaction C). The former reacts with the amino acid component to give the peptide derivative and the free amino acid (Fig. 4.21, reaction D). An undesired side reaction is the base-catalyzed acyl migration from the isourea oxygen to nitrogen; this results in an N-acylurea **71** that does not undergo further aminolysis (Fig. 4.21, reaction B). This O → N acyl migration is catalyzed not only by an excess of base, but also by the basic amino component or the carbodiimide. Polar solvents additionally favor this reaction pathway.

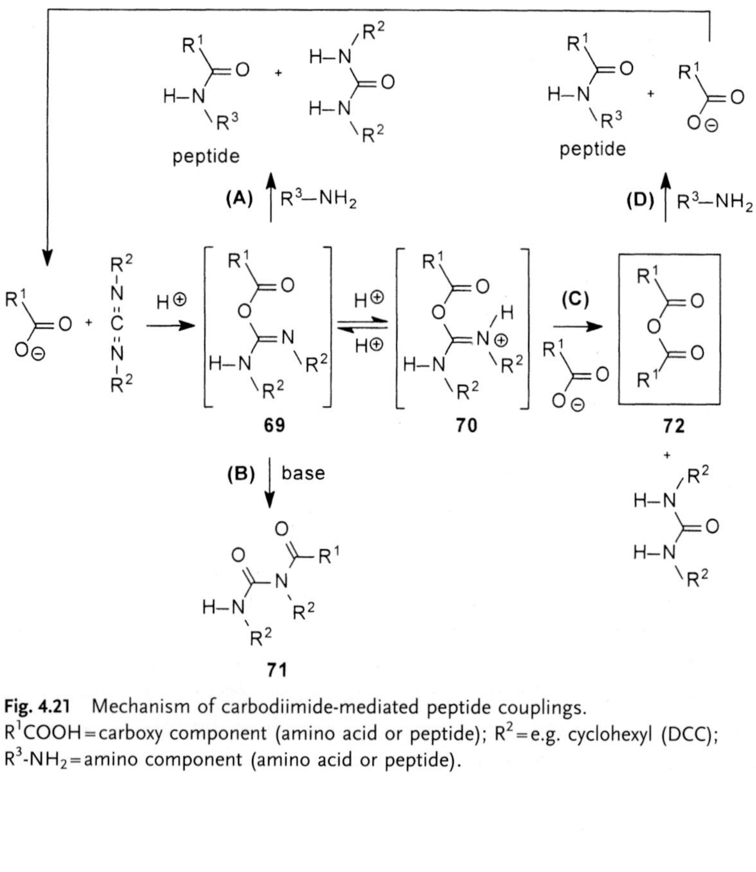

Fig. 4.21 Mechanism of carbodiimide-mediated peptide couplings. R^1COOH = carboxy component (amino acid or peptide); R^2 = e.g. cyclohexyl (DCC); R^3-NH_2 = amino component (amino acid or peptide).

Furthermore, the *O*-acylisourea may undergo cyclization to give the 5(4*H*)-oxazolone **74**, again with liberation of the *N,N'*-dialkylurea as a good leaving group. This intramolecular side reaction explains the high propensity of the carbodiimide method towards racemization, when applied to N-acyl-protected amino acids and peptides. 5(4*H*)-Oxazolones are themselves acylating agents, but are prone to racemization (cf. Section 4.4.2).

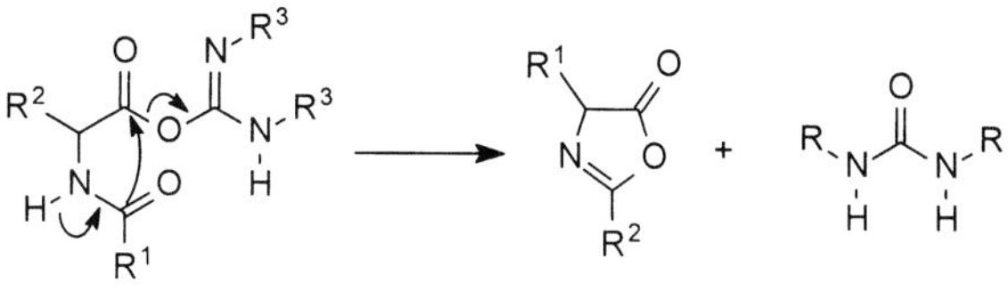

73 **74**

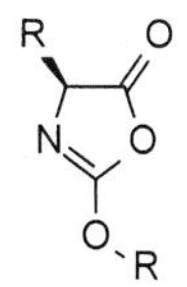

75

Derivatives with N-protecting groups of the urethane type are the only exception. Contrary to earlier speculations, 2-alkoxy-substituted 5(4*H*)-oxazolones **75** are real intermediates that are formed during this reaction from the *O*-acylisourea, but are converted immediately with free carboxy component to give the symmetrical anhydride **72** [243]. All three postulated intermediates are attacked by nucleophiles such as amino groups, with the reactivity decreasing in the series *O*-acyl isourea > 2-alkoxy-5(4*H*)-oxazolone > symmetrical anhydride. Which of the three activated carboxy derivatives is involved in the peptide bond-forming step depends on the presence, availability, and reactivity of the nucleophilic amino component. In cases where no amino component is present (e.g., in ester syntheses) or where the activated carboxy component is added in excess (in SPPS), the symmetrical anhydride is formed predominantly.

The oxazolone-forming cyclization is favored especially upon application of water-soluble 1-ethyl-3-(3-dimethylaminopropyl)-carbodiimide hydrochloride (EDC, WSCI). Aminolysis of β-methyl-substituted amino acid derivatives of type **69** is not favored. Consequently, cyclization and rearrangements may compete efficiently with aminolysis after activation of these derivatives. Different reactivities of 2-alkoxy-5(4*H*)-oxazolones, depending on the amino acid side chain R^1 have been observed, especially with respect to hydrolysis.

The DCC method is a reliable variant for the stepwise assembly of peptides using urethane-protected amino acids (Z, Boc, Fmoc) in both solid-phase and solution-phase peptide synthesis, as well as for segment condensations. However, only peptide segments containing C-terminal glycine or proline residues should be used as acyl components, because of the risk of epimerization at the C-terminal position. The N-acyl urea formation mentioned above may be a prominent side reaction that can be suppressed by keeping the reaction at low temperatures, or by using nonpolar solvents in which the dicyclohexyl urea should, nevertheless, dissolve. Free dimethylamine, which is present in low concentrations even in highly pure DMF or N,N-dimethylacetamide also favors the O → N acyl migration to produce N-acyl urea. When activated with DCC, unprotected glutamine or asparagine side chains may undergo dehydration of the carboxamide group to yield the nitrile. Therefore, a global (maximum) protection strategy for the side-chain functionalities should be applied when DCC couplings are involved.

Besides N-acyl urea formation, a serious problem in peptide synthesis when using the carbodiimide method – and especially in the synthesis of difficult sequences – is imposed by incomplete separation of the *N,N'*-dialkyl urea, and especially of *N,N'*-dicyclohexyl urea. Several carbodiimide derivatives which have been substituted with different tertiary amino or quaternary ammonium groups have been introduced. The ureas formed during peptide synthesis with these modified reagents are easily separated from the reaction mixture because of their increased solubility in aqueous or acidic aqueous solutions. Reagents of this type, for example EDC (WSCI) or 1-cyclohexyl-3-(3-trimethylammoniopropyl) carbodiimide iodide, are appropriate for peptide couplings even in alcohol [244] (or aqueous solution as required) for coupling reactions involving proteins or peptide cyclizations. In addition, polymeric carbodiimides have been developed which allowing for straightforward separation of the urea formed during the reaction. Unsymmetrically substituted carbodiimides have been reported to be suitable for suppression of the O → N acyl migration forming the undesired N-acyl urea, as was shown for 1-benzyl-3-ethylcarbodiimide [245]. Simultaneously, a lower tendency towards racemization has been observed.

The application of appropriate additives was investigated in a systematic manner in order to suppress or diminish both N-acyl urea formation and racemization. Additives intercept highly reactive intermediates, such as *O*-acylisourea, thereby forming an active ester which has lower reactivity but is still sufficiently potent to allow for rapid amide bond formation. The DCC-additive protocol was introduced into peptide chemistry in 1966 during the course of the syntheses of glucagon and calcitonin [246, 247], whereby one equivalent of DCC and two equivalents of *N*-hydroxysuccinimide (HOSu) were applied simultaneously (Wünsch-Weygand protocol). Subsequently, the search for further possible additives continued because of the internally competing reactions between DCC and HOSu [248] that were observed during the coupling of sterically hindered peptides. The side reaction mentioned does not occur upon addition of *N*-hydroxy-5-norbornene-2,3-dicarboxamide (HONdc), which proved to be advantageous in challenging syntheses.

Doubtlessly, the König-Geiger protocol belongs to the DCC-based coupling methods that have found the most widespread application [249]. In this variant, 1-hydroxybenzotriazole (HOBt, Tab. 4.6) is mainly preferred as the additive. 3,4-Dihydro-3-hydroxy-4-oxo-1,2,3-benzotriazine (HODhbt) is more efficient, but is also prone to side reactions [250]. Like all additives of this type, 1-hydroxybenzotriazole (HOBt) reacts rapidly with the *O*-acylisourea initially formed to produce the corresponding active ester derivative. Benzotriazolyl esters (Fig. 4.22) undergo aminolysis up to three orders of magnitude faster than succinimidyl or related esters.

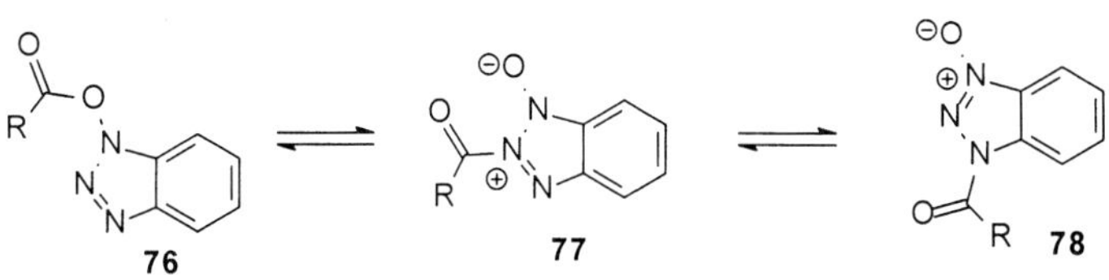

Fig. 4.22 Amino acid benzotriazolyl esters.

Thus, rearrangements, cyclization, and the formation of symmetrical anhydrides are efficiently suppressed. Dehydration of asparagine or glutamine residues is also not observed. Likewise, 2-alkoxy-5(4*H*)-oxazolones react immediately with HOBt or HOSu, respectively, to produce the active esters that undergo peptide coupling with amino components present in the reaction mixture [251].

The N^{α}-protected amino acid HOBt esters **76** are thought to exist in solution in an equilibrium with the N-oxides **77** and **78** [249]. In the crystalline state, structure **78** has been observed using X-ray analysis [252].

Many other N-hydroxy compounds besides HOSu and HOBt have been suggested as potential additives for the DCC method, for example ethyl 1-hydroxy-1H-1,2,3-triazole-4-carboxylate (HOCt) [253] or other hydroxy-substituted triazole and tetrazole derivatives [254]. Interestingly, zinc halogenides ($ZnCl_2$, ZnF_2 [255]) have also been applied successfully as additives. $CuCl_2$ was found to be even more superior in terms of suppressing racemization, and led (especially when applied in combination with HOBt) to a virtually complete retention of the configuration, as was shown in several highly sensitive test systems [256]. Liquid crystal-forming compounds have also been reported as being efficient additives [257–259]. The most efficient additive is currently the 7-aza analogue of HOBt, commonly referred to as 1-hydroxy-7-azabenzotriazole [260–263]. This combines the features of HOBt and a tertiary base in the same molecule, and is reported to accelerate reaction rates significantly as well as provide concomitant optimal coupling yields and minimal racemization [262]. Nowadays, phosphonium and uronium salts derived from this compound (cf. Sections 4.3.6 and 4.3.7) are also available as excellent coupling reagents.

4.3.4
Active Esters

Peptide bond formation by ester aminolysis proceeds analogously to ester saponification (X=O, S, Se) according to a $B_{Ac}2$ mechanism. Nucleophilic attack of the amino component **79** to the carboxy group of **80** leads, in the rate-determining step of this reaction, to the tetrahedral intermediate **81**, the formation of which is favored by electron-withdrawing groups XR^3. The rate of peptide bond formation correlates with the leaving group capacity of XR^3. For example, a good leaving group is the conjugate base of a relatively strong acid, and this is the case when the leaving group anion $^{-}XR^3$ is well-stabilized by inductive or mesomeric effects which facilitate cleavage of the (C=O)–XR^3 bond.

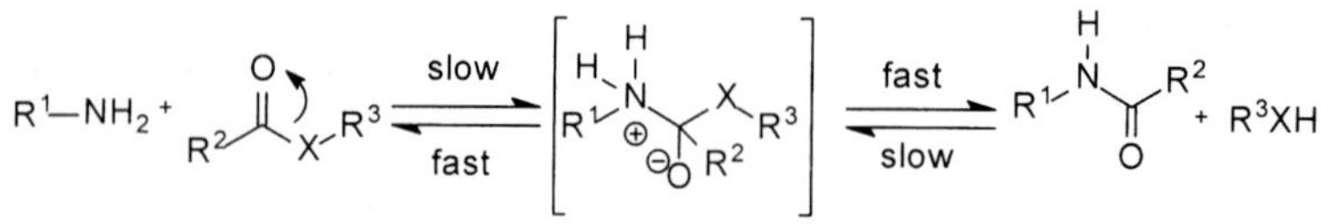

79 80 81

The initial application of thiophenyl-, cyanomethyl-, 4-nitrophenyl-, and pentafluorophenyl esters triggered further development of a broad variety of active esters (Tab. 4.6) [264].

A new type of active ester was suggested, the aminolysis reaction of which was characterized by a slightly different mechanistic pathway compared to the $B_{Ac}2$ mechanism of substituted alkyl esters, unsubstituted or substituted aryl esters, or their thio and seleno analogues. These N-hydroxypiperidinyl esters described by Young et al. [265] and 8-quinolyl esters described by Jakubke and Voigt [266] featured an additional proton-accepting group stabilizing a hydrogen-bonded transition state during aminolysis. This led to high aminolytic activity and extensively decreased the racemization-prone oxazolone formation. This general principle is outlined in Fig. 4.23 for the aminolysis of 8-quinolyl esters as an example. Comparison of the pK_a values of the isomeric 3-hydroxyquinoline (8.06), 6-hydroxyquinoline (8.88), and 8-hydroxyquinoline (9.89) leads to the conclusion that the leaving group ability of the corresponding bases should decrease in the same order. The opposite holds true; 8-quinolyl esters display significantly higher aminolysis reactivity because nucleophilic attack of the amino group to the carboxy group of the ester is favored by the anchimeric assistance of the quinoline nitrogen. This nitrogen atom acts as a basic catalyst and "delivers" the attacking amino group to the reactive center. As the reaction proceeds, the tetrahedral intermediate is further stabilized by an intramolecular hydrogen bond. Racemization of 8-quinolyl esters **82** is disfavored, because the amide oxygen in **83** cannot intramolecularly attack the weakly electrophilic ester group to give oxazolone **84**. Hence, efficient discrimination between aminolysis and oxazolone-based racemization (cf. Section 4.4.2) is possible.

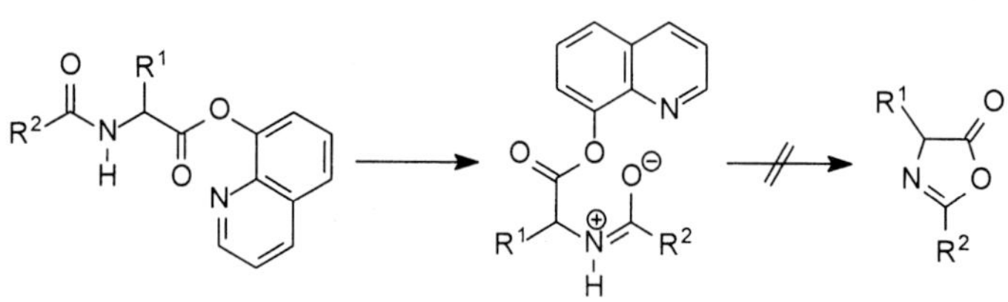

82 83 84

The same situation applies to other active esters of high practical importance, such as derivatives of *N*-hydroxysuccinimide (HOSu) **85** [267], 1-hydroxybenzotriazole (HOBt) esters **86** [249], and *N*-hydroxy-5-norbornene-2,3-dicarboximide (HONdc) [268]. The 7-aza analogue of HOBt, commonly referred to as 1-hydroxy-7-azabenzotriazole **87** (HOAt; the correct IUPAC name is 1-hydroxy-1,2,3-triazo-

Tab. 4.6 Active esters of the type X-(C=O)-R.

Group	*Symbol*	*X*
Thiophenyl	SPh	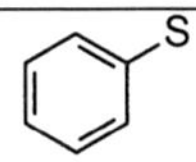
Cyanomethyl	OCm	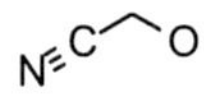
Nitrophenyl		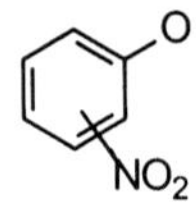
2-Nitrophenyl	O2Np	
4-Nitrophenyl	ONp	
2,4-Dinitrophenyl	O2,4Np	
Chlorphenyl		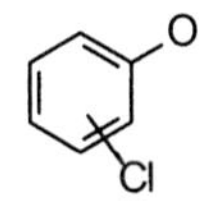
2,4,5-Trichlorophenyl	OTcp	
Pentachlorophenyl	OPcp	
Pentafluorophenyl	OPfp	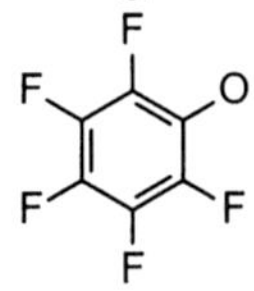
N-Hydroxypiperidinyl	OPip	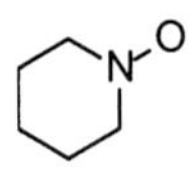
8-Quinolyl	OQ	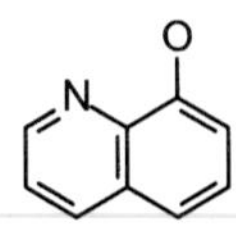
N-Hydroxysuccinimidyl	OSu	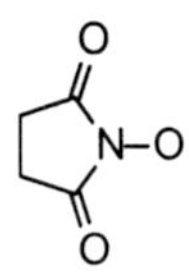
1-Hydroxybenzotriazolyl	OBt	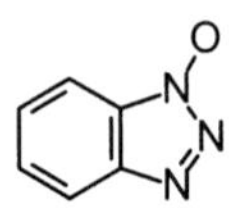
7-Aza-1-hydroxybenzotriazolyl	OAt	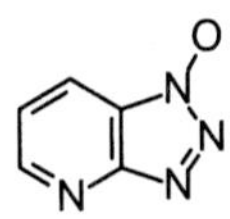
N-Norbornene-2,3-dicarboximidooxy	ONdc	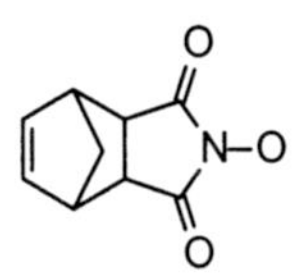
Ethyl-1-hydroxy-1H-1,2,3-triazole-4-carboxylate	OCt	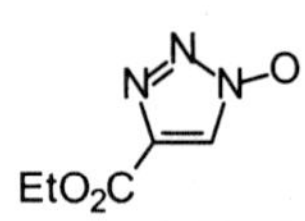

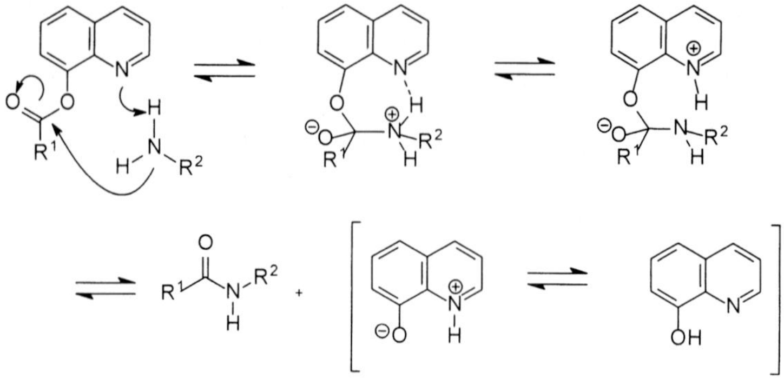

Fig. 4.23 Intramolecular base catalysis during aminolysis of acylamino 8-quinolyl esters. R^1COOH = carboxy component (protected amino acid or peptide); R^2NH_2 = amino component (amino acid or peptide).

lo[5,4-b]pyridine), has been recommended as a highly efficient additive for DCC coupling reactions [260]. As mentioned above for the hydroxyquinoline derivatives, 7-aza-1-hydroxybenzotriazole is most efficient [269]. Its active esters are characterized by very little racemization, this being explained by mechanistic considerations analogously as described for 8-quinolyl esters. HOAt as an additive is superior to HOBt for stepwise peptide couplings and segment condensations, especially with respect to reaction rate and the degree of racemization.

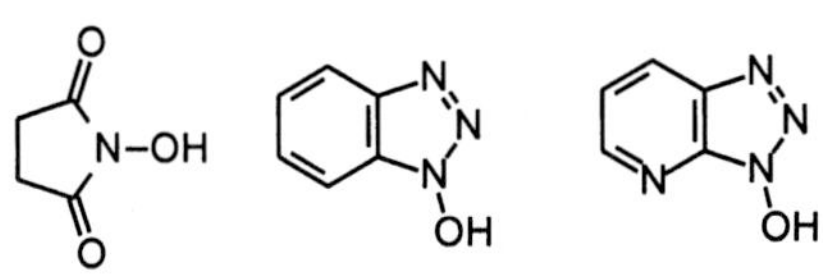

HOSu **85** HOBt **86** HOAt **87**

Active esters of protected amino acids can be synthesized by the mixed anhydride method, the carbodiimide method, the carbonate method, and other more specialized variants. Protected amino acid active esters can also be obtained starting from amino acid salts in a "one-pot" reaction, provided that the carbonate component has been carefully chosen. Treatment of amino acid salts with benzyl (4-nitrophenyl)carbonate or *tert*-butyl (4-nitrophenyl)carbonate, respectively, gives the corresponding Z- or Boc-protected derivatives that, upon acidification and further addition of a carbodiimide, yield the amino-protected (4-nitrophenyl)esters. These are crystalline compounds that may be stored in the dark at room temperature for longer periods of time and which display high reactivity in aminolysis reactions. Aminolysis may even be further enhanced after the addition of catalysts such as azoles or N-hydroxy compounds. Separation of the 4-nitrophenol formed during aminolysis may sometimes be tedious, though precipitation with DMF/water or DMF/ether, adsorption on neutral alumina, or complexation with pyridine at pH 6.5 may be successful. Incomplete separation often disturbs subsequent reactions such as hydrogenolysis.

Likewise, the crystalline *N*-hydroxysuccinimidyl esters are characterized by high reactivity with amines and low sensitivity towards hydrolysis in aqueous/organic solvent mixtures (water/ethanol, water/dioxan, water/tetrahydrofuran, etc.) that makes peptide syntheses feasible in such media. The water-solubility of the *N*-hydroxysuccinimide formed upon aminolysis is a further advantage of this type of active ester.

Halogen-substituted phenyl esters usually are crystalline compounds with high aminolysis activity. Separation of the halogenated phenols liberated during reaction of these active esters with amines often is less problematic compared to the corresponding 4-nitrophenols. The "backing off" protocol is a suitable method for the synthesis of optically pure N-protected activated peptidyl esters. Moreover, a halogenated phenol (X=Cl, F) may be reacted with dicyclohexyl carbodiimide to give a crystalline *O*-arylisourea/phenol complex. For instance, the *O*-pentafluorophenyl-*N,N'*-dicyclohexylisourea/pentafluorophenol complex **88** possesses high acylation potential and may be used directly for segment condensations.

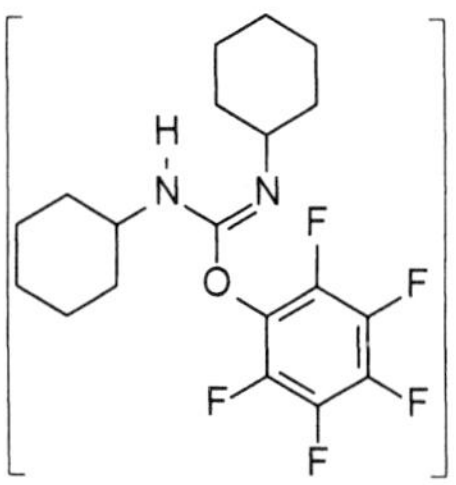

88

For mechanistic reasons, the isoxazolium method [270] (Fig. 4.24) can be discussed in the context of this chapter, because the *N*-ethyl-5-phenylisoxazolium-3′-

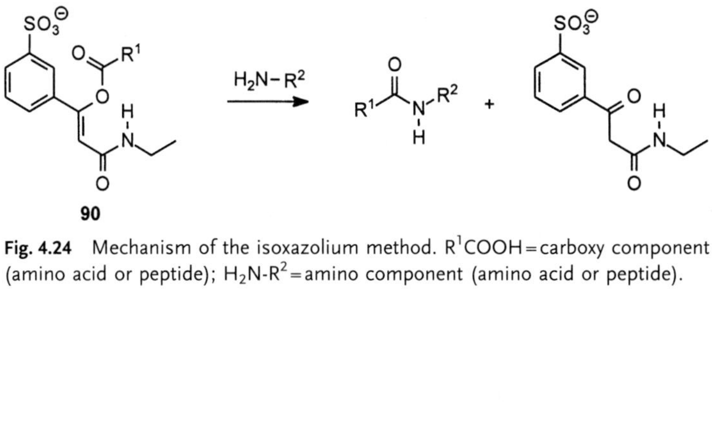

Fig. 4.24 Mechanism of the isoxazolium method. R^1COOH=carboxy component (amino acid or peptide); H_2N-R^2=amino component (amino acid or peptide).

sulfonate **89** (Woodward reagent K) forms an N-protected enol ester **90** with N-protected amino acids and peptides under mildly basic conditions. This enol ester couples *in situ* with amino components and forms, besides the desired product, a water-soluble aryl sulfonate that is easy to separate.

4.3.5
Acyl Halides [271]

By analogy to organic chemistry, an obvious method of activating a carboxy group for amide formation at room temperature or below would utilize the acid chloride. However, amino acid chlorides have until recently [62, 272] rarely been used as they are reputed to be "overactivated" and prone to numerous side reactions, including racemization.

Despite this, because the Fmoc group is highly stable under acidic conditions and does not readily undergo S_N1 or S_N2 displacement reactions at the 9-fluorenylmethyl residue, the Fmoc-protected amino acids have emerged as ideal substrates for conversion into the corresponding acid chlorides. Consequently, these derivatives have experienced a renaissance in peptide synthesis. All common amino acids lacking polar side chains, as well as a number of those bearing benzyl- or allyl-based side-chain protection, can be converted to stable, crystalline acid chlorides. For chemical reasons, amino acids bearing *tert*-butyl-protected side chains cannot generally be accommodated. A hindered base such as 2,6-di-*tert*-butylpyridine is used routinely as hydrogen chloride acceptor, as conversion to oxazolone is slow with such bases. Ammonium or potassium salts of *N*-hydroxybenzotriazole (HOBt) have also been applied as bases in solid-phase syntheses involving Fmoc amino acid chlorides.

Fmoc amino acid chlorides have been used as stable, easily accessible derivatives for rapid coupling reactions, without any racemization. Their general application is, however, somewhat limited as not all Fmoc-protected amino acid derivatives are accessible. This holds true especially for trifunctional amino acids with side chain-protecting groups of the *tert*-butyl type (Boc, *tert*-butyl ester, *tert*-butyl ether).

In contrast, the corresponding acid fluorides do not suffer from such limitations. Urethane-protected amino acid fluorides of trifunctional amino acids with a side-chain protection of the *tert*-butyl or trityl type can be obtained and are sufficiently stable.

Further advantages of the acid fluorides relative to the chlorides include their greater stability toward water, including moisture in the air, and their relative lack of conversion to the corresponding oxazolones on treatment with tertiary organic bases. Fmoc-protected amino acid fluorides were found to be efficient carboxy-activated derivatives suited both for solution syntheses and for SPPS [273–275]. The highly reactive Fmoc amino acid fluorides are especially recommended for SPPS of complicated longer peptides, and also mainly for the coupling of sterically hindered amino acid building blocks [276]. The advantage of amino acid fluorides in the coupling of sterically hindered amino acid components, compared to other coupling methods, is clearly based on an efficient stabilization of tetrahedral transition state by the highly polarized C–F dipole and the relatively small size of the fluoride leaving group. Acylations with Fmoc amino acid fluorides also occur in the absence of a tertiary base – a fact which is important with respect to minimal racemization [277]. Fmoc amino acid fluorides are usually stable crystalline derivatives and can be synthesized from the N-protected amino acid with 2-diethylamino sulfur trifluoride (DAST), cyanuric fluoride, or tetramethyl fluoroformamidinium hexafluorophosphate $[(Me_2N)_2CF]^+$ PF_6^- (TFFH). The latter reagent is a nonhygroscopic, stable salt that is also suitable for generating the acyl fluoride *in situ* on treatment with ethyl diisopropyl amine and is, therefore, also regarded as a coupling reagent for solution- or solid-phase syntheses [278].

4.3.6 Phosphonium Reagents

Coupling reagents allowing for the in-situ generation of active esters are of considerable importance in peptide chemistry. The HOBt-derived benzotriazol-1-yloxy-tris(dimethylamino)phosphonium hexafluorophosphate (BOP) **91** developed by Castro et al. [279] is a very efficient reagent which has also been applied successfully in SPPS. The main disadvantage of this reagent however is that the highly toxic and carcinogenic HMPA is formed during the course of the reaction, whilst an unacceptably high risk of racemization has also been observed [280]. The PyBOP coupling reagent **92** [281] and its HOAt analogue PyAOP [282], where the dimethylamino groups are replaced by pyrrolidine substituents, represent viable alternatives to BOP that do not produce the hazardous by-product. Sterically hindered amino acids often can be coupled successfully using bromo-tris(pyrrolidino)phosphonium hexafluorophosphate PyBroP **93** [283]. These reagents do not react with α-amino groups, and so may be added directly to a mixture of the amino and carboxy component to be coupled. A tertiary amine is usually added in order to form the anion of the carboxy component. Nucleophilic attack of the carboxylate leads to a highly reactive acylphosphonium species that immediately is transformed into the HOBt ester in the presence of HOBt.

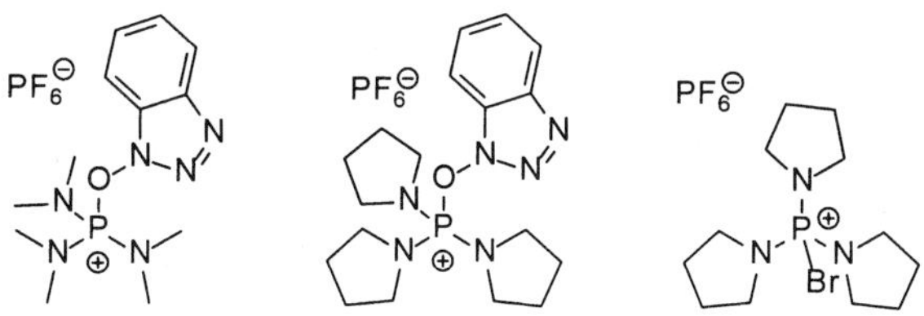

BOP **91** PyBOP **92** PyBroP **93**

Recently, the application of 3-(diethoxyphosphoryloxy)-1,2,3-benzotriazin-4(3H)-one (DEPBT) as a coupling reagent with superior performance has been published by the group of Murray Goodman [284].

4.3.7
Uronium Reagents

Several uronium reagents have found widespread application as coupling reagents in SPPS, for segment condensations, and other purposes of peptide and protein chemistry. Among them are TBTU (**94**, X=BF_4^-) and HBTU (**94**, X=PF_6^-) [285, 286], which are based on *N*-hydroxybenzotriazole HOBt **86** [249] as well as the 7-aza-benzotriazole HOAt **87** [260, 261] (Tab. 4.7). Considering the fact that HBTU and HATU are synthesized from HOBt or HOAt, respectively, they have been erroneously classified – by structural analogy to BOP – as uronium reagents. However, X-ray crystallographic studies revealed that they exist, at least in the solid state, as zwitterions **95** comprising a triazole-*N*-oxide and a guanidinium moiety. It is very likely that other uronium reagents display the same behavior. According to the IUPAC nomenclature, the forms of HATU must be named as 1-[Bis-(dimethylamino)methyliumyl]-1H-1,2,3-triazolo[4,5-b]pyridine-3-oxide hexafluorophosphate or as 1-[(dimethylamino)-(dimethyliminium)methyl]-1H-1,2,3-triazolo[4,5-b]pyridine-3-oxide hexafluorophosphate. It is plausible that the HBTU/TBTU forms **94** and **95** exist in an equilibrium in solution. The same applies to other derivatives.

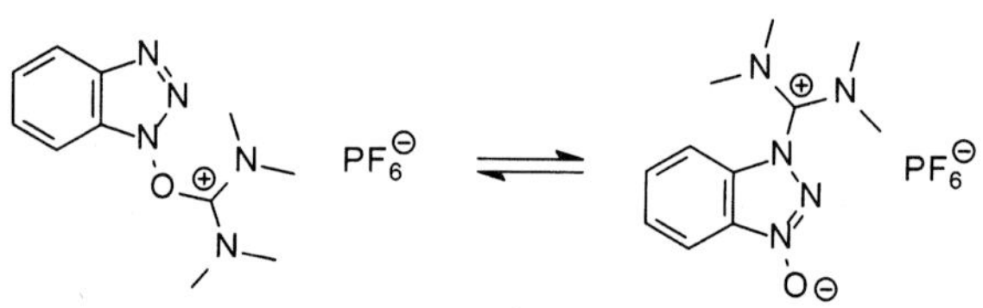

94 HBTU: $X^\ominus = PF_6^\ominus$ TBTU: $X^\ominus = BF_4^\ominus$ **95**

Mechanistically, the uronium reagents function in a similar way as the phosphonium reagents mentioned above. Nucleophilic attack of the carboxylate results in an *O*-acyluronium species that further reacts to give the HOBt (or HOAt) active ester. If a reaction of the guanidinium form of HOBt (or HOAt) is taken into consideration, it may be attacked by the carboxylate ion and then undergo a further acyl migration to give the HOBt/HOAt active ester.

Unlike carbodiimides or phosphonium salt reagents, they may undergo a reaction with free amino groups present. Consequently, the formation of tetramethylguanidinium derivatives of free amines (e.g., N-terminal amino groups) has been observed [287, 288]. The replacement of the dimethylamino moieties by pyrrolidino groups has also been suggested [289], and other active ester types, such as in PfPyU, have been considered [290].

All phosphonium and uronium salt reagents based on HOBt (or even better, on HOAt) serve as direct coupling reagents. A comparative study [288] and theoreti-

Tab. 4.7 HOAt-derived "uronium-type" coupling reagents.

Group	*Structure*
HATU	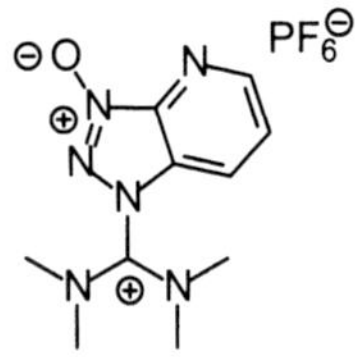
HAPyU	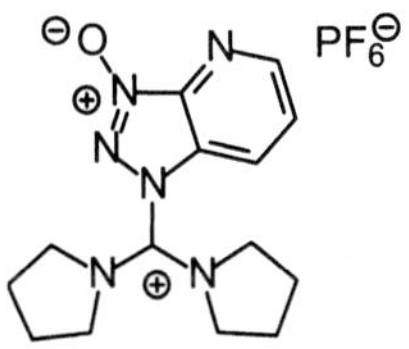
HAPipU	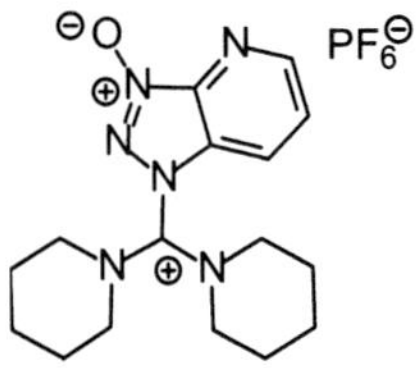
HAMDU	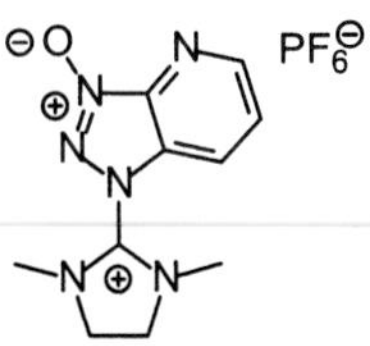
HAMTU	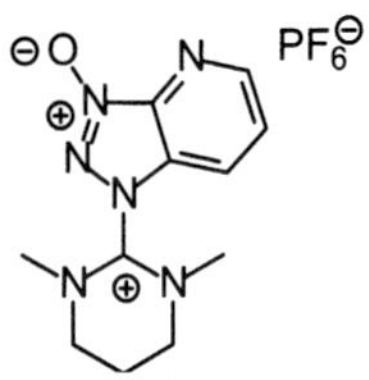

cal investigations [291] on the mechanism have been published. BOP **91**, HATU, HAPyU, and HAPipU (Tab. 4.7) are especially suited for peptide cyclizations [292]. 2-Trifluoroacetylthiopyridine-HOBt as a highly efficient reagent for segment couplings must also be mentioned in this context [293].

In addition, HOBt- or HOAt-derived iminium salts have been proposed [294, 295].

4.3.8
Further Special Methods

Comparison of biochemical peptide and protein synthesis with the chemical methods justifies the application of phosphorus compounds for peptide synthesis, as adenosine triphosphate serves as an activating reagent in amino acid and protein biosynthesis. The elimination of pyrophosphate leads to an enzyme-bound mixed anhydride involving the amino acid and the phosphate group of the nucleotide which are able to transfer the aminoacyl component to the accepting group of the transfer RNA (tRNA).

In the classical phosphorazo-method [296] and other related protocols, phosphorus compounds – in many cases mixed anhydrides of phosphate derivatives and the amino acid – are used for peptide bond formation.

Mechanistically, the Mukaiyama reaction is an oxidation-reduction condensation reaction where a disulfide reduction combined with a phosphine oxidation provides the driving force for the formal dehydratation, and also accepts the water hydrogen and oxygen atoms [297]. The combination of triphenylphosphine and 2,2′-dipyridyldisulfide was found to be highly favorable as the pyridine-2-thiol formed initially tautomerizes spontaneously into the stable pyridine-2-thione; hence the addition of a thiol capture reagent is unnecessary (Fig. 4.25). This method may also be applied in solid-phase synthesis.

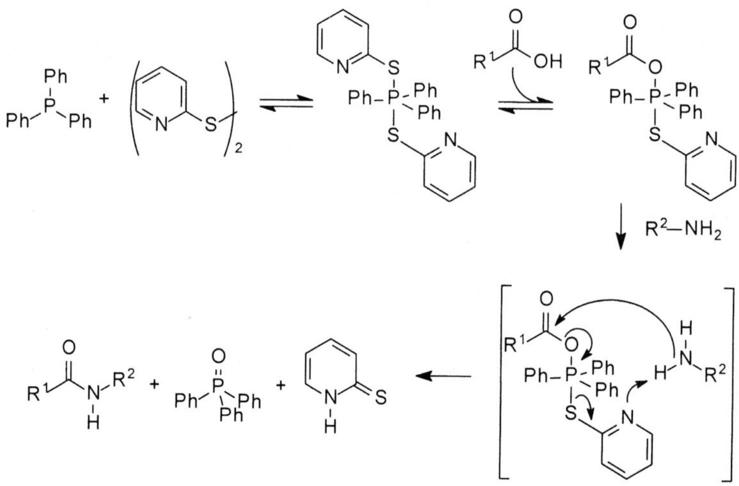

Fig. 4.25 Redox condensation according to Mukaiyama. R^1COOH = carboxy component (amino acid), R^2-NH_2 = amino component (amino acid or peptide).

4.4
Racemization During Synthesis

An inherent risk of racemization is imposed on all reactions involving functional groups directly connected to the stereogenic center of an α-amino acid. This may lead to a partial or total loss of the stereochemical information. While the expression "racemization" in organic chemistry is used for the complete conversion of a single enantiomer into the racemate, it is often used in peptide chemistry also for partial or total epimerization at one chiral center [298], irrespective of whether a mixture of diastereomers or enantiomers is formed in the course of the process. The optical purity and stereochemical integrity of a synthetic peptide consequently depends very much on the degree of partial epimerization during coupling steps or other types of reactions sensitive toward racemization. If only 1% epimerization occurs on each amino acid residue during peptide synthesis of a 10-peptide, the mixture of diastereomers contains only 90.4% of the peptide with the correct stereochemistry. In the case of a 50-peptide, only 60.5% of the desired diastereomer are obtained in theory. As the biological activity of a peptide or a protein depends strongly on the configuration of the chiral centers of the building blocks, racemization must be minimized during peptide synthesis [298, 299].

4.4.1
Direct Enolization

Although free amino acids are usually configurationally stable, the situation may be different when carboxy-activated derivatives **96** are involved. In an equilibrium, reversible carbanion formation **97** leading to a partial loss of the stereochemical integrity may be observed, because re-protonation occurs without facial selectivity.

The acidity of the proton H^{α} of an amino acid depends on the nature of the substituents. Only a base-catalyzed enolization mechanism (Fig. 4.26 A) plays a significant role under peptide coupling conditions. The rate of racemization depends on the electron-withdrawing effect of the groups R, Y, and X, on the solvent, temperature, and on the chemical properties of the base B used. Base-catalyzed racemization has been observed under different coupling conditions, during ester saponification, etc., while an acid-catalyzed enolization (Fig. 4.26B) results in the racemization of N-substituted N-methyl amino acids upon reaction with HBr/HOAc.

4.4.2
5(4*H*)-Oxazolone Mechanism

Racemization during coupling reactions is mainly caused during amino acid activation by the formation of stereochemically labile 5(4*H*)-oxazolones (2-oxazolin-5-ones, azlactones) D/L-**100** as reaction intermediates (Fig. 4.27).

The propensity toward 5(4*H*)-oxazolone formation strongly correlates with the activation potential of the activating group X in **98**, and with the electronic proper-

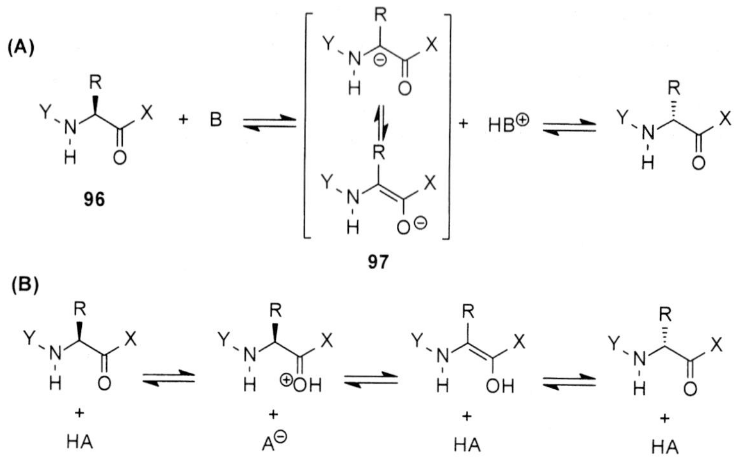

Fig. 4.26 Mechanisms of base-catalyzed (A) and acid-catalyzed epimerization (B) of activated amino acid and peptide derivatives. X=activating group; Y=N$^{\alpha}$-protecting group of peptide chain; B=base; HA=acid; R=amino acid side chain.

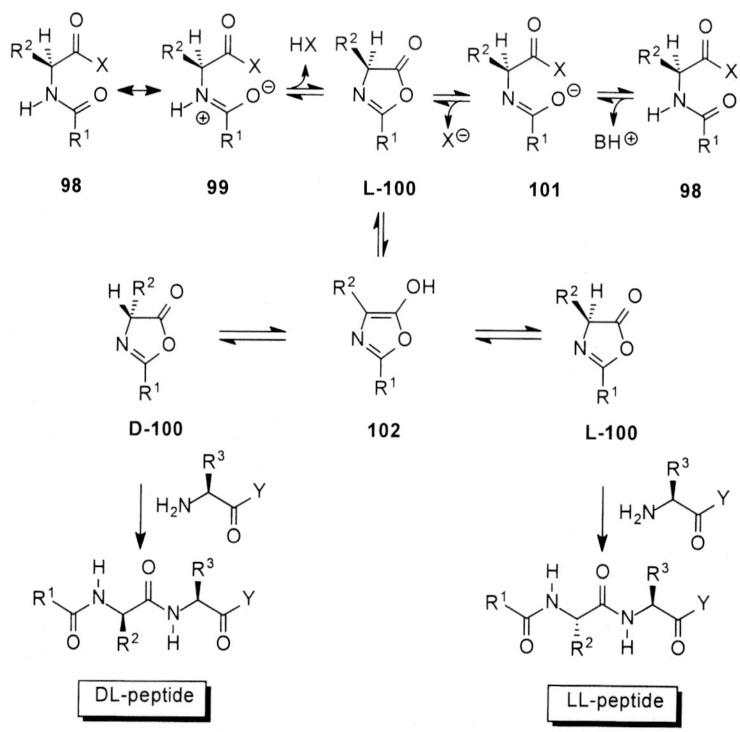

Fig. 4.27 Racemization via 5(4H)-oxazolone formation. R^1=alkyl, aryl or alkoxy residue; R^2, R^3=amino acid side chain; B=base; X=activating group.

ties of the N-acyl residue R^1-CO-. Acyl residues such as an acetyl, benzoyl, or trifluoroacetyl favor oxazolone formation due to the nucleophilic potential of the carbonyl oxygen. The mesomeric electron donation by the amino group enhances the nucleophilic attack of the oxygen atom to the activated carboxy group (Fig. 4.27). The oxazolone L-**100** formed by this process from an L-configured amino acid, is prone to epimerization in the tautomeric equilibrium between L-**100**, **102**, and D-**100**. An oxazolone of type **100** may easily racemize under coupling conditions in a base-catalyzed reaction via the aromatic intermediate **102**, giving rise to a mixture of both enantiomers L-**100** and D-**100**. In the past, oxazolone formation was not supposed to occur with N-alkoxycarbonyl-protected amino acid derivatives. If at all, activated N-alkoxycarbonyl amino acids **103** were believed to undergo cyclization giving N-carboxy anhydrides **104**, but not to suffer from 2-alkoxy-5(4*H*)-oxazolone (**105**) formation. The synthetic tactics of peptide chain assembly starting from the C-terminus using urethane-type protecting groups, which is regarded to be devoid of racemization, relies on this assumption. However, it was shown in 1977 that 2-alkoxy-5(4*H*)-oxazolones may in fact be formed upon activation of the carboxy group [233, 234, 251].

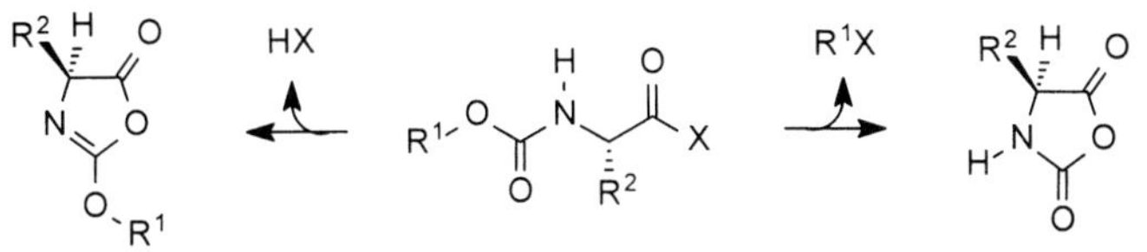

105 **103** **104**

These derivatives seem to be much less prone to racemization than the 2-alkyl or 2-aryl counterparts, and also undergo aminolysis much more rapidly [243, 299]. Although the reasons for that behavior have not yet been fully clarified, the decreased acidity of a urethane-protected NH group (R-O-CO-NH) compared to an amide group (R-CO-NH) seems to be a crucial factor. As depicted in Fig. 4.27, the base abstracts the amide proton in a rapid reaction step. The rate-determining ring closure is initiated by nucleophilic attack of the oxygen atom of the anionic amide group to the activated carboxy group (**101**). Hence, a decrease in the NH acidity – as is the case in the urethane-protected derivatives – lowers the rate of oxazolone formation. Preactivation of amino acids with coupling reagents is sometimes accompanied by racemization, and hence should be avoided when, for example, racemization-prone building blocks are to be incorporated.

Despite their tendency to racemize, the oxazolones still represent carboxy-activated amino acid derivatives, and so are incorporated into the peptide. The degree of racemization observed after peptide coupling reactions depends on both the propensity for oxazolone formation and the rate of oxazolone aminolysis. The nucleophilicity/basicity relationship of the amino component is also a crucial parameter.

Asymmetric induction may also be involved in this kind of epimerization process. In 1966, Weygand et al. reported that so-called positive diastereomers (L,L-peptides) are predominantly formed upon reaction of racemic 5(4*H*)-oxazolones with L-configured amino acid esters [300]. This phenomenon relies on the reaction of a configurationally labile molecule present in two enantiomeric forms L-**100**, D-**100** with a chiral compound and may, in a positive sense suppress or, in a negative sense enhance racemization in peptide synthesis. Consequently, the degree of epimerized product observed after a peptide synthesis is the result of both the racemization according to the oxazolone mechanism and asymmetric induction.

4.4.3
Racemization Tests: Stereochemical Product Analysis

Determination of the degree of racemization occurring in one discrete coupling step, and evaluation of the diastereomeric purity of peptide products, is a necessary measure because of the imminent risk of racemization. Subsequent to the proof of sequence homogeneity of the product using different analytical techniques (e.g., high-performance liquid chromatography (HPLC), capillary electrophoresis, mass spectrometry), the absence of undesired diastereomers in the final product has to be proven. This may be done by complete proteolytic degradation, as reported by Finn and Hofmann [301]. Total hydrolysis of the peptide, followed by chiral resolution of all amino acid components with analytical techniques is a viable alternative, but is impeded by the uncertainty of partial epimerization during hydrolysis. The application of isotope-labeled hydrolytic agents (deuterium- or tritium-labeled HCl) circumvents this disadvantage [298, 302]. The configurational integrity of amino acids may be determined by incubation with L-amino acid oxidase, the transformation into diastereomeric dipeptides, or gas-liquid chromatography (GLC) on enantiomerically pure stationary phases.

Information on the degree of racemization during a coupling reaction may be obtained by the synthesis of model peptides, applying either an activated N-acyl amino acid or an activated N-protected peptide. Sometimes, the additional chiral center of isoleucine is utilized in a simple test method, as the D-allo-isoleucine residue formed by racemization at the isoleucine C^{α} is a diastereomer of L-isoleucine and hence can easily be separated. The variety of methods described has been summarized in review articles [298, 299]. Despite the value of such model systems [261, 303–305] in the development of new coupling methods, a generalization of the results is commonly not possible, because racemization is also sequence-dependent. A HPLC-based racemization test utilizing fluorescence detection has been published by Griehl et al. (Fig. 4.28) [304]. In this investigation, the coupling of Z-Ala-Trp-OH to H-Val-OMe was examined using different coupling methods. The diastereomeric tripeptides L,L,L-Z-Ala-Trp-Val-OMe and L,D,L-Z-Ala-Trp-Val-OMe were separated by reversed-phase (RP) HPLC. Fluorimetric detection provides 100 times higher sensitivity compared to the UV-detection commonly used.

In summary, the problem of racemization still impedes the chemical synthesis of peptides.

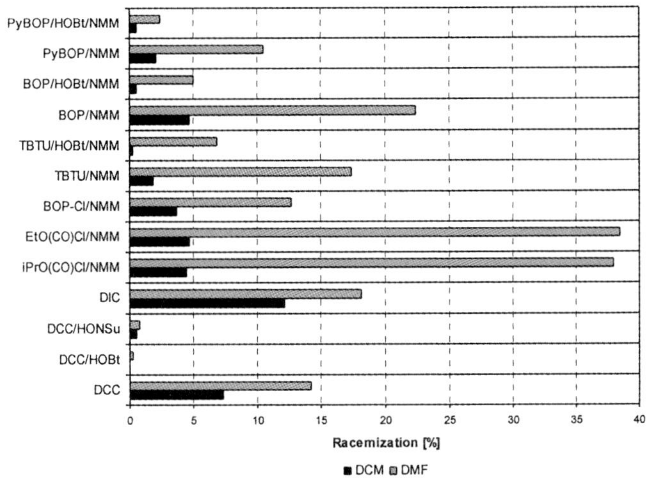

Fig. 4.28 Percentage of racemization upon coupling of Z-Ala-Trp-OH to H-Val-OMe with different coupling reactions.

4.5 Solid-Phase Peptide Synthesis (SPPS)

Ever since the early days of peptide chemistry, reactions involving peptide synthesis have been traditionally performed in solution, and many impressive results – including the total synthesis of small proteins – have been achieved using this conventional technique. In 1981, ribonuclease A – which consists of 124 amino acids – was synthesized in crystalline form by Yajima and Fujii [306] and shown to have a high specific activity. Unfortunately, peptide synthesis in solution is highly labor-intensive and requires extensive knowledge with regard to the strategy and tactics in choosing protecting groups and coupling methods, as well as solving problems of solubility.

One major advantage of solution-based synthesis is the high purity of the final product, though this is clearly dependent upon the purification of intermediate compounds. Despite these restrictions, highly specialized and experienced research teams have successfully completed many ambitious syntheses following this technique.

The ingenious concept of peptide synthesis on a solid support – which is now known as SPPS – was developed by Robert Bruce Merrifield in 1963 [307], and provided a major breakthrough in peptide chemistry. Merrifield was awarded the Nobel prize in 1984 for this unique invention that has revolutionized organic chemistry during the past 20 years. Today, the concept has been extended and generalized to organic synthesis on polymeric supports, which includes not only heterogeneous reactions involving an insoluble polymer, but also the application of soluble polymeric materials which allow homogeneous reactions (liquid-phase peptide synthesis) to be conducted. A combination of these two variants is also possible, being referred to as alternating solid-liquid-phase peptide synthesis.

In SPPS the peptide chain is assembled in the usual manner, starting from the C-terminus. The amazingly simple concept is that the first amino acid of the peptide to be synthesized is connected via its carboxy group to an insoluble polymer that may be easily separated from either reagents or dissolved products by the use of filtration. The general principle is shown in Fig. 4.29.

A necessary prerequisite is that anchoring groups (linkers) are introduced into the polymeric material (Fig. 4.29, step 1). An amino acid which is protected at N^{α} is then reacted with the functional group of the linker (step 2). Subsequently, the temporary protecting group is removed (step 3) and the next amino acid compo-

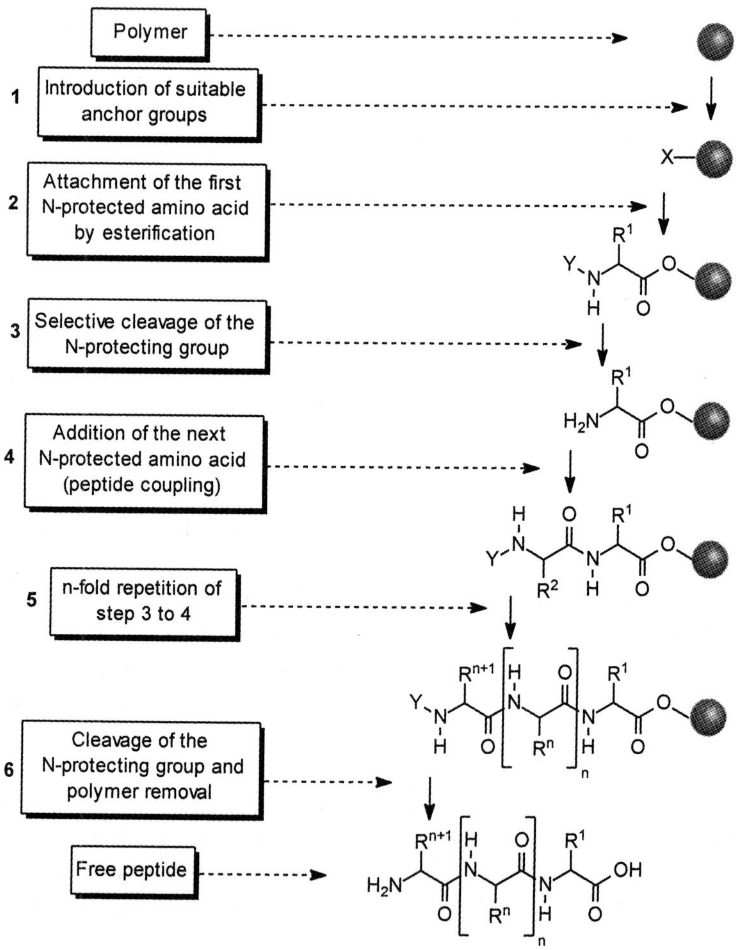

Fig. 4.29 Process of solid-phase peptide synthesis. Y=temporary protecting group; R^1, R^2, R^n, R^{n+1}=amino acid side chains, if necessary, protected with semipermanent protecting group.

nent is coupled (step 4). Steps 3 and 4 are then repeated (step 5) until the required peptide sequence has been assembled. Finally, the covalent bond between the linker moiety and the peptide chain is cleaved. In many cases the semipermanent side chain-protecting groups may be simultaneously removed (step 6). The insoluble polymeric support is then separated from the dissolved product by filtration.

When syntheses are carried out on a polymeric support there is no need to perform the tedious and time-consuming isolation and purification of all intermediates, as would be necessary in solution synthesis. The product of all the reactions (the growing peptide chain) remains bound to the support during steps 3 to 5, and excess reagents and by-products are removed by filtration. The simple technical operation and the potential for automation initially led to the euphoric conclusion that the chemical synthesis of polypeptides and proteins had, in principle, been solved by the solid-phase concept. Clearly, the Merrifield synthesis had a major influence on chemical peptide and protein synthesis as well as on solid-phase organic synthesis. The concept of chemical synthesis on a polymeric support also promoted the chemical synthesis of many other biomolecules, including as oligo- and polynucleotides.

Unfortunately, despite the high expectations that were initially imposed on the Merrifield method, it does suffer from several limitations:

- The final product of a synthesis carried out on a polymeric support is only a homogeneous compound, if all deprotection and coupling steps proceed quantitatively.
- A large excess of each amino acid component is required in the corresponding coupling reaction in order to achieve complete conversion.
- There is a permanent risk of undesirable side reactions during activation, coupling, and deprotection.
- Monitoring the reaction progress and analysis of complete conversion are difficult to perform in heterogeneous reaction systems, and are hampered by experimental error.
- Swelling properties of the polymeric resin and diffusion of the reagents are important parameters for the success of a solid-phase synthesis.
- Aggregation phenomena of the growing peptide chain may complicate the synthesis.
- On occasion, drastic conditions required to cleave the peptide from the polymer may also damage the final product.

Although all methods described for synthesis of peptides and organic molecules on polymeric support rely on the basic principle introduced by Merrifield, the expression "Merrifield synthesis" has mainly been coined for SPPS. More specialized information is available in a series of review articles and monographs [308–318].

4.5.1
Solid Supports and Linker Systems

The polymeric support must inevitably be chemically inert, mechanically stable, completely insoluble in the solvents used, and easily separated by filtration. It must contain a sufficient number of reactive sites where the first amino acid of the peptide chain to be synthesized can be attached. Interactions between the peptide chains bound to the resin should be minimal. As with a carboxy-protecting group, the resin must be stable under the chemical conditions of the synthetic procedures, and cleavage under mild conditions must be possible. Furthermore, it has to display mechanical stability in order for the necessary separation procedures to be carried out.

Initially, a copolymer of polystyrene with 1–2% divinyl benzene as cross-linker was used in SPPS. The dry resin beads (Fig. 4.30) are normally 20–80 μm in diameter, and are able to swell to five- or six-fold volume in the different organic solvents used for peptide synthesis (e.g., dichloromethane; Tab. 4.8). Consequently, the polymeric support – when suspended in these solvents – is not a static solid matrix but a well-solvated gel with mobile polymeric chains. This facilitates diffusional access of the reagents to all reaction sites. Polystyrene with a low-

Fig. 4.30 Polystyrene/divinylbenzene cross-linked resin.

Tab. 4.8 Swelling properties (based on dry volume) of polystyrene/1% divinylbenzene resin beads in different solvents.

Solvent	*Swelling factor*	*Solvent*	*Swelling factor*
Tetrahydrofuran	5.5	N,N-Dimethylacetamide	3.4
Dichloromethane	5.1	Diethylether	2.5
Dioxan	4.6	Acetonitrile	2.0
Toluene	4.5	Ethanol	1.05
N,N-Dimethylformamide	3.5	Methanol	0.95

er degree of cross-linking (1% divinylbenzene rather than 2%) was found to have a superior performance, mainly because of the improved swelling properties.

By using autoradiography with tritium-labeled building blocks, it has been shown that the peptides are distributed uniformly within the gel matrix (Fig. 4.31). This finding contradicts the previous assumption that the reactions proceeded only on the surface of the resin beads. Approximately 10^{12} polypeptide chains can locate on one polystyrene/divinylbenzene bead of 50 μm diameter, and with a resin loading of 0.3 mmol peptide per gram resin.

The heterogeneous reactions used in SPPS are usually two- or three-fold slower than homogeneous reactions. The reaction sites are uniformly distributed within the polymeric matrix, and the reaction rates of the early coupling and deprotection steps are usually comparable with those at the end of a synthesis. An optimum degree of resin loading with the first amino acid ranges from 0.2 to 0.5 mmol per gram resin. On the basis of this assumption, it can be calculated that the peptide, after coupling of 15 amino acids, contributes more than 50% of the total mass. The growing mass ratio of peptide/resin material does not usually reduce the efficiency of further chain elongations, although the swelling behavior in nonpolar solvents is significantly reduced. The addition of higher percentages of DMF or urea is reported to be advantageous in such cases. Support materials that are more polar (e.g., polyamide-based resins) have also been recommended for the synthesis of longer peptides.

The classical Merrifield resin complies well with the requirements for suitable polymeric support materials, but it is not optimal with respect to mechanical stability, loading capacity, diffusional problems, and differences in solvation between the polymer and the peptide.

Cross-linked poly(dimethylacrylamide) was introduced as a more hydrophilic support by Atherton and Sheppard [312]. This forms a gel and is highly solvated

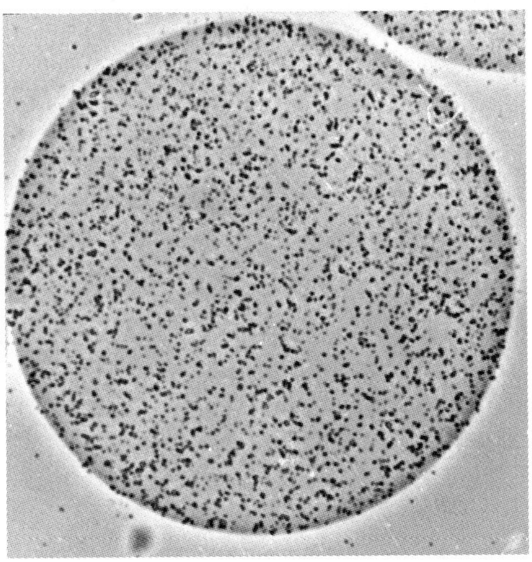

Fig. 4.31 Distribution of ^{3}H-labeled peptides in a single polystyrene/divinylbenzene bead, as revealed by autoradiography.

by solvents appropriate for peptide synthesis (e.g., DMF) which favors peptide synthetic reactions. The swelling properties of this polymer are superior to those of polystyrene-based resins, and this allows improved accessibility of the functional groups present on the polymer matrix, even in the micropores. This type of support has also been used in further developments such as polymerization in the pores of a macroporous rigid silica (kieselguhr) matrix, providing improved mechanical stability. Hence, syntheses can be performed under moderate pressure in a column, while the reaction progress can be monitored continuously by spectrophotometry (see Fig. 4.35).

Hydrophilic tentacle polymers are obtained by grafting polyethylene glycol (PEG) chains with an arbitrary degree of polymerization onto polystyrene beads. Grafted copolymers using porous polystyrene are suited best for this purpose [313], with the high solvation of the PEG chains conferring high mobility on the peptides bound to the copolymer. The term "tentacle polymer gel" (TentaGel) was coined for these materials, which allow solid-state NMR techniques such as magic angle spinning (MAS) ^{1}H-NMR or even solution-phase ^{13}C-NMR techniques to be used to monitor reaction progress. Different types of conformation have been found in relaxation time measurements that distinguish between solvated tentacles and nonsolvated random coils. While the latter are characterized by broad lines, the former give rise to sharp NMR absorptions, even though they are present in the solid state. The swelling factors in different solvents are quite similar; therefore, these polymers are stable to moderate pressure and may be used in columns for continuous-flow peptide synthesis.

Other types of polymers have been tested for applications in synthesis. The polystyrene resin itself has been chemically modified, but structurally different polymers were also studied. Nonswelling, highly cross-linked polystyrene/divinylbenzene matrices are rigid and have a high inner surface area; pellicular and brush-type resins have also been developed. While the former are obtained by the polymerization of a thin polymeric layer on the surface of inert glass beads, the latter consist of a linear polystyrene brush polymer grafted onto a polyethylene film surface. These present the anchoring linker moieties in the form of a brush on the surface. Polymers in the shape of strips, films, and fibers have also been tested in automated syntheses.

Sintered polyethylene, cellulose, silica, controlled pore glass (CPG), and chitin have also been used as planar support materials.

Nowadays, in line with the growing development of combinatorial chemistry and solid-phase bioassays [319], further requirements must be imposed on resin materials for synthetic purposes. Among these are biocompatibility, good swelling properties in either aqueous or buffered aqueous solutions, and a reduced nonspecific binding affinity to biomolecules.

The introduction of anchoring groups (linkers) on a polymeric support is the precondition for application in peptide synthesis, and the many different linker systems currently available for SPPS and solid-phase organic synthesis have been recently reviewed [320]. Additional handles in the form of bifunctional linker moieties may be attached to the polymer matrix. One of the functional groups fulfils the requirements for a protecting group allowing for cleavage under mild conditions, while the other is used for attachment to the resin. This concept provides improved control of the level of resin loading, and also allows greater freedom in the choice of polymer matrix type (e.g., polystyrene, polyacrylamides, CPG, chitin, cellulose). Handles often contain internal reference amino acids (IRAA); these improve the monitoring of reaction progress, provide an exact yield determination, and help to control the integrity of the resin-bound peptide chain [312].

Because linkers must be considered as protecting groups for the C-terminal carboxy group of the peptide to be synthesized, their design and implementation must consider especially their chemical properties. Depending on whether the C-terminus of the desired peptide is required as a carboxylate, a carboxamide, a hydrazide (for backing-off strategy), an ester, a thioester [321], or an alcohol, the linker system must be chosen appropriately. In most cases, the first amino acid is connected to the resin via an ester bond, though amide or hydrazide bonds may also be involved.

Correct selection of the handle allows yield optimization with respect to the attachment and cleavage steps, as well as suppression of racemization during attachment of the first amino acid and diketopiperazine formation on the dipeptide stage. For certain applications (e.g., peptide backbone cyclizations) the linker must behave orthogonally to the other semipermanent and temporary protecting groups.

Meanwhile, many different variants for the chemical functionalization of polymers have been identified, and the number is steadily increasing as the solid-phase synthesis of peptides and organic molecules undergoes rapid development.

Chloromethyl resin (Merrifield resin) [322, 323] The chloromethyl group in **106** is the classical anchoring moiety present in the Merrifield resin, and it is introduced into polystyrene/divinylbenzene resins by Friedel-Crafts-type chloromethylation with an alkoxy-substituted chloromethane in the presence of tin(IV)chloride. Attachment of the first amino acid to the resin to give **108** is performed as a nucleophilic substitution reaction of chloride by the amino acid carboxylate **107** under reflux conditions in an organic solvent such as ethanol, tetrahydrofuran, or dioxan. Often, cesium salts or alkyl ammonium salts are used. Other bases such as tertiary amines or anhydrous potassium fluoride may also be applied.

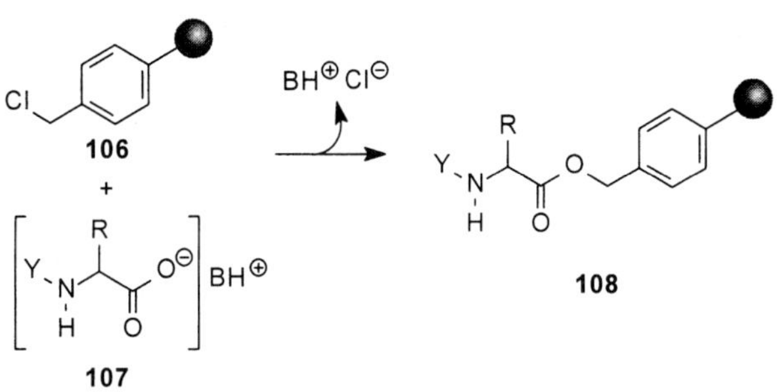

This protocol prevents racemization that might otherwise occur with an activated amino acid derivative. Therefore, the chloromethyl resin is preferred over the corresponding hydroxymethyl resin. The resulting benzyl-type peptidyl ester can be cleaved on completion of the peptide chain assembly only with very strong acids such as liquid HF, TFMSA, or HBr/TFA. Consequently, this type of resin is used in combination with Boc as a temporary protecting group and semipermanent TFA-stable side chain protection of the benzyl or cyclohexyl type to obtain peptides with a free carboxy group at the C-terminus (peptide acids). One disadvantage of the Merrifield resin is that ~1–2% of the growing peptide is cleaved from the resin during each of the repetitive acidolytic deprotection steps.

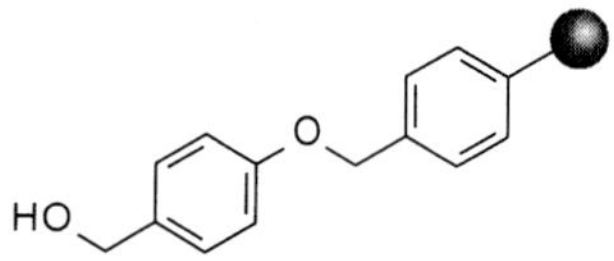

109

4-Benzyloxybenzyl alcohol resin (Wang resin) [324] The Merrifield resin can easily be converted into the Wang resin **109** by etherification with methyl 4-hydroxybenzoate, followed by reduction of the methyl ester with $LiAlH_4$.

This anchoring group comprises a 4-alkoxy-substituted benzyl alcohol moiety, which confers increased acid-lability onto the linker. Hence, the Wang resin is used routinely in batch Fmoc chemistry.

Active esters (e.g., pentafluorophenyl esters) or carbodiimides (e.g., DCC or carbonyldiimidazol) are used for the direct attachment of the first amino acid on the resin. 4-(Dimethylamino)pyridine often is added as an acylation catalyst in carbodiimide couplings, but in these cases racemization may occur and the addition of HOBt is recommended. Cleavage of the final product proceeds smoothly with 95% TFA, and yields peptide acids with concomitant removal of *tert*-butyl-type side chain-protecting groups.

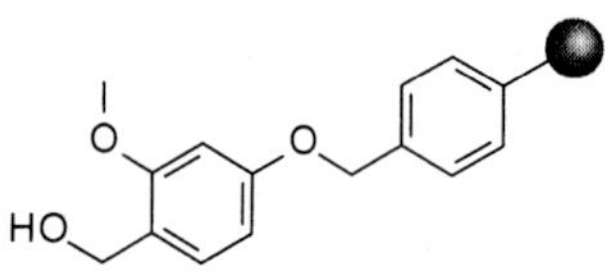

110

2,4-Dialkoxybenzyl alcohol resin (super acid-sensitive resin, SASRIN®) [325] The introduction of one more methoxy group in the 2-position of the benzyl alcohol moiety (**110**) results in a further increased sensitivity towards acid, because the benzyl cation intermediate of acidolysis is highly stabilized by both electron-donating alkoxy groups in 2- and 4-positions. Cleavage occurs even in the presence of 0.5–1% TFA in dichloromethane, or by treatment with the acidic alcohol 1,1,1,3,3,3-hexafluoroisopropanol in dichloromethane. This property makes the resin very well suited for the synthesis of peptide acids that are fully protected at the side chains. Attachment of the first amino acid component is performed analogously, as discussed for the Wang resin. Racemization during this esterification may be a severe problem.

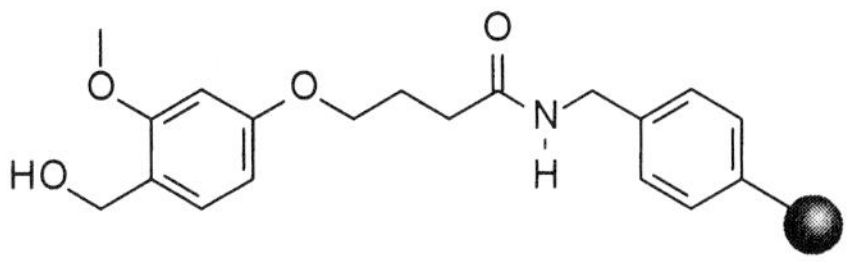

111

4-Hydroxymethyl-3-methoxyphenoxybutyric acid resin (HMPB resin) The HMPB handle type **111** is chemically related to the 2,4-dialkoxybenzyl alcohol resin, and displays similar cleavage characteristics. Detachment of the final product, even in the form of fully protected peptides, is achieved with dilute acid (1% TFA/dichloromethane) to yield peptide acids.

4-(Hydroxymethyl)phenylacetamidomethyl resin (PAM resin) [326, 327] The PAM resin (Fig. 4.32) was developed as an alternative to the Merrifield resin, where loss of peptides during acidolytic Boc deprotections and insufficient reactivity in the nucleophilic displacement of the linker benzyl chloride by the first amino acid carboxylate is sometimes observed.

Aminomethyl groups can also be attached to the aromatic moieties of polystyrene. The electron-withdrawing carboxamidomethyl group renders the PAM resin about 100 times more stable towards treatment with TFA. It is introduced by acylation of aminomethyl polystyrene with (4-hydroxymethylphenyl)acetic acid. The first amino acid is introduced by carbodiimide or active ester chemistry forming a resin-bound ester.

Alternatively, a PAM anchoring group, where the first amino acid residue is already attached to it, may be introduced by the reaction sequence displayed in Fig. 4.32 (pre-formed handle strategy). The bromide substituent in 4-(bromomethyl)phenylacetic acid phenacyl ester undergoes nucleophilic displacement upon treatment with the cesium carboxylate of the corresponding amino acid. Subsequently, the phenacyl ester is cleaved by reduction with zinc in acetic acid and the loaded PAM anchoring group is attached to aminomethyl polystyrene.

The PAM resin is fully compatible with the Boc/Bzl protection scheme, and is cleaved with strong acids such as liquid HF, TFMSA, or HBr/TFA to give peptides with a free acid located C-terminally.

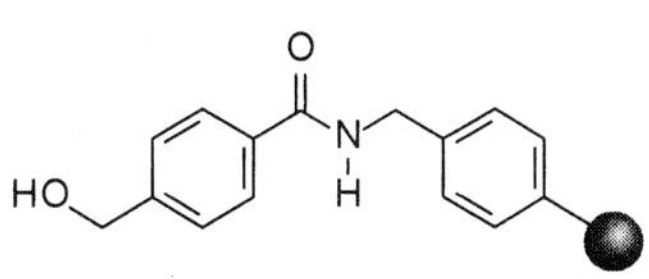

112

Hydroxymethylbenzoic acid resin (HMBA resin) [328] The hydroxymethylbenzoic acid resin **112** is closely related to the PAM resin. The only difference is the miss-

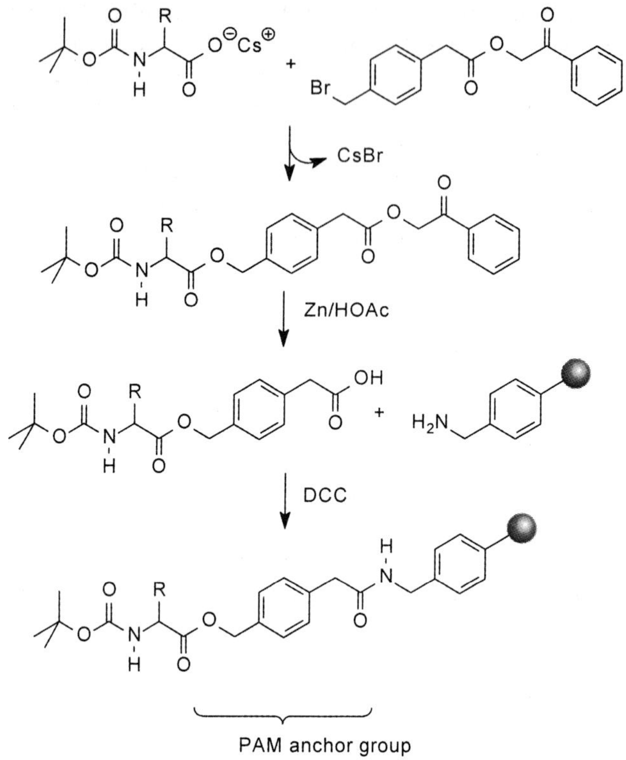

Fig. 4.32 Coupling of the PAM linker, loaded with the first Boc-protected amino acid, to aminomethylated polystyrene. PAM = phenylacetamidomethyl.

ing benzylic CH_2 group, but this results in a completely altered chemical reactivity. It is obtained by acylation of aminomethyl polystyrene with 4-(hydroxymethyl)benzoic acid and is completely resistant towards treatment with acids (even liquid HF). Consequently, on-resin side-chain deprotection can be accomplished. The first amino acid is introduced using carbodiimide or active ester chemistry. The highly versatile HMBA linker is cleaved by a variety of nucleophiles such as hydroxide ions (to give peptide acids), alcohols (to give peptide esters), ammonia or amines (to give peptide amides), hydrazine (to give peptide hydrazides), or $LiBH_4$ (to give peptide alcohols).

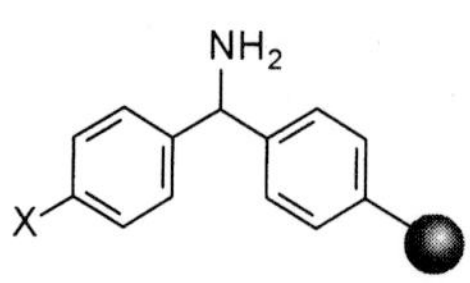

113 X = H (BHA)
114 X = CH_3 (MBHA)

Benzhydrylamine resin (BHA resin) and 4-methylbenzhydrylamine resin (MBHA resin) [329, 330] The acid lability of a benzyl ester can be increased not only by the introduction of additional electron-donating groups on the aromatic ring, but also by the attachment of one (BHA resin **113**, MBHA resin **114**, Rink amide resin **115**) or two additional aromatic rings (Barlos resin **116**) to the benzylic carbon atom. The benzhydrylamine resin (BHA resin) was designed accordingly. Acidolysis with HF, TFMSA, or HBF_4/TFA yields the peptide amide. This fact renders the BHA resin suitable for the Boc/Bzl protection scheme. Increased acid lability is observed after introduction of an electron-donating methyl substituent in the 4-position of the second aromatic ring (MBHA resin **114**).

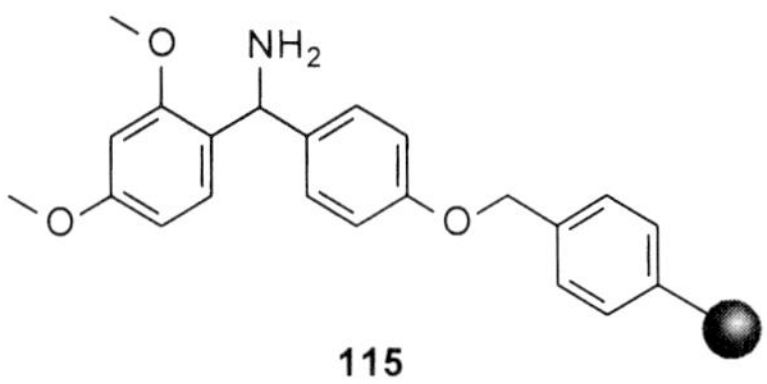

115

(2,4-Dimethoxy)benzhydrylamine resin (Rink amide resin) [331] The combination of a benzhydrylamine group with two additional methoxy substituents on the second aromatic ring renders the (2,4-dimethoxy)benzhydrylamine resin **115** [(4-(4′-dimethoxyphenyl)aminomethyl)phenoxymethylpolystyrene] even more labile towards acidolysis, and allows cleavage of peptide amides with 95% TFA.

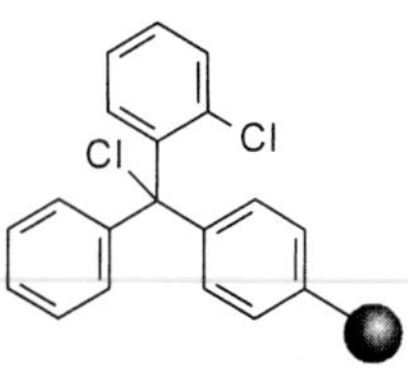

116

o-Chlorotritylchloride resin (Barlos resin) [332]
Attachment of the first amino acid to the *o*-chlorotritylchloride resin **116** is accomplished upon reaction with an amino acid carboxylate salt (e.g., diisopropylethylammonium or triethylammonium salts). This procedure does not suffer from racemization, as the amino acid acts as the nucleophile and not as an electrophilic species. The steric constraints of the *o*-chlorotrityl group impedes diketopiperazine formation on the dipeptide stage [333]. The trityl group has already been discussed in the context of side chain-protecting groups (cf. Section 4.2.4). *o*-Chloro substitution slightly increases stability towards acids, but the *o*-chlorotrityl linker remains very sensitive to acid. Cleavage occurs upon treatment with 0.5% TFA in dichloromethane or with 1,1,1,3,3,3-hexafluoroisopropanol in dichloromethane. The *o*-chlorotrityl handle is also suitable for the attachment of C-terminal alcohols, thiols, and amines.

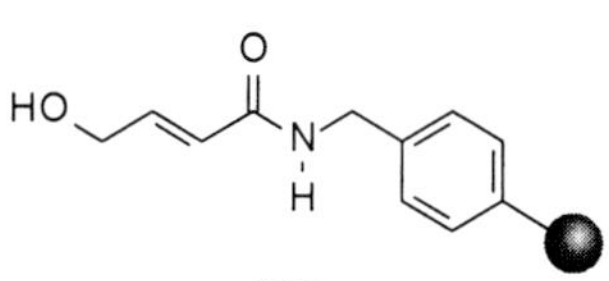

117

Hydroxycrotonoyl aminomethyl resin (Hycram resin) [334] Allyl esters are completely orthogonal to other types of protecting groups, because they are completely stable towards acids or bases usually employed in peptide synthesis. Consequently, an allyl-based handle represents a useful alternative to other linker types. The Hycram resin **117**, loaded with the first amino acid component is obtained upon acylation of aminomethyl polystyrene with esters of, for example, Boc- or Fmoc-protected 4-hydroxycrotonyl derivatives. Cleavage of the final product from the resin is accomplished under neutral conditions by treatment with a Pd(0) catalyst that effects allyl transfer to a suitable nucleophile. Fully protected peptides or very sensitive derivatives (e.g., glycopeptides) can be synthesized according to this concept.

Photolabile linkers Photocleavage provides an additional dimension in orthogonality. Similarly to photolabile protecting groups, linker moieties (e.g., **118**, **119**) are available that allow for cleavage by photolysis [335].

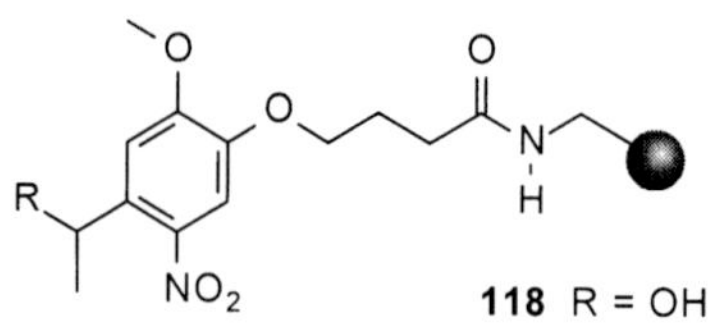

118 R = OH
119 R = NH_2

4.5.2
Safety-Catch Linkers

Safety-catch linker strategies [320, 336] rely on a special linker moiety that connects the polymeric support and the growing peptide chain. This linker is stable under the conditions of peptide synthesis but is finally activated for cleavage by a discrete chemical modification. In addition, the activation step usually provides an activated C-terminus of the peptide which may react with diverse nucleophiles. Intramolecular reaction of this activated position with the free N-terminus of the peptide leads to cyclic peptides and simultaneously liberates the target compound from the solid support. Two major types of safety-catch linkers may be distinguished:

- Those characterized by a hidden orthogonality of the stable and labile form, which means that cleavage mechanisms of the two forms are different.
- Those which rely on different kinetics of the cleavage reaction in both forms.

Kenner's sulfonamide safety-catch linker tethers a carboxylic acid to a solid support (e.g., aminomethyl polystyrene, **120**, **121** or MHBA, **122**, **123**), and is stable towards acidic and basic reaction conditions. After activation by treatment with diazomethane, the secondary sulfonamide **124** can be cleaved by aminolysis, peptidolysis, hydrazinolysis, or saponification to provide the primary amides, peptides, hydrazides, and carboxylic acids, respectively [337]. However, relatively poor loading capacity, racemization during resin loading, and poor reactivity of the N-methylated peptidyl sulfonamide obtained upon diazomethane activation are major limitations of this method. The application of iodoacetonitrile as alkylating agent led to a significant improvement because the *N*-cyanomethyl *N*-acylsulfonamide **125** is extremely reactive in nucleophilic displacement reactions under mild conditions. Alkanesulfonamide-type linkers (**121**, **123**) are superior compared to arenesulfonamide linkers (**120**, **122**) because of the greater nucleophilicity of the alkanesulfonamide nitrogen [338, 339]. Coupling conditions have been developed to load Boc- or Fmoc-protected amino acids with high efficiency and minimal racemization. Near-racemization-free displacement of the activated linker moiety with various nucleophiles has been demonstrated.

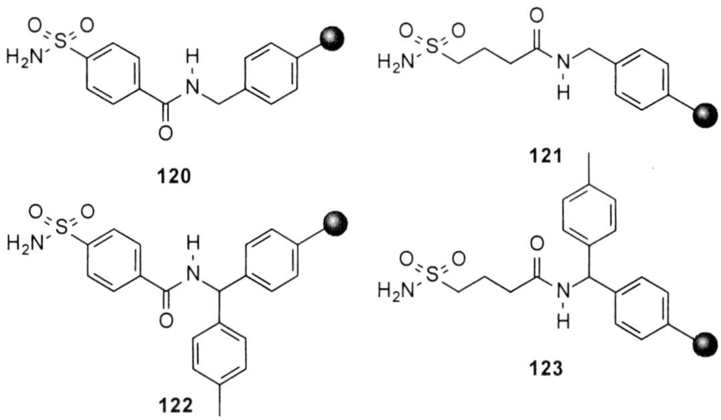

120 **121** **122** **123**

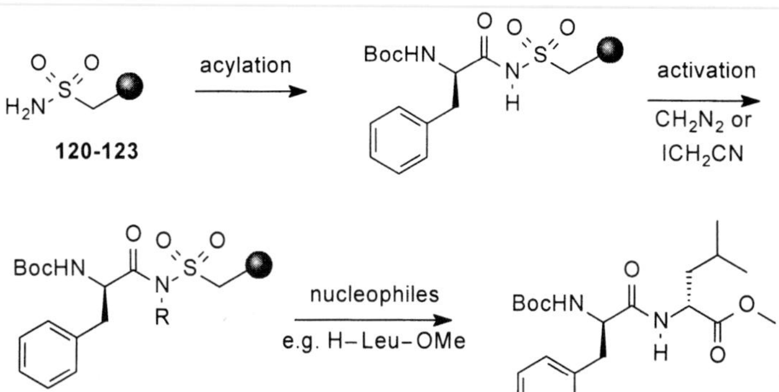

124 R = CH_3 (CH_2N_2)
125 R = CH_2CN (ICH_2CN)

Even C-terminal thioesters, that may be subsequently used for ligation purposes, can be obtained on thiolysis of peptides bound to a safety-catch linker [340].

Safety-catch linkers relying on oxidative activation have been employed in the synthesis of linear and cyclic peptides [341, 342], but their application is limited to the synthesis of peptides that do not contain any oxidation-sensitive amino acid. The sulfoxide-containing safety-catch linker DSA **126** resists treatment with acid but undergoes acidolysis after reduction of the sulfoxide group; this generates a third electron-releasing group on the aromatic system [343].

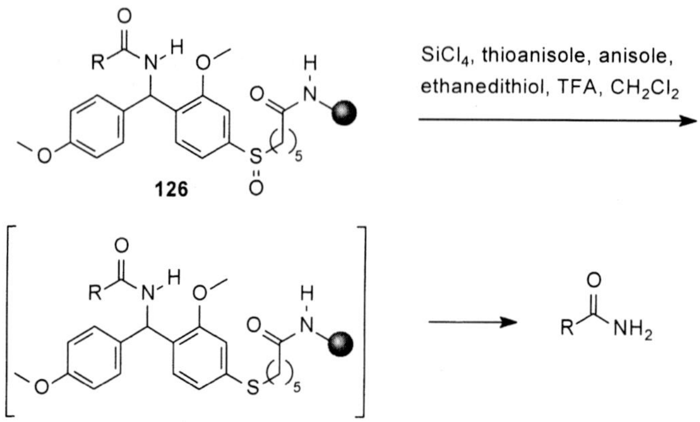

The DSA linker is structurally related to the safety-catch amide linker (SCAL) developed by Lebl [344]. In the oxidized form **127**, SCAL is stable to a wide range of chemical conditions (TFA, HF, thiophenol, reducing agents, basic conditions, and Hg^{2+}). Upon reduction of the two aryl sulfoxide groups to give the corresponding aryl sulfide **128** by $PPh_3/Me_3SiCl/CH_2Cl_2$ or $(EtO)_2P(S)SH/DMPU$, SCAL becomes labile towards treatment with TFA. SCAL has also been used for SPPS by chemical ligation [345].

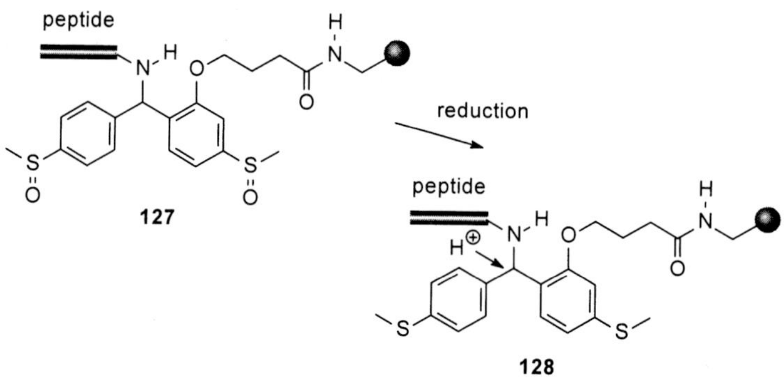

Another concept for a safety-catch linker **129**, developed by Geysen et al. [346–348] relies on the triggered formation of peptide with a C-terminal diketopiperazine **130** that is accompanied by cleavage from the resin.

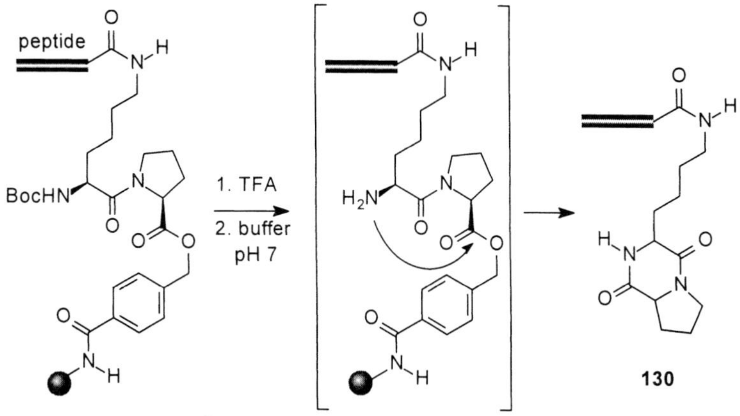

129

130

A similar linker for TentaGel or “large” PS-PEG resin has been designed by Atrash and Bradley; this combines diketopiperazine formation with subsequent fragmentation. Cleavage of the Boc group at proline in **131** triggers diketopiperazine **132** formation, which in turn releases a peptidyloxymethylphenolate **133** that spontaneously undergoes fragmentation, liberating the peptide assembled on the resin as the free carboxylate [349]. These linkers have been shown to be suitable for a direct biological evaluation of a peptide released from the resin *in situ*.

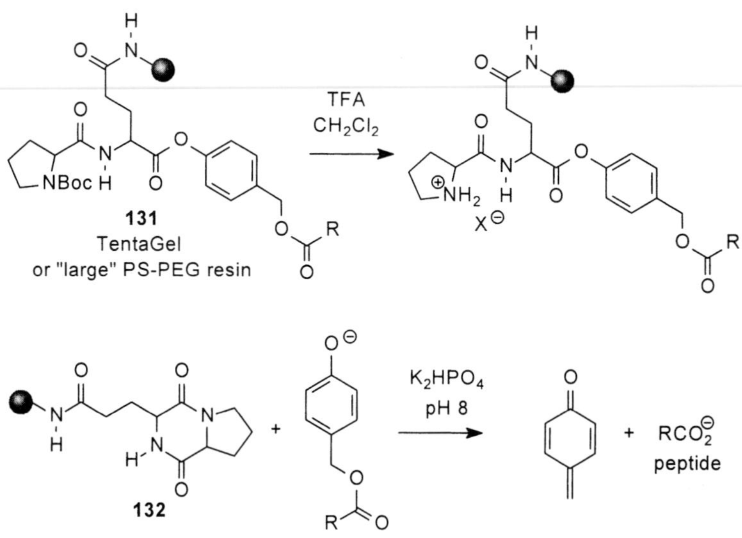

131
TentaGel
or "large" PS-PEG resin

132

133

A concept involving linker cleavage by neighboring group participation has been presented by Frank et al. [350, 351].

4.5.3 Protection Schemes

Each of the repetitive synthesis cycles (deblocking – coupling) is initiated by selective cleavage of the N^{α}-protecting group. The anchoring group (linker) and the semipermanent amino acid side chain-protecting groups must be correctly chosen with respect to the chemical properties of the temporary N^{α}-protecting group, which is usually held constant for all amino acids used throughout the synthesis. The selection and optimization of protecting group chemistry doubtlessly holds a key position in the success of a peptide synthesis.

In SPPS, as in liquid-phase synthesis, either the acid-labile *tert*-butyloxycarbonyl group or the base-labile fluorenyl-9-methyloxycarbonyl group are applied preferably as temporary protecting groups (cf. Section 4.2.1.1).

4.5.3.1 Boc/Bzl-protecting Groups Scheme (Merrifield Tactics)

The standard Merrifield system is based on Boc-protecting group tactics, which rely on selective acidolytic cleavage. The *tert*-butyloxycarbonyl group is used as the temporary N^{α}-protecting group in combination with benzyl-type semipermanent side-chain protection, as Boc is usually cleaved with TFA (20–50%). Fig. 4.33 shows an example of the protecting group combination for a tripeptide amide synthesis (H-Asp-Gly-Tyr-NH_2) on MBHA resin.

The semipermanent benzyl-type groups must be stable under the conditions of repetitive Boc cleavage. In order to safeguard the stability of the side-chain protection, benzyl-type groups with electron acceptors may be applied (e.g., 2,6-dichlorobenzyl, 2,6-Dcb). Cyclohexyl esters often are applied preferentially for the protection of side-chain carboxy groups. The semipermanent protecting groups and the MBHA

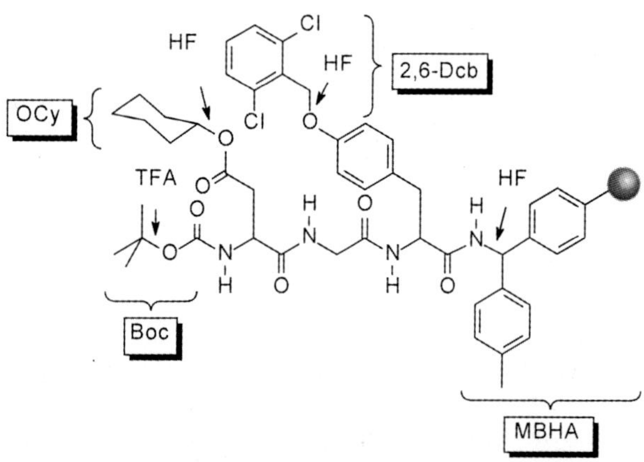

Fig. 4.33 Boc/Bzl protecting group tactics in solid-phase peptide synthesis according to Merrifield. MBHA = 4-methylbenzhydrylamine resin.

linker group are cleaved simultaneously by treatment with liquid HF on completion of the synthesis, as shown in Fig. 4.33. Peptide assembly on the PAM resin would result in the peptide with C-terminal acid under similar cleavage conditions.

As discussed previously, scavengers such as anisole, thioanisole, dimethyl sulfide, or triisopropylsilane must be added to the deprotection reaction in order to avoid side reactions of the intermediate carbenium ions. Other scavengers have especially been recommended in the context of the two-stage, low–high HF cleavage method that usually provides improved product purity [352, 353].

Absolute stability of the side chain-protecting groups is generally required for the success of a peptide synthesis. In cases where N^{α}-deprotection relies on acidolysis, this is especially critical because cleavage of the temporary and semipermanent protecting groups as well as of the linker proceed mechanistically in a very similar manner. Fine-tuning of the cleavage kinetics is required.

Special attention must be paid to selection of the side-chain protection. As any protected amino acid residue employed in the synthesis must be strictly compatible with the protecting group tactics chosen for the synthesis of a particular peptide, some protecting group schemes are more appropriate than others. The following semipermanent protecting groups are very compatible with Boc/Bzl tactics (final cleavage with liquid HF):

- Arg(Tos), Arg(Mts)
- Asp(OBzl), Asp(OCy), Glu(OBzl), Glu(OCy)
- Cys(Acm), Cys[Bzl(4-Me)], Cys(Mob)
- His(Bom), His(Dnp) – pre-cleavage treatment is necessary, His(Z), His(Tos)
- Lys[Z(2-Cl)]
- Ser(Bzl), Thr(Bzl)
- Trp(For) – pre-cleavage treatment is necessary
- Tyr[Z(2-Br)].

4.5.3.2 **Fmoc/tBu-protecting Groups Scheme (Sheppard Tactics)**

The Fmoc-protecting group tactics makes use of the base lability of the fluorenyl-9-methyloxycarbonyl group (Fmoc). It is a widely applied alternative to the Boc/Bzl scheme with two-dimensional orthogonality (Fig. 4.34) [312]. Fmoc is cleaved by base-catalyzed elimination where the secondary amine (piperidine) also traps the dibenzofulvene initially formed in the reaction. The semipermanent side chain-protecting groups are mostly of the *tert*-butyl type, and can be cleaved under relatively mild reaction conditions with TFA. Linker moieties displaying comparable acid lability are mainly used.

Although introduced into peptide chemistry as early as 1970, the Fmoc/tBu scheme has been widely applied in SPPS only since 1978. Many different coupling methods are compatible with the Fmoc group, among which are the highly reactive amino acid chlorides and amino acid fluorides [277]. Fig. 4.35 shows a schematic view of a semi-automatic continuous-flow peptide synthesizer developed by Atherton and Sheppard [312]. Pressure-stable resin material is required for application in a column reactor.

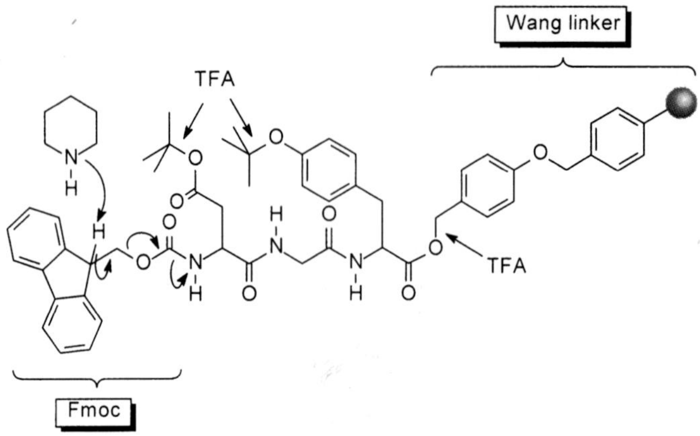

Fig. 4.34 Fmoc/tBu protecting group tactics in solid-phase peptide synthesis according to Sheppard.

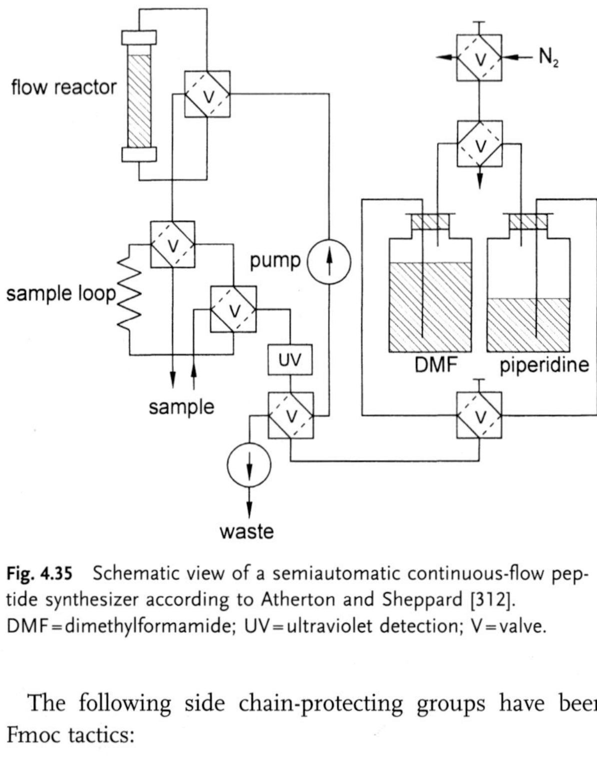

Fig. 4.35 Schematic view of a semiautomatic continuous-flow peptide synthesizer according to Atherton and Sheppard [312]. DMF = dimethylformamide; UV = ultraviolet detection; V = valve.

The following side chain-protecting groups have been used preferentially in Fmoc tactics:

- Arg(Pmc), Arg(Pbf)
- Asp(OtBu), Glu(OtBu)
- Asn(Trt), Asn(Tmb), Gln(Trt), Gln(Tmb)
- Cys(Trt)

- His(Trt)
- Lys(Boc)
- Ser(tBu), Thr(tBu)
- Trp(Boc)
- Tyr(tBu).

4.5.3.3 Three- and More-Dimensional Orthogonality

As alternatives, completely orthogonal protection schemes are available [1, 308]. These are characterized by a set of protecting groups where deprotection reactions are chemically independent. The third dimension of orthogonality relies on the application of protecting groups or linkers labile to, for example, photolysis, thiolysis (Dts), hydrazinolysis, or cleavage with transition metal catalysts (Aloc) [41, 42].

The acid-stable N^{α}-dithiasuccinoyl group (Dts) represents an elegant example of three-dimensional orthogonality in protection schemes of SPPS [308], and can be cleaved by thiolysis under neutral reaction conditions. A possible protecting group combination is shown in Fig. 4.36. While free thiols cleave the intramolecular disulfide bond of the temporary protecting group Dts, the semipermanent side-chain protection is removed by acidolysis. The linker containing a photolabile 2-nitrobenzyl moiety can be cleaved photolytically at 350 nm. The sequence of cleavage reactions can be chosen arbitrarily, and as a consequence fully or partially protected peptide acids can be obtained.

4.5.4 Chain Elongation

4.5.4.1 Coupling Methods

The coupling reaction is a process of eminent importance for the success of a SPPS, because complete conversion is the basic precondition for the formation of a homogeneous final product. Coupling reagents are, therefore, applied in excess

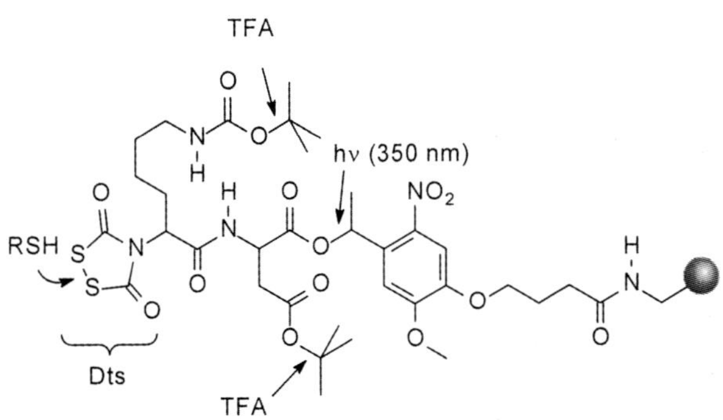

Fig. 4.36 Completely orthogonal protecting group tactics (three-dimensional orthogonality) in solid-phase peptide synthesis [309].

(usually three-fold); this increases the conversion but may also give rise to several undesired side reactions. Initially carbodiimides – especially DCC, but also DIC, because of the higher solubility of the corresponding urea – were used preferentially as coupling reagents. Peptide couplings in solid-phase syntheses are usually performed in dichloromethane, sometimes with addition of DMF, and at ambient temperature. The carbodiimide/HOBt method again prevails due to its major advantages, and not only with respect to the detrimental dehydration of Asn or Gln carboxamide side chains. Symmetrical anhydrides likewise proved to be efficient acylating agents. Active esters, especially of the pentafluorophenyl type, have also become very popular. The rather new coupling reagents of the uronium or phosphonium type, such as TBTU, HBTU, BOP, and PyBOP, UNCA derivatives, and Fmoc-protected amino acid fluorides encompass high coupling yields in nonpolar solvents and minimize or even exclude racemization and other side reactions.

4.5.4.2 **Undesired Problems During Elongation**

Incomplete conversion in solid-phase synthesis leads to a mixture of mismatch and core sequences, as shown in Fig. 4.37.

Considering the synthesis of a pentapeptide A-B-C-D-E, incomplete coupling may lead to four core sequences and three mismatch sequences. The latter arise when acylation or deprotection are incomplete and one or more amino acid components are skipped in the chain elongation. Mismatch sequences may also be formed when the amino acid sequence is correct, but acylation of nucleophilic side-chain functionalities, for example, occurs after partial deblocking.

The separation of undesired side products from the target peptide on completion of the synthesis is very tedious, and often impossible on a preparative scale. Hence, all possible preventative measures must be considered to safeguard any quantitative heterogeneous reaction. Double coupling steps represent one of these measures.

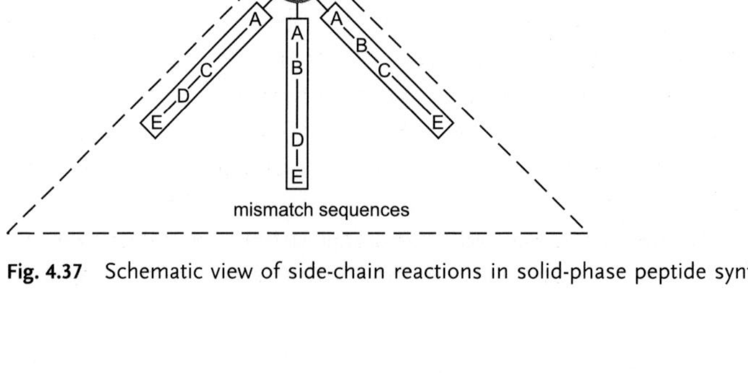

Fig. 4.37 Schematic view of side-chain reactions in solid-phase peptide synthesis.

The application of large excesses of reagents used may, however, result in other side reactions, such as aminoacyl insertions. If an acylation reaction is found to remain incomplete even after multiple repetition of a single coupling step, capping is a viable alternative. Capping means acylation of the unreacted amino groups with acetic anhydride, *N*-acetyl imidazole, or other acylating agents in order to avoid further chain elongation of the mismatch sequence at a later stage. These capped core sequences can usually easily be separated from the final product.

Peptides with a chain-terminating pyroglutamyl residue may be formed during N^{α}-deprotection of glutamine or glutamate residues located at the N-terminus.

The permanent and irreparable risk of side reactions on all synthetic steps is, despite all advantages, one of the peculiarities of SPPS. Mismatch and incomplete core sequences are caused by incomplete conversions in each of the repetitive steps of the synthesis cycle. If the yield of each cycle is hypothetically assumed to be 99%, the desired product is formed after 10 cycles in a maximum total yield of 90%, while after 100 cycles only 37% of the product is present in the mixture. An individual yield of 95% per cycle causes the total product yield after 10 or 100 cycles to decrease to 60% or 0.6%, respectively. Hence, the desired product may easily be contaminated by a series of structurally and chemically very similar compounds such as diastereomers, formed by epimerization, or mismatch and incomplete core sequences. Isolation and purification of one product from such a microheterogeneous mixture is a very difficult, though not unsolvable, problem. Despite all attempts that have been undertaken during past years to investigate most of the possible side reactions and to elucidate the mechanisms, single-cycle yields exceeding 99.5% do not appear attainable. Consequently, the remaining 0.5% adds up to a cumulative product contamination.

Mismatch and incomplete core sequences may be identified and quantified by a modified Edman degradation, also known as preview analysis [354]. In a successful synthesis, these are present in an amount below 0.3%. Mismatch or even wrong sequences are often the result of harmful or sometimes destructive reaction conditions during the repetitive deblocking steps or the final deprotection reaction.

Furthermore, N^{α}-Boc derivatives from commercial sources, for example, might in the past have been contaminated with ~0.3% of the corresponding N^{α}-*sec*-butyloxycarbonyl derivative. Once incorporated into the growing peptide chain, these derivatives resist acidolytic deprotection and so terminate the peptide chain.

Fmoc-protected amino acids formerly obtained by the classical methods might contain 1–4% of Fmoc-protected dipeptides as a consequence of competing mixed anhydride activation of the amino acid during the protection step. Incorporation of one or more of these contaminating dipeptides leads to mismatched elongated peptides. Nowadays, most N-protected amino acid derivatives are commercially available with chemical and optical purities >99.5%. Indeed, it is essential that all reagents and solvents, as well as the amino acid building blocks, be of high purity when used for peptide synthesis.

The application of N^{α}-urethane-protected amino acids usually maintains the degree of racemization below the analytical limits of the test systems employed

(<0.03%). Consequently, in normal cases this process does not significantly impede the outcome of a synthesis. Nevertheless, the esterification of amino acids, for example in the course of an attachment to a hydroxymethyl linker moiety or the activation of the C-terminus of a peptide, may lead to racemization of between 0.2 and 10%.

Diketopiperazine formation may efficiently compete with the coupling of the third amino acid to a resin-bound dipeptide. The diketopiperazine is cleaved from the resin during the attack of the free amino terminus to the C-terminal ester bond. Some amino acids such as glycine, proline, D-amino acids, and N-alkyl amino acids in the second position especially favor this process.

4.5.4.3 **Difficult Sequences**

One major precondition for complete peptide couplings is diffusion of the acyl component through the matrix, which only occurs to a satisfactory extent if the peptidyl-resin is sufficiently solvated. The most serious problem in SPPS is associated with sequence-related incomplete acylation reactions. In this context, random and nonrandom "difficult sequences" may be distinguished. Both types are characterized by reproducible incomplete acylation, limited improvement by repetitive coupling or the introduction of capping steps, and even inferior results when sterically hindered amino acids or higher resin loading are applied. Random difficult aminoacylation is associated with the incorporation of sterically hindered amino acids, for example β-branched derivatives or building blocks with bulky side chain-protecting groups. Nonrandom problems occur due to the sequence-specific formation of stable β-sheets that tend to aggregate and prevent further acylation [355, 356]. Secondary structures stabilized by intermolecular hydrogen bonds impede the diffusion and accessibility of the reactive N-terminus.

Merrifield et al. reported that a SPPS proceeded well up to the eleventh amino acid, but provided low yields at residues 12 to 17 [357]. According to Kent [355], such sequences may occur at a distance of 5–15 residues from the C-terminal residue at the resin. Despite the considerable research carried out in order to understand this phenomenon [358–360], it remains a major obstacle in peptide and protein synthesis.

Nonrandom difficult sequences may be predicted to a certain extent [361]. The propensity of the amino acids to occur in a random coil structure, facilitating synthesis, or in ordered secondary structures, hampering acylation, has been expressed by P_c^* values. The inherent tendency of a peptide chain to form a random coil conformation is then expressed by the average of the single P_c^* values of the n amino acids:

$$\langle c^* \rangle = \left(\sum P_c^*\right)/n$$

Sequences with $\langle P_c^* \rangle$ values >1.0 are associated with near-quantitative acylation reactions, while sequences with $\langle P_c^* \rangle$ values of 0.9–1.0 require longer reaction times or multiple couplings. $\langle P_c^* \rangle$ values <0.9 indicate that persistent acylation problems

Tab. 4.9 Conformational parameters of proteinogenic amino acids.

Amino acid	P_c	*Amino acid*	P_c
Ala	0.75	Leu	0.75
Arg	0.96	Lys	0.90
Asn	1.26	Met	0.70
Asp	1.07	Phe	0.76
Cys	1.08	Pro	1.64
Gln	0.83	Ser	1.21
Glu	0.80	Thr	0.98
Gly	1.47	Trp	0.79
His	0.96	Tyr	0.98
Ile	0.74	Val	0.73

may occur ("nonrandom difficult sequences"). β-Sheet formation can be monitored using FT-IR spectroscopy [362].

Aggregation is not a severe problem in Boc chemistry, because it is abolished by the repetitive treatment with TFA during cleavage of the temporary protecting group.

Fmoc chemistry does not encompass this advantage, but sustains β-sheet formation. This usually leads to prolonged reaction times, and either an increase in the reaction temperature or solvent changes often becomes necessary; this in turn may favor racemization [229, 363] and premature Fmoc cleavage [364, 365].

The β-sheet aggregates can be disintegrated by the application of:

- solvent additives that break up stable secondary structures, such as 2,2,2-trifluoroethanol [366], 1,1,1,3,3,3-hexafluoroisopropanol [366, 367], DMSO [368], or ethylene carbonate and Triton X [369];
- chaotropic salts in aprotic solvents [370, 371];
- temporary backbone amide-protecting groups (cf. Section 4.2.3) such as *N*-phenylthiomethyl [372], Hmb [91, 92, 373], Hnb [93], SiMb [94], or serine or threonine-derived pseudoproline dipeptide building blocks Fmoc-Xaa-Ser(Ox)-OH and Fmoc-Xaa-Thr(Ox)-OH [95–98];
- HOAt-based coupling reagents;
- polymeric supports based on polyoxyethylene-polystyrene copolymers [374], that diminish interaction between growing peptide chains;
- in-situ neutralization in the context of a modified Boc/Bzl synthesis scheme [375, 376];
- introduction of a pre-sequence [377];
- sonication [378];
- reduced resin loading; and
- coupling at elevated temperature [379].

4.5.4.4 On-Resin Monitoring

Stringent control of the coupling reactions is another indispensable prerequisite in SPPS. The ninhydrin test is one of the simplest and most frequently used methods for this purpose [380]. A positive color reaction, performed with a small aliquot of the resin material, indicates unconverted amino groups. This so-called Kaiser test is simple, reliable (in most cases), and requires only minutes to perform. Titrimetric methods are also quite easy to perform. Further monitoring methods that rely on color reactions include the TNBS test [381], the acetaldehyde/chloroanil test [382], and the bromophenolblue test [383]. Dorman developed a protocol involving the determination of chloride ions after protonation of unconverted amino groups of the peptide with pyridinium chloride and subsequent elution of bound chloride with triethylamine [384]. Free amino groups may also be directly titrated with 0.1 N $HClO_4$, for example. Although these methods could be integrated in an automated process, they are too time-consuming. A similar situation applies to gel-phase NMR spectroscopy, which allows a direct determination of resin-bound groups.

In the Fmoc synthesis scheme the coupling of the N^{α}-Fmoc-protected amino acids can be monitored directly and quantitatively by the decrease in absorbance of the reaction solution at 300 nm. Similarly, cleavage of the Fmoc group with piperidine can be monitored by the increase in absorbance at the same wavelength.

Cleavage and total hydrolysis of an aliquot of resin-bound peptide after a certain number of synthetic steps is also an important and sensible tool to check the integrity of the product. Analysis may also be performed on each synthetic step by mass spectroscopy (MALDI-ToF MS) [385]. Exact yield determination and control of the integrity of the peptides bound to the resin can reliably be performed using the IRAA technique [312, 386, 387]. Dual linker analytical constructs comprising additional analytical components for analysis have also been developed [388].

4.5.5
Automation of the Process

The first automated peptide synthesizer was developed by Merrifield in 1966 for batchwise peptide synthesis (Fig. 4.38). The original machine, which is now in the Smithsonian Museum, consists of a reactor unit and a controlling unit. The former comprises the reaction vessel and valve systems for solvents, reagents, and amino acid derivatives as well as reservoirs for all these components. Nowadays, different types of synthesizers are commercially available.

Solid-phase peptide syntheses may also be performed in a continuous-flow mode using resin-filled columns. In some instances this method has proven to be superior compared to the batchwise synthesis developed by Merrifield. Major advantages lie in the reduced reagent and solvent consumption, and in the very short coupling cycles (1–2 min for TentaGel polymers of size 8 μm). The reaction progress may be monitored on a real-time basis by recording the conductivity of the solution [389]. However, the usual types of polymeric support material cannot be used, because of their insufficient pressure stability and the solvent-dependent

Fig. 4.38 Robert Bruce Merrifield and the first-generation solid-phase peptide synthesizer.

swelling properties. Hence, chemically modified silica gel [390] and kieselguhr/polydimethylacrylamide hybrid materials have been successfully applied to automated continuous-flow synthesis [391, 392].

TentaGel-type polymers are also appropriate solid support materials for continuous-flow syntheses [313] as they are chemically and physically inert, and also stable towards moderate pressure. Their uniform spherical shape and homogenous swelling behavior are further advantages.

4.5.6
Special Methods

The application of soluble polymeric support materials for peptide synthesis is characterized by the advantage of homogeneous reaction conditions. This may overcome some disadvantages of SPPS. Although the use of non-cross-linked soluble polystyrene (M_r ~200 kDa) as a solid support allows synthesis to be carried out in homogeneous phase [393], the excess reagents present after each coupling step can only be separated by quite tedious precipitation operations.

The use of polyethyleneglycol (PEG) as a polymeric support and C-terminal-protecting group for the growing peptide chain led to the important improvement of liquid-phase peptide synthesis [394]. In this method, excess low-molecular weight coupling reagents can be separated by ultrafiltration. Further progress was achieved in the development of the so-called crystallization method of Bayer and Mutter [395]. The addition of a suitable organic solvent (diethylether) rapidly induces a spontaneous and quantitative crystallization of the peptidyl-PEG by the formation of helical structures. Low-molecular weight starting materials and reagents are usually not co-precipitated, and may easily be separated. The principle of liquid-phase peptide synthesis is shown in Fig. 4.39. Although coupling times correspond to those of conventional synthesis, an excess of acylating agents and double or multiple couplings are necessary in order to obtain a maximum conversion. The growing polypeptide chain influences the solubility, and viscous solutions may result even when DMF is used as solvent. The preferred area for the application of liquid-phase peptide synthesis is for polypeptides containing fewer

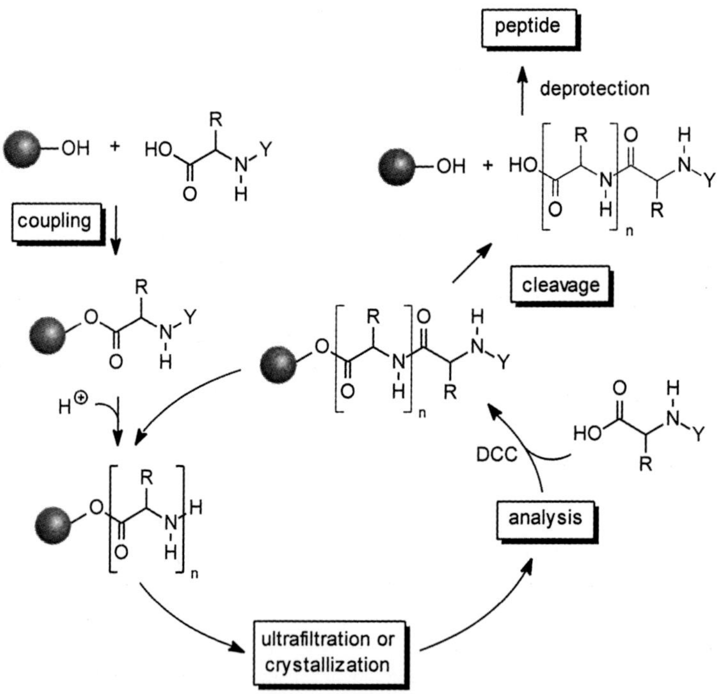

Fig. 4.39 Scheme of liquid-phase peptide synthesis on a soluble polymer. Y=amino-protecting group.

than 30 amino acid building blocks. The clear disadvantages of liquid-phase peptide synthesis are the long time taken for the operation, and the difficulties in its automation.

An alternating solid-phase/liquid-phase peptide synthesis [396] combines the strategic aspects of both methods, and of the polymer reagent peptide synthesis. The basic idea comprises a physical modification of the growing polypeptide chain during the coupling reaction in order to facilitate separation of the starting material from the product. This interesting concept has not yet received very much practical application, however.

The polymer reagent peptide synthesis does not utilize synthesis on polymeric support, because the growing polypeptide chain is retained permanently in solution. Instead, the polymer-bound reagents – and especially the activated carboxy components – are reacted with amino components with liberation of the polymer and of the protected peptide derivative. The polymer-bound reagents can be applied in excess, and the peptide formed may be separated – usually without difficulty. Wieland and Birr [397] developed a continuous process for peptide synthesis by applying polymer-bound active esters. The columns contain a polymer cross-linked with *p,p'*-dihydroxydiphenylsulfone where the active esters of N-protected amino acids are attached. However, the repetitive acidolytic deblocking reagents with subsequent neutralization steadily increase the salt concentration of the reaction solution. Photolabile N^{α}-amino-protecting groups have also been recommended for this column operation. Following synthesis, these are cleaved simply by photolysis, after which the resulting peptide with the free amino-terminus may be subjected to the next synthetic cycle.

4.5.7 Peptide Cleavage from the Resin

Final cleavage of the peptide from the polymeric support usually represents the last step of SPPS. As the anchoring group (linker) usually is chosen to be compatible with the peptide synthetic operations, selective cleavage between the C-terminus of the peptide and the solid support occurs upon treatment with reagents that may concomitantly effect partial or complete deprotection of the peptide side chains. In most cases, acidolytic methods are used, though other methods (transition metal catalysis for allyl-based linkers, photolysis [398, 399]) are available for the synthesis of sensitive compounds or special applications.

4.5.7.1 Acidolytic Methods

The Boc/Bzl scheme often relies on final cleavage by anhydrous liquid HF at 0 °C. This reaction must be carried out in a special apparatus, because HF decomposes glassware. Scavengers such as anisole or others must be added. Both the anchoring group and all nitrogen- or oxygen-protecting groups based on the *tert*-butyl type or benzyl type are cleaved. The nitro group of nitroarginine and the methoxybenzyl group of cysteine for example are also labile towards these conditions.

Other functionalities, such as N^{im}-benzyl groups, N^{im}-2,4-dinitrophenyl groups (His), S-alkylthio groups (Cys), and 4-nitrobenzylesters (Asp, Glu) are resistant to these cleavage conditions. The two-step, high–low HF protocol often provides the final product at greater purity.

Other acidolytic cleavage reactions do not require a special apparatus and may be performed in the reactor that is used for solid-phase synthesis. Cleavage reactions with TFMSA in TFA as the solvent can be improved considerably by the addition of scavenging sulfur compounds such as thioanisole or methionine [400, 401]. Several trialkylsilyl compounds have been selected as cleavage reagents on the basis of the HSAB (*h*ard and *s*oft *a*cids and *b*ases) concept developed by Pearson. Hard Lewis acids such as trimethylsilyl trifluoromethane sulfonate (TMSOTf) or trimethylsilyl bromide (TMSBr) are combined in TFA as the solvent with soft bases (e.g., thioanisole) as a highly efficient deblocking system both for solution-phase and for solid-phase purposes [402]. Cleavage with this "cocktail" (e.g., 1 M TMSOTf/thioanisole in TFA; 0 °C for 60 min) is much faster than the conventional method applying 1 M TFMSA/thioanisole in TFA. The different methods for acidolytic sulfide-assisted cleavage in nonaqueous solvents have been reviewed with respect to mechanism and applications [403].

The Fmoc/tBu scheme allows for much milder final deprotection conditions, for example with TFA or acetic acid in dichloromethane as the solvent (cf. Section 4.5.1).

In the case of special linker systems (e.g., HMBA), cleavage with different nucleophiles is possible, and results in the formation of C-terminal peptide acids (cleavage with aqueous or alcoholic hydroxide solutions, thiophenolate in DMF, or solvated cyanide), peptide esters (alcohols in the presence of tertiary amines), peptide amides (ammonia or amines), peptide hydrazides (hydrazine), or even peptide aldehydes and alcohols (reductive cleavage with $LiBH_4$).

Without doubt, the final cleavage reaction is the most critical step of a SPPS, though on occasion even the separation of the two steps, linker cleavage and side-chain deprotection, may be advantageous.

4.5.7.2 **Side Reactions**

Sensitive peptide bonds such as Asp-Pro or Aib-Pro may be cleaved under the strongly acidic cleavage conditions. Asp-Gly sequences are often prone to succinimide formation and concomitant isoaspartyl peptide ($\alpha \rightarrow \beta$ shift) formation, especially when treated with strong acids such as HF or TFMSA. Similar reaction conditions facilitate an N → O acyl shift involving serine or threonine residues. The products formed are thermodynamically labile, and the process may be reversed in aqueous bicarbonate solution (pH 7.5).

Tryptophan or tyrosine residues may, under certain circumstances, suffer from oxidative degradation. Tyrosine is prone to electrophilic C-alkylation, which may be prevented by the 2-bromobenzyl protecting group in Boc tactics. Arginine residues are reported to undergo deguanylation to give ornithine residues. As mentioned previously, methionine and cysteine residues are susceptible to oxidation and alkylation. In Fmoc tactics, most side reactions (e.g., side-chain alkylation or lactam formation) and others are sufficiently reduced by appropriately protected amino acid building blocks, for example Fmoc-Arg(Pbf)-OH [404], Fmoc-Arg(Pmc)-OH [405], Fmoc-Trp(Boc)-OH [406], and Fmoc-Asn(Trt)-OH [407].

4.5.7.3 Advantages and Disadvantages of the Boc/Bzl and Fmoc/tBu Schemes

Although, as mentioned previously, one advantage of the Boc tactics lies in the termination of β-sheet structures during repetitive acidolysis of the temporary Boc protection, the major drawbacks are to be seen in the harsh reaction conditions for the final deprotections that give rise to different side reactions (e.g., side-chain alkylation, cleavage of sensitive peptide bonds, rearrangements, cf. Section 4.5.7.2). Moreover, special equipment is needed for the final HF cleavage reaction.

Fmoc tactics are characterized by the possibility of synthesizing side chain-protected fragments for segment condensations and peptide cyclizations, the relatively mild cleavage conditions of the side chain-protecting groups and of the linker, and by UV-spectroscopic monitoring of the reaction progress. A detrimental feature is the favored formation of β-sheets and the propensity for diketopiperazine formation on the dipeptide stage. Spontaneous cleavage of Fmoc has also been observed [364, 365].

Systematic investigations slightly favor the Fmoc/tBu scheme over the Boc/Bzl scheme [408]. However, this conclusion does not in the main rely on the synthetic performance, but rather on the greater practicability and easy technical application. Rapid continuous-flow protocols [409] and optimized support materials [374] are regarded as beneficial.

Nonetheless, a general decision regarding which protecting group tactics would be preferred cannot be made. All significant parameters in the synthesis of longer peptides or peptides containing difficult sequences should be carefully analyzed, and the selected synthesis scheme should first be investigated in a small-scale synthesis (preview synthesis) combined with appropriate analytical methods in order to identify potential weak points in the synthetic plan.

4.5.8 Examples of Syntheses by Linear SPPS

Since its invention in 1963, the Merrifield concept has been applied to the synthesis of thousands of biologically active peptides and peptide analogues, and in this respect, a large number of improved variants have been used besides the original protocol. It must be mentioned that this synthetic concept has had an enormous impact on the development of peptide chemistry and organic chemistry during the past few decades. Indeed, the manifold approaches to combinatorial chemistry would not have been developed without the foundations having been set by Bruce Merrifield.

SPPS underwent extreme activity between 1968 and 1972, when the commercialization of automated synthesizers fostered its widespread application. In those early days, the chemical synthesis of small proteins had its limitations, and an attempted synthesis of lysozyme resulted in a complex mixture with a specific activity of only 0.5–1%. The total synthesis of ribonuclease A [410] proved to be a benchmark in that it was the first synthesis of an enzyme with a satisfactory activity of 3.5%. The application of HPLC methods in the purification of synthetic peptides during the early 1980s again led to an enormous increase in solid-phase synthesis activities that continues to this day. Initially, the solid-phase syntheses of molecules such as conotoxin G1, thymosin α, glucagon, β-endorphin, parathyrin, β-lipotropin, interleukin-3, HIV protease, and green fluorescent protein provided an impressive announcement of the vast potential of this synthetic method. Today, the commercial large-scale production of peptides includes the preparation of compounds such as corticotropin, atriopeptin, parathyrin, and calcitonin on a scale up to 200 g [310].

4.6 Biochemical Synthesis

Cells are capable of assembling amino acids and modifying peptides and proteins in a highly regulated and chemically efficient manner (cf. Section 3.2). These in-vivo reactions occur under mild conditions, and the catalyzing enzymes are the keys to the specificity and efficiency of the process. Before the ribosomal synthesis pathway had been elucidated, general peptide and protein biosynthesis was believed to be performed by a number of simple enzymatic processes. Moreover, despite the fact that the ribosomal cycle of protein biosynthesis (cf. Section 3.2.1) is now known to comprise complex machinery, the formation of a peptide bond was originally thought to be catalyzed by an enzyme that was tentatively named peptidyl transferase (nowadays known to be a ribozyme).

Regiospecificity and stereospecificity are two major advantages of biocatalysts. In chemical peptide synthesis, the obligatory protection and deprotection of the α amino function, the carboxy group and the various side-chain functionalities before and after a synthetic step are the most fundamental time-consuming operations. Furthermore, as many of the synthetic operations take place at groups adjacent to a center of chirality, a permanent risk of racemization must be taken into consideration. Therefore, the chemical synthesis of peptides continues to be a formidable chemical effort.

Although there are more than 150 chemical methods for peptide bond formation, an ideal coupling method should allow for a rapid reaction, without racemization or other side reactions, and result in quantitative coupling when the carboxy and amino components are applied stoichiometrically.

Recombinant DNA techniques, genetic engineering, genome sequencing, biochemical protein ligation (cf. Section 5.3.4), and the application of enzymes in chemical synthesis to control stereo- and regiospecificity have led to the so-called "biotechnological revolution". For this reason, new biochemically inclined methodologies are required at the synthetic front, both for the functional analysis of peptides and proteins, and the discovery of new therapeutic agents.

4.6.1
Recombinant DNA Techniques

The nucleotide sequence of the human genome has now been elucidated. Despite the existence in human cells of about 150 000 different genes each containing approximately 5×10^9 base pairs, the isolation of gene sequences is a realistic approach. The methods used for sequence analysis of DNA according to Maxam and Gilbert (1977) or Sanger (1977) can be applied reliably. Today, the sequence analysis of an isolated gene is more efficient than protein sequence analysis, and consequently protein sequences are often derived from DNA sequences.

Only a limited number of specialized peptides (cyclic peptides, peptide antibiotics) are synthesized nonribosomally by multi-enzyme complexes. A preparative procedure for the ribosomal synthesis of polypeptides and proteins is based on the principles of gene technology. Proteins and peptides can be expressed on a large scale by the recombination and expression of genetic material, for example in bacteria, and pharmacologically active peptides and proteins form a special focal point of interest in this context [411]. Indeed, based on their endogenous functions in the human body, these compounds may find application as therapeutic drugs, as well as influencing both pathophysiology and healing processes (cf. Chapter 9).

4.6.1.1 Principles of DNA Technology

Since the elucidation of the molecular structure of DNA by Watson and Crick, genetics and molecular biology have experienced almost unbelievable development. There is no species barrier for DNA, which serves as the "blueprint" for protein biosynthesis. For example, bacteria such as *Escherichia coli* may not only accept genetic material from other microorganisms and stably transmit it to their successors, but may also may serve as recipients of genetic material from either plants or animals.

Gene expression comprises synthesis of the corresponding mRNA (transcription) and synthesis of the protein (translation). The biosynthesis of a foreign gene product (protein) in an organism relies on a recombination of the genetic material of the microorganism with the DNA fragment encoding for the desired protein.

Eukaryotic genes contain noncoding elements (introns) in addition to the sequence information encoding for a protein. The precursor messenger RNA (mRNA) formed after transcription is subsequently processed into mature mRNA by splicing, at which point the introns are removed from the sequence.

Consequently, the genomic DNA of eukaryotes is not appropriate for direct transfection; instead, DNA which is complementary to the mature mRNA (cDNA) must be synthesized and used for transfection. The formation of cDNA is a prerequisite in order to obtain the correct protein sequence derived from a human gene, for example. Mature mRNA isolated from the donor cells is used as a template to synthesize cDNA *in vitro* with the enzyme reverse transcriptase, which occurs naturally in retroviruses. This enzyme requires a synthetic primer (of about 12 nucleotides) that is complementary to the sequence of the 3'-end of mRNA. In eukaryotic cells, this nucleotide sequence is often a poly-adenosine tail, but it may also encode for the C-terminal amino acid sequence of the protein. The hybrid composed of cDNA and mRNA is then subjected to alkaline hydrolysis, which destroys the mRNA. The cDNA is subsequently replicated and amplified by polymerase I. Blunt ends are formed by S1-nuclease, and synthetic oligonucleotide sequences containing cleavage points for restriction endonucleases are attached using DNA ligase for insertion into the vector. Alternatively, homopolymeric sequence are added using terminal transferase. The procedure is shown schematically in Fig. 4.40. Finally, the encoding DNA may be sequenced for verification purposes and subsequently be inserted into expression vectors of suitable host cells.

The process comprises the following steps (Fig. 4.41):

1. Isolation of the encoding DNA fragment from the donor organism.
2. Insertion of the DNA into a vector.
3. Transfection of the vector into the host organism.
4. Cultivation of the host organism (cloning), which leads to gene amplification, mRNA synthesis, and protein synthesis.
5. Isolation of the recombinant protein.

Originally, the term "cloning" meant the amplification of cells, with formation of identical daughter cells, but nowadays the term is also used for the amplification of DNA. Initially, minute amounts of the protein are identified using biological as-

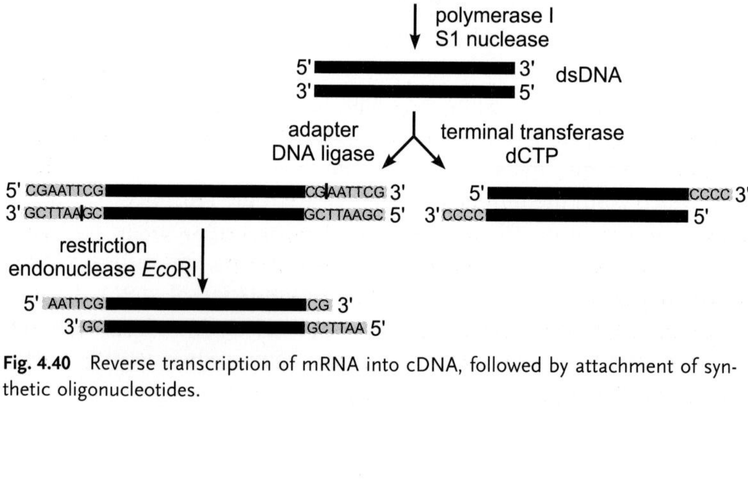

Fig. 4.40 Reverse transcription of mRNA into cDNA, followed by attachment of synthetic oligonucleotides.

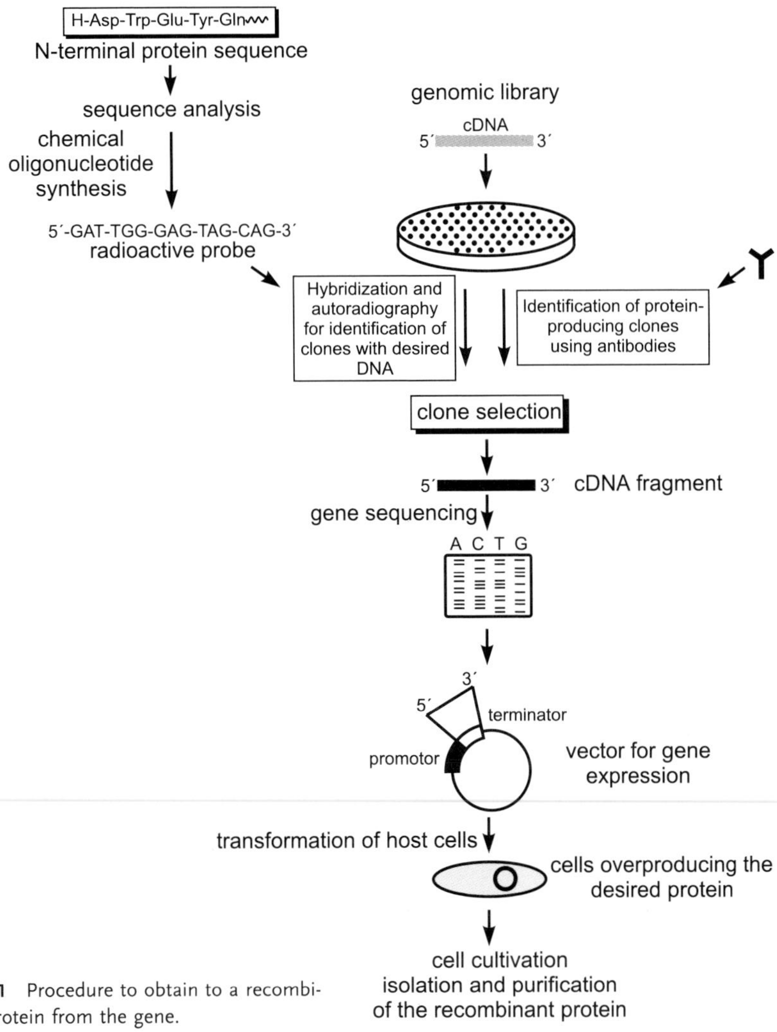

Fig. 4.41 Procedure to obtain to a recombinant protein from the gene.

says. Often, even microgram amounts are sufficient in order to raise antibodies or to perform partial sequencing, starting from the N-terminus. By using this sequence information, the corresponding genes can be identified and either isolated or obtained from gene banks.

The DNA fragments containing the genetic material to be transfected can be introduced into bacteria or other host organisms using so-called vectors. Usually, either plasmids (circular double-stranded DNA molecules that exist extrachromosomally in bacteria) or phages (viruses that infect bacteria) are used as vectors. The discovery and isolation of restriction endonucleases capable of cleaving double-stranded DNA at well-defined positions was the major precondition in the development of gene technology.

The DNA encoding for a protein can be obtained by isolation of the corresponding gene, by enzymatic synthesis from mRNA, or, in special cases by chemical synthesis. Chemical DNA synthesis [412] can be performed on commercial automatic synthesizers more efficiently compared to peptide synthesis. Genes of different proteins (e.g., the genes of human interferons) that contain more than 500 base pairs, or the DNA sequences encoding for the two insulin chains, have been synthesized chemically. The chemical synthesis of oligonucleotides containing special cleavage sequences for restriction endonucleases is also of eminent importance. Such synthetic linker groups are attached to the ends of a DNA molecule (Fig. 4.40) in order to allow the insertion of this DNA fragment into the vector. The DNA to be amplified and expressed, as well as the vector DNA, are cut by an appropriate restriction endonuclease, and the desired DNA fragment is inserted. The phosphate linkage between the DNA portions is formed by ligases to give the intact recombinant vector DNA molecule, which then is introduced into the host organism. Successful transfection can be monitored with labeled oligonucleotides (in-situ hybridization), or on the basis of the protein product using antibodies against this specific protein.

The successful expression of a recombinant protein in a host system requires not only the DNA sequence that encodes for the corresponding protein, but also additional regulatory sequences on the DNA inserted into a plasmid or another vector. Usually, highly efficient regulatory DNA domains are inserted preceding the DNA fragment encoding for the gene product in order to provide a high protein yield.

The selection of new host organisms for recombinant DNA is a major task, because *E. coli* as a host suffers from several disadvantages. Very often, eukaryotic proteins are modified post-translationally (e.g., by glycosylation), and hence contain carbohydrate chains of different lengths. Microorganisms do not modify proteins post-translationally, and usually do not export proteins into the cell culture medium. Furthermore, *E. coli* produces endotoxins that must be separated from the desired protein before its use as a pharmacologically active compound.

In the meantime, recombinant techniques for *Bacillus, Rhizobium* and *Streptomyces* have been described. Eukaryotic cells (from yeasts, insects, or mammals) can also be used as host organisms. Baker's yeast (*Saccharomyces cerevisiae*) or other yeasts are ideal expression systems that also allow for post-translational glycosylation. Stable plasmids are known for baker's yeast. Yeast applied in biotechnological processes usually is allowed to reproduce asexually, which safeguards stable transmission of the genetic material. The genes for interferon and insulin have been expressed in yeast.

Complex post-translational modifications such as glycosylation patterns specific for humans, or γ-carboxylations on glutamate residues, require mammalian cells as expression host systems.

An interesting concept to expand the genetic code has been introduced by Schultz et al. [413]. This uses chemically synthesized 3'-aminoacyl suppressor tRNA that allows for the incorporation of nonproteinogenic amino acids of any type into a protein. The codon UAG on mRNA usually associates with nonacylated suppressor tRNA, and leads to protein chain termination. If the suppressor tRNA has been either chemically or enzymatically [414, 415] loaded with a nonproteinogenic aminoacyl residue, this residue is incorporated into the peptide chain.

4.6.1.2 Examples of Synthesis by Genetic Engineering

As outlined in the preceding section, microorganisms may be genetically programmed to produce human hormones (e.g., insulin, somatotropin), enzymes, or vaccines. The potential of this area of applied gene technology has quickly been recognized, and many new applications are currently under investigation in medicinal chemistry, biotechnology, and agriculture. Currently, many therapeutic peptides and proteins are obtained by the fermentation of recombinant cells (cf. Chapter 9).

Somatostatin was the first peptide to be obtained by the use of recombinant DNA techniques [416]. The DNA encoding for the amino acid sequence of somatostatin was synthesized chemically using the amino acid triplet codons used most frequently in *E. coli*, after which the DNA sequence was inserted into an expression plasmid. The relatively small peptide was protected against proteolysis by fusion to the protein β-galactosidase. Cleavage of the 14-peptide from the hybrid protein with bromocyan was only possible because somatostatin does not contain methionine.

Interestingly, the genes encoding for the interferons from human leukocytes do not contain any introns, and so interferons can be produced in a straightforward manner in bacteria.

The production of human growth hormone using recombinant DNA techniques represents a major milestone in the substitution therapy of dwarfism, as before 1985 this hormone could be obtained only from human hypophyses at autopsy. All analogues produced natively by other mammalian organisms have different amino acid sequences, and the potential for viral contamination has also been a point of criticism. The gene family of the human growth hormone was discovered during molecular biology studies. The gene contains four introns, and the regulatory N-terminal pre-sequence is composed of 26 amino acids, which is typical for an export protein. This sequence is cleaved during secretion by a signal peptidase. Expression of the cDNA in *E. coli* originally provided the human growth hormone at a yield of 2.4 mg L^{-1}, which is far from optimum [417]. Protropin® (Genentech), which was approved for therapeutic use in 1985, additionally contains one nonsequence-specific N-terminal methionine residue, while the peptide with the original sequence was marketed as Humatrop® (Eli Lilly) [418].

Two alternative procedures are described for the overexpression of insulin. Eli Lilly expressed human insulin by separate overexpression of the A and B chains in the form of fusion proteins in *E. coli* (Fig. 4.42A) [419]. Following the isolation of both chains, they were connected by reductive thiolysis and subsequent oxidation (in air) to produce the active hormone. In the second variant (Fig. 4.42B), proinsulin was synthesized analogously to biosynthesis as a fusion protein in *E. coli* and transformed enzymatically into active insulin [420].

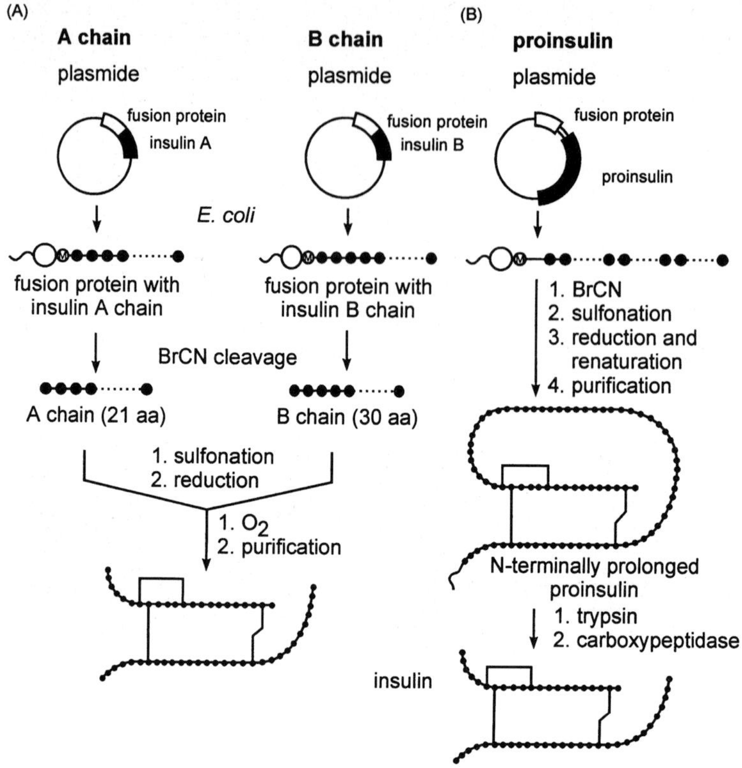

Fig. 4.42 Biotechnological insulin synthesis. (A) Separate synthesis of A and B chain, followed by oxidation to give active insulin. (B) Synthesis of proinsulin with subsequent enzymatic processing.

Currently, insulin analogues designed to improve the therapeutic properties of the hormone have been produced by recombinant DNA technology [421]. The first clinically available insulin analogue, lispro, was marketed in the U.S. in 1996 [422].

Insulin glargine (HOE 901, Lantus®) is produced using a nonpathogenic genetically engineered strain of *E. coli* and has found application as a long-acting agent for the treatment of type 1 and 2 diabetes mellitus (cf. Section 9.3.1). A typical DNA technology production plant shows Fig. 4.43.

4.6.1.3 **Cell-free Translation Systems**

As shown above, the biotechnological synthesis of natural products is mainly based on cellular systems. Methods for the expression of foreign genetic material in living cells often suffers from limitations however, and this is especially true for the production of polypeptides of a comparatively short length (up to 50–60 amino acid residues). Since most bioactive peptides and polypeptides are released

Fig. 4.43 Production plant for industrial gene technological insulin synthesis (Photo: Walter Kloos/Hoechst AG).

after the complex processing of biosynthetic precursors, it is difficult to create vectors that result in the formation of active peptide products when expressed in either prokaryotic or eukaryotic cells. Even when an expression system has been successfully created, the peptide products may be unstable or toxic to the cells. Other problems also result from insoluble peptide products forming aggregates and inclusion bodies. These general limitations may be avoided when translation is performed in cell-free systems.

Zamechnik et al. [423] described the first systems for cell-free protein synthesis, and demonstrated that peptide synthesis proceeds on ribosomes, and requires ATP, GTP, and tRNA. The coat protein of coliphage F2 was the first protein synthesized in a cell-free translation system using an extract from *E. coli* [424]. The main disadvantage of cell-free translation systems is the low peptide yield, as under these conditions only two to three polypeptide chains are formed per mRNA molecule used [425, 426].

A continuous cell-free translation system capable of producing polypeptides in high yield was described by Spirin et al. [427]. This continuous-flow, cell-free (CFCF) system allows for a continuous feeding of amino acids, ATP, and GTP to the reaction mixture containing the cell extract and mRNA, and a continuous removal of the polypeptide product (product removal proved to be essential to maintain prolonged ribosomal activity). The simplest device for performing a continu-

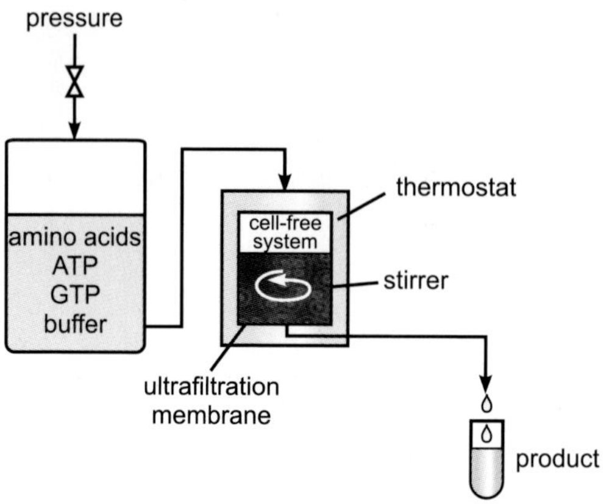

Fig. 4.44 Schematic diagram of a bioreactor for continuous-flow translation, fitted with an ultrafiltration membrane.

ous-flow translation contained an Amicon 8 MC microfiltration system which was installed in a thermostatically controlled chamber (Fig. 4.44).

Later, a thermostatically controlled chromatography column supplied with an ultrafiltration membrane at the output and a standard column adapter at the input was used. These continuous-flow bioreactors have been applied to cell-free translation using both prokaryotic (*E. coli* lysate) and eukaryotic (wheat embryos, *Triticum* sp.) systems. The mRNA of human [Val^8]calcitonin was obtained by in-vitro SP6 phage polymerase transcription of the synthetic calcitonin gene inserted into a plasmid under SP6 promoter, and was used for human [Val^8]calcitonin synthesis. Between 150 and 300 copies of calcitonin were synthesized per mRNA in both types of translation systems during 40 h. Pulsed bioreactors fitted with a hollow fiber ultrafiltration membrane and with a flat ultrafiltration membrane, respectively, have also been used. The inclusion of the translation mixture in granules (e.g., alginate) or polysaccharide vesicles was found to be advantageous. In this configuration, the translation of calcitonin mRNA proceeded for 100 h and produced 250 μg calcitonin from 5 mL of the incubation mixture. The CFCF system enables the production of rather pure polypeptides based on the fact that the ultrafiltration membrane prevents leakage of most proteins present in the reaction mixture. The CFCF system has doubtlessly opened new fields of application, and many of the limitations originally imposed on cellular systems have now been overcome using cell-free gene expression. An ongoing problem is that the productivity and efficiency of cell-free systems are still low, and newer bioreactors that incorporate efficient systems to monitor the translation course are required. Moreover, strict maintenance of ATP and GTP concentration, and the development of efficient regeneration systems for both cofactors, are indispensable. Several promising approaches to increase the efficiency of the CFCF system have been suggested, however [428–430]. A coupled replication-translation system might be developed which would solve the problems of RNA template stability and the efficiency of its translation.

4.6.2
Enzymatic Peptide Synthesis

4.6.2.1 Introduction

Although enzymes have in general become valuable tools in medium- to large-scale synthetic organic chemistry [431], a universal C–N ligase with high catalytic efficiency for all possible combinations of the 21 proteinogenic amino acids to be coupled has yet to be developed. Indeed, the demands on specificity are so severe that not even nature could solve this problem! Hence, protein biosynthesis (cf. Section 3.2.1) has been developed as a stepwise strategy, starting from the N-terminus of the growing peptide chain. The process is catalyzed by the ribosomal peptidyl transferase, and followed by limited proteolysis and other post-translational modifications of the precursor molecules (cf. Section 3.2.2). Despite intensive investigations during the past years, details of the nature and the basic mechanism of the peptidyl transferase reaction within the ribosome remain largely unknown. According to studies by the Nobel laureate Thomas R. Cech et al. [432], an in-vitro-selected ribozyme is capable of catalyzing the same type of peptide bond formation as a ribosome, and its sequence and secondary structure appear strikingly similar to the "helical wheel" portion of 23S rRNA involved in the activity of ribosomal peptidyl transferase. These findings provided evidence for the feasibility of the "RNA world" hypothesis, by demonstrating that RNA itself is capable of catalyzing peptide bond formation. In comparison to the specificity requirements of proteases, peptidyl transferase appears to be an old, unspecific ribozyme, which is in accordance with its function in evolution.

Even if it were possible to separate ribozyme activity from the ribosome, or to isolate an in-vitro-selected ribozyme that catalyzes peptide bond formation like a ribosome, such a biocatalyst does not seem suitable for simple practical use in peptide synthesis. In principle, this conclusion is also valid for the nonribosomal poly- or multienzymes involved in the biosynthesis of peptide antibiotics (cf. Section 3.2.3).

Consequently, only those enzymes which usually act as hydrolases, catalyzing the cleavage of peptide bonds, should be considered as peptide ligases. The reaction of proteolysis is irreversible, as peptide bond hydrolysis is exergonic. Although the ionic hydrolysis products are thermodynamically more stable, the fundamental suitability of proteases to catalyze the formation of peptide bonds is based on the principle of microscopic reversibility that was predicted by J. H. van't Hoff in 1898. Despite this, about 40 years elapsed before the first experimental proof of van't Hoff's prediction was produced, this being based on the clear-cut protease-catalyzed synthesis of an amide bond [433]. Unfortunately, another 40 years elapsed before this approach gained any industrial importance, though during the past few decades considerable effort has been made to determine the optimum conditions for protease-catalyzed peptide synthesis (for reviews, see [434–438]).

4.6.2.2 Approaches to Enzymatic Synthesis

According to the short notation of protease-substrate interactions [439], proteases have binding grooves ($S_n...S_3$-S_2-$\mathbf{S_1}$-$\mathbf{S'_1}$-S'_2-$S'_3..S'_n$) on both sides of the scissile bond -$\mathbf{P_1}$-$\mathbf{P'_1}$- of the substrate, whereas substrate binding ($P_n...P_3$-P_2-$\mathbf{P_1}$-$\mathbf{P'_1}$-P'_2-$P'_3..P'_n$) in proteolysis is mostly reduced to the P_1 amino acid residue. Peptide bond formation is, in contrast to proteolysis, a two-substrate reaction that not only requires a specificity-dependent insertion of the carboxy component into the S subsites of the active site, but also optimal binding of the amino component in the S′ region. Therefore, the S′–P′ interactions essentially determine the outcome of the C–N ligation reaction.

As mentioned above, the equilibrium of a protease-catalyzed reaction normally lies on the side of the thermodynamically more stable cleavage products. Several mechanistically different manipulations are possible to shift the equilibrium in favor of peptide bond formation. Approaches towards protease-catalyzed peptide bond formation are generally classified according to the type of carboxy component used (Fig. 4.45).

In the equilibrium-controlled approach (Fig. 4.45A), the carboxy component contains a free carboxy group, while in the kinetically controlled approach (Fig. 4.43B) the carboxy component is employed in activated form, mainly as an alkyl ester. Both strategies are fundamentally different due to the energy required for the conversion of the starting components into the peptide products.

Equilibrium-controlled synthesis This equilibrium-controlled or thermodynamic approach (Fig. 4.45A) represents the direct reversal of proteolysis. Consequently, all proteases can be used – independent of their mechanisms. Besides this advantage, the large amount of enzyme required and the low reaction rate are draw-

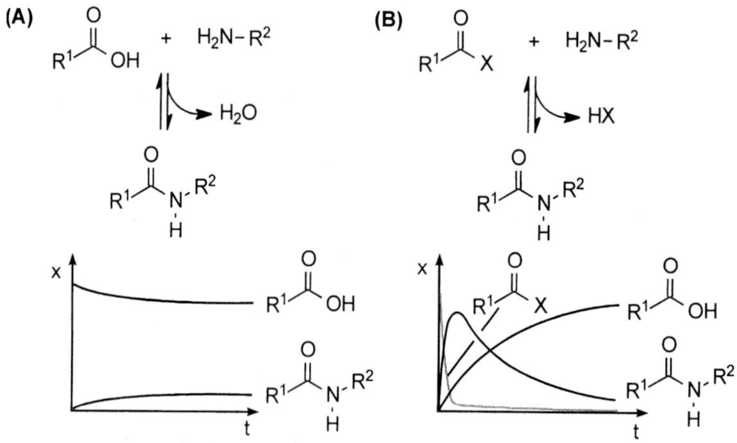

Fig. 4.45 Comparison of the equilibrium- (A) and the kinetically controlled approach (B) of protease-catalyzed peptide synthesis. R^1-COOH = carboxy component; R^1-CO-X = slightly activated carboxy component; H_2N-R^2 = amino component; HX = leaving group.

backs of this approach. The ionization equilibrium must be manipulated to shift the equilibrium towards peptide bond formation. Addition of water-miscible organic solvents to the aqueous reaction mixture is a viable approach. However, the catalytic activity of the enzyme may decrease under such conditions. The lower dielectric constant of the medium reduces the acidity of the carboxy group and, to a smaller extent, the basicity of the amino group of the nucleophilic amino component. In addition to the manipulations mentioned above, reverse micelles, anhydrous media containing minimal water concentrations, water mimics, and reaction conditions promoting product precipitation are often employed.

Kinetically controlled synthesis In contrast to the equilibrium-controlled approach, the kinetically controlled protease-catalyzed peptide synthesis [440] requires much less enzyme, the reaction time to obtain maximum product yield is significantly shorter, and the product yield depends both on the properties of the enzyme used and its substrate specificity. Whereas the equilibrium-controlled approach ends up in equilibrium (Fig. 4.45A), the concentration of the product formed in the kinetic approach passes a maximum before the slower hydrolysis of the product becomes important. Amidase activity of most proteases is lower than esterase activity. The product inevitably is hydrolyzed, if the reaction is not stopped once the acyl donor ester is consumed. Fig. 4.46 indicates the kinetic model of a protease-catalyzed acyl transfer reaction.

The kinetic approach requires an acyl donor ester (Ac-X) as carboxy component, and is limited to proteases that rapidly form an acyl enzyme intermediate (Ac-E), for example serine and cysteine proteases. The enzyme (EH) acts as a transferase catalyzing the transfer of the acyl moiety to the nucleophilic amino component (HN). The amino component reacts in competition with water with the acyl enzyme to form the peptide (Ac-N). The ratio between aminolysis and hydrolysis is of decisive importance for successful preparative peptide synthesis. However, serine and cysteine proteases are not perfect acyltransferases, and undesired reactions may take place due to their limited specificity (e.g., hydrolysis of the acyl en-

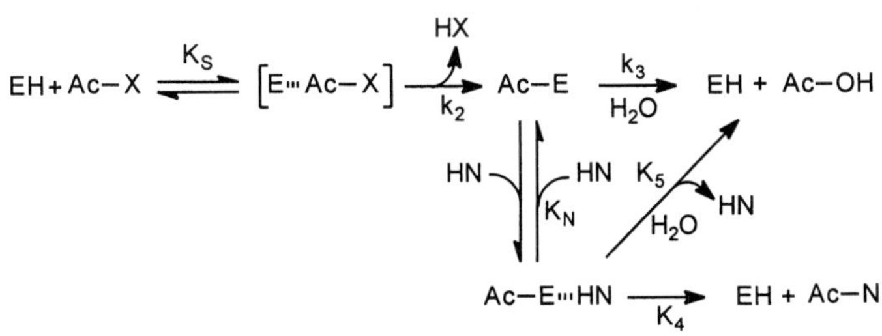

Fig. 4.46 Kinetic model for protease-catalyzed acyl transfer reaction. EH = free enzyme; Ac-X = acyl donor (carboxy component); HX = leaving group; Ac-E = acyl enzyme complex; Ac-OH = hydrolysis product; HN = acyl acceptor (amino component); Ac-N = aminolysis product (synthesized peptide).

zyme, secondary hydrolysis of the ligation product and proteolytic cleavage of other protease-labile peptide bonds present in the segments to be ligated). An understanding of the molecular interactions between the acyl enzyme and the attacking nucleophilic amino component allows for an optimization of the acyl transfer efficiency. Efficient nucleophilic attack of the amino component essentially depends on optimal binding within the active site by S′-P′ interactions. More information on the specificity of the S′ subsites of serine and cysteine proteases has been obtained by systematic acyltransfer studies with libraries of nucleophilic amino components, known as S′ subsite mapping [440].

Esterase activity can be further positively manipulated by varying the kind of leaving group. For preparative peptide synthesis, such manipulation is very important as it allows complete conversion of the acyl donor ester before the product will be hydrolyzed. There is no doubt that the course of kinetically controlled protease-catalyzed peptide synthesis can be more efficiently influenced than the equilibrium approach.

In contrast to the chemical approach, peptide synthesis using stepwise enzymatic chain assembly may start either from the N-terminus (cf. Fig. 5.2) or the C-terminus, because enzyme-catalyzed couplings are devoid of racemization. However, stepwise protease-catalyzed assembly should not be directly compared to automatic solid-phase technique (cf. Section 4.5). Selected di- and tripeptides (cf. Section 9.2), which are important intermediates in many pharmaceutical products, can be synthesized enzymatically on a large scale even in a continuous process, using solubilizing protecting groups [441–443].

In addition, the application of cross-linked enzyme crystals (CLECs) of thermolysin [444] and subtilisin [445] has been recommended for the synthesis of small peptides and peptide alkylamides. As will shown below, the solid-to-solid conversion has been proven as a very useful method for the synthesis of selected short peptides that comply with the requirements for this special synthetic procedure. Methodological support for practical application of protease-catalyzed peptide synthesis is available [446].

Although the kinetic approach should be preferred, the final decision depends on the overall synthesis concept. The largest industrial-scale application of the equilibrium approach is most likely the enzymatic synthesis of Z-Asp-Phe-OMe, the precursor of the peptide sweetener aspartame [447]. Transpeptidation technology on a large industrial scale is well documented for the conversion of porcine insulin into human insulin by trypsin [448] – a process that had widely been used before the gene technology production of insulin.

4.6.2.3 **Manipulations to Suppress Competitive Reactions**

The most important factors limiting the widespread routine application of proteases in kinetically controlled peptide synthesis are hydrolysis of the acyl donor ester and proteolysis within the starting fragments to be coupled, or the final condensation product. These undesired reactions can be minimized by various manipulations of the reaction medium, the enzyme, the substrate, and on mechanistic features of the process.

Medium engineering is based on the replacement of water by organic solvents, and uses mono- or biphasic aqueous-organic and organic monophasic mixtures or microaqueous systems [449]. Competing hydrolysis can be minimized by decreasing the water concentration in the reaction medium of enzymatic synthesis in microaqueous systems that do not employ organic solvents, termed solid-to-solid conversion [450–452], and in frozen aqueous systems [453].

Enzyme engineering describes a range of techniques, from deliberate chemical modification to genetic remodeling of a wild-type enzyme [454]. A mutant of subtilisin BNP′, termed "subtiligase", was obtained by Jackson et al. [455] by protein design and used in a further synthesis of RNase A (cf. Section 5.2.2.2, Fig. 5.9).

4.6.2.4 Irreversible C–N Ligations by Mimicking Enzyme Specificity

The substrate-mimetic approach represents a very specific and efficient form of substrate engineering. Despite the development of various possibilities to suppress competing reactions, proteolytic cleavage of the peptide bond formed cannot be completely suppressed. In this context, the fact must be considered that the newly formed peptide can be cleaved during the course of the catalytic process by the same enzyme. The peptide bond-forming step and the proteolytic cleavage step do not differ in substrate specificity requirements. Since the specificity of proteases is manifested by the side chain of the P_1 amino acid residue, an irreversible peptide bond formation does not seem possible.

In ribosomal peptide synthesis (cf. Section 3.2.1), the selection of the amino acid residues to be coupled is performed as a specific recognition process prior to the acyl transfer reaction, catalyzed by the unspecific ribozyme peptidyl transferase. An irreversible C–N ligation in a biomimetic fashion would be possible also in enzymatic peptide synthesis if the specificity selection could be separated from the actual peptide bond-forming step. This proved to be feasible by transferring the moiety that determines the specificity from the P_1 amino acid side chain to the leaving group of the acyl donor ester. Hence, the enzyme is capable of recognizing, in a kinetically controlled acyl transfer reaction, the acyl donor ester by the specificity determinant present in the leaving group. The specificity-bearing leaving group is released upon formation of the acyl enzyme followed by peptide bond-forming aminolysis. Consequently, the newly formed peptide bond can no longer be cleaved by the enzyme, because the moiety responsible for recognition by the protease is then lacking.

This irreversible synthesis concept was for the first time confirmed in a model peptide synthesis catalyzed by trypsin involving various nonspecific N^{α}-protected amino acid 4-guanidinophenyl esters (OGp) as acyl donors, and various amino acid and peptide derivatives as nucleophilic acyl acceptors. Trypsin accepts the guanidino group of the 4-guanidinophenyl ester as a substitute for the arginine residue, that is usually recognized by the S_1 subsite [456]. This concept was exploited in further examples by another group, who named this type of acyl donor "ester inverse substrate" [457]. The more correct term "substrate mimetic" was introduced by Bordusa et al. [458], together with an extension of this approach to irreversible peptide segment condensations with other proteases. The mechanism of the substrate mimetic concept has been elucidated [459–461]. As shown in Fig. 4.47, a common acyl donor **A** binds its acyl residue to the S-binding site, where the leaving group is present at the S'_1 subsite.

The scissile ester bond may be attacked by Ser^{195} in the case of trypsin, followed by acylation of the enzyme. However, applying the same binding principle, a substrate mimetic with the leaving group located either in S'_1 (**B**) or in S_1 (**C**) leads to catalytically unproductive binding. Recognition of the specificity determinant in the leaving group by the S_1 subsite fails in binding arrangement **B**, whereas in the binding position **C** the scissile ester bond would be far away from the active Ser^{195}, and neither hydrolysis nor acylation would occur. Hence, the binding of a substrate mimetic and the subsequent acylation leads to an acyl enzyme having the acyl residues at the S′-subsites (**D**). Prior to the deacylation step, the acyl residue has to flip to the S-subsite of the enzyme. The common kinetic model has to be extended by a rearrangement step between the two acyl enzyme species E-Ac, with the acyl moiety in the S′ subsite and the common acyl enzyme Ac-E. Peptide nucleophiles that bind specifically to the S′ subsite should be able to remove the acyl moiety from the S′ region more efficiently than water, which decreases unwanted hydrolysis of the acyl enzyme.

In summary, cationic, anionic, and hydrophobic types of substrate mimetics have been designed, characterized and applied for various peptides synthesis, including segment condensations [462]. Consequently, the substrate mimetic strategy should be useful for combining the enzymatic ligation of segments with suitable SPPS methods, especially for the preparation of acyl component peptide segments in the form of appropriate esters (cf. Section 5.2.2.2, Fig. 5.10).

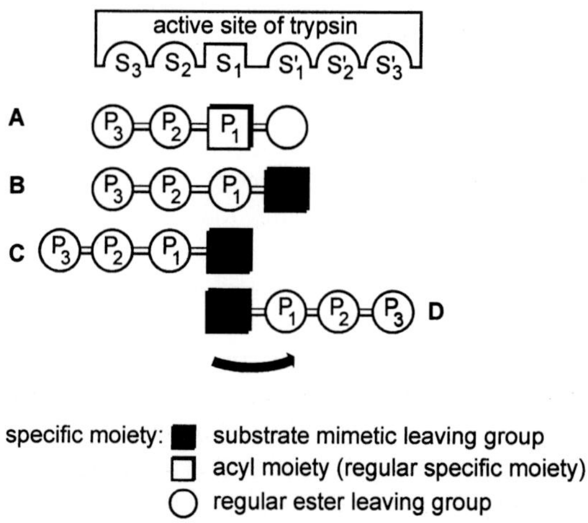

Fig. 4.47 Comparison of the binding of a common acyl donor ester (A) and a substrate mimetic in different binding positions (B–C) to the active site of trypsin with the hypothetical binding mode of a substrate mimetic (D).

Enzymatic methods have several advantages over chemical procedures, but at present more peptides are synthesized by chemical synthesis than in protease-catalyzed processes. The use of peptide synthesizers, in addition to recent new developments in the field of chemical ligation procedures (cf. Section 5.4), still favors chemical methods over the enzymatic approach. However, the enzymatic methods have undoubted advantages, such as the prevention of racemization, no requirement for time-consuming and expensive protection/deprotection procedures of side-chain functions, a reduced use of problematic (toxic) solvents and reagents, and possible re-use of the biocatalysts. Hence, the question should not be whether to use chemical or enzymatic approaches in peptide synthesis. Rather, an ingenious combination of chemical and enzymatic steps should promote general progress in peptide synthesis. Indeed, it might be demonstrated that once the synthesis conditions have been optimized, kilogram quantities of biologically active peptides and analogues could be obtained using enzymatic coupling methods. The semisynthetic synthesis of human insulin and the large-scale production (hundreds to thousands of tons) of the aspartame precursor underline the industrial importance of the enzymatic approach. Ultimately, even proteins may be synthesized via this approach, as shown for the synthesis of RNase A.

The C–N ligation strategy based on the substrate mimetic concept allows, for the first time, an irreversible peptide bond formation catalyzed by proteases. When combined with frozen-state enzymology (as demonstrated in model reactions [463]) or with protease mutants lacking amidase activity, the substrate mimetics C–N ligation approach will undoubtedly contribute to significant progress in enzymatic peptide synthesis. Recombinant polypeptide thioesters as substrate mimetics may be coupled in fragment condensations with chemically synthesized or recombinant fragments. This specific programming of enzyme specificity by molecular mimicry corresponds in practice to a conversion of a protease into a C–N ligase, and results in a biocatalyst which has not been developed evolutionarily by nature.

4.6.3
Antibody-catalyzed Peptide Bond Formation

According to Linus Pauling [464], the action of an enzyme depends to a large extent on the complementarity between the active site of the enzyme and the transition state of the enzyme-catalyzed reaction, as shown schematically for the peptide bond-forming step (Fig. 4.48).

In 1969, it was proposed by Jencks [465] that antibodies could be produced to function as catalysts. According to this idea, structures that mimic the stereoelectronic properties of transition states might be capable of eliciting antibodies that accelerate the corresponding chemical reaction. Consequently, if an antibody could bind a substrate already in the transition state conformation, it might act as an enzyme catalyzing the reaction to which the transition state conformation is predisposed. For this reason, catalytic antibodies have been also termed "abzymes". Investigations concerning requirements for the catalysis of peptide bond

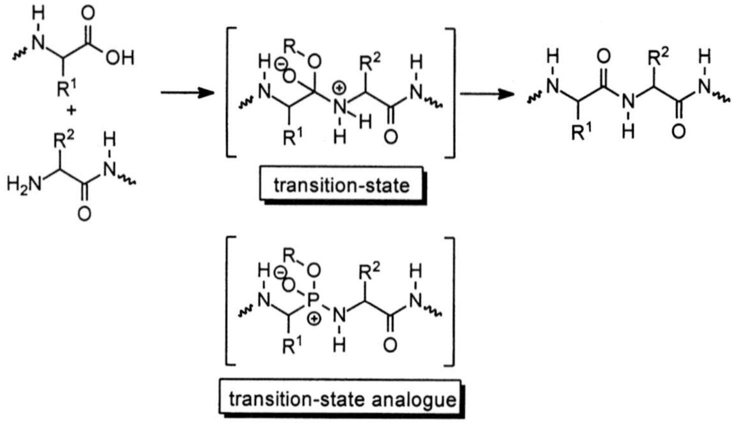

Fig. 4.48 Comparison of the transition-state structure of peptide synthesis by ester aminolysis with the structure of a designed transition-state analogue.

formation using catalytic antibodies was the consequence of progress in antibody catalysis for organic transformations [466, 467].

The normal approach to generate catalytic antibodies requires the design of haptens that are topologically and structurally near-identical to the transition state of the appropriate reaction, as shown in Fig. 4.48. In particular, peptide synthesis should require a large number of catalytic antibodies in order to consider the specificity requirements of the different amino acid residues in the C-terminal position of the carboxy component, and the N-terminal position of the amino component to be coupled.

As shown in Fig. 4.49, Hirschmann et al. [468] synthesized the phosphonamide shown in Fig. 4.49 A as a transition-state analogue (cf. Section 7.2.5) which was used to generate antibodies for catalyzing the model peptide synthesis (Fig. 4.49 B). The cyclohexyl moiety of the transition- state analogue should create a binding site in the antibody that would accommodate various hydrophobic side-chain moieties of the carboxy component. The *p*-nitrobenzyl residue in the phosphonate was designed in order to facilitate the dissociation of both the leaving group and peptide from the antibody. The authors described antibody-catalyzed dipeptide syntheses starting from all possible stereoisomeric combinations of the ester and amide substrates in yields between 44% and 94%. Two monoclonal antibodies (16G3 and 18C10) were found to cause a ~220-fold acceleration of the reaction. Further studies with the so-called antibody ligase 16G3 have demonstrated that this abzyme catalyzes not only the synthesis of dipeptides but also the coupling of an activated amino acid with a dipeptide to form a tripeptide, as well as a simple (2+2) segment condensation with average yields of 80% within a reaction time of 20 min [469].

Jacobsen and Schultz [470] created an antibody (Fig. 4.50) against a neutral phosphonate diester transition-state analogue (**A**) that was capable of catalyzing

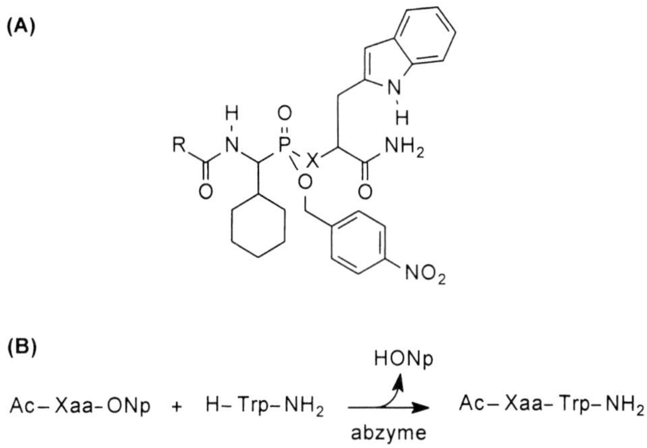

Fig. 4.49 Structural variants of the transition-state analogue (A) for an abzyme-catalyzed model peptide synthesis (B). Xaa=Val, Leu, Phe; X=NH; R=-$(CH_2)_3COOH$/X=O; R=-$(CH_2)_3COOH$/X=O; R=CH_3.

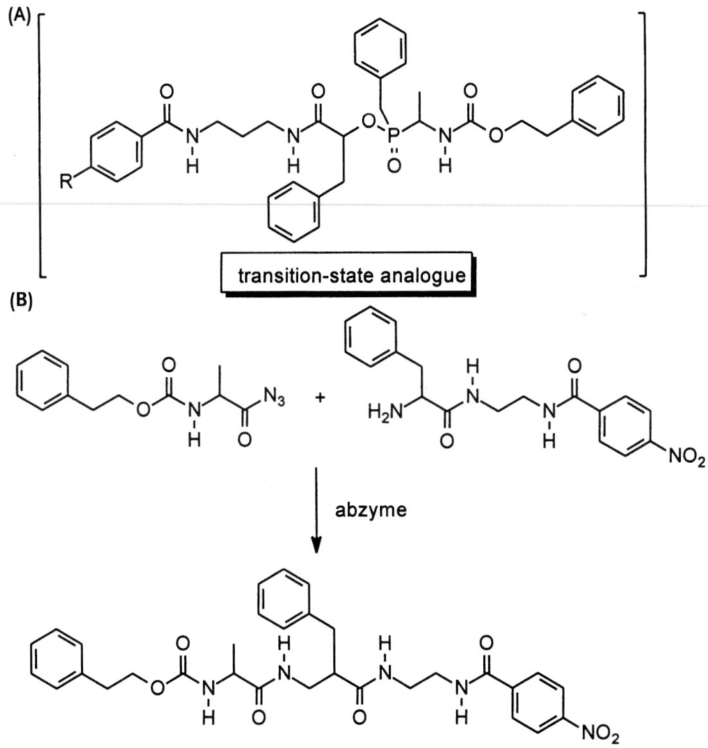

Fig. 4.50 Transition-state analogue (A) and model peptide synthesis reaction (B) catalyzed by a catalytic antibody. R=NH_2 or NO_2.

the reaction (**B**) between *N*-phenylethoxycarbonyl-L-alanyl azide with *N*-(4-nitrobenzoyl)-*N'*-L-phenylalanyl-1,3-diaminopropane at a 10,000-fold rate relative to the uncatalyzed reaction, and in 65% yield.

Interestingly, in both cases the catalytic antibodies do not catalyze the hydrolysis of both the active ester and the azide to an appreciable degree. Product hydrolysis and racemization have not been observed. As mentioned earlier, the main disadvantage of the catalytic antibody approach is the requirement of a large number of antibody catalysts to accommodate the wide specificity pattern of amino acids in coupling reactions. On the other hand, these results provide an impetus for producing further generations of antibodies capable of catalyzing the ligation of longer unprotected fragments, combined with a general strategy for the development of sequence-specific antibody ligases.

4.7 References

1 F. Albericio, *Biopolymers* **2000**, *55*, 123
2 M. Bergmann, L. Zervas, *Chem. Ber.* **1932**, *65*, 1192
3 S.C. McKay, N.F. Albertson, *J. Am. Chem. Soc.* **1957**, *79*, 4686
4 F. Weygand, E. Nintz, *Z. Naturforsch.* **1965**, *20b*, 429
5 A. Patchornik, B. Amit, R.B. Woodward, *J. Am. Chem. Soc.* **1970**, *92*, 6333
6 D.T. Gish, F.H. Carpenter, *J. Am. Chem. Soc.* **1953**, *75*, 950
7 F.H. Carpenter, D.T. Gish, *J. Am. Chem. Soc.* **1952**, *74*, 3818
8 R.A. Boissonnas, G. Preitner, *Helv. Chim. Acta* **1955**, *36*, 875
9 L. Kisfaludy, S. Dualszky, *Acta Chim. Acad. Sci. Hung.* **1960**, *24*, 301
10 K. Blaha, J. Rudinger, *Coll. Czech. Chem. Commun.* **1965**, *30*, 985
11 N. Izumiya, K. Noda, S. Terada, *Bull. Chem. Soc. Japan* **1970**, 43, 1883
12 B.W. Erickson, R.B. Merrifield, *J. Am. Chem. Soc.* **1973**, *95*, 3757
13 J.W. Chamberlin, *J. Org. Chem.* **1966**, *31*, 1658
14 C. Birr, W. Lochinger, G. Stahnke, P. Lang, *Ann. Chem.* **1972**, *763*, 162
15 R. Schwyzer, *Angew. Chem.* **1959**, *71*, 742
16 G.R. Matsueda, J.M. Stewart in: *Peptides: Chemistry, Structure and Biology*, R. Walter, J. Meienhofer (Eds.), Ann Arbor Scientific Publishers, Ann Arbor, **1975**
17 D.S. Kemp, C.F. Hoyng, *Tetrahedron Lett.* **1975**, 4625
18 P. Sieber, B. Iselin, *Helv. Chim. Acta* **1968**, *51*, 622
19 D. Wildemann, M. Drewello, G. Fischer, M. Schutkowski, *J. Chem. Soc., Chem. Commun.* **1999**, 1809
20 A. Tun-Kyi, R. Schwyzer, *Helv. Chim. Acta* **1976**, *59*, 1642
21 D.F. Veber, W.J. Paleveda, Jr., Y.C. Lee, R. Hirschmann, *J. Org. Chem.* **1977**, *42*, 3286
22 G.W. Anderson, A.C. McGregor, *J. Am. Chem. Soc.* **1957**, *79*, 6180
23 E. Wünsch, R. Spangenberg, *Chem. Ber.* **1971**, *104*, 2427
24 H. Eckert, M. Listl, I. Ugi, *Angew. Chem.* **1978**, *90*, 388; *Angew. Chem. Int. Ed. Engl.* **1978**, *17*, 361
25 W.L. Haas, E.V. Krumkalns, K. Gerzon, *J. Am. Chem. Soc.* **1966**, *88*, 1988
26 H. Kalbacher, W. Voelter, *Angew. Chem.* **1978**, *90*, 998; *Angew. Chem. Int. Ed. Engl.* **1978**, *17*, 944
27 G. Jäger, R. Geiger in: *Peptides 1971*, H. Nesvadba (Ed.), North-Holland, Amsterdam, **1971**, 78
28 L.A. Carpino, *Acc. Chem. Res.* **1987**, *20*, 401
29 L.A. Carpino, G.Y. Han, *J. Am. Chem. Soc.* **1970**, *92*, 5748
30 B. Henkel, E. Bayer, *J. Peptide Sci.* **2001**, *7*, 152

31 A.T. Kader, C.J.M. Stirling, *J. Chem. Soc.* **1964**, 258

32 E.T. Wolters, G.I. Tesser, R.J.F. Nivard, *J. Org. Chem.* **1974**, *39*, 3388

33 C.G.J. Verhart, G.I. Tesser, *Recl. Trav. Chim. Pays-Bas* **1988**, *107*, 621

34 V. V. Samukov, A.N. Sabirov, P.I. Pozdnyakov, *Tetrahedron Lett.* **1994**, *35*, 7821

35 A.N. Sabirov, Y.D. Kim, H.J. Kim, V.V. Samukov, *Protein Peptide Lett.* **1997**, *4*, 307

36 L.A. Carpino, M. Philbin, *J. Org. Chem.* **1999**, *64*, 4315

37 L.A. Carpino, M. Philbin, M. Ismail, G.A. Truran, E.M.E. Mansour, S. Iguchi, D. Ionescu, A. El-Faham, C. Riemer, R. Warrass, M.S. Weiss, *J. Am. Chem. Soc.* **1997**, *119*, 9915

38 L.A. Carpino, M. Ismail, G.A. Truran, E.M.E. Mansour, S. Iguchi, D. Ionescu, A. El-Faham, C. Riemer, R. Warrass, *J. Org. Chem.* **1999**, *64*, 4324

39 L.A. Carpino, E.M.E. Mansour, *J. Org. Chem.* **1999**, *64*, 8399

40 C.M. Stevens, R. Watanabe, *J. Am. Chem. Soc.* **1950**, *72*, 725

41 A. Loffet, H.X. Zhang, *Int. J. Peptide Protein Res.* **1993**, *42*, 346

42 F. Guibé, *Tetrahedron* **1998**, *54*, 2967

43 L.A. Carpino, A.C. Sam, *J. Chem. Soc., Chem. Commun.* **1979**, 514

44 B.H. Lipshutz, P. Papa, J.M. Keith, *J. Org. Chem.* **1999**, *64*, 3792

45 D. Stevenson, G.T. Young, *J. Chem. Soc., Chem. Commun.* **1967**, 900

46 P. Wang, R. Layfield, M. Landon, R.J. Mayer, R. Ramage, *Tetrahedron Lett.* **1998**, *39*, 8711

47 L.A. Carpino, H.N. Parameswaran, R. K. Kirkley, J.W. Spiewak, E. Schmitz, *J. Org. Chem.* **1970**, *35*, 3291

48 H.A. Moynihan, W.P. Yu, *Synth. Commun.* **1998**, *28*, 17

49 R. Pascal, R. Sola, *Tetrahedron Lett.* **1998**, *39*, 5031

50 A.R. Brown, S.L. Irving, R. Ramage, G. Raphy, *Tetrahedron* **1995**, *51*, 11815

51 R. Ramage, L. Jiang, Y.-D. Kim, K. Shaw, J.-L. Park, H.-J. Kim, *J. Peptide Sci.* **1999**, *5*, 195

52 C. Carreno, M.E. Mendéz, Y.D. Kim, H.-J. Kim, S.A. Kates, D. Andreu, F. Albericio, *J. Peptide Res.* **2000**, *56*, 63

53 P.M. Balse, H.J. Kim, G. Han, V.J. Hruby, *J. Peptide Res.* **2000**, *56*, 70

54 P. Gómez-Martínez, M. Dessolin, F. Guibé, F. Albericio, *J. Chem. Soc., Perkin Trans. I* **1999**, 2871

55 N. Thieriet, P. Gómez-Martínez, F. Guibé, *Tetrahedron Lett.* **1999**, *40*, 2505

56 G. Barany, R.B. Merrifield, *J. Am. Chem. Soc.* **1977**, *99*, 7363

57 G. Barany, F. Albericio, *J. Am. Chem. Soc.* **1985**, *107*, 4936

58 M. Planas, E. Bardaji, K.J. Jensen, G. Barany, *J. Org. Chem.* **1999**, *64*, 7281

59 S.C. Miller, T.S. Scanlan, *J. Am. Chem. Soc.* **1997**, *119*, 2301

60 E. Vedejs, C. Kongkittingam, *J. Org. Chem.* **2000**, *65*, 2309

61 S.C. Miller, T.S. Scanlan, *J. Am. Chem. Soc.* **1998**, *120*, 2690

62 L.A. Carpino, D. Ionescu, A. El-Faham, P. Henklein, H. Wenschuh, M. Bienert, M. Beyermann, *Tetrahedron Lett.* **1998**, *39*, 241

63 T. Fukuyama, C.K. Jow, M. Cheung, *Tetrahedron Lett.* **1995**, *36*, 6373

64 T. Fukuyama, M. Cheung, C.K. Jow, Y. Hidai, K. Toshiyuki, *Tetrahedron Lett.* **1997**, *38*, 5831

65 L. Zervas, D. Borovas, E. Gazis, *J. Am. Chem. Soc.* **1963**, *85*, 3660

66 E. Wünsch, R. Spangenberg, *Chem. Ber.* **1972**, *105*, 740

67 A. Tun-Kyi, *Helv. Chim. Acta* **1978**, *61*, 1086

68 G.B. Bloomberg, G. Askin, A.R. Gargaro, M. J. Tanner, *Tetrahedron Lett.* **1993**, *36*, 4709

69 P. Dumy, I.M. Eggleston, S. Cervigni, U. Sila, X. Sun, M. Mutter, *Tetrahedron Lett.* **1995**, *36*, 1255

70 L.A. Chhabra, B. Hothi, D.J. Evans, P.D. White, B.W. Bycroft, W.C. Chan, *Tetrahedron Lett.* **1998**, *39*, 1603

71 B. Kellam, B.W. Bycroft, W.C. Chan, S.R. Chhabra, *Tetrahedron* **1998**, *54*, 6817

72 M.G. Ryadnov, L.V. Klimenko, Y.V. Mitin, *J. Peptide Res.* **1999**, *53*, 322

73 N. Thieriet, F. Guibé, F. Albericio, *Organic Lett.* **2000**, *2*, 1815

74 A. Joansson, E. Åkerblom, K. Ersmark, G. Lindeberg, A. Hallberg, *J. Comb. Chem.* **2000**, *2*, 496

75 B. Löhr, S. Orlich, H. Kunz, *Synlett* **1999**, 1136

76 H. Kessler, R. Siegmeier, *Tetrahedron Lett.* **1983**, *24*, 281
77 F. Albericio, E. Nicolás, J. Rizo, M. Ruiz-Gayo, E. Pedroso, E. Giralt, *Synthesis* **1990**, 119
78 P. Sieber, *Helv. Chim. Acta* **1977**, *60*, 2711
79 H. Gerlach, *Helv. Chim. Acta* **1977**, *60*, 3039
80 M. Wagner, H. Kunz, *Synlett* **2000**, 400
81 W.C. Chan, B. W. Bycroft, D.J. Evans, P.D. White, *J. Chem. Soc., Chem. Commun.* **1995**, 2209
82 Z.L. Carpino, H.-G. Chao, S. Ghassemi, E.M.E. Mansour, C. Riemer, R. Warrass, D. Sadat-Aalaee, G.A. Truran, H. Imazumi, A. El-Faham, D. Ionescu, M. Ismail, T.L. Kowaleski, C.H. Han, *J. Org. Chem.* **1995**, *60*, 7718
83 M. Goodman, A. Felix, L. Moroder, C. Toniolo, *Synthesis of Peptides and Peptidomimetics* in: Houben-Weyl, Methoden der organischen Chemie, Vol. E22a, K.H. Büchel (Ed.), Thieme, Stuttgart, **2002**
84 H.-D. Jakubke, H. Jeschkeit, *Aminosäuren, Peptide, Proteine,* Verlag Chemie, Weinheim, **1982**
85 E. Gross, J. Meienhofer, *The Peptides: Analysis, Synthesis, Biology,* Academic Press, New York, **1979**
86 M. Bodanszky, A. Bodanszky, *The Practice of Peptide Synthesis,* Springer Verlag, Berlin, **1994**
87 G.N. Müller, H. Waldmann, *Tetrahedron Lett.* **1999**, *40*, 3549
88 F. Weygand, W. Steglich, J. Bjarnason, *Chem. Ber.* **1968**, *101*, 3642
89 M. Narita, M. Doi, H. Sugasawa, K. Ishikawa, *Bull. Chem. Soc. Jpn.* **1985**, *58*, 1473
90 R. Bartl, K.-D. Klöppel, R. Frank in: *Peptides 1992,* C.H. Schneider, A.N. Eberle (Eds.), Escom, Leiden, **1993**
91 C. Hyde, T. Johnson, D. Owen, M. Quibell, R.C. Sheppard, *Int. J. Peptide Protein Res.* **1994**, *43*, 431
92 T. Johnson, M. Quibell, R.C. Sheppard, *J. Peptide Sci.* **1995**, *1*, 11
93 L.P. Miranda, W.D.F. Meutermans, M.L. Smythe, P.F. Alewood, *J. Org. Chem.* **2000**, *65*, 5460
94 P. May, M. Bradley, D.C. Harrowven, D. Pallin, *Tetrahedron Lett.* **2000**, *41*, 1627
95 T. Haack, M. Mutter, *Tetrahedron Lett.* **1992**, *33*, 1589
96 T. Wöhr, M. Mutter, *Tetrahedron Lett.* **1995**, *36*, 3847
97 T. Wöhr, F. Wahl, A. Nefzi, B. Rohwedder, T. Sato, X. Sun, *J. Am. Chem. Soc.* **1996**, *118*, 9218
98 W.R. Sampson, H. Patsiouras, N.J. Ede, *J. Peptide Sci.* **1999**, *5*, 403
99 S. Bajusz, *Acta Chim. Acad. Sci. Hung.* **1965**, *44*, 23
100 G. Jäger, R. Geiger, *Chem. Ber.* **1970**, *103*, 1727
101 H. Rink, P. Sieber, F. Raschdorf, *Tetrahedron Lett.* **1984**, *25*, 621
102 O. Nishimura, M. Fujino, *Chem. Pharm. Bull.* **1976**, *24*, 1568
103 H. Yajima, M. Takeyama, J. Kanaki, K. Mitani, *J. Chem. Soc., Chem. Commun.* **1978**, 482
104 M. Fujino, O. Nishimura, M. Wakimazu, C. Kitada, *J. Chem. Soc., Chem. Commun.* **1980**, 668
105 R. Ramage, J. Green, A.J. Blake, *Tetrahedron* **1991**, *47*, 6353
106 L.A. Carpino, H. Shroff, S.A. Triolo, E.-S.M.E. Mansour, H. Wenschuh, F. Albericio, *Tetrahedron Lett.* **1993**, *34*, 7829
107 C.G. Fields, G.B. Fields, *Tetrahedron Lett.* **1993**, *34*, 6661
108 H.B. Arzeno, D.S. Kemp, *Synthesis* **1988**, 32
109 T. Johnson, R.C. Sheppard, R. Valerio in: *Peptides 1990,* E. Giralt, D. Andreu (Eds.), Escom, Leiden, **1991**, 34
110 M. Bodanszky, J.C. Tolle, S.S. Bdeshmane, A. Bodanszky, *Int. J. Peptide Protein Res.* **1978**, *12*, 57
111 R. Dölling, M. Beyermann, J. Haenel, F. Kernchen, E. Krause, M. Brudel, M. Bienert, *J. Chem. Soc., Chem. Commun.* **1994**, 853
112 M. Quibell, D. Owen, L.C. Packman, T. Johnson, *J. Chem. Soc., Chem. Commun.* **1994**, 2343
113 J.L. Lauer, C.G. Fields, G.B. Fields, *Lett. Peptide Sci.* **1994**, *1*, 197
114 J.-L. Fauchére, R. Schwyzer in: *The Peptides: Analysis, Synthesis, Biology,* Volume 3, E. Gross, J. Meienhofer (Eds.), Academic Press, New York, **1981**
115 A.H. Karlström, A.E. Unden, *Tetrahedron Lett.* **1995**, *36*, 3909

116 R. G. Hiskey in: *The Peptides: Analysis, Synthesis, Biology*, Volume 3, E. Gross, J. Meienhofer (Eds.), Academic Press, New York, **1981**, 137

117 V. Du Vigneaud, L. F. Audrieth, H. S. Loring, *J. Am. Chem. Soc.* **1930**, *52*, 4500

118 N. Fujii, A. Otaka, S. Funakoshi, K. Bessho, T. Watanabe, K. Akaji, H. Yajima, *Chem. Pharm. Bull.* **1987**, *35*, 2339

119 S. Akabori, S. Sakakibara, Y. Shimonishi, Y. Nobuhara, *Bull. Chem. Soc. Japan* **1964**, *37*, 433

120 M. C. Munson, C. Garcia-Echeverria, F. Albericio, G. Barany, *J. Org. Chem.* **1992**, *57*, 3013

121 L. Zervas, I. Photaki, *Chimia* **1960**, *14*, 375

122 L. Zervas, I. Photaki, *J. Am. Chem. Soc.* **1962**, *84*, 3887

123 G. Amiard, R. Heymés, L. Velluz, *Bull. Soc. Chim. Fr.* **1956**, *192*, 698

124 I. Photaki, J. Taylor-Papadimitriou, C. Sakarellos, P. Mazarakis, L. Zervas, *J. Chem. Soc., Chem. Commun.* **1970**, 2683

125 P. Sieber, B. Kamber, A. Hartmann, A. Jöhl, B. Riniker, W. Rittel, *Helv. Chim. Acta* **1974**, *57*, 2617

126 S. N. McCurdy, *Peptide Res.* **1989**, *2*, 147

127 A. Chimiak in: *Peptides 1963*, G. T. Young (Ed.), Pergamon Press, Oxford, **1963**, 37

128 D. F. Veber, J. D. Milkowski, R. G. Denkewalter, R. Hirschmann, *Tetrahedron Lett.* **1968**, 3057

129 D. F. Veber, J. D. Milkowski, S. L. Varga, R. G. Denkewalter, R. Hirschmann, *J. Am. Chem. Soc.* **1972**, *94*, 5456

130 Y. Kiso, M. Yoshida, Y. Fujiwara, T. Kumura, M. Shimokura, K. Akaji, *Chem. Pharm. Bull.* **1990**, *38*, 673

131 E. Wünsch, R. Spangenberg in: *Peptides 1969*, E. Scoffone (Ed.), North-Holland, Amsterdam, **1971**, 30

132 U. Weber, P. Hartter, *Hoppe-Seylers Z. Physiol. Chem.* **1970**, *351*, 1384

133 E. Atherton, M. Pinori, R. C. Sheppard, *J. Chem. Soc., Perkin Trans I* **1985**, 2057

134 R. Eritja, J. P. Ziehler-Martin, P. A. Walker, T. D. Lee, K. Legesse, F. Albericio, B. Kaplan, *Tetrahedron* **1987**, *43*, 2675

135 R. Matsueda, R. Walter, *Int. J. Peptide Protein Res.* **1980**, *16*, 392

136 A. M. Kimbonguila, A. Merzouk, F. Guibé, A. Loffet, *Tetrahedron* **1999**, *55*, 6931

137 Y. Han, G. Barany, *J. Org. Chem.* **1997**, *62*, 3841

138 H. J. Musiol, F. Siedler, D. Quarzago, L. Moroder, *Biopolymers* **1994**, *34*, 1553

139 Y. Han, F. Albericio, G. Barany, *J. Org. Chem.* **1997**, *62*, 4307

140 T. Kaiser, G. J. Nicholson, H. J. Kohlbau, W. Voelter, *Tetrahedron Lett.* **1996**, *37*, 1187

141 T. Kaiser, G. J. Nicholson, H. J. Kohlbau, W. Voelter, *Tetrahedron Lett.* **1996**, *37*, 1187

142 Y. X. Han, *J. Org. Chem.* **1997**, *62*, 4307

143 J. J. Pastuszak, A. Chimiak *J. Org. Chem.* **1981**, *46*, 1868

144 B. Kamber, W. Rittel, *Helv. Chim. Acta* **1968**, *51*, 2061

145 R. G. Hiskey, F. I. Carroll, R. M. Babb, J. O. Bledsoe, R. T. Puckett, B. W. Roberts, *J. Org. Chem.* **1961**, *26*, 1152

146 B. Kamber, *Helv. Chim. Acta* **1973**, *56*, 1370

147 H. Lamthanh, C. Roumestand, C. Deprun, A. Menez, *Int. J. Peptide Protein Res.* **1993**, *41*, 85

148 P. Sieber, *Helv. Chim. Acta* **1977**, *60*, 2711

149 H. Lamthanh, H. Virelizier, D. Frayssinhes, *Peptide Res.* **1995**, *8*, 316

150 M. Weinert, D. Brandenburg, H. Zahn, *Hoppe-Seylers Z. Physiol. Chem.* **1969**, *350*, 1556

151 H. Immer, I. Eberle, W. Fischer, E. Moser in: *Peptides 1988*, G. Jung, E. Bayer (Eds.), de Gruyter, Berlin, **1989**, 94

152 S. Sakakibara, T. Fujii, *Bull. Chem. Soc. Japan*, **1969**, *42*, 1466

153 V. DuVigneaud, O. K. Behrens, *J. Biol. Chem.* **1937**, *117*, 27

154 S. Shaltiel, *Biochem. Biophys. Res. Commun.* **1967**, *29*, 178

155 J. Jones, *The Chemical Synthesis of Peptides*, Clarendon Press, Oxford, **1991**, 89

156 E. Schnabel, H. Herzog, P. Hoffmann, E. Klauke, I. Ugi, *Ann. Chem.* **1968**, *716*, 175

157 L. Moroder, A. Hallet, E. Wünsch, O. Keller, G. Wersin, *Hoppe-Seyler's Z. Physiol. Chem.* **1976**, *357*, 1655

158 K. Barlos, D. Papaioannou, D. Theodoropolous, *J. Org. Chem.* **1982**, *47*, 1324
159 P. Sieber, B. Riniker, *Tetrahedron Lett.* **1987**, *28*, 6031
160 G. Losse, U. Krychowski, *J. Prakt. Chem.* **1970**, *312*, 1097
161 S. Sakakibara, T. Fujii, *Bull. Chem. Soc. Japan* **1970**, *43*, 3954
162 K. Kitagawa, K. Kitade, Y. Kiso, T. Akita, S. Funakoshi, N. Fujii, H. Yajima, *J. Chem. Soc., Chem. Commun.* **1979**, 955
163 R. Colombo, F. Colombo, J.H. Jones, *J. Chem. Soc., Chem. Commun.* **1984**, 292
164 A.M. Kimbonguila, S. Boucida, F. Guibé, A. Loffet, *Tetrahedron* **1997**, *53*, 12525
165 S.J. Harding, J.H. Jones, *J. Peptide Sci.* **1999**, *5*, 399
166 S.J. Harding, J.H. Jones, A.N. Sabirov, V.V. Samukov, *J. Peptide Sci.* **1999**, *5*, 368
167 P. Sieber, *Tetrahedron Lett.* **1987**, *28*, 6147
168 T. Brown, J.H. Jones, J.D. Richards, *J. Chem. Soc., Perkin Trans. I* **1982**, 1553
169 R. Pipkorn, B. Ekberg, *Int. J. Peptide Protein Res.* **1986**, *27*, 583
170 K. Okawa, A. Tani, *J. Chem. Soc. Japan* **1950**, *75*, 1197
171 B.W. Erickson, R.B. Merrifield, *J. Am. Chem. Soc.* **1973**, *95*, 3757
172 L. Lapatsanis in: *Peptides 1978*, I.Z. Siemion, G. Kupryszewski (Eds.), University Press, Wroclaw, **1979**, 105
173 Y. Nishiyama, S. Shikama, K.-I. Morita, K. Kurita, *J. Chem. Soc., Perkin Trans. I* **2000**, 1949
174 D. Yamashiro, C.H. Li, *J. Org. Chem.* **1973**, *38*, 591
175 J.P. Tam, R.B. Merrifield in: *The Peptides*, Volume 9, S. Udenfriend, J. Meienhofer (Eds.), Academic Press, New York, **1987**, 185
176 F.M. Callahan, G.W. Anderson, R. Paul, J.E. Zimmermann, *J. Am. Chem. Soc.* **1963**, *85*, 201
177 F. Weygand, W. Steglich, I. Lengyel, F. Fraunberger, A. Maierhofer, W. Oettmeier, *Chem. Ber.* **1966**, *99*, 1944
178 P.M. Pojer, S.J. Angyal, *Tetrahedron Lett.* **1976**, 3067
179 A.D. Morley, *Tetrahedron Lett.* **2000**, *41*, 7401
180 T. Mizoguchi, G. Levin, D.W. Wooley, J.M. Stewart, *J. Org. Chem.* **1968**, *33*, 903
181 E. Wünsch, G. Fries, A. Zwick, *Chem. Ber.* **1958**, *91*, 542
182 H. Sajiki, K. Hirota, *Tetrahedron* **1998**, *54*, 13981
183 E. Schröder, *Ann. Chem.* **1964**, *670*, 127
184 A. Di Fenza, M. Tancredi, C. Galoppini, P. Povero, *Tetrahedron Lett.* **1998**, *39*, 8529
185 H. Kunz, *Chem. Ber.* **1976**, *109*, 2670, 3693
186 H. Huang, D.L. Rabenstein, *J. Peptide Res.* **1999**, *53*, 548
187 M. Girard, F. Cavelier, J. Martinez, *J. Peptide Sci.*, **1999**, *5*, 457
188 M. Ohno, S. Tsukamoto, S. Sato, N. Izumiya, *Bull. Chem. Soc. Japan* **1973**, *46*, 3280
189 D. Yamashiro, C.H. Li, *J. Org. Chem.* **1973**, *38*, 2594
190 P. White in: *Peptides, Chemistry, Structure and Biology*, J.A. Smith, J.E. Riviere (Eds.), Escom, Leiden, **1992**, 537
191 Y. Kiso, T. Kumura, M. Shimokura, T. Narukami, *J. Chem. Soc., Chem. Commun.* **1988**, 287
192 H.M. Franzen, K. Nagren, L. Grehn, B. Langström, U. Ragnarsson, *J. Chem. Soc., Perkin Trans. I* **1988**, 497
193 H.A. Guillaume, J.W. Perrich, J.D. Wade, G.W. Tregear, R.B. Johns, *J. Org. Chem.* **1989**, *54*, 5731
194 S. Capasso, P. Di Cerbo, *J. Peptide Res.* **2000**, *56*, 382
195 M. Xie, J. Aubé, R. T. Borchardt, M. Morton, E.M. Topp, D. Vander Velde, R.L. Schowen, *J. Peptide Res.* **2000**, *56*, 165
196 R. Li, A.J. D'Souza, B.B. Laird, R.L. Schowen, R.T. Borchardt, E.M. Topp, *J. Peptide Res.* **2000**, *56*, 326
197 P. Sieber, B. Riniker, *Tetrahedron Lett.* **1991**, *32*, 739
198 B. Riniker, A. Flörsheimer, H. Fretz, P. Sieber, B. Kamber, *Tetrahedron* **1993**, *49*, 9307
199 R.W. Holley, *J. Am. Chem. Soc.* **1955**, *77*, 2552
200 J.-D. Glass in: *The Peptides: Analysis, Synthesis, Biology*, Volume 9, E. Gross, J. Meienhofer (Eds.), Academic Press, New York, **1987**, 167
201 P. Hermann, *Biomed. Biochim. Acta* **1991**, *50*, 19

202 T. Pathak, H. Waldmann, *Curr. Opin. Chem. Biol.* **1998**, *2*, 112

203 F. Eisele, D. J. Owen, H. Waldmann, *Bioorg. Med. Chem.* **1999**, *7*, 193

204 M. Schelhaas, H. Waldmann, *Angew. Chem.* **1996**, *108*, 2192; *Angew. Chem. Int. Ed. Engl.* **1996**, *35*, 2056

205 H. Waldmann, D. Sebastian, *Chem. Rev.* **1994**, *94*, 911

206 T. Kappes, H. Waldmann, *Liebigs Ann./Recueil* **1997**, 803

207 B. Sauerbrei, T. Kappes, H. Waldmann, *Top. Curr. Chem.* **1997**, *186*, 65

208 T. Kappes, H. Waldmann, *Carbohydr. Res.* **1998**, *305*, 341

209 A. G. Gum, T. Kappes-Roth, H. Waldmann, *Chem. Eur. J.* **2000**, *6*, 3714

210 H. Waldmann, E. Nägele, *Angew. Chem.* **1995**, *107*, 2425; *Angew. Chem. Int. Ed. Engl.* **1995**, *34*, 2259

211 E. Nägele, M. Schelhaas, N. Kuder, H. Waldmann, *J. Am. Chem. Soc.* **1998**, *120*, 6889

212 T. Pohl, H. Waldmann, *Angew. Chem.* **1996**, *108*, 1829; *Angew. Chem. Int. Ed. Engl.* **1996**, *35*, 1720

213 P. Braun, H. Waldmann, H. Kunz, *Bioorg. Med. Chem.* **1993**, *1*, 197

214 P. Braun, H. Waldmann, H. Kunz, *Synlett* **1992**, 39

215 H. Kunz, D. Kowalczyk, P. Braun, G. Braun, *Angew. Chem.* **1994**, *106*, 353; *Angew. Chem. Int. Ed. Engl.* **1994**, *33*, 336

216 J. Eberling, P. Braun, D. Kowalczyk, M. Schultz, H. Kunz, *J. Org. Chem.* **1996**, *61*, 2638

217 F. Eisele, J. Kuhlmann, H. Waldmann, *Angew. Chem.* **2001**, *113*, 382; *Angew. Chem. Int. Ed. Engl.* **2001**, *40*, 369

218 J. Sander, H. Waldmann, *Chem. Eur. J.* **2000**, *6*, 1564

219 M. Schelhaas, S. Glomsda, M. Hänsler, H.-D. Jakubke, H. Waldmann, *Angew. Chem.* **1996**, 108, 82; *Angew. Chem. Int. Ed. Engl.* **1996**, *35*, 106

220 J. Sander, H. Waldmann, *Angew. Chem.* **1999**, *111*, 1337; *Angew. Chem. Int. Ed. Engl.* **1999**, *38*, 1250

221 M. Schelhaas, E. Nägele, N. Kuder, B. Bader, J. Kuhlmann, A. Wittinghofer, H. Waldmann, *Chem. Eur. J.* **1999**, *5*, 1239

222 A. Cotté, B. Bader, J. Kuhlmann, H. Waldmann, *Chem. Eur. J.* **1999**, *5*, 922

223 S. Flohr, V. Jungmann, H. Waldmann, *Chem. Eur. J.* **1999**, *5*, 669

224 J. Meienhofer in: *The Peptides: Analysis, Synthesis, Biology*, Volume 1, E. Gross, J. Meienhofer (Eds.), Academic Press, New York, **1979**, 197

225 T. Shioiri, K. Ninomiya, S. Yamada, *J. Am. Chem. Soc.* **1972**, *94*, 6203

226 J. Honzl, J. Rudinger, *Coll. Czech. Chem. Commun.* **1961**, *26*, 2333

227 J. Meienhofer in: *The Peptides: Analysis, Synthesis, Biology*, Volume 1, E. Gross, J. Meienhofer (Eds.), Academic Press, New York, **1979**, 263

228 F. M. F. Chen, R. Steinauer, N. L. Benoiton in: *Peptides 1984*, U. Ragnarsson (Ed.), Almqvist & Wiksell, Stockholm, **1984**

229 F. M. F. Chen, Y. Lee, R. Steinauer, N. L. Benoiton, *Can. J. Chem.* **1987**, *65*, 613

230 F. M. F. Chen, N. L. Benoiton, *Can. J. Chem.* **1987**, *65*, 619

231 N. L. Benoiton, Y. Lee, F. M. F. Chen, *Int. J. Peptide Protein Res.* **1988**, *31*, 443

232 N. L. Benoiton, Y. C. Lee, F. M. F. Chen, *Int. J. Peptide Protein Res.* **1993**, *42*, 278

233 J. H. Jones, M. J. Witty, *J. Chem. Soc., Chem. Commun.* **1977**, 281

234 J. H. Jones, M. J. Witty, *J. Chem. Soc., Perkin Trans. I* **1979**, 3203

235 R. Ramage, D. Hopton, M. J. Parrot, R. S. Richardson, G. W. Kenner, G. A. Moore, *J. Chem. Soc., Perkin Trans. I* **1985**, 461

236 J. Cabré, A. L. Palomo, *Synthesis* **1984**, 413

237 R. Hirschmann, R. G. Denkewalter, H. Schwam, R. G. Strachan, T. E. Beesley, D. F. Veber, E. F. Schoenewaldt, H. Barkemeyer, W. J. Paleveda, J. T. A. Jacob, *J. Am. Chem. Soc.* **1966**, *88*, 3163

238 R. Hirschmann, R. G. Denkewalter, *Naturwissenschaften* **1970**, *57*, 145

239 W. D. Fuller, M. P. Cohen, M. Shabankarech, R. Blair, M. Goodman, F. R. Naider, *J. Am. Chem. Soc.* **1990**, *112*, 7414

240 T. B. Sim, H. Rapoport, *J. Org. Chem.* **1999**, *64*, 2532

241 D. H. Rich, J. Singh in: *The Peptides: Analysis, Synthesis, Biology*, Volume 1, E.

Gross, J. Meienhofer (Eds.), Academic Press, New York, **1979**, 241
242 J. Izdebski, D. Kunce, *J. Peptide Sci.* **1997**, *3*, 141
243 N. L. Benoiton, F. M. F. Chen in: *Peptides: Chemistry and Biology*, G. R. Marshall (Ed.), Escom, Leiden, **1988**, 152
244 S. Nozaki, *J. Peptide Res.* **1999**, *54*, 162
245 H. Ito, N. Takamatsu, I. Ichikizaki, *Chem. Lett.* **1975**, 577
246 E. Wünsch, F. Dress, *Chem. Ber.* **1966**, *99*, 110
247 F. Weygand, D. Hoffmann, E. Wünsch, *Z. Naturforsch. B* **1966**, *21*, 426
248 H. Gross, L. Bilk, *Tetrahedron* **1968**, *24*, 6935
249 W. König, R. Geiger, *Chem. Ber.* **1970**, *103*, 788
250 W. König, R. Geiger, *Chem. Ber.* **1970**, *103*, 2024
251 N. L. Benoiton, *Biopolymers*, **1996**, *40*, 245
252 K. Barlos, D. Papaioannou, S. Voliotis, R. Prewo, J. H. Bieri, *J. Org. Chem.* **1985**, *50*, 696
253 N. Robertson, L. Jiang, R. Ramage, *Tetrahedron* **1999**, *55*, 2713
254 J. C. Spetzler, M. Meldal, J. Felding, P. Vedsø, M. Begtrup, *J. Chem. Soc., Perkin Trans. I*, **1998**, 1727
255 J. Przybylski, B. Sokolowska, *Pol. J. Chem.* **1984**, *58*, 455
256 Y. Nishiyama, M. Tanaka, S. Saito, S. Ishizuka, T. Mori, K. Kurita, *Chem. Pharm. Bull.* **1999**, *47*, 576
257 J. Przybylski in: *Peptides: Structure and Biological Function*, E. Gross, J. Meienhofer (Eds.), Pierce, Rockford, **1979**, 393
258 H. Jeschkeit, M. Strube, J. Przybylski, H. Miecznikowska, G. Kupryszewski, *J. Prakt. Chem.* **1984**, *326*, 638
259 C. Griehl, H. Jeschkeit, *J. Prakt. Chem.* **1994**, *336*, 77
260 L. A. Carpino, *J. Am. Chem. Soc.* **1993**, *115*, 4397
261 L. A. Carpino, A. El-Faham, F. Albericio, *Tetrahedron Lett.* **1994**, *35*, 2279
262 L. A. Carpino, A. El-Faham, C. A. Minor, F. Albericio, *J. Chem. Soc., Chem. Commun.* **1994**, 201
263 L. A. Carpino, A. El-Faham, *Tetrahedron* **1999**, *55*, 6813
264 M. Bodanszky in: *The Peptides: Analysis, Synthesis, Biology*, Volume 1, E. Gross, J. Meienhofer (Eds.), Academic Press, New York, **1979**, 105
265 S. M. Beaumont, B. O. Handford, G. T. Young, *Acta Chim. Sci. Hung.* **1965**, *44*, 37
266 H.-D. Jakubke, A. Voigt, *Chem. Ber.* **1966**, *99*, 2944
267 G. W. Anderson, J. E. Zimmerman, F. M. Callahan, *J. Am. Chem. Soc.* **1963**, *85*, 3039
268 D. S. Kemp, M. Trangle, K. Trangle, *Tetrahedron Lett.* **1974**, *15*, 2695
269 L. A. Carpino, H. Imazumi, B. M. Foxman, M. J. Vela, P. Henklein, A. El-Faham, J. Klose, M. Bienert, *Organic Lett.* **2000**, *2*, 2253
270 R. B. Woodward, R. A. Olofson, *J. Am. Chem. Soc.* **1961**, *83*, 1010
271 L. A. Carpino, M. Beyermann, H. Wenschuh, M. Bienert, *Acc. Chem. Res.* **1996**, *29*, 268
272 E. Falb, T. Yechezkel, Y. Salitra, C. Gilon, *J. Peptide Res.* **1999**, *53*, 507
273 L. A. Carpino, B. Cohen, K. E. Stephens, D. Sadat-Aalaee, J. H. Tien, D. C. Langridge, *J. Org. Chem.* **1986**, *51*, 3732
274 L. A. Carpino, D. Sadat-Aalaee, H. G. Chao, R. H. De Selms, *J. Am. Chem. Soc.* **1990**, *112*, 9651
275 H. Wenschuh, M. Beyermann, S. Rothemund, L. A. Carpino, M. Bienert, *Tetrahedron Lett.* **1995**, *36*, 1247
276 H. Wenschuh, M. Beyermann, R. Winter, M. Bienert, D. Ionescu, L. A. Carpino, *Tetrahedron Lett.* **1996**, *37*, 5438
277 H. Wenschuh, M. Beyermann, A. El-Faham, S. Ghassemi, L. A. Carpino, M. Bienert, *J. Chem. Soc. Chem., Commun.* **1995**, 669
278 L. A. Carpino, A. El-Faham, *J. Am. Chem. Soc.* **1995**, *117*, 5401
279 B. Castro, J. R. Dormoy, G. Evin, C. Selve, *Tetrahedron Lett.* **1975**, 1219
280 R. Steinauer, F. M. F. Chen, N. L. Benoiton, *Int. J. Peptide Protein Res.* **1989**, *34*, 295
281 J. Coste, D. Le-Nguyen, B. Castro, *Tetrahedron Lett.* **1990**, *31*, 205

282 F. Albericio, M. Cases, J. Alsina, S.A. Triolo, L.A. Carpino, S.A. Kates, *Tetrahedron Lett.* **1997**, *38*, 4853

283 C. Auvin-Guette, E. Frerot, J. Coste, S. Rebuffat, P. Jouin, B. Bodo, *Tetrahedron Lett.* **1993**, *34*, 2481

284 H. Li, X. Jiang, Y.-H. Ye, C. Fan, T. Romoff, M. Goodman, *Organic Lett.* **1999**, *1*, 91

285 V. Dourtoglou, J.-C. Ziegler, B. Gross, *Tetrahedron Lett.* **1978**, 1269

286 R. Knorr, A. Trzeciak, W. Bannwarth, D. Gillessen, *Tetrahedron Lett.* **1989**, *30*, 1927

287 H. Gausepohl, M. Kraft, R.W. Frank, *Int. J. Peptide Protein Res.* **1989**, *34*, 287

288 F. Albericio, J. M. Bofill, A. El-Faham, S. A. Kates, *J. Org. Chem.* **1998**, *63*, 9678

289 S. Chen, J. Xu, *Tetrahedron Lett.* **1992**, *33*, 647

290 J. Habermann, H. Kunz, *Tetrahedron Lett.* **1998**, *39*, 265

291 J.M. Bofill, F. Albericio, *Tetrahedron Lett.* **1999**, *40*, 2641

292 A. Ehrlich, S. Rothemund, M. Brudel, M. Beyermann, L.A. Carpino, M. Bienert, *Tetrahedron Lett.* **1993**, *34*, 4781

293 U. Schmidt, H. Grieser, *J. Chem. Soc. Chem. Commun.* **1993**, 1461

294 P. Li, J.-C. Xu, *Tetrahedron Lett.* **2000**, *41*, 721

295 P. Li, J.-C. Xu, *J. Peptide Res.* **2000**, *55*, 110

296 S. Goldschmidt, F. Obermeier, *Ann. Chem.* **1954**, *588*, 24

297 T. Mukaiyama, R. Matsueda, M. Ueki in: *The Peptides: Analysis, Synthesis, Biology*, Volume 2, E. Gross, J. Meienhofer (Eds.), Academic Press, New York, **1980**, 383

298 D.S. Kemp in: *The Peptides: Analysis, Synthesis, Biology*, Volume 1, E. Gross, J. Meienhofer (Eds.), Academic Press, New York, **1979**, 315

299 N.L. Benoiton in: *The Peptides: Analysis, Synthesis, Biology*, Volume 5, E. Gross, J. Meienhofer (Eds.), Academic Press, New York, **1983**, 217

300 F. Weygand, W. Steglich, X. Barocio de la Lama, *Tetrahedron Suppl.* **1966**, *8*, 9

301 F.M. Finn, K. Hofmann in: *The Proteins*, Volume 2, H. Neurath, R.L. Hill (Eds.), Academic Press, New York, **1976**, 183

302 M. Manning, E. Coy, W.H. Sawyer, *Biochemistry* **1970**, *9*, 3925

303 M. Goodman, P. Keogh, H. Anderson, *Bioorg. Chem.* **1977**, *6*, 239

304 C. Griehl, J. Weight, H. Jeschkeit, *J. High. Res. Chrom.* **1994**, *17*, 700

305 C. Griehl, J. Weight, H. Jeschkeit, Z. Palacz in: *Peptides 1992*, C.H. Schneider, A.N. Eberle (Eds.), Escom, Leiden, **1993**, 459

306 H. Yajima, N. Fujii, *J. Chem. Soc., Chem. Commun.* **1980**, 115

307 R.B. Merrifield, *J. Am. Chem. Soc.* **1963**, *85*, 2149

308 G. Barany, R.B. Merrifield in: *The Peptides: Analysis, Synthesis, Biology*, Volume 2, E. Gross, J. Meienhofer (Eds.), Academic Press, New York, **1980**, 1

309 R.B. Merrifield, *Angew. Chem. Int. Ed. Engl.* **1985**, *24*, 799; *Angew. Chem.* **1985**, *97*, 801

310 G. Barany, N. Kneib-Cordonier, D.G. Mullen, Int. J. Peptide Protein Res. **1987**, *30*, 705

311 J.M. Stewart, J.D. Young, *Solid-Phase Peptide Synthesis*, Pierce, Rockford, **1984**

312 E. Atherton, R.C. Sheppard, *Solid Phase Peptide Synthesis: A Practical Approach*, IRL Press, Oxford, **1989**

313 E. Bayer, *Angew. Chem.* **1991**, *103*, 117; *Angew. Chem. Int. Ed. Engl.* **1991**, *30*, 113

314 M.W. Pennington, B.M. Dunn, *Peptide Synthesis Protocols*, Humana Press, Totowa, **1994**

315 G.B. Fields, *Solid-Phase Peptide Synthesis*, Academic Press, New York, **1997**

316 G.A. Grant, *Synthetic Peptides. A User's Guide*, W.H. Freeman & Co, New York, **1992**

317 P. Lloyd-Williams, F. Albericio, E. Giralt, *Chemical Approaches to the Synthesis of Peptides and Proteins*, CRC Press, Boca Raton, **1997**

318 W.C. Chan, P.D. White, *Fmoc Solid Phase Peptide Synthesis: A Practical Approach*, Oxford University Press, Oxford, **2000**

319 M. Meldal, I. Svendsen, K. Breddam, F. Auzanneau, *Proc. Natl. Acad. Sci. USA* **1994**, *91*, 3314

320 I.W. James, *Tetrahedron* **1999**, *55*, 4855

321 D. Swinnen, D. Hilvert, *Org. Lett.* **2000**, *2*, 2439
322 J.H. Jones, W.I. Ramage, *J. Chem. Soc., Chem. Commun.* **1978**, 472
323 J.H. Jones, A.V. Stachulsky, *J. Chem. Soc., Perkin Trans. I* **1979**, 2261
324 W. D. Fuller, M. Goodman, F.R. Naider, Y.F. Zhu, *Biopolymers* **1996**, *40*, 183
325 M. Mergler, R. Tanner, J. Gosteli, P. Grogg, *Tetrahedron Lett.* **1988**, *29*, 4005
326 A.R. Mitchell, B.W. Erickson, M.N. Ryabtsev, R.S. Hidges, R.B. Merrifield, *J. Am. Chem. Soc.* **1976**, *98*, 7357
327 A.R. Mitchell, S.B.H. Kent, M. Engelhardt, R.B. Merrifield, *J. Org. Chem.* **1978**, *43*, 2845
328 R.C. Sheppard, B.J. Williams, *Int. J. Peptide Protein Res.* **1982**, *20*, 451
329 P.G. Pietta, P.F. Cavallo, K. Takahashi, G.R. Marshall, *J. Org. Chem.* **1974**, *39*, 44
330 G.R. Matsueda, J.M. Stewart, *Peptides* **1981**, *2*, 45
331 H. Rink, *Tetrahedron Lett.* **1987**, *28*, 3787
332 K. Barlos, O. Chatzi, D. Gatos, G. Stavropoulos, *Int. J. Peptide Protein Res.* **1991**, *37*, 513
333 M. Gairi, P. Lloyd-Williams, F. Albericio, E. Giralt, *Int. J. Peptide Protein Res.* **1995**, *46*, 119
334 H. Kunz, B. Dombo, *Angew. Chem.* **1988**, *100*, 732; *Angew. Chem. Int. Ed. Engl.* **1988**, *27*, 711
335 C.P. Holmes, D.G. Jones, *J. Org. Chem.* **1995**, *60*, 2318
336 M. Pátek, M. Lebl, *Biopolymers* **1998**, *47*, 353
337 G.W. Kenner, J.R. McDermott, R.C. Sheppard, *J. Chem. Soc., Chem. Commun.* **1971**, 636
338 B.J. Backes, A.A. Virgilio, J.A. Ellman, *J. Am. Chem. Soc.* **1996**, *118*, 3055
339 B.J. Backes, J.A. Ellman, *J. Org. Chem.* **1999**, *64*, 2322
340 R. Ingenito, E. Bianchi, D. Fattori, A. Pessi, *J. Am. Chem. Soc.* **1999**, *121*, 11369
341 D.L. Marshall, I.E. Liener, *J. Org. Chem.* **1970**, *35*, 867
342 E. Flanigan, D.L. Marshall, *Tetrahedron Lett.* **1970**, 2403
343 T. Kimura, T. Fukui, S. Tanaka, K. Akaji, Y. Kiso, *Chem. Pharm. Bull.* **1997**, *45*, 18
344 M. Patek, M. Lebl, *Tetrahedron Lett.* **1991**, *32*, 3891
345 A. Brik, E. Keinan, P.E. Dawson, *J. Org. Chem.* **2000**, 65, 3829
346 N.J. Maeji, A.M. Bray, H.M. Geysen, *J. Immunol. Methods* **1990**, *134*, 23
347 N.J. Maeji, A.M. Bray, R.M. Valerio, M.A. Seldon, J.-X. Wang, H.M. Geysen, *Peptide Res.* **1991**, *4*, 142
348 A.M. Bray, N.J. Maeji, R.M. Valerio, R.A. Campbell, H.M. Geysen, *J. Org. Chem.* **1991**, *56*, 6659
349 B. Atrash, M. Bradley, *J. Chem. Soc., Chem. Commun.* **1997**, 1397
350 S. Hoffmann, R. Frank, *Tetrahedron Lett.* **1994**, *35*, 7763
351 G. Panke, R. Frank, *Tetrahedron Lett.* **1998**, *39*, 17
352 J.P. Tam, W.F. Health, R.B. Merrifield, *J. Am. Chem. Soc.* **1983**, *105*, 6442
353 W.F. Health, J.P. Tam, R.B. Merrifield, *Int. J. Peptide Protein Res.* **1986**, *28*, 498
354 G.W. Tregear, J. van Rietschoten, R. Sauer, H.D. Niall, H.T. Keutmann, *Biochemistry* **1977**, *16*, 2817
355 S.B.H. Kent, *Annu. Rev. Biochem.* **1988**, *57*, 959
356 J. Bedford, C. Hyde, T. Johnson, W. Jun, D. Owen, M. Quibell, R.C. Sheppard, *Int. J. Peptide Protein Res.* **1992**, *40*, 300
357 R.B. Merrifield in: *Recent Progress in Hormone Research*, Volume 23, G. Pincus (Ed.), Academic Press, New York, **1967**, 451
358 M. Beyermann, M. Bienert, *Tetrahedron Lett.* **1992**, *33*, 3745
359 V. Krchnak, Z. Flegelova, J. Vagner, *Int. J. Peptide Protein Res.* **1993**, *42*, 450
360 M. Narita, S. Ouchi, *J. Org. Chem.* **1994**, *52*, 686
361 R.C. de L. Milton, S.C. F. Milton, P.A. Adams, *J. Am. Chem. Soc.* **1990**, *122*, 6039
362 B. Henkel, E. Bayer, *J. Peptide Sci.* **1998**, *4*, 461
363 R. Nagaraj, P. Balaram, *Tetrahedron* **1981**, *37*, 2001

364 M. Bodanszky, S.S. Deshmane, J. Martinez, *J. Org. Chem.* **1979**, *44*, 1622

365 H. Wenschuh, M. Beyermann, E. Krause, M. Brudel, R. Winter, M. Schümann, L.A. Carpino, M. Bienert, *J. Org. Chem.* **1994**, *59*, 3275

366 M. Narita, S. Honda, H. Umeyama, S. Obana, *Bull. Chem. Soc. Japan* **1988**, *61*, 281

367 H. Kuroda, Y.-H. Chen, T. Kumura, S. Sakakibara, *Int. J. Peptide Protein Res.* **1992**, *40*, 294

368 E. Oliveira, L. Miranda, F. Albericio, D. Andreu, A.C.M. Paiva, C.R. Nakaie, M. Tominaga, *J. Peptide Res.* **1997**, *49*, 300

369 L. Zhang, C. Goldammer, B. Henkel, G. Panhaus, F. Zühl, G. Jung, E. Bayer in: *Innovation and Perspectives in Solid Phase Synthesis*, R. Epton (Ed.), Intercept, Andover, **1994**, 711

370 D. Seebach, A. Thale, A.K. Beck, *Helv. Chim. Acta* **1989**, *72*, 857

371 M.W. Pennington, I. Zaydenberg, M.E. Byrnes, R.S. Norton, W.R. Kem, *Int. J. Peptide Protein Res.* **1994**, *43*, 463

372 R. Bartl, K.-D. Klöppel, R. Frank in: *Peptides 1992*, C.H. Schneider, A.N. Eberle (Eds.), Escom, Leiden, **1993**, 277

373 T. Johnson, M. Quibell, D. Owen, R.C. Sheppard, *J. Chem. Soc., Chem. Commun.* **1993**, 369

374 W. Rapp, L. Zhang, R. Häbich, F. Bayer in: *Peptides 1988*, G. Jung, E. Bayer (Eds.), de Gruyter, Berlin, **1989**

375 M. Schnölzer, P. Alewood, A. Jones, D. Alewood, S.B.H. Kent, *Int. J. Peptide Protein Res.* **1992**, *40*, 180

376 M. Schnölzer, S.B.H. Kent, *Science* **1992**, *256*, 221

377 B.D. Larsen, A. Holm, *J. Peptide Res.* **1998**, *52*, 470

378 H.G. Chao, M.S. Bernatowicz, G.R. Matsueda, *J. Org. Chem.* **1993**, *58*, 2640

379 L.M. Varanda, M.T.M. Miranda, *J. Peptide Res.* **1997**, *50*, 102

380 E. Kaiser, R.L. Colescott, C.D. Bossinger, P.I. Cook, *Anal. Biochem.* **1970**, *34*, 595

381 W.S. Hancock, J.E. Battersby, *Anal. Biochem.* **1976**, *71*, 260

382 T. Voikovsky, *Peptide Res.* **1995**, *8*, 236

383 V. Krchnák, J. Vágner, M. Lebl, *Int. J. Peptide Protein Res.* **1988**, *32*, 415

384 L.C. Dorman, *Tetrahedron Lett.* **1969**, 2319

385 G. Talbo, J.D. Wade, N. Dawson, M. Manoussios, G.W. Tregear, *Lett. Peptide Sci.* **1999**, *4*, 121

386 E. Atherton, D.L.J. Clive, R.C. Sheppard, *J. Am. Chem. Soc.* **1975**, *97*, 6584

387 F. Albericio, G. Barany, *Int. J. Peptide Protein Res.* **1993**, *41*, 307

388 G.M. Williams, R.A.E. Carr, M.S. Congreve, C. Kay, S.C. McKeown, P.J. Murray, J.J. Scicinski, S.P. Watson, *Angew. Chem.* **2000**, *112*, 3431; *Angew. Chem. Int. Ed. Engl.* **2000**, *39*, 3293

389 S.A. Salisbury, E.J. Treemer, J. W. Davies, D.E.I.A. Owen, *J. Chem. Soc., Chem. Commun.* **1990**, 538

390 E. Bayer, G. Jung, I. Halasz, I. Sebastian, *Tetrahedron Lett.* **1979**, 4503

391 R. Frank, H. Gausepohl in: *Modern Methods in Protein Chemistry*, Volume 3, H. Tschesche (Ed.), de Gruyter, Berlin, **1988**, 41

392 A. Dryland, R.C. Sheppard, *J. Chem. Soc., Chem. Commun.* **1986**, 125

393 M.N. Shemyakin, Y.A. Ovchinnikov, A.A. Kiryushkin, I.V. Kozhevnikova, *Tetrahedron Lett.* **1965**, 2323

394 E. Bayer, M. Mutter, *Nature* **1972**, *237*, 512

395 M. Mutter, E. Bayer, *Angew. Chem.* **1974**, *86*, 101; *Angew. Chem. Int. Ed. Engl.* **1974**, *13*, 88

396 H. Frank, H. Hagenmair, *Experientia* **1975**, *31*, 131

397 T. Wieland, C. Birr, *Angew. Chem.* **1966**, *78*, 303; *Angew. Chem. Int. Ed. Engl.* **1966**, *5*, 310

398 N. Kneib-Gordonier, F. Albericio, G. Barany in: *Peptides: Chemistry, Structure and Biology*, J.E. Rivier, G.R. Marshall (Eds.), Escom, Leiden, **1990**, 895

399 D.L. Whitehouse, S.N. Savinov, D.J. Austin, *Tetrahedron Lett.* **1997**, *38*, 7851

400 H. Irie, N. Fujii, H. Ogawa, H. Yajima, M. Fujino, S. Shinagawa, *J. Chem. Soc., Chem. Commun.* **1976**, 922

401 Y. Kiso, S. Nakamura, K. Ito, K. Kawa, K. Kitagawa, T. Akita, H. Moritoki, *J. Chem. Soc., Chem. Commun.* **1979**, 971

402 N. Fujii, A. Otaka, O. Ikemura, K. Akaji, S. Funakoshi, Y. Hayashi, Y. Kuro-

da, H. Yajima, *J. Chem. Soc., Chem. Commun.* **1987**, 274
403 J. P. Tam in: *Peptide Chemistry 1987*, T. Shiba, S. Sakakibara (Eds.), Protein Research Foundation, Osaka, **1988**, 199
404 L. A. Carpino, H. Shroff, S. A. Triolo, E. M. E. Mansour, H. Wenschuh, F. Albericio, *Tetrahedron Lett.* **1993**, *34*, 7829
405 R. Ramage, J. Green, *Tetrahedron Lett.* **1987**, *28*, 2287
406 P. White in: *Peptides: Chemistry and Biology*, J. A. Smith, J. E. Rivier (Eds.), Escom, Leiden, **1992**, 537
407 P. Sieber, B. Riniker, *Tetrahedron Lett.* **1991**, *32*, 739
408 A. J. Smith, J. D. Young, S. A. Carr, D. R. Marshak, L. C. Williams, K. R. Williams in: *Techniques in Protein Chemistry III*, R. H. Angeletti (Ed.), Academic Press, Orlando, **1992**, 219
409 W. Rapp in: *Peptides 1992*, C. H. Schneider, A. N. Eberle (Eds.), Escom, Leiden, **1993**, 243
410 B. Gutte, R. B. Merrifield, *J. Am. Chem. Soc.* **1969**, *91*, 501
411 D. Blohm, C. Bollschweiler, H. Hillen, *Angew. Chem.* **1988**, *100*, 213; *Angew. Chem. Int. Ed. Engl.* **1988**, *27*, 207
412 J. W. Engels, E. Ullmann, *Angew. Chem.* **1989**, *101*, 733; *Angew. Chem. Int. Ed. Engl.* **1989**, *28*, 716
413 D. Mendel, V. W. Cornish, P. G. Schultz, *Annu. Rev. Biophys. Biomol. Struct.* **1995**, *24*, 435
414 D. L. Liu, T. J. Magliery, M. Pastranak, P. G. Schultz, *Proc. Natl. Acad. Sci. USA* **1997**, *94*, 10092
415 D. A. Dougherty, *Curr. Opin. Chem. Biol.* **2000**, *4*, 645
416 K. Itakura, T. Hirose, R. Crea, A. D. Riggs, H. L. Heyneker, F. Bolivar, H. W. Boyer, *Science* **1977**, *198*, 1056
417 D. V. Goeddel, H. L. Heyneker, T. Hozumi, R. Arentzen, K. Itakura, D. G. Yansura, M. J. Ross, G. Miozzari, R. Crea, P. H. Seeburg, *Nature* **1979**, *281*, 544
418 H. M. Hsiung, N. G. Mayne, G. W. Becker, *Bio/Technology* **1986**, *4*, 991
419 D. V. Goeddel, D. G. Kleid, F. Bolivar, H. L. Heynecker, D. G. Ansura, R. Crea, T. Horose, A. Kraszewski, K. Itakura, A. D. Riggs, *Proc. Natl. Acad. Sci. USA* **1979**, *76*, 106
420 L. Villa-Komaroff, A. Efstratiadis, S. Broome, P. Lomedico, R. Tizard, S. P. Nabes, W. L. Chick, W. Gilbert, *Proc. Natl. Acad. Sci. USA* **1978**, *75*, 3727
421 A. H. Barnett, D. R. Owens, *Lancet* **1997**, *349*, 47
422 B. Zinman, H. Tildesley, J. Chiasson, E. Tsui, F. Strack, *Diabetes* **1997**, *46*, 440
423 M. B. Hoagland, M. L. Stephenson, J. F. Scott, L. I. Hecht, P. C. Zamechnik, *J. Biol. Chem.* **1958**, *231*, 241
424 D. Nathans, G. Notani, J. H. Schwatz, N. D. Zinder, *Proc. Natl. Acad. Sci. USA* **1962**, *48*, 1424
425 B. E. Roberts, B. M. Paterson, *Proc. Natl. Acad. Sci. USA* **1973**, *70*, 2330
426 T. Suryanarayana, A. R. Subramanian, *Biochemistry* **1983**, *22*, 2715
427 A. S. Spirin, V. I. Baranov, L. A. Ryabova, S. Y. Ovodov, Y. B. Alakhov, *Science* **1988**, *242*, 1162
428 Y. Wu, D. Y. Zhang, F. R. Kramer, *Proc. Natl. Acad. Sci. USA* **1992**, *89*, 11789
429 I. Y. Morozov, V. I. Ugarov, A. B. Chetverin, A. S. Spirin, *Proc. Natl. Acad. Sci. USA* **1993**, *90*, 9325
430 L. A. Ryabova, E. V. Volianik, O. Kurnosov, A. S. Spirin, Y. Wu, F. R. Kramer, *J. Biol. Chem.* **1994**, *269*, 1501
431 W. Gerhartz, *Enzymes in Industry*, VCH, Weinheim, **1990**
432 B. Zhang, T. R. Cech, *Chem. Biol.* **1998**, *5*, 539
433 M. Bergmann, H. Fraenkel-Conrat, *J. Biol. Chem.* **1937**, *119*, 707
434 W. Kullmann, *Enzymatic Peptide Synthesis*, CRC Press, Boca Raton, **1987**
435 H.-D. Jakubke in: *The Peptides: Analysis, Synthesis, Biology*, Volume 9, *Chapter 3*, S. Udenfriend, J. Meienhofer (Eds.), Academic Press, New York, **1987**, 103
436 C.-H. Wong, K.-T. Wang, *Experientia* **1991**, *47*, 1123
437 J. Bongers, E. P. Heimer, *Peptides* **1994**, *15*, 183
438 F. Bordusa, H.-D. Jakubke, in: *Houben-Weyl: Synthesis of Peptides and Peptidomimetics*, Volume E22a, M. Goodman, A. Felix, L. Moroder, C. Toniolo (Eds.), Georg Thieme Verlag, Stuttgart, **2002**, 642
439 I. Schechter, A. Berger, *Biochem. Biophys. Res. Commun.* **1967**, *27*, 157

440 V. Schellenberger, H.-D. Jakubke, *Angew. Chem.* **1991**, *103*, 1440; *Angew. Chem. Int. Ed. Engl.* **1991**, *30*, 1437
441 A. Fischer, A. Bommarius, K. Drauz, C. Wandrey, *Biocatalysis* **1994**, *8*, 289
442 A. Flörsheimer, M.-R. Kula, H. J. Schütz, C. Wandrey, *Biotechnol. Bioeng.* **1989**, *33*, 1400
443 G. Hermann, A. Schwarz, C. Wandrey, G. Knaup, K. Drauz, H. Berndt, *Biotechnol. Appl. Biochem.* **1991**, *13*, 346
444 R. A. Persichetti, N. L. S. Clair, J. P. Griffith, M. A. Navia, A. L. Margolin, *J. Am. Chem. Soc.* **1995**, *117*, 2732
445 Y.-F. Wang, K. Yakolevsky, A. L. Margolin, *Tetrahedron Lett.* **1996**, *37*, 5317
446 D. Ullmann, H.-D. Jakubke in: *Proteolytic Enzymes, Tools and Targets*, E. S. Sterchi, W. Stöcker (Eds.), Springer, Berlin, **1999**, 312
447 T. Harada, S. Irino, Y. Kunisawa, K. Oyama, EP 768 384, **1997**, *Chem. Abstr.* **1997**, *126*, 316
448 J. Markussen, *Human Insulin by Tryptic Transpeptidation of Porcine Insulin and Biosynthetic Precursors*, MTP Press, Lancaster, **1987**
449 J. S. Dordick, *Enzyme Microb. Technol.* **1989**, *11*, 194
450 P. J. Halling, U. Eichhorn, P. Kuhl, H.-D. Jakubke, *Enzyme Microb. Technol.* **1995**, *17*, 601
451 U. Eichhorn, A. S. Bommarius, K. Drauz, H.-D. Jakubke, *J. Peptide Sci.* **1997**, *3*, 261
452 M. Erbeldinger, X. W. Ni, H. J. Halling, *Enzyme Microb. Technol.* **1998**, *23*, 141
453 M. Hänsler, H.-D. Jakubke, *Amino Acids* **1996**, *11*, 379
454 C.-H. Wong, *Chimia* **1993**, *47*, 127
455 D. Y. Jackson, J. Vurnier, C. Quan, M. Stanley, J. Tom, J. A. Wells, *Science* **1994**, *266*, 243
456 V. Schellenberger, H.-D. Jakubke, N. P. Zapevalova, Y. V. Mitin, *Biotechnol. Bioeng.* **1991**, *38*, 104
457 H. Sekizaki, K. Itoh, E. Toyota, K. Tanizawa, *Tetrahedron* Lett. **1997**, *38*, 1777
458 F. Bordusa, D. Ullmann, C. Elsner, H.-D. Jakubke, *Angew. Chem.* **1997**, *109*, 2583; *Angew. Chem. Int. Ed. Engl.* **1997**, *36*, 2473
459 F. Bordusa, *Braz. J. Med. Biol. Res.* **2000**, *33*, 469
460 M. Thormann, S. Thust, H.-J. Hofmann, F. Bordusa, *Biochemistry* **1999**, *38*, 6056
461 R. Günther, S. Thust, H.-J. Hofmann, F. Bordusa, *Eur. J. Biochem.* **2000**, *267*, 3496
462 V. Cerovsky, F. Bordusa, *J. Peptide Res.* **2000**, *55*, 325
463 N. Wehofsky, S. W. Kirbach, M. Hänsler, J.-D. Wissmann, F. Bordusa, *Organic Lett.* **2000**, *2*, 2027
464 L. Pauling, *Am. Sci.* **1948**, *36*, 51
465 W. P. Jencks, *Catalysis in Chemistry and Enzymology*, McGraw-Hill, New York, **1969**
466 R. A. Lerner, S. J. Benkovics, P. J. Schultz, *Science* **1991**, *252*, 659
467 D. Hilvert, G. MacBeath, J. A. Shin in: *Bioorganic Chemistry: Peptides and Proteins*, S. M. Hecht (Ed.), Oxford University Press, Oxford, **1998**
468 R. Hirschmann, A. B. Smith, III, C. M. Taylor, P. A. Benkovic, S. D. Taylor, K. M. Yager, P. A. Sprengeler, S. J. Benkovic, *Science* **1994**, *265*, 234
469 D. B. Smithrud, P. A. Benkovic, S. J. Benkovic, C. M. Taylor, K. M. Yager, J. Witherington, B. W. Philips, P. A. Sprengler, A. B. Smith, III, R. Hirschmann, *J. Am. Chem. Soc.* **1997**, *119*, 278
470 J. R. Jacobsen, P. G. Schultz, *Proc. Natl. Acad. Sci. USA* **1994**, *91*, 5888

5
Synthesis Concepts for Peptides and Proteins

5.1
Strategy and Tactics

Before synthesizing larger peptides (and even proteins), one of the most important prerequisites to be performed is a thorough planning with regard to the order of peptide bond-forming steps, and also the choice of protection scheme(s) to be used. It may also be useful to consider the general method of peptide chain assembly. Bodanszky made his first recommendations for the design of schemes for peptide synthesis in 1966, but it was not until 15 years later that he suggested the terminology used – and especially the terms "strategy" and "tactics" – may have caused some "blurring" of his original concept [1].

Originally, the term "strategy" was coined for the overall planning of peptide synthesis, whilst "tactics" was thought to comprise the selection of protecting groups and coupling methods. There is, however, no consensus among peptide chemists concerning strategy.

In principle, peptides can be synthesized either by stepwise chain assembly, starting preferentially from the C-terminus, or by the condensation of peptide segments. The Merrifield method in particular has often been referred to as a "solid-phase strategy", although this technique in its original form is simply a variant of the linear (stepwise) coupling of amino acids in the $C \rightarrow N$ direction. In addition, the two major protection schemes in solid-phase polypeptide synthesis (SPPS) – the Boc/Bzl and Fmoc/tBu approaches – have occasionally been raised to the status of a strategy, despite their being only categories of protecting group tactics. In general, there is no advantage in exaggerating the discussions regarding terminology; rather, it is more important to circumvent unforeseen problems of peptide and protein synthesis by utilizing not only the extreme flexibility of the global concept, but also the experience and creativity of peptide chemists in the laboratory.

5.1.1
Linear or Stepwise Synthesis

This strategy is defined as the incremental addition of each amino acid in the sequence, until the entire sequence of the target peptide has been assembled. Inde-

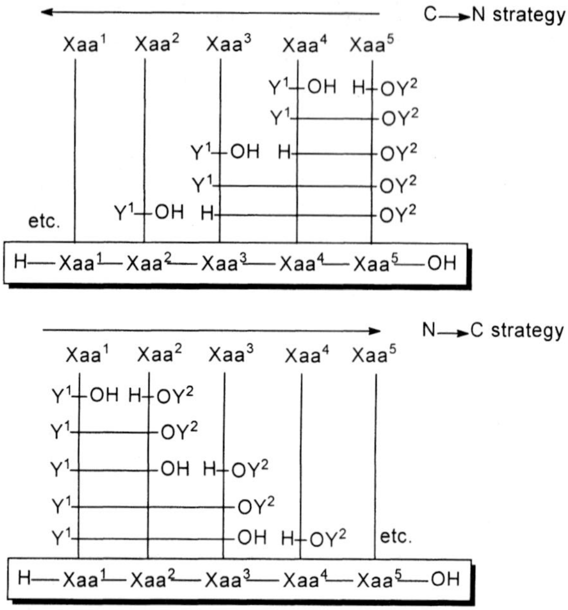

Fig. 5.1 Strategy of linear (step-wise) chain elongation. $Y^1 = N^\alpha$-amino-protecting group; $Y^2 = C^\alpha$-carboxy-protecting group.

pendently of the medium in which a synthesis is performed, the strategy is considered to be "linear" when stepwise chain elongation starts at either the C-terminal or N-terminal residue (Fig. 5.1).

The stepwise chain elongation starting at the N-terminal residue may occasionally be referred to as "inverse peptide synthesis" (Fig. 5.1), and is in agreement with Nature's way of synthesizing proteins (cf. Sect. 3.2.1). Moreover, it has the advantage that only the N-terminal amino acid residue requires a N^α-protecting group, and this is removed on the completion of chain assembly. The permanent risk of racemization (cf. Sect. 4.4) on the dipeptide stage remains a serious drawback of chemical coupling procedures, but this disadvantage can be completely circumvented using enzymatic coupling procedures (cf. Sect. 4.6.2), as shown in Fig. 5.2.

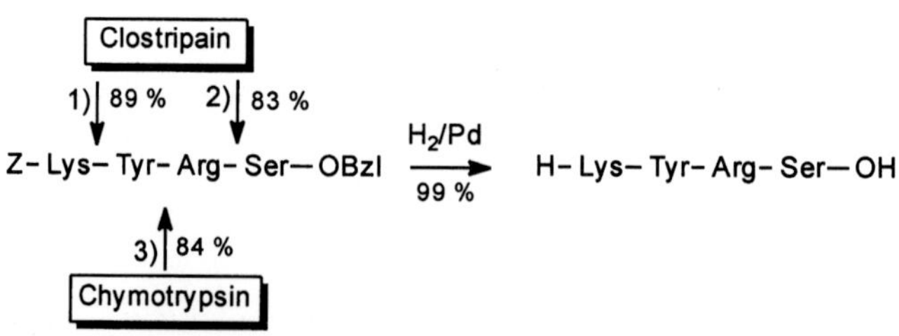

Fig. 5.2 Enzymatic synthesis of a tetrapeptide in the N → C direction without any side-chain protection [2].

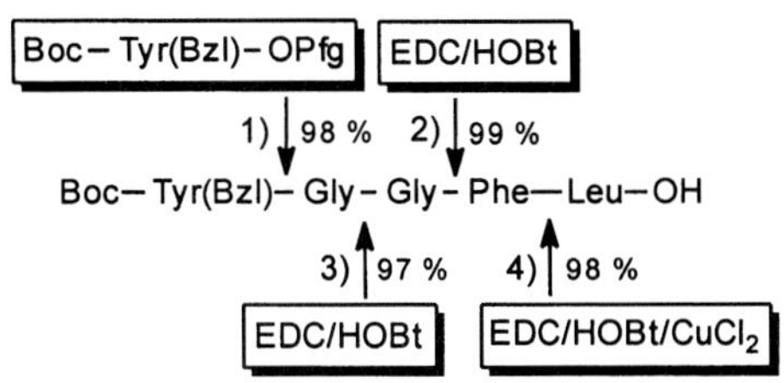

Fig. 5.3 Synthesis of [Leu]enkephalin by linear N → C strategy using mainly free amino acids as amino components [3].

Bordusa et al. [2] synthesized the tetrapeptide Z-Lys-Tyr-Arg-Ser-OBzl as a model system for stepwise enzymatic synthesis in the N → C direction, without any side chain protection, and starting with Z-Lys-SBzl and repetitive enzymatic coupling (Fig. 5.2, reactions 1 to 3) of the following three amino acid alkyl esters. The terminal blocking groups were cleaved by catalytic hydrogenation, and the free tetrapeptide H-Lys-Tyr-Arg-Ser-OH could be obtained in an overall yield of 62% and purity >98% (by HPLC).

Stepwise chemical peptide synthesis in the N → C direction has been described both in solution and as a solid-phase variant. An interesting approach to synthesis in solution using free amino acids as amino components has been described by Mitin et al. [3, 4]. As shown in Fig. 5.3, protected [Leu]enkephalin was synthesized in a model reaction starting with Boc-Tyr(Bzl)-OPfg followed by coupling of the next three free amino acids (Gly, Phe and Leu) using 1-ethyl-3-(3-dimethylamino-propyl) carbodiimide (EDC) in the presence of HOBt. This coupling agent achieves high yields in every coupling step, and in the case of the last coupling reaction $CuCl_2$ as an additional additive suppresses racemization to a significant degree. The problem of the low solubility of free amino acids in aprotic solvents has been solved by the application of a 3 M solution of $Ba(ClO_4)_2$, $Ca(ClO_4)_2$ or $Ca(ClO_4)_2$ in dimethylformamide, yielding 0.4 to 10 M solutions of the amino acids, depending on their structure. The application of this principle in solid-phase synthesis should offer several interesting advantages, provided that the general procedure can be applied for all amino acids without side reactions.

A solid-phase variant of a linear synthesis starting from the N-terminus was described by Letsinger et al. [5] as early as 1963. For this purpose, the N-terminal starting amino acid ester was linked to the polymeric resin via a benzyloxycarbonyl moiety **1**.

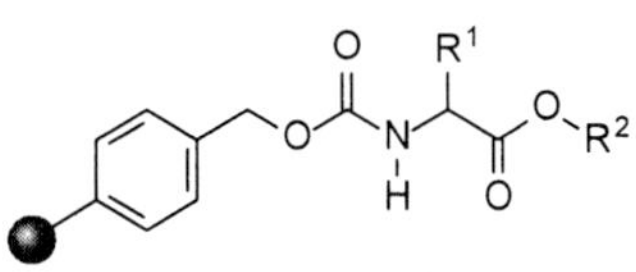

1

After saponification of the ester, the next amino acid residue was coupled via a mixed anhydride. Despite further modifications, this procedure did not find wide application. Recently, a new variant of inverse synthesis using HOBt salts of amino acid 9-fluorenylmethyl esters has been described [6]. The first amino acid was attached via a trityl linker on TentaGel resin. The least degree of racemization was observed using *N,N'*-diisopropylcarbodiimide (DIC) as coupling agent, but the application of TBTU/NMM led to increased racemization. Another attempt of SPPS in the reverse (N → C) direction was based on the application of 2-Cl-trityl resin, allyl ester as the temporary protecting group, and either $Cu(OBt)_2$/DIPCDI or HATU/DIEA as coupling method [7].

Stepwise chain elongation starting with the C-terminal residue (Fig. 5.1) requires the incorporation of each residue in blocked and activated form, and removal of the N^{α}-protecting group after each chain elongation cycle. Although this type of strategy seems to be too laborious for practical purposes, the advantages are doubtlessly that: (i) the application of urethane-type N^{α}protection generally prevents racemization of the amino acid residue it is attached to; and (ii) the use of acylating agent in high excess drives the reaction to near-completion. Bodanszky and du Vigneaud [8] demonstrated the efficiency of this approach in a stepwise synthesis of oxytocin. The application of active esters (*p*-nitrophenyl ester in the case of the oxytocin synthesis) of blocked amino acids for the incorporation offers the advantage that the excess reagent remains mainly unchanged and can be recovered. Despite the successful application of the incremental chain elongation in the syntheses of oxytocin, and even for the 27-peptide porcine secretin [9], the need for long series of coupling and deblocking procedures did not convince the majority of users about such advantages. In practice, under large-scale conditions peptides of up to five or six amino acids in length are generally synthesized by the stepwise approach in solution. However, the consequence has been the development of the solid-phase synthesis by Merrifield (cf. Section 4.5 and 5.3.1). The performance of the stepwise C → N strategy under solid-phase conditions resulted in the method's full appreciation, and its broad application.

5.1.2
Segment Condensation or Convergent Synthesis

Segment condensation (fragment condensation), also termed convergent synthesis, is defined as the construction of the target structure by final assembly of separately synthesized intermediate segments. Assuming, for example, that the formation of each peptide bond could be achieved with a yield of 80%, in the stepwise synthesis of an octapeptide the overall yield is 21%, whereas a segment condensation strategy involving four dipeptides that are coupled pairwise to give two tetrapeptides and final assembly to the target octapeptide results in an overall yield of 51%. Although, based on this formal consideration, the segment condensation approach seems superior, such a conclusion is not valid because other factors also determine the synthesis success.

The convergent synthesis generally allows for greater flexibility in the choice of protecting groups and coupling methods. However, in the synthesis of larger peptides this strategy is sometimes impeded by: (i) an increased risk of racemization; (ii) poor solubility of larger protected intermediate segments; and (iii) low coupling rate, with a concurrent risk of side reactions. In order to exclude or minimize racemization, it is advantageous to have Gly or Pro as the C-terminal residue of a protected peptide segment, or to use either the new additives for DCC couplings, such as DCC/HOBt, DCC/HOAt etc., or, for example, the reagents HBTU [10] or HATU [11].

In convergent synthesis, a judicious dissection of the target sequence will greatly improve the results. As mentioned above, optimum segments should contain C-terminal Gly or Pro residues, but if this is not possible then Ala or Arg should be chosen as C-terminal residues as they are less prone to racemization. With regard to the size of peptides synthesized by linear or convergent synthesis, there is a difference between the laboratory-scale synthesis and the large-scale synthesis performed on an industrial level. As a rule, in large-scale synthesis the segments should (preferably) be no longer than approximately five amino acids, which is about the same size as peptides generally manufactured by linear synthesis [12].

In addition to the conventional coupling of protected peptide segments in solution and minimally protected peptides in aqueous solution, both the solid-phase coupling of protected peptide segments and the chemical and enzymatic ligation of peptide segments have significantly extended the methodology of synthesizing peptides and proteins.

5.1.3 Tactical Considerations

5.1.3.1 Selected Protecting Group Schemes

The tactics of peptide synthesis comprise selection of the optimum combination of protecting groups and the most suitable coupling method that is optimum for each peptide bond-forming step. Even when synthesizing the simplest peptide in a controlled manner, it is essential that certain functional groups are protected. Tactical issues in segment condensation strategies are much more complex. First, it is of great importance to introduce – and, more importantly, to remove – all protecting groups under conditions that do not damage the integrity, and especially the stereochemical purity, of the peptide to be synthesized. As mentioned in Section 4.2 and 4.5.3, a differentiation must be made between temporary amino-protecting groups and semipermanent side chain-protecting groups. The latter must be stable both during the repetitive cleavage reactions of the temporary protecting groups, and during the coupling reactions. The third category of protection applies to the C-terminus of the peptide. The demands on C-terminal protecting groups are very high. On the one hand, they must be stable throughout the whole synthesis route, but on the other hand C-terminal protecting groups are sometimes required which can be removed in the presence of all other protecting groups. This is particularly important if a fully protected peptide segment is required to be coupled at its C-terminal amino acid residue. Especially, in large-scale solution synthesis the most suitable C-terminal protecting group is the free acid itself, though this limits the choice of coupling agents.

The selection of methods for protection and deblocking in linear synthesis requires special attention due to the repetitive character. Acidolysis at two widely different levels of acidity is commonly applied. Accordingly, the removal of temporary *α*-amino-protecting groups is often performed with trifluoroacetic acid (TFA), while the deblocking of side chain-protecting groups with liquid HF. However, this system of differential acidolysis generally presents increasing problems in the synthesis of larger peptides.

A better compatibility of protecting groups is based on the so-called orthogonality of protecting groups, although the term "orthogonal" has nothing to do with the absolute or relative geometries of protecting groups according to the normal meaning of "right-angled" or "situated at right-angles". Orthogonal protecting groups will be completely removed by one reagent, without affecting the other groups. For example, complete chemical selectivity allows for catalytic hydrogenolysis of the Z group in the presence of *tert*-butyl-type side chain-protecting groups, that in turn are cleaved by mild acidolysis. Furthermore, the Boc group can be used as N^{α}-protecting group in combination with benzyl-based side-chain protection. Another orthogonal protecting group combination that relies on two different chemical mechanisms is the Fmoc/tBu scheme characterized by the base-labile Fmoc group as temporary protective group and acid-labile permanent *tert*-butyl side-chain protection. However, the N^{α}-Fmoc protection is also not free of problems due to sequence-dependent Fmoc deprotection inefficiency and premature deprotection in the course of chain assembly [13]. It is of great advantage that several research groups [14–18] have been engaged in the "fine-tuning" of side chain-protecting groups, such that both types of chemistry are undergoing steady refinements.

The solubility of intermediate segments in organic solvents plays a fundamental role in the synthesis of large peptides and proteins. Besides the safe maximum protection scheme with a global masking of all functional groups, the minimum protection tactics has been developed for reducing solubility problems and minimizing the number of synthesis steps. The following arbitrarily selected three segments vary in the degree of side-chain protection. Segment **2** is soluble in DMF, but insoluble in DCM/trifluoroethanol, whereas segment **3** is soluble in DCM/trifluoroethanol, yet insoluble in DMF. In contrast, the water-soluble segment **4**, when dispensed completely from side-chain protection, can only be used in enzymatic segment conden-

Boc–Ala–Ile–Ser(Bzl)–Thr(Bzl)–Asn–Val–Gly–Ser(Bzl)–

Asn–Thr(Bzl)–Tyr[Z(2-Br)]–NH_2

2

Z–pGlu–Phe–Thr(Bzl)–Asp(OCy)–Val–Asp(OCy)–Cys(Acm)–

Ser(Bzl)–Val–Ser(Bzl)–OPac

3

iNoc–Lys–Glu–Thr–Ala–Ala–Ala–Lys–Phe–Glu–Arg–Glu–His–Asp–Ser–

Ser–Thr–Ser–Ala–Ala–OGlyc–Phe–NH_2

4

sation (cf. Sect. 4.6.2). In particular, unprotected ε-amino functions of lysine residues do not allow for any chemical segment condensation.

There is no doubt that the minimal side-chain protection approach is desirable for large-scale synthesis, as it minimizes the number of synthesis steps, though in practice it is not possible to omit side-chain protection completely. Segment **4** is an exception, as unprotected ε-amino functions cannot be used in chemical segment condensations. Furthermore, it is necessary to protect also the thiol function of cysteine, preferentially with the Acm group or a S-benzyl group. The Acm group can be removed by iodine with concomitant disulfide formation, but the benzyl group requires reduction with sodium in liquid ammonia. In addition, the trityl residue is another thiol-protecting group that, in combination with Acm, allows site-directed cyclizations. The carboxy groups of Asp and Glu must also be protected. From a practical point of view, it should be stated that the more hydrophilic properties of minimal side chain-protected peptides do not promote extractive work-up procedures. Generally, there is no doubt that the formation of unexpected side products in the course of chemical operations with minimally protected peptides cannot be excluded. Taking this into consideration, the maximum protection approach seems to be the preferred tactical variant both in the chemical solution-phase synthesis and in solid-phase synthesis.

Selected protecting group schemes used in the maximum protection approach are listed in Tab. 5.1.

Both the Boc/Bzl and Fmoc/tBu protection schemes (also termed Merrifield tactics and Sheppard tactics, respectively) are the preferred tactics used in SPPS. Side chain deprotection and detachment from the resin are performed in a single step under acidic conditions. The Boc/Bzl chemistry requires very strong acids for deprotection – in contrast to Fmoc/tBu chemistry – but the choice of scavengers is critical in both cases. Anhydrous hydrogen fluoride (HF) [19] or trifluoromethane sulfonic acid (TFMSA) [20] are the preferred agents for the final depro-

Tab. 5.1 Selected protecting group schemes for the maximum protection approach.

Protecting groups temporary	***Fmoc/tBu tactics*** ***Fmoc***	***Boc/Bzl tactics*** ***Boc***	***Boc/Bzl/Pac tactics*** ***Boc***
semipermanent			
Asp/Glu	OtBu	OBzl	OCy
Arg	Mtr/Pmc	Tos/Mts	Tos
Lys	Boc	Z(2-Cl)	Z(2-Cl)
His	Boc/Bum/Trt	Tos/Dnp	Bom
Cys	Trt/Tmb	Npys/Fm/Bzl(4-Me)	Acm
Ser	tBu	Bzl	Bzl
Thr	tBu	Bzl	Bzl
Tyr	tBu	Z(2-Br)/2,6-Cl_2Bzl	Z(2-Br)/3-Pn [a)]
Trp	–	For	For,Hoc
Asn/Gln	Trt/Tmb	Xan	Xan [a)]

a) Essential for the HMFS resin method.

tection of benzyl-type protecting groups. From the mechanistic point of view, a distinction can be made between the high HF procedure (S_N1 mechanism) and the low HF procedure (S_N2 mechanism) [21]. The tBu-based protecting groups are removed by acidolysis with TFA [22]. The chemistry of both procedures has different features, and different problems [23]. For a large-scale solution synthesis the Fmoc group is less attractive, because of the lack of volatility and the reactivity of the dibenzofulvene by-product.

The maximum protection tactics (Boc/Bzl/Pac), termed Sakakibara tactics, is characterized by simple deprotection with HF; moreover, the by-products are easily separated. The Acm group is stable to HF and allows for additional purification of a peptide intermediate before folding takes place [24]. The cyclohexyl ester (OCy) group resists aminosuccinyl bond formation of aspartyl residues during coupling reactions [25]. The 4-methylbenzyl [Bzl(4-Me)] moiety as a protecting group for Cys permits the isolation of peptides with free thiol groups after a single treatment with HF [26, 27]. The Z(2-Br) group for the aromatic hydroxy function prevents the benzylation of Tyr during HF cleavage [28], and the Z(2-Cl) group is stable to TFA during Boc deprotection [29].

5.1.3.2 **Preferred Coupling Techniques**

A veritable arsenal of couplings methods exists for the coupling of the individual amino acids, as described in Sect. 4.3. The classical procedures, such as mixed carboxylic-carbonic anhydrides, symmetrical anhydrides, carbodiimides, azides and commercially available pre-activated amino acid derivatives, such as various active esters, *N*-carboxyanhydrides (NCAs) and urethane-protected NCAs, known as UNCAs [30], have not lost their importance as coupling reagents. In particular, the pre-activated derivatives are compatible with unprotected C-terminal amino acid residues in the amino component, and have the benefit that by-products are not generated from the activating agent. More recently, uronium- and phosphonium-based reagents (e.g., HATU, HBTU, TBTU) have found increasing popularity. This has been primarily based on their superior reactivity, convenience and efficiency, as well as their minimal side reactions. The latter reagents have been used especially in segment coupling reactions. The carbodiimide method, in combination with auxiliary nucleophiles such as HOBt and HOAt, has also been used in large-scale segment coupling. The same holds true for the mixed anhydride method (with or without auxiliary nucleophiles) and the classical azide coupling procedure that has advantages due to the low racemization sensitivity. Furthermore, amino components with unprotected C-terminal carboxy functions are compatible with the azide method, though the by-product N_3^- gives rise to safety concerns in large-scale syntheses. HATU is also a very promising coupling agent for chemical protein synthesis. The special features of large-scale peptide synthesis [12] are discussed in Chapter 9, but details with regard to tactical coupling methods are outlined in the following paragraphs.

5.2 Synthesis in Solution

The first synthesis of oxytocin was performed by the Nobel laureate V. du Vigneaud and his co-workers [31] in 1953, and today most peptide pharmaceuticals are prepared commercially using classical synthesis in solution. Indeed, despite the dominance of SPPS, peptide synthesis in solution remains a major approach to the preparation of peptides, and even proteins. Peptide synthesis in solution can be performed by both linear and convergent strategies. In principle, a stepwise synthesis in solution can be applied to small oligopeptides and segments up to about five amino acids, whereas SPPS is much more successful for the synthesis of longer peptides. Various features such as the complexity of the target molecule, the protection scheme chosen, and the economical as well as logistical considerations, determine the strategic route. Convergent peptide synthesis (CPS) in solution has been used for the large-scale production of small to medium-length peptides in quantities of up to hundreds of kilograms, or even metric tons per year; examples include inhibitors of angiotensin-converting enzymes (ACE) and HIV protease, analogues of luteinizing hormone-releasing hormone (LH-RH), oxytocin, desmopressin, and aspartame [12]. Calcitonins obtained from various species and containing 32 amino acid residues are among the longest peptides manufactured for commercial application.

Using the classical solution procedure, the product can be purified and characterized at each step in the reaction. As assembly of the entire molecule can be performed with purified and well-characterized segments, the desired final product is isolated in a highly homogeneous form. Unfortunately, this major advantage requires the investment of a considerable amount of work and time, although when experienced teams of workers can simultaneously synthesize several different segments. Within the context of convergent peptide synthesis in solution, it should be borne in mind that individual segments may also be synthesized using SPPS.

5.2.1 Convergent Synthesis of Maximally Protected Segments

The low solubility of intermediate segments in the usual organic solvents is a major difficulty when synthesizing large peptides in solution. Indeed, Erich Wünsch, a pioneer of segment condensation in solution using fully protected segments, predicted about 30 years ago that this ideal strategy might involve an insolubility problem when synthesizing peptides in the range between 30 and 50 amino acid residues [32]. In 1981, Fujii and Yajima [33, 34] described the solution synthesis of the 124-residue protein ribonuclease A (RNase A) by coupling 30 relatively small-sized segments. The azide procedure was used for segment coupling, as at that time the method was believed to be devoid of racemization. After deprotection and protein folding, the enzyme was isolated in crystalline form and shown, after chromatography, to have full biological activity. The procedure was problematic, however, because during the segment elongation process of every coupling step, a large excess (between 3- and 30-fold) of each carboxy segment was required. In addition to causing serious problems of insolubility of the intermediates, the necessary excess of carboxy segments and separation of by-products formed via Curtius rearrangement greatly increased purification difficulties.

5.2.1.1 The Sakakibara Approach to Protein Synthesis

Following the impressive synthesis of RNase A, and despite the problems of product purification, Sakakibara's group at the Peptide Institute in Osaka began to investigate a general procedure for the solution synthesis of peptides containing more than 100 residues. In a recent review, Sakakibara [35] provided a comprehensive description of this effective synthesis concept, and predicted that linear peptide sequences consisting of more than 200 residues could be synthesized using this approach.

This actual concept of solution synthesis of proteins is based on new types of solvent systems for dissolving fully protected segments, in which segment condensation reactions can be performed smoothly in solution. The entire target molecule is assembled from fully protected segments with a size of about 10 residues. This size fits the requirement for purifying segments with currently available methods (e.g., HPLC, column chromatography), and for characterizing homogeneity. Furthermore, each segment is designed to have a common structure of Boc-peptide-OPac (Pac, phenacyl), and all side chain functions are protected by benzyl-type groups (cf. Tab. 5.1). Pro and Hyp are not suitable as N-terminal residues in the segments, but Val and Ile are still acceptable. However, Gly, Ala, Leu, Pro, and Hyp(Bzl) are suitable C-terminal residues, while Gln, Asn, Lys[Z(Cl)], Glu(OCy), Asp(OCy), Trp(For), and Ser(Bzl) might be acceptable. In contrast, His(Bom), Cys(Acm), Arg(Tos), Tyr[Z(2-Cl)], Tyr(3-Pn), Trp(Hoc), Ile, Val, Phe, and Thr(Bzl) must not occur in this position. An alternative removal of the Boc or OPac group yields the segments, which are coupled using the water-soluble 1-ethyl-3-(3′-dimethylaminopropyl)-carbodiimide (EDC) in the presence of 3,4-dihydro-3-hydroxy-4-oxo-1,2,3-benzotriazine (HODhbt) to give the entire sequence. The general principle of the Sakakibara approach is shown schematically in Fig. 5.4.

A mixture of trifluoroethanol (TFE) in chloroform or dichloromethane (DCM) in a ratio of 1:3 (v/v) displays highly suitable properties as the coupling solvent, especially, when HODhbt is used as additive instead of HOBt.

The C→N strategy for the synthesis of the protected segments is shown schematically in Fig. 5.5. The synthesis of the protected segments starts from Boc-Xaa^1-OPac, and the first coupling reaction is performed after removal of the Boc group.

As mentioned previously, once the Boc-peptide-OPac is obtained, it can be used alternately as an amino component after removal of the Boc group, or as the carboxy component after cleavage of the OPac group. The Pac group can be removed under mild conditions by reduction with zinc in acetic acid at 40 or 50 °C [36], and it is stable to TFA. Interestingly, segment synthesis has sometimes also been

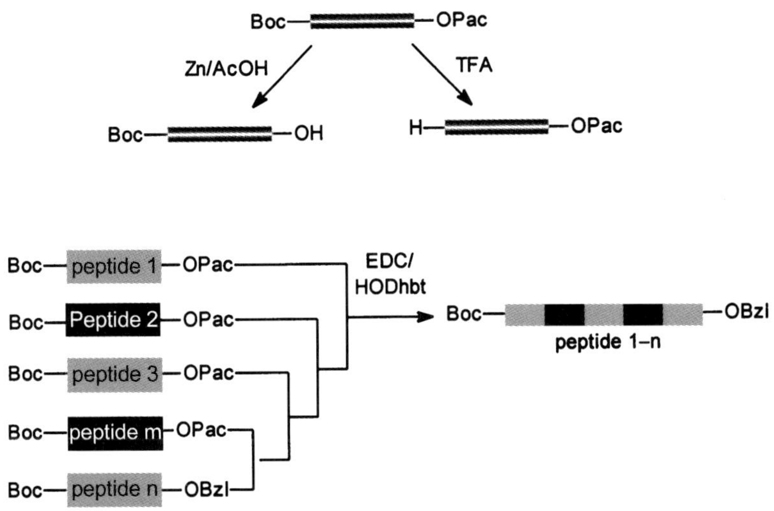

Fig. 5.4 Alternative deblocking of Boc or Pac groups and segment condensation using EDC/HODhbt according to SAKAKIBARA [35].

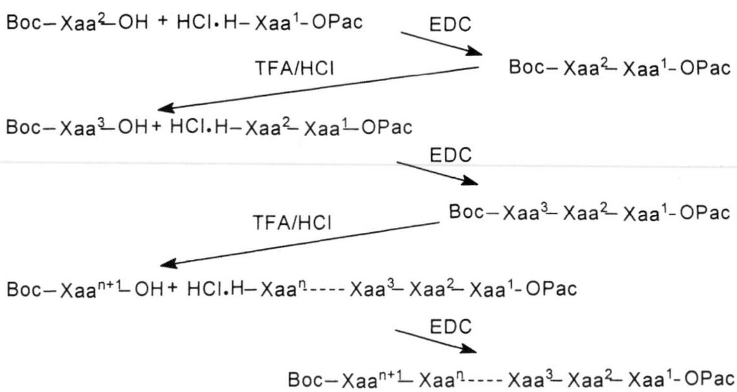

Fig. 5.5 C → N strategy of segment synthesis starting from an amino acid phenacyl ester according to SAKAKIBARA [35].

carried out using SPPS on HMFS-resin, in which *N*-[9-(hydroxymethyl)-2-fluorenyl]succinamic acid (HMFS) is attached to a cross-linked aminomethyl polystyrene [37]. In this case HBTU was used as coupling agent for Boc amino acids, and removal of the protected segment from the resin was performed with 20% morpholine in DMF. Selected biologically active peptides and proteins synthesized by the Sakakibara approach are listed in Tab. 5.2.

Synthesis of the green fluorescent protein (GFP) precursor molecule with 238 residues was the masterly performance of a highly sophisticated synthesis technique,

Tab. 5.2 Selected biologically active peptides or proteins synthesized according to the Sakakibara approach [35].

Name	*Residues*	*Segments*	*Year*
h-PTH	84	10	1982
ω-Conotoxin GVIA	27	4	1982
h-ANP	28	4	1985
C5a Anaphylatoxin	74	9	1988
p-Endothelin I	21	4	1988
h-Angiogenin	123	15	1990
Charybdotoxin	37	5	1990
Calciseptine	60	7	1992
Elafin	57	8	1992
Na^+,K^+-ATPase Inhibitor-1 (SPAI-I)	49	6	1992
μ-Conotoxin GIIIB	22	3	1993
w-Agatoxin IVA	48	6	1993
h-Osteocalsine	49	7	1995
PLTX II	44	6	1995
Conotoxin GS	34	5	1995
Calcicludine	60	9	1996
Human midkine	121	13	1996
Green fluorescent protein (GFP) precursor	238	26	1998
Dendrotoxin-I	60	6	1998
Muscarinic toxin I	66	7	1999
h-Pleiotrophin (PTN)	136	14	1999

and belongs to the state-of-the-art protein syntheses [38]. The entire sequence of GFP was divided into 26 segments, from which only four were synthesized by the classical solution procedure, whilst the major proportion was assembled on HFMS-resin. Further coupling of the four large segments (Boc-(1–51)-OPac, Boc-(52–116)-OPac, Boc-(117–174)-OPac and Boc-(175–238)-OBzl) with preceding removal of the corresponding protecting groups resulted in the two final segments, Boc-(1–116)-OPac and Boc-(117–238)-OBzl. The last segment condensation in chloroform/TFE (3:1) Boc-(1–116)-OH + H-(117–238)-OBzl → Boc-(1–238)-OBzl yielded the fully protected target molecule. Following cleavage of the terminal Boc group with TFA, the remaining product (465 mg) was treated with HF in the presence of Cys · HCl and *p*-cresol at –5 °C for 1 h. Purification by reversed-phase (RP)-HPLC, followed by removal of the two remaining Acm groups, yielded the final product (25 mg). Finally, under these experimental conditions, about 10% of the synthetic GFP precursor was found to fold into the native GFP protein.

5.2.1.2 **Condensation of Lipophilic Segments**

In order to circumvent solubility problems in solution-phase segment condensation, a combined solid-phase/solution approach has been described by Riniker et al. [39–41]. This lipophilic segment coupling approach has been applied to the synthesis of some medium-sized peptides, such as human calcitonin-(1–33), human neuropeptide Y, and the sequence 230–249 of mitogen-activated 70K S6 kinase. Lipophilic protected segments, that are relatively soluble in organic solvents (e.g., chloroform, tetrahydrofuran, ethyl acetate), can be obtained using a special, maximum protection chemistry. For this purpose the side chains of Asn, Gln, and His are always blocked with the Trt group, Arg with the Pmc group, and Trp with the Boc group. All other side-chain functions, with the exception of the thiol group of Cys, are protected with *tert*-butyl-type groups. Cys may also be blocked with the Trt residue, but if acid-stable Cys protection is required, then Acm protection is necessary.

The synthesis of the lipophilic segments is performed by Fmoc/tBu SPPS on a resin bearing the highly acid-labile 4-(4-hydroxymethyl-3-methoxyphenoxy)-butyric acid (HMPB) handle. The length of the segments can vary up to approximately 20 amino acid residues, and preferentially they should contain, if possible, Gly or Pro at the carboxy terminus.

The principle of the lipophilic segment coupling approach is shown schematically in Fig. 5.6. As most peptide segments are soluble in *N*-methylpyrrolidone, the segment coupling reactions have been performed in this solvent using TPTU/

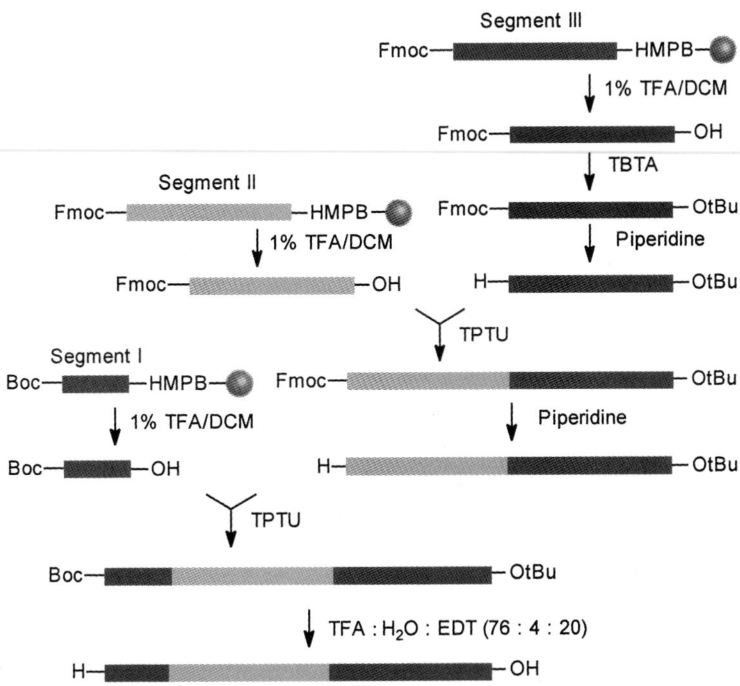

Fig. 5.6 Principle of the lipophilic segment coupling approach according to Riniker et al. [39]. The amino acid side chains are protected as the tBu ethers (Ser, Tyr, Thr), tBu esters (Asp, Glu), Boc derivatives (Lys, Trp), Trt derivatives (Asn, Gln, His, Cys), or Pmc derivatives (Arg).

HOBt. Most segment condensations reach completion within a few minutes at room temperature, and the resultant protected peptides are purified by flash or medium-pressure chromatography on silica gel. Finally, acidolytic deprotection is usually carried out in TFA/H_2O/ethanedithiol (76:4:20).

5.2.2
Convergent Synthesis of Minimally Protected Segments

5.2.2.1 Chemical Approaches

As early as 1969, Hirschmann et al. [42] succeeded in synthesizing ribonuclease S (a protein containing 104 residues), using minimally protected segments, and in which only the side-chain functions of Cys and Lys were blocked. With the exception of Trp, the sequence of the ribonuclease S protein contains all other trifunctional amino acids. For the synthesis of the segments in the range between tri- and nonapeptides, only the NCA/NTA method and HOSu procedure could be used. The azide method was mainly used for assembling the segments to larger ones. Only the final azide coupling of the two large segments with 44 and 60 residues led to a significantly lower yield.

As already observed in the classical ribonuclease S protein synthesis, the chemical coupling of minimally protected segments is associated with a permanent risk of unexpected side reactions. In principle, coupling of two unprotected segments requires the selective activation of the C-terminal carboxylic acid of one of the segments, without affecting any of the other carboxy groups present in either segment. Moreover, all other amino functions except the one involved in peptide bond formation must be blocked or deactivated. Some attempts have been made, though an ideal solution of this very complicated problem has not been found.

The thiocarboxy segment condensation method pioneered by Blake [43] has an interesting feature, in which selective activation of the thiocarboxy group at the C-terminus by silver ions is achieved without affecting side-chain carboxy groups (Fig. 5.7). Both segments to be coupled can be synthesized using Boc/Bzl SPPS.

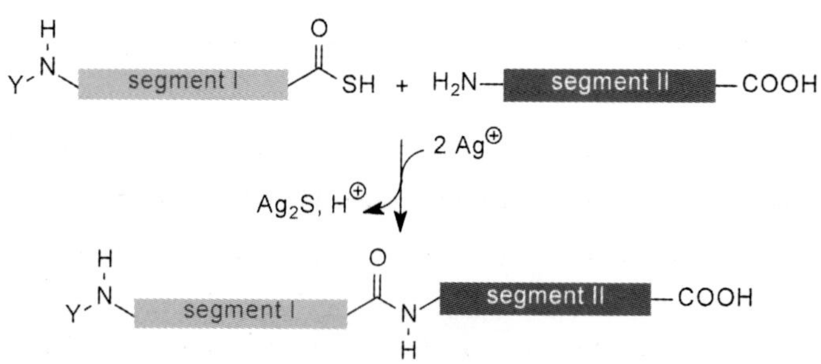

Fig. 5.7 Schematic description of the thiocarboxy segment condensation method according to BLAKE [43].

In general, thioglycine is mainly chosen as the C-terminus in order to avoid racemization during the following segment condensation. After standard SPPS, the segments are detached from the resin by treatment with liquid HF. However, the N-terminus of the carboxy component must bear a protecting group Y that is stable under the strongly acidic cleavage conditions, such as Fmoc, Msc, Troc, iNoc, TFA, or Ac. The protection of other side-chain functions is not necessary, with the exception of free amino groups. Both α-inhibin containing 92 residues [44, 45], and β-lipotropin [46], were synthesized using this method. This elegant method has some limitations, however:

- thiocarboxylic acids are not stable toward oxidation and hydrolysis;
- free side-chain amino groups in the segments can undergo undesired amide bond formation with the activated thiocarboxy group; and
- no procedure could be developed to synthesize cysteine-containing polypeptides based on the thiocarboxy segment condensation approach.

Boc-OSu cannot be used for the essential blocking of side-chain amino groups due to side reactions with the highly nucleophilic thiol moiety of the thiocarboxy group, and the alternative citaconyl group lacks stability, even under conditions of mild acidity.

Aimoto et al. [47] have described a conceptually similar, but different and significantly improved, approach to polypeptide synthesis. The Aimoto thioester approach [48] is characterized by converting an S-alkyl thioester moiety in the presence of a silver salt ($AgNO_3$ or AgCl) into an active ester derived from HOBt or HODhbt, followed by segment condensation of partially protected segments. Peptide segment thioesters can be prepared via a Boc solid-phase method. The insertion of one amino acid between a thioester linker and an MBHA resin diminishes loss of the growing peptide during TFA treatment for Boc deblocking. The segments, except for the N-terminal ones, must be blocked by the iNoc group, this being removable by zinc dust in aqueous acetic acid, even in the presence of Boc groups. Following HF treatment, Boc-OSu was used as the reagent for protecting side-chain amino groups.

H–Ala^{1}–Gln–Val–Ile–Asn^{5}–Thr–Phe–Asp–Gly–Val^{10}–Ala–Asp–Tyr–Leu–
Gln^{15}–Thr–Tyr–His–Lys–Leu^{20}–Pro–Asp–Asn–Tyr–Ile^{25}–Thr–Lys–Ser–
Glu–Ala^{30}–Gln–Ala–Leu–Gly ↓ Trp^{35}–Val–Ala–Ser–Lys–Gly^{40}–Asn–Leu–
Ala–Asp–Val^{45}–Ala–Pro–Gly–Lys–Ser^{50}–Ile–Gly ↓ Gly–Asp–Ile^{55}–Phe–
Ser–Asn–Arg–Glu^{60}–Gly–Lys–Leu–Pro–Gly^{65}–Lys–Ser–Gly–Arg–Thr^{70}–
Trp–Arg–Glu–Ala–Asp^{75}–Ile–Asn–Tyr–Thr–Ser^{80}–Gly ↓ Phe–Arg–Asn–
Ser^{85}–Asp–Arg–Ile–Leu–Tyr^{90}–Ser–Ser–Asp–Trp–Leu^{95}–Ile–Tyr–Lys–
Thr–Thr^{100}–Asp–His–Tyr–Gln–Thr^{105}–Phe–Thr–Lys–Ile–Arg^{110}–OH

Fig. 5.8 Primary structure of barnase. The arrows indicate the segments used in total synthesis.

For the synthesis of barnase (Fig. 5.8), a bacterial RNase containing 110 amino acid residues, the sequence was divided into four partially protected segments: Boc-[Lys(Boc)^{19,27}]Barnase(1–34)-S-C$(CH_3)_2$-CH_2-CO-Nle-NH_2 (A); iNoc-[Lys(Boc)^{39,49}] Barnase(35–52)-S-C$(CH_3)_2$-CH_2-CO-Nle-NH_2 (B); iNoc-[Lys(Boc)^{62,66}] Barnase(53–81)-S-C$(CH_3)_2$-CH_2-CO-Nle-NH_2 (C); and H-[Lys(Boc)^{98,108}]Barnase (82–110)-OH (D), that were coupled stepwise in DMSO using HOSu, $AgNO_3$ and NMM starting with C+D, followed by B+(C–D), and A+(B–C–D), respectively. Finally, barnase was obtained in 11% yield, based on the fragment D. Further examples of polypeptide synthesis are detailed in a review [48].

5.2.2.2 Enzymatic Approaches

A mutant of subtilisin BNP, termed subtiligase, was obtained by Jackson et al. [49] by protein design. Subtiligase was used in a further synthesis of RNase A (cf. Section 5.2.1 and 5.2.2.1). This approach combines chemical solid-phase synthesis of the segments and enzymatic segment condensation. Starting from the C-terminal fragment RNase A(98–124), the further fragments bearing the N-terminal iNoc group (77–97, 64–76, 52–63, 21–51, and 1–20) were chosen such that the appropriate C-terminal amino acid residues (P_1 residues) of the fragments (Tyr^{97}, Tyr^{76}, Val^{63}, Leu^{51}, and Ala^{20}) were closely matching the substrate specificity for large and hydrophobic residues of the protease mutant. The first segment condensation is shown schematically in Fig. 5.9.

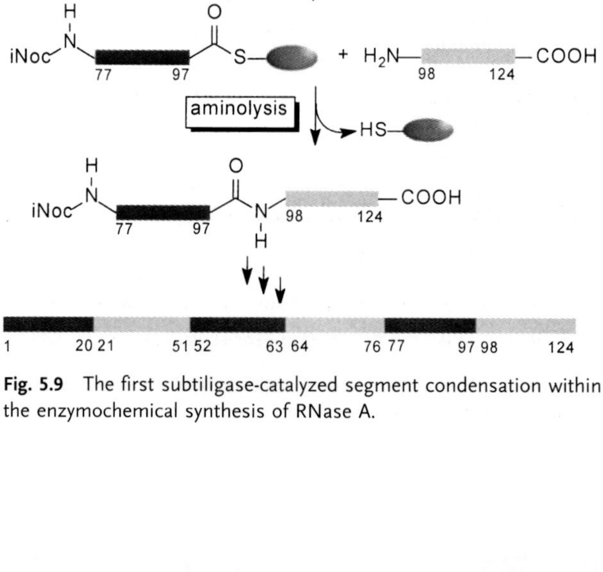

Fig. 5.9 The first subtiligase-catalyzed segment condensation within the enzymochemical synthesis of RNase A.

In particular, an efficient leaving group of the acyl donor ester can provide high reaction rates in combination with a decreasing risk of possible proteolysis of starting segments and the final product. The excellent leaving group of the Phe-NH_2-modified carboxamido methyl ester, which even in unmodified form was shown to be a highly efficient acyl donor moiety [50], and the application of a considerable excess of acyl donor segments ensured that, in the RNase fragment condensations, most of the possible side reactions could be minimized.

It is clear that the substrate mimetic strategy (cf. Sect. 4.6.2.4) should be also useful for combining irreversible enzymatic ligation of segments with suitable methods of SPPS, especially in the preparation of acyl component peptide segments in the form of appropriate esters [51]. This methodology has also been used by Bordusa's group [52] in the semisynthesis of the biologically active 493–515 sequence of human thyroid regulatory subunit anchoring protein Ht 31 via α-chymotrypsin-catalyzed (8+16) segment condensation via phenyl ester and 4-guanidinophenyl ester, respectively (Fig. 5.10).

The resulting 24-peptide represents a minimum region of Ht 31 required to bind to the regulatory subunit of cAMP-dependent protein kinase. The Ht 31-(493–515) peptide inhibited forskolin-dependent activation of a chloride channel

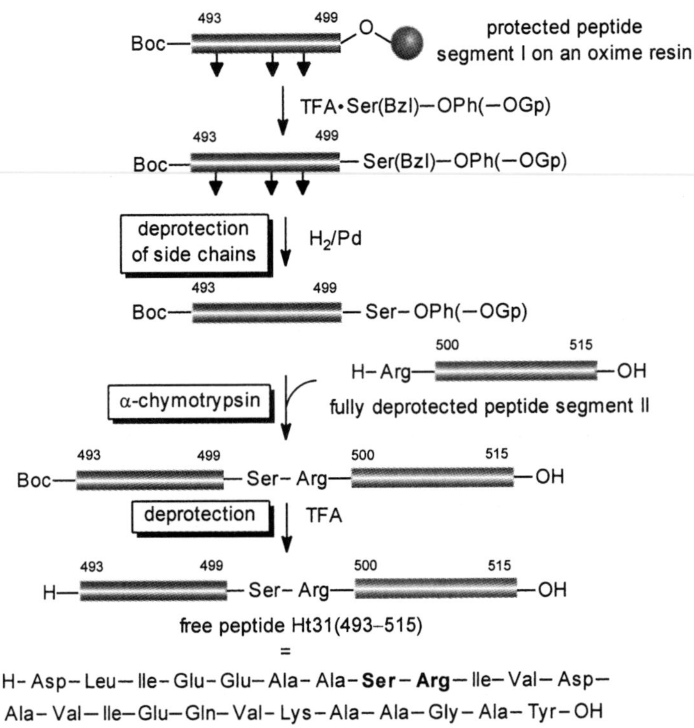

Fig. 5.10 Chemoenzymatic synthesis of the 24-peptide Ht31(493–515) using the substrate mimetic approach. OGp=4-guanidinophenyloxy.

in mammalian heart cells, thus suggesting an involvement of protein kinase A-anchoring proteins in this process. From a synthetic point of view, this example represents the first synthesis of a longer biologically active peptide that was prepared by the substrate mimetics-mediated semisynthetic approach.

An ingenious combination of chemical and enzymatic strategies, as was demonstrated in the synthesis of RNase A, should at present represent the state of the art in this field. However, the C–N ligation strategy based on the substrate mimetic concept allows for the first time the irreversible formation of peptide bond, catalyzed by proteases. When combined with frozen-state enzymology, as shown for model reactions [53], or using protease mutants with either minimal [54] or absent amidase activity, the substrate mimetics C–N ligation approach will contribute towards significant progress in enzymatic segment coupling. Fragment condensations using recombinant polypeptides generated as mimetic-based carboxy components, together with chemically synthesized or recombinant amino components, will doubtlessly broaden the application of this approach in the near future. This specific programming of enzyme specificity by molecular mimicry corresponds in practice to the conversion of a protease into a C–N ligase, resulting in a biocatalyst that nature was incapable of developing evolutionarily.

5.3 Optimized Strategies on Polymeric Support

5.3.1 Stepwise SPPS

The main repeating steps in stepwise elongation of a peptide chain are coupling and deprotection using preformed blocked amino acids. In conventional solution synthesis with equimolar amounts of reactants, the basic operations entail a large number of reactions at each stage, together with purification procedures (e.g., washing, crystallization), and the collection of solid products by filtration or centrifugation, and chromatography.

During the mid-twentieth century the need to facilitate the solution synthesis process had been clear for some time, and it was in 1963 that R. B. Merrifield [55] first introduced stepwise peptide synthesis. This entailed having the C-terminal starting amino acid attached covalently to polystyrene beads in order to simplify the classical synthesis in solution (a detailed description is provided in Chapter 4). The original synthesizer developed by Merrifield (see Fig. 4.38) is now located in the Smithsonian Museum; however, developments in the field have been very rapid, and today a number of commercial instruments are available which perform most of the synthesis steps under computer control.

The mechanization and automation of chain assembly are the ultimate goals of SPPS. Consequently, during the past 40 years a series of continuous developments and improvements have led to a revolution not only in peptide synthesis, but also in organic synthesis – and especially in combinatorial synthesis [56–59].

Tab. 5.3 Relationship between average yield per step and total yield depending on the number of residues of the peptide in SPPS.

Peptide	*Number of residues*	*Total yield [%]*			
		Yield [%] per step [a]			
		95	*98*	*99*	*99.9*
Growth hormone	191	0.006	2.2	15.0	82.8
Ribonuclease	124	0.2	8.3	29.1	88.4
Trypsin inhibitor (bovine)	58	5.4	31.6	56.4	94.5
Insulin A chain	21	35.8	66.8	81.8	98.0

a) Related to the C-terminal amino acid.

In its early stages, Merrifield's concept roused a great degree of skepticism, based mainly on two objections. The first objection was that synthetic intermediates could not be purified during chain assembly, and purification and characterization of the synthesized product would be possible only on completion of the synthesis. The second objection was that nonquantitative reactions in N^{α} deprotection and coupling, in connection with incomplete orthogonality between temporary and permanent protecting groups, would cause the production of truncated and deleted sequences. Since deletion peptides are structurally closely related to the desired peptide, their separation from the target product at the final purification step might be very difficult. The need for yields approaching 100% at every step is demonstrated with data compiled in Tab. 5.3.

Until now, the chemistry involved in SPPS has been refined to such an extent that most of the reactions are performed repetitively and reproducibly in near-quantitative yields. This, together with a parallel refinement of analytical and purification techniques, underlines why at present the vast majority of peptides are synthesized by SPPS. It might be misleading however, to believe that most polypeptides (and even small proteins) can – either now or in the future – be synthesized efficiently and safely using automated synthesizers running standardized reaction protocols. With few exceptions, it can be assumed that carefully implemented stepwise SPPS protocols are effective in producing peptides which are up to 50 residues in length, and which can be further purified using the power of RP-HPLC. In general, this purification technique has been increasingly exploited as a manufacturing tool by the major pharmaceutical companies to produce a range of commercial products, including somatostatin, LH-RH, salmon calcitonin, and other analogues. By contrast, the simple principle of SPPS and its subsequent technical improvements have brought peptide synthesis within the scope of the undergraduate chemist or biochemist, albeit with dangerous consequences. Clearly, many syntheses of longer peptides up to the size of small proteins have been attempted, but the data not published because of failure or inconclusive results. Perhaps it should also be pointed out that, on occasion, commercially prepared custom-synthesized peptides have also been produced using multiple synthesis machines that are not under adequate analytical control. Consequently, it is possible that such products might contain undetected erroneous structures, or be contaminated with undesirable by-products [60, 61].

Despite SPPS being based on a simple principle, its operation requires the knowledge and expertise of an experienced peptide chemist in order to be successful. An example is in the area of the so-called "difficult sequences" [62–66], which clearly demonstrates the intense effort required to tackle SPPS under problematic conditions (cf. Sect. 4.5.4.3).

In summary it is clear that, while stepwise SPPS represents a very useful means of producing small quantities of peptides, solution synthesis is preferable for the preparation of larger quantities of peptides, and especially of the kilogram quantities required in industrial production. The accumulation of by-products that occurs during a long stepwise synthesis may be avoided by subdivision of the target polypeptide or protein into several segments, each of which is synthesized by linear SPPS. Following detachment of the segments from the resin in suitable protected form and subsequent purification, they can be used for assembly of the complete sequence.

5.3.2
Convergent SPPS

In addition to linear SPPS, segment condensation in solution and chemical ligation (cf. Sect. 5.4), convergent solid-phase peptide synthesis (CSPPS) has been developed to circumvent the problems caused by the poor solubility of protected seg-

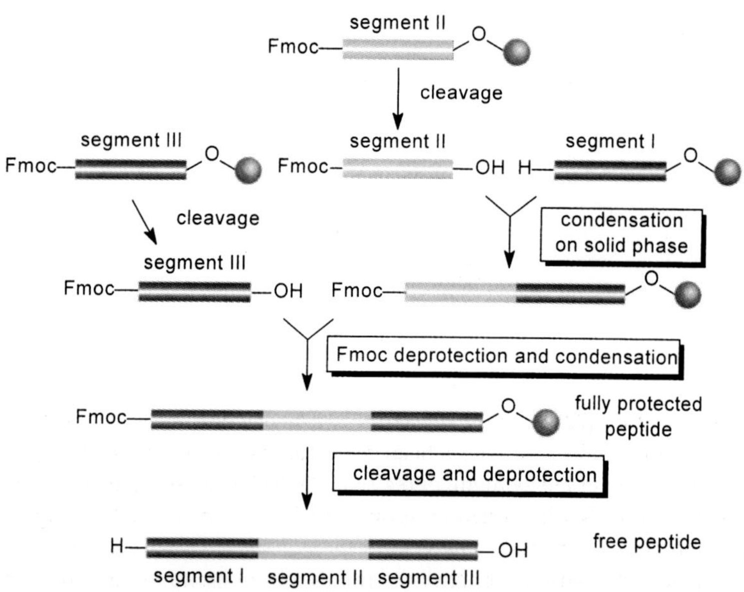

Fig. 5.11 Principle of convergent solid-phase peptide synthesis (CSPPS) starting from the C-terminus.

ments in solution [67–70]. The assembly of segments to produce the target polypeptide or protein can be performed both in the C → N and N → C directions, or it may also start from a middle region and proceed in both directions [71]. However, the C → N direction has until now been the preferred method for the synthesis of proteins, as shown schematically in Fig. 5.11.

Generally, the target protein must be divided into segments bearing (preferentially) Gly and Pro as the C-terminal amino acid in order to minimize the risk of epimerization during the coupling reaction.

5.3.2.1 **Solid-Phase Synthesis of Protected Segments**

The synthesis of protected segments on a solid support requires some modifications of the chemistry involved in comparison to the linear SPPS, since in the latter case it is mainly the free target peptide that results following final detachment from the resin. Hence, procedures are required for detachment of the protected segment from the solid support, under mild conditions, at high yield and without affecting the configurational integrity of the C-terminal amino acid. A high degree of compatibility between the blocking groups of the segment and the peptide resin anchorage is of particular interest in this respect.

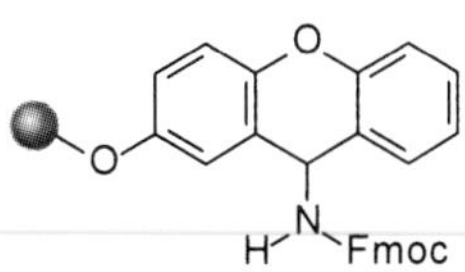

5

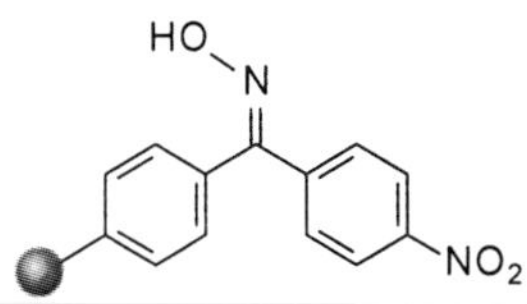

E/Z-**6**

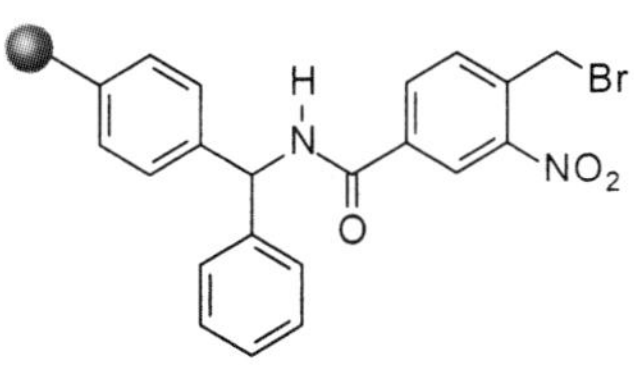

7

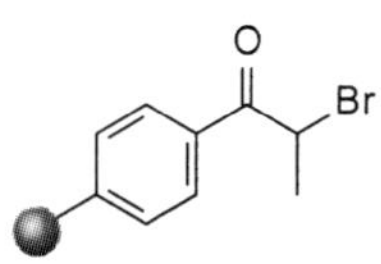

8

Fmoc/tBu-based CSPPS [72] is compatible with highly acid-labile resins and linkers. The 2-chlorotrityl resin [73, 74] (cf. Sect. 4.5.1) allows the detachment of protected segments by treatment with mixtures of AcOH/TFE/DCM, or even in the complete absence of acids with hexafluoroisopropanol in DCM. The latter reagent excludes the contamination of the protected segment with a carboxylic acid, as such contamination can cause capping of the free N amino group of the protected segment in coupling reactions. Furthermore, resins and linkers of the trityl type suppress diketopiperazine formation on the dipeptide stage of the synthesis – an unwanted side reaction that is especially pronounced with C-terminal Pro or Gly residues. A further highly acid-labile resin is known as SASRIN (*S*uper *A*cid *S*ensitive *Res*IN) [75] (see Sect. 4.5.1). The detachment of protected peptides from SASRIN can be performed at high yields using 1% TFA in DCM. The highly acid-labile xanthenyl resin [76] **5** can be used as the C-terminal segment if a protected peptide amide is required. Boc/Bzl CSPPS requires base-labile solid supports, but photolysis and allyl transfer are two further options. The 4-nitrobenzophenone oxime resin **6**, also termed "Kaiser oxime resin" [77, 78], has been used extensively for the synthesis of Boc/Bzl-protected segments. Several methods have been described for the cleavage of protected segments from the resin, including hydrazinolysis, ammonolysis, or aminolysis using suitable amino acid esters. Transesterification of the peptide resin with *N*-hydroxypiperidine, followed by treatment of the hydroxypiperidine ester with zinc in acetic acid, yielding the free carboxylic acid, is the preferred procedure. The only disadvantage is the lability of **6** towards nucleophiles (including the free N amino group of the growing peptide chain) during neutralization with base after the acidolytic deprotection of the Boc group. Photolytic cleavage of protected peptide segments from the solid support is possible using, for example, nitrobenzyl and phenacyl resins. 3-Nitro-4-bromomethyl-benzhydrylamido-polystyrene **7**, known as Nbb-resin [79, 80], is fully compatible with the Boc/Bzl CSPPS, which allows for detachment of the protected segment by irradiation at 360 nm in a mixture of DCM and TFE. The photolabile α-methylphenacyl ester resin **8** [81] is only compatible with the Boc/Bzl approach.

Although the Fmoc/tBu approach very often provides protected segments in high purity after cleavage from the resin, in other cases a thorough purification of the segments is essential before segment condensation. Preparative RP-HPLC using acetonitrile/water systems as the eluant is a suitable method for purification of the Fmoc/tBu-blocked segments [39, 82], as well as silica gel column chromatography. The poor solubility of the segments is generally a serious problem for effective purification; however, the solubility of protected peptide segments may be enhanced by backbone protection (see Sect. 4.5.4.3).

5.3.2.2 **Solid Support-mediated Segment Condensation**

In CSPPS, the resin-bound C-terminal segment is a prerequisite for high efficiency. In principle, it can be synthesized directly on the resin by stepwise SPPS, but to ensure the highest possible purity, the re-attachment of an independently synthesized, purified and characterized segment onto the resin offers a better approach. The loading of the resin should be chosen to be lower than that used in linear SPPS [74, 82, 83]. As a rule, at the end of the segment condensation the weight ratio between the protected target polypeptide and the resin should exceed 1:2, and loading values between 0.04 and 0.2 mmol g^{-1} should fit these requirements. With a higher loading, the resin loses its swelling and polarity properties during peptide chain elongation.

As the condensation of protected segments with the resin-bound C-terminal segment depends to a greater extent on the concentration of the carboxy component than on the excess applied, solutions with the highest possible concentration should be used. The purity of solvents and reagents must be carefully checked for the very often long-lasting segment coupling reactions. There is no significant difference to linear SPPS with respect to coupling methods. Nowadays carbodiimides, usually in the presence of additives such as HOSu, HOBt, HOAt, and reagents based on phosphonium or uronium salts in the presence of HOBt and HOAt, have been used successfully in CSPPS.

There is no doubt that epimerization at the C-terminal-activated amino acid of the segment is the most important side reaction (cf. Sect. 4.4). Placing Gly in this position excludes epimerization, which is also normally minimized in the case of Pro. Epimerization remains the main problem of CSPPS for all other residues, however [84]. Methods to quantify the extent of epimerization that might occur in a given segment coupling have been described [85, 86]. In a simple CSPPS model system, the lowest epimerization was found using DIC/HOBt as coupling reagent [87]. Barlos and Gatos [72] studied the extent of epimerization of the C-terminal residue Glu^{94} during condensation of the prothymosin α (ProTα) segments ProTα-(87–94) and ProTα-(95–109) using various coupling reagents and solvent systems. Epimerization was efficiently suppressed using carbodiimides and acidic additives in DMSO (D-Glu: $<0.2\%$), whereas the use of phosphonium and uronium salts also in DMSO, but in the presence of bases (e.g., diisopropylethylamine), resulted in an unacceptably high degree of epimerization (D-Glu: $\sim 34\%$). The use of DMF as solvent or DCM as cosolvent in DMSO (50:50) for DCC/HOBt segment couplings led to a similar D-Glu percentage. Furthermore, it must be considered that epimerization is also heavily dependent on the reaction time, and so long-lasting segment condensations will be accompanied by a high degree of epimerization, even though model studies conducted under identical conditions and normal reaction times have shown a lack of effect on the chiral integrity of the C-terminal residue of the carboxy component. For this reason, a short double coupling with the carboxy component, followed by acetylation, seems to be more effective than slow and incomplete condensation with high epimerization of the C-terminal residue of the carboxy component.

Monitoring of the coupling reaction during CSPPS can be performed using the Kaiser test, but this is less sensitive with increasing length of the peptide chain. Edman sequencing offers an alternative method to determine the yields of the segment couplings. The removal of an aliquot of resin, and cleavage of the peptide followed by analysis with RP-HPLC, mass spectrometry, capillary electrophoresis, or NMR, should be the most straightforward monitoring strategy.

The above-mentioned 109-residue protein prothymosin *α* is one of the largest target peptides synthesized by CSPPS [83]. The C-terminal 34-residue segment, ProTα-(76–109)-2-chlorotrityl resin was synthesized by linear Fmoc/tBu SPPS, and then assembled with nine segments containing on average 7–10 amino acids each to the fully protected Fmoc-ProTα-(1–109)-2-chlorotrityl resin. Prothymosin *α* was obtained at 11% overall yield after cleavage from the resin, deprotection and purification, and showed identical biological activity as the natural protein. Selected peptides and proteins synthesized by CSPPS are listed in Tab. 5.4.

Tab. 5.4 Selected peptides and proteins synthesized by CSPPS.

Year	*Name*	*Residues*	*Reference*
1988	Rat atrial natriuretic factor (8–33)	26	111
1990	Human prothymosin α	109	106
1990	Human gastrin-I	17	112
1992	Human [Gly^{33}] calcitonin	33	106
1992	β-Amyloid protein	42	113
1995	Antifreeze protein type III	64	114
1995	N-terminal repeat region of γ-zein	48	115
1995	3-Repeat region of human Tau-2	94	105
1996	HIV-1_{Bru}tat(1–72)	72	114
1998	Tyr^0-Atriopeptin (rat)	24	116
1998	Tetanus toxin MUC-1 oligomers	115	117

5.3.3
Phase Change Synthesis

The choice between solid-phase synthesis and synthesis in solution should not be considered as a dogma. Consequently, a number of hybrid approaches have been developed, referred to as phase change synthesis (hybrid approach) [72]. This combined solid-phase and solution approach has already been used in lipophilic segment coupling (see Sect. 5.2.1.2). The synthesis of large, protected segments by SPPS that are subsequently assembled either by solution-phase and by solid-phase procedures has opened a new approach to the manufacture of complex sequences. Although not yet used widely, phase change synthesis shows great promise in the production of large peptides on a commercial scale [12].

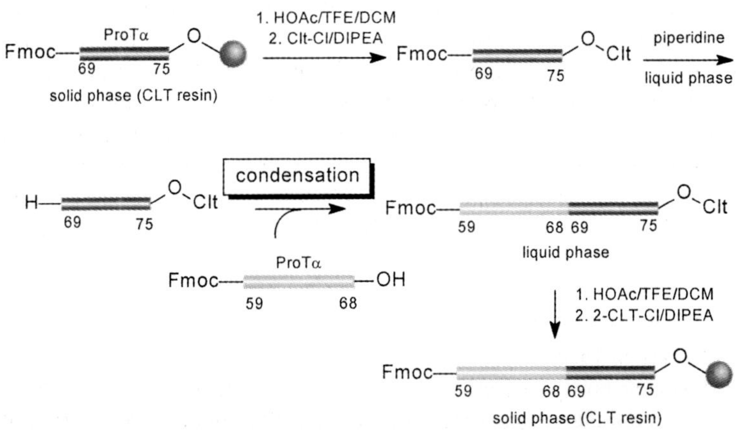

Fig. 5.12 Phase-change synthesis of the protected prothymosin α segment (59–75) according to Barlos and Gatos [72].

A benzyl alcohol linker attached to the xanthenyl resin allows C-terminal-protected esters to be prepared. Otherwise, segments synthesized by SPPS are cleaved from the resin, followed by subsequent esterification at the C-terminal carboxy group.

The phase change synthesis of ProTα-(59–75) according to Barlos and Gatos [72] is shown in Fig. 5.12. After cleavage of the protected segment (69–75) from the 2-chlorotrityl resin, the esterification is performed by treatment with 2-chlorotritylchloride (Clt-Cl) and diisopropylethylamine (DIPEA) in DMF/DCM mixtures [71]. The N-terminal Fmoc group is cleaved with piperidine, yielding the C-terminal segment for the condensation with the segment Fmoc-ProTα(59–68)-OH in solution. The peptide is re-attached to the support after selective cleavage of the 2-chlorotrityl group from Fmoc-ProTα(59–75)-Clt with 1% TFA in DCM, or alternatively with AcOH/TFE/DCM.

5.3.4
Soluble-Handle Approaches

Alternative approaches to simplify solution synthesis, and also to make it less time-consuming, have been examined already before the invention of SPPS. Attempts have been directed using "soluble handles" in order to facilitate isolation and to avoid the full characterization of intermediates at each step.

5.3.4.1 Picolyl Ester Method

In 1968, Young's group introduced an interesting combination of solid-phase synthesis and synthesis in solution [88]. These authors used the picolyl ester moiety (see Tab. 4.2), which is quite resistant to acids, as a semipermanent blocking group for the C-terminal carboxy group. Hence, the Boc group could be used for protection of the *α*-amino function. The protonated picolyl handle increases the solubility of peptide intermediates in aqueous medium, and allows the isolation from the reaction mixture. This can be easily performed by passing the crude reaction mixture through an appropriate ion-exchange resin. The by-products are not retained, while the desired peptide can be recovered by elution with a suitable buffer. Both the coupling reaction and the deprotection are carried out in solution by well-established procedures. The picolyl ester method has been used in several syntheses, including that of complex peptides [89], but was superseded by SPPS, mainly because the process was not readily amenable to mechanization and automation.

5.3.4.2 Liquid-Phase Method

In 1965, Shemyakin's group [90] proposed a polymeric support that is soluble in organic solvents as an alternative resin material for SPPS. High-molecular weight non-cross-linked polystyrene ($M_r \sim 200$ kDa) is chloromethylated in the usual way and esterified with the first Boc amino acid. The stepwise chain elongation from the C-terminal end is achieved with Boc amino acid *N*-hydroxysuccinimide esters. After completion of the necessary steps of one synthesis cycle, the intermediates are separated from solution by dilution with water. The resulting precipitate is intensively washed to remove low-molecular weight by-products. Although effective, this procedure suffers from some disadvantages, mainly related to the properties of the polystyrene.

In 1971, Mutter et al. [91] proposed polyethylene glycol (PEG), HO-CH_2-CH_2-(O-CH_2-CH_2)$_n$-O-CH_2-CH_2-OH, or the monofunctional monoethyl ether of PEG as more efficient polymers that allow for the precipitation of the intermediates by addition of diethyl ether, thereby providing a simple method of purification after the coupling of each amino acid [92]. PEG is soluble in the majority of organic solvents. Deblocking and coupling reactions take place in homogeneous solution (cf. Fig. 4.39). The solubilizing effect of PEG permits even coupling in aqueous solution by water-soluble carbodiimides (EDC), and allows for monitoring of the intermediates by NMR spectroscopy. Although the liquid-phase method is versatile for stepwise and convergent synthesis, it has some inherent drawbacks and hence has not attracted the attention of automated synthesizer manufacturers.

5.4
Ligation of Unprotected Peptide Segments

The phenotype of an organism is determined not only by the genome, but also by the protein equivalent of the genome – first designated a "proteome" by Marc Williams in 1994. Proteomics, the quantitative analysis of proteins present in an organism at a certain time and under certain conditions [93], also requires efficient tools for the synthesis of native proteins for protein chemical characterization of the altered proteins. Furthermore, synthetic access to novel protein analogues for structure-activity relationships is a prerequisite to obtain new insight into the molecular basis of protein function in biological systems. Unfortunately, due to the extremely high specificity requirements, nature has not developed an enzyme for the ligation of unprotected peptide segments, and the currently available methods of enzymatic ligation (see Sect. 5.2.2.2) have not yet found widespread application.

During the past two decades, most investigations of the molecular basis of protein function have been performed using recombinant DNA-based expression of proteins in genetically engineered cells [94, 95]. Protein expression in engineered cells is an extremely powerful technology [96], but has limited scope. It is well documented that the expression of small proteins ($M_r < 30$ kDa) can be carried out more easily than the expression of multidomain proteins, while the folding of the latter molecules may also represent a further challenge. Despite the efforts made to incorporate noncoded amino acids into cell-free translation systems [97, 98], many noncoded amino acids are not compatible with ribosomal synthesis [99]. Until now, the amounts of pure protein obtained from cell-free translation systems remains unsatisfactory [100]. Moreover, the uncontrolled processing of nascent protein in the cell may cause product heterogeneity, while overexpressed proteins that may be toxic to the cell might represent an additional, unwanted problem.

As shown in above, several chemical synthesis strategies have achieved success in the preparation of proteins of 75 to ~150 amino acids. Although approaches have been described that are technologically feasible to synthesize proteins exceeding this range, they have not reached the point of becoming routine, and are not (yet) applicable to larger proteins. As unprotected peptides and proteins of different sizes can be obtained from either chemical or recombinant sources, their ligation to larger polypeptides and proteins should provide a unifying operational strategy both for total and semisynthesis of proteins.

As a consequence, suitable ligation chemistry is required to fill the gap that currently exists in the synthesis of proteins starting from unprotected segments. As shown above, the controlled formation of a peptide bond between two peptide segments normally requires minimal or maximal protection schemes. However, when a nonamide bond is used to link two peptide segments, protecting groups are less essential. Finally, protecting groups can be omitted totally if the reaction is carried out chemoselectively. In this case, both segments must bear unique and complementary functional groups that are mutually reactive with each other, but are nonreactive with all other functionalities present in the segments to be coupled.

The price to be paid for realizing this chemoselective ligation principle is the formation of a nonpeptide bond (non-natural structure) at the ligation site. Formally, in the sense of rational nomenclature, the resulting ligation products should be defined as protein analogues lacking one peptide bond. However, in practice the non-natural structures are often well tolerated within a folded protein, and numerous examples of fully active proteins prepared by such a method underline this assumption. This approach is often called "chemical ligation" according to the proposal of Muir and Kent [101], though this term is somewhat confusing as ligation or chemical ligation describes, without any doubt, the preferential classical condensation of peptide segments joining by peptide bonds. Nonetheless, on the basis of this chemical ligation, Kent and co-workers [102] later extended this term to "native chemical ligation" in order to encompass true peptide bond-forming ligation.

5.4.1 Backbone-engineered Ligation

The first example of backbone-engineered ligation, also termed chemical ligation, was published by Schnolzer and Kent in 1992 [103].

The principle of this thioester-forming ligation is shown schematically in Fig. 5.13. Segment I with a thiocarboxy group at its C-terminus (thiocarboxy glycine) reacts at acidic pH with the N-terminal bromoacetyl group of segment II, forming the condensation product with the thioester group as non-natural bond. Under this condition, all free amino groups are protonated. This approach was first applied to the synthesis of a HIV-1 protease analogue. The protein analogue

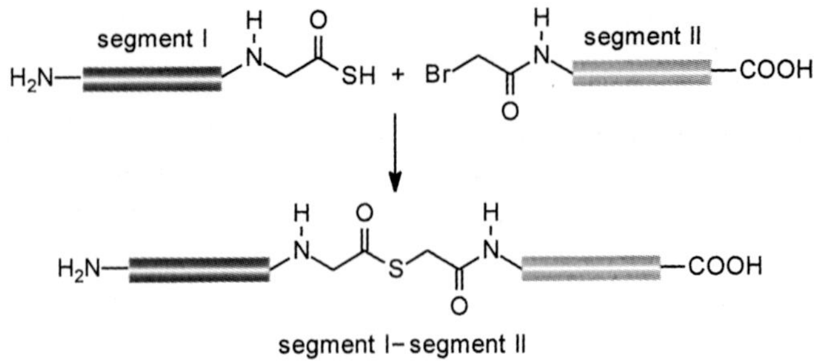

Fig. 5.13 Thioester-forming ligation approach

is stable in the pH range 3–6, but undergoes base-catalyzed hydrolysis above pH 7. Analogously, the mirror image enzyme D-HIV-1 protease was prepared [104]. Other examples for chemoselective ligations using thioesters have been described by Gärtner et al. [105] and Wallace [106]. Thioether-forming ligation [107–109] and directed disulfide formation [110] are further possibilities developed by Kent's group.

Hydrazone-forming ligation [111–113] starts from C-terminal peptide hydrazides that are chemically ligated at pH 4.6 with amino components bearing an aldehyde function at the N-terminus obtained by mild periodate oxidation of N-terminal Ser or Thr residues.

The resulting backbone-engineered peptide hydrazone **9** can be reduced with sodium cyanoborohydride to produce the more stable peptide hydrazide. Oxime-forming ligation [114, 115] is based on the coupling of a O-peptidyl hydroxylamine as amino component with a peptide aldehyde forming the peptide oxime **10**.

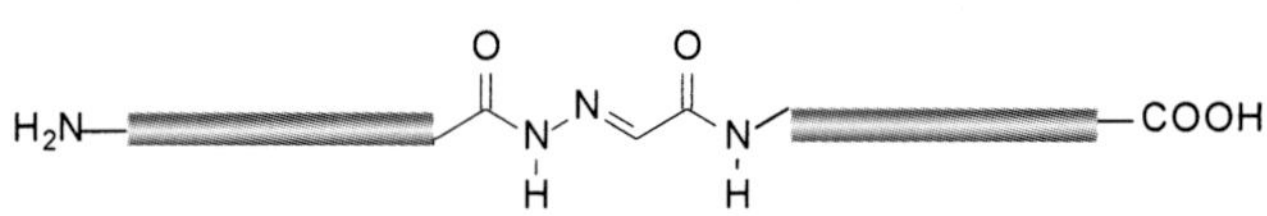

9

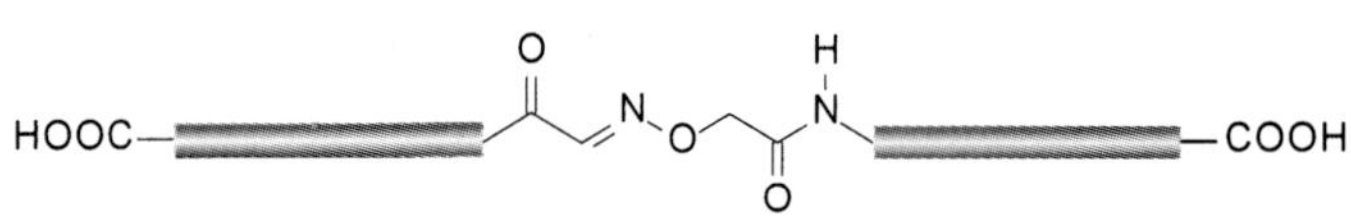

10

Furthermore, the application of more than one type of ligation chemistry allows for the ligation of several unprotected segments in a specific fashion. The synthesis of the cMyc-Max transcription factor-related protein could be performed using both the thioester and oxime approaches [116].

5.4.2
Prior Capture-mediated Ligation

In contrast to the chemoselective backbone-engineered ligation, which results in well-defined artificial protein analogues, the synthesis of tailor-made proteins with a "native" backbone structure requires a peptide bond-forming step at the ligation site. At the very beginning of chemical peptide synthesis Theodor Wieland [117] and Max Brenner [118] independently created the idea to convert the bimolecular reaction between two peptides to be ligated into an intramolecular reaction by bringing together the respective C- and N-termini on a template in order to facilitate intramolecular acyl transfer reaction. A similar mechanism is practically realized in the reverse ligase action of serine and cysteine proteases in which O- or S-acyl intermediates undergo O- or S- to N-acyl transfer reaction forming a peptide bond (cf. Sect. 5.2.2.2). Furthermore, in protein splicing a series of acyl transfer reactions lead to a final covalent O- or S-acyl intermediate resulting in a spontaneous uncatalyzed O- or S- to N-acyl transfer forming the corresponding peptide bond (cf. Sect. 5.5.3). All these chemical and biological advancements might have stimulated several groups to vary this fundamental basic principle in the development of both template-mediated amide bond formations and segment condensation by prior chemical ligation.

5.4.2.1 Template-mediated Ligation

The principle of a template-mediated amide bond formation according to Brenner et al. [118] using a salicylamide template is based on a base-catalyzed intramolecular rearrangement [119]. Demonstrated in a general sense, the carboxy function of salicylic acid **11** is acylated with a peptide alkyl ester, whereas the hydroxyl group is esterified with a protected amino acid.

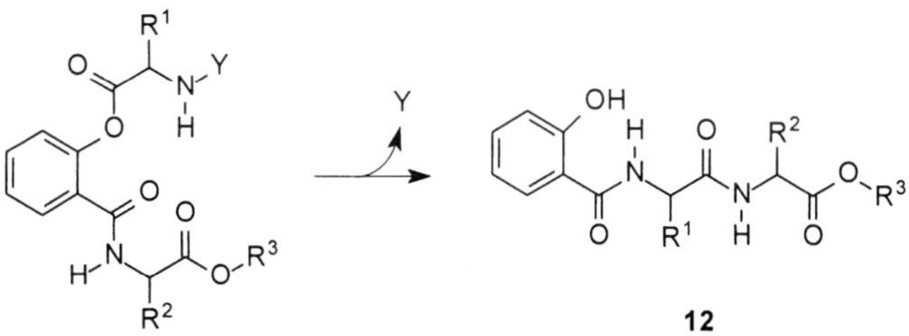

11 **12**

After removal of the N-protecting group (Y), base-catalyzed intramolecular rearrangement gives the *N*-salicoylpeptide alkyl ester **12** that bears the newly incorporated amino acid in the N-terminal position. The next amino acid can then be esterified onto the free phenolic hydroxy group of the template, thereby continuing chain elongation. However, this very interesting concept suffers from certain limitations, namely problems with selective removal of the N-salicoyl group on com-

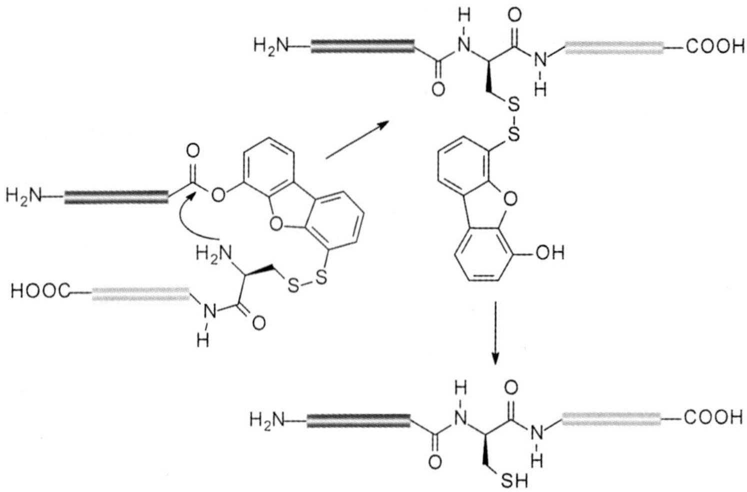

Fig. 5.14 Thiol capture ligation according to KEMP [120].

pletion of the synthesis, and bringing together longer segments on the template. Nevertheless, this inherent template-assisted synthesis concept based on entropic activation facilitates intramolecular acylation reactions.

The thiol capture ligation, introduced by Kemp [120], is an elegant current variation of template-promoted amide bond formation between two segments using 4-hydroxy-6-mercaptodibenzofuran as a template (Fig. 5.14).

The template is attached by disulfide bridge formation with a suitable solid support that is used for Bpoc/tBu-SPPS of the C-terminal peptide segment to be ligated [121, 122]. After deprotection with TFA, Cys residues may be maintained protected (Acm). The second peptide segment, acting as amino component, must bear a N-terminal Cys residue and can be synthesized on Wang resin by standard Fmoc/tBu-SPPS. After appropriate deprotection the latter is coupled to the template by formation of a disulfide bridge, thus yielding the two segments ready for intramolecular acyl transfer reaction. This procedure could be used for the synthesis of a 39-peptide [123]. In contrast to conventional segment condensations, the very rapid intramolecular acyl transfer reaction does not require an excess of one segment, and the solubility of the intermediates in aqueous solvent systems is significantly improved due to mostly unprotected amino acid side-chain functions.

5.4.2.2 **Native Chemical Ligation**

As shown above, the capture step using an external template has been successful in bringing together the respective C- and N-termini of the two segments that allows for peptide bond formation via intramolecular acyl transfer reaction. In 1953, Wieland et al. [117] described a reaction between a thioester and a cysteine deriva-

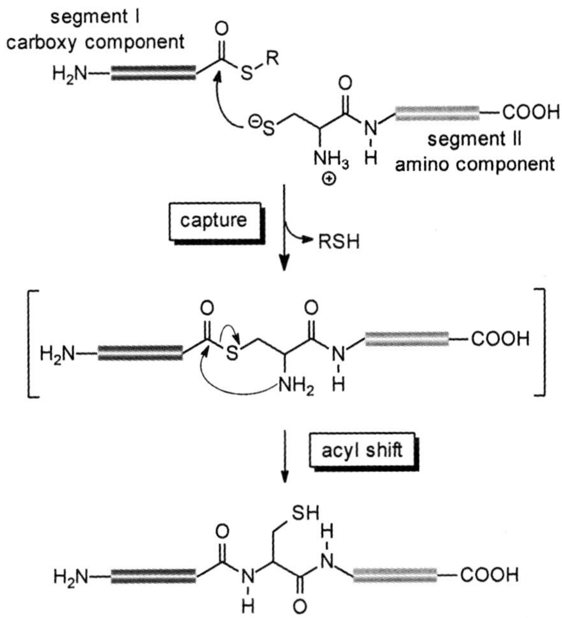

Fig. 5.15 Principle of native chemical ligation.

tive resulting in the formation of a cysteine peptide bond between the two reactants. The N-terminal Cys in the amino component functions as "interior" thioester capture, thus forming a new thioester through transesterification. The covalent S-acyl intermediate undergoes spontaneously S- to N-acyl migration forming a peptide bond via a five-membered ring intermediate.

This principle of segment coupling by prior chemical ligation does not require an external template as the thiol moiety acts as an interior capture. In 1994, Kent and colleagues [102] used this principle (Fig. 5.15) in order to ligate unprotected peptide segments in the presence of an N-terminal Cys residue in the amino component giving a "native" amide (i.e., peptide) bond at the site of ligation. In contrast to original types of ligation chemistry resulting in nonpeptide bonds (cf. Section 5.4.1), these authors named this technique "native chemical ligation" (NCL). The synthesis of human interleukin 8 (Fig. 5.16) was the first example of NCL published by Kent's group [102].

One year later, Tam et al. described an alternative ligation approach, which was termed not only chemoselective capture activation ligation [124], but also intramolecular [125] or biomimetic ligation [126]. Later, Tam et al. [127] proposed the new term "orthogonal ligation" for the two-step reaction sequence of capture and intramolecular acyl transfer in accordance with other chemical concepts that distinguish between two functional sites based on chemoselectivity. Recent reviews of Kent's group [128, 129] and the latest review of Tam et al. [127] have described this new methodology in great depth.

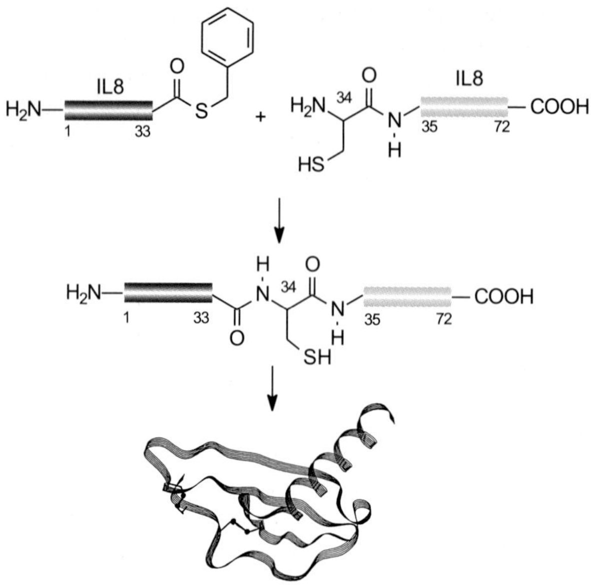

Fig. 5.16 Synthesis of interleukin 8 (IL8) by native chemical ligation.

In all ligation procedures used, the capture reaction brings the reactive sites of the two segments into close proximity, and this results in an intramolecular acyl transfer to form the peptide bond at the ligation site. Only the type of interior capture and the capture reaction vary among the different ligation procedures. In principle, thioester and imine capture can be differentiated [127].

Thioester capture (as shown above) as NCL is based on reversible thiol/thioester exchange between the C-terminal thioester moiety of the acyl donor and an unique N-terminal Cys of the acyl acceptor, giving a thioester-linked intermediate that undergoes spontaneous rearrangement via intramolecular nucleophilic attack forming a peptide bond at the ligation site. This reaction is independent of additional internal Cys residues present in either segment to be ligated, since the α-amino group of the N-terminal Cys facilitates trans-thioesterification as a general base. No protecting groups are required, and practically no racemization was detected in the ligation product above a limit of $<1\%$ D-amino acid content [130]. Most ligation procedures are carried out under aqueous conditions buffered at pH 7–8. Although ligations can be performed under nondenaturing conditions, the uncertainty of the conformational behavior of large segments routinely favors strongly denaturing conditions of 6–7 M guanidinium chloride [102, 131]. In order to prevent N,S-bisacylated by-product and disulfide formation, the maintenance of reducing conditions by addition of, for example, thiophenol, 2-mercaptoethane-sulfonate, or a combination of tri(2-carbonylethyl)phosphin has been proposed.

The synthesis of a C-terminal α-thioester can be performed by Boc-SPPS, yielding peptide segments with C-terminal thioester [124] or thioacid functionalities

[102], depending on the starting thiol-modified resin. In addition, the progress in the development of suitable safety-catch thioester linkers for Fmoc-SPPS [132–134] has provided the possibility of the synthesis of glycopeptide α-thioesters leading to the first total synthesis of a glycoprotein [133]. Further improved methods for the rapid, cost-effective synthesis of large numbers of unprotected peptide thioesters can be expected to lead to widespread use in the ligation approach [135].

The solid-phase chemical ligation (SPCL) [136] of unprotected peptide segments in aqueous solution can be performed in both $N \rightarrow C$ and $C \rightarrow N$ directions, and has special advantages in relay ligation procedures. The C-terminal segment of the target protein is attached via a cleavable linker to a water-compatible solid support. This strategy requires an N-terminal-protecting group to prevent cyclization/polymerization reactions with the internal (N-terminal Cys and thioester-containing) segment during the condensation reaction. A wide range of N-terminal protecting-groups which are removable by acidic, basic, photolytic, reductive, or oxidative conditions can be used for this purpose. The omission of otherwise necessary chromatographic separations and lyophilization steps results in a saving of time, and also considerably decreases the yield. By using this approach, it was possible to assemble some unprotected segments of 35–50 residues to the target polymer-bound protein by consecutive ligation on a water-compatible, cellulose-based polymer support. A complementary approach utilizing a highly stable, safety catch acid-labile (SCAL) linker [137] (cf. Sect. 4.5.2) producing an amide C-terminus has been described for a three-segment synthesis of vMIP I, a 71-amino acid chemokine [138]. Despite the advantages of this procedure, the relatively low chemical stability of cellulose-based supports toward HF and prolonged TFA treatment, for example, is a drawback that may be circumvented using more robust resins, perhaps of the superpermeable organic combinatorial chemistry resin (SPOCC) type [139].

Tam et al. [124] described a reverse thioester ligation to form a Cys bond by nucleophilic attack of the thiocarboxy group of a peptide on the bromomethyl group of a N-terminal β-bromoalanine residue of the amino component (Fig. 5.17).

The resulting covalent thioester undergoes rearrangement, giving the Cys at the ligation site. The Cys-perthioester ligation, developed also by Tam's group [140], requires thiolytic reduction of the resulting hydrodisulfide intermediate in order to give a Cys residue at the ligation site.

With respect to ligation sites, current native ligation chemistry requires a Cys residue at the site of peptide bond formation joining two unprotected peptide segments. For this reason, it would be desirable to have the option to apply thioester-supported ligation at residues other than Cys.

The first example for a potential solution to this sequence restriction was based on an N^{α}-linked removable thiol moiety. This principle could be later improved using the removable N^{α}-(1-phenyl-2-mercaptoethyl) auxiliary [141]. Further approaches to extend the principle of thioester ligation to noncysteinyl residues are, for example, Met ligation [126] and His ligation [142]. For ligation of noncysteine-containing segments, a specific Ala residue in the parent protein is replaced

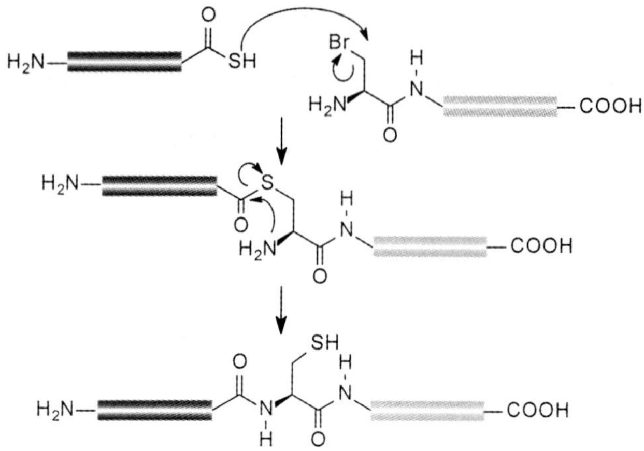

Fig. 5.17 Reverse thioester ligation approach.

with Cys in order to facilitate the ligation procedure. After assembly of the two segments, selective desulfurization with H_2/metal reagents converts the Cys residue to Ala and Met to Abu [143].

Studies to extend the applicability of procedures to noncysteine ligation sites are in progress [144]. In comparison to ligation under denaturing conditions that requires more than several hours, similar condensation reactions under folding conditions proceed conformationally assisted in <2 min [145]. Surprisingly, in suitable systems even the requirement for a Cys residue at the ligation site does not seem to be necessary by mixing a peptide-α-thioester under folding conditions with another segment lacking the N-terminal Cys. From the mechanistic point of view, the intramolecular rearrangement normally forming the peptide bond at the ligation site is not an essential prerequisite in this case. Doubtlessly, the capability of chemically ligating peptide segments between any of the 21 DNA-encoded amino acids without racemization will be an ultimate goal in near future, but will not be easy to reach. The concept of an auxiliary assisted ligation strategy is a challenge. Basic model studies such as the so-called traceless Staudinger ligation procedure [146] to generate amide bonds from the reaction of an azide and a functionalized phosphine [147] are encouraging, but require refinement and adaptation to the much more complex protein ligation systems.

Imine capture is based on the reaction of an acyl-aldehyde with a suitable N-terminal amino acid (Cys, Ser, Thr, His, Trp, or Asn) of an appropriate amine segment leading to an imine intermediate that undergoes rapid ring chain tautomerization followed by O,N-acyl migration [124, 127].

The principle of the Tam approach is shown schematically in Fig. 5.18. Several methods have been developed to prepare the ester aldehyde segments acting as carboxy component. At the ligation site, the N-terminal amino acid residues Cys (R=H, X=S), Ser (R=H, X=O), and Thr ($R=CH_3$, X=O) of the amino component form five-membered thiazoline or oxazolidine rings with the glycolaldehyde

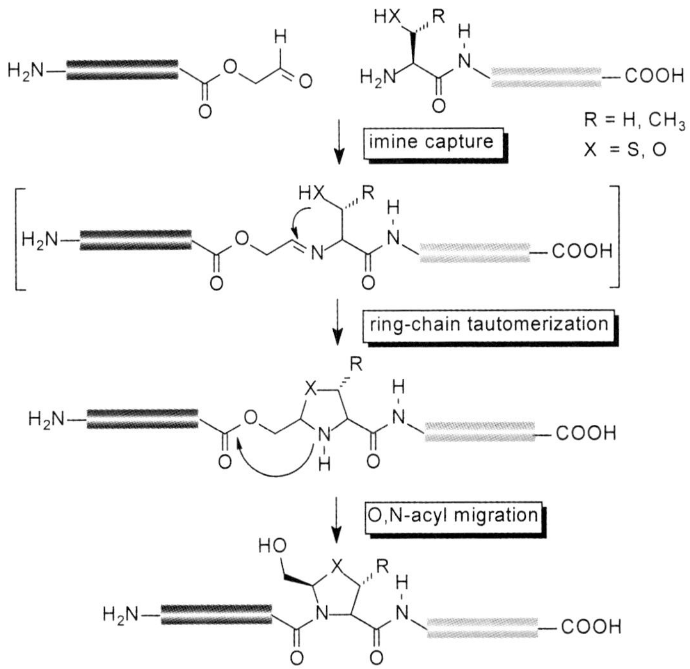

Fig. 5.18 Imine capture ligation approach. N-terminal Cys (X=S, R=H), Ser (X=O, R=H), and Thr (X=O, R=CH_3) form pseudoprolines.

ester. Both segments are now linked by an amide bond based on 2-hydroxymethyl thiaproline in the case of Cys, and 2-hydroxymethyl oxaproline if Ser or Thr act as N-terminal residues. Reactions with amino components bearing N-terminal Trp, His or Asn yield much more complex heterocycles. The effectiveness of the thiaproline ligation has been confirmed in the synthesis of analogues of TGF, and a HIV-1 protease analogue with substituted pseudoproline (SPro) instead of Pro at the ligation site with unchanged enzyme activity [127].

Until now, hundreds of biologically active polypeptides and proteins have been successfully synthesized using the new protein ligation strategy [127–129]. With the help of these synthetic proteins, the elucidation of gene function, the discovery of new biology, the systematic development of proteins with enhanced potency and specificity as therapeutic candidates, and the determination of protein structure by NMR and X-ray crystallography might be realized.

Difficulties associated with SPPS of polypeptides longer than approximately 50 residues lead to the consequence that the native chemical/orthogonal ligation of just two segments in this range becomes increasingly difficult for target molecules of >100 building blocks. As mentioned above, multiple ligation step strategies provide an alternative to overcome this limitation, but the inherent technically very demanding requirements have led only to the application of a limited number of target proteins. An interesting solution to this problem should be the linkage of suitable recombinant polypeptides and proteins with synthetic peptides (containing also unnatural amino acid buildings blocks) but, furthermore, also with recombinant proteins as amino components in order to obtain much larger target proteins. Methods used specifically to generate reactive groups on bacterially expressed proteins necessary for protein ligation, together with some applications, are described In the next paragraph.

5.4.3
Biochemical Protein Ligation

In 1998, a semisynthesis approach was described independently by two groups that allows peptides or proteins to be chemically ligated (cf. Sect. 5.4.2.2) to the C-terminus of recombinant proteins via a normal peptide bond. Muir and co-workers [148, 149] termed this procedure "expressed protein ligation" (EPL), whereas Xu and co-workers [150, 151] named this new method of protein engineering "intein-mediated protein ligation" (IPL). The crunch point of this process is the generation of a recombinant protein α-thioester that is capable of participating in subsequent native chemical ligation reactions. A newly discovered form of post-translational processing, named protein splicing, provided the solution to the problem. In protein splicing, which was first discovered in yeast by Kane et al. [152] in 1990, a proprotein undergoes an intramolecular rearrangement; this results in the extrusion of an internal sequence, the intein, followed by concomitant fusion of the two flanking protein regions (exteins) via a native peptide bond [153–157]. It is of interest to note that native chemical ligation [102] was developed simultaneously with the elucidation of the mechanism of protein splicing [153]. Furthermore, the high chemical similarities between the two mechanisms, especially with regard to the involvement of thioester-linked intermediates and the final step of an S- or O- to N-acyl shift to form the peptide bond, have initiated the development of a new route to the formation of recombinant protein α-thioester for the application in native chemical ligation. Xu and co-workers [158] demonstrated in a series of elegant experiments that protein splicing can be halted at the thioester/ester intermediate stage by mutation of the key Asn in the intein to an Ala residue. Such an engineered intein led to the development of on-column protein purification systems, and the same protocols were then applied to generate specifically a C-terminal thioester [159] or an N-terminal Cys [151] on bacterially expressed proteins for EPL/IPL procedures.

Protein splicing in a simplified form (Cys residues may also be replaced by Ser or Thr) is shown schematically in Fig. 5.19. The biochemical protein ligation procedure (Fig. 5.20) uses a genetically engineered intein and a chitin binding domain (CBD) as fusion partners in order to express a protein or protein segment of interest.

CBD allows for the separation of the target protein of interest by binding to a chitin resin. Incubation of the chitin-bound protein with a suitable thiol reagent results in the cleavage between the target protein and the intein, yielding the ap-

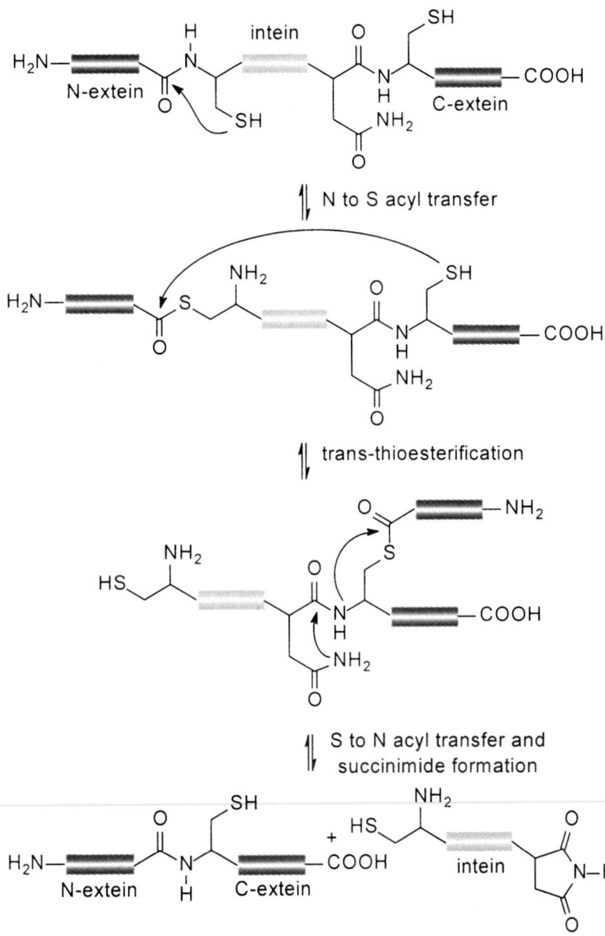

Fig. 5.19 Protein splicing.

propriate α-thioester. The latter can be ligated with chemically synthesized peptide segments, which allows also unnatural amino acids to be site-specifically introduced into proteins [160]. In addition, according to a modified protein splicing reaction developed by Xu and co-workers [151], specially generated recombinant proteins bearing an N-terminal cysteine residue can be used.

EPL/IPL represents a novel semisynthetic approach to extend greatly the native chemical ligation strategy. This procedure found application, for example, in the isolation of cytotoxic proteins [150], the incorporation of noncoded amino acids into a protein sequence [160], or of isotope-labeled protein sequences for NMR analysis [161], and investigations on protein structure-function relationships. Furthermore, the circularization and polymerization of bacterially expressed polypeptides and proteins has been described using a versatile two-intein (TWIN) sys-

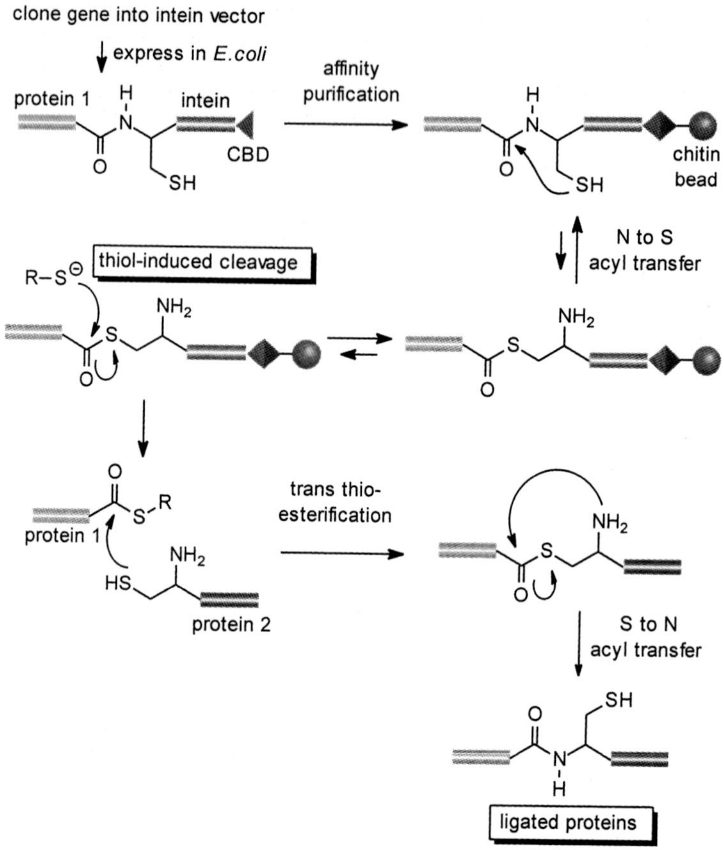

Fig. 5.20 Expressed protein ligation.

tem. Additionally, it has been suggested that in the future modified inteins may be applied to control the fusion of three or more protein segments in succession [162].

5.5 References

1 M. Bodanszky in: *Perspectives in Peptide Chemistry*, A. Eberle, R. Geiger, T. Wieland (Eds.), Karger, Basel, **1981**, 15

2 F. Bordusa, D. Ullmann, H.-D. Jakubke, *Angew. Chem.* **1997**, *109*, 1125; *Angew. Chem. Int. Ed. Engl.* **1997**, *36*, 1099

3 Y.V. Mitin, M.G. Ryadnow, *Protein Peptide Lett.* **1999**, *6*, 87

4 M.G. Ryadnov, L.V. Klimenko, Y.V. Mitin, *J. Peptide Res.* **1999**, *53*, 322

5 R.L. Letsinger, M.J. Kornet, *J. Am. Chem. Soc.* **1963**, *85*, 3045

6 B. Henkel, L. Zhang, E. Bayer, *Liebigs Ann./Recueil* **1997**, 2161

7 N. Thieriet, F. Guibe, F. Albericio, *Organic Lett.* **2000**, *2*, 1815

8 M. Bodanszky, V. du Vigneaud, *J. Am. Chem. Soc.* **1959**, *81*, 5688

9 M. Bodanszky, M.A. Ondetti, S.D. Levine, N.J. Williams, *J. Am. Chem. Soc.* **1967**, *89*, 6753

10 V. Dourtoglou, B. Gross, V. Lambropoulo, C. Zioudrou, Synthesis **1984**, 572

11 L.A. Carpino, *J. Am. Chem. Soc.* **1993**, *115*, 4397

12 L. Andersson, L. Blomberg, M. Flegel, L. Lepsa, B. Nilsson, M. Verlander, *Biopolymers* **2000**, *55*, 227

13 B.D. Larsen, A. Holm, *Int. J. Peptide Protein Res.* **1994**, *43*, 1

14 A.H. Karlstrom, A. Unden, *Tetrahedron Lett.* **1995**, *36*, 3909

15 A.H. Karlstrom, A. Unden, *Int. J. Peptide Protein Res.* **1996**, *48*, 305

16 A.H. Karlstrom, A. Unden, *Chem. Commun.* **1996**, 959

17 M. Royo, J. Alsina, E. Giralt, U. Slomcyznska, F.J. Albericio, *J. Chem. Soc. Perkin Trans. I* **1995**, 1095

18 M.C. Munson, C. Garcia Echeverria, F. Albericio, G.F. Barany, *J. Org. Chem.* **1992**, *57*, 3013

19 S. Sakakibara, Y. Shimonishi, *Bull. Chem. Soc. Japan* **1965**, *38*, 1412

20 H. Yajima, N. Fujii, M. Shimokura, K. Akaji, S. Kiyama, M. Nomizu, *Chem. Pharm. Bull.* **1983**, *31*, 1800

21 J.P. Tam, W.F. Health, R.B. Merrifield, *J. Am. Chem. Soc.* **1983**, *105*, 6442

22 R. Schwyzer, H. Kappeler, *Helv. Chim. Acta* **1963**, *46*, 1550

23 S.B.H. Kent, *Annu. Rev. Biochem.* **1988**, *57*, 957

24 D.F. Veber, J.D. Milkowski, R.G. Denkewalter, R. Hirschmann, *Tetrahedron Lett.* **1968**, 3057

25 J.P. Tam, T.W. Wong, M.W. Riemen, F.S. Tjoeng, R.B. Merrifield, *Tetrahedron Lett.* **1979**, 4033

26 B.W. Erickson, R.B. Merrifield, *J. Am. Chem. Soc.* **1973**, *95*, 3750

27 Y. Nakagawa, Y. Nishiuchi, J. Emura, S. Sakakibara in: *Peptide Chemistry 1980*, K. Okawa (Ed.), Protein Research Foundation, Osaka, **1981**, 41

28 D. Yamashiro, C.H. Li, *J. Org. Chem.* **1973**, *38*, 591

29 B.W. Erickson, R.B. Merrifield, *Isr. J. Chem.* **1974**, *12*, 79

30 W.D. Fuller, M. Goodman, F. Naider, Y.-F. Zhu, Biopolymers **1996**, *40*, 183

31 V. du Vigneaud, C. Ressler, J.M. Swan, C.W. Roberts, P.G. Katsoyannis, S. Gordon, *J. Am. Chem. Soc.* **1953**, *75*, 4879

32 E. Wünsch, *Angew. Chem.* **1971**, *10*, 773; *Angew. Chem. Int. Ed. Engl.* **1971**, *10*, 786

33 N. Fujii, H. Yajima, *J. Chem. Soc., Perkin Trans. I*, **1981**, 831

34 N. Fujii, H. Yajima, *Chem. Pharm. Bull.* **1981**, *29*, 600

35 S. Sakakibara, *Biopolymers* **1999**, *51*, 279

36 J.B. Hendrickson, C. Kandall, *Tetrahedron Lett.* **1970**, 343

37 F. Rabanal, E. Giralt, F. Albericio, *Tetrahedron* **1995**, *51*, 1449

38 Y. Nishiuchi, T. Inui, H. Nishio, J. Bodi, T. Kimura, F.I. Tsuji, S. Sakakibara, *Proc. Natl. Acad. Sci. USA* **1998**, *95*, 13549

39 B. Riniker, A. Flörsheimer, H. Fretz, P. Sieber, B. Kamber, *Tetrahedron* **1993**, *49*, 9307

40 B. Riniker, A. Flörsheimer, H. Fretz, B. Kamber in: *Peptides 1992*, C.H. Schneider, A.N. Eberle (Eds.), Escom, Leiden, **1993**

41 B. Kamber, B. Riniker in: *Peptides, Chemistry and Biology*, Proceedings, 12th American Peptide Symposium, J.A. Smith, J.E. Rivier (Eds.), Escom, Leiden, **1992**, 525

42 R. Hirschmann, R. Nutt, D.F. Veber, R.A. Vitali, S.L. Varga, T.A. Jacob, F.W. Holly, R.G. Denkewalter, *J. Am. Chem. Soc.* **1969**, *91*, 507

43 J. Blake, *Int. J. Peptide Protein Res.* **1981**, *17*, 273

44 J. Blake, D. Yamashiro, K. Ramasharma, C.H. Li, *Int. J. Peptide Protein Res.* **1986**, *28*, 468

45 D. Yamashiro, C.H. Li, *Int. J. Peptide Res.* **1988**, *31*, 322

46 J. Blake, C.H. Li, *Proc. Natl. Acad. Sci. USA* **1983**, *80*, 1556

47 S. Aimoto, N. Mizoguchi, H. Hojo, S. Yoshimura, *Bull. Chem. Soc. Japan* **1989**, *62*, 524

48 S. Aimoto, *Biopolymers* **1999**, *51*, 247

49 D.Y. Jackson, J. Vurnier, C. Quan, M. Stanley, J. Tom, J.A. Wells, *Science* **1994**, *266*, 243

50 P. Kuhl, U. Zacharias, H. Burckhardt, H.-D. Jakubke, *Monatsh. Chem.* **1986**, *117*, 1195

51 V. Cerovsky, F. Bordusa, *J. Peptide Res.* **2000**, *55*, 325

52 V. Cerovsky, J. Kockskämper, H. Glitsch, F. Bordusa, *Chembiochem.* **2000**, *2*, 126

53 N. Wehofsky, S.W. Kirbach, M. Hänsler, J.-D. Wissmann, F. Bordusa, *Organic Lett.* **2000**, *2*, 2027

54 R. Grünberg, I. Domgall, R. Günter, K. Rall, H.-J. Hofmann, F. Bordusa, *Eur. J. Biochem.* **2000**, *267*, 7024

55 R.B. Merrifield, *J. Am. Chem. Soc.* **1963**, *85*, 2149

56 E. Atherton, R.C. Sheppard, *Solid Phase Peptide Synthesis: A Practical Approach*, Oxford University Press, Oxford, **1989**

57 G.B. Fields, R.L. Noble, *Int. J. Peptide Protein Res.* **1990**, *35*, 161

58 R.B. Merrifield in: *Peptides, Synthesis, Structures, and Applications*, B. Gutte (Ed.), Academic Press, New York, **1995**, 94

59 P. Lloyd-Williams, F. Albericio, E. Giralt, *Chemical Approaches to the Synthesis of Peptides and Proteins*, CRC Press, New York, **1997**

60 J.D. Fontenot, J.M. Ball, M.A. Miller, C.M. David, R.C. Montelaro, *Peptide Res.* **1991**, *4*, 19

61 A.J. Smith, J.D. Young, S.A. Carr, D.R. Marshak, L.C. Williams, K.R. Williams in: *Techniques in Protein Chemistry III*, R.H. Angeletti (Ed.), Academic Press, New York, **1992**, 219

62 R.B. Merrifield in: *Recent Progress in Hormone Research*, Volume 23, G. Pincus (Ed.), Academic Press, New York, **1967**, 451

63 S.B.H. Kent, *Annu. Rev. Biochem.* **1988**, *57*, 957

64 M. Beyermann, M. Bienert, *Tetrahedron Lett.* **1992**, *33*, 3745

65 V. Krchnak, Z. Flegelova, J. Vagner, *Int. J. Peptide Protein Res.* **1993**, *42*, 450

66 M. Narita, S. Ouchi, *J. Org. Chem.* **1994**, *52*, 686

67 E. Pedrose, A. Grandas, M.A. Saralegui, E. Giralt, C. Granier, J. van Rietschoten, *Tetrahedron* **1982**, *38*, 1183

68 P. Lloyd-Williams, F. Albericio, E. Giralt, *Tetrahedron* **1993**, *49*, 11065

69 H. Benz, *Synthesis* **1994**, 337

70 F. Albericio, P. Lloyd-Williams, E. Giralt, *Methods Enzymol.* **1997**, *289*, 313

71 P. Athanassopoulos, K. Barlos, O. Hatzi, D. Gatos, C. Tzavara in: *Solid Phase Synthesis and Combinatorial Libraries*, R. Epton (Ed.), Mayflower Scientific, Birmingham, **1996**, 243

72 K. Barlos, D. Gatos, *Biopolymers* **1999**, *51*, 266

73 K. Barlos, D. Gatos, S. Kapolos, G. Papaphotiu, G. Schäfer, Y. Wenqing, *Tetrahedron Lett.* **1989**, *30*, 3947

74 K. Barlos, D. Gatos, G. Papaphotiou, W. Schäfer, *Liebigs Ann. Chem.* **1993**, 215

75 M. Mergler, R. Nyfelder, R. Tanner, J. Gosteli, P. Grogg, *Tetrahedron Lett.* **1988**, *29*, 4005, 4009

76 P. Sieber, *Tetrahedron Lett.* **1987**, *28*, 2107

77 W. DeGrado, E.T. Kaiser, *J. Org. Chem.* **1982**, *47*, 3258

78 E.T. Kaiser, *Acc. Chem. Res.* **1989**, *22*, 47

79 D.H. Rich, S.K. Gurwara, *J. Chem. Soc., Chem. Commun.* **1973**, 610

80 N. Kneib-Cordonier, F. Albericio, G. Barany, *Int. J. Peptide Protein Res.* **1990**, *35*, 527

81 S.S. Wang, *J. Org. Chem.* **1976**, *41*, 3258

82 M. Quibell, L.C. Packman, T.J. Johnson, *J. Am. Chem. Soc.* **1995**, *117*, 11656

83 K. Barlos, D. Gatos, W. Schäfer in: *Peptides 1990*, E. Giralt, D. Andreu (Eds.), Escom, Leiden, **1991**

84 D.A. Tomalia, A.M. Naylor, W.A. Goddard III, *Angew. Chem.* **1990**, *102*, 119; *Angew. Chem. Int. Ed. Engl.* **1990**, *29*, 138

85 M. Goodman, P. Keogh, H. Anderson, *Bioorg. Chem.* **1977**, *6*, 239

86 R. Steinauer, F.M.F. Chen, N.L. Benoiton, *J. Chromatogr.* **1985**, *325*, 111

87 M. Quibell, L.C. Packman, T.J. Johnson, *J. Chem. Soc. Perkin Trans. I*, **1996**, 1219

88 R. Camble, R. Garner, G.T. Young, *Nature* **1968**, *217*, 247

89 R. Macrae, G.T. Young, *J. Chem. Soc. Perkin Trans. I*, **1975**, 1185

90 M. M. Shemyakin, Y. A. Ovchinnikov, A. A. Kiryushkin, I. V. Kozhevnikova, *Tetrahedron Lett.* **1965**, 2323

91 M. Mutter, H. Hagenmaier, E. Bayer, *Angew. Chem.* **1971**, *83*, 883; *Angew. Chem. Int. Ed. Engl.* **1971**, *10*, 811

92 M. Mutter, E. Bayer, *Angew. Chem.* **1974**, *86*, 101; *Angew. Chem. Int. Ed. Engl.* **1974**, *13*, 88

93 F. Lottspeich, *Angew. Chem.* **1999**, *111*, 2630; *Angew. Chem. Int. Ed. Engl.* **1999**, *38*, 2476

94 B. W. Matthews, *Annu. Rev. Biochem.* **1993**, *62*, 139

95 M. Smith, *Angew. Chem.* **1994**, *106*, 1277; *Angew. Chem. Ed. Int. Ed. Engl.* **1994**, *33*, 1214

96 J. L. Cleland, C. S. Craik, *Protein Engineering: Principles and Practice*, Wiley, New York, **1996**

97 S. M. Hecht, *Acc. Chem. Res.* **1992**, *25*, 545

98 D. Mendel, V. W. Cornish, P. G. Schultz, *Annu. Rev. Biophys. Biomol. Struct.* **1995**, *24*, 435

99 V. W. Cornish, D. Mendel, P. G. Schultz, *Angew. Chem.* **1995**, *107*, 677; *Angew. Chem. Int. Ed. Engl.* **1995**, *34*, 621

100 J. E. Barret, C. M. Lucero, P. G. Schultz, *J. Am. Chem. Soc.* **1999**, *121*, 7965

101 T. W. Muir, S. B. H. Kent, *Curr. Opin. Biotechnol.* **1993**, *4*, 420

102 P. E. Dawson, T. W. Muir, I. Clark-Lewis, S. B. H. Kent, *Science* **1994**, *266*, 776

103 M. Schnolzer, S. B. H. Kent, *Science* **1992**, *256*, 221

104 R. deLisle-Milton, S. C. F. Milton, M. Schnolzer, S. B. H. Kent in: *Techniques in Protein Chemistry IV*, R. H. Angeletti (Ed.), Academic Press, New York, **1993**, 251

105 H. Gärtner, K. Rose, R. Cotton, D. Timms, R. Camble, R. E. Offord, *Bioconj. Chem.* **1992**, *3*, 262

106 C. J. A. Wallace, *FASEB J.* **1993**, *7*, 505

107 F. A. Robey, R. L. Fields, *Anal. Biochem.* **1989**, *177*, 373

108 Y.-A. Lu, P. Clavijo, M. Galatino, Z.-Y. Shen, W. Liu, J. P. Tam, *Mol. Immunol.* **1991**, *6*, 623

109 D. R. Englebretsen, B. G. Garnham, D. A. Bergman, P. F. Alewood, *Tetrahedron Lett.* **1995**, *36*, 8871

110 M. Baca, T. W. Muir, M. Schnolzer, S. B. H. Kent, *J. Am. Chem. Soc.* **1995**, *117*, 1881

111 T. P. King, S. W. Zhao, T. Lam, *Biochemistry* **1986**, *25*, 5774

112 I. Fisch, G. Künzi, K. Rose, R. Offord, *Bioconj. Chem.* **1992**, *3*, 147

113 H. F. Gaertner, R. E. Offord, R. Cotton, D. Timms, R. Camble, K. Rose, *J. Biol. Chem.* **1994**, *269*, 7224

114 S. Pochon, F. Buchegger, A. Pelegrin, J.-P. Mach, R. E. Offord, J. E. Ryser, K. Rose, *Int. J. Cancer* **1989**, *43*, 1188

115 K. Rose, *J. Am. Chem. Soc.* **1994**, *116*, 30

116 L. E. Canne, A. R. Ferre-D'Amare, S. K. Burley, S. B. H. Kent, *J. Am. Chem. Soc.* **1995**, *117*, 2998

117 T. Wieland, E. Bokelmann, L. Bauer, H. U. Lang, H. Lau, *Liebigs Ann. Chem.* **1953**, *583*, 129

118 M. Brenner, J. P. Zimmermann, J. Wehrmüller, P. Quitt, A. Hartmann, W. Schneider, U. Beglinger, *Helv. Chim. Acta* **1957**, *40*, 1497

119 P. L. Russel, R. M. Topping, D. E. Tutt, *J. Chem. Soc.* (B) **1971**, 657

120 D. S. Kemp, *Biochemistry* **1981**, *20*, 1793

121 D. S. Kemp, N. G. Galakatos, *J. Org. Chem.* **1986**, *51*, 1821

122 D. S. Kemp, N. G. Galakatos, B. Bowen, K. Tam, *J. Org. Chem.* **1986**, *51*, 1829

123 D. S. Kemp, R. I. Carey, *J. Org. Chem.* **1993**, *58*, 2216

124 J. P. Tam, Y.-A. Lu, C. F. Liu, J. J. Shao, *Proc. Natl. Acad. Sci. USA* **1995**, *92*, 12485

125 C.-F. Liu, C. Rao, J. P. Tam in: *Peptides: Chemistry, Structure and Biology*, P. T. P. Kaumaya, R. S. Hodges (Eds.), Mayflower Worldwide, Kingswinford, UK, **1996**, 835

126 J. P. Tam, Q. Yu, *Biopolymers* **1998**, *46*, 319

127 J. P. Tam, Q. Yu, Z. Miao, *Biopolymers* **1999**, *51*, 311

128 T. Muir, P. E. Dawson, S. B. H. Kent, *Methods Enzymol.* **1997**, *289*, 266

129 P. E. Dawson, S. B. H. Kent, *Annu. Rev. Biochem.* **2000**, *69*, 923

130 W.Y. Lu, M.A. Quasim, S.B.H. Kent, *J. Am. Chem. Soc.* **1996**, *118*, 8518

131 P.E. Dawson, M.J. Chirchill, M.R. Ghadiri, S.B.H. Kent, *J. Am. Chem. Soc.* **1997**, *119*, 4325

132 S. Futaki, K. Sogowa, J. Maruyama, T. Asahara, M. Niwa, *Tetrahedron Lett.* **1997**, *38*, 6237

133 Y. Shin, K.A. Winans, B.J. Backes, S.B.H. Kent, J.A. Ellman, C.R. Bertozzi, *J. Am. Chem. Soc.* **1999**, *121*, 11684

134 A.B. Clippingdale, C.J. Barrow, J.D. Wade, *J. Peptide Sci.* **2000**, *6*, 225

135 R. Ingenito, E. Bianchi, D. Fattori, A. Pessi, *J. Am. Chem. Soc.* **1999**, *121*, 11369

136 L.E. Canne, P. Botti, R.J. Simon, Y. Chen, E.A. Dennis, S.B.H. Kent, *J. Am. Chem. Soc.* **1999**, *121*, 8720

137 M. Patek, M. Lebl, *Tetrahedron Lett.* **1991**, *32*, 3891

138 A. Brik, E. Keinan, P.E. Dawson, *J. Org. Chem.* **2000**, *65*, 3829

139 J. Rademann, M. Grotli, M. Meldal, K. Bock, *J. Am. Chem. Soc.* **1999**, *121*, 5459

140 C.F. Liu, C. Rao, J.P. Tam, *Tetrahedron Lett.* **1996**, *37*, 933

141 D.W. Low, M.G. Hill, M.R. Carrasco, S.B.H. Kent, P. Botti, *Proc. Natl. Acad. Sci. USA* **2001**, *98*, 6554

142 L. Zhang, J.P. Tam, *Tetrahedron Lett.* **1997**, *38*, 3

143 L.Z. Yan, P.E. Dawson, *J. Am. Chem. Soc.* **2001**, *123*, 526

144 T.M. Hacking, J.H. Griffin, P.E. Dawson, *Proc. Natl. Acad. Sci. USA* **1999**, *96*, 10068

145 G.S. Beligere, P.E. Dawson, *J. Am. Chem. Soc.* **1999**, *121*, 6332

146 E. Saxon, C.R. Bertozzi, *Science* **2000**, *287*, 2007

147 E. Saxon, J.I. Armstrong, C.R. Bertozzi, *Organic Lett.* **2000**, *2*, 2141

148 T.W. Muir, D. Sondhi, P.A. Cole, *Proc. Natl. Acad. Sci. USA* **1998**, *95*, 6705

149 V. Severinov, T.W. Muir, *J. Biol. Chem.* **1998**, *273*, 16205

150 T.C. Evans, Jr., J. Benner, M.-Q. Xu, *Protein Sci.* **1998**, *7*, 2256

151 T.C. Evans, Jr., J. Benner, M.-Q. Xu, *J. Biol. Chem.* **1999**, *274*, 3923

152 P.M. Kane, C.T. Yanashiro, D.F. Wolczyk, N. Neff, M. Goebl, T.H. Stevens, *Science* **1990**, *250*, 651

153 F.B. Perler, E.O. Davis, G.E. Dean, F.S. Gimble, W.E. Jack, N. Neff, C.J. Noren, J. Thorner, M. Belfort, *Nucleic Acids Res.* **1994**, *22*, 1125

154 M.-Q. Xu, F.B. Perler, *EMBO J.* **1996**, *15*, 5146

155 Y. Shao, M.-Q. Xu, H. Paulus, *Biochemistry* **1996**, *35*, 3810

156 S. Chong, Y. Shao, H. Paulus, J. Benner, F.B. Perler, M.-Q. Xu, *J. Biol. Chem.* **1996**, *271*, 22159

157 C.J. Noren, J. Wang, F.B. Perler, *Angew. Chem.* **2000**, *112*, 458; *Angew. Chem. Int. Ed. Engl.* **2000**, *39*, 450

158 S. Chong, F.B. Mersha, D.G. Comb, M.E. Scott, D. Landry, L.M. Vence, F.B. Perler, J. Benner, R.B. Kucera, C.A. Hirvonen, J.J. Pelletier, H. Paulus, M.-Q. Xu, *Gene* **1997**, *192*, 271

159 F.B. Perler, M.-Q. Xu, H. Paulus, *Curr. Opin. Chem. Biol.* **1997**, *1*, 292

160 B. Ayers, U.K. Blaschke, J.A. Camarero, G.J. Cotton, M. Holford, T.W. Muir, *Biopolymers* **1999**, *51*, 343

161 R. Xu, B. Ayers, D. Cowburn, T.W. Muir, *Proc. Natl. Acad. Sci. USA* **1999**, *96*, 388

162 T.C. Evans, Jr., M.-Q. Xu, *Biopolymers* **1999**, *51*, 333

6
Synthesis of Special Peptides and Peptide Conjugates

6.1
Cyclopeptides

Cyclopeptides form a large class of naturally occurring or synthetic compounds with a variety of biological activities [1]. In addition, cyclopeptides have gained enormous importance as models for turn-forming peptide and protein structures. Cyclopeptides can be classified chemically as either homodetic or heterodetic compounds (cf. Section 2.2). Cyclic depsipeptides contain α-hydroxy acids, and are members of the latter class because the peptide backbone also contains ester bonds. Cyclopeptides, where a disulfide bond is involved in the ring-closure reaction, are also classified as heterodetic cyclopeptides, and will be discussed in Section 6.2. Results gained from several biological studies involving cyclopeptides have led to the conclusion that this class of compounds is often characterized by increased metabolic stability, improved receptor selectivity, controlled bioavailability, and improved profiles of activity. Moreover, the constrained geometry of cyclopeptides is a favorable precondition for conformational and molecular modeling studies on key secondary structure elements. Cyclopeptides and cyclodepsipeptides can be metabolized in mammalian organisms only to a very limited extent because of their resistance towards enzymatic degradation; however, they are excreted via hepatic clearance more easily than are open-chain analogues because of their increased lipophilicity.

Cyclopeptides display manifold biological activities in the form of hormones, antibiotics, ion carrier systems, antimycotics, cancerostatics, and toxins. Several representatives of this class of compound had been isolated, their structures elucidated, and the compounds produced by total synthesis by the mid-twentieth century. However, during the past four decades there has been a rapid increase in the number of cyclopeptides with unusual structures isolated from plants, fungi, bacteria, and marine organisms. Cyclopeptides and cyclodepsipeptides often contain unusual amino acids, among them D-amino acids, β-amino acids, and α,β-didehydro amino acids. Consequently, ribosomal synthesis can be excluded for most cases. Biosynthesis usually proceeds via an activation of the amino acids as thioester in multienzyme complexes (thiotemplate mechanism, cf. Section 3.2.3). For some classes, such as the lantibiotics, a ribosomal precursor synthesis with subsequent processing by decarboxylations, dehydrations, dehydrogenations, and formation of the cyclic sulfide has been proven.

The fact that cyclopeptides also display decreased flexibility and restriction of conformation in biologically important moieties allows for tailored interaction with receptors, and important conclusions on the structural requirements of a receptor-ligand interaction can be derived from these studies. Several reviews on the natural occurrence, biological importance, design and application – including synthetic aspects – of the cyclopeptides are available [2–6], and in this context the studies of Kessler et al. [6–8], which relate to the cyclic analogues of antamanide, thymopoietin, somatostatin, as well as cyclic RGD and LDT peptides, are of eminent importance.

Interest in synthetic cyclopeptides is motivated by different objectives that range from the development of synthetic methodology, through physico-chemical studies to investigations into the role of different structural factors, for example ring size, building block configuration, and influence of the side-chain functionalities. Structure-activity relationship studies on bioactive cyclopeptides remain the focus of general interest, and include conformational investigations with respect to peptide-receptor interaction. Cyclic peptides serve as models of protein-recognition motifs, and are used to mimic β-sheets, β-turns, or γ-turns.

Total synthesis is the best method to obtain structural variation and to optimize the biological properties. In most cases, sufficient quantities of the peptide for use in biological investigations can only be provided by synthetic means, as these naturally occurring highly bioactive compounds occur in only minute amounts and their isolation is extremely difficult. Whilst an exhaustive presentation of all aspects of cyclopeptides is beyond the scope of this book, some principles regarding their synthesis will be discussed, using selected examples, in the following paragraphs.

In addition to backbone head-to-tail cyclization, in principle there are three further general topologies for peptide cyclization (Fig. 6.1). Side chain-to-side

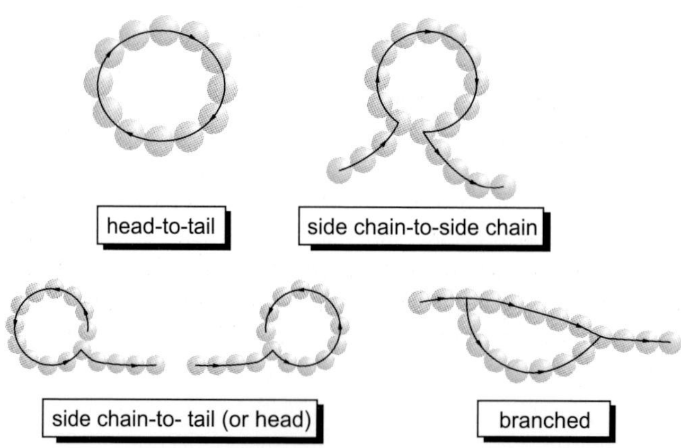

Fig. 6.1 General topologies for peptide cyclization.

chain ring closure may be performed by disulfide or amide bond formation between suitable functional groups. Head-to-side chain or side chain-to-tail cyclizations are performed in a similar manner.

6.1.1
Backbone Cyclization (Head-to-Tail Cyclization)

As is known from organic chemistry, the formation of medium-size rings with 9–12 atoms is extremely difficult, with carbocyclic ring systems of medium size being energetically disfavored because of ecliptic and transannular interactions. However, the situation changes when heteroatoms are present, as for example in macrocyclic natural products, as these minimize – or even exclude – ecliptic or transannular strain.

The synthesis and conformation of backbone-cyclized cyclopeptides containing more than nine backbone atoms has undergone extensive examination. In principle, all methods suitable for the formation of a peptide bond can be used for the cyclization of linear peptide sequences involving amide bonds. Ring-closure reactions usually proceed much more slowly compared to normal peptide bond formations, and side reactions [9, 10], such as undesired intermolecular peptide bond formation leading to cyclodimerization of linear peptide precursors, may predominate. This may be suppressed by performing the cyclization reaction under high dilution conditions in 10^{-4}–10^{-3} M solution. The problem of cyclo-oligomerization in the case of small peptides is attributed to the predominantly trans-configured peptide bonds favoring an extended conformation of the linear precursors.

A sometimes unacceptably high risk of epimerization at the C-terminal amino acid (racemization) must be taken into account for cyclization reactions involving carboxy activation of chiral amino acid building blocks, especially because of the prolonged reaction times [11]. Hence, precursors with C-terminal glycine or proline residues that are less prone to racemization should be chosen whenever possible.

Head-to-tail cyclizations of 7- to 10-peptides are usually not impeded by sequence-specific problems. For shorter sequences, the reaction rate of the ring closure depends on the presence of turn-inducing elements such as D-amino acids, proline, glycine, or N-alkyl amino acids, that favor turns or *cis*-peptide bonds. Purely L-configured tetra- and pentapeptides that do not contain any glycine, proline or D-amino acid residues are usually very difficult to cyclize.

Cyclic dipeptides (diketopiperazines) **1** are very easily formed by intramolecular aminolysis of dipeptide esters. These cyclodipeptides, which comprise a six-membered ring, are usually stable towards proteolysis and are often formed as unwanted side products in solution- or solid-phase synthesis of linear peptides on the dipeptide stage. Diketopiperazines may serve as scaffolds for combinatorial synthesis in drug discovery [12–14].

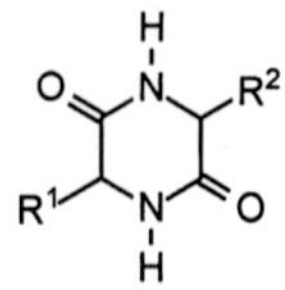

1

Cyclic tripeptides (nine-membered rings) usually are not formed when ring closure is attempted starting from carboxy activated tripeptides **2**. In most cases, the reaction results in cyclodimerization, forming the corresponding hexapeptides **3**. Only cyclic tripeptides containing N-substituted glycine, proline, or hydroxyproline residues are known as exceptions. Basic investigations by Rothe et al. [15] revealed that stable cyclotripeptides must contain at least two secondary amino acid building blocks.

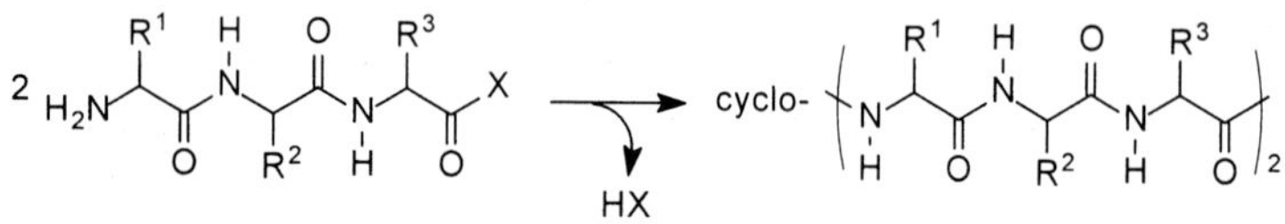

2 **3**

Preorganization of the linear precursor seems to be a major precondition for successful ring closure. The pseudoproline-containing tripeptide **4** is cyclized to give **5** without oligomerization, even at 0.1 M concentration [16].

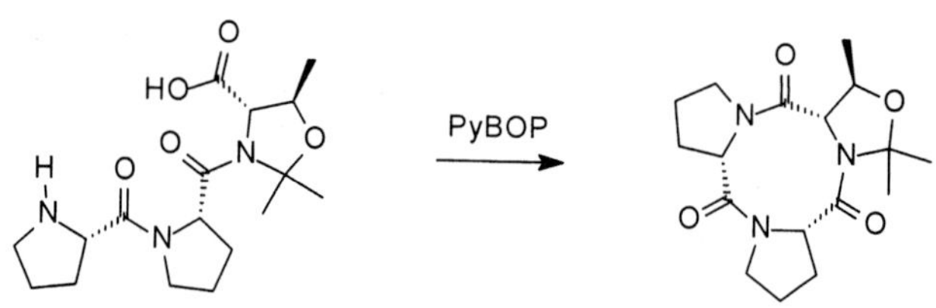

4 **5**

Cyclic tetrapeptides (12-membered rings) that are assembled exclusively from L-amino acids are also not readily accessible. The conformation of the linear precursor influences the tendency towards cyclization [17]. If the linear precursor lacks N-alkyl amino acids, usually only traces of product are formed [18]. Polycondensation efficiently competes with ring closure. Cyclotetrapeptides containing N-alkyl amino acids such as sarcosine are formed more easily.

When the usual coupling methods are employed, cyclic pentapeptides are obtained in acceptable to good yields, but competing cyclodimerization, by which the corresponding cyclodecapeptides are formed, may also be observed.

In the case of the cyclodepsipeptides the formation of a peptide bond during cyclization is preferred. Cyclization yields generally are rather low, and a yield of 50% is usually regarded as high.

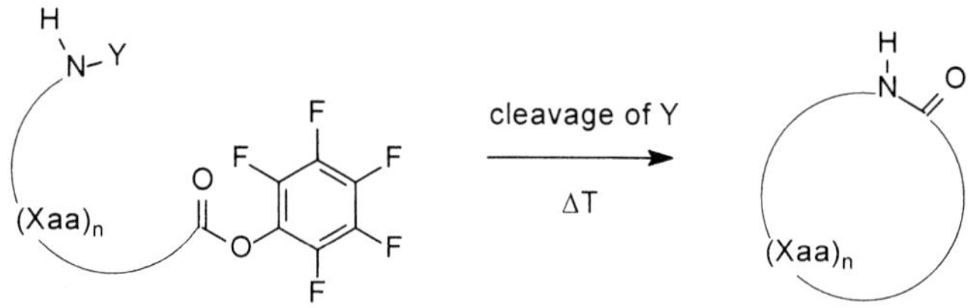

Fig. 6.2 Pentafluorophenylester cyclization of Z-, Fmoc-, or Boc-protected peptides. Cleavage conditions: H_2/Pd (Y=Z), TFA, then base (Y=Boc), piperidine (Y=Fmoc).

The active ester method was found to be a straightforward method for peptide cyclization. The pentafluorophenylester ring-closure reactions performed by Schmidt et al. proceed smoothly in the presence of DMAP as an acylation catalyst. The principle of the reaction is shown schematically in Fig. 6.2 [19].

The linear N-terminally protected peptides are transformed into the pentafluorophenylester, for example by treatment with a carbodiimide and pentafluorophenol. These protected and activated intermediates are not stable for an unlimited period of time. Three variants are possible depending on the N-terminal N^{α}-amino protective group of the linear peptide pentafluorophenylester. In the case of an N-terminal Z group, the protected peptide pentafluorophenylester is added dropwise slowly to a hot dioxane solution (95 °C) containing Pd/C, some alcohol, and dimethylaminopyridine under a hydrogen atmosphere. The Z group is cleaved under these conditions, and ring closure proceeds on the palladium surface. Yields of 70–80% have been obtained in a series of cyclotetrapeptides containing three L-configured and one D-configured amino acid, or in a series of peptide alkaloids.

The 23-membered didemnines **6** have been obtained in 70% yield using the Boc/OPfp protocol [20]. The cyclotetrapeptides of the chlamydocine group can be synthesized in high yields by this procedure. Chlamydocine **7** and WF3161 **8**,

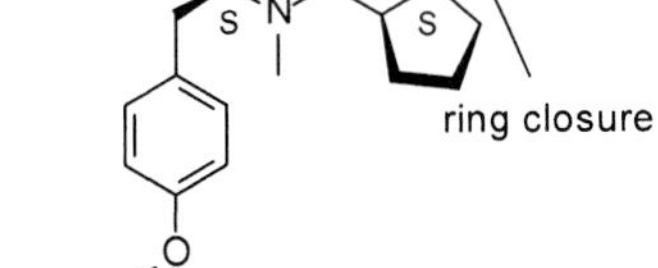

6

Didemnin A: R = H-(R)-MeLeu

Didemnin B: R = H-Lac-Pro-(R)-MeLeu

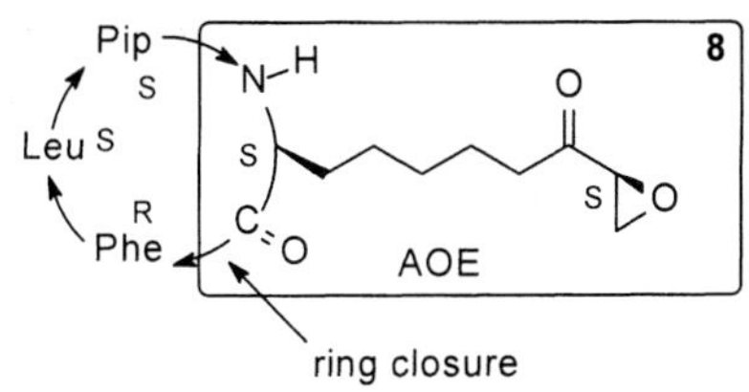

both of which belong to the most potent cancerostatics *in vitro*, have been obtained in 95 and 70% yields, respectively. The non-natural amino acid (*S*)-2-amino-8-oxo-(*S*)-9,10-epoxidecanoic acid (AOE) present in **7** and **8** was assembled stereoselectively in a seven-step synthesis.

Peptides containing an amino acid building block at the C-terminus which is prone to racemization may be cyclized using the azide method or related variants such as the diphenylphosphoryl azide (DPPA) protocol [21], though these methods are characterized by prolonged reaction times [10]. The coupling reagents BOP and TBTU permit rapid cyclization reactions to be carried out, but suffer from significant C-terminal racemization [11]. Better results have been obtained using the azide method or DPPA compared to BOP or HBTU in ring-closure reactions of D-Pro-containing β-casomorphin tetra- and pentapeptides [10].

Comparative studies regarding the efficiency of peptide cyclizations using different new coupling reagents have revealed that reagents based on HOAt gave best results with respect to yields and minimization of racemization. C-terminal D-amino acid residues favor the formation of pentapeptide rings [22]. However, the application of HATU, for example, sometimes suffers from undesired N-guanylation if the cyclization reaction is too slow. This problem may be overcome by the addition of HOAt [23].

Cyclopeptides may be synthesized by solution- or solid-phase methods (or by a combination of these) where a linear sequence is synthesized on polymeric support and cyclization is performed in solution after cleavage from the resin. The side chain-protecting groups are cleaved in the final step. The major obstacles of classical cyclization reactions in solution (cyclo-oligomerization and cyclo-dimerization) must be avoided by high dilution, as discussed previously. A viable route towards cyclopeptides comprises the assembly and cyclization of a peptide while it is still bound to the resin. The so-called pseudo-dilution, a kinetic phenomenon favoring intramolecular reactions of resin-bound peptides over intermolecular side reactions, presents a major advantage [24–27], provided that the resin loading is not too high. The additional application of a larger excess of soluble reagent may favor the cyclization reaction. As the desired product is bound to the resin, simple washing and filtration processes facilitate the whole process and provide the potential for automation. One precondition for on-resin cyclization is the attachment of the first amino acid to the polymeric support via a side-chain functional group (Fig. 6.3, Tab. 6.1).

This strategy requires one further orthogonal protective group for the C-terminal carboxy group, and the synthesis of cyclo-(-Gly-His-)$_3$ in 42% purified yield was a first successful example of this [36]. Such types of cyclization reaction can be performed using orthogonal protecting groups such as Fmoc/*tert*-butyl/allyl type [37]. Alternatively, resin attachment is possible via a backbone amide linker (Fig. 6.4) [38].

The success of the peptide cyclization reaction either in solution or on solid support depends primarily on the conformational preferences of the linear target sequence, and also on the solvent, the bases applied, the concentration, and the temperature.

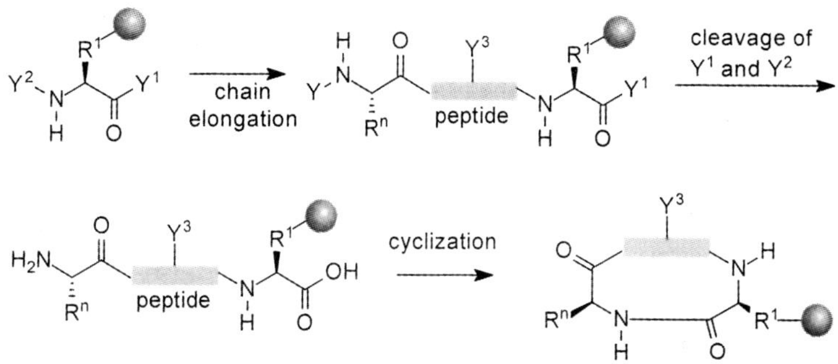

Fig. 6.3 On-resin cyclization of peptides with side chain resin attachment. Y^1 = carboxy-protecting group; Y^2 = temporary amino-protecting group; Y^3 = semipermanent side chain-protecting groups.

Tab. 6.1 Selected protecting group/linker tactics for backbone cyclization.

Amino acid	*N^{α}-protection*	*Resin*	*Orthogonal CO-protecting group*	*Reference(s)*
Asp	Fmoc	Wang	ODmb	28, 29
Asp	Fmoc	PAC/PAL	OAl	30
Asp/Glu	Fmoc	Pepsyn-K	ODmab	31
Asp/Glu	Boc	Merrifield (HM-PS)	OFm	32
Asp/Glu	Boc	MBHA	OFm	32
Lys/Orn/Dab	Boc	HM-PS, linked via urethane group	OFm	32
Ser	Boc	AM-PS, linked via succinyl group	ONbz	32
Tyr	Boc	Merrifield (HM-PS)	ONbz	32
Lys	Fmoc	Wang (modified)	OAl	33
Ser/Thr/Tyr	Boc	Merrifield (HM-PS, modified)	OAl	34
His	Fmoc	Trt	OAl	35

The first solid-phase head-to-tail cyclization by intramolecular aminolysis of a resin bound *o*-nitrophenyl peptide ester was described as early as 1965 by Patchornik et al. The concept of combined cyclization and cleavage from the resin does not require side chain or backbone anchoring, but makes use of special linkers for C-terminal attachment (oxime resin [39, 40], thioester resin [41]). The acid-stable *p*-nitrobenzophenone oxime resin was found to be the most suitable [42]. Whilst in the former case tyrocidin A was obtained in 30% yield after purification, the synthesis of cyclo-(-Arg-Gly-Asp-Phg-), a cyclic tetrapeptide with inhibitory activity against cell adhesion, was achieved with a yield exceeding 50%.

This strategy experienced a further renaissance with the development of the so-called safety-catch resins (cf. Section 4.5.2). During the course of this process, the

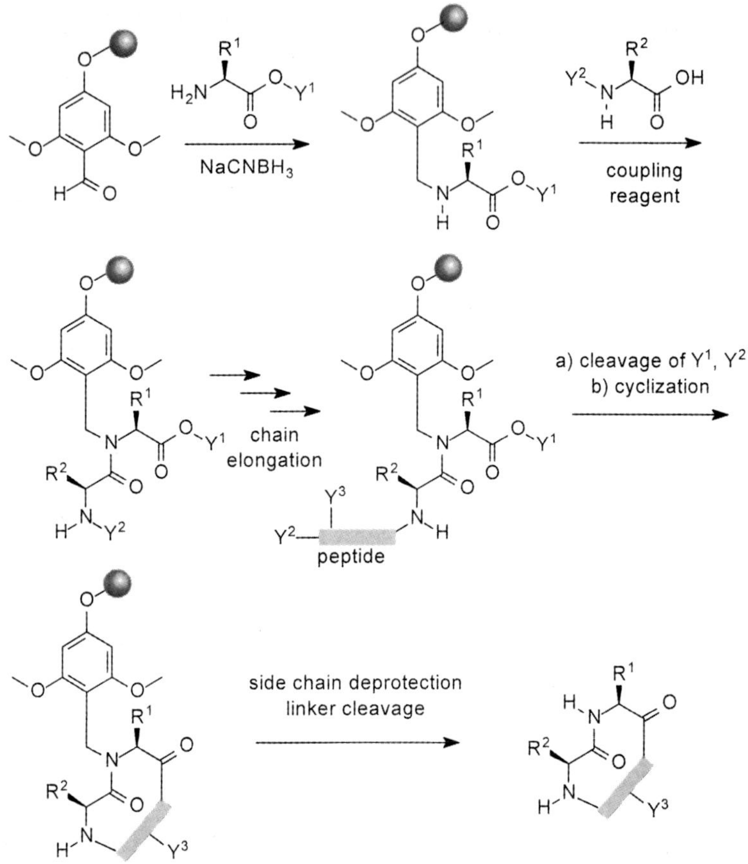

Fig. 6.4 On-resin cyclization of peptides with resin attachment via backbone amide linker. Y^1 = carboxy-protecting group; Y^2 = temporary amino-protecting group; Y^3 = semipermanent side chain-protecting groups.

linker moiety is activated so that the bound peptide is able to undergo intramolecular aminolysis (Fig. 6.5) [43].

Thiol-mediated backbone cyclization of cysteine-containing peptides (cf. Section 5.4) has also been described [44, 45]. Peptide thioesters may be cyclized enzymatically using a thioesterase [46, 47] and the in-vivo synthesis of cyclic peptides has also been described [48].

Macrocyclization between two additional functional groups connected to backbone amides may also be regarded as backbone cyclization, or as a side chain-to-side chain cyclization. The cyclization may occur either via amide bond formation [49] or via ring-closing metathesis, where two N-allyl, N-homoallyl or longer derivatives react intramolecularly under transition metal catalysis to give peptide cyclization [50, 51]. Cyclopeptide-derived macrocycles have been obtained by ring-clos-

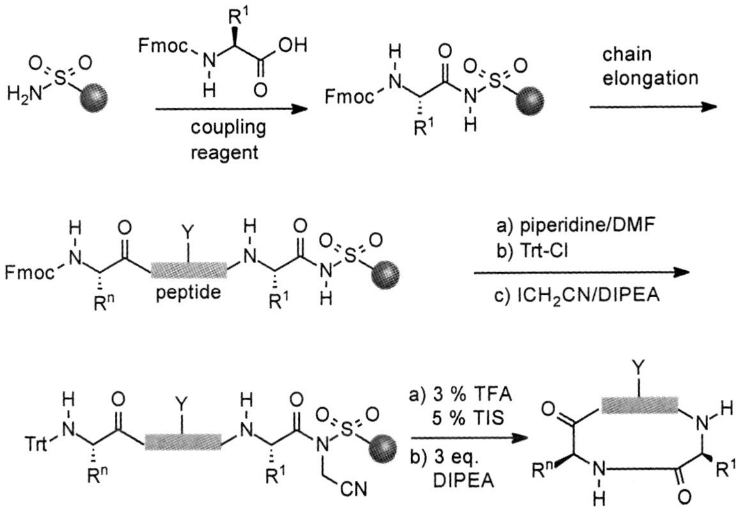

Fig. 6.5 Peptide cyclization with simultaneous cleavage from safety-catch resin. Y = semipermanent side chain-protecting groups.

ing metathesis with Grubbs catalyst using a strategy that involved simultaneous metathesis macrocyclization and resin cleavage [52].

6.1.2
Side Chain-to-Head and Tail-to-Side Chain Cyclizations

Side chain-to-head or tail-to-side chain cyclizations may proceed via macrolactamization between a side-chain amino group and the C-terminus, or between a side-chain carboxy group and the N-terminus. On-resin cyclizations have been reported [53–55], and lactone formation involving side-chain hydroxy groups is also possible. Furthermore, intramolecular thioalkylation of an N-terminal bromoacetyl group with a cysteine residue of the peptide chain is feasible and results in the formation of a cyclic thioether [56, 57]. Head-to-side chain cyclizations and branched cyclopeptides have been reviewed in [1].

6.1.3
Side Chain-to-Side Chain Cyclizations

Macrocyclic disulfides (cf. Section 6.2), thioethers [58], lactams [54, 59, 60] and lactones are obtained upon reactions between appropriate functional groups present in the side chains. Ring-closure reactions between side-chain functional groups were described for the first time by Schiller et al. [61], and have been examined with respect to methodology by Felix et al. [62]. Subsequently, many different types of cyclopeptides have been synthesized since that early investigative period.

6.2
Cystine Peptides [63–66]

Disulfide bridges, which are formed upon oxidative coupling of two thiol groups of cysteine residues, frequently occur in peptides and proteins. As discussed in Chapter 2, disulfide bonds contribute greatly to the formation and stabilization of distinct three-dimensional structures in proteins and peptides.

A distinction can be made between intrachain and interchain disulfide bridges. In the former type, a disulfide bridge links two cysteine residues of one peptide, chain leading to a cyclic peptide; in contrast, interchain disulfide bridges lead to a covalent connection of two peptide molecules. Disulfide bridge-containing peptides can be classified as shown in Tab. 6.2.

Tab. 6.2 Symmetrical and unsymmetrical cystine peptides.

Type	*Occurrence*
Symmetrical cystine peptides 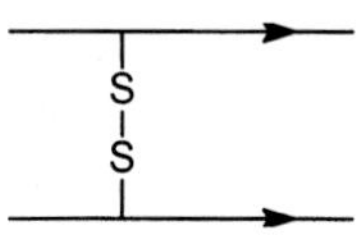	This motif does not occur very frequently in nature, but it is found for instance in oxidized glutathione
Unsymmetrical cystine peptides – with one intrachain disulfide bridge 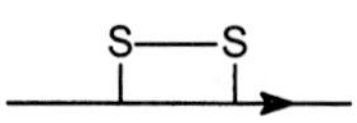	These cyclic heterodetic structures are found for instance in oxytocin, vasopressin, and somatostatin. Furthermore, this motif has often been used for the design of conformationally constrained cyclic peptides
– with two or more intrachain disulfide bridges 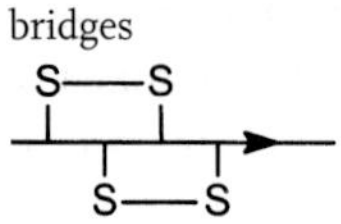	These polycyclic heterodetic structures occur, e.g., in proinsulin, epidermal growth factor, insulin-like growth factors I and II, and some toxins
– with one interchain disulfide bridge 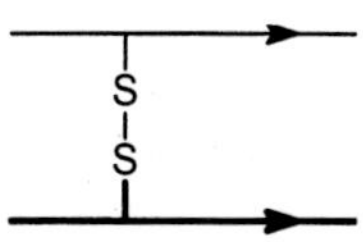	Two different peptide chains are connected via a cystine bridge. This motif is observed in some enzymatic protein digests
– with two or more interchain disulfide bridges 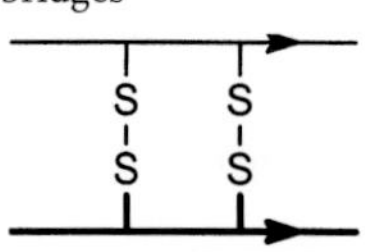	This motif leading to cyclic heterodetic structures is found, e.g., in insulin and relaxin. In this case the two different peptide chains may be arranged in a parallel or antiparallel manner

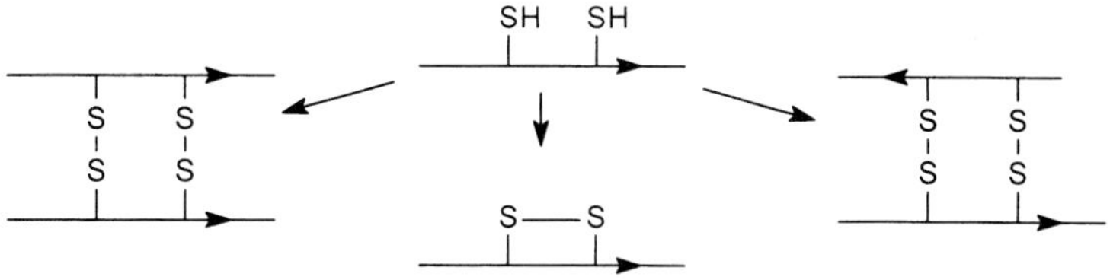

Fig. 6.6 Side reactions in peptide cyclization via disulfide bond formation.

The synthesis of simple monocyclic cystine-containing peptides is usually not difficult. The linear peptide sequence is assembled either in solution or on solid phase, with both thiol groups being protected by the same protecting group. Following cleavage of either all side chain-protecting groups or, alternately, of only the cysteine side chain-protecting groups, cyclization is performed in solution under high dilution (10^{-3}–10^{-5} M). Oxygen (air), iodine, potassium, hexacyanoferrate(III), azodicarboxylates, pyridylsulfenylchlorides, alkoxycarbonylsulfenylchlorides, alkyltrichlorosilane/sulfoxide, thallium(III) trifluoroacetate, and DMSO, respectively, are used as oxidizing agents for the cyclization. Cyclodimerization and oligomerization may be observed as an undesired side reaction (Fig. 6.6). Likewise, selenocysteine peptides may be oxidized to produce diselenide-bridged compounds [67].

Some protecting groups for the side-chain thiol function of cysteine allow for simultaneous cleavage and disulfide formation (cf. Section 4.2.4.4). Treatment of peptides containing Cys(Acm) or Cys(Trt) residues with iodine simultaneously cleaves the protecting group and oxidizes the free thiols to give the disulfide. Accordingly, peptides containing a single Cys residue are converted to the symmetrical dimer, while peptides containing two cysteine residues may be cyclized under high dilution conditions.

The synthesis of unsymmetrical cystine peptides, especially of those containing more than one disulfide bridge, is far more difficult. Two basic strategies may be distinguished: (i) a cystine derivative, which already contains the desired disulfide bridge is used for the assembly; or (ii) appropriately protected cysteine-containing peptides are assembled and subsequently transformed into the cystine-containing peptide [66]. The former variant requires correct selection of the amino- and carboxy-protecting groups. In the latter variant, the complexity of the synthesis and of the protecting group strategy increases exponentially with the number of cysteine residues present [68].

Sieber et al. [69] succeeded in the total synthesis of crystalline human insulin by controlled formation of the disulfide bonds. This synthesis was based on the unsymmetrical cystine peptide **9** obtained by Kamber, where one of the two interchain disulfide bridges of insulin had been already formed.

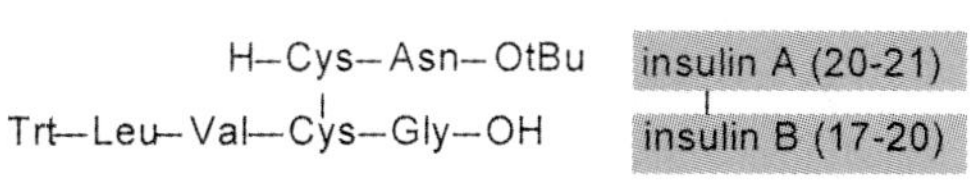

9

Unsymmetrical cystine peptides may be obtained by a statistical oxidation of two different cysteine-containing peptides. This route was used in the first total synthesis of insulin, where three disulfide bridges must be formed. Although the yield of insulin initially was less than 4%, optimization of the reaction conditions led to a significant increase in yield. This method was also used in the final step of the biotechnological synthesis of human insulin.

The second variant – the assembly of linear, suitably protected cysteine-containing peptide, with subsequent oxidative cystine bridge formation – requires selective orthogonal and pairwise deprotection of cysteine residues. Both Fmoc and Boc tactics may be employed for the peptide chain assembly. Disulfide bond formation is performed either in solution, or while the peptide is still bound to the resin [70]. Selective unsymmetrical interchain disulfide bridge formation requires the activation of one thiol component by attachment of a good leaving group in order to achieve selective formation of the disulfide bond. For these activation purposes, 2,2′-dipyridyldisulfide or azodicarboxylates may be used. The thiol group present in one peptide chain is thus activated with one of these reagents, and subsequently reacted with the free thiol group of the second peptide chain [69].

6.3 Glycopeptides

The various properties and functions of glycoproteins were discussed in Chapter 3 (Section 3.2.2.4). In order to perform biological studies and therapeutic evaluations, a large amount of material is needed, and consequently the synthesis of glycopeptides as models for glycoproteins is currently of great interest. This synthetic task represents a major challenge however, because in addition to the requirements of peptide synthesis, the methods of carbohydrate chemistry must also be considered.

Glycopeptide synthesis and the so-called glycopeptide remodeling are of major importance because many glycoproteins are of pharmaceutical interest; examples include juvenile human growth hormone, CD4, and the tissue plasminogen activator. Glycopeptide remodeling requires modification of glycoproteins by the removal or addition of carbohydrate units. Without doubt, enzymes are very much suited to manipulations of the oligosaccharide moiety of glycoproteins, the sensitivity and polyfunctionality of which requires high selectivity. Unfortunately, however, many of the glycosyltransferases that might be used for this purpose are not yet available. It should be noted that, in addition to remodeling, the synthesis of new protein oligosaccharide conjugates is of major importance [71, 72]. As well as the problems of peptide synthesis that have been discussed previously, the chemical synthesis of glycopeptides imposes high requirements with regard to the reversible and highly selective protection of additional functional groups, and also to the stereoselective formation of glycosidic bonds [73–82]. The development of the methodology for glycopeptide synthesis requires special consideration of the additional complexity and lability of the carbohydrate moiety [72]. The protecting groups must be chosen correctly to ensure that their removal is selective, and does not interfere with the acid- and base-labile oligosaccharide groups.

The formation of a glycosidic bond between the carbohydrate moiety and the peptide unit is the crucial step in glycopeptide synthesis and, in general, two strategies for this can be distinguished:

1. The most common approach uses glycosylated amino acid building blocks which are appropriately protected for the stepwise synthesis of the glycopeptide.
2. The block glycosylation approach utilizes conjugation of the final carbohydrate unit to the full-length peptide. Unfortunately this approach is hampered by side reactions such as aspartimide formation in the case of N-glycosides, or difficulties in forming stereoselective O-glycosylation reactions with complex targets, not to mention the complex protecting groups strategies necessary.

The hydroxy functions present in the carbohydrates are usually reversibly blocked by either benzylether, acyl, or acetal protecting groups. In particular, benzyl ester and benzylidene groups exclude the simultaneous application of the benzyloxycarbonyl group and of benzyl esters as temporary protecting groups in the peptide moiety. The high chemical lability of the glycosidic bonds greatly complicates peptide synthesis. Glycosidic bonds are usually hydrolyzed under acidic conditions, and a permanent risk of β-elimination exists for all glycosyl serine and threonine derivatives, even under weakly basic conditions [83].

The β-N-glycosidic bond between N-acetylglucosamine and asparagine is formed by carbodiimide coupling of N-protected aspartic acid α-benzyl ester **10** with 2-acetamido-3,4,6-tri-*O*-acetyl-2-deoxy-β-glucopyranosylamine **11**, the latter molecule being obtained from the corresponding glucosyl chloride via the azide. The azide is selectively reduced to the amine by catalytic hydrogenation on Raney-nickel, without cleaving the benzyl-type protecting groups. The conjugate **12** is formed in 49% yield.

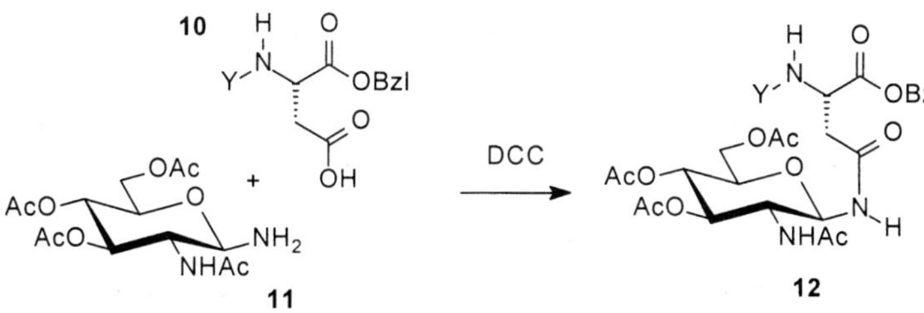

Y: N$^{\alpha}$-protecting group

Glycosyl amines regularly undergo anomerization under acidic conditions, where the amine is protonated. Equilibration favors the β-anomer as a result of the reverse anomeric effect [76].

The formation of an α-O-glycosidic bond between N-acetylgalactosamine and serine or threonine very often makes use of the azido group as a masked 2-acetamido moiety. The azido group does not exert neighboring group participation like the 2-acetamido group, which would lead to predominant formation of the β-anomer. Most often, the 1-halogeno carbohydrates are employed in Königs-Knorr-type glycosylations of Fmoc-protected serine and threonine esters. Highest α-selectivities are obtained upon application of an insoluble promoter such as silver perchlorate. Subsequently, the 2-azido group of **13** can be reduced and acetylated to give the 2-acetylamido group of **14**. Selective reduction of the azido group in complex glycopeptides is achieved using nickel tetrahydridoborate.

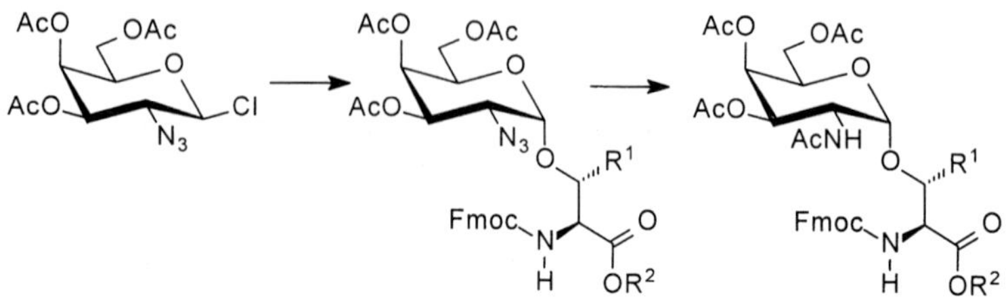

13 **14**

The synthesis of a β-O-glycosidic bond between N-acetylglucosamine and serine or threonine makes use of the anchimeric assistance of the 2-acetamido group in **15**. $BF_3 \cdot Et_2O$ activation of glycosyl acetates clearly improves the yields of N-protected glycosyl amino acid derivatives [84, 85]. The application of readily available anomeric acetates usually does not require protection of the amino acid carboxy group, thus rendering protecting group manipulation unnecessary.

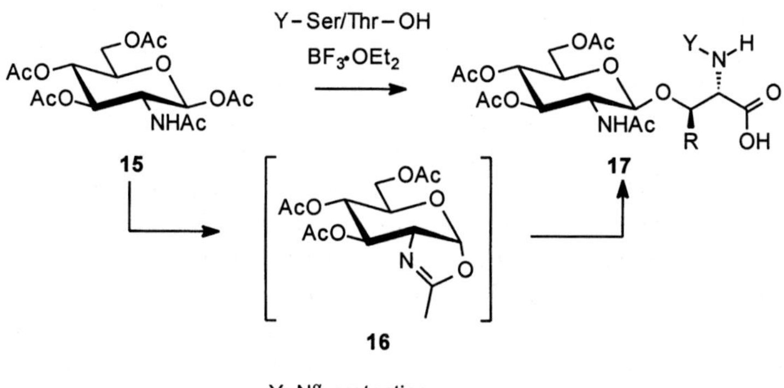

Y: N^{α}-protecting group
R: H, CH_3

Both the anomeric effect and anchimeric assistance of the acetyl group present on the 2-hydroxy function are exploited for the selective synthesis of α-mannosyl-serine **18** (R=H) and threonine **19** ($R=CH_3$) under Königs-Knorr-type conditions [86].

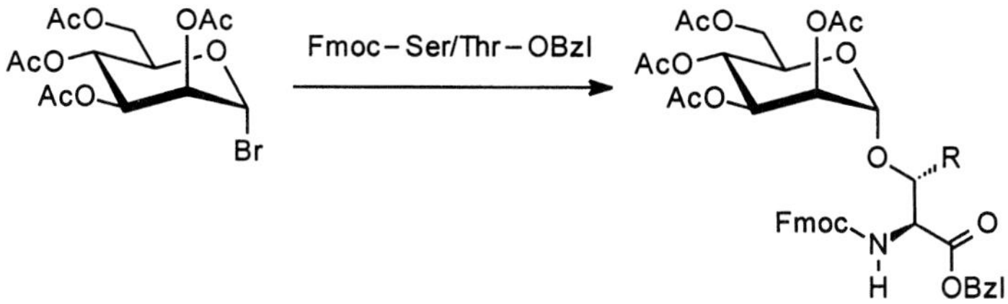

18 R = H
19 R = CH_3

In principle, glycopeptides may be synthesized either in solution or on solid phase. An N-glycopeptide cluster containing two sialyl-Lewis^x moieties has been synthesized in solution by Kunz et al. [87]. First, the asparagine residue is protected as the allyl ester, and with the Boc group; the glycosylated amino acid is then selectively deprotected at the amino group, without hampering the acetyl-protecting groups of the carbohydrate moiety. The fully protected glycopeptide is then assembled using preformed glycosylated amino acid building blocks. Allyl ester protection of the carboxy group allows for selective deblocking by rhodium(I)-catalyzed deallylation. Finally, the *tert*-butyl ester can be cleaved using formic acid, after which the acetyl groups of the carbohydrate moieties are removed using highly diluted sodium methoxide.

A very useful protocol for the synthesis of glycopeptides makes use of enzymatic reactions. Glycosyl amino acids are incorporated into oligopeptides by enzymatic coupling steps, and further elaboration of the oligosaccharide part can be achieved using glycosyltransferases [88]. This enzymatic strategy allows for a synthesis in aqueous solution and requires only minimum protection. The synthesis of a glycopeptide using a thermostable thiolsubtilisin mutant (S221C, M50F, N76D, G169S) is shown in Fig. 6.7. This mutant favors aminolysis over hydrolysis by a factor of approximately 10000. Glycosyltransferases, as well as exo- and endoglycosidases, are valuable catalysts for the formation of specific glycosidic linkages. The glycosyltransferases transfer a given carbohydrate from the corresponding sugar nucleotide donor to a specific hydroxy group of the acceptor sugar. Currently, a large number of eukaryotic glycosyltransferases (e.g., β1–4-galactosyl transferase, β1–4-GalT; Fig. 6.7) have been cloned that exhibit exquisite linkage and substrate specificity [74].

Solid-phase peptide synthesis (SPPS) can also be adapted to the construction of glycopeptides. Basically, the same linker systems as already described for SPPS (see Section 4.5.1) can be used for the assembly of glycopeptides. Enzymatic glycosylation may also be performed on solid phase, using an aminopropyl-modified silica gel support [89]. An enzyme-labile linker molecule was used in the synthesis shown in Fig. 6.8.

This elegant strategy allows for a rapid and iterative formation of peptide and glycosidic bonds in either organic or aqueous solvents, including enzymatic deprotection of the glycopeptide from the solid support [90]. The selective removal of glycosyl amino acid or glycopeptide protecting groups has remained an unsolved

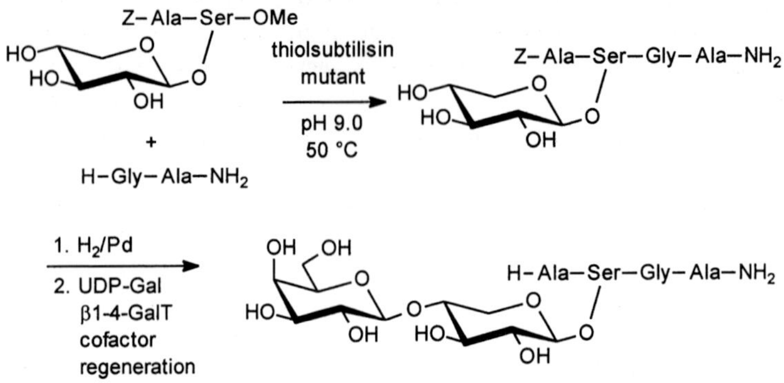

Fig. 6.7 Enzymatic synthesis of an O-glycopeptide using a subtilisin mutant for peptide coupling and a galactosyl transferase.

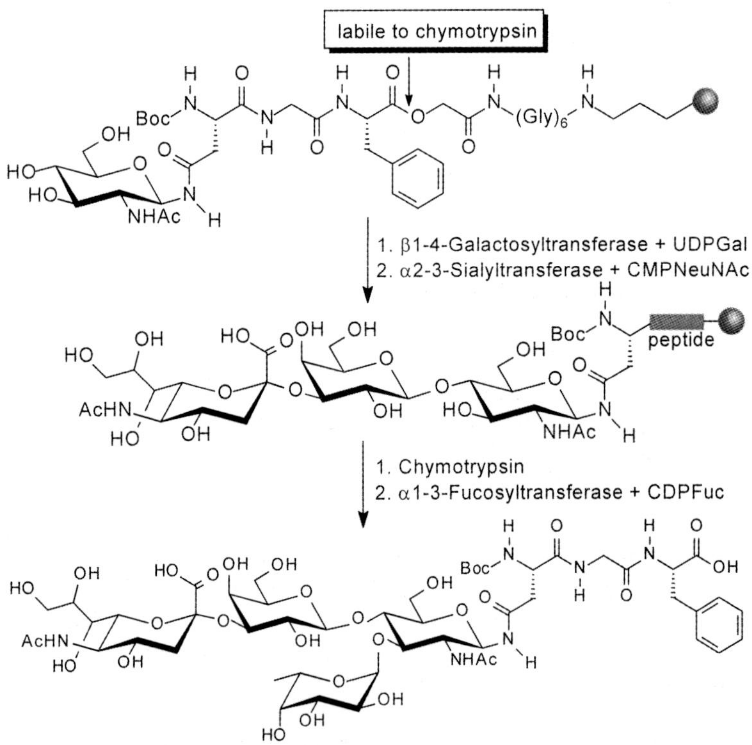

Fig. 6.8 Solid-phase glycopeptide synthesis involving enzymatic glycosyl transfer and a chymotrypsin-labile resin anchor.

problem for quite some time, having first been tackled during the late 1970s. A Z group in β-glycosyl amino acid derivatives cannot be cleaved using HBr/AcOH without destroying glycosidic bonds. Although cleavage of the Boc residue is possible under certain reaction conditions, acidolytic deprotection reactions should be used in the synthesis of glycopeptides only after careful consideration. As mentioned, the base sensitivity of glycosylserine or glycosylthreonine derivatives (β-elimination) further restricts the repertoire of deblocking reactions. Consequently, protecting groups must allow for cleavage under either mild or neutral reaction conditions.

The two-step protecting groups fulfil both requirements, namely stability during synthesis, and lability on deblocking. The 2-pyridylethoxycarbonyl (2-Pyoc) group **21** and its 4-pyridyl analogue (4-Pyoc) are stable under both acidic and basic conditions, but can be converted into labile derivatives upon methylation at the pyridine nitrogen atom. The alkylated derivatives are cleaved under very mild conditions (morpholine/CH_2Cl_2). Glycosylation of 2-Pyoc-Thr-OBzl **21** starting from the glucopyranosylbromide **20** according to the silver triflate procedure in the presence of an equimolar amount of trifluoromethanesulfonic acid provides access to the glycosyl amino acid **23**. The labilizing reactions mentioned above can, however, only be applied if no other amino acids that readily would undergo alkylation reactions are present in the target sequence.

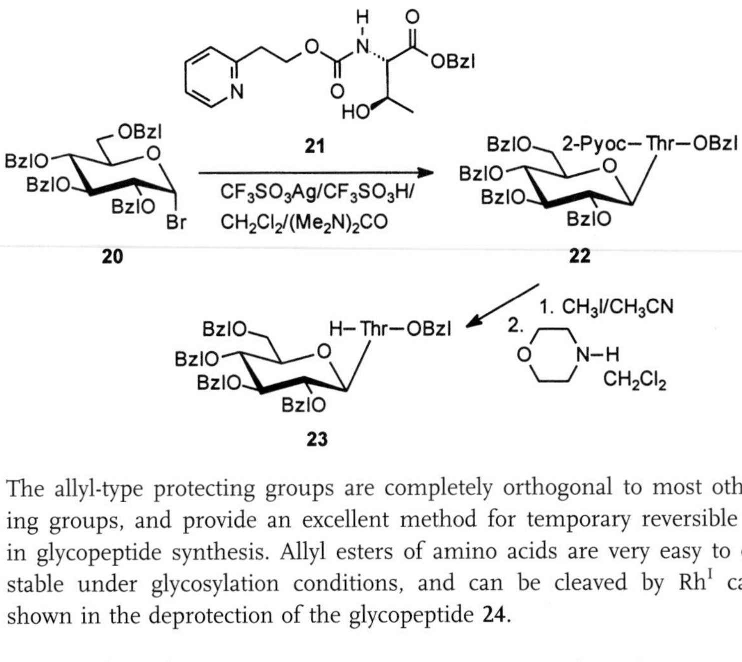

The allyl-type protecting groups are completely orthogonal to most other protecting groups, and provide an excellent method for temporary reversible protection in glycopeptide synthesis. Allyl esters of amino acids are very easy to obtain, are stable under glycosylation conditions, and can be cleaved by Rh^I catalysis, as shown in the deprotection of the glycopeptide **24**.

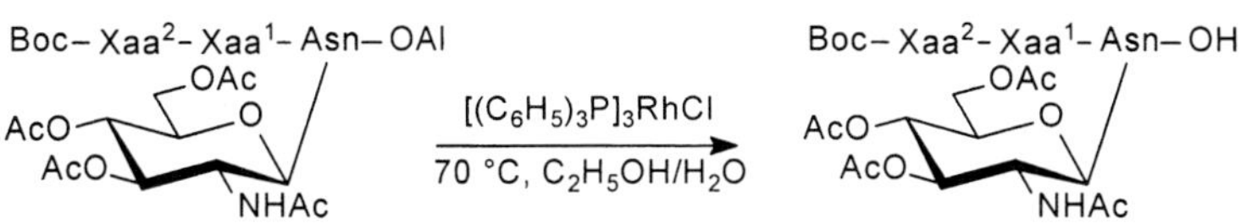

The allyl ester is isomerized under these reaction conditions to produce a vinyl ester (propenyl ester) that immediately undergoes hydrolysis under the reaction conditions.

An even milder method for the selective cleavage of allyl esters utilizes palladium(0)-catalyzed allyl transfer to morpholine. This principle is especially suited for the sensitive, elimination-prone O-glycosylserine and O-glycosylthreonine derivatives, and also allows for a smooth cleavage of the allyloxycarbonyl (Aloc) group in peptides and glycopeptides [91]. The Aloc residue is stable towards treatment with trifluoroacetic acid (TFA). Cleavage of the Aloc protecting group from the glycosylasparagine derivative **26** results in the formation of **27** in 90% yield. The allyl group in this case is transferred to the 1,3-diketone as the nucleophile.

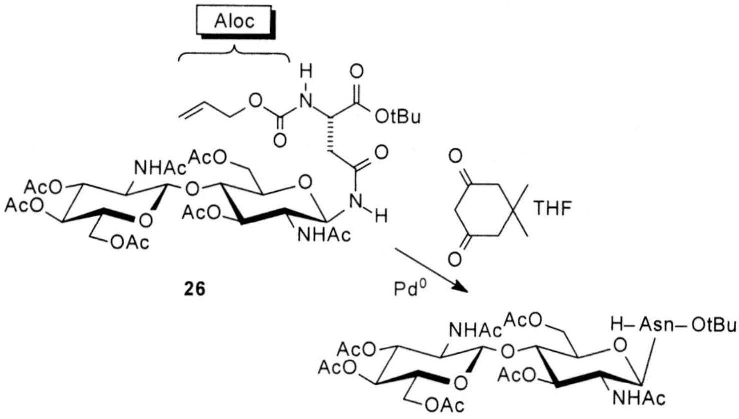

26 **27**

Allyl-type linker moieties are also appropriate for the solid-phase synthesis of complex glycopeptides [92]. Flexible polar oligo-ethyleneglycol spacers are often used to minimize steric hindrance and associations with the hydrophobic polystyrene matrix.

The HYCRON linker **28** was applied successfully in a synthesis of peptide T, an O-glycosylated tetrapeptide which occurs in the sequence of the HIV envelope protein gp120, and of an O-glycosylated glycopeptide of the mucine type [93]. Details of further studies in the solid-phase synthesis of glycopeptides are outlined in several references [73, 75–77, 94].

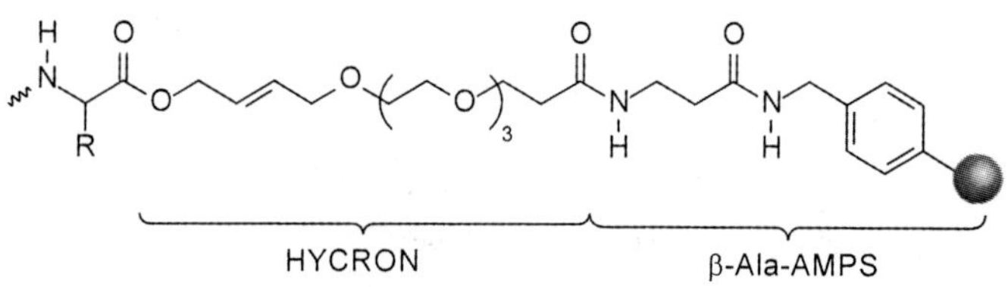

28

6.4
Phosphopeptides

In the view of the importance of protein phosphorylation (cf. Section 3.2.2.5), it is clear that phosphorylated peptides are highly valuable tools for the study of protein phosphorylation and dephosphorylation, as well as the recognition of phosphorylated proteins. Phosphorylated peptides for example can be used to determine the specificity of protein phosphatases [95], and may be synthesized via two fundamentally different routes: (i) global phosphorylation; or (ii) a building block approach. Whilst the former method – which is also known as post-assembly phosphorylation – makes use of selectively side chain-deprotected serine, threonine or tyrosine residues that are being phosphorylated on completion of the synthesis, the building block approach utilizes phosphorylated amino acids. This, of course, imposes further problems with respect to protecting group strategy. Both solution-phase and solid-phase syntheses of phosphorylated peptides have been reported. Different types of protecting groups have been used for the phosphate group: allyl, methyl, benzyl, *tert*-butyl, and 2,2-dichloroethyl. Appropriate phosphoserine, phosphothreonine, or phosphotyrosine derivatives for the building block approach can be synthesized by two different methods.

The first of these methods involves phosphorylation with a dialkyl- or diallylchlorophosphate under alkaline conditions (Fig. 6.9A). One advantage of this method is that the phosphorus atom is already in the correct oxidation state P(V). Alternately, Fmoc-Tyr(PO(NMe_2)$_2$)-OH may for example be applied in peptide synthesis [96].

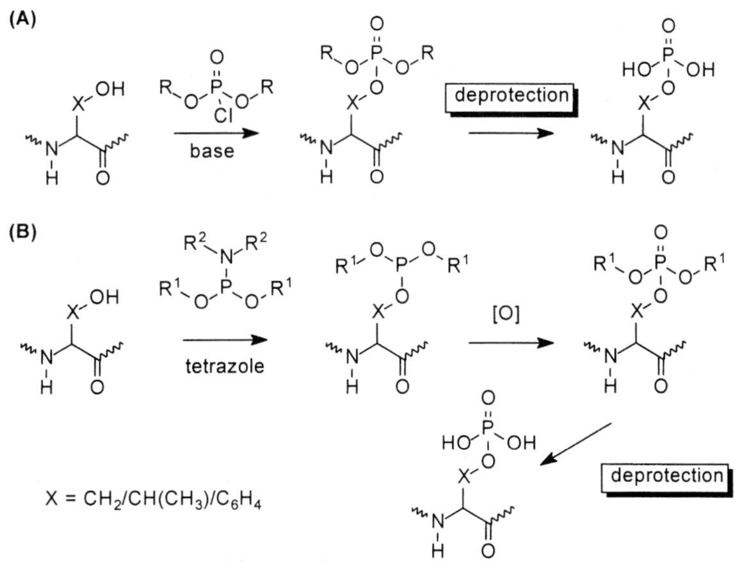

Fig. 6.9 Phosphorylation with phosphorus(V) and phosphorus(III) reagents.

The second method uses phosphorus(III) compounds, as they are used in oligonucleotide synthesis (Fig. 6.9B). A phosphoramidite is reacted with a suitably protected amino acid under mildly acidic conditions (tetrazole) to form a phosphite. Subsequently, an oxidation is performed using iodine, meta-chloroperbenzoic acid, or *tert*-butylhydroperoxide. Phosphorylation is usually performed on N^{α}-protected amino acids where the carboxy group additionally is blocked. Following phosphorylation of the side-chain hydroxy group, the carboxy group protection is selectively removed.

Peptide synthesis methodology is somewhat limited for serine and threonine derivatives, because these compounds readily undergo piperidine-mediated β-elimination to produce dehydroamino acid derivatives; hence, the N^{α}-Fmoc protection scheme is not universally applicable. The N^{α}-Aloc protection scheme is highly compatible with the synthesis of phosphopeptides. β-Elimination is suppressed when only one hydroxy function of the phosphate group is protected. Hence, under basic conditions this phosphate group is deprotonated and, consequently converted into a poor leaving group. Under these conditions, Fmoc tactics can be applied to the synthesis of phosphopeptides with Fmoc-Ser(PO(OBzl)OH)-OH [97, 98] and Fmoc-Thr(PO(OBzl)OH)-OH [98]. Partially protected phosphoamino acids, however, suffer from both pyrophosphate formation [99] and incompatibility with PyBOP and carbodiimide coupling reagents [100].

Enzyme-labile protecting group techniques provide in these cases an interesting and advantageous alternative (cf. Section 4.2.5). The selectively phosphorylated pentapeptide (Fig. 6.10) has been synthesized by a combination of enzymatic and chemical protecting group operations. In this sequence, both heptyl esters (cleavage with lipases) and allyl esters (cleavage with Pd^0) have been employed as pro-

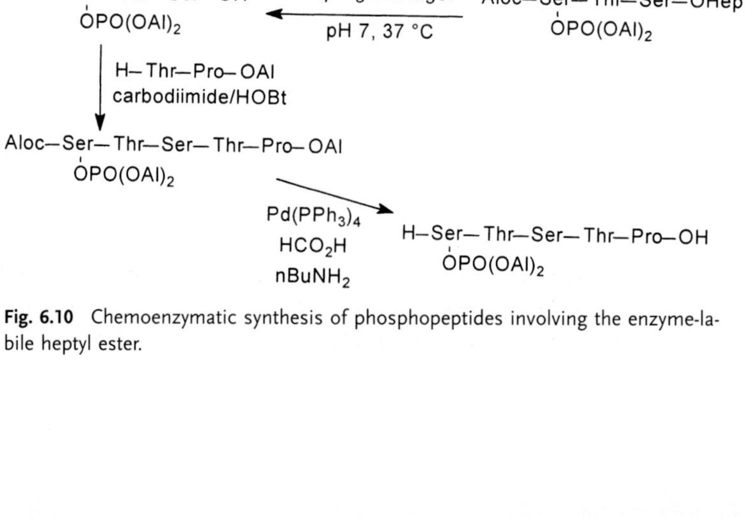

Fig. 6.10 Chemoenzymatic synthesis of phosphopeptides involving the enzyme-labile heptyl ester.

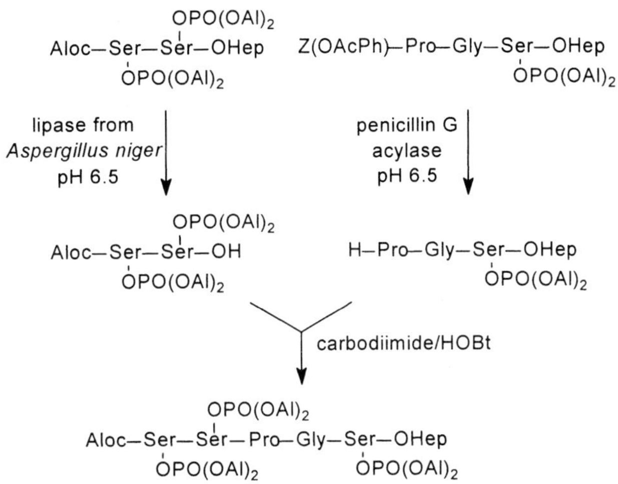

Fig. 6.11 Chemoenzymatic synthesis of phosphopeptides involving the enzyme-labile heptyl ester and Z(OAcPh)-protecting groups.

tecting groups for the C-terminus. The N-terminus may be protected with an allyl-type group (Aloc) or, alternately, with an enzyme-labile protecting group such as Z(OAcPh) (Fig. 6.11).

Treatment with piperidine not only leads to β-elimination, but may also cleave one alkyl group from dimethyl- or dibenzyl-protected phosphotyrosine. This dealkylation can be reduced by applying an alternative Fmoc deprotection reagent such as 2% 1,8-diazabicyclo[5.4.0]undec-7-ene (DBU) in dimethylformamide (DMF).

Most coupling reagents are compatible with phosphorylated building blocks. If the phosphate group is unprotected, then amino acid active esters may be used. Benzyl and *tert*-butyl protecting groups on the phosphate moiety are labile to TFA-based cleavage conditions typically used in Fmoc tactics. Treatment of pTyr-containing peptides with liquid HF causes significant dephosphorylation. This also holds true for phosphoserine/phosphothreonine-containing peptides and treatment with HBr/AcOH. Special procedures must be applied when phosphorylated peptides are to be synthesized according to Boc tactics [101].

Several nonhydrolyzable phosphotyrosine mimetics have been developed and applied to peptide synthesis [102] (Fig. 6.12).

6.5 Lipopeptides

Membrane-bound proteins are frequently modified covalently by lipid residues (cf. Section 3.2.2.7) [73, 103, 104]. G-protein-coupled receptors are S-palmitoylated at a C-terminal cysteine residue. The α-subunits of heterotrimeric G-proteins and nontyrosine receptor kinases contain N-myristoylated N-terminal glycine residues

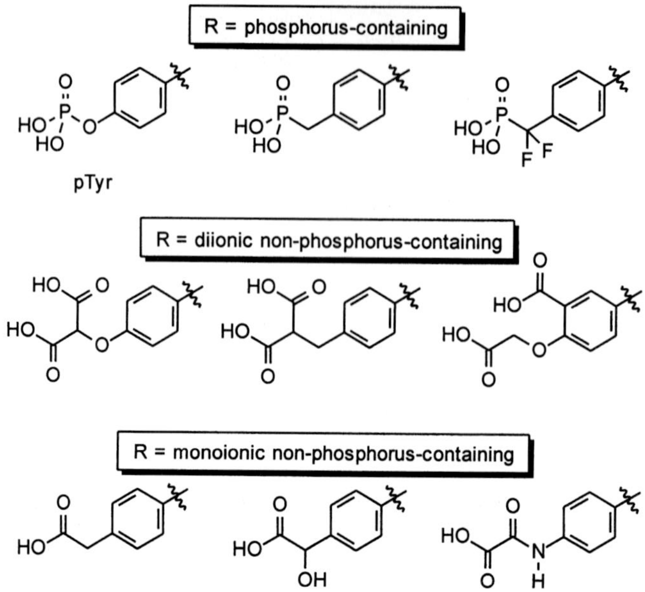

Fig. 6.12 Phosphotyrosine (pTyr) mimetics.

together with S-palmitoylation of a neighboring cysteine residue. The γ-subunits of G-proteins are S-farnesylated or S-geranyl-geranylated at cysteine residues.

The lipid moieties are necessary to recruit and anchor proteins to the membrane. Furthermore, it has been postulated that lipidation of proteins represents an event involved in signal transduction [105].

Lipid-modified peptides can be used as a tool in order to study the biosynthesis and processing of lipid-modified proteins, as well as their biological role. A single N-myristoylation or S-farnesylation is not sufficiently hydrophobic to achieve stable membrane attachment of peptides. Only double or multiple lipid modifications lead to stable association of the peptides to membranes [106–108].

Lipid-modified peptides are chemically very sensitive; the thiol ester moiety present in S-palmitoyl derivatives hydrolyzes spontaneously at pH >7 in aqueous solution, and β-elimination might also be problematic during synthesis. The alkene groups of farnesyl residues, for example, may undergo side reactions upon treatment with acids, and consequently all coupling and deprotection reactions in the synthesis of lipid-modified peptides must be carried out under very mild conditions [109].

Enzyme-labile protection groups are especially appropriate in the synthesis of these highly sensitive derivatives (cf. Section 4.2.5).

Synthesis of the S-palmitoylated and S-farnesylated lipohexapeptide of the C-terminus of human N-Ras protein could be accomplished by the application of a choline ester as carboxy-protecting group, which can be cleaved by butyrylcholine esterase at near-neutral pH, without any risk of β-elimination.

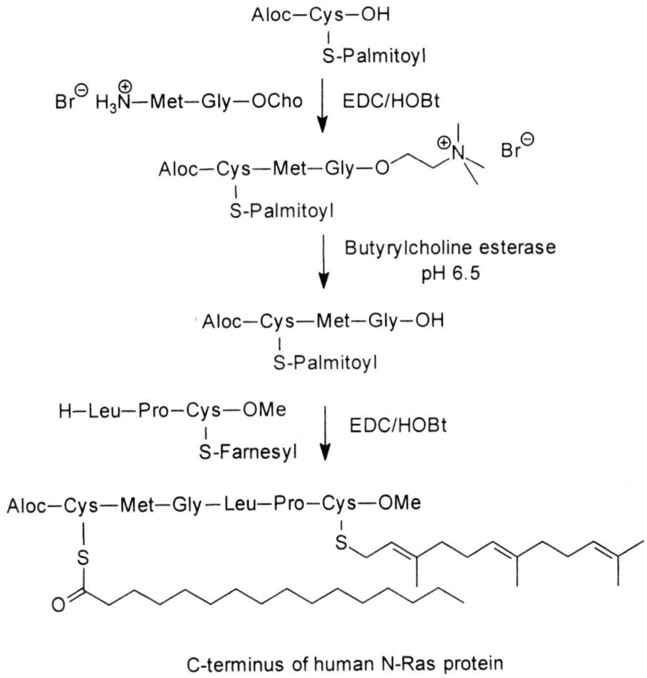

Fig. 6.13. Chemoenzymatic synthesis of lipidated peptides enzyme-labile choline esters. OCho = choline

Classical methods can be applied only to a limited extent to the assembly of such sensitive molecules. If only S-palmitoylation of cysteine residues is required in the target peptide, then Boc tactics may be used as the thioester is acid stabile. On the other hand, lipopeptides containing only the acid-labile S-farnesyl cysteine can be synthesized using modified Fmoc tactics. In addition, allyl-type protecting groups have proven useful in the synthesis of sensitive derivatives [106, 110].

6.6 Sulfated Peptides

The chemical synthesis of peptides containing tyrosine-O-sulfate (cf. Section 3.2.2.9) is a difficult task for peptide chemistry, mainly because the sulfate moiety is intrinsically labile. When using Fmoc tactics, the final cleavage and deprotection with acid represents the critical step in the synthesis because of the acid sensitivity of the tyrosine sulfate.

The solution synthesis of tyrosine-O-sulfate-containing peptides, applying Z tactics, was described during the early 1980s [111, 112].

The acid lability of amino acid O-sulfate esters has been evaluated [113, 114]. Post-assembly sulfatation has also been reported as an alternative to the building block approach, where barium or sodium salts of suitably N-protected tyrosine-O-sulfate are used, though the former strategy suffers greatly from the nonselectivity of the sulfation reaction. The N-terminus, as well as the side chains of Lys, Ser, and Thr, must be correctly protected, and the synthesis of peptides containing multiple tyrosine sulfate residues remains one of the most challenging problems in peptide chemistry [115]. Tyrosine may be sulfated using chlorosulfonic acid/pyridine, $Me_3N \cdot SO_3$ or $DMF \cdot SO_3$ complex.

Analogues of the C-terminal nonapeptide amide cholecystokinin-(2–33) have been synthesized in solution using Z as the temporary protecting group, together with side-chain protection of the *tert*-butyl type, with the tyrosine-*O*-sulfate building block being used as the barium salt [111]. The 34-peptide human big gastrin-2, as well as CCK-33, and CCK-39 have recently been synthesized by Fmoc-based solid-phase synthesis on 2-chlorotrityl resin with Fmoc-Tyr(SO_3Na)-OH as the tyrosine-O-sulfate building block. Final cleavage/deprotection with TFA is possible; this does not cause destruction of the O-sulfate group if the reaction mixture temperature is kept low, and no sulfur-containing scavengers are added [114].

6.7 References

1 S.A. Kates, N.A. Solé, F. Albericio, G. Barany in: *Peptides: Design, Synthesis, and Biological Activity*, C. Basava, G.M. Anantharamaiah (Eds.), Birkhäuser, Boston, **1994**, 39

2 V.J. Hruby, F. Al-Obeidi, W. Kazmierski, *Biochem. J.* **1990**, *268*, 249

3 U. Schmidt, *Pure Appl. Chem.* **1986**, *58*, 295

4 Y.A. Ovchinnikov, V.T. Ivanov in: *The Proteins*, H. Neurath, R.L. Hill (Eds.), Academic Press, New York, **1982**, 307

5 J. Rizo, L.M. Gierasch, *Annu. Rev. Biochem.* **1992**, *61*, 387

6 R. Haubner, D. Finsinger, H. Kessler, *Angew. Chem.* **1997**, *109*, 1440; *Angew. Chem. Int. Ed. Engl.* **1997**, *36*, 1374

7 H. Kessler, *Angew. Chem.* **1982**, *94*, 509; *Angew. Chem. Int. Ed. Engl.* **1982**, *21*, 512

8 J. Boer, D. Gottschling, A. Schuster, M. Semmrich, B. Holzmann, H. Kessler, *J. Med. Chem.* **2001**, *44*, 2586

9 N. Izumiya, T. Kato, M. Waki, *Biopolymers* **1981**, *20*, 1785

10 R. Schmidt, K. Neubert, *Int. J. Peptide Protein Res.* **1991**, *37*, 502

11 N.L. Benoiton, Y.C. Lee, R. Steinauer, F.M.F. Chen, *Int. J. Peptide Protein Res.* **1992**, *40*, 559

12 A.K. Szardenings, T.S. Burkoth, H.H. Lu, D.W. Tien, D.A. Campbell, *Tetrahedron* **1997**, *53*, 6573

13 Y. Hamuro, W. J. Marshall, M.A. Scialdone, *J. Comb. Chem.* **1999**, *1*, 163

14 E. Giralt, R. Eritja, J. Josa, C. Kuklinski, E. Pedroso, *Synthesis* **1985**, 181

15 M. Rothe, J. Haas in: *Peptides 1990*, E. Giralt, D. Andreu (Eds.), Escom, Leiden, **1991**, 212

16 T. Rückle, P. De Lavallaz, M. Keller, P. Dumy, M. Mutter, *Tetrahedron* **1999**, *55*, 11281

17 D. Besser, R. Olender, R. Rosenfeld, O. Arad, S. Reissmann, *J. Peptide Res.* **2000**, *56*, 337

18 M. El Haddadi, F. Cavelier, E. Vives, A. Azmani, J. Verducci, J. Martinez, *J. Peptide Sci.* **2000**, *6*, 560

19 U. Schmidt, A. Lieberknecht, H. Griesser, J. Talbiersky, *J. Org. Chem.* **1982**, *47*, 3261

20 U. Schmidt, M. Kroner, H. Griesser, *Tetrahedron Lett.* **1988**, *29*, 3057

21 T. Shioiri, K. Ninomiya, S. Yamada, *J. Am. Chem. Soc.* **1972**, *94*, 6203

22 A. Ehrlich, H.-U. Heyne, R. Winter, M. Beyermann, H. Haber, L. A. Carpino, M. Bienert, *J. Org. Chem.* **1996**, *61*, 8831

23 F. Albericio, J.M. Bofill, A. El-Faham, S.A. Kates, *J. Org. Chem.* **1998**, *63*, 9678

24 G. Barany, R.B. Merrifield in: *The Peptides: Analysis, Synthesis, Biology*, Volume 2, E. Gross, J. Meienhofer (Eds.), Academic Press, New York, **1980**, 1

25 L.T. Scott, J. Rebek, L. Ovsyanko, L. Sims, *J. Am. Chem. Soc.* **1977**, *99*, 626

26 A.W. Mott, U. Slomczynska, G. Barany in: *Forum Peptides le Cap d'Agde*, B. Castro, J. Martinez (Eds.), Les Impressions Dohr, Nancy, **1984**, 321

27 S. Mazur, P. Jayalekshmy, *J. Am. Chem. Soc.* **1979**, *101*, 677

28 A. Müller, F. Schumann, M. Koksch, N. Sewald, *Lett. Peptide Sci.* **1997**, *4*, 275

29 J.S. McMurray, *Tetrahedron Lett.* **1991**, *32*, 7679

30 S.A. Kates, N.A. Solé, C.R. Johnson, D. Hudson, G. Barany, F. Albericio, *Tetrahedron Lett.* **1993**, *34*, 1549

31 M. Cudic, J.D. Wade, L. Otvos, Jr., *Tetrahedron Lett.* **2000**, *41*, 4527

32 P. Romanovskis, A.F. Spatola, *J. Peptide Res.* **1998**, *52*, 3565

33 J. Alsina, F. Rabanal, E. Giralt, F. Albericio, *Tetrahedron Lett.* **1994**, *35*, 9633

34 J. Alsina, C. Chiva, M. Ortiz, F. Rabanal, E. Giralt, F. Albericio, *Tetrahedron Lett.* **1997**, *38*, 883

35 G. Sabatino, M. Chelli, S. Mazzucco, M. Gianneschi, A.M. Papini, *Tetrahedron Lett.* **1999**, *40*, 809

36 S.S. Isied, C.G. Kuehn, J.M. Lyon, R.B. Merrifield, *J. Am. Chem. Soc.* **1982**, *104*, 2632

37 A. Trzeciak, W. Bannwarth, *Tetrahedron Lett.* **1992**, *33*, 4557

38 J. Alsina, K.J. Jensen, F. Albericio, G. Barany, *Chem. Eur. J.* **1999**, *5*, 2787

39 G. Osapay, A. Profit, J.W. Taylor, *Tetrahedron Lett.* **1990**, *31*, 6121

40 M. Xu, N. Nishino, H. Mihara, T. Fujimoto, N. Izumiya, *Chem. Lett.* **1992**, 191

41 L.S. Richter, J.Y.K. Tom, J.P. Brunier, *Tetrahedron Lett.* **1994**, *35*, 5547

42 N. Nishino, M. Xu, H. Mihara, T. Fujimoto, Y. Ueno, H. Kumagai, *Tetrahedron Lett.* **1992**, *33*, 1479

43 L. Yang, G. Morriello, *Tetrahedron Lett.* **1999**, *40*, 8197

44 L. Zhang, J.P. Tam, *J. Am. Chem. Soc.* **1997**, *119*, 2363

45 Y. Shao, W. Lu, S.B.H. Kent, *Tetrahedron Lett.* **1998**, *39*, 3911

46 M.A. Marahiel, T. Stachelhaus, H.D. Mootz, *Chem. Rev.* **1997**, *97*, 2651

47 J.W. Tranger, R.M. Kohll, H.D. Mootz, M.A. Marahiel, C.T. Walsh, *Nature* **2000**, *407*, 215

48 C.P. Scott, E. Abel-Santos, M. Wall, D.C. Wahnon, S.J. Benkovic, *Proc. Natl. Acad. Sci. USA* **1999**, *96*, 13638

49 G. Gellerman, A. Elgavi, Y. Salitra, M. Kramer, *J. Peptide Res.* **2001**, *57*, 277

50 J.N. Lambert, J.P. Mitchell, K.D. Roberts, *J. Chem. Soc., Perkin Trans. I* **2001**, 471

51 J.F. Reichwein, C. Versluis, R. Liskamp, *J. Org. Chem.* **2000**, *65*, 6187

52 J. Pernerstorfer, M. Schuster, S. Blechert, *J. Chem. Soc., Chem. Commun.* **1997**, 1949

53 M. Lebl, V.J. Hruby, *Tetrahedron Lett.* **1984**, *25*, 2067

54 V. Cavallaro, P. Thompson, M. Hearn, *J. Peptide Sci.* **1998**, *4*, 335

55 S. Plane, *Int. J. Peptide Protein Res.* **1990**, *35*, 510

56 F.A. Robey, R.L. Fields, *Anal. Biochem.* **1989**, *177*, 373

57 K.D. Roberts, J.N. Lambert, N.J. Ede, A. M. Bray, *Tetrahedron Lett.* **1998**, *39*, 8357

58 L. Yu, Y. Lai, J.V. Wade, S.M. Coutts, *Tetrahedron Lett.* **1998**, *39*, 6633

59 P.W. Schiller, T.M.-D. Nguyen, J. Miller, *Int. J. Peptide Protein Res.* **1985**, *31*, 231

60 P. Grieco, P.M. Gitu, V.J. Hruby, *J. Peptide Res.* **2001**, *57*, 250

61 P.W. Schiller, T.M.-D. Nguyen, J. Miller, *Int. J. Peptide Protein Res.* **1985**, *25*, 171

62 A.M. Felix, C.-T. Wang, E.P. Heimer, A. Fournier, *Int. J. Peptide Protein Res.* **1988**, *31*, 231

63 D. Andreu, F. Albericio, N.A. Solé, M.C. Munson, M. Ferrer, G. Barany

in: *Peptide Synthesis Protocols*, M.W. Pennington, B.M. Dunn (Eds.), Humana Press, Totowa, **1994**
64 P. Lloyd-Williams, F. Albericio, E. Giralt, *Chemical Approaches to the Synthesis of Peptides and Proteins*, CRC Press, Boca Raton, **1997**
65 F. Albericio, I. Annis, M. Royo, G. Barany in: *Fmoc Solid Phase Peptide Synthesis: A Practical Approach*, W.C. Chan, P.D. White (Eds.), Oxford University Press, Oxford, **2000**
66 M. Pons, F. Albericio, M. Royo, E. Giralt, *Synlett* **1999**, 172
67 D. Besse, L. Moroder, *J. Peptide Sci.* **1997**, *3*, 442
68 J. Ottl, L. Moroder, *Tetrahedron Lett.* **1999**, *40*, 1487
69 P. Sieber, B. Kamber, A. Hartmann, A. Jöhl, B. Riniker, W. Rittel, *Helv. Chim. Acta* **1974**, *57*, 2617
70 J.A. Camarero, E. Giralt, D. Andreu, *Tetrahedron Lett.* **1995**, *36*, 1137
71 B.D. Livingston, E.M.D. Robertis, J.C. Paulson, *Glycobiology* **1990**, *1*, 39
72 O. Seitz, I. Heinemann, A. Mattes, H. Waldmann, *Tetrahedron* **2001**, *57*, 2247
73 M. Meldal, P.M. St Hilaire, *Curr. Opin. Chem. Biol.* **1997**, *1*, 552
74 K.M. Koeller, C.-H. Wong, *Chem. Rev.* **2000**, *100*, 4465
75 J.B. Jones, *Synlett* **1999**, 1495
76 C.M. Taylor, *Tetrahedron* **1998**, *54*, 11317
77 O. Seitz, *Chembiochem.* **2001**, *1*, 214
78 T. Bielfeldt, S. Peters, M. Meldal, K. Bock, H. Paulsen, *Angew. Chem.* **1992**, *104*, 881; *Angew. Chem. Int. Ed. Engl.* **1992**, *31*, 857
79 F.E. McDonald, S.J. Danihefsky, *J. Org. Chem.* **1992**, *57*, 7001
80 S.T. Cohen-Anisfeld, D.T. Langbury, Jr., *J. Am. Chem. Soc.* **1993**, *115*, 10531
81 H. Kunz, *Pure Appl. Chem.* **1993**, *65*, 1223
82 A.L. Handlon, B. Fraser-Reid, *J. Am. Chem. Soc.* **1993**, *115*, 3796
83 P. Sjölin, J. Kihlberg, *J. Org. Chem.* **2001**, *66*, 2957
84 O. Seitz, C.-H. Wong, *J. Am. Chem. Soc.* **1997**, *119*, 8766
85 T. Pol, H. Waldmann, *J. Am. Chem. Soc.* **1997**, *119*, 6702
86 H. Vegad, C.J. Gray, P.J. Somers, A. S. Dutta, *J. Chem. Soc., Perkin Trans. I* **1997**, 1429
87 K. von dem Bruch, H. Kunz, *Angew. Chem.* **1994**, *106*, 87; *Angew. Chem. Int. Ed. Engl.* **1994**, *33*, 101
88 C.-H. Wong, M. Schuster, P. Wang, P. Sears, *J. Am. Chem. Soc.* **1993**, *116*, 5893
89 M. Schuster, P. Wang, J.C. Paulson, C.-H. Wong, *J. Am. Chem. Soc.* **1994**, *116*, 1135
90 C.-H. Wong, R.L. Halcomb, Y. Ichikawa, T. Kajimoto, *Angew. Chem.* **1995**, *107*, 569; *Angew. Chem. Int. Ed. Engl.* **1995**, *34*, 521
91 H. Kunz, C. Unverzagt, *Angew. Chem.* **1984**, *96*, 426; *Angew. Chem. Int. Ed. Engl.* **1984**, *23*, 436
92 W. Kosch, J. März, H. Kunz, *React. Polym.* **1994**, *22*, 181
93 O. Seitz, H. Kunz, *Angew. Chem.* **1995**, *107*, 901; *Angew. Chem. Int. Ed. Engl.* **1995**, *34*, 803
94 H. Herzner, T. Reipen, M. Schultz, H. Kunz, *Chem. Rev.* **2000**, *100*, 4495
95 J.S. McMurray, D.R. Coleman, IV, W. Wang, M.L. Campbell, *Biopolymers* **2001**, *60*, 3
96 H.-G. Chao, B. Leiting, P.D. Reiss, A.L. Burkhardt, C.E. Klimas, J.B. Bolen, G.R. Matsueda, *J. Org. Chem.* **1995**, *60*, 7710
97 T. Wakamiya, K. Saruta, J. Yasuoka, S. Kusunoto, *Chem. Lett.* **1994**, 1099
98 T. Vorherr, W. Bannwarth, *Bioorg. Med. Chem. Lett.* **1995**, *5*, 2661
99 E.A. Ottinger, Q. Xu, G. Barany, *Peptide Res.* **1996**, *9*, 223
100 J.W. Perich, N.J. Ede, S. Eagle, A.M. Bray, *Lett. Peptide Sci.* **1999**, *6*, 91
101 A. Otaka, K. Miyoshi, M. Kaneko, H. Tamamura, N. Fujii, M. Nomizu, T.R. Burke, P.P. Roller, *J. Org. Chem.* **1995**, *60*, 3967
102 T.R. Bruke, Jr., Z.-J. Yao, D.-G. Liu, J. Voigt, Y. Gao, *Biopolymers* **2001**, *60*, 32
103 P.J. Casey, *Science* **1995**, *268*, 221
104 G. Milligan, M. Parenti, A.I. Magee, *Trends Biochem. Sci.* **1995**, *20*, 181
105 S. Moffet, B. Mouillac, H. Bonin, M. Bouvier, *EMBO J.* **1993**, *12*, 349
106 H. Waldmann, M. Schelhaas, E. Nägele, J. Kuhlmann, A. Wittinghofer,

H. Schröder, J. R. Silvius, *Angew. Chem.* **1997**, *109*, 2334, *Angew. Chem. Int. Ed. Engl.* **1997**, *36*, 2238

107 H. Schröder, R. Leventis, S. Rex, M. Schelhaas, E. Nägele, H. Waldmann, J. R. Silvius, *Biochemistry* **1997**, *36*, 13102

108 H. Schröder, R. Leventis, S. Shahinian, P. A. Walton, J. R. Silvius, *J. Cell Biol.* **1996**, *134*, 647

109 D. Kadereit, P. Deck, I. Heinemann, H. Waldmann, *Chem. Eur. J.* **2001**, *7*, 1184

110 T. Schmittberger, A. Cotté, H. Waldmann, *J. Chem. Soc., Chem. Commun.* **1998**, 937

111 L. Moroder, L. Wilschowitz, M. Gemeiner, W. Göhring, S. Knof, R. Scharf, P. Thamm, J. D. Gardner, T. E. Solomon, E. Wünsch, *Hoppe Seylers Z. Physiol. Chem.* **1981**, *362*, 929

112 E. Wünsch, L. Moroder, L. Wilschowitz, W. Göhring, R. Scharf, J. D. Gardner, *Hoppe Seylers Z. Physiol. Chem.* **1981**, *362*, 143

113 B. Penke, F. Hajnal, J. Lonovics, G. Holzinger, T. Kadar, G. Telegdy, J. Rivier, *J. Med. Chem.* **1984**, *27*, 845

114 K. Kitagawa, C. Aida, H. Fujiwara, T. Yagami, S. Futaki, M. Kogire, J. Ida, K. Inoue, *J. Org. Chem.* **2001**, *66*, 1

115 A. Leppänen, P. Mehta, Y.-B. Ouyang, T. Ju, J. Helin, K. L. Moore, I. van Die, W. M. Canfield, R. P. McEver, R. D. Cummings, *J. Biol. Chem.* **1999**, *27*, 24838

7
Peptide and Protein Design, Pseudopeptides, and Peptidomimetics

Native peptides can be directly applied as pharmacologically active compounds only to a very limited extent. The major disadvantages of the application of a peptide in a biological system – for example, rapid degradation by proteases, hepatic clearance, undesired side effects by interaction of conformationally flexible peptides with different receptors [1–7], and low membrane permeability due to their hydrophilic character – are in most cases detrimental to oral application. Furthermore, most peptides are not able to pass body barriers such as the blood-brain barrier, and also suffer from rapid processing and excretion. The high structural flexibility, especially of linear peptides, has the consequence that the receptor-bound, biologically active conformation represents only one member of a large ensemble of conformers.

Peptide chemistry however may contribute considerably to drug development. The interaction of a peptide or a protein epitope with a receptor or an enzyme is the initial event based on molecular recognition [8], and generally elicits a biological response. Likewise, ligand or substrate binding to a protein relies on the surface complementarity of the binding partners.

Chemically modified peptides with improved bioavailability and metabolic stability may be directly used as drugs (cf. Chapter 9). Alternately, peptidic receptor ligands or enzyme substrates (peptide leads) may serve as starting points for the development of nonpeptide drugs [9]. The first step of any such investigation is to identify the amino acid side-chain residues responsible for receptor-ligand interaction. Subsequently, the topography of these functional (pharmacophoric) groups is reproduced by similar nonpeptidic functionalities on a rigid scaffold [10]. Many efforts have been made to develop peptide-based, pharmacologically active compounds, including peptide modification and the design of peptidomimetics. Whilst modified peptides (by definition) contain nonproteinogenic or modified amino acid building blocks, peptidomimetics (cf. Section 7.3) are nonpeptidic compounds that imitate the structure of a peptide in its receptor-bound conformation and – in the case of agonists – also the biological mode of action on the receptor level.

According to the definition by Ripka and Rich [11], three different types of peptidomimetics may be distinguished:

- Type I: these are peptides modified by amide bond isosteres (cf. Section 7.2.2) and secondary structure mimetics (cf. Section 7.2.4). These derivatives are usually designed to closely match the peptide backbone.
- Type II: these are small nonpeptide molecules that bind to a receptor or enzyme (functional mimetics, cf. Section 7.3). However, despite being often presumed to serve as structural analogues of native peptide ligands, these nonpeptide antagonists often bind to a different receptor subsite and, hence, do not necessarily mimic the parent peptide.
- Type III: these may be regarded as ideal mimetics, because they are nonpeptide compounds and contain the functional groups necessary for the interaction of the native peptide with the corresponding protein (pharmacophoric groups) grafted onto a rigid scaffold.

The design of all three types of peptidomimetics may be assisted by X-ray crystallographic or nuclear magnetic resonance (NMR) data [12], computational de-novo design ("in-silico screening") [13], and combinatorial chemistry (cf. Chapter 8) [14].

Furthermore, the construction of novel bioactive polypeptides and artificial protein structures or miniaturized enzymes currently form the focal point of intensive investigations [15]. Although the secondary structure elements (cf. Section 2.5) of a polypeptide chain with known primary structure can be predicted to limited extent, our knowledge on protein folding is still incomplete. Therefore, reliable predictions on the composition of secondary structure elements to produce a three-dimensionally folded biologically active peptide or protein are, at present, beyond reach.

7.1 Peptide Design

The development of a drug (peptide drug) may start with the identification and design of the peptide sequence essential for biological activity. Such structure-activity relationships for example rely in the first instance on a systematic substitution of each amino acid residue of a native peptide by a simple amino acid such as alanine ("alanine scan"). Studies on the receptor selectivity or on the agonism or antagonism of a peptide hormone require structure-activity relationships to be performed, using many synthetic peptide analogues. In the past, this approach was highly labor-intensive, but today it is much more straightforward following the application of combinatorial chemistry and automation (cf. Chapter 8).

A viable design procedure is outlined in Fig. 7.1. A peptide sequence **1** is identified as being responsible for interaction with a receptor. In the example shown, the peptide contains the sequence Arg-Gly-Asp (RGD) that is known as a universal recognition motif for cell-cell and cell-matrix interactions. Some peptides which contain the RGD sequence in a well-defined conformation efficiently inhibit the binding of extracellular proteins (e.g., fibrinogen, vitronectin) to cellular receptors (integrins). The corresponding protein-receptor interaction is involved, for

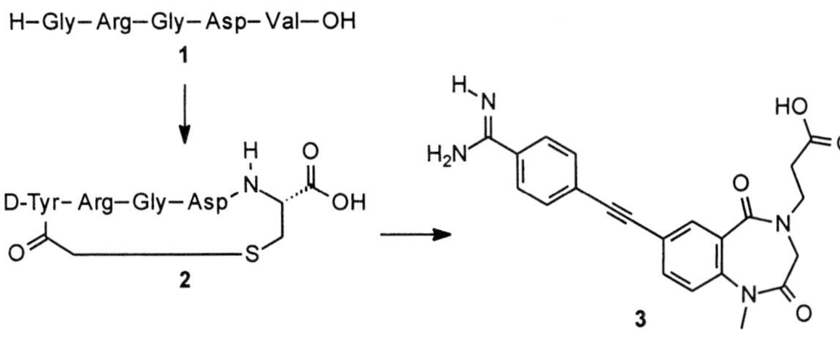

Fig. 7.1 Conversion of a peptide sequence into a nonpeptide lead.

example, in blood platelet aggregation, tumor cell adhesion, angiogenesis, and osteoporosis [16]. The peptide conformation subsequently is constrained, perhaps by cyclization (**2**) or by sterically hindered amino acids (type I peptidomimetics). If the biological activity is retained, it is likely that the receptor-bound conformation is still accessible [17]. In the next step, the three-dimensional array of the pharmacophoric groups can be deduced and a nonpeptide lead compound **3** (type III peptidomimetic) may be constructed, where the pharmacophoric groups are attached to a rigid scaffold and are presented in the appropriate three-dimensional array [18, 19].

In-vitro or in-vivo biological activity is not sufficient for a drug candidate, however, and many further investigations of toxicology, pharmacokinetics (ADME: *a*bsorption, *d*istribution, *m*etabolism, *e*xcretion), pharmacology, and clinical trials are required before a drug candidate is approved by national authorities for the treatment of patients. Considerable effort is needed to develop a biologically active compound (lead compound) as a drug candidate, and finally to register it as a pharmacologically active drug that can be marketed. Vast research expense, long development times, and high risks are imposed on innovative drugs where, in contrast to the so-called "me-too" products, no other substances with a similar mode of action are being marketed concurrently. In most cases, the development pathway is aimed towards a specific drug, and starts with a pathological phenomenon (biological target). Extensive biochemical and medical knowledge is required to understand the physiological and pathological events involved, and in this respect the recent advances made in biochemistry, molecular biology, gene technology, protein purification, protein crystallography, protein NMR, genomics, proteomics, and computational methods will provide the necessary platform for drug development.

Originally, the term "drug design" was coined for the rational design of a molecule that interacts appropriately on a structural or mechanistic basis with the target (enzyme, receptor, or DNA sequence) in a complementary manner, and that ultimately exerts the desired pharmacological effect by activation or blocking of the corresponding target. The complex interactions in drug design are shown schematically in Fig. 7.2 [20].

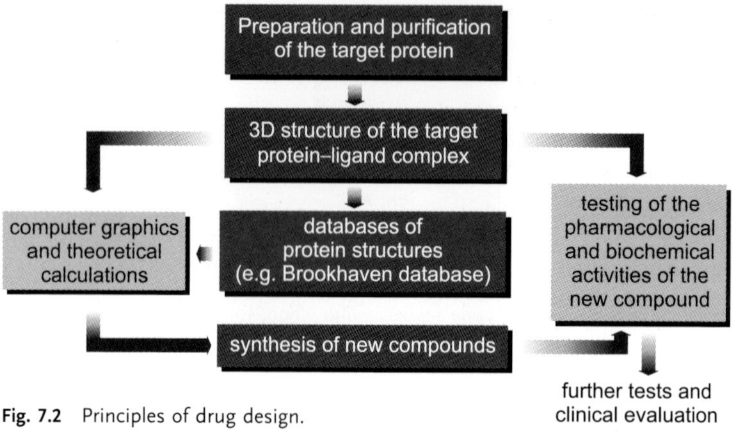

Fig. 7.2 Principles of drug design.

Determination of the three-dimensional (3D) structure of the target protein is a crucial precondition for the rational design of small molecule agonists, antagonists or enzyme inhibitors [21]. In many cases, protein engineering facilitates structure determination because the corresponding protein can be obtained in larger amounts by overexpression (cf. Section 4.6.1). Methods of gene technology may also provide single domains of a protein for structure determination by X-ray crystallography or NMR methods. The 3D structure often results in suggestions for site-directed mutagenesis, and/or the exchange of single or multiple amino acid residues using molecular biology methods in order for example to optimize the stability, substrate specificity, or pH-optimum of an enzyme-catalyzed reaction.

Methods of molecular modeling are used to visualize the structures of receptors and peptide drugs on a computer, and then to simulate their interaction. However, as the torsion angles of the peptide backbone and of the side-chain residues may adopt different values, numerous conformations usually result for an unbound peptide ligand that are not biased for the most efficient conformation with respect to ligand-receptor interaction. Force-field calculations have been developed as a strategy to predict conformations by comparison of their relative potential energy [22]. The single atoms of a protein are considered as hard spheres, and classical mechanics calculations are used to simulate their time-dependent motion. Even solvent molecules may be explicitly included.

Statistical methods (Monte Carlo methods) may also be used to produce a huge number of conformations and to calculate subsequently their potential energy in the local minima [23]. Finally, the conformation with the lowest energy or a family of low-energy conformations are further investigated. This procedure, which initially is based on statistics, can be supplemented by molecular dynamics calculations and may result in an evaluation of the conformational flexibility of molecular structures by force-field methods [22, 24]. In computer-aided molecular design (CAMD) [25], the complementary fitting of a peptide or nonpeptide drug to a receptor plays a crucial role, though sensible application of this concept requires knowledge of the 3D structure of the receptor. During recent years much effort has been imposed on developing new methods and algorithms (e.g., neuronal network methods, genetic algorithms, machine learning, and graph theoretical methods) in order to predict molecular structures [25–28].

If the site of action of a biologically active compound is known, then direct computer-aided drug design is used [29]. Tailor-made molecular structures are constructed to optimally fulfil the requirements for binding to the receptor or the active site of an enzyme [30]. In this context, protein–ligand docking is an important method, and many docking programs based on a series of searching algorithms such as surface complementary matching [31, 32], fragment growing [33, 34], random sampling (Monte Carlo), simulated annealing [35–37], and genetic algorithms [38, 39] have been developed.

One NMR-based method has been described to identify and optimize small organic molecules binding to proximal subsites of a single protein in order to produce from these moieties ligands with high affinity to the protein. Di- or multivalent protein-ligand interactions are obtained by covalently linking these pharmacophoric groups. This approach experimentally provides structure-activity relationship data, as obtained from NMR [40,41]. The method relies on ^{15}N-labeled proteins for instance, because changes of the ^{15}N or ^{1}H amide chemical shift are observed in ^{15}N heteronuclear single quantum correlation spectra (^{15}N HSQC) upon addition of a ligand to the labeled protein (Fig. 7.3).

If no structural data of a target protein are available, the known 3D structure of another, sequence-related protein may be used as the starting point for the so-called homology modeling.

Indirect computer-aided drug design (indirect CADD) is a very useful protocol in the process of rational drug design when the molecular structure of the target is unknown. It allows for the evaluation of structural changes upon modification of a lead structure. Systematic substituent variation of a lead structure is correlated with biological data, and this results in the identification of crucial structural elements of the ligand that are required for high-affinity receptor binding (receptor mapping). This finally allows for the creation of a hypothetical receptor model, and is intensively applied in modern drug design.

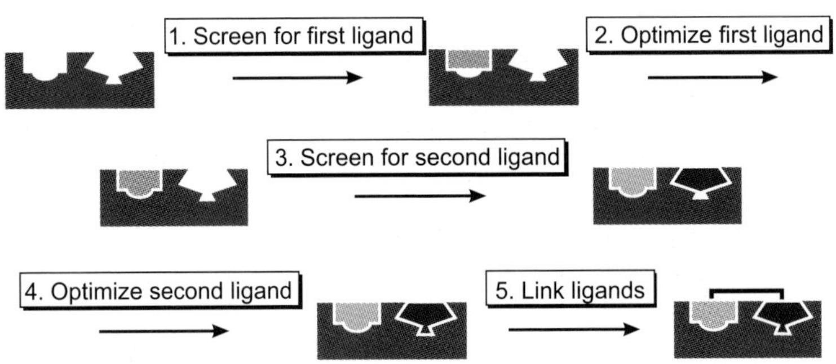

Fig. 7.3 Optimization process in establishing structure-activity relationships (SAR) by NMR.

The synthesis of chemically modified peptides often employs the incorporation of conformationally constrained amino acids (proline, proline analogues, N-methyl amino acids, C^{α}-dialkyl amino acids, etc.). This type of modification results in rather rigid peptide analogues where the conformational freedom is reduced. A further viable route for the design of peptides with reduced conformational freedom relies on peptide cyclization (cf. Section 6.1).

The design of a peptide depends primarily not only on the targeted application, but also on synthetic considerations [42, 43]. In general, several steps are necessary to obtain a peptide suitable for biochemical or biomedical applications, and several optimization cycles are often required. There is no such thing as a general routes for the design, although the procedure is guided by the desired function.

Conformational analysis is a valuable tool in the design of new peptide drugs. Linear peptide chains are usually characterized by high flexibility which renders determination of the 3D structure nontrivial. The basics of conformational analysis were established by Ramachandran et al. [44], Scheraga et al. [45], and Blout et al. [46]. These authors established basic concepts for the classification and description of peptide structures (cf. Section 2.5), and contributed extensively to an improved understanding of peptide and protein conformation and of protein-folding processes. Conformational transitions in polypeptides have been studied using several different methods, for example by Goodman et al. [47]. Cyclic peptides offer the advantage of limited conformational freedom, and this makes them particularly suited for conformational studies. Furthermore, many cyclic peptides (Section 6.1) display interesting biological profiles of activity. Peptide conformations can be very efficiently analyzed using NMR spectroscopy (Section 2.5.3) as reviewed in [48, 49].

Sequence-based peptide design relies not only on the deletion, the exchange, or the modification of amino acid-building blocks, but also on structural elements of the peptide backbone. Besides the 21 proteinogenic amino acids (including selenocysteine), at present more than 1000 nonproteinogenic amino acids are known. Some of these occur naturally in the free form, or are bound in peptides.

The truncation of a sequence by deletion of amino acid-building blocks that practically do not contribute to biological activity is an important approach in the design of pharmacologically active peptides starting from longer bioactive peptides. Many peptide hormones and also other bioactive peptides very often contain a minimum sequence responsible for binding to the receptor and the biological activity. Of course, it is sometimes not very easy to identify the essential partial sequence. Often, the spatial arrangement of those residues responsible for activity depends on the 3D peptide structure where non-neighboring amino acids, with respect to the primary structure, come into close vicinity by peptide folding. By contrast, many peptides are known where the residues essential for the biological activity are localized in well-conserved partial sequences. Messenger segments, responsible for the triggering of physiological events, are located C-terminally in the tachykinins. The C-terminal N^{α}-acetylated 8-peptide sequence Ac-His-Trp-Ala-Val-Gly-His-Leu-Met-NH_2 of gastrin-releasing peptide displays higher relative activity compared to the native sequence. Furthermore, this C-terminal sequence is highly homologous to the 14-peptide bombesin isolated from frog skin. Melanin-concentrating hormone (MCH) is a 17-peptide with a disulfide bridge between Cys^5 and Cys^{14}; only the partial sequence 5–15 is necessary for full biological activity, with Trp^{15} being of eminent importance for interaction with the MCH receptor.

7.2 Modified Peptides

7.2.1 Side-Chain Modification

The most straightforward approach for peptide modification is to introduce changes into the side chains of single amino acids. On this level, a multitude of possibilities for the synthesis of nonproteinogenic amino acids already exists, and useful preparative routes for the asymmetric synthesis of many derivatives have been developed. This strategy allows for the incorporation of amino acids with side chains that do not occur naturally in peptides or proteins, with the aim being to introduce special functional groups, to restrict the conformational flexibility of a peptide, or to enhance its metabolic stability. Furthermore, D-configured amino acids, N^{α}-alkylated amino acids, or C^{α}-dialkylated amino acids may be employed.

Stereochemical substitutions provide information about the steric requirements in the formation of distinct secondary structures and their influence on peptide-receptor interaction. Furthermore, the incorporation of D-amino acids usually confers higher metabolic stability on the peptide. [1-deamino,D-Arg^8]Vasopressin (DDAVP) **4** is a modified peptide applied as a selective antidiuretic agent with long-term activity suitable for the treatment of diabetes insipidus. DDAVP acts as a V_2-receptor agonist in the kidney, with the missing α-amino group increasing proteolytic stability. According to investigations of Manning et al. [50], the exchange of Gln^4 by Val results in the potent V_2-receptor agonist [Val^4]vasopressin **5** that simultaneously is a weak V_2-receptor antagonist.

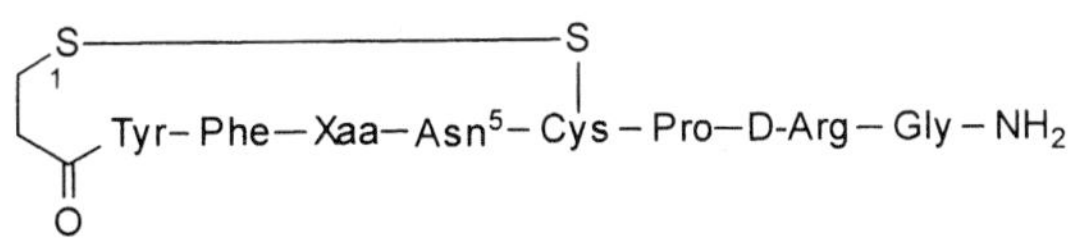

4 Xaa = Gln
5 Xaa = Val

Isosteric and other substitutions similarly provide valuable information on peptide-receptor interactions. Often, methionine is exchanged in this context by norleucine (Nle); indeed, such a substitution in the 14-peptide α-MSH resulted in [Nle^4]α-MSH displaying two-fold increased activity compared to the native peptide. Oxytocin modification through an exchange of Tyr^2 by Phe resulted in only weak agonist activity, while [4-$MePhe^2$]oxytocin is an antagonist.

Many substitutions have been performed in gonadoliberin (gonadotropin-releasing hormone, GnRH) that can be used either for diagnostic purposes or as a drug against infertility. Interestingly, GnRH superagonist analogues result in receptor down-regulation. Such long-term activity reduces LH and FSH liberation, and in males this provides therapy for prostate carcinoma.

Conformational and topographical restrictions are particularly suited as manipulations for peptide design targeted towards an increase of receptor selectivity, metabolic stability, and the development of highly potent agonists or antagonists. The incorporation of N-methyl amino acids influences the *cis/trans* ratio of the peptide bond by lowering the relative energy for the *cis*-isomer. Consequently, the torsion angle ω can also be influenced by the incorporation of special amino acids. Pseudoprolines **6** are synthetic proline analogues that can be obtained by a cyclocondensation reaction of the amino acids cysteine, threonine or serine with aldehydes or ketones [51]. These derivatives, which have been mentioned in the context of the synthesis of difficult sequences (cf. Section 4.5.4.3), also influence the *cis/trans* peptide bond ratio [52, 53]. The introduction of different substituents at C^2 of the pseudoprolines can be used to predetermine the *cis/trans* peptide bond ratio. In particular, the C^2-disubstituted analogues **6** ($R^3 = R^4 = CH_3$) induce up to 100% *cis*-configured peptide bonds in di- or tripeptides [52–54].

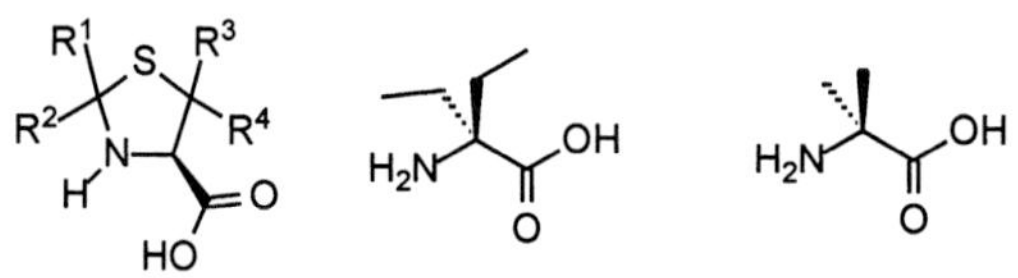

6 **7** **8**

The influence of certain naturally occurring amino acids on the secondary structure of proteins and peptides has been discussed in Chapter 2 (Section 4). This information on the conformational bias of certain amino acids can be utilized to guide the design of peptide analogues. Besides their ability to influence the *cis/trans* peptide bond ratio, N-alkyl-substituted amino acids can favor the formation of turn structures because they very often occur in position i+2 of a β-turn [55]. Similarly, D-amino acids may also be used for the rational design of peptides with defined secondary structure because they stabilize a βII′-turn where they occur in position i+1. The same position i+1 is often favored by proline in βI- or βII-turns. In these cases, a *trans*-peptide bond is involved. Proline with a *cis*-peptide bond usually occupies position i+2 in a βVIa- or βVIb-turn; this situation is also often observed for N-alkyl amino acids.

The dipeptide sequence D-Pro-Gly especially favors a β-turn, and also promotes hairpin nucleation [56]. Strategies for the stabilization of peptide conformations by secondary structure mimetics will be discussed in Section 7.2.4.

The [*N*-MeNle^3]CCK-8 analogue displays higher specificity for the CCK-B receptor in comparison to the CCK-A receptor. This preference is due to a *cis*-peptide bond present in the [Nle^3]CCK-8 analogue, but not in the [N-Me-Nle^3]CCK-8 derivative as has been shown by NMR spectroscopy [57].

Some C$^\alpha$-dialkyl amino acids, such as diethylglycine (Deg) **7**, α-aminoisobutyric acid (Aib) **8**, and isovaline (Iva, 2-amino-2-methylbutyric acid) occur naturally. These derivatives are often incorporated into peptides to investigate the conformational requirements of receptors [58]. They also play an important role as building blocks for the stabilization of short peptides in a well-defined conformation, depending on the nature of the two substituents attached to C$^\alpha$-carbon. 3_{10}-Helices are stabilized by the incorporation of Aib and other C$^\alpha$-dialkyl-substituted amino acids [59, 60].

Compounds **9–13** are constrained derivatives of phenylalanine. The introduction of a sterically demanding functional group at C$^\beta$, as in **9**, mainly constrains conformations around the side-chain dihedral angle χ^1. This angle, in combination with the backbone dihedrals ϕ, ψ, and ω, determines the 3D positioning of side-chain functional groups (e.g., as pharmacophores). Consequently, amino acids with additional substituents at C$^\alpha$, C$^\beta$, or cyclic amino acids are suitable building blocks for the introduction of a restriction in "χ-space" [9, 61, 62].

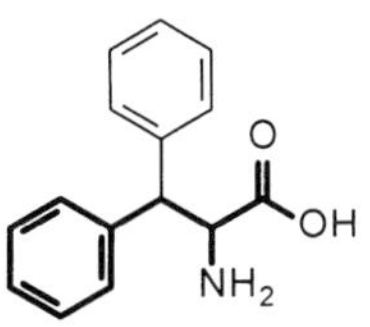

9

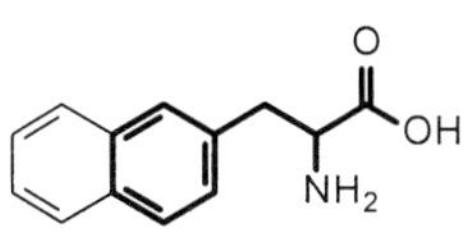

10

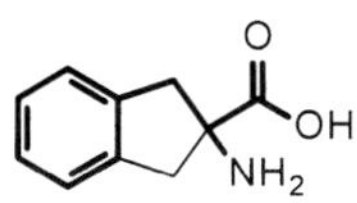

11

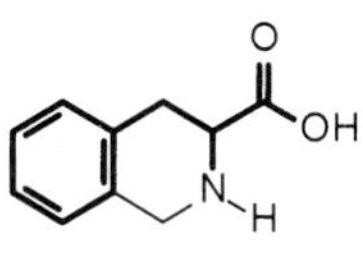

12

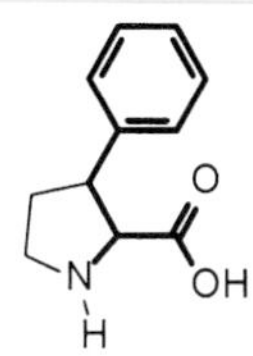

13

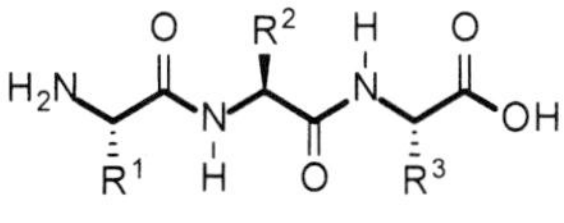

14
(unmodified sequence)

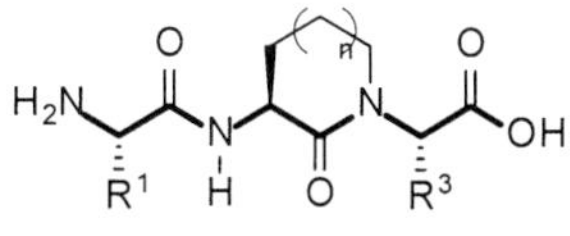

15

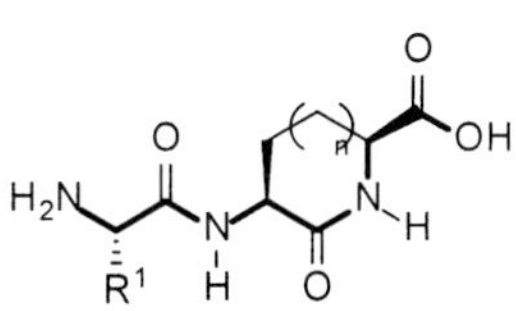

16

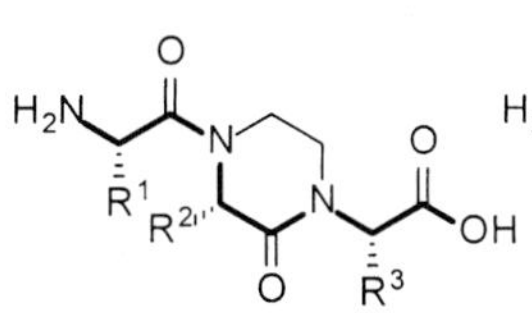

17

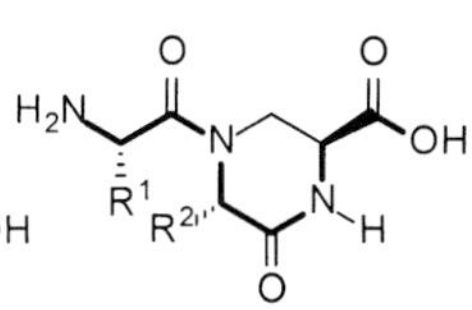

18

2-Naphthylalanine **10** is a sterically demanding derivative that may also lead to improved interaction of the aromatic system with hydrophobic areas of a protein receptor. Compound **11** is a representative of C^{α}-dialkyl amino acids. Tetrahydroisoquinoline carboxylic acid **12** (Tic) is a phenylalanine analogue where the torsion angle χ^1 is limited to very small range of values. 3-Phenylproline **13** may be regarded as a chimera containing both proline and phenylalanine substructures.

The introduction of bridges between two adjacent amino acid residues leads to the formation of dipeptide mimetics.
Bridges can, for example, be formed between C^{α} and N^{α} (**15**, **18**), between two C^{α} (**16**), or between two N^{α} (**17**). This type of modification also makes the peptide more rigid.

7.2.2
Backbone Modification

The modification of the peptide chain (backbone) comprises, for instance, the exchange of a peptide bond by amide analogues (e.g., ketomethylene, vinyl, ketodifluoromethylene, amine, cyclopropene). Some basic types of peptide backbone modifications are displayed in Fig. 7.4 [63].

The NH group of one or more amino acids within a peptide chain may be alkylated [64] or exchanged by an oxygen atom (depsipeptide), a sulfur atom (thioester), or a CH_2 group (ketomethylene isostere) [65]. However, peptidylthioesters are quite unstable. The CH moiety (C^{α}) may be exchanged in a similar manner by a nitrogen atom (azapeptide) [66, 67], by a C-alkyl group (C^{α}-disubstituted amino acid) [58], or by a BH^- group (borapeptide). The carbonyl groups may be replaced by thiocarbonyl groups (endothiopeptide) [68], CH_2 groups (reduced amide bond) [69], SO_n groups (sulfinamides, n=1; sulfonamides, n=2), POOH groups (phosphonamides), or B-OH groups. The peptide bond of one or more amino acid residues in **19** (cf. Fig. 7.5) may be inverted (NH-CO), giving retro peptides. In order to maintain the original side-chain orientation, the retro modification (**20**, **21**) has to be accompanied by an appropriate stereochemical compen-

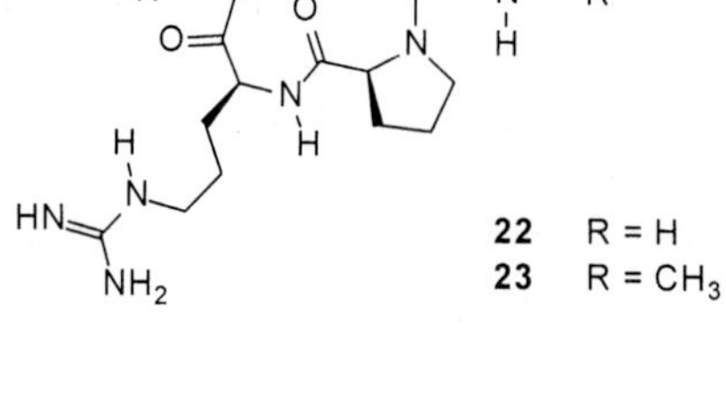

22 R = H
23 R = CH_3

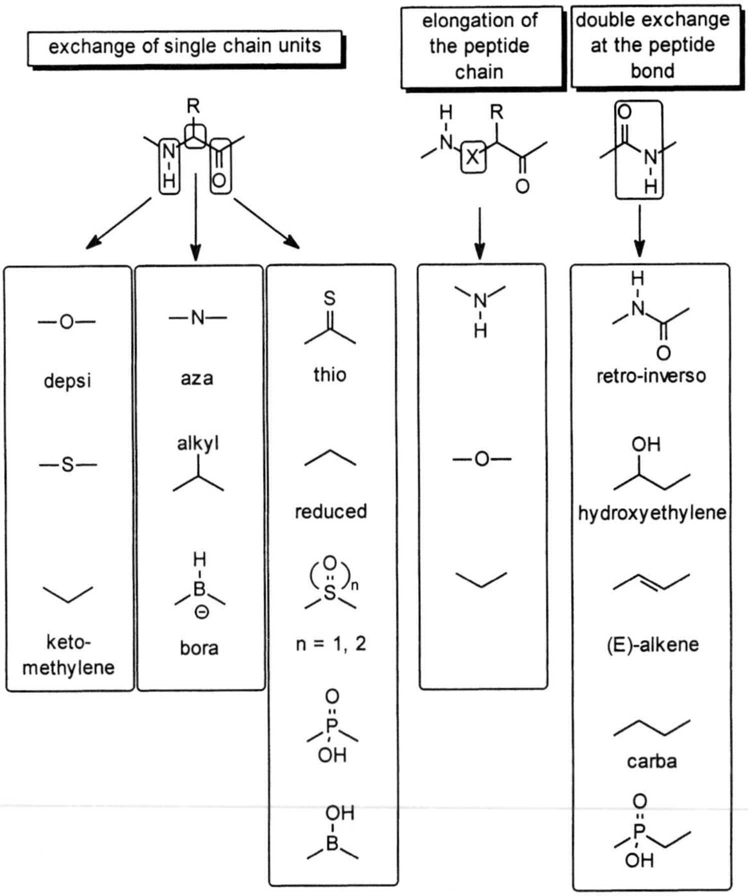

Fig. 7.4 Selected types of peptide backbone modification.

sation (inversion of the configuration at C^{α}). The consequence is formation of the so-called retro-inverso peptides (Fig. 7.5) [70].

Peptides, where an amide bond has been replaced to give hydroxyethylene [71], E-alkene [72, 73], or alkane structures (carbapeptides), have also been described. Besides these isosteric replacements, the peptide chain may be extended by one atom (O: aminoxy acid [74]; NH: hydrazino acid [75, 76]; CH_2: β-amino acid [77–79]; Section 7.4).

Vinylogous peptides contain amino acids extended by a vinyl group as building blocks, which is also not an isosteric replacement (Section 7.4). Such peptide analogues have been found in nature, as in the compounds cyclotheonamide A **22** (R=H) and B **23** (R=Me) that occur in the sponge *Theonella*. Besides vinylogous tyrosine, these peptides further contain the nonproteinogenic amino acid β-amino-α-oxohomoarginine [80]. They act as thrombin inhibitors and are of interest in medicinal chemistry as potential antithrombotic agents.

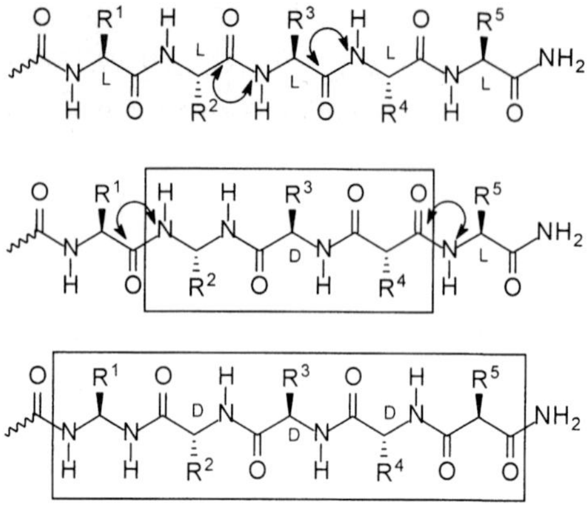

19

20

21

Fig. 7.5 Retro-inverso peptide modification.

7.2.3
Combined Modification (Global Restriction) Approaches

The introduction of global restrictions in the conformation of a peptide is achieved by limiting the flexibility of the peptide through cyclization (Section 6.1). This may eventually lead to a stabilization of peptidic secondary structures, and consequently this is an important approach toward drug development. Turns (or loops) are important conformational motifs of peptides and proteins, besides α-helix or β-sheet structures. Reverse turns comprise a diverse group of structures with a well-defined 3D orientation of amino acid side chains. β-Turns constitute the most important subgroup, and are formed by four amino acids (see Section 2.5.1.3). Turns lead to a reversion of the peptide chain, and may be mimicked by cyclic peptides with well-defined conformation. Depending on the type of cyclization (Sections 6.1 and 6.2) and on the ring size, the angles ψ, ϕ, ω, and χ will have limited flexibility. This reduction in the decrease of freedom may eventually lead to receptor binding with high affinity because of entropic reasons, provided that the receptor-bound conformation is still accessible to the modified peptide.

In heterodetic cyclic peptides where the cyclization is obtained by cysteine disulfide bridges, cysteine residues may also be replaced by sterically hindered cysteine analogues such as penicillamine (3-mercapto-valine, Pen). Such an exchange may lead to a differentiation between agonistic and antagonistic analogues, as has been shown by Du Vigneaud et al. for [Pen^1]oxytocin **24** and by Hruby et al. for cyclic enkephalin analogues [61]. [D-Pen^1,D-Pen^5]enkephalin (DPDPE), a cyclic disulfide, is a highly potent and selective δ-opioid receptor antagonist.

Ring size and the formation of stable conformations play important roles in the design of cyclic peptide analogues. This has been intensively investigated in the oxytocin and vasopressin series. In these cases cyclization to produce a 20-membered ring is essential for biological activity. However, this cyclization does not necessarily have to occur by disulfide formation, because the carba analogues,

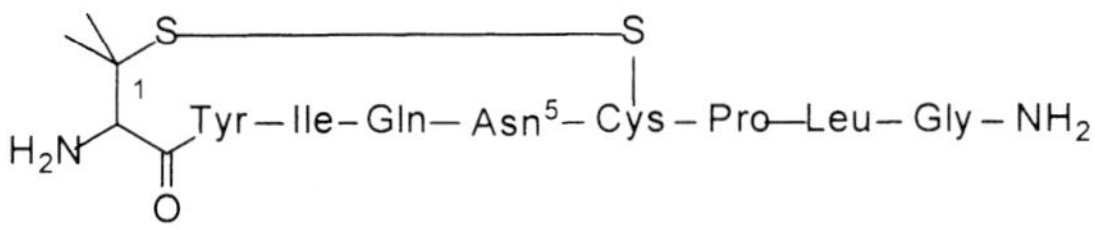

24

25

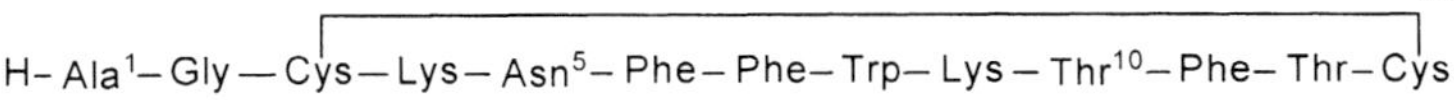

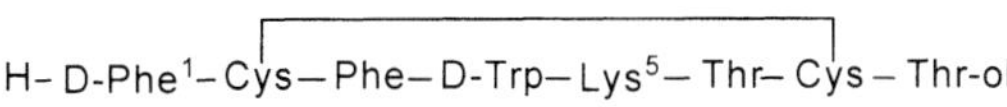

26

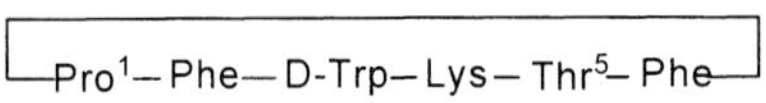

27

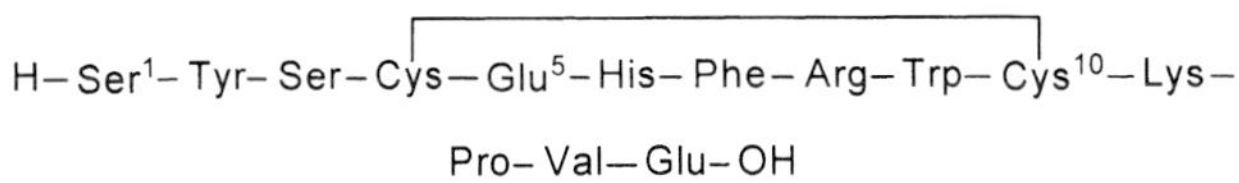

28

where the sulfur atoms are replaced by CH_2 groups, also display biological activity.

Somatostatin **25**, which is a regulatory peptide with a broad spectrum of activity, likewise contains an intrachain disulfide bridge. Potential applications have been envisaged in the treatment of acromegaly or of retinopathy in diabetes because **25** suppresses the liberation of growth hormone and glucagon, an increased level of which contributes especially to organ damage in diabetes. Native somatostatin is rapidly metabolized and therefore is not suited to the treatment of juvenile diabetes; hence, metabolically stable analogues have to be developed. The excision of six amino acid residues from the ring in **25** produced the 20-membered somatostatin analogue **26** developed by Bauer et al. [81]. This peptide drug, named octreotide, has been approved for the treatment of acromegaly and of patients with metastasizing carcinoid and vasoactive tumors. Finally, the ring size may be further reduced to produce an 18-membered cyclic hexapeptide **27** that induces the liberation of growth hormone, insulin and glucagon with the same potency as somatostatin [82].

Peptides that natively are linear sequences can also be modified by the incorporation of cyclic structural elements in order to improve receptor interaction. In this context, the superagonist **28** of the α-melanocyte-stimulating hormone should be mentioned, where a β-turn predicted for the linear peptide may be stabilized by the replacement of Met^4 and Gly^{10} with two cysteine residues and disulfide bridge formation [83].

Kessler et al. [16] first established the concept of "spatial screening", whereby small libraries of stereoisomeric peptides with conformational constraints (e.g., cyclic peptides) are used for different 3D presentations of pharmacophoric side-chain groups (Section 6.1). The correlation of biological activity with peptide conformation provides useful information about the peptide conformation with the best fit to the corresponding target receptor [16].

7.2.4
Modification by Secondary Structure Mimetics

It is desirable to have a repertoire of building blocks that reliably induce a desired conformation in a peptide. Consequently, numerous efforts have been undertaken to synthesize secondary structure mimetics that induce a well-defined secondary structure [84]. The desired conformation should be imitated as closely as possible, and the synthetic route for the secondary structure mimetic should permit the introduction of appropriate side chains onto the mimetic scaffold. While certain α-amino acids also exert conformational bias on the peptide chain (see Sections 2.5.1.3 and 6.1), this section focuses on synthetic nonproteinogenic modules that induce and stabilize secondary structure. A large variety of conformationally constrained dipeptide mimetics that may be introduced into a strategic position in a peptide, using methods of peptide synthesis, in order to induce a defined secondary structure has been compiled in reviews [85–87]. Derivatives of this type may also be used as scaffolds for peptidomimetics (Section 7.3).

The β-turn is a structural motif that is often postulated to occur in the biologically active form of linear peptides [88]. Consequently, most secondary structure mimetics imitate a β-turn. Compounds **29–37** are examples of β-turn mimetics (**29** [89], **31** [90], **32** [91], **33** [92], **34** [93], **35** [94], **36** [95], **37** [96]).

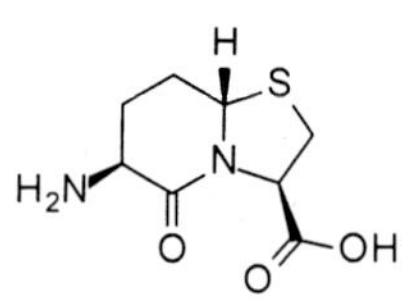

29

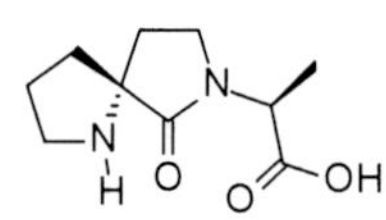

30

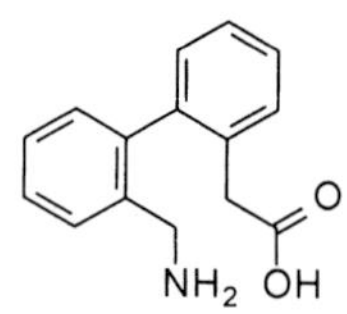

31

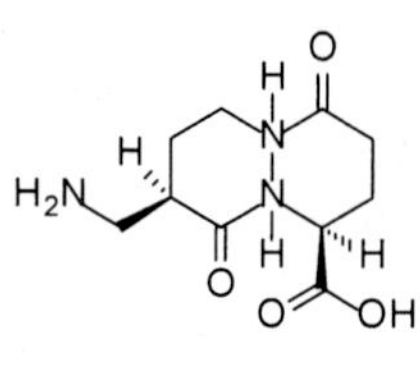

32

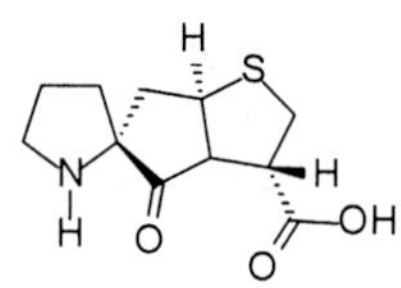

33

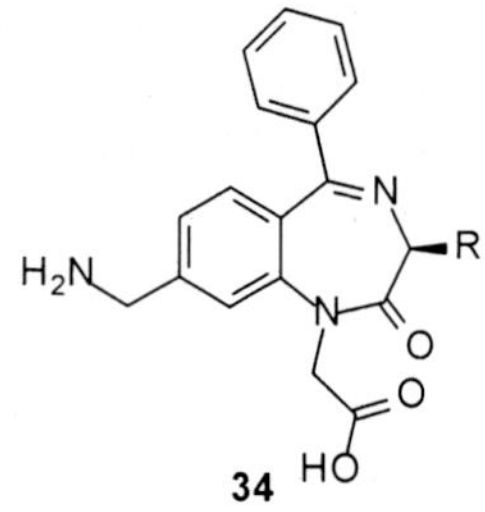

34

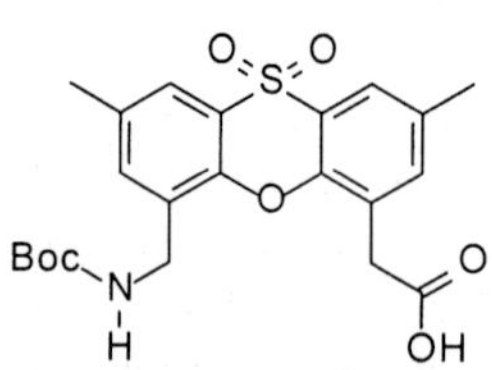

35

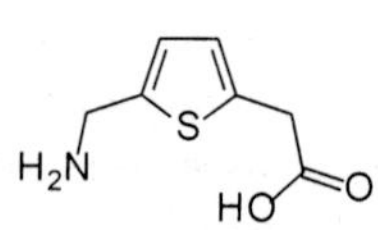

36

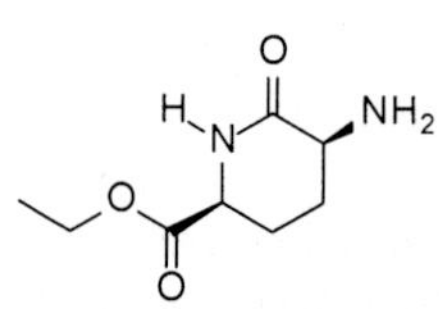

37

However, as has been shown by both experimental [97] and theoretical methods [98], not all scaffolds designed as β-turn mimetics actually display the desired properties.

γ-Turns, which occur less frequently than β-turns in proteins and peptides, may also be mimicked by a series of compounds. β-Amino acids are most likely the simplest γ-turn mimetics; they induce stable conformations, for example in cyclic tetra- and pentapeptides composed of four α-amino acids and one β-amino acid [79]. Other γ-turn mimetics are compounds **38** [99], **39** [100], and **40** [101].

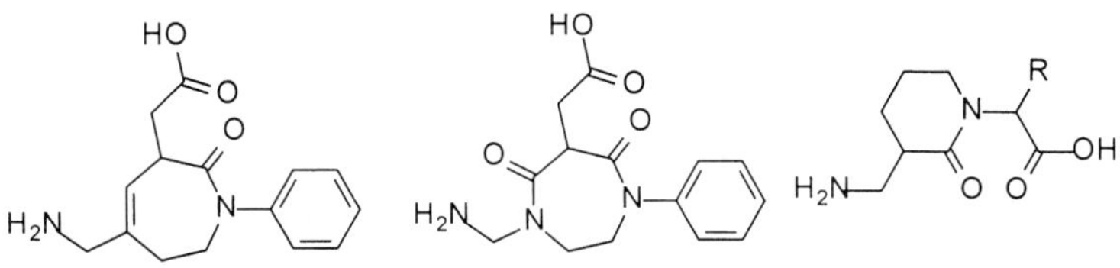

38 **39** **40**

α-Helix initiators [102, 103] and ω-loop mimetics have also been described [104]. ω-Loops are larger reverse turns in a peptide chain comprising 6–16 amino acid residues.

7.2.5
Transition State Inhibitors

Structural complementarity is a key issue in enzyme-substrate interactions, and both substrates and substrate analogues compete for binding to the active site and formation of the enzyme–substrate complex. It was first recognized in the 1920s that the affinity of a substrate towards a catalyst increases concomitantly with the distortion of the reactant towards its structure in the transition state. The transition state of an enzyme reaction can be assumed to differ significantly from the substrate ground state conformation. For example, a transition state with a tetrahedral configuration is formed in nucleophilic displacement reactions starting from a sp^2-hybridized carboxy group, as is present in peptide bond hydrolysis. Pauling [105] suggested that a potent enzyme antagonist might be developed that mimics the enzyme-substrate transition state despite being unreactive (transition state analogue, Fig. 7.6). Substrate and substrate analogues are relatively weakly bound in their ground state, because tight binding would be counterproductive to efficient catalysis, while the substrate in its transition state is bound much more closely [106].

Based on this concept, transition state inhibitors of enzyme reactions (e.g., protein or peptide cleavage by a protease) have been developed (Fig. 7.6) [5, 107, 108]. Nature once more provided a prototype for this kind of inhibitors with pepstatine **41**, an inhibitor of aspartyl proteases. This peptide (isovaleryl-Val-Val-Sta-Ala-Sta-OH), which is isolated from culture filtrates of *Streptomyces* species, contains – besides the N-terminal acyl moiety – the unusual amino acid statine (Sta), which is

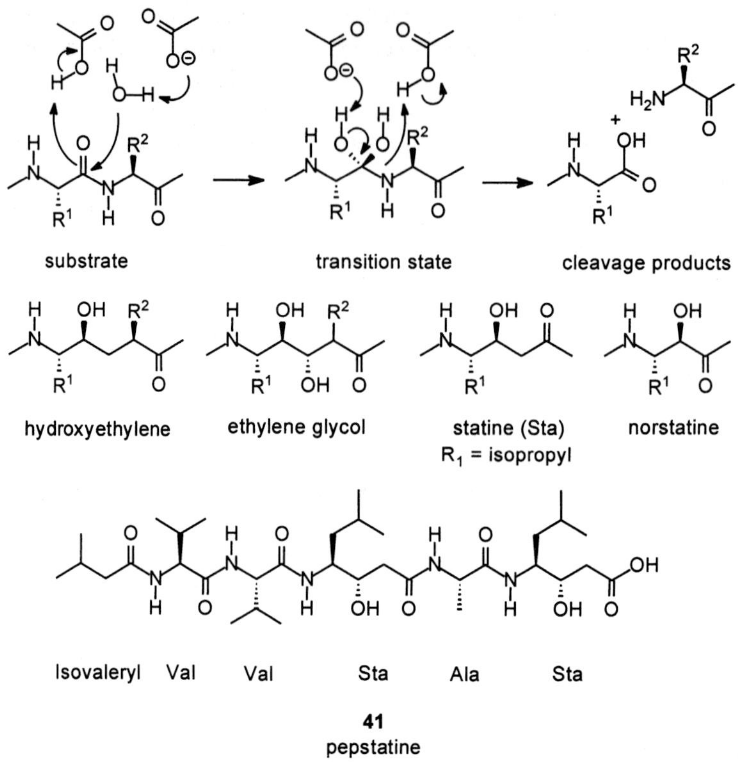

Fig. 7.6 Mechanism of proteolysis by an aspartyl protease and structures of some transition state inhibitors.

regarded as an amino acid analogue mimicking the transition state of peptide bond cleavage. Hydroxyethylene, ethylene glycol, or norstatine represent other building blocks for transition state inhibitors. This amino acid and its analogues have been applied in the development of many mechanism-based inhibitors of proteases and other enzymes [5]. The concept has also been utilized for the design of suitable haptens in the creation of catalytic antibodies (see Section 4.6.3) [109, 110].

7.3 Peptidomimetics

As the therapeutic application of peptides is limited because of certain disadvantages (see above), small organic molecules remain in most cases the most viable approach for the identification and optimization of potential drugs [111]. A major goal of modern medicinal chemistry is to find rational first principles for systematically transforming the information present in a natural peptide ligand into a nonpeptide compound of low molecular weight (type II and type III peptidomimetics). Chemical structures designed to convert the topographical information present in a peptide into small nonpeptide structures are referred to as peptidomimetics. Today, these compounds – which combine bioavailability and stability superior to that of bioactive peptides with increased receptor selectivity – are the subject of major interest by pharmaceutical companies. Peptidomimetics range from peptide-isosteric molecules to compounds where similarities to peptides can barely be identified. Information obtained from the structure-activity relationships and conformational properties of peptide structures can enhance the design of nonpeptide compounds [112]. The design of a peptidomimetic relies primarily on knowledge of the conformational, topochemical, and electronic properties of the native peptide and its corresponding receptor. Structural effects, such as a favorable fit to the binding site, the stabilization of a conformation by introducing rigid elements, and the placement of structural elements (functional groups, polar or hydrophobic regions) into strategic positions favoring the required interactions (hydrogen bonds, electrostatic bonds, hydrophobic interactions) must be taken into account.

The major objective in the development of small molecules displaying pharmacological activity is to mimic the complex molecular interactions between natural proteins and their ligands. Often, the mode of action of a biologically active peptide on the receptor level can be imitated by a small molecule (agonism), or can be blocked (antagonism) [1, 4, 9, 113]. A peptidomimetic may also be designed as an enzyme inhibitor.

The design of peptidomimetics exceeds the mere application of peptide modifications, and is targeted towards nonpeptidic compounds, characterized by a high degree of structural variation. Both random screening and design using molecular modeling have proved to be helpful in drug development. Peptidomimetic receptor antagonists or enzyme inhibitors usually imitate the binding of nonadjacent peptide substructures (pharmacophoric groups) to a protein, with this interaction often also relying on hydrophobic protein-ligand interactions. However, small organic molecules containing hydrophobic moieties may undergo a conformational transition in aqueous solutions because of an undesired intramolecular hydrophobic interaction (hydrophobic collapse). Consequently, the pharmacophoric groups should be presented on a rather rigid scaffold in order to avoid hydrophobic collapse of the bioactive conformation in an aqueous environment [5].

Peptidomimetics may be identified or discovered by extensive screening of natural or synthetic product collections (compound libraries). The combinatorial synthesis (Chapter 8) of a multitude of different peptidic or nonpeptidic compounds, combined with careful evaluation of receptor binding, are promising approaches for the discovery of new lead structures that subsequently must be further optimized with respect to their pharmacological properties.

Asperlicin **42** was discovered during a screening of fungal metabolites, and was found to be a lead structure as a cholecystokinin A (CCK-A) antagonist. **42** is structurally very similar to diazepam, which consequently was combined with a D-tryptophan structural motif; this eventually led to the design and synthesis of a se-

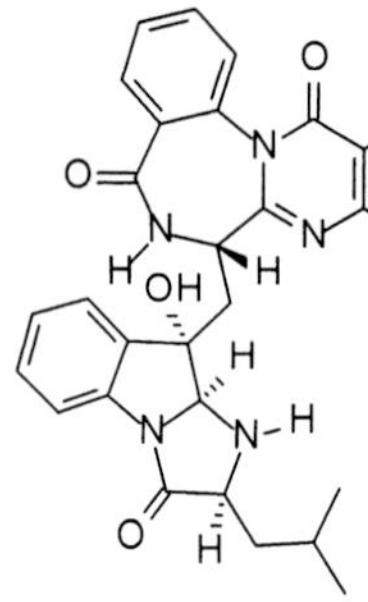

42

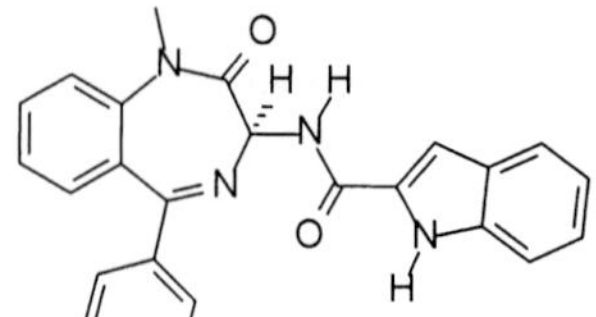

43

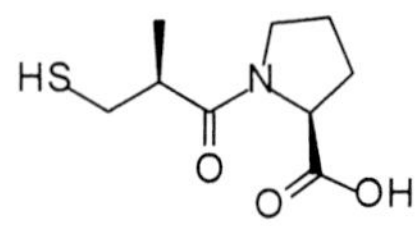

44

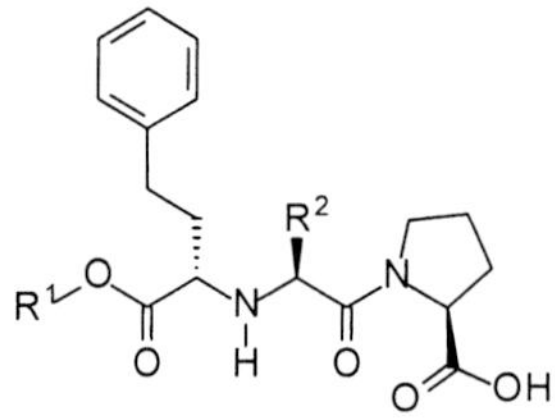

45 R^1 = H, R^2 = $(CH_2)_4NH_2$
46 R^1 = CH_2CH_3, R^2 = CH_3

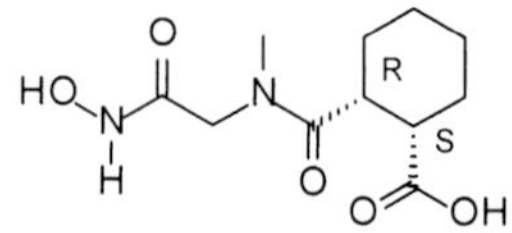

47

lective orally administered peptidomimetic antagonist of the peptide hormone cholecystokinin **43** [114].

The angiotensin-converting enzyme (ACE) inhibitor captopril **44** [115] was the first peptidomimetic compound to have been developed by rational design. Wolfenden introduced (*R*)-2-benzylsuccinic acid as an inhibitor for carboxypeptidase A (CPA), the enzyme mechanism of which is closely related to that of ACE [116, 117]. (*R*)-2-Benzylsuccinic acid belongs to the class of transition state analogues (Section 7.2.5).

The 9-peptide teprotide Pyr-Trp-Pro-Arg-Pro-Gln-Ile-Pro-Pro-OH isolated from snake venoms possesses bradykinin-potentiating activity that is based on the inhibition of ACE. While the modified C-terminal dipeptide sequence Ala-Pro displayed only weak inhibition of ACE, exchange of the N-terminal amino group by a carboxy group resulted in a more potent ACE inhibitor. This modification was based on the finding by Wolfenden concerning the inhibitory activity of benzylsuccinic acid on CPA. Replacement of the carboxy group by a thiol function, which strongly coordinates metal ions (e.g., Zn^{2+}), resulted in captopril, which is an inhibitor of the Zn-dependent metalloprotease ACE, and has been approved as an orally administered drug. The highly potent analogues lisinopril **45** and enalapril **46** have been synthesized by variation of different regions of captopril [118]. The hydroxamate **47** displays extremely low toxicity and an activity for ACE inhibition which is comparable to that of **44** (IC_{50}=7 nM) [119].

Losartan **48** is highly active angiotensin II antagonist that was optimized by molecular design. It is the first nonpeptide angiotensin II antagonist, and is currently used for the treatment of hypertonia.

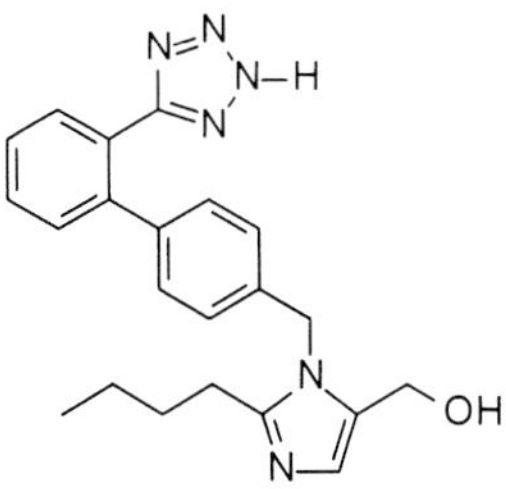

48

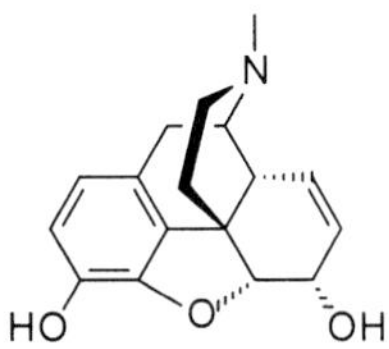

49

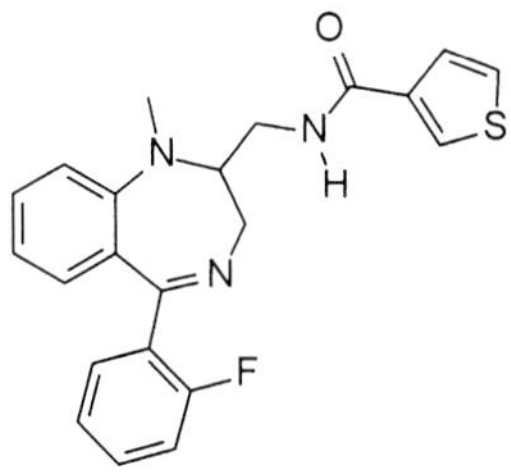

50

Morphine **49**, as a representative of the opioid alkaloids, is the classic example of a nonpeptidic compound that was found to be a mimetic of endogenous peptides. Morphine imitates the biological effect of β-endorphin on the corresponding receptor. The endogenous opioids resemble morphine very much from a pharmacological point of view, even in the side effects such as addictive potential and respiratory depression. Important developments following the discovery of opioid peptides included evidence of the existence of three different opioid receptor classes (μ, δ, κ), each with various subclasses (see Section 3.3.2.1). High expectations were imposed on this field of research, which was aimed towards the development of an opioid compound with high analgesic potency but without any adverse side effects, and as a result the field of opioid peptide and peptidomimetics became popular with regard to the application of new design principles. Concepts such as conformational restriction, peptide bond replacements, the incorporation of turn mimetics, and library screening for the identification of novel ligands were extensively tested in the area of opioids. However, despite many interesting developments, the original aim has not (yet) been achieved. For example, the benzodiazepine derivative tifluadom **50** is an agonist of the κ-opioid receptor and an antagonist of the CCK-A receptor. Although animal experiments have shown to be devoid of addictive potential and respiratory depression [120], its administration is associated with locomotor incoordination [121].

7.4
Pseudobiopolymers

Antisense oligonucleotides form one of the first classes of non-natural biopolymers [122]. Several other types of pseudobiopolymers have been described, and these are composed of nonpeptidic molecules in a similar, repetitive manner as peptides. Moreover, some pseudobiopolymers fold to give stable, reproducible sec-

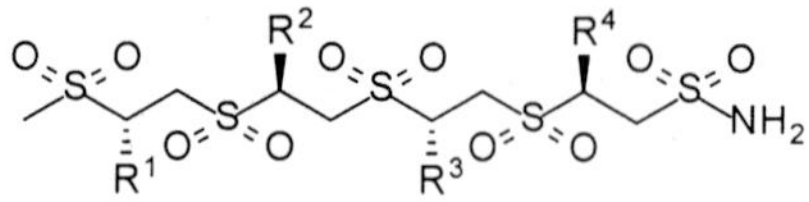

51

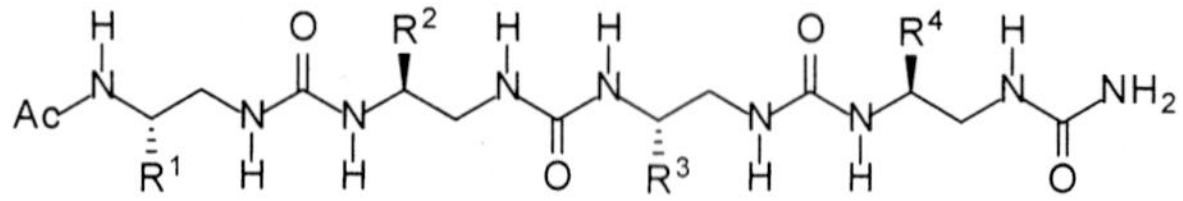

52

ondary structures that mimic protein secondary structures such as helices, sheets, or turns [84, 123–125]. They often are characterized by metabolic stability, resemble peptidic structures, and some of them may be used therapeutically. The peptoids (Section 7.4.1) [126], peptide nucleic acids (Section 7.4.2) [127], and β-peptides (Section 7.4.3) [128–133] are important representatives of this class of compounds. Oligomers of aza-amino acids (hydrazine carboxylates) have been named azatides [134, 135]. Oligosulfonamides [136–138], hydrazinopeptides, composed of α-hydrazino acids [75, 139] and aminoxypeptides, composed of α-aminoxy acids [74], may be regarded as β-peptide analogues with respect to the backbone atom pattern (Section 7.4.3). Oligocarbamates (Section 7.4.4) [140], oligopyrrolinones (Section 7.4.5) [141], oligosulfones **51**, γ-peptides [142, 143], oligoureas **52** [125], vinylogous peptides [144], vinylogous oligosulfonamides [145], and carbopeptides (oligomers of carbohydrate-derived amino acids) [146, 147] form further classes of pseudobiopolymers. Gellman first coined the term foldamers for any polymer that reproducibly adopts a specific ordered conformation [123]. Some representatives of these classes will be briefly introduced in the following sections.

7.4.1 Peptoids

Peptoids are oligomers composed of N-substituted glycine building blocks [126, 148, 149]. The side chain of each amino acid in the peptides formally is shifted in the peptoids by one position from C^{α} to the amino group nitrogen. Comparison of the peptide chain **53** with the peptoid chain **54** shows that the direction of the peptide bond in **54** should be reversed (retro-sequence) in order to provide the

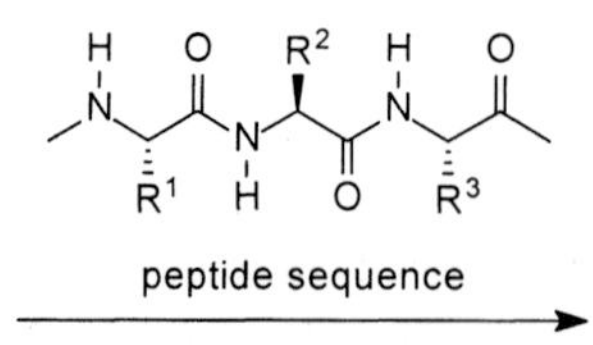

53

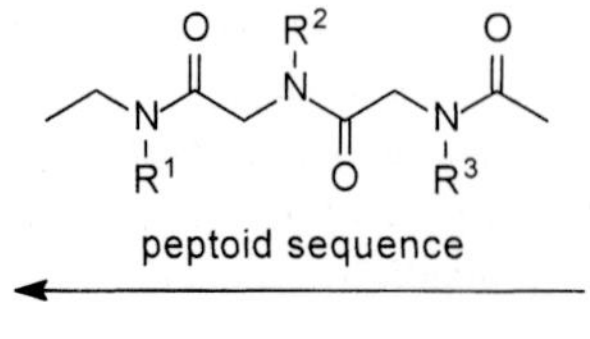

54

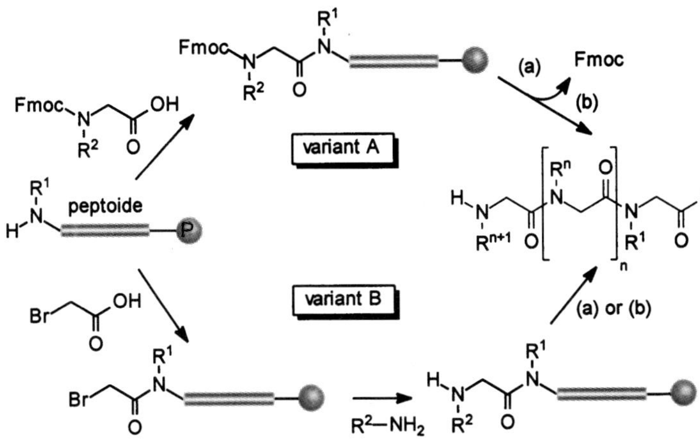

Fig. 7.7 Solid-phase synthesis of peptoids. Variant A, building block approach; variant B, submonomer approach. (a) = chain elongation (n repetitions); (b) = deprotection and cleavage from solid support; $R^1 \ldots R^n$ = side-chain residues; X = NH_2, OH.

same relative arrangement of side-chain residues R and carbonyl groups. Peptoids are achiral, and resemble the structures of the retro-inverso-peptidomimetics [70]. Helical peptoid structures are favored when bulky N-alkyl side chains are present [150]. Interestingly, these helices are not stabilized by hydrogen bonds, and the helicity depends on side-chain chirality [151] and oligomer length. N-Substituted glycine derivatives can be synthesized easily. The principle of solid-phase peptoid synthesis is shown in Fig. 7.7 [7].

Among the different types of polymeric support which may be used, the Rink amide resin has provided good results. Variant A utilizes the Fmoc/tBu tactics where the suitably protected N-substituted glycine derivatives are coupled to polymer-bound NHR groups using PyBOP or PyBroP [126]. The submonomer solid-phase synthesis (variant B) [148] does not apply the N-substituted glycine monomer. Instead, bromoacetyl building blocks are coupled to the growing peptoid chain and then subjected to a nucleophilic substitution reaction of the bromide substituent by a primary amine carrying the side-chain substituent. Additional protection of the amino group is not necessary, and no different monomeric building blocks have to be synthesized.

Conformational studies revealed that peptoids display higher conformational flexibility compared to peptides. They resist proteolysis and are, therefore, metabolically stable. Preliminary studies show that peptoids with structural resemblance to biologically active peptides have for example a similar activity as enzyme inhibitors compared to the native peptide sequence. The peptoid analogue of the hepatitis A 3C-protease inhibitor Ac-Glu-Leu-Ala-Thr-Gln-Ser-Phe-Ser-NH_2 displays, in the retro-sequence, a similar inhibition to the original peptide. An additional important finding is that peptoid analogues of peptide sequences from HIV-Tat protein bind to the corresponding RNA. This class of peptoids represents a promising concept in the search for new lead structures to influence pathological events. Peptoid libraries may be synthesized in a similar manner to peptide libraries, and this facilitates high-throughput screening for the identification of novel nonpeptide drugs.

7.4.2
Peptide Nucleic Acids (PNA)

Peptide nucleic acids are nonionic analogues of oligonucleotides containing a peptide backbone [127, 152–154]. They are mimetics of DNA that can be employed as antisense nucleotides. PNA originally were designed to mimic an oligonucleotide binding to double-stranded DNA in the major groove, but they have eventually been shown as good structural analogues of nucleic acids. The antisense concept is based on the application of oligonucleotides to hybridize with DNA or RNA in order to prevent transcription by the formation of a triple helix, or to bind mRNA by duplex formation. However, the first examples – which were described as DNA/RNA backbone analogues [155, 156] – did not display any ability to hybridize with oligonucleotides. Today, this concept is of therapeutic importance in cases where diseases can be treated on the DNA or mRNA level. The nonionic oligonucleotides of the PNA type are totally inert towards degradation by nucleases, and can also be synthesized in larger amounts by solid-phase synthesis.

The desoxyribose-phosphate backbone of an oligodesoxynucleotide **55** is replaced in the PNA **56** by a pseudopeptidic 2-(aminoethyl)glycine unit. Besides the 2-(aminoethyl)glycine moiety, other monomers may be used for PNA synthesis [127].

The N-protected monomer unit **57** (Y=Boc or Fmoc) comprises an amino acid [2-(aminoethyl)glycine] that is connected via a methylenecarbonyl spacer to the corresponding nucleobase B (thymine, cytosine, adenine, or guanine).

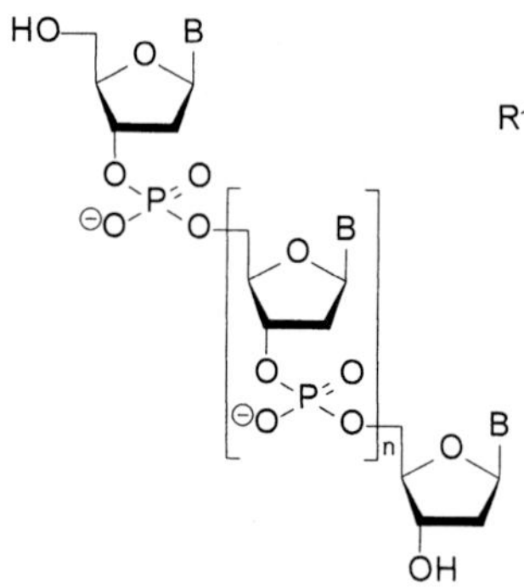

55

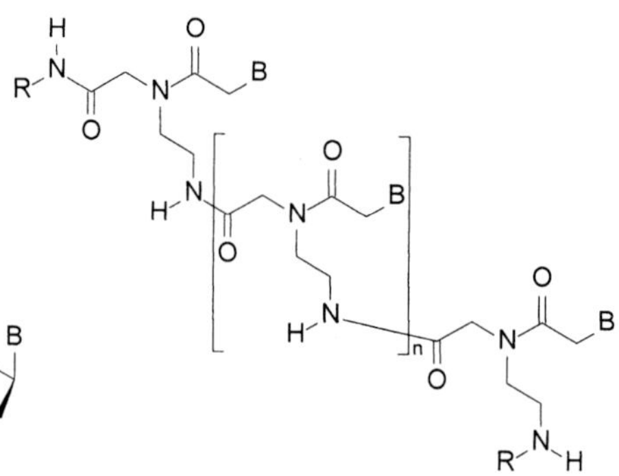

56

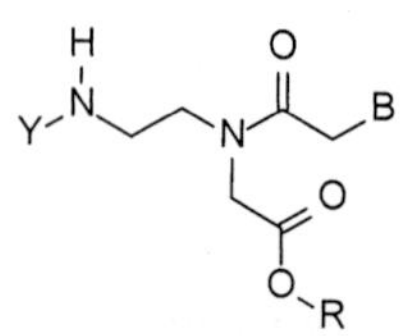

57

In principle, most of the coupling reagents used in peptide chemistry can be applied to the oligomerization of PNA. The active ester **57** (R=Pfp) may be used. The building blocks of type **57** can be obtained by alkylation of the corresponding base (B) with methyl-2-bromoacetate in DMF in the presence of K_2CO_3. After saponification of the methyl ester and conversion of the free carboxylic acid to the pentafluorophenylester, the nucleobase derivative is reacted with N-(N′-Boc-aminoethyl)glycine (N′-Boc-Aeg), which again is subsequently converted into the pentafluorophenylester **57** using DCC. An alternative synthesis of Fmoc-PNA-OPfp **57** (Y=Fmoc) has been described [157].

7.4.3
β-Peptides, Hydrazino Peptides, Aminoxy Peptides, and Oligosulfonamides

One characteristic feature of β-amino acids and β-homoamino acids (where an additional C_1 unit is inserted between the carboxy group and C^{α} of a natural amino acid) is the additional carbon atom between the amino group and the carboxy group. Side chains may be attached either to C^{α} (β^2-homoamino acid, **59**) or to C^{β} (β^3-homoamino acid, **58**), or to both of them. This fact considerably influences the secondary structure of the corresponding oligomer. The first structural investigations on homo-oligomers of β-amino acids were performed over 30 years ago.

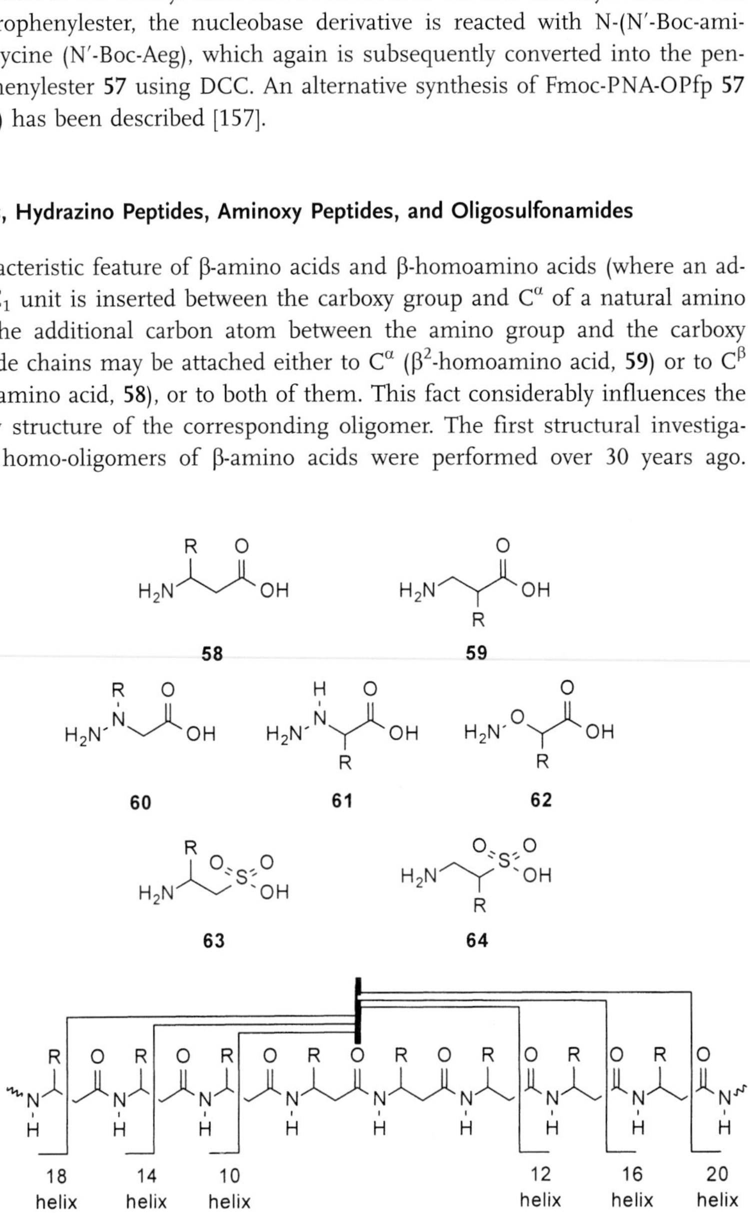

Fig. 7.8 β-Amino acids, β-amino acid analogues and hydrogen bond patterns in β-peptide helices.

Kovacs et al. postulated in 1965 that poly-(L-β-aspartic acid) forms a 3.4_{14}-P-helix [right-handed helix with 3.4 amino acid residues per turn and a 14-membered hydrogen bonded "ring" from NH (i) to CO (i+2)] [158]. Homopolymers of poly-(L-β-aspartic acid) have been investigated by Muñoz-Guerra et al. [128, 159], while Gellman et al. [123, 129, 130] and Seebach et al. [131, 132] profoundly examined the conformational behavior of a series of different well-defined β-peptides. All these compounds have been found to adopt predictable and reproducible helical conformations with hydrogen bond patterns depending on the substituent position (Fig. 7.8). Indeed, many β-peptides prefer 3_{14}-helices even in aqueous solutions [132, 160–162]. Helix formation occurs already in β-peptide hexamers, while in the case of α-amino acids usually more than 10–12 amino acid residues are required to form a stable helix. In molecular dynamics simulations, β-peptide helices have been shown to refold spontaneously after thermal defolding [163]. An alternating combination of β^2- (**59**) and β^3- (**58**) amino acids leads to an irregular helix with 10- and 12-membered hydrogen bond "rings" [164]. Both Boc- [165, 166] and Fmoc-protected β-amino acids [77] are available for peptide synthesis.

Hydrazino peptides [75, 139] and aminoxy peptides [74], composed of α-hydrazino acids (**60, 61**) or α-aminoxy acids (**62**) are structurally related to β-peptides, because they also contain an additional skeleton atom between amino and carboxy group. Aminoxy peptides form helices stabilized by hydrogen bonds in eight-membered "rings" [74, 136–138].

7.4.4
Oligocarbamates

Oligocarbamates **65**, which form another class of pseudobiopolymers (biopolymer mimetics), contain a carbamate moiety instead of a peptide bond [140]. The oligocarbamates may be considered as γ-peptide analogues, with three skeleton atoms

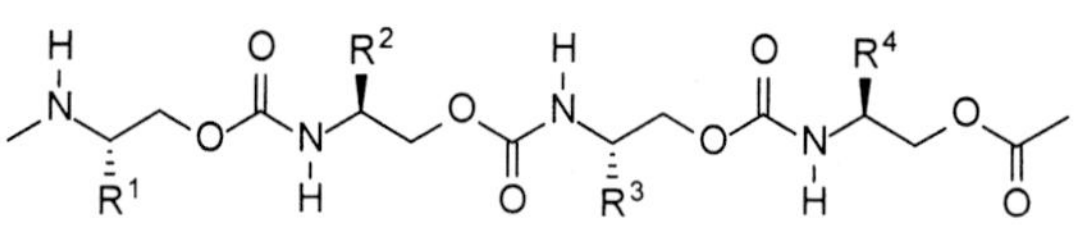

65

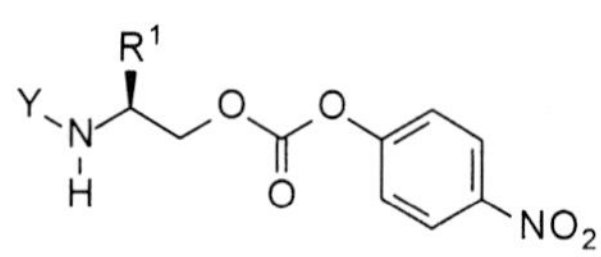

66

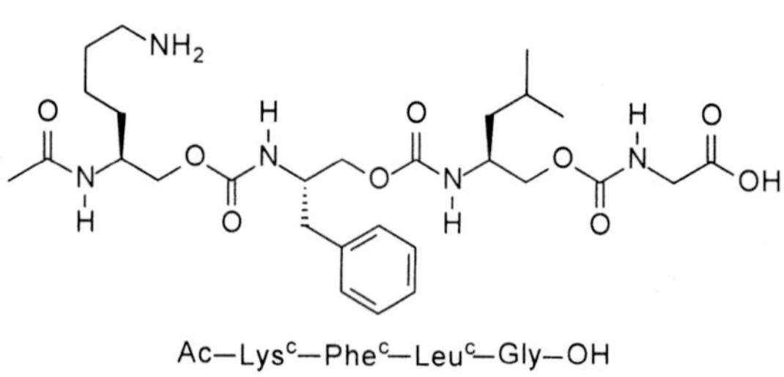

67

being found between the amino group and the carboxy group of one monomer. Oligocarbamates are stable towards proteolysis and are more hydrophobic compared to the peptides; hence, they may be able to permeate body barriers such as the blood-brain barrier.

The monomeric N-protected aminoalkylcarbonates are accessible starting from the corresponding amino alcohols that are converted into a carbamate active ester (e.g., *p*-nitrophenylester **66**). The oligocarbamates may be obtained in solid-phase synthesis using either the base-labile Fmoc group (**66**, Y=Fmoc) or the photolabile nitroveratryloxycarbonyl group (**66**, Y=Nvoc). Cho et al. [140] succeeded in synthesizing a library containing 256 different oligocarbamates, with **67** being one representative. The superscript "c" indicates the carbamate.

7.4.5
Oligopyrrolinones

Oligopyrrolinones contain a novel peptidomimetic principle where a strongly modified peptide backbone is integrated in cyclic structures also containing vinylogous amino acids [141]. A structural comparison between the oligopyrrolinone **71**

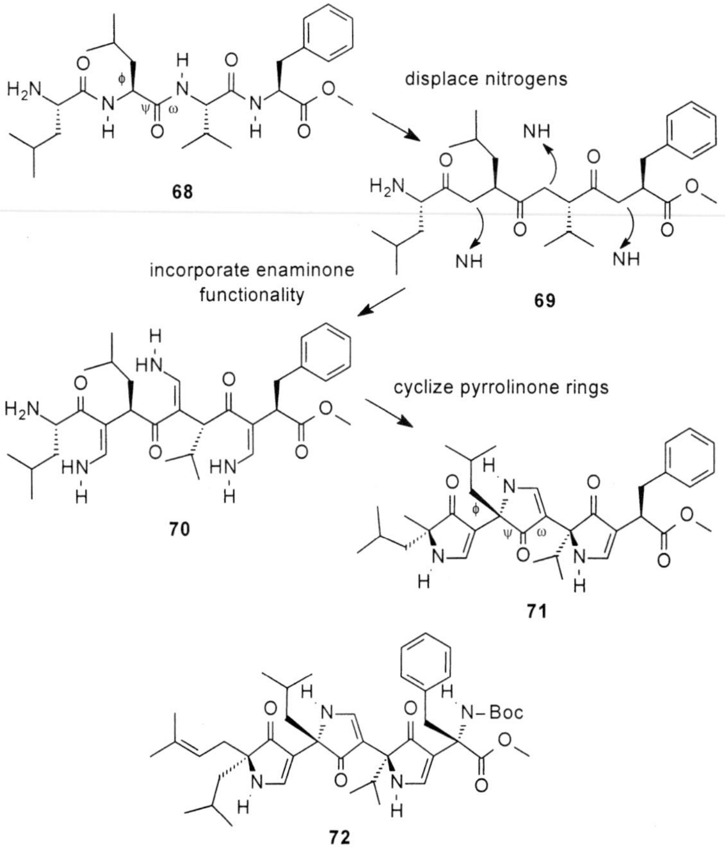

and the tetrapeptide **68** with a parallel β-sheet structure clearly shows that **71** comprises comparable dihedral angles φ, ψ, and ω as well as a similar orientation of the side chains. Formally, the amide nitrogen is displaced (**69**), incorporated in an enaminone (**70**), and then cyclized (**71**). Preconditions for the formation of hydrogen bonds, as in a β-sheet, are fulfilled. The 3D similarity of both the side chains and carbonyl groups in **68** and **72** has been proven by X-ray crystallography. **72** is present in an antiparallel β-sheet structure.

7.5 Macropeptides and De-novo Design of Peptides and Proteins

The de-novo design of peptides and proteins has emerged as a challenging approach to study the relationship between the structure and function of a protein. Solving the protein folding problem remains as the "Holy Grail" for computational structural biology, and this is still out of reach. De-novo design can be regarded as an alternate route to tackle the problem of protein folding, the term having been coined for approaches that involve the construction of a protein with a well-defined 3D structure that does not consist of a sequence directly related to that of any natural protein.

7.5.1 Protein Design

Proteins are characterized by an impaired variety of shape and function. Consequently, protein design is not easy to realize compared to the design of smaller model compounds, for example the active sites of enzymes. Protein design represents a major challenge, and provides valuable information on the complex interplay between the structure and function of proteins. The principles and methods for the design of proteins has been the topic of several reviews [167–172].

Most de-novo design strategies rely on knowledge obtained from an increasing amount of crystallographic protein structure data. These data provide basic information on the propensity of a single amino acid (or of a short amino acid sequence) to form a particular secondary structure. The assembly of supersecondary and tertiary folds relies on further hydrophobic or electrostatic interactions. Recent success in protein design is not merely the result of a simple hierarchic procedure, but has relied heavily on computer-based strategies [173–175].

Although the assembly of proteins by use of chemical synthesis, chemoenzymatic strategies and gene technological approaches is feasible (see also Chapters 4 and 5), it is not a straightforward task. Different concepts have been developed that mainly follow two important strategies [176]:

- The assembly of a larger peptide with known secondary structure motifs that forms a well-defined tertiary structure.
- The design of smaller peptide moieties of known secondary structure followed by assembly on a template.

De-novo protein design comprises the arrangement of simple secondary structures (β-turn, α-helix, β-sheet) to tertiary structures (folds) of increasing complexity (β-hairpin [177, 178], helix-turn-helix [179], coiled coils, helix bundles; Section 2.4) [167, 168, 180–182].

The design of non-natural redox proteins according to the former strategy has been realized by a sensible combination of complex chemistry and protein chemistry [168].

An interesting concept utilizes metal ion-induced self-association for the difficult organization of small amphiphilic peptide subunits in aqueous solution to give topologically well-determined protein structures. Such a de-novo design of functional protein requires the assembly of molecular scaffolds with predetermined structure, where the desired functional groups of an active site of an enzyme can be introduced. Using this approach, Ghadiri et al. [183] obtained a Ru^{II} metalloprotein with an exactly defined metal coordination site. Within this metalloprotein, three histidine residues were located proximal to the C-terminus of the three-helix protein bundle and formed a coordinating site for a metal, thereby favoring the incorporation of Cu^{II} to give the $Ru^{II}Cu^{II}$ protein **73**.

Two different metal ions and three peptide subunits assemble in a highly selective process of self-aggregation to a heterodinuclear metalloprotein with three parallel helices. This achievement is one of the preconditions for the design of potentially redox-active metalloproteins and of artificial light-collecting metalloproteins.

DeGrado et al. [184] synthesized helical 31-peptides that assembled to give dimeric bundles of four helices that are connected pairwise via disulfide bridges. The model peptide **74** contains two histidine building blocks in each subunit (His^{10}, His^{24}), dimerizes upon addition of Fe^{II}-heme to give four redox centers at the putative binding sites, and also serves as a mimetic for the cytochrome b subunit of cytochrome bc_1. Furthermore, miniaturized metalloproteins have been designed and synthesized *de novo* [168, 169, 185].

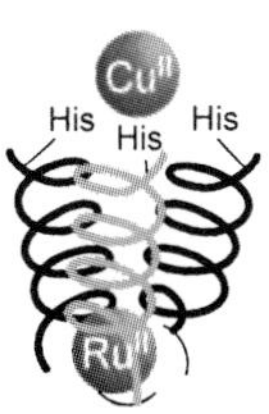

73

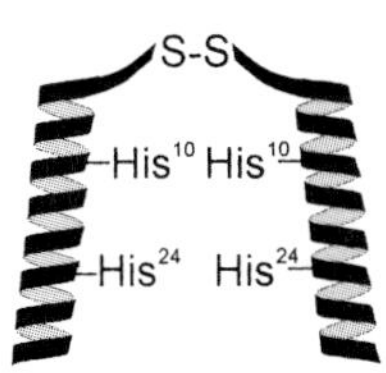

74

The Ghadiri group has also rationally designed tubular peptide nanostructures (peptide nanotubes) and transmembrane ion channels (Fig. 7.9). This concept is based on the assumption that cyclopeptides with alternating D- and L-configured amino acid building blocks form a flat ring where the planes of the amide groups in the scaffold are arranged perpendicularly to the ring plane [186].

Furthermore, it was postulated that this special arrangement is energetically favored under certain conditions, leading to the formation of tubular associates that

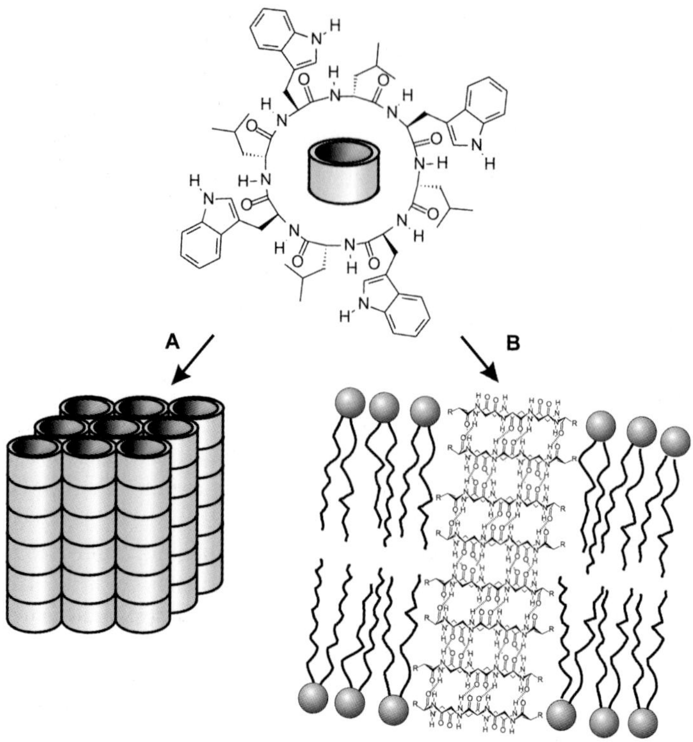

Fig. 7.9 Assembly of peptide nanotubes (A) and transmembrane ion channels (B) starting from cyclic peptides with alternating D- and L-amino acids.

are open at both ends. This process is favored by intermolecular hydrogen bonds together with interactions leading to a stacking of the cyclopeptide rings. Eventually, a selectively N-methylated cyclic peptide, cyclo-[(-Phe-D-N^2-Me-Ala-$)_4$] could be synthesized that assembles in apolar organic solvents by self-organization to produce a discrete, soluble, cylinder-shaped associate. In the solid state, the peptide forms by self-organization a crystal that is characterized by a pore structure in the shape of an ordered parallel arrangement of water-filled channels with a diameter of 7–8 Å.

The rational design of such structures is of considerable interest in the investigation into molecular transport processes, for the chemistry of inclusion compounds, and for catalytic processes, as well as the manufacture of new optical and electronic units.

The preparation of new artificial proteins and enzymes with predetermined 3D structure and tailor-made chemical, biological, and catalytic properties represents the ultimate goal for peptide and protein chemists.

The second strategy referred to at the beginning of this chapter is the template-associated synthetic proteins (TASP) concept, developed by Mutter et al. [187, 188]. These TASP with predetermined 3D structure are obtained by synthesis. The de-

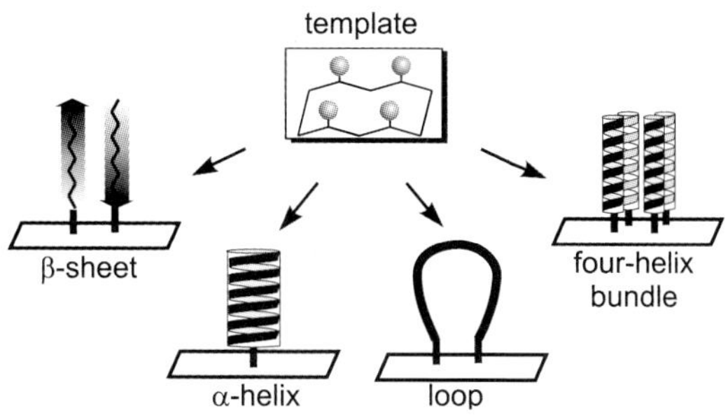

Fig. 7.10 Structural motifs of a template-assembled synthetic protein (TASP).

novo design of polypeptide sequences is limited by the protein folding problem, though this can be avoided by constructing proteins of non-natural chain architecture with predetermined intramolecular folding. TASP molecules combine structural features of natural proteins such as peptide moieties with predetermined secondary structure and synthetic elements such as topological templates; the result is branched structures with different possible folding topologies [170–172] (Fig. 7.10).

By selecting the correct template, TASP with for example βαβ, α-helix bundle, or β-barrel tertiary structures can be obtained, whilst protein-like folding motifs may be constructed from a "molecular kit". The individual building blocks (α-helix, β-sheet, loops, turns) are assembled in a regioselective manner by chemoselectively addressable groups at both ends of each building block [171, 189]. The schematic view of a TASP molecule **75** shows that the peptide chains attached to a cyclic octapeptide template via the lysine side chains form α-helical structures, and also associate intramolecularly to a four α-helix bundle that closely resembles protein structures.

The computational design of a TASP helical bundle, for example, requires the following steps:

- The design of amphiphilic helical peptide blocks based on the principles for secondary structure formation.
- Self-associations of the peptide blocks by optimal intramolecular interaction.
- The design of tailor-made templates and functional attachment of the peptide building blocks.
- Stabilization of hypothetical TASP structure by minimization of the conformational energy.

Solid-phase synthesis is used for the assembly of the peptides. The nonconsecutive mode of connection of the subunits excludes the application of synthetic methods by recombinant DNA techniques. Regioselectively addressable functionalized templates (RAFT) are used for chemoselective ligation [170, 190], and allow for the attachment of four different peptide building blocks or for the assembly of binding loops as receptor mimetics. The antiparallel four-helix bundle TASP T_4-($2\alpha_{14}\downarrow,\alpha_{14}\uparrow$) **76** was designed as a mimetic for the lysozyme epitope mAb HyHel-

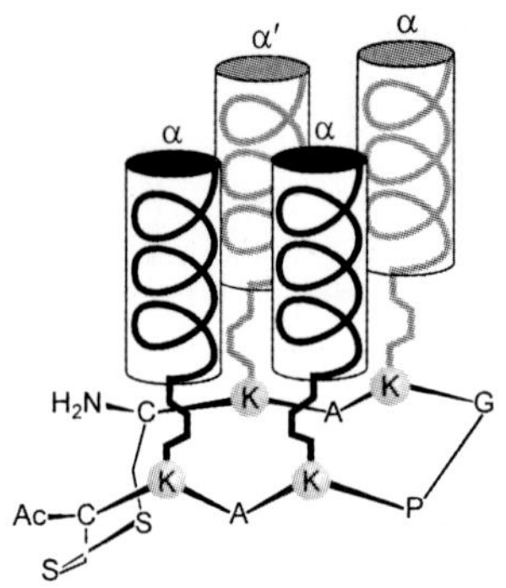

75

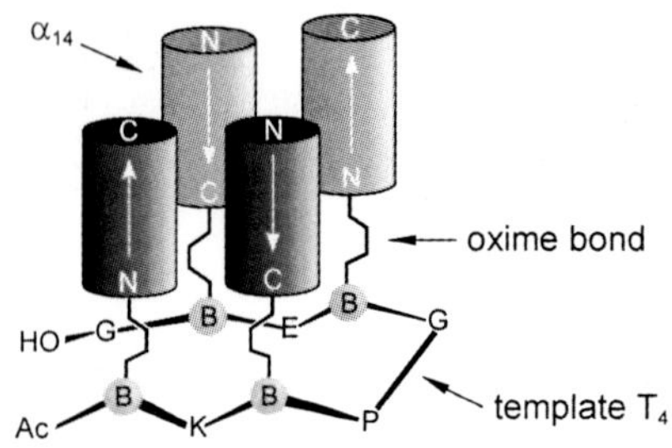

76

10 [191]. The TASP concept has also been used for the de-novo synthesis of four-helix bundle metalloproteins in a combinatorial approach [192].

The α-helical coiled coil is another simple but highly versatile protein folding motif. Approximately 2–3% of all protein residues form coiled coils [193], the latter consisting of two to five amphipathic α-helices that are wrapped around each other in a left-handed twist with a seven-residue periodicity. The stability of this fold is mainly achieved by a packing of apolar side chains into a hydrophobic core ("knobs-into-holes") [194]. In the seven-residue repeat the first and fourth positions are occupied by hydrophobic uncharged (U) residues forming a hydrophobic core at the interhelical interface. The fifth and seventh positions of the heptad repeat are often occupied by charged residues (C) (Fig. 7.11). The residues 2,3,6 in the outer boxes (X) do not contribute to interhelical contacts (Fig. 7.11).

This α-helical coiled coil motif is often regarded as ideal model system for investigations into the principles of protein stability and for de-novo design [195–199]. Many studies have elucidated the influence of sequence variations on structure, orientation and aggregation state of coiled coils [200–203]. The design principle of coiled coil proteins has also been used for the concept of a de-novo-designed peptide ligase [204].

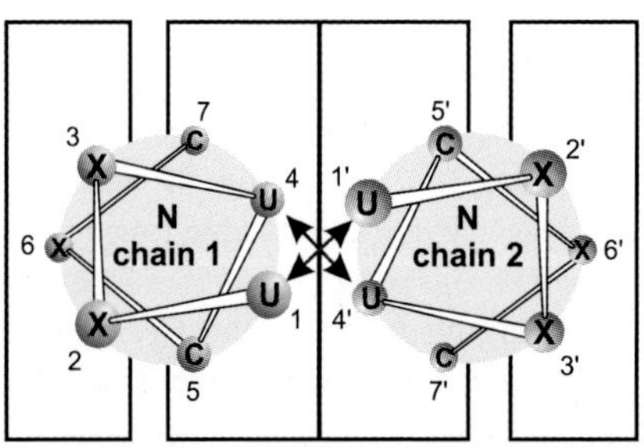

Fig. 7.11 Interactions in coiled coils. U = uncharged amino acid; C = charged amino acid; X = any amino acid.

7.5.2
Peptide Dendrimers

Dendrimers are highly ordered, highly branched polymers with a wide variety of potential chemical applications. Dendrimers bearing peptide moieties with predetermined secondary structure are useful tools in the de-novo design of proteins that may be applied as antigens and immunogens, for serodiagnosis, and for drug delivery [205]. Dendrimers are formed by successive reactions of polyfunctional monomers around a core, and consequently have many terminal groups.

Dendrimeric macromolecules differ from normal polymers as they are constructed from AB_n monomers in an iterative fashion. They may be synthesized either by a divergent or by a convergent strategy (Fig. 7.12). The divergent strategy makes use of the assembly starting from the initiator core to the periphery. Each new layer of monomer units attached to the growing molecule is called a "generation". In the convergent strategy, the assembly of dendrimers starts from the periphery and continues towards the central core [206].

Tam et al. [207] utilized branched oligolysine cores as synthetic carriers of fully synthetic antigens. The multiple antigen peptides (MAP) developed by Tam et al. comprise a simple amino acid such as glycine or β-alanine as an internal standard, an inner oligolysine core, and multiple copies of the synthetic peptide antigen (Fig. 7.13).

One advantage of MAP is that they have a well-defined and high antigen:carrier ratio. The derivatives are stable, and no conjugation to an antigenic protein is necessary; hence, undesired epitopes are not present.

Dendrimers may be synthesized either in solution or on solid phase. While the former approach is often very challenging and requires long reaction times and nontrivial purification steps, the application of solid-phase strategies provides the possibility of driving the reactions to completion by use of a large excess of reagents. Furthermore, a distinction may be made between direct and indirect approaches for the synthesis (Fig. 7.14).

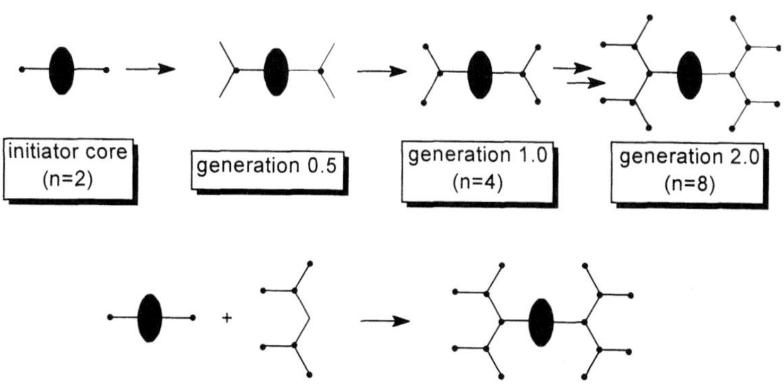

Fig. 7.12 Divergent and convergent synthesis of dendrimers.

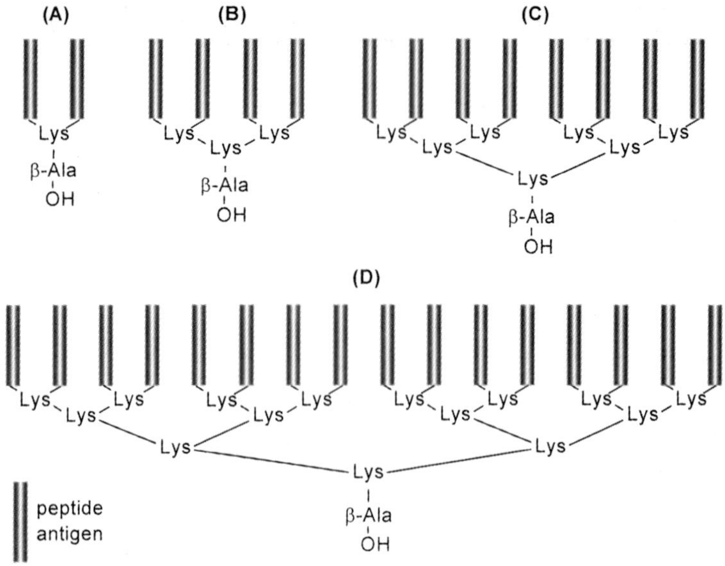

Fig. 7.13 Schematic structure of (A) di-, (B) tetra-, (C) octa-, and (D) hexadecavalent multiple antigen peptides (MAP).

In the direct approach, multiple copies of the peptide antigen are synthesized by stepwise solid-phase peptide synthesis (SPPS) on a solid-phase-bound dendrimer core. Both Fmoc and Boc tactics have been successfully applied [208, 209]. Low resin loading is necessary in the synthesis of peptide dendrimers in order to minimize interchain interaction that may limit coupling efficiency [210]. The efficiency must be monitored, and when necessary repetitive couplings should be performed using different activating agents. Similarly, galactose has been used as a core moiety, where four identical peptide strands were assembled by solid-phase synthesis [211].

The indirect approach (Fig. 7.14) is characterized by the separate synthesis of a suitably functionalized core to where the full-length peptide is ligated in the final step. A classical fragment condensation in solution using fully protected peptides may be applied, but this often suffers from low yield, slow coupling reactions, poor solubility, and problems of racemization. On the other hand, the possibility of purifying the peptide segments before ligation excludes the occurrence of mismatched sequences. Alternately, unprotected or partially protected peptide epitopes, prepared by SPPS, may be conjugated to the dendrimer core in aqueous solution under mild conditions. In this approach, either thiol chemistry, hydrazone chemistry, or oxime chemistry [212] may be applied (for further information, see [205, 213]).

Using this approach, either lipidated or glycosylated peptide dendrimers may also be synthesized [205, 214, 215]. Cationic dendrimers may also be used for gene delivery because they form stable complexes with DNA. Polylysine dendri-

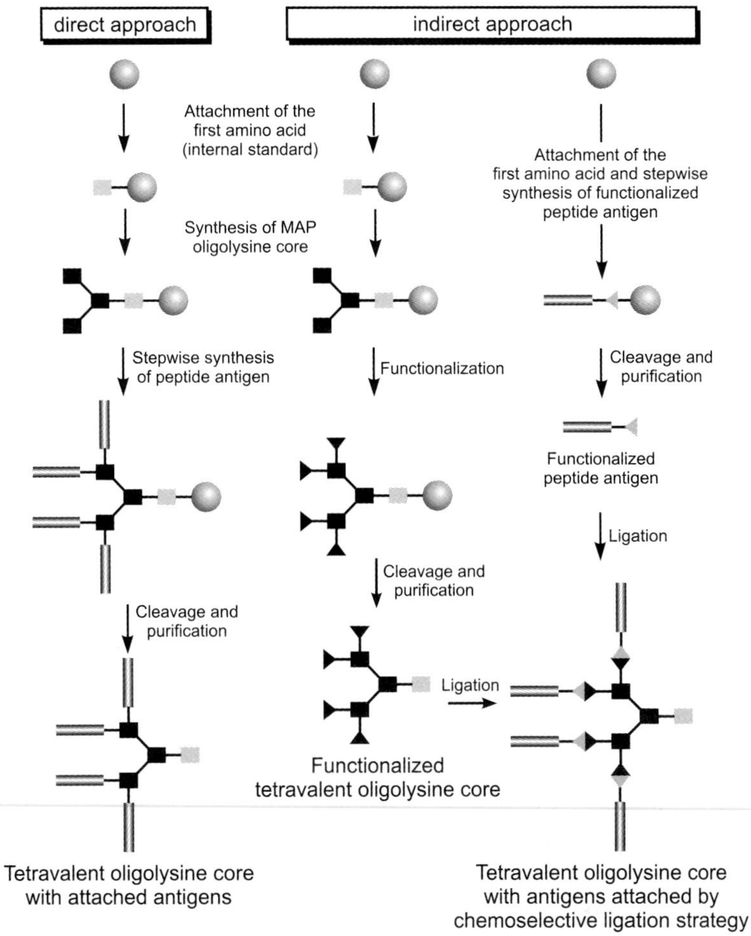

Fig. 7.14 Methods for the preparation of multiple antigen peptides (MAP).

mers containing a polyethylene glycol moiety have been reported to form a spherical water-soluble complex with DNA [216].

Peptide dendrimers containing a porphyrin core form another important group of functional dendrimers that may be used as either chemosensors or catalysts [217–219].

7.5.3 Peptide Polymers

As oligo- or polypeptides are polymers by definition, three other types of polymers will be discussed in the following section. The first of these comprises homopolymers composed of only one type of amino acid. In particular, polylysine or polyarginine, both of which are present under physiological conditions as polycations, have a unique ability to cross the plasma membrane of cells [220]. Consequently, they may be used to transport a variety of biopolymers and small molecules into cells [221]. For example, polylysine interacts electrostatically with the negatively charged phosphate backbone of DNA and may, therefore, be used for gene transfer [222].

Polylysine is synthesized by polymerization of the N-carboxy-anhydride of lysine, and this is subsequently fractionated and characterized with respect to the average molecular weight [223]. The heterogeneity of commercially available polylysine in terms of degree of polymerization is a major obstacle in the preparation of reproducible, stable formulations however. Therefore, rationally designed synthetic peptides have been suggested as DNA delivery systems.

A second type of peptide-based polymer (also called protein-based polymers or sequential polypeptides) is formed by compounds composed of repeating peptide sequences. In contrast to other polymers, these can be prepared either by chemical synthesis (solution or solid phase) or alternately by recombinant technology.

Peptide-based polymers can be transformable hydrogels, elastomers, regular thermoplasts, or inverse thermoplasts that are postulated to be molecular machines [224].

The third type of peptide polymers comprises peptide drugs that are chemically conjugated to nanoscale polymer particles such as hydroxypropylmethacrylamide or polyethyleneglycol; this may be carried out using a formulation approach in order to enhance drug stability and targeting possibilities. It has been shown that a vinylacetate derivative of luteinizing hormone-releasing hormone (LHRH) could be co-polymerized with butylcyanoacrylate to form particles of a mean size of 100 nm that are stable *in vitro*. In this way, the average half-life of LHRH in blood could be increased from 2–8 min (peptide) to up to 12 h [225].

Co-polymers of peptides and polyethyleneglycol or polyacrylates have an intermediate position between the second and third types referred to above [226, 227].

7.6 References

1 B. A. Morgan, J. A. Gainor, *Annu. Rep. Med. Chem.* **1989**, *24*, 243

2 R. M. J. Liskamp, *Recl. Trav. Chim. Pays-Bas* **1994**, *113*, 1

3 V. J. Hruby, *Biopolymers* **1993**, *33*, 1073

4 A. Giannis, T. Kolter, *Angew. Chem.* **1993**, *105*, 1303; *Angew. Chem. Int. Ed. Engl.* **1993**, *32*, 1244

5 R. A. Wiley, D. H. Rich, *Med. Res. Rev.* **1993**, *13*, 327

6 A. F. Spatola in: *Methods in Neurosciences*, Volume 13, P. M. Conn (Ed.), Academic Press, San Diego, **1993**, 19

7 J. Gante, *Angew. Chem.* **1994**, *106*, 1780; *Angew. Chem. Int. Ed. Engl.* **1994**, *33*, 1699

8 M. W. Peczuh, A. D. Hamilton, *Chem. Rev.* **2000**, *100*, 2479

9 V. J. Hruby, P. M. Balse, *Curr. Med. Chem.* **2000**, *7*, 945

10 J. Boer, D. Gottschling, A. Schuster, B. Holzmann, H. Kessler, *Angew. Chem.* **2001**, *113*, 3988; *Angew. Chem. Int. Ed. Engl.* **2001**, *40*, 3870

11 A. S. Ripka, D. H. Rich, *Curr. Opin. Chem. Biol.* **1998**, *2*, 441

12 R.E. Babine, S.L. Bender, *Chem. Rev.* **1997**, *97*, 1359

13 R.S. Bohacek, C. McMartin, *Curr. Opin. Chem. Biol.* **1997**, *1*, 157

14 K.S. Lam, M. Lebl, V. Krchnák, *Chem. Rev.* **1997**, *97*, 411

15 K.H. Mayo, *Trends Biotechnol.* **2000**, *18*, 212

16 R. Haubner, D. Finsinger, H. Kessler, *Angew. Chem.* **1997**, *109*, 1440; *Angew. Chem. Int. Ed. Engl.* **1997**, *36*, 1374

17 J.B. Ball, R.A. Hughes, P.F. Alewood, P.R. Andrews, *Tetrahedron* **1993**, *49*, 3467

18 R.S. McDowell, T.R. Gadek, P.L. Barke, D.L. Burdick, K.S. Chan, C.L. Quan, N. Skelton, M. Struble, E.D. Thorsett, M. Tischler, J.Y. Tom, T.R. Webb, J.P. Burnier, *J. Am. Chem. Soc.* **1994**, *116*, 5069

19 R.S. McDowell, B.K. Blackburn, T.R. Gadek, L.R. McGee, T. Rawson, M.E. Reynolds, K.D. Robarge, T.C. Somers, E.D. Thorsett, M. Tischler, R.R. Webb II, M.C. Venuti, *J. Am. Chem. Soc.* **1994**, *116*, 5077

20 W.G.J. Hol, *Angew. Chem.* **1986**, *98*, 765; *Angew. Chem. Int. Ed. Engl.* **1986**, *25*, 767

21 L.M. Amzel, *Curr. Opin. Biotechnol.* **1998**, *9*, 366

22 W. Wang, O. Donini, C.M. Reyes, P.A. Kollman, *Annu. Rev. Biophys. Biomol. Struct.* **2001**, *30*, 211

23 R.E. Bruccoleri, M. Karplus, *Biopolymers* **1987**, *26*, 137

24 W.F. van Gunsteren, *Protein Eng.* **1988**, *2*, 5

25 T. J. Marrone, J.M. Briggs, J.A. McCammon, *Annu. Rev. Pharmacol. Toxicol.* **1997**, *37*, 71

26 G. Böhm, *Biophys. Chem.* **1996**, *59*, 1

27 P.J. Whittle, T.L. Blundell, *Annu. Rev. Biophys. Biomol. Struct.* **1994**, *23*, 349

28 H.J. Böhm, *Prog. Biophys. Mol. Biol.* **1996**, *66*, 197

29 K. Gubernator, H.-J. Böhm (Eds.), *Structure-based Ligand Design*, Wiley-VCH, Weinheim, **1998**

30 C. Quan, N.J. Skelton, K. Clark, D.Y. Jackson, M.E. Renz, H.H. Chiu, S.M. Keating, M.H. Beresini, S. Fong, D.R. Artis, *Biopolymers* **1998**, *47*, 265

31 D.J. Bacon, J. Moult, *J. Mol. Biol.* **1992**, *225*, 849

32 V. Sobolev, R.C. Wade, G. Vriend, M. Edelman, *Proteins* **1996**, *25*, 120

33 M. Rarey, B. Kramer, T. Lengauer, G. Klebe, *J. Mol. Biol.* **1996**, *261*, 470

34 M. Rarey, B. Kramer, T. Lengauer, *J. Comput. Aid. Mol. Des.* **1997**, *11*, 369

35 S.Y. Yue, *Protein Eng.* **1990**, *4*, 177

36 G.M. Morris, D.S. Goodsell, R. Huey, A.J. Olson, *J. Comput. Aid. Mol. Des.* **1996**, *10*, 293

37 M. Liu, S. Wang, *J. Comput. Aid. Mol. Des.* **1999**, *13*, 435

38 C.M. Oshiro, I.D. Kuntz, J.S. Dixon, *J. Comput. Aid. Mol. Des.* **1995**, *9*, 113

39 G. Jones, P. Willett, R.C. Glen, A.R. Leach, R. Taylor, *J. Mol. Biol.* **1997**, *267*, 727

40 S.B. Shuker, P.J. Hajduk, R.P. Meadows, S.W. Fesik, *Science* **1996**, *274*, 1531

41 T. Diercks, M. Coles, H. Kessler, *Curr. Opin. Chem. Biol.* **2001**, *5*, 285

42 M.L. Moore in: *Synthetic Peptides: A User's Guide*, G. A. Grant (Ed.), Freeman, New York, **1992**, 9

43 D. Ward (Ed.), *Peptide Pharmaceuticals, Approaches to the Design of Novel Drugs*, Open University Press, Milton Keynes, **1991**

44 G.N. Ramachandran, V. Sasisekharan, *Adv. Protein Chem.* **1968**, *23*, 283

45 H.A. Scheraga, *Chem. Rev.* **1971**, *71*, 195

46 S.M. Bloom, G.D. Fasman, C. Deloze, E.R. Blout, *J. Am. Chem. Soc.* **1961**, *84*, 458

47 M. Goodman, R.P. Saltman, *Biopolymers* **1981**, *20*, 1929

48 H. Kessler, *Angew. Chem.* **1982**, *94*, 509; *Angew. Chem. Int. Ed. Engl.* **1982**, *21*, 512

49 W.R. Croasmun, R.M.K. Carlson, *Two-Dimensional NMR-Spectroscopy: Application for Chemists and Biochemists*, VCH, Weinheim, **1994**

50 M. Manning, L. Balaspiri, M. Agosta, *J. Med. Chem.* **1973**, *16*, 975

51 T. Haack, M. Mutter, *Tetrahedron Lett.* **1992**, *33*, 1589

52 P. Dumy, M. Keller, D.E. Ryan, B. Rohwedder, T. Wöhr, M. Mutter, *J. Am. Chem. Soc.* **1997**, *119*, 918

53 M. Keller, C. Sager, P. Dumy, M. Schutkowski, G. S. Fischer, M. Mutter, *J. Am. Chem. Soc.* **1998**, *120*, 2714
54 A. Wittelsberger, M. Keller, L. Scarpellino, P. Patiny, H. Acha-Orbea, M. Mutter, *Angew. Chem.* **2000**, *112*, 1153; *Angew. Chem. Int. Ed. Engl.* **2000**, *39*, 1111
55 Y. Takeuchi, G. R. Marshall, *J. Am. Chem. Soc.* **1998**, *120*, 5363
56 H. E. Stanger, S. H. Gellman, *J. Am. Chem. Soc.* **1998**, *120*, 4236
57 V. J. Hruby, S. Fang, R. Knapp, W. M. Kazmierski, G. K. Lui, H. I. Yamamura, *Int. J. Peptide Protein Res.* **1990**, *35*, 566
58 P. Balaram, T. S. Sudha, *Int. J. Peptide Protein Res.* **1983**, *21*, 381
59 I. L. Karle, *Biopolymers* **1996**, *40*, 157
60 B. Pispisa, L. Stella, M. Venzani, A. Palleschi, F. Marchiori, A. Polese, C. Toniolo, *Biopolymers* **2000**, *53*, 169
61 V. J. Hruby, G. Li, C. Haskell-Luevano, M. Shenderovich, *Biopolymers* **1997**, *43*, 219
62 S. E. Gibson, N. Guillo, M. J. Tozer, *Tetrahedron* **1999**, *55*, 585
63 A. F. Spatola in: *Chemistry and Biochemistry of Amino Acids, Peptides and Proteins*, B. Weinstein (Ed.), Marcel Dekker, New York, **1983**, *7*, 267
64 P. Manavalan, F. A. Momany, *Biopolymers* **1980**, *19*, 1943
65 J. V. N. Vara Prasad, D. H. Rich, *Tetrahedron Lett.* **1990**, *31*, 1803
66 J. Gante, *Synthesis* **1989**, 405
67 J. Magrath, R. H. Abeles, *J. Med. Chem.* **1992**, *35*, 4279
68 K. Clausen, M. Thorsen, S. O. Lawesson, A. F. Spatola, *J. Chem. Soc., Perkin Trans. I* **1984**, 785
69 L. El Masdouri, A. Aubry, C. Sakarellos, E. J. Gomex, M. T. Cung, M. Marraud, *Int. J. Peptide Protein Res.* **1988**, *31*, 420
70 M. Chorev, M. Goodman, *Acc. Chem. Res.* **1993**, *26*, 266
71 M. Szelke, D. M. Jones, B. Atrash, A. Hallett, B. J. Leckie in: *Peptides: Structure and Function*, V. J. Hruby, D. H. Rich (Eds.), Pierce, Rockford, **1980**, 579
72 M. M. Hann, P. G. Sammes, P. D. Kennewell, J. B. Taylor, *J. Chem. Soc., Chem. Commun.* **1980**, 234
73 M. Kranz, H. Kessler, *Tetrahedron Lett.* **1996**, *37*, 5359
74 D. Yang, J. Qu, B. Li, F.-F. Ng, X.-C. Wang, K. K. Cheung, D.-P. Wang, Y.-D. Wu, *J. Am. Chem. Soc.* **1999**, *121*, 589
75 A. Aubry, J.-P. Mangeot, J. Vidal, A. Collet, S. Zerkout, M. Marraud, *Int. J. Peptide Protein Res.* **1994**, *43*, 305
76 A. Cheguillaume, F. Lehardy, K. Bouget, M. Baudy-Floc'h, P. Le Grel, *J. Org. Chem.* **1999**, *64*, 2924
77 A. Müller, C. Vogt, N. Sewald, *Synthesis* **1998**, 837
78 A. Müller, F. Schumann, M. Koksch, N. Sewald, *Lett. Peptide Sci.* **1997**, *4*, 275
79 F. Schumann, A. Müller, M. Koksch, G. Müller, N. Sewald, *J. Am. Chem. Soc.* **2000**, *122*, 12009
80 M. Hagihara, S. L. Schreiber, *J. Am. Chem. Soc.* **1992**, *114*, 6570
81 W. Bauer, U. Boiner, W. Haller, R. Huguenin, R. Marbach, P. Petcher, J. Pless in: *Peptides* **1982**, K. Blaha, P. Malon (Eds.), de Gruyter, Berlin, **1983**
82 D. F. Veber, R. M. Freidinger, D. S. Prelow, W. J. Poleveda, F. M. Holly, R. G. Strachan, R. F. Nutt, B. H. Arison, C. Homnick, W. C. Randall, M. S. Glitzer, R. Saperstein, R. Hirschmann, *Nature* **1981**, *292*, 55
83 T. K. Sayer, V. J. Hruby, P. S. Darman, M. E. Hadley, *Proc. Natl. Acad. Sci. USA* **1982**, *79*, 1751
84 K. D. Stigers, M. J. Soth, J. S. Nowick, *Curr. Opin. Chem. Biol.* **1999**, *3*, 714
85 S. Hanessian, G. McNaughton-Smith, H.-G. Lombart, W. D. Lubell, *Tetrahedron* **1997**, *53*, 12789
86 P. Gillespie, J. Cicariello, G. L. Olson, *Biopolymers* **1997**, *43*, 191
87 L. Halab, F. Gosselin, W. D. Lubell, *Biopolymers* **2000**, *55*, 101
88 G. D. Rose, L. M. Gierasch, J. A. Smith, *Adv. Protein Chem.* **1985**, *37*, 1
89 C. Nagai, K. Sato, *Tetrahedron Lett.* **1982**, *23*, 3759
90 V. Brandmeier, M. Feigel, *Tetrahedron* **1989**, *45*, 1365
91 D. S. Kemp, W. E. Stites, *Tetrahedron Lett.* **1988**, *29*, 5057
92 M. J. Genin, R. L. Johnson, *J. Am. Chem. Soc.* **1992**, *114*, 8778

93 W.C. Ripka, G.V. DeLucca, A.C. Bach II, R.S. Pottorf, J.M. Blaney, *Tetrahedron* **1993**, *49*, 3593

94 M. Feigel, *Liebigs Ann. Chem.* **1989**, 459

95 M. Feigel, G. Lugert, C. Heichert, *Liebigs Ann. Chem.* **1987**, 367

96 D.S. Kemp, E.T. Sun, *Tetrahedron Lett.* **1982**, *23*, 3759

97 R. Haubner, W. Schmitt, G. Hölzemann, S.L. Goodman, A. Jonczyk, H. Kessler, *J. Am. Chem. Soc.* **1996**, *118*, 7881

98 G. Müller, G. Hessler, H.Y. Decornez, *Angew. Chem.* **2000**, *112*, 926; *Angew. Chem. Int. Ed. Engl.* **2000**, *39*, 894

99 J.F. Callahan, J.W. Bean, J.L. Burgess, D.S. Eggleston, S.M. Hwang, K.D. Kopple, P.F. Koster, A. Nichols, C.E. Peishoff, J.M. Samanen, J.A. Vasko, A. Wong, W.F. Huffman, *J. Med. Chem.* **1992**, *35*, 3970

100 J.F. Callahan, K.A. Newlander, J.L. Burgess, D.S. Eggleston, A. Nichols, A. Wong, W.F. Huffman, *Tetrahedron* **1993**, *49*, 3479

101 M. Sato, J.Y. H. Lee, H. Nakanishi, M.E. Johnson, R.A. Chrusciel, M. Kahn, *Biochem. Biophys. Res. Commun.* **1992**, *187*, 1000

102 D.S. Kemp, T.P. Curran, J.G. Boyd, T.J. Allen, *Biochem. Biophys. Res. Commun.* **1991**, *56*, 6683

103 D.S. Kemp, T.P. Curran, W.M. Davis, J.G. Boyd, C. Muendel, *J. Org. Chem.* **1991**, *56*, 6672

104 R. Sarabu, K. Lovey, V.S. Madison, D.C. Fry, D.W. Greeley, C.M. Cook, G.L. Olson, *Tetrahedron* **1993**, *49*, 3629

105 L. Pauling, *Chem. Eng. News* **1946**, *24*, 1375

106 R. Wolfenden, *Bioorg. Med. Chem.* **1999**, *7*, 647

107 H.-J. Böhm, G. Klebe, H. Kubinyi, *Wirkstoffdesign*, Spektrum Verlag, Heidelberg, Berlin, Oxford, **1996**

108 M.L. Moore, G.B. Dreyer, *J. Comput. Aided Mol. Des.* **1993**, 85

109 K.D. Janda, D. Schloeder, S.J. Benkovic, R.A. Lerner, *Science* **1988**, *241*, 1188

110 J. D. Stewart, P.A. Benkovic, *Chem. Soc. Rev.* **1993**, 213

111 T. Kieber-Emmons, R. Murali, M.I. Green, *Curr. Opin. Biotechnol.* **1997**, *8*, 435

112 D.C. Horwell, *Trends Biotechnol.* **1995**, *13*, 132

113 D.F. Veber in: *Peptides: Chemistry and Biology*, J.A. Smith, J.E. Rivier (Eds.), Escom, Leiden, **1992**, 3

114 B.E. Evans, M.G. Bock, K.E. Rittle, R.M. DiPrado, W.L. Whitter, D.F. Veber, P.S. Anderson, R. M. Freidinger, *Proc. Natl. Acad. Sci. USA* **1986**, *83*, 4918

115 M.A. Ondetti, B. Rubin, D.W. Cushman, *Science* **1977**, *196*, 441

116 L.D. Byers, R. Wolfenden, *J. Biol. Chem.* **1972**, *247*, 606

117 L.D. Byers, R. Wolfenden, *Biochemistry* **1973**, *12*, 2070

118 D.H. Rich, *Compr. Med. Chem.* **1990**, *2*, 400

119 L. Turbanti, G. Cerbai, C. Di Bugno, R. Giorgi, G. Garzelli, M. Criscuoli, A.R. Renzetti, A. Subissi, G. Bramante, *J. Med. Chem.* **1993**, *36*, 699

120 D. Römer, H.H. Büscher, R.C. Hill, R. Maurer, T.J. Petcher, H. Zeugner, W. Benson, E. Finner, W. Milkowski, P.W. Thies, *Nature* **1982**, *298*, 759

121 L.A. Dykstra, D.E. Gmerek, G. Winger, J.H. Woods, *J. Pharmacol. Exp. Ther.* **1987**, *242*, 413

122 A. Peyman, *Chem. Rev.* **1990**, *90*, 543

123 S.H. Gellman, *Acc. Chem. Res.* **1998**, *31*, 173

124 K. Kirshenbaum, R.N. Zuckermann, K.A. Dill, *Curr. Opin. Struct. Biol.* **1999**, *9*, 530

125 J.S. Nowick, *Acc. Chem. Res.* **1999**, *32*, 287

126 R.J. Simon, R.S. Kania, R.N. Zuckermann, V.D. Huebner, D.A. Jewell, S. Banville, S. Wang, S. Rosenberg, C.K. Marlowe, D.C. Spellmeyer, R. Tan, A.D. Frankel, D. V. Santi, F.E. Cohen, P.A. Bartlett, *Proc. Natl. Acad. Sci. USA* **1992**, *89*, 9367

127 P.E. Nielsen, *Acc. Chem. Res.* **1999**, *32*, 624

128 J.M. Fernández-Santín, J. Aymamí, A. Rodríguez-Galán, S. Muoz-Guerra, J.A. Subirana, *Nature* **1984**, *311*, 53

129 D.H. Appella, L.A. Christianson, I.L. Karle, D.R. Powell, S.H. Gellman, *J. Am. Chem. Soc.* **1996**, *118*, 13071

130 D.H. Appella, L.A. Christianson, D.A. Klein, D.R. Powell, X. Huang, J.J. Barchi, Jr. S.H. Gellman, *Nature* **1997**, *387*, 381

131 D. Seebach, P.E. Ciceri, M. Overhand, B. Jaun, D. Rigo, L. Oberer, U. Hommel, R. Amstutz, H. Widmer, *Helv. Chim. Acta* **1996**, *79*, 2043

132 D. Seebach, J. L. Matthews, *J. Chem. Soc., Chem. Commun.* **1997**, 2015

133 W. F. DeGrado, J. P. Schneider, Y. Hamuro, *J. Peptide Res.* **1999**, *54*, 206

134 H. Han, K.D. Janda, *J. Am. Chem. Soc.* **1996**, *118*, 2539

135 H. Han, J. Yoon, K.D. Janda, *Bioorg. Med. Chem. Lett.* **1998**, *8*, 117

136 C. Gennari, B. Salom, D. Potenza, A. Williams, *Angew. Chem.* **1994**, *106*, 2181; *Angew. Chem. Int. Ed. Engl.* **1994**, *33*, 2067

137 W. J. Moree, G. A. van der Marel, R.J. Liskamp, *J. Org. Chem.* **1995**, *60*, 5157

138 C. Gennari, M. Gude, D. Potenza, U. Piarulli, *Chem. Eur. J.* **1998**, *4*, 1924

139 R. Günther, H. J. Hofmann, *J. Am. Chem. Soc.* **2001**, *123*, 247

140 C. Y. Cho, E. J. Moran, S. R. Cherry, J.C. Stephans, S. P.A. Fodor, C.L. Adams, A. Sundaram, J. W. Jacobs, P.G. Schultz, *Science* **1993**, *261*, 1303

141 A. B. Smith, III, T. P. Keenan, R.C. Holcomb, P.A. Sprengeler, M.C. Guzman, J. L. Wood, P.C. Carroll, R. Hirschmann, *J. Am. Chem. Soc.* **1992**, *114*, 10672

142 T. Hintermann, K. Gademann, B. Jaun, D. Seebach, *Helv. Chim. Acta* **1998**, *81*, 983

143 S. Hanessian, X. Luo, R. Schaum, *Tetrahedron Lett.* **1999**, *40*, 4925

144 M. Hagihara, N. J. Anthony, T. J. Stout, J. Clardy, S.L. Schreiber, *J. Am. Chem. Soc.* **1992**, *114*, 6568

145 C. Gennari, C. Longari, S. Ressel, B. Salom, U. Piarulli, S. Ceccarelli, A. Mielgo, *Eur. J. Org. Chem.* **1998**, 2437

146 M. D. Smith, T. D.W. Claridge, G.E. Tranter, M.S.P. Sansom, G.W.J. Fleet, *J. Chem. Soc., Chem. Commun.* **1998**, 2041

147 E. Locardi, M. Stöckle, S. Gruner, H. Kessler, *J. Am. Chem. Soc.* **2001**, *123*, 8189

148 R.N. Zuckermann, J. M. Kerr, S.B.H. Kent, W. H. Moos, *J. Am. Chem. Soc.* **1992**, *114*, 10646

149 H. Kessler, *Angew. Chem.* **1993**, *105*, 572; *Angew. Chem. Int. Ed. Engl.* **1993**, *32*, 543

150 K. Kirshenbaum, A. E. Barron, R.A. Goldsmith, P. Armand, E. K. Bradley, K.T.V. Truong, K.A. Dill, F.E. Cohen, R.N. Zuckermann, *Proc. Natl. Acad. Sci. USA* **1998**, *95*, 4303

151 P. Armand, K. Kirshenbaum, R.A. Goldsmith, S. Farr-Jones, A.E. Barron, K.T.V. Truong, K.A. Dill, D.F. Mierke, F.E. Cohen, R.N. Zuckermann, E.K. Bradley, *Proc. Natl. Acad. Sci. USA* **1998**, 95, 4309

152 P.E. Nielsen, M. Egholm, R.H. Berg, O. Buchardt, *Science* **1991**, *25*, 1497

153 M. Egholm, O. Buchardt, P.E. Nielsen, R.H. Berg, *J. Am. Chem. Soc.* **1992**, *114*, 1895

154 C. Meier, J.W. Engels, *Angew. Chem.* **1992**, *104*, 1039; *Angew. Chem. Int. Ed. Engl.* **1992**, *31*, 1008

155 H. De Koning, U.K. Pandit, *Recl. Trav. Chim. Pays-Bas* **1971**, *90*, 1069

156 J.D. Buttrey, A.S. Jones, R.T. Walker, *Tetrahedron* **1975**, *31*, 73

157 S.A. Thomson, J.A. Josey, R. Cadilla, M.D. Gaul, C.F. Hassman, M.J. Luzzio, A.J. Pipe, K.L. Reed, D.J. Ricca, R.W. Wiethe, S.A. Noble, *Tetrahedron* **1995**, *51*, 6179

158 J. Kovacs, R. Ballina, R. Rodin, O. Balasubramanian, J. Applequist, *J. Am. Chem. Soc.* **1965**, *87*, 119

159 F. López-Carrasquero, M. García-Alvarez, J.J. Navas, C. Alemán, S. Muoz-Guerra, *Macromolecules* **1996**, *29*, 8449

160 D.H. Appella, J.J. Barchi, S.R. Durell, S.H. Gellman, *J. Am. Chem. Soc.* **1999**, *121*, 2309

161 P.I. Arvidsson, M. Rueping, D. Seebach, *J. Chem. Soc., Chem. Commun.* **2001**, 649

162 R.P. Cheng, W.F. DeGrado, *J. Am. Chem. Soc.* **2001**, *123*, 5162

163 X. Daura, W.F. van Gunsteren, D. Rigo, B. Jaun, D. Seebach, *Chem. Eur. J.* **1997**, *3*, 1410

164 D. Seebach, K. Gademann, J.V. Schreiber, J.L. Matthews, T. Hintermann, B.

JAUN, L. OBERER, U. HOMMEL, H. WIDMER, *Helv. Chim. Acta* **1997**, *80*, 2033
165 E. JUARISTI (Ed.), *Enantioselective Synthesis of β-Amino Acids*, Wiley-VCH, New York, **1997**
166 J. PODLECH, D. SEEBACH, *Liebigs Ann. Chem.* **1995**, 1217
167 P. BALARAM, *J. Peptide Res.* **1999**, *54*, 195
168 W. F. DEGRADO, C. M. SUMMA, V. PAVONE, F. NASTRI, A. LOMBARDI, *Annu. Rev. Biochem.* **1999**, *68*, 779
169 R. B. HILL, D. P. RALEIGH, A. LOMBARDI, W. F. DEGRADO, *Acc. Chem. Res.* **2000**, *33*, 745
170 G. TUCHSCHERER, L. SCHEIBLER, P. DUMY, M. MUTTER, *Biopolymers* **1998**, *47*, 63
171 G. TUCHSCHERER, G. GRELL, M. MATHIEU, M. MUTTER, *J. Peptide Res.* **1999**, *54*, 185
172 J. FERNANDEZ-CARNEADO, D. GRELL, P. DURIEUX, J. HAUERT, T. KOVACSOVICS, G. TUCHSCHERER, *Biopolymers* **2000**, *55*, 451
173 P. KOEHL, M. LEVITT, *J. Mol. Biol.* **1999**, *293*, 1161
174 P. KOEHL, M. LEVITT, *J. Mol. Biol.* **1999**, *293*, 1183
175 D. N. WOOLSON, *Curr. Opin. Struct. Biol.* **2001**, *11*, 464
176 H.-B. KRAATZ, *Angew. Chem.* **1994**, *106*, 2143; *Angew. Chem. Int. Ed. Engl.* **1994**, *33*, 2055
177 M. FAVRE, K. MOEHLE, L. JIANG, B. PFEIFFER, J. A. ROBINSON, *J. Am. Chem. Soc.* **1999**, *121*, 2679
178 L. JIANG, K. MOEHLE, B. DHANAPAL, D. OBRECHT, J. A. ROBINSON, *Helv. Chim. Acta* **2001**, *83*, 3097
179 J. M. TRAVINS, F. A. ETZKORN, *J. Org. Chem.* **1997**, 8387
180 A. BANERJEE, S. RAGHOTHAMA, P. BALARAM, *J. Chem. Soc., Perkin Trans. 2* **1997**, 2087
181 C. K. SMITH, L. REGAN, *Acc. Chem. Res.* **1997**, *30*, 153
182 C. MICKLATCHER, J. CHMIELEWSKI, *Curr. Opin. Chem. Biol.* **1999**, *3*, 724
183 M. R. GHADIRI, M. A. CASE, *Angew. Chem.* **1993**, *105*, 1663; *Angew. Chem. Int. Ed. Engl.* **1993**, *32*, 1594
184 D. E. ROBERTSON, R. S. FARID, C. C. MOSER, J. L. URBAUER, S. E. MULHOLLAND, R. PIDIKITI, J. D. LEAR, A. J. WAND, W. F. DEGRADO, P. L. DUTTON, *Nature* **1994**, *368*, 425
185 A. LOMBARDI, C. M. SUMMA, S. GEREMIA, L. RANDACCIO, V. PAVONE, W. F. DEGRADO, *Proc. Natl. Acad. Sci. USA* **2000**, *97*, 6298
186 M. R. GHADIRI, *Adv. Mater.* **1995**, *7*, 675
187 M. MUTTER, G. VUILLEUMIER, *Angew. Chem.* **1989**, *101*, 551; *Angew. Chem. Int. Ed. Engl.* **1989**, *28*, 535
188 G. TUCHSCHERER, M. MUTTER, *J. Biotechnol.* **1995**, *41*, 197
189 G. TUCHSCHERER, C. LEHMANN, M. MATHIEU, *Angew. Chem.* **1998**, *110*, 3160; *Angew. Chem. Int. Ed. Engl.* **1998**, *37*, 2990
190 K. ROSE, L. A. VILASECA, R. WERLEN, A. MEUNIR, I. FISCH, R. M. L. JONES, R. E. OFFORD, *Bioconj. Chem.* **1991**, *2*, 154
191 O. NYANGUILE, M. MUTTER, G. TUCHSCHERER, *Lett. Peptide Sci.* **1994**, *1*, 9
192 R. SCHNEPF, P. HÖRTH, E. BILL, K. WIEGHARDT, P. HILDEBRANDT, W. HAEHNEL, *J. Am. Chem. Soc.* **2001**, *123*, 2186
193 P. BURKHARD, J. STETEFELD, S. V. STRELKOV, Trends Cell Biol. **2001**, *11*, 82
194 F. C. H. CRICK, *Acta Crystallogr.* **1953**, *6*, 689
195 J. A. TALBOT, R. S. HODGES, *Acc. Chem. Res.* **1982**, *15*, 224
196 N. E. ZHOU, B.-Y. ZHU, C. M. KAY, R. S. HODGES, *Biopolymers* **1992**, *32*, 419
197 J. G. ADAMSON, N. E. ZHOU, R. S. HODGES, *Curr. Opin. Biotechnol.* **1993**, *4*, 428
198 W. D. KOHN, C. M. KAY, R. S. HODGES, *Protein Sci.* **1995**, *4*, 237
199 K. S. THOMPSON, C. R. VINSON, E. FREIRE, *Biochemistry* **1993**, *32*, 5491
200 R. B. HILL, W. F. DEGRADO, *J. Am. Chem. Soc.* **1998**, *120*, 1138
201 N. E. ZHOU, C. M. KAY, R. S. HODGES, *J. Mol. Biol.* **1994**, *237*, 500
202 D. M. ECKERT, V. N. MALASHKEVICH, P. S. KIM, *J. Mol. Biol.* **1998**, *284*, 859
203 H. WENDT, C. BERGER, A. BAICI, R. M. THOMAS, H. R. BOSSHARD, *Biochemistry* **1995**, *34*, 4097
204 A. J. KENNAN, V. HARIDAS, K. SEVERIN, D. H. LEE, M. R. GHADIRI, *J. Am. Chem. Soc.* **2001**, *123*, 1797
205 P. VEPŘEK, J. JEŽEK, *J. Peptide Sci.* **1999**, *5*, 5

206 N.J. Wells, A. Basso, M. Bradley, *Biopolymers* **1998**, *47*, 381

207 J.P. Tam, *Proc. Natl. Acad. Sci. USA* **1988**, *85*, 5409

208 A. Basak, A. Boudreault, A. Chen, M. Chretien, N.G. Seidah, C. Lazure, *J. Peptide Sci.* **1995**, *1*, 385

209 B. Nardelli, Y.-A. Lu, R.D. Shiu, C. Delpiere-Defoort, A.T. Profy, J.P. Tam, *J. Immunol.* **1992**, *148*, 914

210 J.P. Tam, Y.A. Lu, *J. Am. Chem. Soc.* **1995**, *117*, 12058

211 K.J. Jensen, G. Barany, *J. Peptide Res.* **2000**, *56*, 3

212 J.P. Mitchell, K.D. Roberts, J. Langley, F. Koentgen, J.N. Lambert, *Bioorg. Med. Chem. Lett.* **1999**, *9*, 2785

213 P. Vepřek, J. Ježek, *J. Peptide Sci.* **1999**, *5*, 203

214 A.T. Florence, T. Sakthivel, I. Toth, *J. Controlled Release* **2000**, *65*, 253

215 G. Purohit, T. Sakthivel, T. Florence, *Int. J. Pharm.* **2001**, *214*, 71

216 J.S. Choi, E.J. Lee, Y.H. Choi, Y. J. Jeong, J.S. Park, *Bioconj. Chem.* **1999**, *10*, 62

217 M. Sakamoto, A. Ueno, H. Mihara, *Chem. Eur. J.* **2001**, *7*, 2449

218 S.A. Vinogradov, D.F. Wilson, *Chem. Eur. J.* **2000**, *6*, 2456

219 M. Sakamoto, T. Kamachi, I. Okura, A. Ueno, H. Mihara, *Biopolymers* **2001**, *59*, 103

220 H. Ryser, R. Hancock, *Science* **1965**, *150*, 501

221 D.J. Mitchell, D.T. Kim, L. Steinman, C.G. Fathman, J.B. Rothbard, *J. Peptide Res.* **2000**, *56*, 318

222 A.D. Miller, *Angew. Chem.* **1998**, *110*, 1862; *Angew. Chem. Int. Ed. Engl.* **1998**, *37*, 1768

223 L.C. Smith, J. Duguid, M.S. Wadhwa, M.J. Logan, C.-H. Tung, V. Edwards, J.T. Sparrow, *Adv. Drug Deliv. Rev.* **1998**, *30*, 115

224 D.W. Urry, *Biopolymers* **1998**, *47*, 167

225 E. Allémann, J.-C. Leroux, R. Gurny, *Adv. Drug Deliv. Rev.* **1998**, *34*, 171

226 N. Belcheva, S.P. Baldwin, W.M. Saltzman, *J. Biomater. Sci. Polym. Ed.* **1998**, *9*, 207

227 K. Kataoka, G.S. Kwon, M. Yokoyama, T. Okano, Y. Sakurai, *J. Controlled Release* **1993**, *24*, 119

8
Combinatorial Peptide Synthesis

During the mid-1980s, several different methods for the synthesis of compound libraries emerged that were inspired by the concept of selection, as verified by nature. This branch of preparative chemistry has been termed combinatorial chemistry [1–7]. Combinatorial chemistry represents an alternative to traditional approaches in pharmaceutical drug development and optimization. Compound libraries of different types [8] eventually introduced a complete change of paradigms in pharmaceutical chemistry.

Combinatorial peptide synthesis and combinatorial organic synthesis – which can be seen as a further development of the former – nowadays have gained widespread acceptance and application. They aim towards generating high synthetic and structural diversity [9]. Modern medicinal chemistry relies on different types of compounds (e.g., peptides, natural products, or small organic molecules) for biological testing, and one source of these is synthesis. The development track proceeds sequentially from test compounds over "lead compounds" to eventually a "drug candidate" (Fig. 8.1).

Formerly, most of the compounds had to be tested in animal models at a very early development stage. However, the past 20 years have witnessed enormous advances in biochemistry and biology (including structural biology) that have led to an improved understanding of the role of proteins and their ligands in pathologic processes. A drug candidate often is a surrogate for a natural ligand that possesses the appropriate structural complementarity such that it may selectively mimic the natural ligand with a therapeutically beneficial result. Consequently, nowadays protein-based primary screening methods have replaced testing in animal models [10].

Usually, several thousands of different compounds have to be synthesized and biologically tested in order to identify a pharmacological lead compound. Therefore, the pace of synthesis had to be increased considerably to satisfy the demand in structural diversity of chemical compounds. The fascination of combinatorial chemistry basically lies in its potential to synthesize several hundreds or thousands of test compounds, and to subject them to biological testing in order to significantly reduce the time necessary for the development of a drug. High-throughput screening enables scientists to test thousands of compounds per day in a biological assay. Combinatorial syntheses of peptides (peptide libraries) and organic

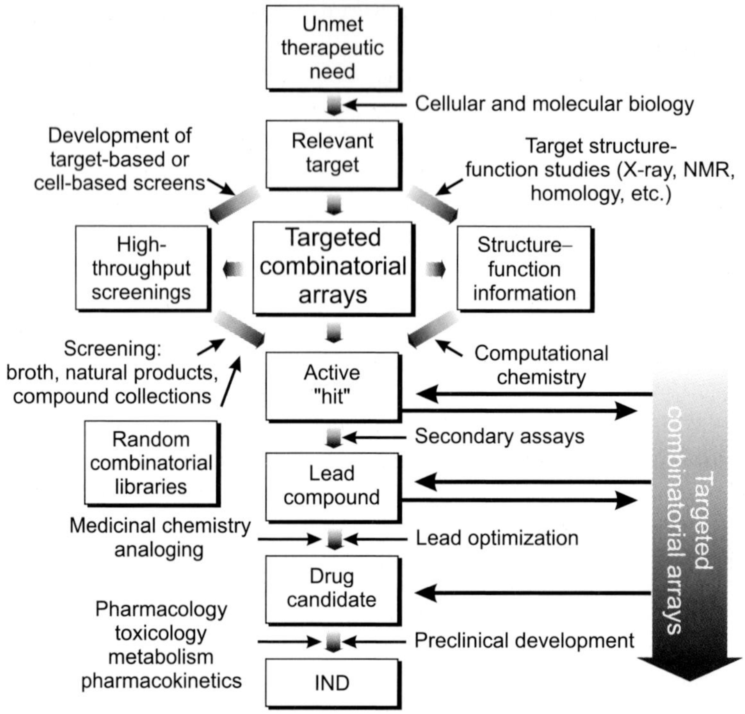

Fig. 8.1 The role of combinatorial technology in the drug discovery process (IND: investigational new drug; adapted from [10]).

compounds (diversomer libraries) are especially suited for the development of lead structures in pharmaceutical research. The expression "library" has originally been coined exclusively for compound mixtures, whilst collections of single compounds are often called "arrays".

Methods for combinatorial synthesis were developed independently by several groups on the basis of different concepts following to two major strategies:

1. Parallel (multiple) synthesis: this comprises the targeted simultaneous multiple peptide synthesis (SMPS). Single substances (peptides) are obtained in a parallel manner, for instance as analogues of a biologically active peptide [11].
2. The synthesis of complex mixtures of compounds (libraries) [12].

The synthetic peptide combinatorial libraries (SPCL) are synthesized and tested as mixtures. Many bioassays used for the development of pharmacologically active compounds display sufficient sensitivity and an inherent biological specificity for active compounds present in a mixture. Arguments against this combinatorial synthesis of complex mixtures with respect to the demanding chemical criteria of purity have been opposed by this fact. However, it has been postulated that, for example, the antagonist activity of one compound may override an agonist activity of another compound present in the mixture, leading to a "null" (false-negative) response. Furthermore, synergism between different library members (false-positive response) cannot be excluded.

While parallel syntheses usually are limited to arrays comprising several hundreds to thousands of compounds, depending on the reaction scale required and the method used, the diversity of a mixture-based combinatorial library is, in principle, only restricted by the detection limit of the biological assay and by the total mass of substance to be produced. Full peptide libraries often are constructed using 19 of the proteinogenic amino acids (Cys normally is omitted). However, many nonproteinogenic or even non-natural amino acids nowadays are commercially available, leading to a high diversity. The numbers (V^P) of different peptides composed of 19 L-amino acids (19 variables V) present in a library, depending on the number of positions P to be varied, are listed in Tab. 8.1.

It can easily be calculated that, for a high number of variable positions or a high number of different building blocks (variables), the total mass of the product mixture (including the resin mass in the case of a solid-phase synthesis) increases exponentially. Consequently, if the reaction scale is limited, such combinatorial mixtures may remain incomplete. Moreover, mixture-based libraries may suffer from nonequimolar incorporation of building blocks because of differences in reactivities. Methods to overcome this problem will be discussed in Section 8.2.

Synthetic strategy is another criterion to distinguish between two different approaches in combinatorial chemistry: solid-phase synthesis versus solution synthesis.

In solid-phase combinatorial synthesis, reagents can be used in excess without separation problems in order to attain complete conversion. Facile purification and automation of the process represent further advantages that have already been discussed in Chapter 4. While solid-phase peptide synthesis (SPPS) nowadays is well developed, the solid-phase synthesis of organic molecules imposes different requirements, additional reaction steps, and often involves further development, for example of the linker or protecting group chemistry [13]. Monitoring of

Tab. 8.1 Library diversity correlates with the number of variable position and the number of diversity elements (e.g., 19 amino acids).

Number of variable positions	*Number of peptides*	*Average mixture weight [1 pmol per peptide]*
2	361	92 ng
3	6859	2.58 µg
4	130321	64.76 µg
5	2476099	1.53 mg
6	47045881	34.64 mg
7	893871739	765.25 mg
8	16983563041	16.57 g
9	322687697779	353.52 g
10	6131066257801	7.45 kg

the reaction progress in most cases is difficult. The solution-phase synthesis of organic molecules [14] may utilize all organic reactions known, without the need for special linker systems and for any adaptation of known reaction conditions. However, reagents can usually not be used in excess, and process automation is much more difficult [15]. Solid-supported reagents and scavengers very much facilitate solution-phase combinatorial synthesis [16].

Which strategy and which degree of diversity (library size) has to be chosen depends on the problem to be investigated. If nothing is known about a target, then maximum diversity is desirable, and the highest possible number of different compounds should be screened in such a case. However, the increasing knowledge of biochemical concepts and biopolymer structures also contributes considerably to lead structure detection. Structure-based molecular design can be employed to guide the lead-finding process. Structural data on a target protein are applied in a rational approach using computational methods to identify potential low-molecular mass ligands. In such a case a smaller, focused library (e.g., an array of single compounds) may be preferred, which is called the targeted diversity approach [10]. Once a lead compound has been identified, the diversity requirements change fundamentally, and structure-activity relationships are performed by substituent variation to optimize the compound properties with the aim of developing a drug candidate. Consequently, a sensible combination of different methodologies and technologies will eventually succeed [17].

8.1 Parallel Synthesis

Multiple peptide synthesis (MPS) means the simultaneous (parallel) synthesis of a multitude of peptide sequences, irrespective of the chain length and amino acid composition.

Peptide chemists first recognized that compound libraries could be synthesized in a combinatorial manner. Many strategies of multiple peptide synthesis (or multiple synthesis in general) are special variants of synthesis on a polymeric support [7]. Combinations of protection groups and activation methods of solid-phase strategy relevant to peptide chemistry can be applied to multiple synthesis. The methodological arsenal ranges from peptide synthesis in the so-called teabags developed by Houghten et al. [18], multipin synthesis (polyethylene rods as polymeric support) developed by Geysen et al. [19, 20], spot synthesis developed by Frank [21], to methods with pipetting robots and fully automated synthesis. Furka et al. very much contributed to the development of combinatorial chemistry with the development of the "split and combine" technique [22, 23]. This technique provides a library in the form of a spatially resolved compound mixture ("one bead – one compound") and will, therefore, be detailed in Section 8.2.

There are differences between the variants of multiple peptide synthesis with respect to the polymeric support used, the number of different peptides to be synthesized with this method, and the amount of products formed in the synthesis. Besides the classical Merrifield support polystyrene/divinylbenzene (DVB), polystyrene/polyethylene films, polyethylene rods and cellulose sheets are also used. A full account on combinatorial methods is far beyond the scope of this book, but some of the different methods will be introduced briefly.

8.1.1
Synthesis in Teabags [18]

A certain amount of the polymeric support is secluded in a porous polypropylene bag marked with solvent-resistant ink. Usually, about 100 mg of polystyrene cross-linked with 1% divinylbenzene are used per teabag [24]. Other support materials may also be used, but the particle size of the polymeric resin should always be bigger than the porosity of the polypropylene bag (64–74 μm). The number of different compounds to be synthesized is equal to the number of teabags used. Cleavage reactions of, for example the α-amino-protecting groups may be performed in one common vessel for all teabags. The number of teabags where the same amino acid component has to be coupled is combined in one reaction vessel and reacted with the activated amino acid component, followed by washing. Vigorous shaking is a necessary precondition for a complete reaction. After each coupling step, the teabags are removed from the reaction containment, pooled, and the N-protecting groups are then cleaved. The teabags are then sorted again, combined appropriately, and treated with the next activated amino acid in separate reaction vessels. In order to remove excess reagents, the teabags are washed and combined again for the next deblocking reaction. Coupling, washing, and deprotection are the repetitive steps (as discussed for SPPS in Chapter 4). Before the last coupling step, the teabags are carefully washed and dried, after which cleavage of the peptides from the polymeric support is performed. The teabag method is appropriate for the Boc tactics using, for example *N,N'*-diisopropylcarbodiimide as the coupling reagent, and employing cleavage from the resin with liquid HF [18, 24], but also for the Fmoc tactics [25], as demonstrated in Fig. 8.2.

The teabag method usually is suitable for the synthesis of amounts between 30 and 50 mg, and a peptide length of more than 10 amino acids. In contrast to other methods of parallel synthesis, purification and characterization prior to biological testing is possible. This protocol is characterized by a high variability and relatively low costs, but it is highly labor-intensive because of the multitude of manual steps. Computer-assisted calculation of the synthetic cycles and the sorting of the resin bags is essential. Automation of the washing steps by using a combination of the teabag method with suitable semi-automatic peptide synthesizers is advantageous [25]. Peptides synthesized by the teabag method have been applied to the investigation of structure-activity relationships, antigen-antibody interactions, and for the conformational mapping of proteins.

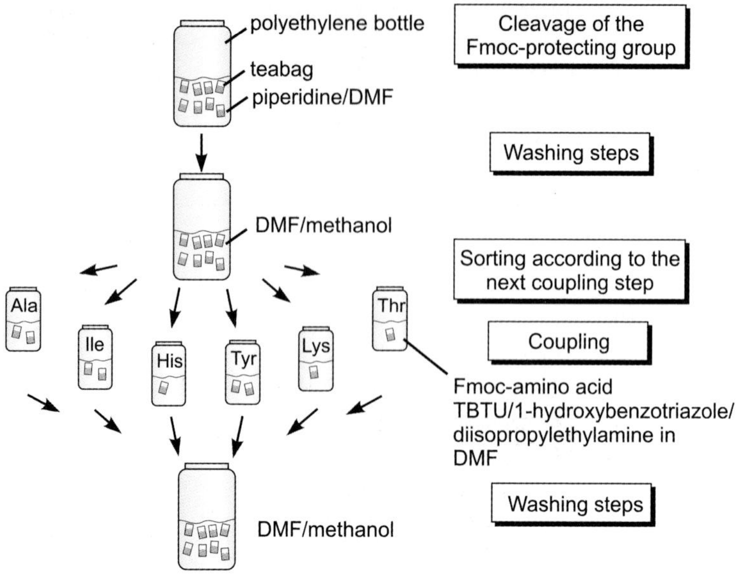

Fig. 8.2 Multiple peptide synthesis according to the teabag method.

8.1.2
Synthesis on Polyethylene Pins (Multipin Synthesis)

Parallel peptide synthesis on polyethylene rods [19] according to the multipin method does usually not provide larger amounts of peptides. In many cases, after the synthesis only the protecting groups are cleaved and the peptides remain bound to the polyethylene rods, for instance for antibody binding studies. Peptide assembly is performed using amino functionalized polyethylene rods (pins) about 4 mm in diameter and 40 mm in length (Fig. 8.3). However, larger rods are also commercially available. Usually, 96 of these pins are assembled on one block in eight rows, with 12 pins each. This arrangement allows for the application in enzyme-linked immunosorbent assay (ELISA) after the synthesis is complete, because the pins fit exactly into the wells of the microtiter plates usually applied in this method. Computer software allows for the calculation of the amounts and distribution of the protected amino acid components into the different wells. Peptide coupling reactions are performed in parallel in the wells of microtiter plates. The activated amino acid component is filled into the corresponding well, and the coupling reaction is started by dipping the whole array of pins mounted onto one plate into the complementary array of wells on the microtiter plate. Cleavage and washing steps usually can be performed in one bulk solution. Instead of pins permanently grafted to the pin holder, removable "crowns" of different sizes can be used [26]. This method is shown schematically in Fig. 8.3. A pin is equipped with Fmoc-protected linker moieties (Fig. 8.3 A) and, after Fmoc cleavage, the first amino acid is attached. An array of pins uses a complementary array of microtiter plate wells as reactors (Fig. 8.3 B).

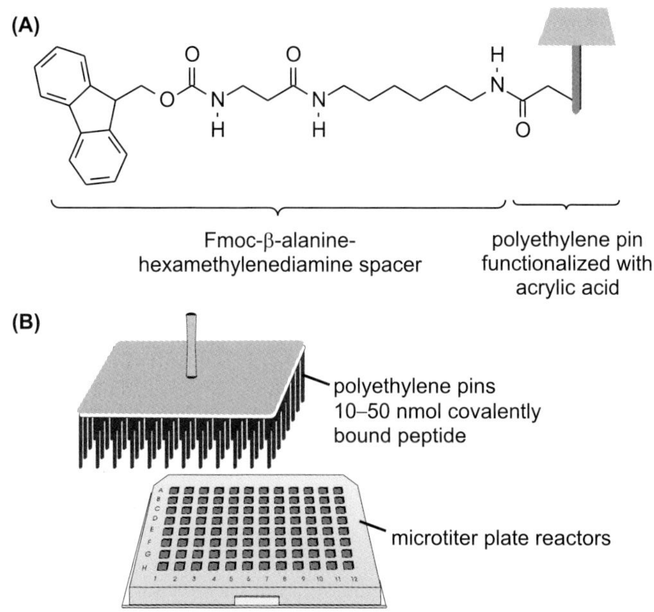

Fig. 8.3 Schematic view of parallel peptide synthesis on polyethylene rods (multipin synthesis). (A) Polyethylene rod (4×40 mm) with Fmoc-protected β-Ala-hexamethylene spacer. (B) Polyethylene rods are mounted on a plate. The parallel reactions are performed in a complementary microtiter plate.

This protocol, which usually provides peptides without any information on the purity, is mainly oriented towards immune analysis. Epitope mapping, which means the systematic screening of partial amino acid sequences of a protein, can be efficiently performed using multipin technology. For example, with this method the amino acid sequence required to be recognized by an antibody can be determined. Between 10 nmol and 2 µmol of a single peptide can be synthesized on each pin, depending on the pin size; hence, this procedure is virtually impractical for preparative synthesis. The pin method also allows the synthesis either of peptides permanently bound to the pins, or of free peptides after cleavage of the final product from the polyethylene rod.

8.1.3
Parallel Synthesis of Single Compounds on Cellulose or Polymer Strips

The spot synthesis [27] is, like the pin method, suited for the synthesis of peptides in minute amounts, and in most cases is restricted to the final products remaining anchored to the matrix. The spot synthesis utilizes a planar sheet of cellulose as the polymeric support. The first (C-terminal) amino acid is connected via an ester bond and a linker molecule to the hydroxy groups of cellulose. Residual hydroxy functions of cellulose are subsequently deactivated. The amino-protecting group is then cleaved, and the next Fmoc amino acid component is attached to the free amino group. Amino acids and coupling reagents are distributed by a pipetting robot in 1 μL aliquots onto the cellulose sheet at a distance of about 1 cm or less. All washing operations, the capping reaction with acetic anhydride, and the cleavage of the Fmoc protecting group is performed by dipping the whole sheet of cellulose into the appropriate reagent solutions. On completion of the synthesis, side-chain protecting groups are cleaved with 20% trifluoroacetic acid/1% triisobutylsilane. This low-cost technology can also be used for screening of biological activity. The peptides are usually not cleaved from the cellulose sheet, and consequently the sheet with the bound peptides can be treated and tested like Western blot membranes. An array of between 96 and several thousands of peptides can be created on a single cellulose sheet in a parallel manner. Despite the fact that only 100 nmol of peptide per spot can be synthesized, free peptides may be obtained. For this purpose, a potential cleavage site (usually in the form of a cleavable linker) must be incorporated before attaching the first amino acid [28]. The single spots can be punched out and the peptides may be cleaved in separate vessels.

Even before the development of spot synthesis an interesting procedure for the multiple peptide synthesis on cellulose as polymeric support had been developed [29, 30], based on previous experience in oligonucleotide synthesis. Small cellulose discs are numbered and used for the semi-automatic synthesis of a peptide. Before each coupling, the cellulose discs (1.55 cm diameter) are appropriately sorted with respect to common building blocks to be coupled, and are piled up to form columns. Each column is then separately subjected to the coupling of the amino acid component. Correct sorting is assisted by computer software, which also minimizes the number of synthesis cycles. The peptides are obtained in crude yields of 8–10 mg, and can also be cleaved from the support when suitable anchoring groups have been used. Immunological tests may be performed without cleavage from the solid support.

Paper strips (Whatman 540) have also been used as solid support. Loading of up to 1–2 $\mu mol\ cm^{-2}$ has been described [31]. Peptides can be synthesized using this approach according to the Boc or Fmoc tactics, and can be cleaved from the support. The peptides bound to the paper may also be subjected after side-chain deprotection directly to antibody binding assays in an ELISA. The process is somewhat related to the teabag strategy in terms of methodology, but cutting of the paper is easier than filling and properly closing the teabags. One obvious disadvantage is the low mechanical stability and the polyfunctionality of the cellulose support.

Cotton-wool displays higher mechanical stability, and therefore discs made from cotton with a diameter of 3 cm proved advantageous as a support in multiple synthesis [32, 33]. In this case, the solvent can be removed after the washing steps by centrifugation.

Chemically functionalized polystyrene-polyethylene films may also be utilized as solid support material in the form of small pieces (1.5×3 cm). The films can be subjected to deblocking reactions and washing steps altogether while being separated into different flasks for the coupling reaction [34]. Although the peptide yields are similar to those obtained using the teabag technology, this variant represents a clear simplification with respect to handling because no tedious filling and closing of the teabags is necessary. The copolymer polystyrene-polyethylene also safeguards minimal loss of resin.

Further automation of multiple peptide synthesis is accompanied by a significant increase in efficiency. The different variants of multiple peptide synthesis described above are clearly characterized by a high percentage of manual operations. On the other hand, repetitive steps such as deprotections, washing and coupling steps offer possibilities for automation. On the basis of using software-controlled pipetting robots, fully automatic multiple peptide synthesis machines have been developed [35, 36]. The number of tasks in manual multiple peptide synthesis may also be reduced by using semiautomatic machines.

8.1.4
Light-Directed, Spatially Addressable Parallel Synthesis

A photolithographic technique, which is well established in microchip technology, enables the parallel synthesis of a compound library on a glass surface. As one certain peptide sequence corresponds to a discrete XY coordinate on the disc, peptides binding with high affinity to a protein can be directly identified in the biological test. Further selective and sensitive analytical methods are not required in this case. The light-directed, spatially addressable parallel chemical synthesis is based on a combination of photolithographic techniques with solid-phase synthesis using photolabile protecting groups such as 6-nitroveratryloxycarbonyl Nvoc **1** [37]. Amino-functionalized glass plates are used as the solid support for this technique [38, 39]. Using suitable masks, photodeprotection of the amino terminus is possible with a spatial resolution of 50×50 μm. Light of wavelength 365 nm is used for cleavage (Fig. 8.4), and subsequently the next Nvoc-protected amino acid is coupled. An array of 2^n different compounds may be synthesized using n different photolithographic masks, each covering half of the surface area and allowing for deprotection in the uncovered area.

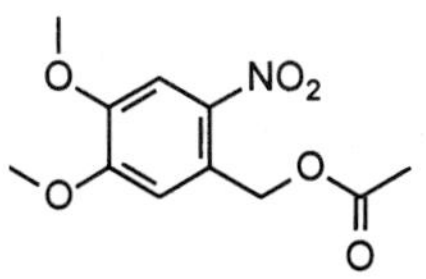

1

An appropriate choice and layout of the lithographic mask enables the synthesis of up to 10000 different peptides per cm^2. The photolithographic technique allows for an exact positioning. Screening with biological test systems can be performed, using fluorescence microscopy or laser confocal fluorescence detection and fluorescently labeled antibodies.

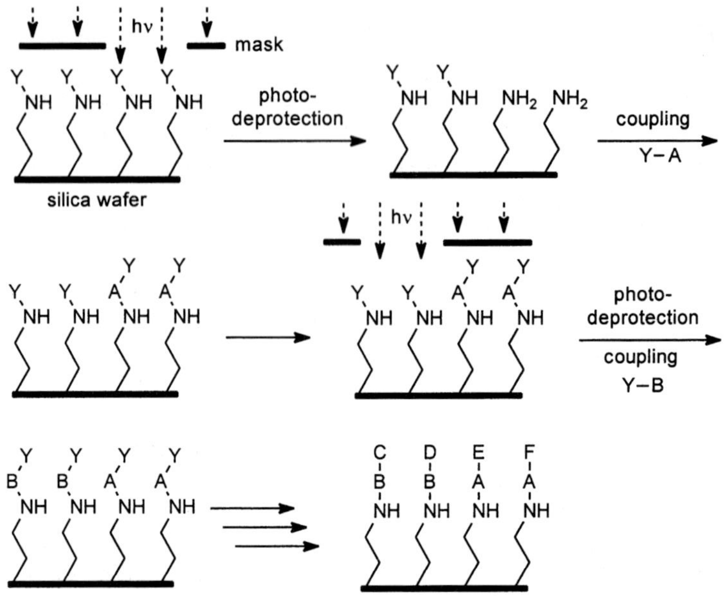

Fig. 8.4 Light-directed, spatially addressable multiple synthesis. Y = photolabile N^{α}-protecting group; A, B, C, D, E, F = amino acids.

8.1.5
Liquid-Phase Synthesis using Soluble Polymeric Support

The principle of liquid-phase peptide synthesis on soluble polymeric support, which was first described by Mutter and Bayer in 1974 [40] (c.f. Section 5.3.4.2), was applied to combinatorial synthesis in 1995 by Janda, and has been termed liquid-phase combinatorial synthesis [14, 41]. Polyethyleneglycol monomethylester with a molecular mass of ~ 5 kDa was used as the soluble polymeric support. This material proved useful in peptide, oligonucleotide, and oligosaccharide synthesis. The crystallization method according to Bayer and Mutter is utilized most favorably on each step of the combinatorial synthetic process. The addition of a suitable organic solvent leads to the crystallization/precipitation of the peptidyl polymer with formation of helical structures, thus avoiding inclusions. A further advantage of the synthesis in solution is the application of a recursive deconvolution strategy [42] in order to obtain and identify the most interesting library. Furthermore, the synthetic progress during this combinatorial synthesis can be monitored using NMR spectroscopy.

8.2
Synthesis of Mixtures

Peptide libraries (mixtures) [11] are obtained using special methods of combinatorial peptide synthesis. In contrast to classical peptide synthesis, it is not the carefully purified and characterized single peptide that is the focal point in this case. A library should be a mixture of peptides with an optimum degree of heterogeneity. Such peptide libraries can be composed of from hundred thousands to millions of different peptides. The advantages of combinatorial synthesis have to be seen in the relatively small number of synthetic steps. The application of suitable testing methods allows the identification and isolation of peptides with a specific mode of action. Sequencing and characterization of these compounds provides the precondition for the synthesis of this single compound on a larger scale. Combinatorial synthesis also permits the assembly of organic compounds that no longer resemble the original peptidic structure, hence providing perspectives for the synthesis on nonoligomeric compound libraries and screening for lead structures.

Suitable biochemical or biological testing methods in combination with synthetic peptide libraries enable investigations to be made on diverse biochemical issues in a much more efficient manner compared to the methods used previously. Epitope mapping of proteins of differing size for diagnostics, as well as the development of vaccines, is of major importance in this context. Furthermore, the systematic variation of amino acid building blocks allows for a rapid and exact study of the folding tendency of longer peptides. Peptide libraries have also contributed significantly to the development of enzyme inhibitors, as well as to screening for new enzymes or the characterization of the specificity of known enzymes. Peptide-specific monoclonal antibodies can be obtained starting from lipopeptide-antigen conjugates containing complete protein sequences in the form of overlapping peptide epitopes. These conjugates allow targeted screening for epitopes of, for example, B-, T-helper, and T-killer cells.

The combination of only 20 proteinogenic amino acids provides 20^6, or 64 million hexapeptides. The synthesis of such a diverse library has been made possible with the development of modern synthesis robots, whilst the testing of such diverse compounds, using high-throughput screening has opened up new dimensions in drug research compared to the classical procedure. One problem might be that the components of a library are not usually checked in terms of their purity, and so the library may be incomplete. In biological testing, the active compound is identified and characterized; subsequently its structure must be elucidated in order to validate the potential lead structure.

8.2.1
Reagent Mixture Method

The "reagent mixture" method allows a mixture of building blocks to be incorporated into the molecule anywhere within the reaction scheme. This method uses a mixture of reagents with predefined ratio in excess with respect to the second reactand to achieve nearly equimolar incorporation of each building block (mixture member) at the position of diversity [43–46]. Equal incorporation of each diversity element (building block) requires profound knowledge of the mechanism and kinetics involved in the specific reaction. Equimolar incorporation of the diversity elements may also be achieved by applying a large excess of each building block where the molar ratio of the building blocks is adjusted according to their different reactivity (isokinetic mixture) [43]. Consequently, a building block with higher reactivity will be applied in the mixture at a lower concentration compared to a building block with lower reactivity. Another method uses double couplings of equimolar building block mixtures, without considering the different reactivities. Consequently, some sequences may be formed in greater amounts than others.

8.2.2
Split and Combine Method

Special procedures are required for the assembly of peptide libraries (mixtures) that contain more than 1000 peptides. The split and combine method [22, 47–49] uses chemically functionalized polymeric beads applied in solid-phase synthesis. Both Boc and Fmoc tactics may be applied. The method is based on the separation of the beads into equal portions prior to coupling, whereas deprotection of the N^{α}-protecting groups and the washing steps are performed together. The resin is divided into a number of portions of about equal size and corresponding to the

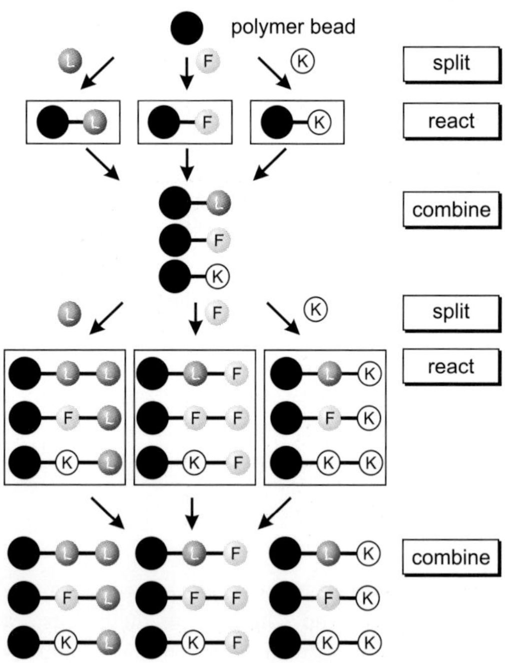

Fig. 8.5 "Split and combine" strategy.

number of different amino acids to be coupled in this position. One amino acid component is coupled to each of these portions in separate containers. The number of reaction vessels is equal to the number of different building blocks to be coupled in one position. Then, all portions are combined and the N^{α}-protecting group is cleaved. Following this, the whole amount of beads is evenly split again into a new number of separate portions. The principle of the split and combine method is shown schematically in Fig. 8.5.

If, for instance, all possible tripeptides comprising Leu, Phe, and Lys are to be synthesized, the resin is split into three portions. The first amino acid is coupled to each portion, the portions are combined, deprotected, washed and split again. Then in each container, the second amino acid component is coupled, the three portions are combined again, protecting groups are cleaved, and the bulk of the resin is split into three portions again. According to this method, all 27 possible tripeptide sequences are synthesized. One single bead contains only peptides with the same sequence ("one bead, one compound"; Fig. 8.6). There may be up to 10^{13} identical molecules on one single bead, which is 100 μm in diameter.

Care must be taken to use the appropriate amount of resin in the synthesis to ensure a statistical representation of all compounds in the library; otherwise, the library may be incomplete. Furthermore, sequence mismatches depending on problems that occur during synthesis cannot be excluded. Although each bead

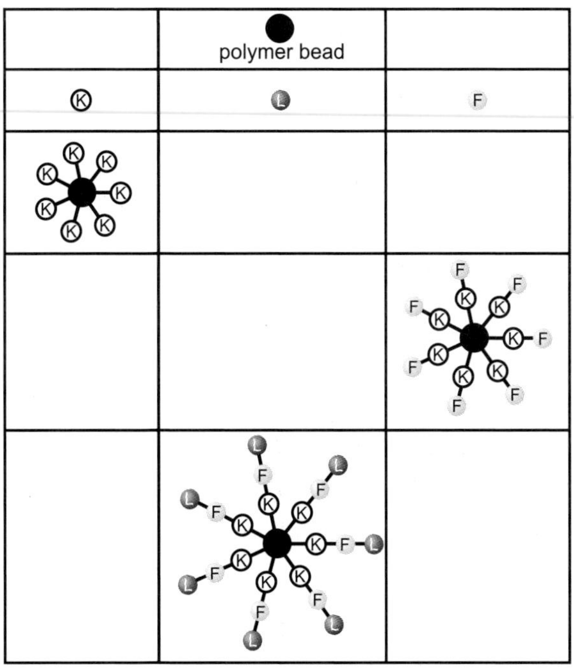

Fig. 8.6 The "split and combine" method provides single compounds on single beads.

contains peptides with the same sequence, the primary structure of these peptides is not known. Peptides libraries synthesized according to this protocol may be used for screening purposes with resin-bound peptides, and the peptides displaying the desired biochemical activity can be identified using suitable bioassays. Special enzyme-labeled antibodies can be used to catalyze color reactions, and consequently this method may be used for antibody epitope mapping. The peptide sequence that binds with high affinity to an antibody will give rise to an intensive color of the bead after incubation with the enzyme-labeled antibody. The labeled resin beads are subsequently separated, the peptide is cleaved, and the sequence is then determined by microsequencing. As the split and combine method is somewhat limited with respect to the size of the library (variability), sublibraries are often created with fixed (defined) amino acid residues in certain positions.

The three repetitive tasks in the split and combine method (coupling, mixing, and splitting) can be easily automated. For the combination of three different building blocks to one peptide, only nine separate synthetic steps are required in combinatorial synthesis to obtain the $3^3=27$ possible tripeptides. This compares favorably with a linear synthesis, where 81 steps are required.

About 50000 cyclic disulfide peptides with random sequences have been synthesized and tested as antagonists for the platelet glycoprotein receptor GP IIb/IIIa (integrin $\alpha_{IIb}\beta_3$). This receptor binds the tripeptide sequence RGD present in the corresponding proteins. The cyclic peptide with the sequence CRGDC was found to be the most active compound, with a binding affinity of 1 μM [50]. In a different study, 2×10^6 peptides have been tested for affinity to the SH3-domain of phosphatidylinositol-3-kinase (PI3K). The peptides were still resin-bound, and the SH3-domain was labeled with a fluorescent marker [51]. The peptides which bound with highest affinity to the protein were consequently present on the brightest beads when examined by fluorescence microscopy. These beads were then selected and the primary structure of the peptide was established by sequencing. The peptide RKLPPRPRR showed the highest affinity towards the SH3-domain of the kinase (binding affinity 7.6 μM). A tetradecameric peptide library eventually provided a thrombin inhibitor with an inhibitory constant (K_i) of 20 nM [51]. The fluorescence-activated cell sorting (FACS) technology or special bead-sorting machines which are based on this principle greatly facilitate the high-speed selection of fluorescent beads, and consequently accelerate the detection of high-affinity ligands.

8.2.3 Encoding Methods

The advantages gained in the synthesis of a peptide library and in the testing of the compound mixture are compensated by the necessary deconvolution in order to identify an active compound. Peptide libraries obtained according to the split and combine method may be analyzed using Edman microsequencing. Binary encoding of compound libraries provides one possible means of overcoming this problem [52]. Brenner and Lerner first described a concept for the generation of an encoded compound library by application of the split and combine method which comprised two consecutive combinatorial synthetic steps. Earlier methods used peptides or oligonucleotides for encoding purposes, but these suffered from limited chemical stability.

Still et al. [53–55] developed nonsequential binary encoding of peptide libraries. In this technique, the encoding molecules are not connected with each other in a sequential manner. Each coupling step of an amino acid component is preceded by the tagging step which is necessary to identify the peptide sequence present on the beads (Fig. 8.7).

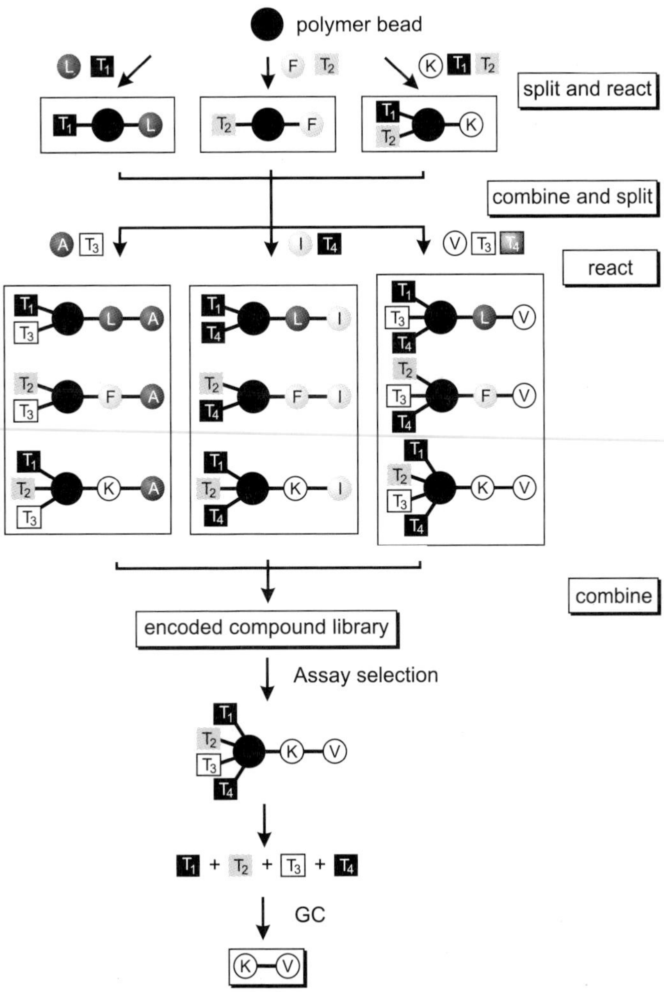

Fig. 8.7 Binary encoding using tag molecules (circles: amino acids, one letter-code; squares: tagging molecules).

Tab. 8.2 Binary code of compound libraries.

Educt [a)]	*Step*	*Binary code*	*Code molecule* [a)]
L	1	10	T_1
F	1	01	T_2
K	1	11	T_1+T_2
A	2	10	T_3
I	2	01	T_4
V	2	11	T_3+T_4

a) Educts are symbolized by the one letter-code, e.g. L, F, K, A, I, V (cf. Fig. 8.7), code molecules by T_1 to T_4.

One disadvantage in the screening of peptides bound to a solid support might be a negative influence of the polymeric matrix on the binding of the peptide to its receptor. Therefore, orthogonal linker systems for the library member and the tag often are required that allow for independent cleavage of the library member and the tag from the solid support for an assay in solution [56]. Decoding relies on the formal assignment of a binary code for single encoding molecules and their mixtures, as shown in Tab. 8.2.

$(2^x–1)$ building blocks in the library can be encoded with x code molecules. Considering a reaction sequence with n steps, the encoding of a library of $(2^x–1)^n$ components requires x^n encoding molecules using combinatorial principles. For example, 20 code molecules are necessary to tag 923 521 different compounds obtained from a four-step reaction sequence. Encoding is usually performed before the coupling, as shown schematically in Fig. 8.7. Only a small amount of encoding molecules and of free amino groups of the solid support are used for the attachment. A photolabile *o*-nitrobenzylester group may be used for example as a linker for the attachment of the encoding molecules such as **2** to safeguard their detachment independently of the library members.

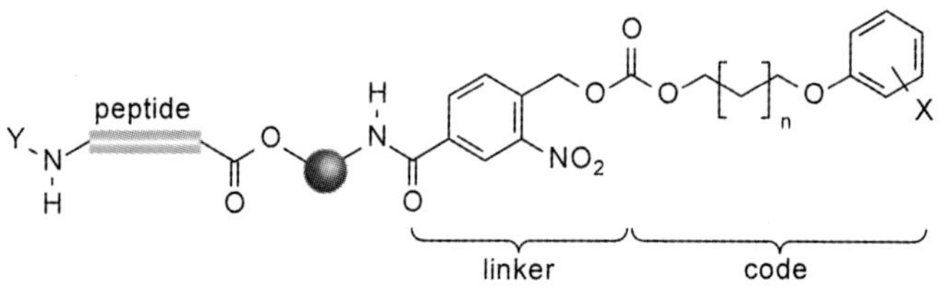

2
X = F or Cl

Photolabile linkers, for example, may be combined with linkers allowing for oxidative cleavage. For the identification of the compound present on a bead, the tagging molecules must be detached and analyzed, and should also usually be volatile in order to allow for gas chromatographic separation (if required). Halogenated aromatic compounds containing alkoxy spacers of varying length and a cleavable linker are coupled to the bead by Rh(II)-catalyzed carbene insertion. The ratio of library member:tag molecule is usually between 200:1 and 100:1. Once bioactivity has been detected in an enzyme assay or receptor binding studies, single active beads are selected and oxidative or photolytic cleavage of the encoding molecule is performed. The resulting alcohols **3** are silylated, and the volatile silyl derivatives are separated by gas chromatography (electron capture detection provides high sensitivity). By using this tagging concept, the identity of a bioactive compound present on a bead may be recognized by selective, highly sensitive methods without the need for peptide sequencing, as has been shown for several examples (e.g. [57, 58]).

Binary encoding (Fig. 8.8) is also very efficient for the synthesis of compound libraries of organic molecules. These have been named as "diversomers" in order to distinguish them from the oligomer libraries (peptide libraries) that can be sequenced.

Although peptide libraries have been useful in lead structure finding, the transformation of a biologically active peptide into a nonpeptide drug often incurs high costs for the necessary modifications. Furthermore, oligomer libraries contain repetitive backbone bonds (peptide bonds, nucleotide bonds) that are detrimental to the concept of diversity. One target in the development of diversomers is that of lead structure finding. Diversomer libraries can be subjected to automatic biological screening systems (high-throughput screening); for example, the synthesis of nonpeptide, nonoligomeric libraries of hydantoins and benzodiazepines with a high degree of chemical diversity has been performed automatically [59]. Clearly, the number of organic molecules obtained in this way is small when compared to the diversity achieved with solution-phase organic chemistry. Nonetheless, the

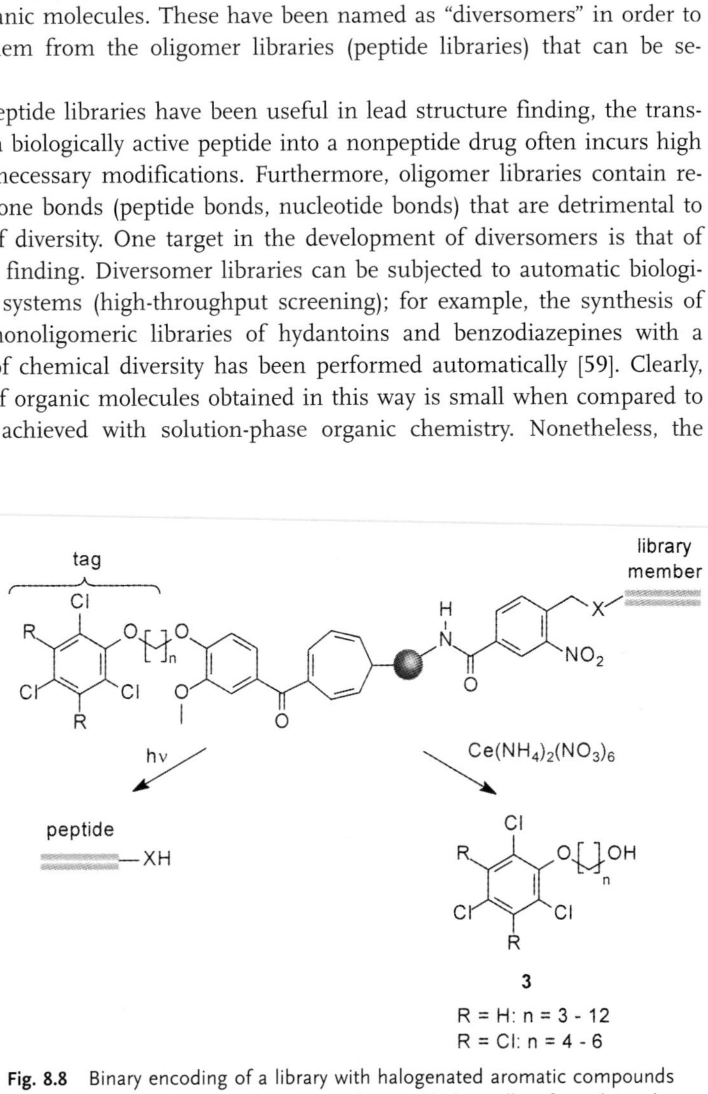

Fig. 8.8 Binary encoding of a library with halogenated aromatic compounds bearing different alkoxy spacers. The orthogonal linkers allow for independent cleavage of the library member and the tag, respectively.

adaptation of further organic chemical reactions to this method will ultimately lead to a high number of diversomer libraries.

Recently, nonchemical encoding methods have also been developed [60–62] in which each "giant" resin bead contains a microchip to store and read all synthetic steps performed with this particular bead, using high-frequency signals. Further developments of this technique have been reviewed [63].

8.2.4
Peptide Library Deconvolution

The deconvolution of a compound library is the process where it must be determined which discrete substance (library member) has the desired property observed in the biological test. Two major strategies are available for this purpose: (i) the iteration method; and (ii) positional scanning.

The iteration method (Fig. 8.9) relies on the creation of sublibraries, once the desired property (e.g., biological activity) has been detected in a mixture library. If a tetrapeptide library has been created where each position is varied by five different amino acid building blocks, the total library consists of $5^4 = 625$ compounds. In the first deconvolution round this total library is re-synthesized in the form of 25 sublibraries, each containing 25 different compounds. In each library, two posi-

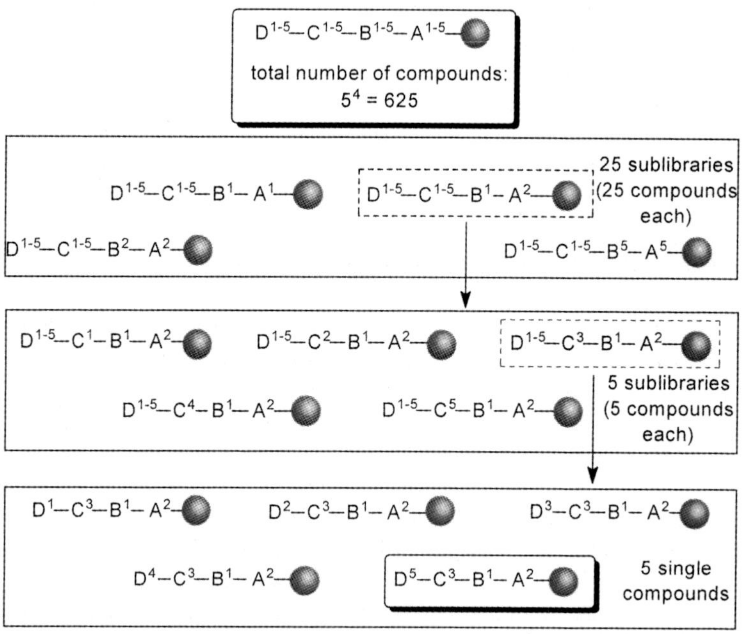

Fig. 8.9 Identification of the most active library component by iteration deconvolution. Sublibraries of the most active library (dashed frame) are synthesized until finally the most active single compound (D^5-C^3-B^1-A^2) is identified.

tions are defined and two positions are varied. Once the sublibrary displaying the highest biological activity has been revealed, the second deconvolution round follows where five further sublibraries are generated on the basis of this result. Each sublibrary comprises five different sequences. Three positions of the tetrapeptide are now fixed, and one position is varied using the five different amino acid building blocks. Again, the biological test is performed and the sublibrary with the highest biological activity is subjected to further deconvolution. Finally, five single compounds are synthesized corresponding to the variation of the fourth tetrapeptide position and the active component should be revealed.

One disadvantage of this iterative procedure is that the most active compound may not necessarily be detected. This protocol is also applicable to the "split and combine" method where the sublibraries, for example of the first deconvolution round, are generated by omitting the last mixing step at the end of the synthesis.

Recursive deconvolution has been described by Janda [41, 42]. After each cycle in the split synthesis an aliquot is retained, the major advantage of this approach being that no new syntheses are necessary because the aliquots secured after each split synthesis step correspond to the necessary sublibraries.

The noniterative method of positional scanning has been suggested by both Houghten et al. [64, 65] and Furka et al. [66]. The partial libraries of positional scanning represent first-order sublibraries in each of which one position is kept invariant, while all other positions are varied (Fig. 8.10). In the tetrapeptide example used above, this would mean that 20 first-order sublibraries would be generated, each containing 125 compounds, and corresponding to the variation of three positions by five different building blocks. The sublibrary with the highest biological activity reveals which amino acid residue in the fixed position has the highest contribution to biological activity.

Omission libraries and amino acid test mixtures have been developed by Furka [67].

8.2.5
Biological Methods for the Synthesis of Peptide Libraries

Biological methods for the synthesis of peptide libraries already have been developed [68–70]. Antibody libraries with a high number of binding specificities are of major importance as they enable a more complete therapeutic and diagnostic application of antibody technology, without being dependent on the immunization of animals or the application of eukaryotic cells. Proof of this principle has been shown in investigations on the specific recognition of double-stranded DNA by semisynthetic antibodies [71]. However, the antibodies selected by panning do not yet satisfy the requirement for technical and medical application. In general, tailor-made antibodies can be used for the recognition of certain DNA sequences [72], a finding which may be important for the synthesis of artificial biocatalysts for the neutralization of viruses or toxins, for the identification of certain DNA sequences in gene diagnostics, or in genome assignments as the antibodies block certain DNA sequences and prevent access to other enzymes. Furthermore, the

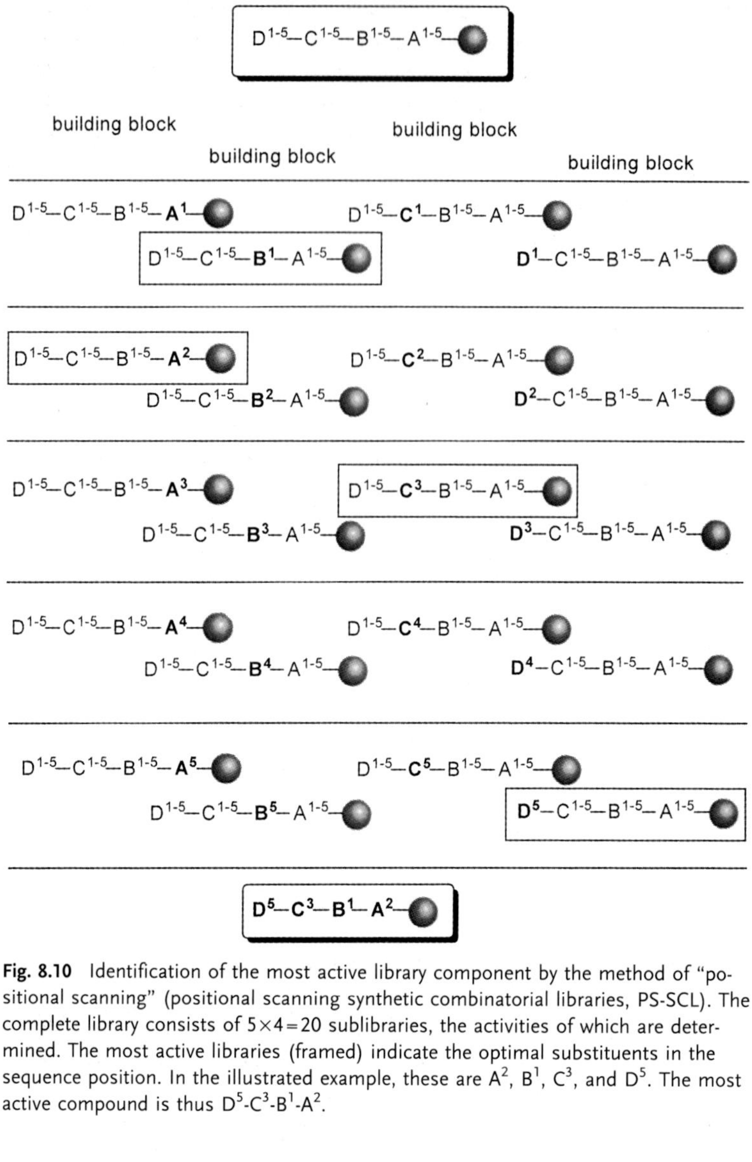

Fig. 8.10 Identification of the most active library component by the method of "positional scanning" (positional scanning synthetic combinatorial libraries, PS-SCL). The complete library consists of $5 \times 4 = 20$ sublibraries, the activities of which are determined. The most active libraries (framed) indicate the optimal substituents in the sequence position. In the illustrated example, these are A^2, B^1, C^3, and D^5. The most active compound is thus D^5-C^3-B^1-A^2.

application may be useful in the selection and improvement of catalytic antibodies [73–75].

Peptide libraries with thousands of different peptides can also be obtained using techniques of molecular biology. Screening is performed by selection and subcloning of phages, and recognition by antibodies. The identification of the active peptide sequence is performed on the DNA level by sequencing of the phage genome [68–70].

Biological compound libraries – and especially phage display libraries – are characterized by the advantage that each library member is able to replicate itself, and that each member carries unique encoding (DNA sequence) [68, 70, 76–78].

Phage display is a technology which links the phenotype of a peptide displayed on the surface of a bacteriophage with the genotype encoding for this peptide. This molecular biology technique permits the generation of a peptide library by site-directed mutagenesis. Some 10^7–10^9 different oligopeptide sequences can be generated and expressed on the surface of phages, whilst subsequent screening using affinity-based techniques allows the selection of phage that present high-affinity peptides on the surface and which may subsequently be amplified. The uniqueness of a filamentous phage is that three of its five virion proteins tolerate the insertion of foreign peptides. The most studied phages are f1, fd, and M13, all of which infect *E. coli* cells through their f-pili. Among the five coat proteins, pIII, pVI, and pVIII have been used for displaying peptides on the phage surface. The peptide cDNA sequences are inserted into the phage genomes in such a way that they join, or neighbor the gene for the structural protein pIII. Consequently, the peptides are presented at the N-terminus of pIII after infection of *E. coli* with the phage. pIII is the longest coat protein, and will tolerate the insertion of larger, foreign polypeptides. Unlike pIII, pVIII is a small protein of 50 amino acids where the length of the peptides displayed is limited to five to six amino acids [79]. In early examples of phage display, polypeptides were fused to the amino terminus of either pIII or pVI on the viral genome level [80], though in several cases the displayed peptides were shown to affect coat protein function. This problem was solved by the development of phagemid display systems.

8.3 References

1 M.C. Densai, R.N. Zuckermann, W.H. Moos, *Drug Dev. Res.* **1994**, *33*, 174
2 R.L. Liu, P.G. Schultz, *Angew. Chem.* **1999**, *111*, 36; *Angew. Chem. Int. Ed. Engl.* **1999**, *38*, 36
3 M.C. Pirrung, *Chem. Rev.* **1997**, *97*, 473
4 M. Lebl, *Biopolymers* **1998**, *47*, 397
5 A. Nefzi, J.M. Ostresh, R.A. Houghten, *Chem. Rev.* **1997**, *97*, 449
6 E.M. Gordon, R.W. Barrett, W.J. Dower, S.P.A. Fodor, M.A. Gallop, *J. Med. Chem.* **1994**, *37*, 1385
7 R.M.J. Liskamp, *Angew. Chem.* **1994**, *106*, 661, *Angew. Chem. Int. Ed. Engl.* **1994**, *33*, 633
8 P.C. Andres, D.M. Leonard, W.L. Cody, T.K. Sayer in: *Multiple and Combinatorial Peptide Synthesis*, B.M. Dunn, M.W. Pennington (Eds.), Humana Press Inc., Totowa, **1994**
9 L. Weber, *Curr. Opin. Chem. Biol.* **2000**, *4*, 295
10 J.C. Hogan, Jr., *Nature Biotechnol.* **1997**, *15*, 328
11 G. Jung, A.G. Beck-Sickinger, *Angew. Chem.* **1992**, *104*, 375; *Angew. Chem. Int. Ed. Engl.* **1992**, *31*, 367
12 R.A. Houghten, C. Pinilla, J.R. Appel, S.E. Blondelle, C.T. Dooley, J. Eichler, A. Nefzi, J.M. Ostresh, *J. Med. Chem.* **1999**, *42*, 3743
13 C. Blackburn, *Biopolymers* **1998**, *47*, 311
14 D.J. Gravert, K.D. Janda, *Chem. Rev.* **1997**, *97*, 489
15 F. Balkenhohl, C. von dem Bussche-Hünnefeld, A. Lansky, C. Zechel, *An-*

gew. Chem. **1996**, 108, 2436; *Angew. Chem. Int. Ed. Engl.* **1996**, *35*, 2288
16 R.J. Booth, J.C. Hodges, *Acc. Chem. Res.* **1999**, *32*, 18
17 H.-J. Böhm, M. Stahl, *Curr. Opin. Chem. Biol.* **2000**, *4*, 283
18 R.A. Houghten, *Proc. Natl. Acad. Sci. USA* **1985**, *82*, 5131
19 H.M. Geysen, R.H. Meloen, S.J. Barteling, *Proc. Natl. Acad. Sci. USA* **1984**, *81*, 3998
20 H.M. Geysen, S.J. Barteling, R.H. Meloen, *Proc. Natl. Acad. Sci. USA* **1985**, *82*, 178
21 R. Frank, *Tetrahedron* **1992**, *48*, 9217
22 Á. Furka, F. Sebestyén, M. Asgedom, G. Dibó in: *Highlights of Modern Biochemistry*, Proceedings of the 14th International Congress of Biochemistry, Volume 5, VSP, Utrecht, **1988**
23 Á. Furka, F. Sebestyén, M. Asgedom, G. Dibó, *Int. J. Peptide Protein Res.* **1991**, *37*, 487
24 R.A. Houghten, *Trends Biotechnol.* **1987**, *5*, 322
25 A.G. Beck-Sickinger, H. Dürr, G. Jung, *Peptide Res.* **1991**, *4*, 88
26 N.J. Maeji, A.M. Bray, R.M. Valerio, W. Wang, *Peptide Res.* **1995**, *8*, 33
27 R. Frank, S. Güler, S. Krause, W. Lindemaier in: *Peptides 1990*, E. Giralt, D. Andreu (Eds.), Escom, Leiden, **1991**, 151
28 A.M. Bray, N.J. Maeji, H.M. Geysen, *Tetrahedron Lett.* **1990**, *31*, 5811
29 B. Blankenmeyer-Menge, R. Frank, *Tetrahedron Lett.* **1988**, *29*, 5871
30 R. Frank, R. Döring, *Tetrahedron* **1988**, *44*, 6031
31 J. Eichler, M. Beyermann, M. Bienert, M. Lebl in: *Peptides 1988*, G. Jung, E. Bayer (Eds.), de Gruyter, Berlin, **1989**
32 J. Eichler, M. Bienert, A. Stierandova, M. Lebl, *Peptide Res.* **1991**, *4*, 296
33 M. Schmidt, J. Eichler, J. Odarjuk, E. Krause, M. Beyermann, M. Bienert, *Bioorg. Med. Chem. Lett.* **1993**, *3*, 441
34 R.H. Berg, K. Almdal, W. Batsberg Pedersen, A. Holm, J.P. Tam, R.B. Merrifield, *J. Am. Chem. Soc.* **1989**, *111*, 8024
35 G. Schnorrenberg, H. Gerhardt, *Tetrahedron* **1989**, *45*, 7759
36 H. Gausepohl, M. Kraft, C. Boulin, R.W. Frank in: *Peptides: Chemistry, Structure and Biology*, J.E. Rivier, G.R. Marshall (Eds.), Escom, Leiden, **1990**, 1003
37 A. Patchornik, B. Amit, R.B. Woodward, *J. Am. Chem. Soc.* **1970**, *92*, 6333
38 S.P. A. Fodor, J.L. Read, M.C. Pirrung, L. Stryer, A.T. Lu, D. Solas, *Science* **1991**, *251*, 767
39 S.P.A. Fodor, R.P. Rava, X.C. Huang, A.C. Pease, C.P. Holmes, C.L. Adams, *Nature* **1993**, *364*, 555
40 M. Mutter, E. Bayer, *Angew. Chem.* **1974**, *86*, 101; *Angew. Chem. Int. Ed. Engl.* **1974**, *13*, 88
41 H. Han, M.M. Wolfe, S. Brenner, K.D. Janda, *Proc. Natl. Acad. Sci. USA* **1995**, *92*, 6419
42 E. Erb, K.D. Janda, S. Brenner, *Proc. Natl. Acad. Sci. USA* **1994**, *91*, 11422
43 J.M. Ostresh, J.H. Winkle, V.T. Hamashin, R.A. Houghten, *Biopolymers* **1994**, *34*, 1681
44 K. Burgess, A.I. Liaw, N. Wang, *J. Med. Chem.* **1994**, *19*, 2985
45 H.M. Geysen, S.J. Rodda, T.J. Mason, *Mol. Immunol.* **1986**, *23*, 709
46 C. Pinilla, J.R. Appel, P. Blanc, R.A. Houghten, *Biotechniques* **1992**, *13*, 901
47 K.S. Lam, E. Salmon, E.M. Hersh, V.J. Hruby, W.M. Kazmierski, R.J. Knapp, *Nature* **1991**, *354*, 82
48 A. Furka, W.D. Bennett, *Comb. Chem. High Throughput Screen.* **1999**, *2*, 105
49 K.S. Lam, M. Lebl, V. Krchnák, *Chem. Rev.* **1997**, *97*, 411
50 S.E. Salmon, K.S. Lam, M. Lebl, A. Kandola, P.S. Khattri, S. Wade, M. Pátek, P. Kočiš, V. Krchnák, D. Thorpe, S. Felder, *Proc. Natl. Acad. Sci. USA* **1993**, *90*, 11708
51 J.K. Chen, W.S. Lane, A.W. Brauer, A. Tanaka, S.L. Schreiber, *J. Am. Chem. Soc.* **1993**, *115*, 12591
52 S. Brenner, R.A. Lerner, *Proc. Natl. Acad. Sci. USA* **1992**, *89*, 5381
53 M.H.J. Ohlmeyer, R.N. Swanson, L.W. Dillard, J.C. Reader, G. Asouline, R. Kobayashi, M. Wigler, W.C. Still, *Proc. Natl. Acad. Sci. USA* **1993**, *90*, 10922
54 A. Borchardt, W.C. Still, *J. Am. Chem. Soc.* **1994**, *116*, 373

55 H. P. Nestler, P. A. Bartlett, W. C. Still, *J. Org. Chem.* **1994**, *59*, 4723

56 C. Chen, L. A. A. Randall, R. B. Millen, A. D. Jones, M. J. Kurth, *J. Am. Chem. Soc.* **1994**, *116*, 2661

57 J. Nielsen, S. Brenner, K. D. Janda, *J. Am. Chem. Soc.* **1993**, *115*, 9812

58 J. M. Kerr, S. C. Banville, R. N. Zuckermann, *J. Am. Chem. Soc.* **1993**, *115*, 2529

59 S. Hobbs De Witt, J. S. Kiely, C. J. Stankovic, M. C. Schröder, D. M. Reynolds Cody, R. R. Pavia, *Proc. Natl. Acad. Sci. USA* **1993**, *90*, 6909

60 R. L. Affleck, *Curr. Opin. Chem. Biol.* **2001**, *5*, 257

61 K. C. Nicolaou, X.-Y. Xiao, Z. Parandoosh, A. Senyei, M. P. Nova, *Angew. Chem.* **1995**, 107, 2476; *Angew. Chem. Int. Ed. Engl.* **1995**, *34*, 2289

62 E. J. Moran, S. Sarshar, J. F. Cargill, M. M. Shahbaz, A. Lio, A. M. M. Mjalli, R. W. Armstrong, *J. Am. Chem. Soc.* **1995**, *117*, 10787

63 C. Barnes, S. Balasubramanian, Curr. Opin. Chem. Biol. **2000**, 4, 346

64 R. A. Houghten, J. R. Appel, S. E. Blondelle, J. H. Cuervo, C. T. Dooley, C. Pinilla, *Biotechniques* **1992**, *13*, 412

65 C. T. Dooley, R. A. Houghten, *Life Sci.* **1993**, *52*, 1509

66 Á. Furka, F. Sebestyén, WO93/24517 1993

67 Á. Furka, *Drug Dev. Res.* **1994**, *33*, 90

68 S. E. Cwirla, E. A. Peters, R. W. Barrett, W. J. Dower, *Proc. Natl. Acad. Sci. USA* **1990**, *87*, 6378

69 J. A. Wells, H. B. Lowman, *Curr. Opin. Biotechnol.* **1992**, *3*, 355

70 J. K. Scott, G. P. Smith, *Science* **1990**, *249*, 386

71 S. M. Barbas, P. Ghazal, C. F. Barbas, III, D. R. Burton, *J. Am. Chem. Soc.* **1994**, *116*, 2161

72 M. Famulok, D. Faulhammer, *Angew. Chem.* **1994**, *106*, 1911; *Angew. Chem. Int. Ed. Engl.* **1994**, *33*, 1827

73 Y. C. J. Chen, T. Danon, L. Sastry, M. Mubaraki, K. D. Janda, R. A. Lerner, *J. Am. Chem. Soc.* **1993**, *115*, 357

74 B. Posner, J. Smiley, I. Lee, S. Benkovic, *Trends Biochem. Sci.* **1994**, *19*, 145

75 J. D. Stewart, S. J. Benkovic, *Int. Rev. Immunol.* **1993**, *10*, 229

76 M. B. Zwick, J. Shen, J. K. Scott, *Curr. Opin. Biotechnol.* **1998**, *9*, 427

77 J. J. Devlin, L. C. Panganiban, P. E. Devlin, *Science* **1990**, *249*, 404

78 G. P. Smith, V. A. Petrenko, *Chem. Rev.* **1997**, *97*, 391

79 S. Cabilly, *Mol. Biotechnol.* **1999**, *12*, 143

80 S. S. Sidhu, *Curr. Opin. Biotechnol.* **2000**, *11*, 610

9
Application of Peptides and Proteins

Almost all biological processes in living cells are controlled by various types of molecular recognition. Regulatory processes comprise initiation or inhibition via specific protein-protein complex formation. In particular, peptides and proteins possess an enormous potential for diversity, and therefore these compounds are well suited for such complicated control functions.

Further progress in synthesis methodology, for example Merrifield's solid-phase technique (see Section 4.5), recombinant DNA technology (Section 4.6.1), and rapid advances in more recently developed methods for the production of peptides and proteins in transgenic animals [1] and plants [2] have paved the way to increase the availability and reduce the cost of peptide and protein production.

Although, initially, the pharmaceutical community was excited about the market potential of peptides and proteins for diagnostic and therapeutic purposes, some decades after the first chemical synthesis of a peptide hormone very little advantage has been taken of the potential of peptides and proteins as pharmaceuticals and tools for application in basic and clinical research. Only relatively few peptides have been approved as drugs, most likely because the application of proteins in therapeutic use may be hampered by factors such as antigenicity, immunogenicity, and stability of the protein when administered. A further reason has been the lack of an efficient large-scale production technology of proteins and peptides.

9.1
Protein Pharmaceuticals

9.1.1
Importance and Sources

Proteins constitute a major fraction of the biopolymers contained in all organisms with respect to diversity and mass. A huge proportion of these biomolecules have regulatory functions in maintaining biochemical or cellular equilibria in healthy organisms, though they may also be involved both in pathophysiological events and healing processes. Until the late 1970s, the human body was the only source for endogenous proteins such as growth hormone or coagulation factor VIII used for replacement therapy. The selection of a suitable protein source for its isolation is based on such criteria as the ease of obtaining sufficient quantities of the appropriate tissue, the amount of the chosen protein in this tissue, and any properties that would aid in its stabilization and isolation. Preferentially, tissues from domesticated animals, easily obtainable microorganisms and plants are often chosen as sources for the isolation. Isolated proteins, sometimes in form of poorly defined mixtures, have been used in many traditional medicines and in so-called alternative medicine. However, purified active proteins are of great importance for both causal and symptomatic treatment, and for prophylaxis.

Since its introduction in 1977, the use of recombinant DNA technology (Section 4.6.1) has provided a new and highly efficient means of producing large amounts of rare and/or novel proteins. The development of molecular cloning techniques offers a new production method for proteins, and has consequently exerted an enormous medical, industrial and agricultural impact. Once a protein-encoding gene has been isolated from its parent organism, it may be genetically engineered if desired, and overexpressed in either bacteria, yeast, or mammalian cell cultures. The biotechnological isolation of the cloned protein is much easier, as it may constitute up to about 35% of the overproducer's total cell protein.

9.1.2
Endogenous Pharmaceutical Proteins

The developments in molecular biology have led to a therapeutic concept based on the pharmaceutical application of endogenous proteins which includes:

- the discovery and synthesis of proteins with therapeutic potential by gene technology;
- the elucidation of their biological actions *in vitro* and *in vivo*; and
- the development of drugs based on the primary protein lead molecule.

Very important indications for pharmaceutical proteins include cancer, infectious diseases, AIDS-related diseases, heart disease, respiratory diseases, autoimmune disorders, transplantations, skin disorders, diabetes, genetic disorders, digestive disorders, blood disorders, infertility, growth disorders, and eye conditions. In 1998, cancer was by far the most prevalent target, accounting for more than 40% of the total number of new medicines according to disease area. The top five drug types in 1998 were vaccines, monoclonal antibodies, gene therapeutics, growth factors, and interferons [3].

A list of proteins of general pharmaceutical interest is provided in Tab. 9.1.

Most of these proteins were cloned during the 1980s [4], at which time examples of approved protein-based products included epidermal growth factor (EGF), Factor VIII, tissue plasminogen activator (tPA), insulin, hepatitis B vaccine, various interferons, monoclonal antibodies, and growth hormone.

Many of the early protein drug candidates failed in clinical trials due to their immunogenicity, short half-life, or low specificity. It has been estimated that up to 1998/99, about 100 drugs produced by biotechnology had been approved, but that

Tab. 9.1 Selected pharmaceutical proteins.[a)]

Protein (abbreviation)	***AA***	***Isolated from***	***Application/mode of action***
Albumin (HSA)	585	Liver (1975)	Plasma expander
Angiogenin (TAF)	123	Bowel cancer cells (1985)	Wound healing; tumors
α_1-Antitrypsin (AAT)	394	Blood (1978)	Anticoagulant
Antithrombin III (AT3)	432	Liver (1979)	Anticoagulant
Erythrocyte differentiation factor (EDF)	110	Leukemia cells (1987)	Tumors
Erythropoietin (EPO)	165	Urine (1977)	Aplastic anemia
Factor VII	406	Plasma (1980)	Blood clotting
Factor VIII	2332	Liver (1983)	Hemophilia A
Factor IX	416	Plasma (1975)	Hemophilia B
Factor XIII	1372	Plasma (1971)	Surgical adhesive
Fibroblast growth factor (basic) (bFGF)	146	(1986)	Wound healing, tumors
Fibronectin (FN)	96	(1970)	Wound healing
Granulocyte colony stimulating factor (G-CSF)	174–177	Tumor cells (1986)	Leukemia, other tumors
Granulocyte macrophage colony stimulating factor (GM-CSF, CSF-2)	127	T cell (1984)	Anemia, tumors
Hepatitis B surface antigen (HBS, HbsAg)	226	Virions (1977)	Hepatitis vaccine
Human collagenase inhibitor (HCI, TIMP)	184	Fibroblasts (1983)	Arthritis
Interferon-α (IFN-α)	166	Leucocytes (1979)	Hairy cell leukemia, tumors
Interferon-β (IFN-β)	166	Fibroblasts (1979)	Keratitis, hepatitis B
Interferon-γ (IFN-γ; MAF)	146	Lymphocytes (1981)	Tumors, arthritis
Interleukin-1 (IL-1, ETAF, LAF)	152	Neutrophils (1984)	Tumors
Interleukin-2 (IL-2, TCGF)	133	T cells (1980)	Tumors
Interleukin-3 (IL-3, Multi-CSF, BPA, MCGF)	133		Leukemia, other tumors
Interleukin-4 (IL-4, BSF-1, BCGF-1)	129		Leukemia, infections
Interleukin-5 (Il-5, TRP, BCGF-II)	112		Autoimmune diseases
Interleukin-6 (IL-6, BSF2, IFN-β2, BCDF)	184	T cells (1985)	Leukemia
Lipase	135	Microbial	Digestive disturbances
Lipomodulin, lipocortin (AIP)	346		Arthritis, allergies
Lung surfactant protein (LSP, PSF)	248	Sputum (1986)	Emphysema, pulmonary
Lymphotoxin (LT, TNF-β)	171	Lymphocytes (1984)	Tumors
Macrophage inhibitory factor (MIF)	Het.	Lymphocytes (1981)	
Macrophage colony stimulating factor (CSF-1, M-CSF)	224	Urine (1982)	Leukemia, tumors

Tab. 9.1 (continued)

Protein (abbreviation)	*AA*	*Isolated from*	*Application/mode of action*
Monoclonal antibody OKT3, Orthoclone OKT3		Hybridoma (1979)	Transplantation
Nerve growth factor (NGF-β)	118		Injuries
Platelet-derived growth factor (PDGF)	241	Platelets (1983)	Wound healing
Plasminogen activator (PAI I)	376–379	Lymphosarcoma (1984)	Blood clotting
Protein C (PC)	262	Plasma (1979)	Anticoagulant
Protein S	635		Anticoagulant
Streptokinase	416	*Streptoccocus*	Myocardial infarction, thromboses
Superoxide dismutase (SOD)	153	Placenta (1972)	After-treatment of myocardial infarction
Tissue plasminogen activator (tPA)	527	Uterus (1979)	Myocardial infarction, embolism
Transforming growth factor-α (TGF-α)	50	Tumor cells (1982)	Wound healing
Transforming growth factor-β (TGF-β)	112	Kidney tumor (1983)	Wound healing, tumors
Tumor necrosis factor (TNF-α, cachectin, DIF)	157	Tumor (1985)	Tumors
Urokinase (UK)	366	Urine (1982)	Thromboses, embolism
Uromodulin, Tamm-Horsfall protein	616	Urine (1985)	Inflammations

a) Based on data published by Blohm et al. [4].

approximately 350 biotechnology drugs are currently under development. Initially, many pharmaceutical proteins were of nonhuman origin, and caused immune responses against the drug itself. Others suffered from suboptimal affinity or poor half-life, resulting in poor efficacy.

9.1.3
Engineering of Therapeutic Proteins

The engineering of therapeutic proteins [5] provides a valuable means of circumventing some of the disadvantages mentioned above. The goals of protein engineering techniques along this line are to:

- minimize the immunogenicity of protein drugs;
- improve pharmacokinetics;
- improve effector functions; and
- improve affinity.

9.1.3.1 Peptide-Based Vaccines

In some countries, especially in south-east Asia and Africa, a relatively high percentage of the population has been infected with the highly infectious hepatitis B virus. This causes jaundice and, as a late consequence of chronic infection, even gives rise to liver tumors. A hepatitis B vaccine, isolated from the blood of virus carriers, has been available since 1982. Some years later, the first vaccine based on a recombinant protein was described containing the pure viral surface antigen. Peptide vaccines may be designed based on the subunit of a pathogen, either with naturally occurring immunogenic peptides or synthetic peptides corresponding to highly conserved regions required for the pathogen's function. The aim of this strategy is vaccination with a minimal structure that consists of a well-defined antigen and elicits effectively a specific immune response, without potentially hazardous risks. About 10 years ago, the groups of Wiesmüller [6] and Jung [7] first described a completely synthetic vaccine. A $CD8^+$ cytotoxic T-cell (CTL) epitope of influenza virus NP was conjugated to the general immune enhancer Pam_3Cys-Ser-Ser. This relatively simple construct resulted in efficient priming of virus-specific cytotoxic T cells when injected into mice, without any additional adjuvant. During the past few years, tremendous progress has been made in the development of fully synthetic vaccines [8, 9]. A successful vaccination for HIV is complicated for several reasons; for example, as the virus attacks and destroys T-helper lymphocytes and undergoes antigenic variations. HPG-30, a HIV-1 p17 synthetic peptide vaccine, was evaluated in Phase I clinical trials [10]. Four epitopes from the HIV-1 regulatory protein Rec [11], and the V3 loop epitope gp 120 [12] have been tested for the in-vitro immunization of human lymphocytes. Immunization protocols using single peptides related to a CTL epitope derived from cancer without added adjuvant have led to hopefully beneficial clinical effects [13].

9.1.3.2 Monoclonal Antibodies

Monoclonal antibodies (MAb) can be considered as a group of natural drugs as they mimic their natural function in an organism, but without inherent toxicity. Orthoclone (OKT3) was the first monoclonal antibody marked for therapeutic purposes. This was launched in 1986 by Ortho Biotech for acute kidney transplant rejection, but first-dose reactions and antimurine antibodies remain drawbacks in its clinical application. Generally, the use of rodent MAb as therapeutic agents is hampered because the human organism recognizes them as foreign. Near-human clinical MAb, called "chimerized" or "humanized" antibodies have been created by fusing mouse variable domains to human constant domains in order to retain binding specificity while simultaneously reducing the portion of the mouse sequence [14]. The first example of an approved chimeric antibody was ReoPro (Abciximab) from Centocor, an anticoagulant, which was registered at the end of 1994 in the USA. Zenapax is a complementary determining region (CDR) grafted MAb targeted to the interleukin-2 (IL-2) receptor on T cells for application in preventing transplant rejection. Further examples for chimeric/humanized MAb-based therapeutics approved during 1997–2001 are listed in Tab. 9.2.

Tab. 9.2 Selected therapeutic monoclonal antibodies on the market.[a)]

Product name (substance name)	*Target antigen*	*Therapeutic use*	*Registered (region)*	*Company*
Mabthera/Rituxan (Rituximab)[b)]	CD 20 surface antigen of B lymphocytes	Treatment of non-Hodgkin's lymphoma	1997 (US) 1998 (EU)	Genentech/IDEC Hoffmann La-Roche (EU)
Zenapax (Daclizumab)[c)]	α-chain of the IL-2 receptor (CD 25)	Prevention of acute kidney transplant rejection	1997 (US) 1999 (EU)	Hoffmann La-Roche
Herceptin (Trastuzumab)[c)]	Human EGF-like receptor 2 (HER-2)	Treatment of metastatic breast cancer over-expressing HER-2 protein	1998 (US)	Genentech
Synagis (Palivizumab)[c)]	Undisclosed epitope on the surface of respiratory syncytial virus	Prophylaxis of lower respiratory disease caused by respiratory syncytial virus in children	1998 (US) 1999 (EU)	Medimmune (US) Abbot (EU)
Simulect (Basilixmab)[b)]	α-chain of the IL-2 receptor (CD 25)	Prophylaxis of acute organ rejection in allogeneic renal transplantation	1998 (EU)	Novartis
Remicade (Infliximab)[b)]	TNF-α	Treatment of Crohn's disease Treatment of rheumatoid arthritis	1998 (US) 1999(EU) 1999 (US) 2000(EU)	Centocor
Campath (Alemtuzumab)[c)]	B lymphocytes	Treatment of B-cell chronic lymphocytic leukemia	2001 (US)	Millenium Pharm. Inc. ILEX Oncology Inc.

a) Slightly changed according to van Dijk and van den Winkel [16].
b) Chimerized mAb.
c) Humanized mAb.

More than 30 CDR grafted mAb are currently undergoing in clinical trials. The mAb currently available commercially are produced in mammalian cell cultures, but this is an expensive process. Economic alternatives may be the development of transgenic animals (goats or cows) that have been genetically engineered to produce mAb in their milk [15]. Besides recent developments to reduce murine components, fully human antibodies will be the next-generation therapeutics [16]. Indeed, various techniques already exist for the development of 100% human antibodies, such as the direct isolation of human antibodies from phage display libraries [17] and transgenic mice containing human antibody genes and disrupted endogenous immunoglobulin loci [18]. A human anti-TNF-α mAb, designated

D2E7 (BASF/CAT), is currently undergoing Phase III clinical trials for the treatment of rheumatoid arthritis [19]. Antibodies and antibody derivatives constitute about 25% of pharmaceutical proteins currently under development, and it is very clear that the immune system is an excellent target for new therapeutic efforts.

9.1.3.3
Protein Pharmaceuticals with Various Functions

Therapeutic proteins with various biological action including growth factors, interferons, interleukins, tissue plasminogen activators, clotting factors, colony stimulating factors, erythropoietin and others have also been engineered to improve the effectiveness as protein therapeutics in the clinic or clinical development pipeline.

Acute myocardial infarction and other thrombotic obstructions of blood vessels are indications for the recombinant tissue plasminogen activator (tPA). This compound belongs to the 'big' pharmaceutical market products. It might be demonstrated that, in the case of tPA, the removal of natural domains may improve the pharmacokinetics and specificity of the protein drug. Retaplase (Boehringer Mannheim) is an extremely truncated tPA molecule lacking the N-terminal finger domain, the epidermal growth factor domain, and the kringle 1 domains. The resultant drug (Rapilysin), which has an improved half-life compared with Retaplase, is used in the treatment of myocardial infarction.

Pharmaceutical proteins which act on immunological functions include, particularly, the interferons and interleukins as well as the growth and differentiation factors, the latter being highly specific triggers of the differentiation steps in hematopoiesis. Angiogenesis factors such as angiogenin and fibroblast growth factors are normally not distributed throughout the bloodstream, but are formed and act locally, for example in the surroundings of inflammations, injuries or tumors. These compounds are not only involved in wound healing, but are also indirectly associated with tumor therapy, as anti-angiogenic compounds may prevent the vascularization of solid tumors.

In order to overcome the disadvantages associated with the therapeutic application of proteins, engineering efforts have been (and are being) directed towards the design and expression of variants with low toxicity and suitable binding profiles. IL-2, which is an approved therapeutic for advanced metastatic cancer, is a representative example. Despite the great potential initially promised, IL-2 has found limited use due to its systemic toxicity. Proleukin (Chiron) is a mutant of IL-2 in which one (Cys^{125}) of the three Cys residues has been converted to Ser, without affecting the biological activity. However, this minimal alteration safeguards that a greater portion of the recombinant product is produced in the correctly folded form.

An improvement in the serum half-life of protein drugs can also be made by the addition of polyethylene glycol (PEG), and this was proved in the case of IL-2 [20]. In addition to PEG-IL-2, several PEG-protein conjugates (e.g., PEG-α-interferon, PEG-G-CSF, and PEG-hemoglobin) are currently undergoing clinical trials, whilst others such as PEG-adenosine-deamidase and PEG-L-asparaginase are already commercially available.

Recombinant growth hormone is used for the treatment of children suffering from dwarfism, a disorder brought about by deficient endogenous synthesis of this hormone. Interestingly, until 1985 growth hormone was obtainable only from human pituitaries removed at autopsy, as the growth hormone of other species is not active in humans. This early drug was heavily criticized because of suspected viral contamination, however. Protropin® (Genentech), which had an additional Met at the N-terminal end, was approved in 1985, and this was followed 2 years later by Humatrop® (Eli Lilly), which had an identical sequence as the native hormone.

Erythropoietin (EPO) stimulates the production of red blood cells, and is used in the treatment of anemia caused by kidney disease. Epidermal growth factor (EGF) combined with poly(acrylic acid) gels has been shown to be successful in the treatment of corneal epithelial wounds [21].

Finally, it should be mentioned that the cost to develop a successful drug and to take it to the marketplace is, on average, US$ 600 million, this being associated with an average development time of about 10 years [22]. By contrast, improved engineered drugs have been successful both in medical and financial terms. For example, the humanized MAb Herceptin (see Tab. 9.2) generated US$ 188 million during its first year of sales, and is undoubtedly one the most successful anticancer drugs launched to date. More information on the action of selected pharmaceutical proteins is provided in the Glossary.

Despite an almost 80-year history of the use of proteins as therapy – starting with the commercial introduction of insulin in 1923, and followed by the approval of recombinant insulin as the first biotechnological drug in 1982 – interest has increased most significantly during the past two decades. The enormous advances in molecular biology (genetic engineering), cell biology and modern techniques in protein chemistry have promoted this rapid development. At present, most efforts are still directed towards the discovery of new proteins with pharmaceutical potential, and the engineering of therapeutic proteins to provide the clinical benefits as discussed above. It is likely that drug developments of the near future will be characterized as a marriage of selection-based and knowledge-based approaches. Mutagenesis, selection, and high-throughput screening (HTS) techniques will be guided by both structural knowledge, obtained by systematic determination of protein structures, and a better understanding of the biological and molecular mechanisms (molecular medicine).

9.1.3.4 Future Perspectives

Drug research and pharmaceutical treatment stand at the dawn of an entirely new scientific area. In mid-2001, the human genome sequence had (mostly) been completed, indicating a total of 30000 to 40000 unique human genes [23, 24]. Moreover, according to the number of splice variants and functional variants resulting from post-translational modifications, the number of proteins will most likely exceed the number of genes.

The next major challenge is directed towards the human proteome. Proteomics represents a formidable task, and may result ultimately in the characterization of every protein encoded by the human genome. The proteome is defined as the set of proteins expressed by a cell at a certain time, and under certain conditions. As different stages of development or pathological events are reflected in changes in the proteome, proteome analysis is usually carried out using a differential approach, by detecting changes in the protein expression profile [25–27]. The proteome dynamically reflects the state of a biological cell system, and responds to either differentiation states, nutrition, temperature, stress, or pathological conditions.

An understanding of the structure, function, molecular interactions, and regulation of every protein in various cell types is a future goal of the highest relevance. However, despite the magnitude of this task, powerful tools in biochemistry, molecular biology, and bioinformatics – combined with massive automation – will be required to reach this goal. Indeed, knowledge gained of the molecular basis of many human diseases such as diabetes, cancer, arthritis, and Alzheimer's disease might eventually lead to the introduction of new therapeutic strategies.

It has been suggested that protein microarrays might represent the backbone of medical diagnostics during the 21st century, and according to a proposal of Kodadek [28], two types can be distinguished. It is likely that protein microarrays will be the next major manifestation of the revolution in genomics and proteomics.

The protein function array comprises thousands of native proteins that are immobilized in a defined pattern so that each protein present in a cell at a certain time occupies a specific x/y-coordinate on the chip. Such devices will permit ultra-high-throughput activity studies of native proteins. Using a fluorescently labeled protein as a potential interaction partner, fluorescent regions on the chip can be considered to contain immobilized proteins that are excellent binding partners for the analyte protein. These arrays will find useful application for studies of activities and binding profiles of native proteins, and will also be useful in addressing the specificity of small, protein-binding molecules, including drug candidates (see Section 9.3).

A second type of protein array is termed a protein-detecting array, and consists of large numbers of arrayed protein-binding agents. This chip allows for the recognition of target proteins and polypeptides in cell extracts or other complex biological solutions. This approach seems to be useful for monitoring the levels and chemical states of native proteins, and can be considered as the proteomics version of DNA microarrays.

However, two major technical problems must be solved before these sophisticated protein-detecting arrays may be used. First, high-throughput technology is required for the isolation of high-specificity and high-affinity protein ligands. Second, a robust method is needed for the detection of protein binding to the array in a biological sample, preferentially without the need to modify chemically the protein analytes. Without doubt, protein microarrays – when fully developed and optimized – will become powerful tools in the quest for new pharmaceutical proteins.

9.2
Large-Scale Peptide Synthesis

Procedures for the industrial production of peptides differ significantly from the well-known laboratory-scale synthesis methods. The term "large-scale" is relatively subjective however, and may be defined as ranging from kilograms to metric tons, according to the excellent review on this subject by Andersson et al. [29]. The chemistry does not differ markedly between large-scale manufacturing processes and those used under laboratory-scale conditions. However, the development of an economic, efficient and safe procedure which fulfils the requirements imposed by regulatory authorities generally comprises the final goal of large-scale peptide production.

The scale-up development strategy must take into account various technical, economic, and safety aspects. Extreme reaction conditions such as high pressure, extremes of temperature, long reaction times, highly anhydrous conditions, and very specialized equipment must be avoided. Reaction temperatures generally range from –20 °C to +100 °C.

The reactors used for large-scale synthesis (Fig. 9.1) include steel reaction vessels, systems for heating and cooling, and a heterogeneous group of units for filtration, concentration under reduced pressure, and hydrogenation. Intermediates isolated during the course of synthesis should be obtained as solids rather than as oils, and the method of choice for this is either precipitation (crystallization if possible) or chromatography. Environmental and economic aspects also determine the selection of reagents and solvents used in industrial processes. For example, it is necessary to eliminate diethyl ether as a precipitation agent due to the high risk of explosion, and also to substitute the ozone-destroying dichloromethane with other solvents. Corrosive cleavage agents such as trifluoroacetic acid (TFA) and

Fig. 9.1 Modern production plant for solution phase synthesis (Photo: BACHEM AG).

HF, or the toxic hydrazoic acid HN_3 which occurs as the by-product of azide couplings, are highly hazardous and must be avoided. The coupling agent BOP should also be substituted, as the by-product hexamethylphosphoramide (HMPA), which is formed in coupling reactions, is known to be a carcinogen. By contrast, the highly efficient (but very expensive) coupling additive HOAt is normally substituted by the less expensive HOBt because of commercial and economic reasons. For economic reasons, it is detrimental to use more than two equivalents of activated amino acids, and reagents and reactants should be used in amounts close to stoichiometry. The decision of whether the more expensive preactivated amino acids instead of nonactivated amino acids are used is usually made when the necessary development studies have been completed. On occasion, in-situ activation protocols may be more time-consuming and accompanied by lower yields and higher amounts of impurities compared to protocols using the more expensive preactivated starting materials.

Although many protected amino acid derivatives are available commercially, minimum protection schemes are preferred for large-scale synthesis (see Section 5.2.2). In particular, the side-chain protection of arginine is minimized to the inexpensive HCl salts, as shown by the first industrial solution-phase synthesis of ACTH(1–24) [30] (Fig. 9.2).

The large-scale solution-phase manufacture of 1-deamino-8-D-arginine-vasopressin (desmopressin), Mpa-Tyr-Phe-Gln-Asn-Cys-Pro-D-Arg-Gly-NH_2 (disulfide bond: Mpa^1-Cys^6), an antidiuretic used to treat diuresis associated with diabetes insipidus, nocturnal enuresis, and urinary incontinence [31], is performed via a [(3+4)+2] segment coupling strategy. Interestingly, the N-terminal segment Mpa(Acm)-Tyr-Phe-NH-NH_2 is synthesized by chymotrypsin-catalyzed coupling of Mpa(Acm)-Tyr-OEt with H-Phe-NH-NH_2, underlining that enzyme-mediated coupling (cf. Section 4.6.2) is also highly efficient in industrial processes.

As shown in Chapter 4.5 solid-phase peptide synthesis has a lot of advantages over the classical solution procedure like shorter production cycle times and often higher yields and purity. Thus this approach is also attractive for large-scale manufacture of selected peptides and, especially, fragments for phase change synthesis (cf. Section 5.3.3). For this purpose various companies have constructed special equipment which differs significantly from the commercially available lab-scale synthesizer.

As an example, Fig. 9.3 shows a large-scale solid-phase reactor which allows in combination with a SP4000-PPS control unit the synthesis of peptides in kg batches. For example, starting from 2 kg resin the procedure results in about 9 kg peptidyl resin which corresponds to about 1 to 1.5 kg peptide. This equipment fulfils the standard of "Current Good Manufacturing Practice (CGMP)" according to the Federal Regulations of the Food and Drug Administration in the same way as the plant for solution-phase peptide production shown in Fig. 9.1.

Another large-scale solution-phase synthesis with an estimated future annual production scale in the range of 50–100 kg has been described for an oxytocin antagonist, named Atosiban [32], which is used to treat preterm labor and delivery [33]. The synthesis strategy for Atosiban, Mpa-D-Tyr(Et)-Ile-Thr-Asn-Cys-Pro-Orn-

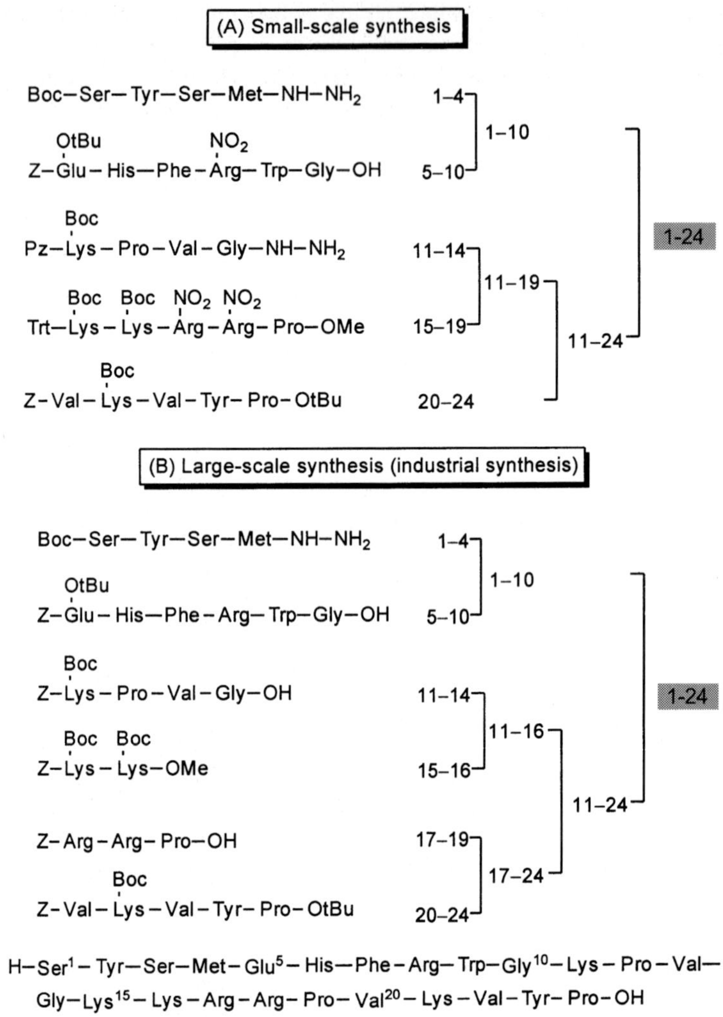

Fig. 9.2 Strategy and tactics of the industrial synthesis of ACTH-(1–24) [30].

Gly-NH_2 (disulfide bond: Mpa^1-Cys^6) is based on a common intermediate for solution-phase and solid-phase syntheses. First, the required quantities of Atosiban for toxicology and early phase clinical studies during drug development were synthesized using the rapid solid-phase method (see Section 4.5). An increasing demand for the peptide in clinical Phase II trials, where defined doses for studies in humans and the determination of a safety profile are required according to the regulations, led to a change in the synthesis protocol and the introduction of a solution-phase scale-up (2+5)+2 strategy. Thus, it was desirable to direct both manufacturing methods to a common intermediate with an identical side-chain protecting group pattern (Fig. 9.4).

Fig. 9.3 Commercially available 60 litres solid-phase reactor unit (Photo: Dr. Streb, Labortech AG).

Under these conditions, the following steps such as deprotection, oxidation, purification, and final isolation are associated with a similar profile of impurities. This combined strategy, leading to a common intermediate, is of general importance in the industrial-scale production of peptides.

Phase-change synthesis (see Section 5.3.3), which is also known as a hybrid approach, has been used in several industrial processes, with synthesis of the 36-peptide T20 being one example [34].

T20 (Fig. 9.5) is derived from the ectodomain of HIV-1 gp41, and is the first representative of a novel family of anti-retroviral agents that inhibit membrane fusion. The major importance of T20 in the development of a drug to treat HIV initiated, in the first instance, a solid-phase manufacturing process based on Fmoc chemistry [35]. However, due to a need for production in the region of metric tons, the strategy was changed at an early stage to a phase change process involv-

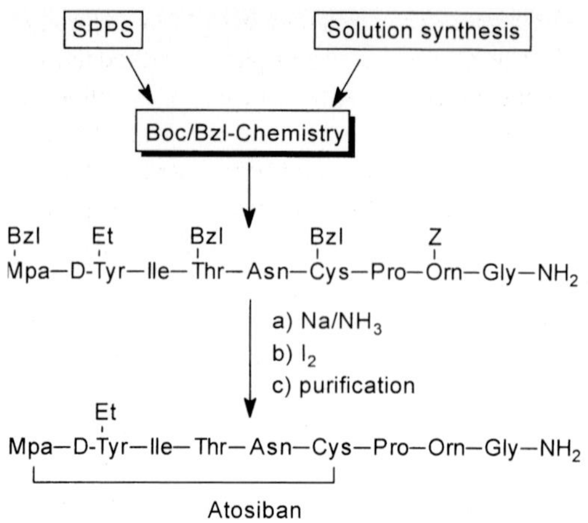

Fig. 9.4 Schematic illustration of a drug development approach to a common intermediate resulting from solid-phase and solution-phase strategy demonstrated for the oxytocin antagonist Atosiban [32].

ing three segments, synthesized on 2-chlorotrityl resin (Section 4.5.1). Initially, Fmoc-T20(27–35)-OH is coupled after cleavage from the resin to H-Phe-NH_2. The N-terminal Fmoc group is cleaved, resulting in H-T20(27–36)-NH_2 which is elongated with Fmoc-T20(17–26)-OH, yielding Fmoc-(17–36)-NH_2. After removal of the N-terminal protecting group, this fragment is coupled to Ac-T20(1–16)-OH, providing the fully protected Ac-T20(1–36)-NH_2. Deprotection with TFA/dithiothreitol/H_2O gives the crude 36-peptide derivative with a relatively high HPLC purity (>70%). This was further purified by preparative reverse-phase-HPLC, and subsequently lyophilized. T20 could thus be synthesized, without scale-up optimization, in batches of >10 kg. Hence, it seems likely that the production of this highly complex peptide on a projected annual scale of metric tons can be realized.

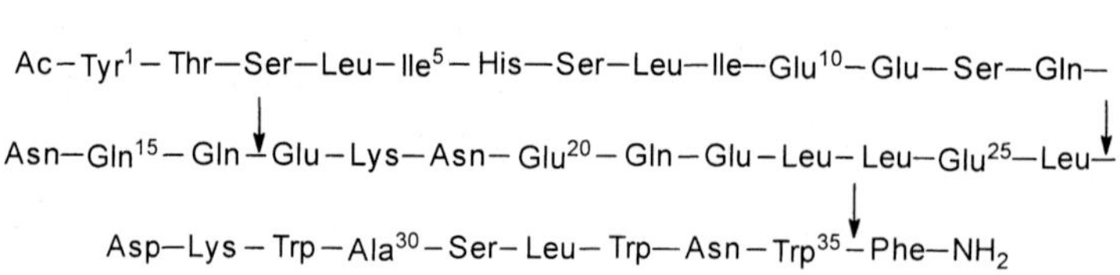

Fig. 9.5 Primary structure of the antiretroviral drug substance T20.

9.3
Peptide Pharmaceuticals

9.3.1
Peptide Drugs and Drug Candidates

Many peptides manufactured by chemical synthesis for commercial applications consist of less than 40 amino acids.

A selection of approved peptide pharmaceuticals, together with their method of manufacture, is provided in Tab. 9.3. The classical solution-phase methodology is the preferred method for the production of small to medium-length peptides in scales up to several hundred of kilograms or even metric tons per year. However, hybrid approaches which combine the advantages of solution-phase and solid-phase techniques are of increasing importance. Many hormones and especially polypeptide pharmaceuticals of more than 40 residues can be produced economically by recombinant techniques not only in bacteria but also in yeast or cultured mammalian cells. Human insulin, human growth hormone, corticotropin (ACTH), relaxin, somatostatin, renin, corticoliberin, atrial natriuretic peptide, β-endorphin, cholecystokinin, chorionic gonadotropin, and follicle-stimulating hormone are examples of regulatory peptides that can be synthesized by genetic engineering. However, not all of these are currently manufactured commercially.

Insulin is one of the oldest biopharmaceuticals approved, and currently more than 2000 kg are marketed each year. Recombinant human insulin was first launched as Huminsulin® by Eli Lilly in 1982, and over the past few decades insulin analogues have been designed with the aim of improving the therapeutic [36]. In order to improve on the pharmacokinetics of insulin, an engineered form of human insulin, termed Liprolog®/Humalog® was produced by Eli Lilly. In this variant, only the partial sequence -Pro^{28}-Lys^{29}- has been reversed [37]. Because of this manipulation, the insulin existed as monomer at physiological concentrations and consequently had a faster onset, but shorter duration of action due to enhanced absorption after subcutaneous administration.

Insulin glargine (Lantus®, formerly known as HOE901), 21^{A}-Gly-30^{B}a-L-Arg-30^{B}b-L-Arg-human insulin, is a long-acting recombinant human insulin analogue produced by DNA technology [38]. The substitution of Asn^{21} of the A chain by Gly, and the N-terminal extension of the B chain by two Arg residues, resulted in a change of the isoelectric point, from 5.4 of the native insulin to 6.7 of insulin glargine. As a result, it is soluble in slightly acidic conditions (pH 4.0) and precipitates at the neutral pH of subcutaneous tissue. In this way, the absorption of insulin glargine is delayed, thereby providing a fairly constant basal insulin supply for about 24 hours.

Even if the peptide sweetener aspartame and peptide antibiotics are included, peptides comprise only about 0.0025% (by mass) of the worldwide annual production of drugs. By contrast, peptide drugs such as the immunoregulator cyclosporin (which nowadays is indispensable in modern organ transplantation) are highly beneficial in therapy, and the current sales of cyclosporin exceed US $ 1 billion each year. The sales values of calcitonin, which has become an important drug in the treatment of hypercalcemia, Paget's disease, osteoporosis and pain (preferentially of patients suffering from bone cancer) are only slightly less.

Without doubt, the therapeutic application of peptides has great potential in various indications such as blood pressure, neurotransmission, growth, digestion, reproduction, and metabolic regulation. The control of almost all biological processes in living cells is carried out by proteins, and involves various types of molecular recognition. Much of this activity is mediated by enzymes, but many regu-

Tab. 9.3 Selected peptides approved as pharmaceuticals, and their methods of production.[a)]

Peptide	*AA*	*Solution-phase synthesis*	*Solid-phase synthesis*	*Recombinant synthesis*	*Extraction*
Oxytocin and analogues					
Oxytocin	9	X			
Atosiban	9	X			
Vasopressin analogues					
Pitressin	9	X			
Lypressin	9	X			
Desmopressin	9	X	X		
Terlipressin	12	X	X		
Adrenocorticotropic hormone	39				
ACTH-(1-24)	24	X			
Insulin[b)]	51	X		X	X
Glucagon	29		X	X	X
Secretin	27				X
Calcitonins					
Human	32	X			
Salmon	32	X	X		
Eel	32	X	X		
Dicarba-Eel (Elcatonin)	31	X	X		
LH-RH and analogues					
LH-RH	10	X	X		
Leuprolide	9	X			
Deslorelin	9		X		
Triptorelin	10		X		
Goserelin	10		X		
Buserelin	9		X		
Parathyroid hormone (PTH)	84				
PTH-(1-34)	34		X		
Corticoliberin (CRH)	41				
Human	41		X		
Ovine	41		X		
Growth hormone-releasing hormone (Somatoliberin)	44		X		
GRH-(1-29)	29				
Somatostatin and analogues					
Somatostatin	14	X	X		
Lanreotide	8		X		
Octreotide	8	X			
Thyroliberin (TRH)	3	X			
Thymosin-α_1	28		X		
Thymopentin (TP-5)	5	X			
Cyclosporin	11				X
Integrilin	7	X			

a) Slightly modified according to Andersson et al. [29].
b) Also manufactured by semisynthesis.

latory processes are initiated by specific protein-protein interactions that still constitute a largely unexploited area of targets in drug discovery and drug design. Unfortunately, the interacting surfaces very often lack the classical features necessary for inhibition with small molecules [39]. The development of peptide and nonpeptide integrin antagonists [40], the design of a platelet-derived growth factor (PDGF) binding molecule with anti-angiogenic and tumor regression properties [41], and the "death antagonist" Bcl-2 [42] have proved that the inhibition of protein-protein interactions is a viable therapeutic strategy. The current focal point of research in this field is directed towards the development of high-affinity protein-protein interaction antagonists and agonists that mimic the binding interface at selected interaction hot spots, based on the "hot spot" concept [43].

Because of their enormous potential for diversity, it is possible that peptides may be uniquely suited for influencing biological control processes based on molecular recognition. As shown in Chapter 8, it is now possible to construct very large libraries of peptides. When discussing peptides as potential pharmaceutical agents, although the final goal is efficacy *in vivo*, the ultimate need for high potency towards the target protein must be linked with few side effects and good (preferentially oral) bioavailability. Unfortunately, one major disadvantage of peptide pharmaceuticals is their putative metabolic instability.

9.3.2
Peptide Drug Delivery Systems

During recent years, progress in the areas of formulation and delivery systems has led to the development of several highly successful peptide drugs. The main goal of these improvements were not only to overcome a lack of oral bioavailability, but also to avoid the need for subcutaneous injection, which often leads to poor patient compliance. Several possibilities of administering peptides and proteins by either oral, pulmonary, mucosal membrane or transcutaneous routes have been reported. These routes of administration very often require specific delivery vehicles and/or permeability enhancers to assist transfer of the drug across the delivery site and into the systemic circulation. Interesting alternatives include nasal sprays for LH-RH (Buserelin®), calcitonin, oxytocin and vasopressin, and rectal suppositories for calcitonin. Ointments are often used for the transdermal application of peptides, but sublingual administration is another possibility. Structure-function studies have led to the design of new peptide derivatives suitable for oral administration, such as the vasopressin analogue, desmopressin. Modified analogues of somatostatin are available which retain the pharmacological properties of the parent hormone but exhibit a significantly prolonged duration of action. Following administration, many peptide and protein biopharmaceuticals exert their intended action in the systemic circulation, and must therefore resist clearance by conventional mechanisms, including molecular filtration by the kidney and clearance by the reticuloendothelial system. PEGylation of peptides and proteins yields polyethylene glycol (PEG)-conjugated derivatives with reduced renal clearance and a more than 50-fold enhancement in circulatory half-life [44, 45].

The recent developments achieved in drug delivery systems for peptide and protein pharmaceuticals will continue to increase the therapeutic application of these materials. In their excellent review, Pettit and Gombotz [46] defined site-specific drug delivery as delivery through a specific site (i.e., the route of administration), as well as delivery to a specific site (i.e., the site of action). The physical and chemical characteristics of both the peptide to be delivered, and of the site to be targeted, must be especially considered in development of the appropriate technology. A synthetic polymer, device or carrier system may be introduced to target the biopharmaceutical to a specific site within the body. Selected examples of site-specific drug delivery are listed in Tab. 9.4.

Radiolabeled tumor-specific peptides have found application for diagnostic purposes. For example, they may be used *in vitro* on tumor sections to obtain information on the so-called receptor status of the cells; alternately, they may be injected into the body in order to locate tumors. ^{125}I is a useful radioligand for in-vitro application, whereas the short-lived ^{123}I is more suitable for in-vivo administration. Nowadays, peptides labeled with radioactive iodine isotopes are increasingly replaced by peptide derivatives equipped with chelators for ^{111}In or ^{99}Tc. Radiolabeled peptides appear to be very useful both for tumor diagnosis in cancer patients, and for tumor therapy. An

Tab. 9.4 Selected examples of site-specific drug delivery according to Pettit and Gombotz [46].

Site targeted	*Remarks*
Route of administration	
Transdermal	Assisted by iontophoresis or ultrasound
Pulmonary	Liquid and dry-powder aerosol delivery
Mucous membranes	Aerosol-mucin charge interactions
Oral/intestinal	Small particles, protein-carrier complexes
Specific tissues or organs	
Tumors	Neovascularization markers are targeted
Lungs	Aerosol, liposomal delivery
Brain	Target the transferrin receptor
Intestines	Protect against proteolysis and acid hydrolysis
Eyes	Mucin charge interactions
Uterine horns	Form biogradable gel *in situ*
Bones	Hydroxyapatite binds bone-promoting growth factor
Skin	Methylcellulose gels
Cellular/intracellular	
Macrophages	Small particles are phagocytosed
Tumor cells	Fusogenic liposomes to deliver intracellular toxins
Molecular targets	
Tumor antigens	Antibody-enzyme conjugates activate prodrugs
Fibrin/site of clot formation	Fusion proteins combine targeting with toxin
Carbohydrate receptors	Mannose and galactose used to target receptor
Systemic circulation	
Injection	Prolong or sustain circulation

interesting approach to cancer chemotherapy is based on the targeting of cytotoxic peptide conjugates to their receptors on tumors. Cytotoxic conjugates are hybrid molecules consisting of a peptide carrier (which binds to the receptors that are up-regulated on tumors) and a suitable cytotoxic moiety. An early example of hormone drug conjugates was that of the DNA intercalator daunomycin linked to the N-terminal amino group of Asp, and also to the ε-amino groups of Lys residues of the β-melanocyte stimulating hormone [47]. Recently, cytotoxic compounds such as doxorubicin linked to LH-RH, bombesin, and somatostatin could be targeted to certain tumors that expressed specific peptide-receptors in higher numbers than would normal cells. Consequently, these conjugates were seen to be especially lethal for cancer cells [48]. Peptide conjugates with daunorubicin or doxorubicin covalently linked to neuropeptide Y have also been described for tumor-specific chemotherapy [49].

Peptide and protein drugs are generally not capable of being transported from the blood to the brain, as the blood-brain barrier (BBB) limits the transfer of soluble peptides and proteins through the brain capillary endothelial wall. However, these compounds may be delivered to the brain using the chimeric peptide strategy for peptide drug delivery [50]. A nontransportable drug is coupled to a BBB drug transport vector such as cationized albumin, transferrin, or transferrin receptor antibodies; this allows the BBB to be bypassed by absorption-mediated and receptor-mediated transcytosis.

The application of drug delivery vectors has also been demonstrated using pegelin and penetratin, both of which may transport doxorubicin across the BBB.

Peptides that target tumor blood vessels have been identified by phage display and coupled to anticancer drugs [51–53]. Tumor-targeting peptide oligonucleotide conjugates have been described for the application of antisense oligodeoxynucleotides as therapeutic agents inhibiting gene expression [54].

9.3.3 Peptides as Tools in Drug Discovery

The use of peptides for affinity labeling of receptors is important in attempts to identify, characterize, and isolate hormone or neurotransmitter receptors. The general approach is to establish a covalent bond between a ligand and its receptor; this can be achieved by chemical affinity labeling and photoaffinity labeling, with the latter technique being the preferred method for receptor identification and isolation. For this purpose, a chemically stable but photolabile moiety is conjugated to a potent ligand. When the modified ligand has bound to its receptor site, photolysis generates highly reactive nitrenes or carbenes that react with chemical functionalities on the receptor molecule, thereby forming a covalent bond between the ligand and the receptor [55].

Synthetic peptides are also used for the delineation of receptor types and subtypes. Receptors for almost all bioactive peptides are expressed by different target cells linking the hormone signal to slightly varying biological effects. Multiple types and subtypes of receptors exist which complicates receptor pharmacology, notably as each subtype plays a particular functional role *in vivo*. Consequently, the design and synthesis of peptides directed toward receptor subtype binding and the determination of the appropriate kinetics are essential aims of current peptide research.

Target-based screening to identify compounds for development is a prerequisite to a powerful methodology in drug discovery research. Conventionally, drugs have been discovered by screening either natural compound collections or small chemical compound libraries. An alternate approach would be the chemical synthesis of compounds based on structural data available for a given target. Unfortunately, all these methods are generally cumbersome and time consuming, and drug companies now routinely assay several hundred thousand compounds against each new drug target by the use of modern HTS techniques (see below). In connection with this, several complementary methods now exist by which large combinatorial peptide libraries may be made available (see Chapter 8).

The importance of peptides as tools in drug discovery has been recently reviewed by Grøn and Hyde-DeRuyscher [56]. The initial step in drug discovery is the selection of a suitable target molecule, and the number of proteins seen as potential targets for drug intervention in order to control human disease or injury has been estimated in the range of 2000 to 5000 [57]. Despite this, the drugs that are currently on the market, together with those which have been discovered during the past 100 years, have been calculated to be directed against not more than 500 target proteins [58]. Interestingly, the term "chemogenomics" has recently been coined as relating to the discovery and description of all possible drugs to all possible drug targets [59]. As mentioned above, the terms genomics and proteomics define the process of identifying and classifying all genes in a genome, as well as the correlation between a gene expression pattern and the phenotype at different stages. Protein modeling forms an integral part of the drug discovery effort [60]. A functional understanding of novel gene products will increase the number of suitable drug targets based on clear synergies of the combination of target structural information with combinatorial chemistry.

9.3.3.1 Peptides Targeted to Functional Sites of Proteins

A functional site of a target protein is characterized as an area where binding of a ligand – a small molecule or a protein – modulates activity. As shown above, most proteins interact with other proteins, but the number of residues critical for binding sometimes is rather low, comprising three to ten amino acids [61]. Thus peptides, for example from combinatorial libraries, may act as "surrogate" ligands. Functional sites are mostly located at grooves in the protein surface [62], and comprise flexible areas where favorable interactions with the ligand support formation of the protein-ligand complex. Often, interactions with single water molecules stabilize the empty functional site of the native protein. One of the driving forces for peptide binding to a target protein is the displacement of water molecules from recesses or cavities in the protein, mainly because of entropic reasons. Target-specific peptides can be used to understand the nature of functional sites and to identify potential binding partners; moreover, they serve as valuable tools in structure-based and HTS drug discovery, as will be shown below.

As mentioned, many peptides have poor pharmacological properties. Consequently, the question remains as to how a peptide ligand that binds to an active site of a target protein can be converted into a drug. Peptides may act immediately as agonists or antagonists under special circumstances, as is the case of cell-surface receptors. Because most small peptides are easily proteolyzed, rapidly excreted and poorly bioavailable, special short-lived peptides are only used for the treatment of acute health problems by intravenous or subcutaneous injection. These limitations have thus necessitated the development of techniques to replace portions of peptides with nonpeptide structures, and this has resulted in nonpeptide therapeutics. Additionally, it is possible to design peptidomimetics (see Chapter 7) that are protease-resistant, readily cross the plasma membrane, and also show desirable pharmacokinetic properties [63].

9.3.3.2 Peptides Used in Target Validation

Target validation is necessary to clarify the function of a protein in a specific biochemical pathway. Peptides may also find application for target validation in the drug discovery process. The increasing amount of genome data, both of the human cell and of selected human pathogens, has provided a rich source of interesting targets. The best candidates for pharmacological interventions can be elucidated by usual target validation tools such as gene knockouts and targeted mutations, in combination with bioinformatics. Unfortunately, genetic knockouts and mutations may result in the complete loss of all functions of the target protein, and the deletion of the protein can therefore be misleading. Target validation with peptides will be faster and can be achieved much more selectively. A peptide normally interferes with only one of several specific functional sites of a protein target, and this resembles the action of a drug. Suitable peptide ligands can either be introduced into a cell or expressed inside cells. They may bind to the target protein, and the physiological effects of binding can be monitored in order to predict the response of a drug binding to the same site. Several means for validating a target with a peptide have been developed. Upon injection of target-specific peptides, for example, for the Src homology 3 (SH3) domain into *Xenopus laevis* oocytes, an acceleration of progesterone-stimulated maturation was observed [64]. This effect might be caused by peptide-induced modulation of the protein, or by a signal transduction pathway. Furthermore, peptide ligands that are specific for the tyrosine kinase Lyn SH3 domain have been transported into mast cells by electroporation, resulting in an inhibition of mast cell activation [65]. The activity of peptidic ligands might be of short duration because of intracellular proteolysis, though this limitation can be overcome using peptides composed of either D-amino acids or β-amino acids [66].

Another delivery route is based on linking peptides to other peptides or protein domains. The resulting peptide conjugates have the capacity to cross the plasma membrane in order to modulate target activity inside cells. In principle, peptides can also be expressed inside cells, either alone or fused to an innocuous reporter protein, for example, to the green fluorescent protein [67] using recombinant DNA [68].

9.3.3.3 **Peptides as Surrogate Ligands for HTS**

A third possibility to utilize peptides in the drug discovery process is the design of in-vitro modular assays suitable for HTS systems of small molecule libraries. Peptides directed to special protein functional sites are used to format an assay where molecules are tested for their capability to displace bound peptide ligands, or to prevent binding. Several competitive binding assays are currently in use to detect inhibitors of peptide binding, and for many targets compounds have been identified to inhibit the activity of the target protein. Several detection formats, including scintillation proximity assays, time-resolved fluorescence (TRF), fluorescence polarization (FP) and fluorescence resonance energy transfer (FRET), have found application for the detection of inhibitors using peptide surrogate ligands. The assays can be performed automatically in a high throughput mode, and it is possible to collect up to 200 000 or more data points per day with appropriate robotic workstations. Any target protein for which a peptide surrogate ligand has been elucidated can be used in the inhibitor screening of large compound libraries.

The homogeneous time-resolved fluorescence (HTRF) assay [69] eliminates many disadvantages associated with some conventional screening assay methodologies like in-plate binding assays and radiometric assays. HTRF is performed in completely homogeneous solution without the need of coating plates, solid supports or time-consuming separation steps. Furthermore, background fluorescence is eliminated and there is no requirement for special handling, monitoring or disposal of reagents. Each microplate is measured in one second. HTRF is based on fluorescence resonance energy transfer between the donor fluorophore europium cryptate (Euk) and the acceptor fluorophore XL665 which is a modified allophycocyanin (Fig. 9.6). A slow signal decay is observed at 665 nm when two biomolecules labeled with the fluorophores are binding to each other. The energy of the laser at 337 nm is absorbed by EuK which transfers its energy to XL665, that emits the fluorescence signal at 665 nm with a slow decay time. HTRF has been proven to be feasible for the detection of protein-protein interaction and receptor binding with a variety of targets like tyrosine kinases, viral proteases and antibodies.

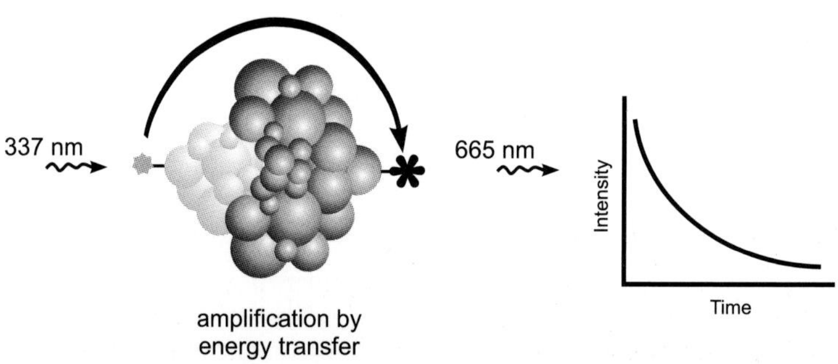

Fig. 9.6 Simplified principle of the homogeneous time-resolved fluorescence (HTRF) screening assay according to Kolb et al. [69].

A further interesting tool for HTS is based on a conformationally dependent binding of peptides to receptors. Traditional assays searching for agonists and antagonists of hormone receptors are based on a binding assay where a labeled natural ligand competes with library constituents. Unfortunately, these assays are not capable of differentiating between an agonist or antagonist; additional cell-based model systems, or even animal models, are then required for the elucidation of the biological effect. In contrast, HTS assays can be formatted for the search of compounds with specific effects on receptor conformation that will contribute to knowledge on biological effects.

HTS technologies underwent a revolution during the late 1990s, with the result that most pharmaceutical companies now use HTS as the primary tool for lead discovery [70–72]. New HTS techniques have significantly increased throughput, and have also reduced assay volumes in offering a new technology for the 21st century [73]. The transition from slow, manual, low-throughput screening to robotic ultrahigh-throughput screening (uHTS) will soon allow screening of more than 200000 samples per day. In addition, new fluorescence methods [74, 75], photoactivatable ligands [76], and miniaturized HTS technologies [77–80], together with key advances both in detection platforms and liquid handling technologies have contributed to the rapid development of uHTS. Modern detection platforms demonstrate significant improvements in sensitivity and throughput, whilst new liquid handling methods allow for the dispensing of compounds and reagents in volumes consistent with miniaturized assay formats. The development from 96-well screening on the microscale towards higher density (for example, 1536-well) nanoscale formats and the advent of homogeneous fluorescence detection technologies serve as benchmarks in HTS development. Ullmann et al. [81] described both new applications and instrumentation for confocal fluctuation fluorescence-based HTS, and new two-dimensional applications of this methodology in which molecular brightness analysis (FIDA) is combined with molecular anisotropy measurements and other reactant principles.

In summary, peptide surrogate ligands play a fundamental role in modern drug discovery programs. Besides other applications, they have been used to format sensitive HTS systems in order to identify compounds that modulate the function of the target protein. Most importantly, they are the starting points for drug leads [82].

9.4 References

1 G. Wright, A. Carver, D. Cottom, D. Reeves, A. Scott, P. Simons, I. Wilmut, I. Garner, A. Colman, *Biotechnology* **1991**, *9*, 830

2 C. Cunningham, A. J. R. Porter, *Methods in Biotechnology*, Volume 3, Humana Press, Totowa, **1997**

3 A. Persidis, *Nature Biotechnol.* **1998**, *16*, 1378

4 D. Blohm, C. Bollschweiler, H. Hillen, *Angew. Chem.* **1988**, *100*, 213; *Angew. Chem. Int. Ed. Engl.* **1988**, *27*, 207

5 J. McCafferty, D. R. Glover, *Curr. Opin. Structural. Biol.* **2000**, *10*, 417

6 K. Deres, H. Schild, K.-H. Wiesmüller, G. Jung, H.D. Rammensee, *Nature* **1989**, *342*, 561

7 H. Schild, K. Deres, K.-H. Wiesmüller, G. Jung, H.G. Rammensee, *Eur. J. Immunol.* **1991**, *21*, 2649

8 K.-H. Wiesmüller, B. Fleckenstein, G. Jung, *Biol. Chem. Hoppe-Seyler* **2001**, *382*, 571

9 W. Zauner, K. Lingnau, F. Mattner, A. von Gabain, M. Buschle, *Biol. Chem. Hoppe-Seyler* **2001**, *382*, 581

10 P.S. Sarin, C.A. Mora, P.H. Naylor, R. Markham, D. Schwartz, J. Kahn, P. Heseltine, B. Gazzard, M. Youle, A. Rios, A. Goldstein, *Cell. Mol. Biol.* **1995**, *41*, 401

11 V. Blazevic, A. Ranki, K.J. Krohn, *AIDS Res. Hum. Retroviruses* **1995**, *11*, 1335

12 G. Gorse, M.C. Keefer, R.B. Belshe, T.J. Mattews, B.D. Forrest, R.H. Hsieh, W.C. Koff, C.V. Hanson, R. Dolin, K. Weinhold et al., *J. Infect. Dis.* **1996**, *173*, 330

13 M. Marchand, N. van Baren, P. Weynants, V. Brichard, B. Dreno, M.-H. Tessier, E. Rankin, G. Parmiani, F. Arienti, Y. Humblet et al., *Int. J. Cancer* **1999**, *80*, 219

14 T.J. Vaughan, J.K. Osbourn, P.R. Tempest, *Nature Biotechnol.* **1998**, *16*, 535

15 D.P. Pollock, J.P. Kutzko, E. Birck-Wilson, J.L. Williams, Y. Echelard, H.M. Meade, *J. Immunol. Methods* **1999**, *231*, 147

16 M.A. van Dijk, J.G.J. van den Winkel, *Curr. Opin. Chem. Biol.* **2001**, *5*, 368

17 J. McCafferty, A.D. Griffiths, G. Winter, D.J. Chiswell, *Nature* **1990**, *348*, 552

18 M. Bruggemann, M.J. Taussig, *Curr. Opin. Biotechnol.* **1997**, *8*, 455

19 J. Kempeni, *Ann. Rheum. Dis.* **1999**, *58*, 170

20 E. Ciszak, J.M. Beals, B.H. Frank, J.C. Baker, N.D. Carter, G.D. Smith, *Structure* **1995**, *3*, 615

21 W.C. Zhou, C. Chen, B. Buckland, J. Aunins, *Biotechnol. Bioeng.* **1997**, *55*, 783

22 M. Hall, *An A to Z of British Medicine Research*, Association of British Pharmaceutical Industry, London, **1998**

23 E.S. Lander, L.M. Linton, B. Birren, C. Nusbaum, M.C. Zody, J. Baldwin, K. Devon, K. Dewar, M. Doyle, W. Fitz-Hugh et al., *Nature* **2001**, *409*, 860

24 J.C. Venter, M.D. Adams, E.W. Myers, P.W. Li, R.J. Mural, G. G. Sutton, H.O. Smith, M. Yandell, C.A. Evans, R.A. Holt et al., *Science* **2001**, *291*, 1304

25 F. Lottspeich, *Angew. Chem.* **1999**, *111*, 2630; *Angew. Chem. Int. Ed. Engl.* **1999**, *38*, 2476

26 P.A. Haynes, S. Gygi, D. Figeys, R. Aebersold, *Electrophoresis* **1998**, *19*, 1862

27 P. James, *Q. Rev. Biophys.* **1998**, *30*, 279

28 T. Kodadek, *Chem. Biol.* **2001**, *8*, 105

29 L. Andersson, L. Blomberg, M. Flegel, L. Lepsa, B. Nilsson, M. Verlander, *Biopolymers* **2000**, *55*, 227

30 M. Feurer in: *Peptides: Proceedings, 5th American Peptide Symposium*, M. Goodman, J. Meienhofer (Eds.), Wiley, New York, **1977**, 487

31 K.-E. Andersson, B. Bengtsson, O. Paulsen, *Drugs of Today* **1988**, *24*, 509

32 C. Johansson, L. Blomberg, E. Hlebowicz, H. Nicklasson, B. Nilsson, L. Andersson in: *Peptides 1994*, H.L.S. Maia (Ed.), Escom, Leiden, **1995**, 34

33 P. Melin, *Baillières Clin. Obstet. Gynaecol.* **1993**, *7*, 577

34 M.K. Lawless, S. Barney, K.I. Guthrie, T.B. Bucy, S.R. Petteway Jr., G. Merutka, *Biochemistry* **1996**, *35*, 13697

35 M.C. Kang, B. Bray, M. Lichty, C. Mader, G. Merutka, US 6015881, **2000**

36 A.H. Barnett, D.R. Owens, *Lancet* **1997**, *349*, 47

37 E. Ciszak, J.M. Beals, B.H. Frank, J.C. Baker, N.D. Carter, G.D. Smith, *Structure* **1995**, *3*, 615

38 R. Hilgenfeld, M. Dörschug, K. Geisen, H. Neubauer, R. Obermeier, G. Seipke, H. Berchtold, *Diabetologia* **1992**, *35* (Suppl.1), A 193

39 A.G. Cochran, *Chem. Biol.* **2000**, *7*, 85

40 H.W. Lark, G.W. Stroup, S.M. Hwang, I.E. James, D.J. Rieman, F.H. Drake, S.N. Bradbeen, A. Mathur, K.F. Erhand, K.A. Newlander et al., *J. Pharmacol. Exp. Ther.* **1999**, *291*, 612

41 M.A. Blaskovich, Q. Lin, F.L. Delarue, J. Sun, H.S. Park, D. Coppolöa, A.D. Hamilton, S.M. Sebti, *Nature Biotechnol.* **2000**, *18*, 1065

42 J.-L. Wang, D. Liu, Z.-J. Zhan, S. Shan, X. Han, S.M. Srinivasula, C.M. Croce, E.S. Alnemri, Z. Huang, *Proc. Natl. Acad. Sci. USA* **2000**, *97*, 7124

43 T. Clackson, J.A. Wells, *Science* **1995**, *267*, 383

44 C. Delgado, G.E. Francis, D. Fisher, *Crit. Rev. Ther. Drug Carrier Syst.* **1992**, *9*, 249

45 D.K. Pettit, T.P. Bonnert, J. Eisenman, S. Srinivasan, R. Paxton, C. Beers, D. Lynch, B. Miller, J. Yost, K.H. Grabstein, W.R. Gombotz, *J. Biol. Chem.* **1997**, *272*, 2312

46 D.K. Pettit, W.R. Gombotz, *Trends Biotechnol.* **1998**, *16*, 343

47 J.M. Varga, *Methods Enzymol.* **1985**, *112*, 259

48 A.V. Schally, A. Nagy, *Eur. J. Endocrinol.* **1999**, *141*, 1

49 M. Langer, F. Kratz, B. Rothen-Rutishauser. H. Wunderli-Allenspach, A.G. Beck-Sickinger, *J. Med. Chem.* **2001**, *44*, 1341

50 U. Bickel, T. Takayoshi, W.M. Pardridge, *Adv. Drug. Delivery Rev.* **2001**, *46*, 247

51 C. Rousselle, P. Clair, J.M. Lefauconnier, M.M. Kaczorek, J.M. Scherrmann, J. Temsamani, *Mol. Pharmacol.* **2000**, *57*, 679

52 D. Derossi, G. Chassaing, A. Prochiantz, *Trends Cell Biol.* **1998**, *8*, 84

53 W. Arap, R. Pasqualini, E. Ruoslahti, *Science* **1998**, *279*, 377

54 W. Mier, R. Eritja, A. Mohammed, U. Haberkorn, M. Eisenhut, *Bioconj. Chem.* **2000**, *11*, 855

55 G. Dormán, G.D. Prestwich, *Trends Biotechnol.* **2000**, *18*, 64

56 H. Grøn, R. Hyde-DeRuyscher, *Curr. Opin. Drug Discov. Dev.* **2000**, *3*, 636

57 J. Drews in: *Human Disease: from Genetic Causes to Biochemical Effects*, J. Drews, S. Ryser (Eds.), Blackwell, Berlin, **1997**

58 J. Drews, S. Ryser, *Nature Biotechnol.* **1997**, *15*, 1318

59 P.R. Caron, M.D. Mullican, R.D. Mashal, K.P. Wilson, M.S. Su, M.A. Murcko, *Curr. Opin. Chem. Biol.* **2001**, *5*, 464

60 M. Norin, M. Sundström, *Curr. Opin. Drug Discov. Dev.* **2001**, *4*, 284

61 J.A. Wells, *Proc. Natl. Acad. Sci. USA* **1996**, *93*, 1

62 R.A. Laskowski, N.M. Luscombe, M.B. Swindells, J.M. Thornton, *Protein Sci.* **1996**, *5*, 2438

63 A.S. Ripka, D.H. Rich, *Curr. Opin. Chem. Biol.* **1998**, *2*, 441

64 A.B. Sparks, L.A. Quilliam, J.M. Thom, C.J. Der, B.K. Kay, *J. Biol. Chem.* **1994**, *269*, 23853

65 T.P. Stauffer, C.H. Martenson, J.E. Rider, B.K. Kay, T. Meyer, *Biochemistry* **1997**, *36*, 9388

66 D.H. Appella, L.A. Christianson, I.L. Karle, D.R. Powell, S.H. Gellman, *J. Am. Chem. Soc.* **1996**, *118*, 13071

67 M. Chalfie, Y. Tu, G. Euskirchen, W.W. Ward, D.C. Prasher, *Science* **1994**, *263*, 802

68 P. Colas, B. Cohen, T. Jessen, I. Grishina, J. McCoy, R. Brent, *Nature* **1996**, *380*, 548

69 A.J. Kolb, P.V. Kaplita, D.J. Hayes, Y.-W. Park, C. Pernell, J.S. Major, G. Mathis, *Drug Discov. Today* **1998**, *3*, 333

70 A.W. Lloyd, *Drug Discov. Today* **1997**, *2*, 397

71 J.J. Burbaum, *Drug Discov. Today* **1998**, *3*, 313

72 S.J. Rhodes, R.C. Smith, *Drug Discov. Today* **1998**, *3*, 361

73 R.P. Hertzberg, A.J. Pope, *Curr. Opin. Chem. Biol.* **2000**, *4*, 445

74 M. Auer, K.J. Moore, F.J. Meyer-Almes, R. Guenther, A.J. Pope, K.A. Stoeckli, *Drug Discov. Today* **1998**, *3*, 457

75 P. Kask, K. Palo, D. Ullmann, K. Gall, *Proc. Natl. Acad. Sci. USA* **1999**, *96*, 13756

76 G. Dorman, G.D. Prestwich, *Trends Biotechnol.* **2000**, *18*, 64

77 K.R. Oldenburg, J. Zhang, T. Chen, A. Maffia, III, K.F. Blom, A.P. Combs, D.Y. Chung, *J. Biomol. Screen.* **1998**, *3*, 55

78 U. Haupts, M. Ruediger, A.J. Pope, *Drug Discov. Today* **2000**, *1*, 3

79 R. TURNER, D. ULLMANN, S. STERRER in: *Handbook of Drug Screening*, R. SEETHALA, P.B. FERNANDES (Eds.), Marcel Dekker, New York, **2001**

80 J. WÖLCKE, D. ULLMANN, *Drug Discov. Today* **2001**, *6*, 637

81 D. ULLMANN, M. BUSCH, T. MANDER, *J. Pharm. Technol.* **1999**, *99*, 30

82 B.A. KENNY, M. BUSHFIELD, D.J. PARRY-SMITH, S. FORGARTY, J.M. TREHEME, *Prog. Drug Res.* **1998**, *51*, 245

Glossary

A

Abrin, a phytotoxic plant protein from the red seeds of *Abrus precatorius*. It consists of an A-chain (M_r ~ 30 kDa) and a B-chain (M_r ~ 35 kDa) joined by disulfide bridges. The A-chain is a highly specific N-glucosidase acting as ribosome-inactivating protein (RIP), whereas the B-chain is responsible for anchoring at the cell surface. A disulfide-cleaving system of the cell releases the A-chain, which enters the cell by endocytosis. RIP cleaves a single adenine residue from the rRNA, resulting in inhibition of protein synthesis followed by cell death. Similar action and structure are possessed by → ricin. The A-chain coupled with a monoclonal antibody directed against a tumor antigen is used in drug targeting [A. J. Cumber et al., *Methods Enzymol.* **1985**, *112*, 207].

Abzyme, *catalytic antibody*, a monoclonal antibody with catalytic activity. An antibody raised against a transition state analogue of a particular reaction can catalyze that reaction. The first abzyme to be generated was capable of catalyzing the hydrolysis of esters. Abzymes have been described that catalyze e.g., acyl transfer, C-C bond cleavage, β-elimination, and C-C bond formation. From X-ray analyses it could be concluded that antibodies bind peptides of various length in elongated grooves using hydrogen bonding, van der Waal's forces, and ionic contacts for recognition. Abzymes are also an interesting choice for peptide bond formation (cf. Section 4.6.3) [R. A. Lerner et al., *Science* **1991**, *252*, 659; R. Hirschmann et al., *Science* **1994**, *265*, 234; D. W. Smithrud et al., *J. Am. Chem. Soc.* **1997**, *119*, 278].

ACE inhibitors, compounds reducing the activity of the → angiotensin converting enzyme (ACE). The beneficial effects of ACE inhibitors such as like → captopril and → enalapril in hypertension and heart failure result primarily from suppression of the renin-angiotensin-aldosterone system. Inhibition of ACE causes a decrease in plasma angiotensin II (→ angiotensins) level which leads to decreased vasopressor activity and to a small decrease in aldosterone secretion. [G. Lawton et al. in: *Advances in Drug Research*, B. Testa (Ed.), Volume 23, p. 161, Academic Press, New York, **1992**].

ACE, acronym of angiotensin-converting enzyme, → angiotensins.

Achatin, a 4-peptide isolated from the ganglia of the African giant snail *Achatina fulica*. The neuroexcitatory peptide *achatin I* (H-Gly-D-Phe-Ala-Asp-OH) contains a D-amino acid (→ dermorphin, → deltorphins, → fulicin) in position 2, whereas *achatin II* with the L-isomer in the same position shows neither physiological nor pharmacological activities. It has been assumed that D-Phe in achatin I is a prerequisite for forming a 15-membered ring with a unique turn conformation structure, which may be the active conformation suitable for interactions with the receptor [Y. Kamatani et al., *Biochem. Biophys. Res. Commun.* **1989**, *160*, 1015; T. Ishida et al., *FEBS Lett.* **1992**, *307*, 253].

ACTH, acronym of adrenocorticotropic hormone, → corticotropin.

Actin, a contractile protein occurring in many eukaryotic cell types. Actin and → myosin are the major components of the muscle. Both proteins account for 60–70% and 20–25% of the total muscle protein, respectively. Besides → thrombomyosin and → troponin, actin is the major constituent of thin filaments. The fibrous *F-actin* forms the core of the thin filament and is formed under physiological conditions by polymerization of the globular *G-actin* (375 aa; $M_r \sim 42$ kDa). The monomeric G-actin consists of two domains, each of which is divided into two subdomains. Actin normally binds one molecule ATP which is hydrolyzed during polymerization to F-actin, and the resulting ADP remains bound to the F-actin monomer unit. ATP and ADP bind in a cleft between the two domains. The F-actin helix (diameter ~ 100 Å) has 2.17 actin monomers per left-handed helix turn (13 subunits in six turns) and a rise per turn of ~ 60 Å. The monomeric unit of each F actin is capable of binding a single myosin S 1 fragment [R.A. Milligan et al., *Nature* **1990**, *348*, 217; P. Sheterline, J.C. Sparrow, *Protein Profiles* **1994**, *1*, 1].

Actinomycins, peptide antibiotics produced by various strains of *Streptomyces*. Actinomycins are orange-red bacteriostatic and cytostatic, but highly toxic chromopeptides. The chromophore *actinocin*, 2-amino-4,6-dimethyl-3-oxo-phenoxazine-1,9-dicarboxylic acid, is linked to two five-membered peptide lactones by the amino groups of two threonine residues. The various naturally occurring and synthetic actinomycins differ mostly in the amino acid sequence of the lactone moieties. *Actinomycin D* (***Formula***) is one of the well-known actinomycins with known 3D structure. Actinomycin D is a useful antineoplastic agent that tightly binds to dsDNA, and in this manner strongly inhibits both transcription and DNA replication. It presumably interferes as an intercalating agent with the passage of RNA polymerase and DNA polymerase, respectively. Actinomycin D is used as a cytostatic in the treatment of the rare types of cancer, e.g., Wilms' carcinoma, chorion carcinoma, and Hodgkin's disease.

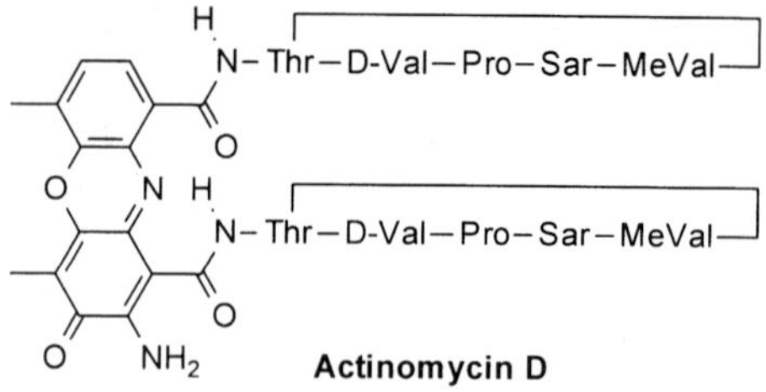

Actinomycin D

Activins, proteins occurring in the follicular fluid from ovaries and in leukemic cells stimulating the release of → follitropin. *Activin A* is a homodimer of the β-chains of → inhibin-A, whereas *activin B* consists of the β-chains of inhibin-A and inhibin-B.

Activity-dependent neurotrophic peptides, ***ADNP***, peptides derived from the neuroprotective protein, named *activity-dependent neuroprotective factor* (*ADNF*). ADNF ($M_r \sim 14$ kDa; pI 8.3) is very potent in preventing neuronal cell death in electrically blocked spinal cord test cultures. *ADNP-*

14, VLGGGSALLR[10]SIPA, is a 14-peptide exhibiting neuroprotection at femtomolar concentration ranges against neurotoxins associated with HIV infection, excitotoxicity, electrical blockade and Alzheimer's disease. *ADNP-8*, H-Asn-Ala-Pro-Val-Ser-Ile-Pro-Gln-OH, is a novel femtomolar-acting peptide that prevents death in enriched neuronal cultures. From these results has been concluded that the 8-peptide can prevent neurotoxicity associated with direct action of the → β-amyloid peptide on neurons [D. E. Brenneman, I. Gozes, *J. Clin. Invest.* **1996**, *97*, 2299; I. Zemlyak et al., *Regul. Peptides* **2000**, *96*, 39].

Adaptins, accessory proteins thought to bind the membrane-spanning receptors for those specific proteins that the coated vesicle (→ clathrin) sequester.

Adhesion molecules, proteins responsible for interactions between cells and their environment, especially, the extracellular matrix and other cells. Several different molecules act as cell adhesion receptors such as → integrins, intercellular adhesion molecules (ICAM), leukocyte LFA-1, Mac-1 and p150/95 molecules, the fibronectin receptor complex (→ fibronectin), → tenascin, and the position-specific (PS) antigens of *Drosophila*.

Adipokinetic hormones, ***AKH***, peptide hormones produced by the African migratory locusts *Locusta migratoria* consisting of *AKH-I*, pGlu-Leu-Asn-Phe-Thr-Pro-Asn-Trp-Gly-Thr-NH_2, *AKH-II*, pGlu-Leu-Asn-Phe-Ser-Ala-Gly-Trp-NH_2, and *AKH-III*, pGlu-Leu-Asn-Phe-Thr-Pro-Trp-Trp-NH_2. The first hormone identified in locusts was AKH-I mobilizing lipids from the fat body. Nearly all the biological effects yet ascribed to AKH appear to be exerted by each of the three family members with varying in potency. In an in-vivo lipid mobilization assay AKH-I is the most potent such hormone. Various analogues have been synthesized [L. Schoofs et al. *Peptides* **1997**, *18*, 145; M. J. Lee et al., *Regul. Peptides* **1997**, *69*, 69].

Adrenocorticotropic hormone, ACTH, → corticotropin.

Adrenomedullin, YRQSMNNFQG[10]LRSFGCRFGT[20]CTVQKL AHQI[30]YQFTDKDKDN[40]VAPRSKISPQ[50] GYa (disulfide bond: C^{16}–C^{21}), a vasoactive 52-peptide amide which is a member of the → calcitonin family and shares 24% sequence homology with → calcitonin gene-related peptide (CGRP). The biological activity profile of adrenomedullin is similar to that of CGRP. Adrenomedullin displays overlapping biological effects with the other members of the calcitonin family owing to their structures and cross-reactivity between receptors. Adrenomedullin was discovered in human pheochromocytoma tissue in 1993 by Kitamura et al. Its gene is situated in a single locus on chromosome 11 in humans. The amino acid sequence is highly conserved across species. The receptor of adrenomedullin has been cloned as an orphan receptor from the rat lung. The cDNA encodes for a protein of 395 amino acids containing seven transmembrane domains with structural homology to other members of the G-protein-linked receptor superfamily including the CGRP-I receptor. The bioactivity of adrenomedullin may be manifested through a variety of second messenger systems, mainly cAMP, but in addition nitric oxide, tissue prostaglandins, and intracellular calcium. From the therapeutic point of view, it may become possible to reduce the rate of degradation of adrenomedullin (thus enhancing its biological activity), or to improve its production. This would be therapeutically beneficial for the treatment of heart failure or hypertension [K. Kitamura et al., *Biochem. Biophys. Res. Commun.* **1993**, *192*, 553; C.J. Charles et al., *Am. J. Hypertens.* **1999**, *12*, 166; S.J. Wimalawansa, *Crit. Rev. Neurobiol.* **1997**, *11*, 167; M. Jougasaki, J.C. Burnett Jr., *Life Sci.* **2000**, *66*, 855].

Adrenorphin, → metorphamide.

Aequorin, a protein from the jellyfish *Aequorea victoria*. This bioluminescent jellyfish produces a greenish luminescence from the margin of its umbrella using aequorin and a chromophore-bearing → green fluorescent protein (GFP). Aequorin is a Ca^{2+}-binding protein (M_r ~21 kDa) and undergoes an intramolecular reaction on binding Ca^{2+}, yielding a blue fluorescent protein in the singlet excited state transferring its energy by resonance to GFP. Aequorin can be used as a Ca^{2+} indicator [H. Morise et al., *Biochemistry* **1974**, *13*, 2656; M. Brini et al., *J. Biol. Chem.* **1995**, *270*, 9896].

Agouti-related protein, *AGRP*, human AGRP(86–132): RCVRLHESCL^{10}GQQVPC CDPC^{20}ATCYCRFFNA^{30}FCYCRKLGTA^{40} MNPCSRT (Disulfide bonds: C^2–C^{17}/C^9–C^{23}/C^{16}–C^{34}/C^{20}–C^{44}/C^{25}–C^{32}), a 47-peptide with some sequence similarity to the agouti protein that was known to affect pigmentation *via* the melanocortin receptor 1 (MC-1). AGRP and the agouti protein bind to distinct types of melanocortin receptors. AGRP prefers the receptors MC-3 and MC-4 that are known to participate in the regulation of feeding, and is acting as a MC-3/4 antagonist stimulating food intake. The appetite-boosting AGRP-(86–132) may be an important tool for elucidating the mechanism of obesity [R.D. Rosenfeld et al., *Biochemistry* **1998**, *37*, 16041; E.J. Bures et al., *Biochemistry* **1998**, *37*, 12172].

Alamethicin, Ac-Aib-Pro-Aib-Ala-Aib-Ala-Gln-Aib-Val-Aib^{10}-Gly-Leu-Aib-Pro-Val-Aib-Aib-Glu-Gln-Phe^{20}-ol, a peptide produced by the fungus *Trichoderma viride* containing eight α-aminoisobutyric acid (Aib) building blocks and L-phenylalaninol (Phe-ol) as C-terminal residue. It is a member of the → peptaibols. Alamethicin is amphiphilic, but of high overall hydrophobicity. It inserts spontaneously into lipid bilayers. From the crystal structure could be revealed that alamethicin is preferentially α-helical with a bend in the helix axis at Pro^{14}. This structure is in agreement with an early model for the mode of action in which a certain number (6–12) of molecules form aggregates like the staves of a barrel. However, there are also other models for channel formation. One model based on a voltage-dependent flip-flop of α-helix dipoles postulates that the membrane-inserted helices attract each other when oriented in anti-parallel fashion. Contrary to the flip-flop model, another model assumes that the gating charge transfer is involved in the opening-closing mechanisms [R.O. Fox, F.M. Richards, *Nature* **1982**, *300*, 325; G. Boheim et al. *Biophys. Struct. Mech.* **1983**, *9*, 181; D.T. Edmonds, *Eur. Biophys. J.* **1985**, *13*, 31; H. Wenschuh et al., *J. Org. Chem.* **1995**, *60*, 405].

Alanyl-histidinato zinc (β), *AHZ*, a zinc-chelated dipeptide (***Formula***) for exogenous administration of zinc. The zinc delivery potential of AHZ is more effective on bone metabolism than zinc sulfate. In-vitro stud-

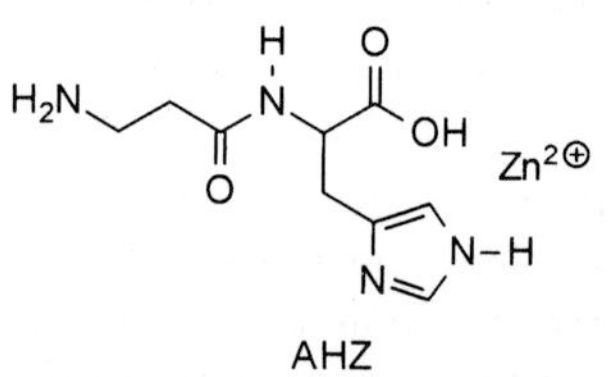

β-Alanyl-histidinato zinc

ies have established that AHZ causes complete inhibition of the decrease of bone calcium in a bone tissue culture system, as well as in the formation of osteoclast-like cells in mouse marrow culture [M. Yamaguchi, *Gen. Pharmacol.* **1995**, *26*, 1179].

Albomycins, nucleoside peptides with antibiotic action against Gram-positive and Gram-negative bacteria isolated from *Streptomyces* sp. *WS 116* (DSM 1692) [G. Benz, *Liebigs Ann. Chem.* **1984**, 1399].

Albumins, a group of water-soluble proteins occurring in body liquids, animal tissues and in some plant seeds. They are rich on both Glu and Asp (20–25%) as well as Leu and Ile (up to 16%). Albumins have low molecular mass, are easily crystallizable, and their isoelectric points are in the weakly acid range. High concentrations of neutral salts are necessary for 'salting out'. *Serum albumin* (*plasma albumin*) (M_r ~ 67.5 kDa) comprise up to 60% of the dry mass of blood serum, and is one of few nonglycosylated proteins in the blood. It has a high binding capacity for Ca^{2+}, Na^+, K^+, fatty acids and drugs, and contributes in maintenance of the osmotic pressure of the blood. Human serum albumin consists of a single peptide chain with 584 residues and 17 disulfide bridges. Bovine and human serum albumin contain 16% nitrogen, and are used as standard proteins for calibration. Further important animal and plant albumins are → lactalbumin, → ovalbumin and → ricin.

Allatostatins, neuropeptides isolated from the brain of the cockroach *Diploptera punctata*. *Type A allostatin I*, APSGAQRLYG^{10}FGLa, and its shortened homologues II–IV, and the *Type B allostatin*, AYSYVSEYKR^{10}LPVYNFGLa, inhibit the synthesis of the juvenile hormone in the *corpora allata* of cockroaches and related insects, but not of flies. Furthermore, besides their allotostatic activity in cockroaches, these peptides have been shown to cause an inhibitory effect on the contractility of a variety of visceral muscles in a broad range of insect species. *Allatostatin-like peptides* have been discovered in the *Drosophila* CNS and in endocrine cells of the gut. Picomolar concentrations of a peptide (H-Ser-Arg-Pro-Tyr-Ser-Phe-Gly-Leu-NH_2) from the head of *Drosophila* activate a fruitfly G-protein-coupled receptor that shows striking sequence similarities to mammalian galanin and somatostatin/opioid receptors [A. P. Woodhead et al., *Proc. Natl. Acad. Sci. USA* **1989**, *86*, 5997; J. G. Yoon, B. Stay, *J. Comp. Neurol.* **1995**, *363*, 475; N. Birgul et al., *EMBO J.* **1999**, *18*, 5892].

Alytensin, pGlu-Gly-Arg-Leu-Gly-Thr-Gln-Trp-Ala-Val^{10}Gly-His-Leu-Met-NH_2, a 14-peptide amide belonging to the → bombesin family. Alytensin was isolated from the skin of the European amphibian *Alytes obstetricans* in 1971. It is structurally very similar to → bombesin and displays similar biological activities when applied to mammals [A. Anastasi et al., *Experientia* **1971**, *27*, 166; V. Erspamer, P. Melchiorri, *Trends Biochem. Sci.* **1980**, *1*, 391].

Alzheimer's disease, ***AD***, → amyloid-β.

Amanitins, a group of toxic components of *Amanita phalloides* (→ amatoxins).

Amatoxins, heterodetic bicyclic 8-peptides from *Amanita* species, but also detected in *Galerina* and *Lepiota* species, which are responsible for the fatal intoxications by the mentioned toadstools. The toxic peptides are readily absorbed by the intestine, and in humans the lethal dose of amatoxin is ~ 0.1 mg kg^{-1} body weight, or even lower. The gut cells of humans seem to be the first cells affected, and the intestinal phase begins about 9 h after administration of the toxins. There is no effective therapy of *amanita* poisoning since the amatoxins inhibit strongly and specifically the eukaryotic RNA-polymerases II (B) in the nuclei. Nine members of the amatoxins have been isolated up to now, from which α-amanitin is the best studied. The individuals are derived from the parent structure (see Fig. 3.5), and differ mainly by the number of hydroxyl groups and by an amide versus carboxy function. Amatoxins contain exclusively L-amino acids and glycine, together with uncommon moieties such as the bridging 6′-hydroxytryptathionine-(*R*)-sulfoxides between residues 4 and 8. The fundamental work on structure elucidation and studying the biochemical effects was carried out by Theodor Wieland and co-workers [Th. Wieland, *Peptides of Poisonous Amanita Mushrooms*, Springer Series in Molecular Biology, Springer-Verlag, New York, Berlin, **1986**].

Amylin, KCNTATCATQ^{10}RLANFLVHSS^{20}NNFGAILSST^{30}NVGSNTYa (human amylin; disulfide bond: C^2–C^7), *islet amyloid polypeptide (IAPP), diabetes-associated peptide (DAP)*, a 37-peptide amide which belongs to the → calcitonin family. Amylin is generated from a gene located on the short arm of chromosome 12 sharing 46% sequence homology with the two → calcitonin gene-related peptides and 20% with human → calcitonin. Amylin was first isolated from an insulinoma and from pancreatic amyloid deposits of patients with noninsulin-dependent (type II) diabetes mellitus (NIDDM). It may be involved in the pathogenesis of type II diabetes by deposition as amyloid within the pancreas which leads to β-cell destruction. Human amylin and other mammalian amylins are synthesized as relatively small precursor proteins. Mature amylin is liberated from the 89-polypeptide precursor by proteolytic processing in a similar way to that of proinsulin and other islet prohormones. Amylin is cosynthesized with insulin, stored in the β-cell secretory granules in analogy to insulin, and secreted with insulin from the islet β cells. Normally, the rate of synthesis is only about 1% of that of insulin, but this can increase after prolonged stimulation *in vivo* with increased concentrations of glucose. This behavior seems to be a link to the contribution of amylin in the pathogenesis of type II diabetes. Amylin has a vasodilatory effect like other members of the calcitonin family, and, also acts as an agonist at the calcitonin receptor. Despite being less potent than insulin, amylin shows growth factor-like effects and inhibits insulin-stimulated incorporation of glucose into muscle glycogen [R. Muffet et al., *Eur. J. Endocrinol.* **1995**, *133*, 17; S.J. Wimalawansa, *Crit. Rev. Neurobiol.* **1997**, *11*, 167].

Amyloid-β, ***Aβ***, β-amyloid, *amyloid*-β peptide, a proteolytic derivative of the amyloid precursor protein (APP). According to the amyloid hypothesis, the deposition of Aβ is a primary event in the pathological cascade for Alzheimer's disease (AD), first described by Alzheimer in 1907. AD is a widespread, neurodegenerative, dementia-inducing disorder of elderly people characterized mainly by amyloid deposits surrounding dying neurons (senile plaques), neurofibrillar degeneration with tangles, and cerebrovascular angiopathies. In 1998, it was estimated that 25 million people worldwide suffered from AD. Amyloid β is the result of the endoproteolytic cleavage of a family of differentially spliced APPs. Two major C-terminal variants of Aβ are known. *Aβ*-(1-42), DAEFRHDSGY^{10}EVHHQKLVFF^{20}AEDVGSNKGA^{30}IIGLMVGGVV^{40}IA, is the major species of amyloid deposits, whereas Aβ-(1-40) is the major component secreted from cultured cells, and occurs in cerebrospinal fluid. Most of the knowledge about APP processing has come from studies with cultured cells. Some fraction of the APP is cleaved at the cell surface within the Aβ sequence by α-secretase, generating the neuroprotective secreted amyloid precursor protein ($APP_{s\alpha}$), which is detected in human plasma and cerebrospinal fluid, and nonamyloidogenic 3 kDa Aβ-secreted products. The β-secretase cleavage product at the N-terminus of the Aβ sequence is termed C100, and comprises the sequence of Aβ; the remaining sequence of the C-terminus of APP_{695} is a physiologically relevant APP fragment in human brain. γ-Secretase (presenilin) cleaves C100 to release Aβ. Mutations in APP can cause either increased overall secretion of Aβ or secretion of "long" (42- to 43-residue) forms of Aβ-(1-42) relative to the shorter Aβ-(1-40) form. Aβ-(1-42) is the major form of brain amyloid deposits in AD. Furthermore, APP has been expressed in various tissues as a family of differentially spliced forms ranging from 695 (APP_{695}) to 770 (APP_{770}) residues. In addition, two longer forms containing an additional 56-peptide domain with homology to Kunitz protease inhibitors have been described. Despite the increasing production of research data, AD remains an enigma. Especially, the heterogeneity of this neuropathology and the lack of screening of AD patients on an early stage and a better delineation of pathological subtypes of AD inhibit significantly the progress in understanding of AD. In order to develop new methods to prevent and treat AD it must be possible to diagnose the preclinical stage of AD using biological markers, before the brain damage becomes irreversible. Suitable markers may include high plasma concentrations of Aβ-(1-42) and findings of hippocampal atrophy on magnetic resonance imaging (MRI) of the brain. Besides the amyloid hypothesis numerous mechanisms for the neuronal cell death in AD have been suggested. For example, it has been proposed that AD may be a dysfunction of the APP that should normally function in the brain as a cell surface signaling molecule. Interestingly, recent reports on transgenic mouse models of AD have demonstrated that Aβ vaccination may prevent memory loss and reduce behavioral impairment and plaques, respectively [J. Kang et al., *Nature* **1987**, *325*, 733; B. Drouet et al., *Cell. Mol. Life Sci.* **2000**, *57*, 705; R.L. Neve et al., *Brain Res.* **2000**, *886*, 54; D. Morgan et al., *Nature* **2000**, *408*, 982; C. Janus et al., *Nature* **2000**, *408*, 979; J. Wiltfang et al., *Gerontology* **2001**, *47*, 65; A.B. Clippingdale et al., *J. Peptide Sci.* **2001**, *7*, 227; K. Fassbender et al., *Naturwissenschaften* **2001**, *88*, 261].

Ancovenin, an analogue of B-Type lantibiotics of the prototype → cinnamycin.

Androctonin, RSVCRQIKIC^{10}RRRGGCYYK C^{20}TNRPY (disulfide bonds: Cys^{4}–Cys^{20}/ Cys^{10}–Cys^{16}), a hydrophilic antimicrobial peptide (→ antimicrobial animal peptides) isolated from the blood of the scorpion *Androctonus australis*. Androctonin inhibits the growth of both Gram-positive and Gram-negative bacteria, and displays a large spectrum of activity against filamentous fungi. In contrast to amphipathic α-helical antimicrobial peptides that bind and permeate negatively charged vesicles, androctonin binds only to negatively charged lipid vesicles; this might explain the selective lytic activity towards bacteria but not to red blood cells [L. Ehret-Sabatier et al., *J. Biol. Chem.* **1996**, *271*, 29537; C. Hetru et al., *Biochem. J.* **2000**, *345*, 653].

Angiotensins, *AT, angiotonins, hypertensins,* tissue peptide hormones occurring both in the periphery and in the brain, with influence on blood pressure. The source of the

ATs is *angiotensinogen*, a plasma protein (M_r ~ 60 kDa) of the α_2-globulin fraction which is initially cleaved by the aspartyl protease → renin yielding the inactive 10-peptide *angiotensin I, AT I*, H-Asp-Arg-Val-Tyr-Ile-His-Pro-Phe-His-Leu^{10}-OH. *Angiotensin II, AT II*, H-Asp-Arg-Val-Tyr-Ile-His-Pro-Phe-OH, is formed primarily in the lungs by proteolytic cleavage of AT I catalyzed by → angiotensin-converting enzyme (ACE). *Angiotensin III, AT III*, H-Arg-Val-Tyr-Ile-His-Pro-Phe-OH, is formed by cleavage of the N-terminal Asp of AT II under the catalysis of aminopeptidase A. *Angiotensin IV, AT IV*, H-Val-Tyr-Ile-His-Pro-Phe-OH, results from cleavage of AT III with aminopeptidase N. Among the main active peptides of the brain renin-angiotensin system, AT II and AT III exhibit the same affinity for type 1 (AT_1) and type 2 (AT_2) angiotensin receptors. AT II strongly increases blood pressure in mammals, mediated via the AT II receptors in a variety of tissues. One-fifth of the adult population suffers from chronic hypertension. The application of AT II inhibition either by → ACE inhibitors or by angiotensin receptor blockade for treatment of hypertension, ischemic heart disease, and heart failure started in the early 1970s but has not lost importance until today. AT II exerts a contracting effect on the vascular smooth muscles of the aorta, myocardium, kidney, intestine and uterus; this results in a potent vasoconstricting and blood pressure-increasing effect. Furthermore, AT II stimulates, similar to AT III, the release of aldosterone and prostaglandin, and also the → pro-opiomelanocortin cascade. AT II is inactivated in the blood by angiotensinase. On the basis of structure-activity relationship studies, it could be established that according to modifications of AT II in positions 8 (type I) and 4 (type II), two classes of antagonists can be synthesized. Type I antagonists show protracted effects on smooth muscle tissue, whereas type II antagonists are competitive antagonists at AT II receptors. Furthermore, cyclic analogues of AT II characterized by their conformational constraint show high activity and selectivity. There are no significant differences in the central effects of AT II and its linear analogs, but cyclic analogues produce sedation or neuroleptic-like activity. AT III is acting as a central regulator of → vasopressin release and blood pressure [J. M. Saavedra, *Endocr. Rev.* **1992**, *13*, 329; R. L. Davisson et al., *Circ. Res.* **1998**, *83*, 1047; Z. Lenkei et al., *Front. Neuroendocrinol.* **1997**, *18*, 383; A. Reaux et al., *Trends Endocrinol. Metab.* **2001**, *12*, 157; H. Gavras, H. R. Brunner, *Hypertension* **2001**, *37* (part 2), 342].

Angiotensin-converting enzyme, ***ACE***, *peptidyl dipeptidase A*, a zinc metallopeptidase that catalyzes the conversion of angiotensin I to the vasoconstrictor substance, angiotensin II (→ angiotensins). The function of ACE in blood pressure control and water and salt metabolism has been elucidated mainly by the use of highly specific → ACE inhibitors [N. M. Hooper, *Int. J. Biochem.* **1991**, *23*, 641].

Anorectin, → somatoliberin.

Antamanide, ***AA***, *"anti-amanita peptide"*, cyclo-(-Val-Pro-Pro-Ala-Phe-Phe-Pro-Pro-Phe-Phe^{10}-), a non-toxic cyclic 10-peptide from *Amanita phalloides*. Antamanide was isolated from the lipophilic part of extracts of *A. phalloides* by chromatographic procedures, structurally characterized and synthesized by Th. Wieland and co-workers in 1968. Administration at 0.5 mg kg^{-1} causes full protection of mice from death by phalloidin (→ phallotoxins) after injection about 1 h before or, at the latest, simultaneously with 5 mg kg^{-1} of the toxin.

Antamanide is a competitive inhibitor of the transport system for phallotoxins and amatoxins in the parenchyma cells of the liver [Th. Wieland, *Peptides of Poisonous Amanita Mushrooms,* Springer Series in Molecular Biology, Springer Verlag, Berlin, New York, **1986**; K. Münster et al., *Biochem. Biophys. Acta* **1986**, *860,* 91].

Anthopleurin-A, GVSCLCDSDG^{10}PSVRGN TLSG^{20}TLWLYPSGCP^{30}SGWHNCKAHG^{40} PTIGWCCKQ, a 49-peptide containing three disulfide bridges from the sea anemone species *Anthopleura.* Aanthopleurin causes, in nanomolar concentration, a positive ionotropic effect [G. Strichartz et al., *Annu. Rev. Neurosci.* **1987**, *10,* 237; M.W. Pennington et al., *Int. J. Peptide Protein Res.* **1994**, *43,* 463].

Antimicrobial animal peptides, *AAP*, antibiotic peptides of the innate immune systems of most multicellular organisms. Animal host defense peptides, either inducible or constitutive, with activity against different types of microorganisms have been found in the last two decades in almost all groups of animals. Several hundreds of structures have been described and can be found in a periodically updated data base on the Internet at http://www.bbcm.univ.tries.it/~tossi. Classification based on chemical-structural criteria defines two broad groups, corresponding to linear and cyclic structures, respectively. The first group can be divided into two subgroups: (a) linear peptides with helical conformation, e.g., *bombolitin,* IKITTMLAKL^{10}GKVLAHVa, from bumblebee (*Megabombus pennsylvanicus*), → bombinin-like peptides (BLP-1), *PGLa,* GMASKA GAIA^{10}GKIAKVALKA^{20}La, from the South African clawed frog (*Xenopus laevis*), → cecropins, → bombinin, → mellitin, → magainins; and (b) linear peptides rich in certain amino acids like Arg, Pro, or Trp, e.g., → PR-39, → Indolicidin, *Apidaecin IA,* GNNR PVYIPQ^{10}PRPPHPRIa, from honeybee (*Apis mellifera).* Within the cyclic cystine-containing peptides also two subgroups can be distinguished: (a) peptides with a single cyclic disulfide, e.g., *bovine dodecapeptide,* RLCRIVVIRV^{10}CR, from ox (*Bos taurus*), *brevenin-1,* FLPVLAGIAA^{10}KVVPALFCKI^{20} TKC, from Japanese frog (*Rana brevipoda porsa*), → ranalexin; and (b) peptides with several internal disulfides, e.g., → protegrins → defensins, → androctonin. Most of the AAP provoke an increase in plasma membrane permeability. Especially, a direct correlation between antibiotic activity and permeabilization capability has been found, for example, for → magainins, → defensins, and → cecropins. Antibiotic peptides have been envisaged for therapeutic applications in clinics in the treatment of bacterial or viral infections, and cancer. The magainin analogue, named *pexiganan acetate,* has obtained approval for the treatment of diabetic foot ulcers [R.E. Hancock, R.I. Lehrer, *Trends Biotechnol.* **1998**, *16,* 82; D. Andreu, L. Rivas, *Biopolymers* **1998**, *47,* 415].

Antisense peptide, *complementary peptide,* a peptide sequence hypothetically deduced from the nucleotide sequence that is complementary to the nucleotide sequence coding for a naturally occurring peptide (*sense sequence*). It could be demonstrated that antisense peptide exerts biological responses through the interaction with the receptors for the sense peptides. Potential applications of antisense peptide lies in the area of biomedical research. For example, antisense peptides (or antibodies against it) may promote the purification of both endogenous ligands as well as receptors. Furthermore, the development of highly selective antisense peptides against tumor cell markers may aid in diagnosis and therapeutic modalities of the appropriate state of the disease. The in-vivo administration of antisense peptides may help in better modulation of biological responses caused by their endogenous sense counterparts. The concept of antisense peptides was suggested by Mekler in 1969, and three years later independently proposed and tested by Jones using a synthetic peptide that was antisense to the C-terminal tetrapeptide of gastrin [L. B. Mekler, *Biophys. USSR (Engl. Trans.)* **1969**, *14*, 613; D. S. Jones, *J. Chem. Soc. Perkin Trans. I* **1972**, 1407; K. L. Bost, J. E. Blalock, *Methods Enzymol.* **1989**, *168*, 16].

Anxiety peptide, → diazepam-binding inhibitor peptide.

Apamin, H-Cys-Asn-Cys-Lys-Ala-Pro-Glu-Thr-Ala-Leu^{10}-Cys-Ala-Arg-Arg-Cys-Gln-Gln-His-NH_2 (disulfide bonds: Cys^1–Cys^{11}/Cys^3–Cys^{15}), a 18-peptide amide of the bee venom (1–3% of the venom) causing neurotoxic effects. The two arginine residues are of essential importance for the biological activity. Similar to the → mast cell-degranulating peptide, apamin blocks selectively Ca^{2+}-dependent K^+ channels in neurons; this results in serious disturbances of CNS function [R. C. Hider, *Endeavour, New Series* **1988**, *12*, 60; E. Moczydlowski et al., *J. Membrane Biol.* **1988**, 105].

Apelin-36, human apelin: LVQPRGSRNG^{10}PGPWQGGRRK^{20}FRRQRPRLSH^{30}KGPMPF, a 36-peptide acting as a ligand for the G protein-coupled orphan receptor APJ [putative receptor protein related to the angiotensin receptor (AT_1)]. The primary structure of apelin was determined by cDNA cloning. The bovine peptide differs in only two residues from the human apelin. Synthetic human bovine *[pGlu]-apelin-13*, corresponding to the carboxy terminal 13 residues with pGlu at the N-terminus of apelin, shows a much higher acidification rate than that of parental apelin. The endogenous functions of apelin are hitherto unknown [K. Tatemoto et al., *Biochem. Biophys. Res. Commun.* **1998**, *251*, 471].

Apidaecin Ia, → antimicrobial animal peptides.

Aprotinin, *bovine pancreatic trypsin inhibitor (Kunitz)*, ***BPTI***, a polypeptide consisting of 58 amino acid residues and three disulfide bridges (M_r ~ 6.5 kDa). Aprotinin belongs to the serpins, and acts as inhibitor for serine proteases, but not for → thrombin and Factor X_a [W. Gebhard et al. in: *Proteinase Inhibitors*, A. J. Barret, G. Salvesen (Eds.), p. 375, Elsevier, Amsterdam, **1986**].

Aquaporins, ***AQP***, member of the major intrinsic protein (MIP) superfamily of integral membrane proteins, which is found throughout nature. AQP facilitate water transport in various eukaryotes and prokaryotes. The archetypal aquaporin, *AQP1* (M_r ~ 28 kDa) forms a partly glycosylated water-selective channel widely expressed in the plasma membranes of various water-permeable epithelial and endothelial cells. The three-dimensional structure (7 Å resolution) of the deglycosylated human erythrocyte AQP1 shows that the structure has an in-plane, intramolecular 2-fold axis of symmetry located in the hydrophobic core of the bilayer. The monomer is composed of six membrane-spanning, tilted α-helices forming a right-handed bundle surrounding a central density. From these results, a model is suggested that identifies the aqueous pore in the AQP1 molecule and indicates the organization of the tetrameric complex in the membrane [J. H. Park, M. H. Saier, *J. Membr. Biol.* **1996**, *153*, 171; T. Walz et al., *Nature* **1997**, *38*, 624; A. Cheng et al., *Nature* **1997**, *38*, 627].

Arrestin, *S-antigen*, *48-kDa protein*, a retinal protein. In the photoreceptor cell, the activation of rhodopsin by a photon initiates signal transduction and signal termination. Arrestin binds selectively to the light-activated rhodopsin in its phosphorylated form, thereby shutting down the phototransduction cascade by blocking transducing activation [V.V. Gurevich, J.L. Benovic, *Methods Enzymol.* **2000**, *315*, 422].

Aschheim-Zondek reaction, → chorionic gonadotropin.

Aspartame, ***APM***, H-Asp-Phe-OMe, a non-nutritive high-intensity sweetener. APM is about 200 times sweeter than sucrose. It was first approved by the FDA in 1981 as a table-top sweetener and an additive, for example, in dry based beverages, dry cereals, chewing gum, gelatins, puddings, instant coffee and tea. APM was discovered accidentally during recrystallization of an intermediate of the synthesis of the C-terminal tetrapeptide of → gastrin at Searle & Co. For commercial synthesis, various methods have been developed including thermolysin-catalyzed synthesis [Y. Isowa et al., *Tetrahedron Lett.* **1979**, *28*, 2611; K. Oyama et al., *J. Org. Chem.* **1981**, *46*, 5242].

Atrial natriuretic peptide, ***ANP***, *atrial natriuretic factor*, ***ANF***, *atriopeptide, atriopeptin, cardionatrin I*, SLRRSSCFGG^{10}RMDRIGA QSG^{20}LGCNSFRY (disulfide bridge: C^7–C^{23}), a 28-peptide hormone isolated from the atrium of the mammalian heart. It belongs to the family of → natriuretic peptides. ANP is a potent hypotensive and natriuretic agent. It is synthesized as prepro-ANP (human: 151 aa), and stored as 126-polypeptide pro-ANP (also termed: atriopeptigen, cardionatrin IV) in specific granules of atrial cardiocytes. The latter is secreted from the atria and then processed proteolytically by a protease of the blood to the human circulating 28-peptide. In the atria, the quantity of ANP is orders of magnitude higher than in extracardiac tissues such as CNS and kidney. Receptors have been found in blood vessels, kidney, and adrenal cortex. ANP mediates in the adrenal cortex the decrease of aldosterone release, and increases glomerular filtration rate, renal blood flow, urine volume and sodium excretion in the kidney [A.J. de-Bold, *Science* **1985**, *230*, 767; K. Kangawa et al., *Nature* **1985**, *313*, 397; G. McDowell et al., *Eur. J. Clin. Invest.* **1995**, *25*, 291].

Avidin, a glycoprotein in the egg whites of many birds and amphibians. Chicken A. (M_r ~ 66 kDa) consists of four identical subunits (without carbohydrate: M_r ~ 14 kDa; 128 aa). Avidin binds four molecules of the vitamin biotin with high affinity (dissociation constant, $K \sim 10^{-15}$ M). The resulting complex is stable against proteolysis, and prevents its intestinal adsorption. Avidin inhibits the growth of microorganisms in egg whites. Together with biotin, avidin is used for the immobilization of ligands in immunoassays and in affinity chromatography.

Azadepsipeptides, a new class of pseudopeptides. Analogously to → azapeptides the α-carbon atom in → depsipeptides is replaced isoelectronically by a trivalent nitrogen. Synthesis and structure evaluation have been demonstrated using a bis-aza analogue of the antiparasitic cyclooctadepsipeptide PF 1022A as a model [H. Dyker et al., *J. Org. Chem.* **2001**, *66*, 3760].

Azapeptides, a class of backbone-modified peptides in which the α-CH of one or more amino acid residues in the peptide chain is isoelectronically replaced by an N atom (further information is given in Chapter 8).

B

Bacilysin, an antibiotic dipeptide (***Formula***) produced from the bacterial genus *Bacillus*. In bacilysin, Ala is linked to anticapsin which is derived from prephenate, the aromatic amino acid precursor. Bacilysin inhibits the microbial glucosamine-6-phosphate synthetase.

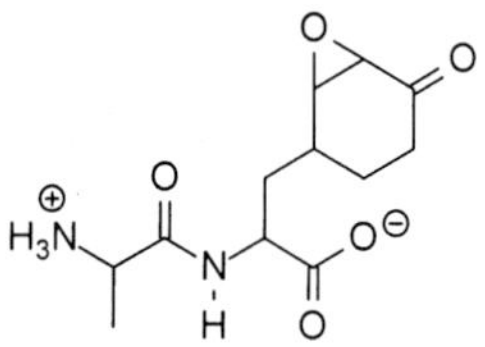

Bacilysin

Bacitracin, a mixture of branched cyclic dodecapeptides from *Bacillus licheniformis*. The most well-known member of this group of peptide antibiotics is *bacitracin A*. The thiazole moiety is formed from the N-terminal Ile and the neighboring Cys (for formula, see Chapter 3, **53**). Biosynthesis is based on the thiotemplate mechanism, and is catalyzed by the bacitracin synthetase. Bacitracin has found considerable medical application, for example, in several antiseptic combinations for the treatment of sore throats and in diagnostic agar formulations to inhibit growth of Gram-positive bacteria. Bacitracin specifically forms a complex with the bacterial bactoprenol C_{55} bisphosphate (undecaprenyl pyrophosphate), which is the membrane-bound carrier for the peptidoglycan and teichoic acid subunits. This interaction inhibits the bacterial cell wall synthesis.

Bactenecin 5, ***Bac5***, RFRPPIRRPP^{10}LRPPF YPPFR^{20}PPIRPPIFPP^{30}IRPPFRPPLR^{40}FP, a member of the → antimicrobial animal peptides with a high content of Pro and Arg. Bac5 was isolated from ox (*Bos taurus*) and is equally active against Gram-negative and Gram-positive bacteria. Although detailed molecular action is missing, it may interact and disorganize bacterial membranes [D. Hultmark, *Trends Genet.* **1993**, *9*, 178].

Bacteriocins, antimicrobial peptides produced by bacteria which are active against closely related strains. Bacteriocins can be simply divided into those produced by Gram-positive bacteria and others produced by Gram-negative bacteria. The *colicins* (→ microcins) produced by various *E. coli* strains were probably the first bacteriocins, and are the prototypes of the first group. In general, they are large, proteinaceous compounds. One domain mediates specific binding to receptor proteins in the outer membrane; another domain mediates the translocation through the peptidoglycan layer and the periplasmic space. Furthermore, the toxic activity resides in a third domain. The bacteriocins of Gram-positive bacteria are principally subdivided into five subtypes from which the first subtype are the → lantibiotics [J. R. Tagg et al., *Bacteriol. Rev.* **1976**, *40*, 722; R. James, C. Lazdunski, F. Pattus (Eds.), *Bacteriocins, Microcins and Lantibiotics*, Springer, Berlin, **1992**].

Baratin, H-Asp-Asn-Ser-Gln-Trp-Gly-Gly-Phe-Ala-OH, a 9-peptide from the brain of the cockroach *Leucophaea maderae*. It appears to play a function in modulation of motor patterns in abdominal ganglia [D. R. Nässel et al., *J. Comp. Neurol.* **2000**, *422*, 267].

Biphenomycins, cyclic peptides containing a biphenyl structure from the culture fluid of *Streptomyces griseorubiginosus* with strong antibiotic activity against Gram-positive, β-lactam-resistant bacteria (***Formula***). *Biphenomycin B* contains with (2*S*,4*R*)-hydroxyornithine and (*S*,*S*)-diisotyrosine two nonproteinogenic amino acids, whereas in *biphe-*

nomycin A, instead of the latter building block, a biphenyl structure consisting of a dimer of (*S*)-2-hydroxy-phenylalanine and (2*S*,3*R*)-2-hydroxyphenylserine is found [U. Schmidt et al., *J. Chem. Soc., Chem. Commun.* **1992**, *13*, 951].

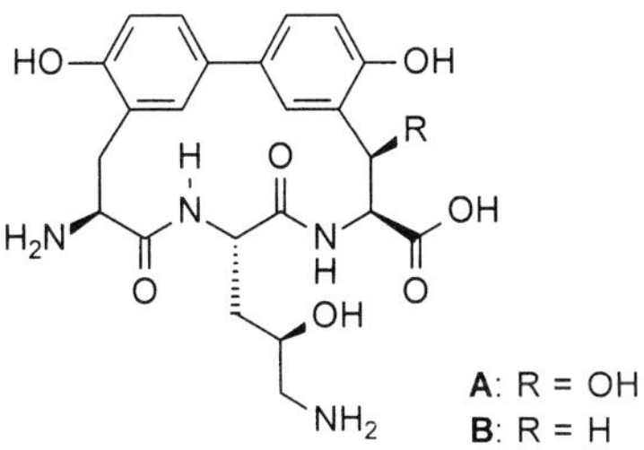

Biphenomycins

BLP-1, acronym for → bombinin-like peptide 1.

Bombesin family, a subfamily of the → bombesin-like family to which belong → bombesin, → alytensin and → gastrin-releasing peptide, a mammalian counterpart for bombesin. The amphibian peptides bombesin and alytensin are structurally very similar and display biological effects when applied to mammals, such as hypertensive action, stimulation of the uterus and digestive tract, hyperglycemic effect, stimulation on the gastric secretion and increase of insulin levels in peripheral blood.

Bombesin, *BN*, <EQRLGNQWAV[10]GHLMa, a 14-peptide amide belonging to the → bombesin family. BN was isolated from the skin of the European amphibian *Bombina bombina*. Applied to mammals it shows, for example, hypertensive action, potent stimulation on the uterus and digestive tract, stimulation of the gastric secretion, hyperglycemic effect and increase of insulin levels in blood. BN is used as a diagnostic aid in the gastric stimulation test [A. Anastasi et al., *Experientia* **1971**, *27*, 166; V. Erspamer, P. Melchiorri, *Trends Biochem. Sci.* **1980**, *1*, 391].

Bombesin-like family, a peptide family comprising bombesin and bombesin-related peptides which are classified into the three subfamilies → bombesin family, → ranatensin family and → phyllolitorin family.

Bombinin, GIGALSAKGA[10]LKGLAKGLKE[20]HFANa, an antimicrobial (→ antimicrobial animal peptides) and hemolytic 24-peptide amide from the yellow-bellied toad (*Bombina variegata*) [A. Csordas, H. Michl, *Monatsh. Chem.* **1970**, *101*, 182].

Bombinin-like peptides, *BLP*, families of antimicrobial peptides related to → bombinin from amphibia (*Bombina variegata* and *Bombina orientalis*). The *B. variegata* bombinin peptides contain 27 residues, bearing a C-terminal amide, differ from each other by only one or a few amino acids, and are characterized by a variable N-terminal sequence and an identical C-terminal region. *BLP-1*, GIGASILSAG[10]KSALKGLAKG[20]LAEHFANa, was isolated from the skin of the Asian toad (*Bombina orientalis*), and shows antibiotic and hemolytic activity. It is more potent in killing bacteria than → magainin. The *B. orientalis* bombinins are 25- and 27-residue peptides, from which the shorter molecules lack the last two C-terminal residues. The precursor contains three peptides, two bombinin-like ones, separated by a 33-mer sequence, followed by a 50-mer spacer region and a C-terminal 17-peptide. The latter sequence belongs to the *type-H bombinins*. The mature product has been subsequently separated in two forms. Interestingly, one of these contains D-Leu in the second position [M. Simmaco et al., *Eur. J. Biochem.* **1991**, *199*, 217; B.W. Gibson et al., *J. Biol. Chem.* **1991**, *266*, 23103].

Bombolitin, → antimicrobial animal peptides.

Bombyxin, an insect brain peptide (M_r ~ 5 kDa) with structural similarity to → insulin. Bombyxin exists in various molecular forms (I–V). The heterodimers of the A and B chains show about 50% and 30% identity to the A and B chains of human insulin. Bombyxin stimulates, together with the prothoracicotropic hormone (PTTH), the synthesis and release of ecdysone [M. Iwami, *Zool. Sci.* **2000**, *17*, 1035].

Bone Gla protein (*BGP*), → osteocalcin, → matrix Gla protein.

Bone morphogenetic proteins, *BMP*, a family of at least eight different polypeptides, termed BMP-1 to BMP-8, regulating cartilage and bone differentiation *in vivo*. Due to the homology of the primary structure BMP-2 to BMP-8 belong to the TGF-β superfamily. It could be established that BMP induce rapid maturation of chondrocytes at the growing stage, transform the cells into rounded cells and induce a marked accumulation of cartilage matrix. Human rBMP may be used for inducing bone formation in clinical practice [J. M. Wozney et al., *Science* **1988**, *242*, 1228; A. J. Celeste et al., *Proc. Natl. Acad. Sci. USA* **1990**, *87*, 9843].

Bovine dodecapeptide, → antimicrobial animal peptides.

Bradykinin, *kinin 9*, H-Arg-Pro-Pro-Gly-Phe-Ser-Pro-Phe-Arg-OH, a member of the → plasma kinins. Bradykinin is a tissue hormone that is released from the precursor kininogen by → kallikrein. It causes dilation of blood vessels resulting in decrease of blood pressure. Bradykinin promotes contraction of the smooth muscles of bronchia, intestine and uterus. The B_1 receptor binds more strongly de-Arg^9-bradykinin than bradykinin, whereas the B_2 receptor shows a greater affinity to the native hormone. Furthermore, B_3 and B_4 receptors have been studied. The degradation of bradykinin occurs primarily in the lungs by kininase II (identical to ACE), but also by other endopeptidases and exopeptidases. Synthetic antagonists of bradykinin are potential agents for the treatment of inflammation, pain, pancreatitis and other disorders. Bradykinin and various analogues such as [Thr^6]bradykinin and C-terminally shortened or lengthened bradykinins have been isolated from the skin of amphibia [N.-E. Rhaleb et al., *Br. J. Pharmacol.* **1990**, *99*, 445; J. E. Taylor et al., *Drug Dev. Res.* **1989**, *16*, 1].

Brain natriuretic peptide, *BNP*, h-BNP: SPKMVQGSGC^{10}FGRKMDRISS^{20}SSGLGCKVLR^{30}RH (disulfide bond: C^{10}–C^{26}), a 32-peptide that belongs to the → natriuretic peptides. It binds to the particulate guanylate cyclase receptor of vascular smooth muscle and endothelial cells, causing increased concentration of cGMP, which serves as a second messenger to dilate veins and arteries, and causes smooth muscle cell relaxation. The precursors of the BNP of various species have been elucidated through the appropriate cDNAs. The sequence of BNP is located in the C-terminal part of the precursor. Human prepro-BNP (134 aa) is proteolytically converted into pro-BNP (108 aa). The natriuretic and hypotensive effects of the BNP are quite similar to these of the → atrial natriuretic peptide. Furthermore, BNP should be involved in the maintenance of body fluid homeostasis via the regulation of the activity of → vasopressin and the angiotensin II-hypothalamic system. BNP was originally isolated from porcine brain as a 26-peptide that resembles strongly ANP [G. P. Vlasuk et al., *Biochem. Biophys. Res. Com-*

mun. **1986**, *136*, 396; J. J. Seilhamer et al., *Biochem. Biophys. Res. Commun.* **1989**, *165*, 650; C. C. Lang et al., *Clin. Sci.* **1992**, *83*, 519; G. McDowell et al., *Eur. J. Clin. Invest.* **1995**, *25*, 291].

Brain-binding peptide, H-Cys-Leu-Ser-Ser-Arg-Leu-Asp-Ala-Cys-OH (disulfide bond: Cys^1–Cys^9), a 9-peptide inhibiting the localization of homologous phages into the brain. After coating onto glutaraldehyde-fixed erythrocytes it causes selective localization of intravenously injected cells into the brain [R. Pasqualini, E. Ruoslahti, *Nature* **1996**, *380*, 364].

Brevenin-1, → antimicrobial animal peptides.

Bufokinin, KPRPDQFYGL^{10}Ma, a substance P-related peptide isolated from an extract of the intestine of the toad *Bufo marinus.* This member of the nonmammalian → tachykinin family shows high binding affinity but low selectivity for mammalian tachykinin receptors [J. M. Conlon et al., *J. Peptide Res.* **1998**, *51*, 210].

C

C peptide, → insulin.

Cachectin, → tumor necrosis factor-α.

Caerin 1 peptides, a family of antibiotic 25-peptide amides from the skin secretion of the Australian tree frog of the genus *Litoria. Caerin 1.1*, for example, has the sequence: GLLSVLGSVA^{10}KHVLPHVVPV^{20} IAEHLa. All members of the family show significant antibiotic activity, especially against Gram-positive microorganisms [S. T. Steinborner et al., *J. Peptide Res.* **1998**, *51*, 121].

Calcineurin, ***CN***, ***PP-2B***, a Ca^{2+}- and calmodulin (CaM)-dependent Ser/Thr protein phosphatase with narrow substrate specificity. It is a heterodimer consisting of a catalytic subunit (CNA) and a regulatory subunit (CNB). CNA genes are highly conserved and encode for a protein consisting of a catalytic domain homologous to other protein phosphatases, and three additional domains (CNB binding domain, CaM binding domain, and autoinhibitory domain). The CNB subunit is highly conserved and shares structural homology with → calmodulin (CaM). Calcineurin is involved in a number of physiological processes such as T-lymphocyte activation, muscle cell differentiation, learning and memory [C. S. Hemenway, J. Heitman, *Cell. Biochem. Biophysiol.* **1999**, *30*, 115; F. Rusnak, P. Mertz, *Physiol. Rev.* **2000**, *80*, 1483].

Calcitonin family, a group of structurally related peptides containing a N-terminal intramolecular disulfide-bridged ring with six to seven amino acid residues and an amidated C-terminus. The calcitonin gene peptide superfamily consists of → calcitonin (CT) → amylin, → adrenomedullin (ADM) and → calcitonin gene-related peptides (CGRP1 and 2). Calcitonin and CGRP are encoded by the CT/CGRP gene which is located on chromosome 11, whereas amylin is generated from a gene encoded on chromosome 12. The latter is thought to be an evolutionary duplication of chromosome 11. Adrenomedullin shares 24% homology with CGRP. A portion of the B-chain of → insulin is strongly homologous to these four members of this family. The peptides of the calcitonin family show overlapping biological effects owing to their structure and cross-reactivity between receptors [S. J. Wimalawansa, *Crit. Rev. Neurobiol.* **1997**,*11*, 167].

Calcitonin gene-related peptide, ***CGRP***, ACDTATCVTH^{10}RLAGLLSRSG^{20}GVVKNN FVPT^{30}NVGSKAFa (h-α-CGRP or hCGRP-I: disulfide bond: C^2–C^7), a 37-peptide belonging to the → calcitonin family. The h-β-CGRP or hCGRP-II differs from the homologous peptide h-α-CGRP by three residues (Asn^3/Met^{22}/Ser^{25}). CGRP was discovered by molecular cloning of the → calcitonin (CT) gene in 1983. It is a product from alternative splicing of the primary RNA transcript of the CT/CGRP gene encoded on chromosome 11. In the CNS, splicing of the CT/α-CGRP gene produces α-CGRP, whereas in the C cells of the thyroid gland CT is formed. The β-CGRP gene, which is also located on chromosome 11 like → insulin, was predicted from cDNA analysis in humans and rats. In contrast to h-β-CGRP, r-α-CGRP differs only in one amino acid from r-β-CGRP. CGRP is widely distributed in the nervous system and in the cardiovascular system. Besides these areas, the receptors of CGRP are also present in the adrenal and pituitary glands, kidney, bone and exocrine pancreas. CGRP shows 19% sequence homology to salmon → calcitonin, and 46% to → amylin. The fully characterized CGRP receptor (M_r ~ 66 kDa; 20% glycosylated) consists of only one subunit. CGRP is primarily a neuropeptide and seems to be involved in modulating sensory neurotransmission, but its presence in nervous tissues and blood vessels demonstrates its role in the regulation of many processes. Besides acting as neuromodulator, a very important function of CGRP is to control blood flow to various organs by its potent vasodilatory activity. It is the most potent endogenous vasodilatory peptide found so far [H.R. Morris et al., *Nature* **1983**, *304*, 129; R. Muff et al., *Eur. J. Endocrinol.* **1995**, *133*, 17; S.J. Wimalawansa, *Crit. Rev. Neurobiol.* **1997**, *11*, 167].

Calcitonin, ***CT***, human CT: $CGNLSTCMLG^{10}TYTQDFNKFH^{20}TFPQTAIGVG^{30}APa$ (disulfide bond: C^1–C^7), a 32 amino acid peptide hormone with important functions in calcium homeostasis and bone remodeling. CT belongs to the → calcitonin family. Sequence and structure of the human hormone was determined in 1968. CT from various species include an N-terminal Cys^1–Cys^7 disulfide bridge and a C-terminal proline amide. Amino acid residues conserved across species are Cys^1, Leu^4, Ser^5, Thr^6, Cys^7, Gly^{28}, and, furthermore, both Leu^9 and Leu^{16} are common to 11 of 12 known sequences. Due to the therapeutic importance of CT, extensive studies on structure-activity relationships have been carried out to understand the structural basis for the activity of this hormone. Interestingly, *elcatonin*, $[Asu^{1-7}]$eel calcitonin, is an analogue of eel calcitonin in which the disulfide bridge is replaced by an ethylene bridge. It shows full biological activity in comparison to human CT. This analogue and salmon CT are widely used clinically because of their superior potency. In humans, CT is synthesized as a large precursor protein (141 aa). The leader sequence (25 aa) is proteolytically eliminated during transport through the ER, providing pro-CT (116 aa). The latter is then cleaved at both ends to yield a 33-peptide bearing a C-terminal glycine which acts as a precursor for the amino group of the 32-amino acid mature CT. In addition, the final processing step generates a N-terminal 57-residue pro-CT peptide and a 21-amino-acid C-terminal peptide, called PDN sequence or → katacalcin. CT is primarily secreted by the parafollicular cells (C cells) of the thyroid. These C cells are scattered throughout the thyroid in mammals, but constitute in submammalian species a distinct organ, the ultimobranchial body. CT affects a variety of tissues and organs including bone, intestine kidney, breast, and the hypothalamopituitary axis. The physiological function of CT is to maintain skeletal mass during periods of calcium stress, as during growth, pregnancy, and lactation. Furthermore, it plays a central role in controlling calcium homeostasis and maintaining serum calcium without significant fluctuations. CT receptors are widely distributed and they exist in different subclasses. It has been shown that CT receptors stimulate phospholipase C and increase intracellular Ca^{2+} through interactions with a G-protein. The main receptor subclass represents the type present on osteoclast cells, but different receptors were found, for instance, in certain brain regions and in the renal cortex. Despite the fact that the CT receptors belong to receptors with seven transmembrane domains, the predicted amino acid sequence of the CT receptor shows no significant homology with other G-protein-coupled receptors. It is also reported that CT acts as a neurotransmitter and brings about receptor-mediated analgesia, possibly caused by release of → β-endorphin. In clinical use, CT and special analogues have found to be safe for the treatment of a number of bone disorders, such as Paget's disease, Sudeck's atrophy, high-turnover osteoporosis, and hypercalcemia caused by malignancy. Especially, second-generation analogues of CT with reduced side effects and new dosage forms (nasal and, potentially, oral) will enhance the usefulness of calcitonin therapy [R. Neher et al., *Nature* **1968**, *220*, 984; C. Basava in: *Peptides: Design, Synthesis, and Biological Activity*, C. Basava, G.M. Anantharamaiah (Eds.), p. 209, Birkhäuser, Boston, **1994**; S. J. Wimalawansa, *Crit. Rev. Neurobiol.* **1997**, *11*, 167; R. Muff et al., *Eur. J. Endocrinol.* **1995**, *133*, 17; P.M. Sexton et al., *Curr. Med. Chem.* **1999**, *6*, 1067; T.J. Martin, *Bone* **1999**, *24*, 63].

Caledonin, a modified peptide isolated from the tunicate *Didemnum rodriguesi*. The amino group of the central Phe residue is linked with (*S*)-3-amino-5-mercaptopentanoic acid, whereas the carboxy function is connected with a cyclic guanine ring system bearing a n-octane side chain. Caledonin is capable of complexing Zn^{2+} (***Formula***), and Cu^{+} ions and may also be involved in ion transport through membranes [M.J. Vazquez et al., *Tetrahedron Lett.* **1995**, *36*, 8853].

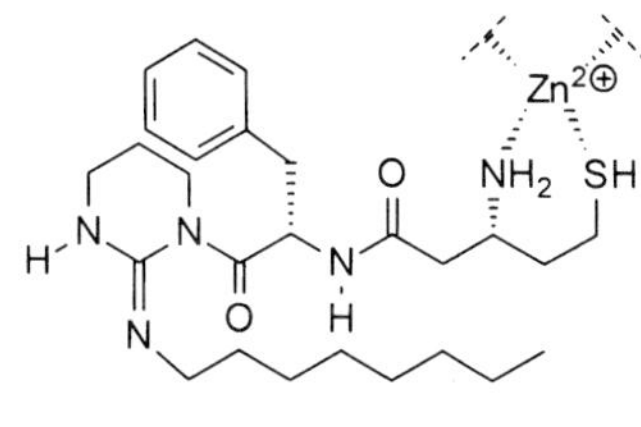

Caledonin

Calmodulin, *calcium-dependent regulator,* ***CDR***, *calcium modulator,* ***CaM***, a Ca^{2+}-binding protein (148 aa; M_r ~ 17 kDa) that mediates various functions in eukaryotes. It contains two similar globular domains separated by a seven-turn α-helix, and two high-affinity Ca^{2+}-binding sites on each of its globular domains. The binding sites are formed by helix-loop-helix motifs named EF hands. Binding of Ca^{2+} causes calmodulin to change conformation, exposing a hydrophobic site that interacts with the protein to be regulated [Y.S. Babu et al., *Nature* **1985**, *315*, 37].

Captopril, 1-[(2*S*)-3-mercapto-2-methylpropionyl]-L-proline, the first orally active inhibitor of the → angiotensin-converting enzyme (ACE) on the market. The positive effects of captopril and other → ACE inhibitors like → enalapril in hypertension and heart failure result primarily from suppression of the renin-angiotensin-aldosterone system. Captopril causes a drop in blood pressure in hypertensive patients.

Carnosine, H-β-Ala-His-OH, a naturally occurring dipeptide found in skeletal muscle in relative high concentration. Carnosine acts as anti-oxidant, free radical scavenger, physiological buffer, histidine source, and immunostimulant. More recently, possible anti-aging actions have been described [A. R. Hipkiss, C. Brownson, *Cell. Mol. Life Sci.* **2000**, *57*, 747].

CART peptides, novel neuropeptides acting as putative neurotransmitters/cotransmitters in the brain and gut. The name CART is an acronym for *c*ocaine and *a*mphetamine *r*egulated *t*ranscript, and refers to the mRNA discovered by Douglass et al. in 1995. CARTs are thought to be involved in feeding, development, stress, endocrine regulation, sensory processing, reward and reinforcement. Human CARTs (hCART) is thought to have neurotrophic properties, as it promotes neuronal development and survival in culture. Furthermore, CART inhibit feeding behavior and show psychostimulant-like effects on locomotor behavior [J. Douglass et al., *J. Neurosci.* **1995**, *15*, 2471; M. J. Kuhar et al., *Trends Neurosci.* **1999**, *22*, 316; L. D. Adams et al., *Brain Res.* **1999**, *848*, 137; M. J. Kuhar et al., *Regul. Peptides* **2000**, *89*, 1].

α-Casein exorphin, → milk protein-derived opioid peptides.

β-Casomorphins, *β-CM*, *β-casorphins*, β-casein-derived opioid peptide receptor ligands which belong to the → exorphins. They are the first → milk protein-derived opioid peptides acting as receptor ligands. β-CM with opioid activity have been found in human, bovine, ovine, and water buffalo β-casein. They can be released from β-casein both in the adult's and the neonate's organism where they might exert opioid activities. *β-casomorphin-7* (β-CM-7), H-Tyr-Pro-Phe-Pro-Gly-Pro-Ile, was the first example of a peptide with morphine-like activity isolated from commercial casein peptone in 1979. This sequence corresponds to bovine β-casein-(60–66). Contrary to moderate potencies and receptor selectivities of the natural β-CMs, some of the synthetic analogues display high agonist potencies and remarkable μ-receptor selectivities. *Morphiceptin*, H-Tyr-Pro-Phe-Pro-NH_2, the N-terminal tetrapeptide amide of β-CM-7, had been synthesized prior to isolation. It is regarded as a standard μ-selective opioid receptor ligand [V. Brantl et al., *Hoppe-Seyler's Z. Physiol. Chem.* **1979**, *360*, 1211; V. Brantl et al., *Peptides* **1982**, *3*, 793; K. Neubert et al. in: *β-Casomorphins and Related Peptides*, F. Nyberg, V. Brantl (Eds.), p. 15, Fyris-Tryk, Uppsala, **1990**].

Casoxins, → milk protein-derived opioid peptides.

Catalytic antibody, → abzyme.

Cecropins, a group of antimicrobial peptides first isolated from *Drosophila* and from the pupae of the giant silk moth *Hyalophora cecropia*, from which the name is derived. *Cecropin A*, KWKLFKKIEK^{10}VGQ NIRDGII^{20}KAGPAVAVVG^{30}QATQIAKa, and *cecropin B* with 35 residues are positively charged linear peptides from *H. cecropia* forming time-variant and voltage-dependent ion channels in planar lipid membranes. Furthermore, homologous peptides were found in several other insect species. It is interesting to note that a porcine cecropin was discovered in the upper intestinal tract, *cecropin P 1*, SWLSKTAKKL^{10}EN SAKKRISE^{20}GIAIAIQGGP^{30}Ra. Cecropins are not lethal for mammalian cells at microbiocidal levels, and have been administered safely to animals. A direct correlation between cationic character and activity has been established for the cecropins, where the less cationic members also show lower activity [H. G. Boman, D. Hultmark, *Annu. Rev. Microbiol.* **1987**, *41*,103; H. G.

Boman et al., *FEBS Lett.* **1989**, *259*, 103; W.F. Broekaert et al., *Crit. Rev. Plant Sci.* **1997**, *16*, 297].

Cell-penetrating peptides, ***CPP***, peptides of different structural classes that are capable of translocating across the plasma membrane of eukaryotic cells and acting as peptidic delivery factors. They have found application for the intracellular delivery of macromolecules with molecular weights several times greater than their own. In order to differentiate from larger proteins that have been shown to function as transporters across biological membranes, CPP on average contain no more than 30 amino acid residues. Until now, with the *protein-derived peptides* and *synthetic* or *chimeric peptides*, two main groups have been described. To the first group belong *penetratin*, RQIKIWFQNR^{10}RMKWKK, *tat fragment(48–60)*, GRKKRRQRRR^{10}PPQ, *signal-sequence-based peptide I*, GALFLGWLGA^{10}AGSTMGAWSQ^{20}PKKKRKV, and *signal-sequence-based peptide II*, AAVALLPAVL^{10}LALLAP, whereas members of the second group are *transportan*, GWTLNSAGYL^{10}KINLKALA^{20}ALAKKIL, and *amphiphilic model peptide*, KLALKLALKA^{10}LKAALKLA. In contrast to endocytosis, the penetration is a seemingly energy-independent mechanism of peptide translocation across the cell membrane which allows addressing of conjugated cargoes into the cytoplasm and the nucleus. However, the mechanism of the cellular translocation by CPPs is still obscure. It seems likely that CPPs from the different groups act by distinct transport mechanisms. Examples of cargoes internalized by CPPs are, for example, the transport of a fibroblast growth factor (FGF) receptor phosphopeptide by penetratin to inhibit FGF receptor signaling in living neurons, and the internalization of the 21-mer galanin receptor antisense by penetratin or transportan in order to regulate galanin receptor levels and modify pain transmission *in vivo* [M. Lindgren et al., *Trends Pharmacol. Sci.* **2000**, 99; S. Calvet et al., *J. Neurosci.* **1998**, *18*, 9751; M. Pooga et al., *Nature Biotechnol.* **1998**, *16*, 857].

Cerulein, *caerulein*, ***CRL***, pGlu^1-Gln-Asp-Tyr(SO_3^-)-Thr^5-Gly-Trp-Met-Asp-Phe^{10}-NH_2, a peptide amide firstly isolated from the skin of the Australian frog *Litoria caerulea* and from the skin of other amphibians. Cerulein belongs as a member of non-mammalian origin to the → gastrin family. Synthetic CRL, named *ceruletide*, was predominantly used for studies of the activity spectrum, and is applied in X-ray diagnostics and in the diagnosis of pancreas function. Activation of the receptor causes, for example, breakdown of phosphoinositides, and mobilization of cellular calcium. In general, it acts in a similar manner as → cholecystokinin. Due to the stimulating effect of the small intestine, ceruletide is used therapeutically in postoperative intestinal atonia and paralytic ileus, and can also be employed for the expulsion of gallstones and in biliary colic.

Ceruletide, → cerulein.

Chaperonins, a family of the → molecular chaperones.

Chemokines, a family of chemotactic → cytokines mediating inflammation. The chemokines are homologous 8- to 10-kDa proteins with 20 to 70% sequence homology. Until now, more than 40 chemokines have been identified. They are subdivided into at least four families based on the relative position of the Cys residues in the mature protein, but only the α- and β-chemokines (both of which contain four Cys residues) have been well characterized. In the *α-chemokines*, the first two Cys residues are separated by a single amino acid residue (Cys-

Xaa-Cys), whereas in the β-chemokines, the first two Cys residues are adjacent to each other (Cys-Cys). α-Chemokines, which contain the sequence Glu-Leu-Arg preceding the Cys-Xaa-Cys sequence, are chemotactic for neutrophils, and those lacking this sequence act on lymphocytes. Lymphotactin and fractalkine represent two additional families that do not fit into the classification. *Lymphotactin* contains only two Cys residues in the mature protein, whereas *fractalkine* is a membrane-bound glycoprotein in which the first two Cys residues are separated by three amino acids (Cys-Xaa-Xaa-Xaa-Cys). Chemokines mediate cell migration and activation by binding to specific G-protein-coupled receptors on the surface of the target cells. The functions of chemokines in the pathophysiology of disease are still being defined, but from studies in animals it can be concluded that the neutralization of chemokine activity may have therapeutic value. Chemokine-receptor antagonists may inhibit allergic, autoimmune, and septic processes. Chemokines may improve the host response to infection, vaccines, and tumors [A.D. Luster, *N. Engl. J. Med.* **1998**, *338*, 436].

Chemotactic peptides, *CP*, peptides acting as potent agonists and antagonists of the neutrophil fMLF receptor. The prototype of CP is For-Met-Leu-Phe-OH (fMLF). N-Ureido-derivatized small peptides are also active compounds. The agonist or antagonist behavior can be achieved depending on the nature of the substituent R of the urea. Benzylureido-Met-Leu-Phe-OH is an antagonist, whereas 4-methoxyphenylureido-Met-Leu-Phe-OH shows agonist activity [S. Aswanikumar et al., *J. Exp. Med.* **1976**, *143*, 1154; R.F. O'Dea et al., *Nature* **1978**, *272*, 462; J.D. Higgins et al., *J. Med. Chem.* **1996**, *39*, 1013].

Cholecystokinin, ***CCK***, formerly named *cholecystokinin/pancreozymin*, ***CCK-PZ***, a peptide synthesized in the small intestinal mucosa and in various regions of the brain. The CCK gene encodes prepro-cholecystokinin, which by differentiated endoproteolytic cleavages is processed to six CCK peptides varying in length from 83 to 8 amino acids: *CCK-8*, H-Asp-Tyr(SO_3^-)-Met-Gly-Trp-Met-Asp-Phe-NH_2, *CCK-22*, *CCK-33*, *CCK-39*, *CCK-58* and *CCK-83*. The six CCK peptides have the same C-terminal octapeptide sequence as shown for CCK-8, in which the tyrosine residue is O-sulfated. CCK-8 has the full activity spectrum of the longer CCK peptides. Decreased feeding in rats is caused by peripherally injected CCK-8. Extensive studies concerning the mechanism of action, behavioral specificity, and physiological significance of this effect have been performed. CCK is a member of the → gastrin family. In 1928, CCK was discovered in small intestine extracts as a gallbladder-emptying hormone, and in the 1940s a stimulator of pancreatic enzyme secretion, named pancreozymin, was found. It took 20 years before it was shown by Jorpes and Mutt that CCK and pancreozymin are identical. Now, only the acronym CCK is used [V. Mutt, J.E. Jorpes, *Eur. J. Biochem.* **1968**, *6*, 156; R.J.L. Deschenes et al., *Proc. Natl. Acad. Sci. USA* **1984**, *81*, 726; G.P. Smith, J. Gibbs, *Ann. NY Acad. Sci.* **1985**, *448*, 417; J.F. Rehfeld, *Physiol. Rev.* **1998**, *78*, 1087].

Chorionic gonadotropin, ***CG***, *human chorionic gonadotropin*, ***hCG***, a α/β two chain glycoprotein hormone synthesized in the syncytiotrophoblasts of the placenta. CG is secreted in the first weeks of pregnancy and stimulates the corpus luteum via the G-protein-coupled hCG receptor in order to continue the secretion of progesterone rather than regressing, and this prevents menstruation. Immunoassay pregnancy tests are based on the detection of CG in blood or urine within a few days after embryo implantation. The α-chain (96 aa) is identical with that of → lutropin (LH), → thyrotropin (TSH) and → follitropin (FSH), whereas the β-chain (145 aa) shows a high degree of sequence identity of those of LH (85%), FSH (45%), and TSH (36%). CG belongs to the → gonadotropins. Human CG (Predalon, Prednesin, Choragon, Primogonyl) is used in the treatment of disturbances of gonadal functions [L. Liu et al., *Endocrinology* **1989**, *124*, 175].

Chorionic mammotropin, ***CM***, *human chorionic somatomammotropin*, ***HCS***, *placenta lactogen*, ***PL***, a single chain proteohormone formed in the placenta during the first trimester of pregnancy. It shows more structural similarities to → somatotropin (STH) than to → prolactin (PRL), but it combines the function of both hormones. CM initially stimulates the development of the breast without influence on milk secretion. With decreasing level of CM after birth, PRL stimulates milk secretion. The growth-stimulating effect of CM is weaker than that of human STH, but CM might play a more important function in the development of the fetus. CM promotes the release of progesterone and estrogen from the corpus luteum, but this effect is additionally potentiated by → chorionic gonadotropin.

Chromogranins, ***Cg***, *secretogranins*, a family of water-soluble acidic glycoproteins. The Cg consist of four distinct members: CgA, CgB (secretogranin I), CgC (secretogranin II) and the neuroendocrine secretory protein 55 (NESP 55). The three Cg are rich in Glu, have calcium-binding capacity and the capability to be glycosylated, phosphorylated and sulfated. The Cg are precursors of various peptides showing biological activities, e.g., NESP 55 produces a tetrapeptide named LSAL which is an endogenous antagonist of the serotoninergic 5-HT1B receptor subtype, and → secretoneurin is a derivative of CgC. *CgA* is the most important member of this family. Human CgA is a single chain glycoprotein (439 aa; M_r ~ 50 kDa) predominantly located in the dense core of granula storing peptide hormones and catecholamine-containing vesicles. Due to the high sensitivity and specificity of CgA, it can be successfully used as a circulating marker in diagnosis, prognosis and follow-up of neuroendocrine tumors. Human CgA is the precursor of human → pancreastatin and → peptide WE-14 [L. Ferrari et al., *Anticancer Res.* **1999**, *19*, 3415].

Cinnamycin, a member of the Type B → lantibiotics. Cinnamycin is characterized by a threefold thioether-based bridging pattern containing an additional lysinoalanine bridge connecting the C-terminus with 2,3-didehydroalanine (Dha) in position 6. Cinnamycin (M_r 2042) is produced from *Streptomyces cinnamoneus*. Since cinnamycin shares considerable sequence similarities with *ancovenin* (M_r 1959; produced from *Streptomyces* ssp.), *duramycin* (M_r 2014; produced from *Streptomyces cinnamoneus*), *duramycin B* (M_r 1951; produced from *Streptoverticillium* ssp.) and *duramycin C* (M_r 2008; produced from *Streptomyces griseolutes*), all these peptides form a group of analogues of the prototype C. In general, only seven amino acids vary, and the substitutions are mainly conservative. Besides their general antimicrobial activity, the members of this group have attracted special interest because of their novel activity directed towards specific enzyme functions, e.g., phospholipase A_2 and → angiotensin-converting enzyme. Furthermore, members of this group of type-B lantibiotics show some antiviral activities, for example against *herpes simplex* type II and other related retroviruses [R.W. Jack, G. Bierbaum, H.-G. Sahl, *Lantibiotics and Related Peptides*, Springer, Berlin, Heidelberg, New York, **1998**].

Cionin, H-Asn-Tyr(SO_3^-)-Tyr(SO_3^-)-Gly-Trp-Met-Asp-Phe-NH_2, a disulfotyrosyl hybrid of → cholecystokinin and → gastrin from the neural ganglion of the protochordate *Ciona intestinalis*. It was concluded that the structure of cionin meets that of a common ancestor for cholecystokinin and gastrin [A.H. Johnsen, F.R. Rehfeld, *J. Biol. Chem.* **1990**, *265*, 3054].

Cobratoxins, polypeptides from the venoms of cobras or poisonous vipers (*Elapidae*). The peptides consist of 60–62 residues and 4–5 disulfide bonds. *α-Cobratoxin* from the venom of the cobra *Naja Naja siamensis* is a 62-peptide containing four disulfide bridges. Cobratoxins show neurotoxic actions; the toxins of Asian cobras are additionally cardiotoxic.

Collagen, an extracellular protein occurring as the major stress-bearing component of connective tissues such as bone, cartilage, teeth, ligament, tendon, and the fibrous matrices of skin and blood vessel. Collagen is composed of three polypeptide chains. At least 30 genetically distinct polypeptide chains comprising 16 collagen variants have been found in mammals. To the most abundant types of collagens belongs *type I collagen* (M_r ~ 285 kDa; width: ~ 14 Å; length: ~ 3000 Å) characterized by the chain composition $[\alpha1(I)]_2\alpha2(I)$ and mainly occurring in bone, tendon, skin, blood, cornea, and blood vessels. Further important members are *type II collagen* $[\alpha1(II)]_3$ found in cartilage and intervertebral disk, and *type III collagen* $[\alpha1(III)]_3$, occurring in blood vessels and fetal skin. The amino acid sequence of collagen α1(I) contains repeating -Gly-Xaa-Yaa- sequence triplets (Xaa often Pro and Yaa often Hyp) over a continuous 1011-residue stretch of its 1042-residue polypeptide chain. In the triple helix of collagen, the left-handed peptide helices are twisted together forming a right-handed superhelical structure. The triple helix has 10 Gly-Xaa-Yaa units per turn and a pitch of 86.1 Å. Collagen is organized into fibrils showing a periodicity of 680 Å and a diameter of 100 to 2000 Å depending on the types of collagen. The molecule contains covalently linked carbohydrates ranging from ~ 0.4 to 12% by weight and consisting mainly of glucose, galactose, and their disaccharides. Collagen fibrils are covalently cross-linked. Specific lysyl and hydroxylysyl residues are oxidized by lysyl oxidase, resulting in the formation of an aldehyde from the side chain amino group. The resulting aldehydes undergo Schiff's base formation and aldol condensations with neighboring side chains, forming a variety of cross-links that contribute to the strength of collagen. A variety of hereditary and environmentally caused disorders are caused by impairment of collagen synthesis. Mutations of Type I collagen usually result in *osteogenesis imperfecta* (brittle bone disease) [M.E. Nimni (Ed.), Collagen, *Volume I*, CRC Press, Boca Raton, FL, **1988**; E.Y. Jones, A. Miller, *J. Mol. Biol.* **1991**, *218*, 209; M. van der Rest, P. Bruckner, *Curr. Opin. Struct. Biol.* **1993**, *3*, 430].

Collectins, a group of mammalian plasma and cell-surface proteins containing collagenous regions and lectin regions. This family includes serum mannan binding protein (MBP), bovine conglutinin, the lung surfactant proteins SP-A and SP-D, and a bovine protein, named collectin-43. Collectins share a very similar modular domain composition and overall three-dimensional structure. The overall domain organization in the collectins is characterized by four distinct regions, of which three are structurally well defined. An independently folded C-type lectin domain is located at the C-terminal end of each polypeptide chain. The latter is preceded by a stretch of about 40 residues, which are capable of forming a three-stranded α-helical coiled-coil when associated with the equivalent regions in two other chains. The third domain is the collagenous region, which is always located directly N-terminal to the α-helical bundle. The biological roles of collectins seem to be the preimmune defense against microorganisms both in serum and lung surfactant. The C-type lectin domains bind to carbohydrate ligands on the cell-surface of pathogens and fulfill a recognition function that can elicit effector functions via the collagen-like region, such as complement activation in the case of serum mannan binding protein, or binding to cell-surface receptors to trigger phagocytosis or oxidative killing [H.-J. Hoppe, K. B. M. Reid, *Protein Sci.* **1994**, *3*, 1143].

Colony stimulating factors, ***CSF***, *hematopoietic growth factors*, glycoprotein growth factors involved in proliferation, differentiation and survival of hematopoietic progenitor cells. Human CSFs are: *granulocyte/macrophage CSF (GM-CSF)*, produced by T lymphocytes, endothelial cells and fibroblasts (M_r ~ 18–20 kDa; monomer, hGM-CSF: 127 aa), *granulocyte CSF (G-CSF)*, produced by macrophages and fibroblasts (M_r ~ 20 kDa; monomer, hG-CSF: 174 aa), *macrophage CSF (M-CSF)*, produced by fibroblasts, macrophages and endothelial cells (M_r ~ 70 kDa; dimer), and *multi-CSF* [also called *interleukin-3 (IL-3)*, → interleukins], produced by T lymphocytes and epidermal cells (M_r ~ 25 kDa, mouse IL-3: 139 aa). The action of the CSFs is completed by the hormone-type CSF → erythropoietin. Some human CSFs are now produced in larger quantities as recombinant proteins, and used clinically to counteract leukocyte death during chemotherapy and to facilitate bone marrow transplantation. Specific receptors for each CSF are coexpressed on granulocytes and monocytes [S. C. Clark et al., *Science* **1987**, *236*, 1229; H. Link, L. Arseniev, *Chemotherapie* **1993**, *2*, 1; Y. Nakagawa et al., *J. Biol. Chem.* **1994**, *269*, 10905].

Complementary peptide, → antisense peptide.

Conantokins, → conotoxins.

Conopressins, → conotoxins.

Conotoxins, toxic peptides from the venom of marine snails (genus *Conus*) comprising on average 9–29 residues and a content of Cys between 22 and 50%. More than 500 species of sea snails exist which live on worms, mollusks, and fish. The fish-hunting sea snails bite the fish, and the simultaneously injected toxin quickly paralyzes the prey. It is assumed that there are potentially ~ 50 000 different conotoxins present in the venoms of living species in the genus *Conus*. Seven various groups of conotoxins have been described: *α, μ-* and *ω-conotoxin, conotoxin K, sleeper-conotoxin, convulsant-conotoxin*, and *conopressins*. *α-Conotoxins* are a family of Cys-enriched peptides occurring in several marine snails from the genus *Conus*. These peptides behave pharmacologically as competitive antagonists of the nicotinic acetylcholine receptor. α-Conotoxins (GIA), ECCNPACGRH^{10}YSCGKa (disulfide bonds: C^2–C^7/C^3–C^{13}), and shortened, slightly sequence-altered forms (GI, GII and MI) block postsynaptic cholinergic receptors in neuromuscular connection. *μ-Conotoxins* are Cys-rich 23-peptide amides (GVIIIA, GVIIIB, GVIIIC) and block muscular Na^+ channels causing paralysis and death. *ω-Conotoxins*, after injection into mice, cause shaking and are therefore named *shaker-peptides*. These toxins block voltage-sensitive Ca^{2+} channels at cholinergic nerve terminals, thus inhibiting the release of acetylcholine. All mentioned conotoxins paralyze and kill fish after injection. These peptides have also been used as model systems in neuroscience research. The *convulsant-conotoxins* containing about 100 residues (M_r ~ 13 kDa), and *conotoxin K* from *Conus magus* (~ 25 aa; three disulfide bridges) are significantly longer, but the CNS effects have not yet been well elucidated. *Conopressins* are basic 9-peptide amides with structural similarities to → vasopressin. Smooth muscle effects result after injection into mice; therefore, they may be involved in the distribution of paralyzing toxins in the guest organism. *[Lys^8]conopressin-G*, H-Cys-Phe-Ile-Arg-Asn-Cys-Pro-Lys-Gly-NH_2 (disulfide bond: Cys^1–Cys^6), and *[Arg^8]conopressin-S*, H-Cys-Ile-Ile-Arg-Asn-Cys-Pro-Arg-Gly-NH_2 (disulfide bond: Cys^1–Cys^6) are typical members of this family. The *conantokins* (also named *sleeper peptides*) are further peptides with neurotoxic effects from sea snails. The name is derived from the Philippine "*antokin*" which means sleepily. *Conantokin-G* (conantokin-GV), H-Gly-Glu-Gla-Gla-Leu-Gln-Gla-Asn-Gln-Gla^{10}-Leu-Ile-Arg-Gla-Lys-Ser-Asn-NH_2, and the conantokins-T and -GS have been identified. It is assumed that the Gla residues in conantokin-G promote the binding of the receptor on cell membranes. The conantokins-G and -T are antagonists of the NMDA receptor which is a subtype of the glutamate receptor. In 2001, Olivera and Cruz reported on the conotoxin story in retrospect, and presented the latest nomenclature of conus peptides, which however has not been used in the text above [W. R. Gray, B. M. Olivera, *Annu. Rev. Biochem.* **1988**, *57*, 665; B. M. Olivera, *J. Comp. Physiol. A* **1999**, *185*, 353; H. R. Arias, M. P. Blanton, *Int. J. Biochem. Cell. Biol.* **2000**, *32*, 1017; J. L. Dutton, D. J. Craik, *Curr. Med. Chem.* **2001**, *8*, 327; B. M. Olivera, L. J. Cruz, *Toxicon* **2001**, *39*, 7].

Contryphans, H-Gly-Cys-Hyp-D-Trp-Xaa-Trp-Cys-NH_2, a D-Trp-containing family of peptides from venoms of fish-hunting cone snails. In the three known mature peptides, seven of eight amino acids are conserved. *Contryphan-R* (Xaa = Glu) from the venom of *Conus radiatus*, *[de-Gly^1]contraphan-R* and *[6-bromo-Trp]contraphan-R* were the members discovered first. With *contryphan-Sm* (Xaa = Gln) from *C. stercusmuscarum* and *contryphan-P* (Xaa = Asp) from *C. purpurascens*, two further members from other *Conus* venoms have been characterized. The contryphans show a remarkable sequence conservation in contrast to the multiply disulfide-bonded → conotoxins [R. Jacobsen et al., *J. Peptide Res.* **1998**, *51*, 173].

Corazonin, QTFQYSRGWT^{10}Na, a cardioexcitatory 11-peptide amide from the *corpora cardiaca* (CC) of the American cockroach, *Periplaneta americana*. CC store and release neurohormones that are produced in neurosecretory cells of the CNS. [His^7]corazonin is capable of inducing dark pigmentation in an albino locust strain [J. A. Veenstra, *Biochem. Biophys. Res. Commun.* **1994**, *204*, 292; A. I. Tawfik et al., *Proc. Natl. Acad. Sci. USA* **1999**, *96*, 7083].

Corticoliberin, *corticotropin-releasing hormone*, ***CRH***, *corticotropin-releasing factor*, ***CRF***, SEEPPISLDL^{10}TFHLLREVLE^{20}MAR AEQLAQQ^{30}AHSNRKLMEI^{40}Ia (human, rat CRH), a 41-peptide amide occurring in the nerve fibers all over the brain stimulating the synthesis and release of → pro-opiomelanocortin and its degradation in the hypophysis and placenta. CRH is formed in the neurons of the hypothalamus, and delivered to the adenohypophysis via a direct circulatory connection. Besides nerve fibers, corticoliberin has also been found, for example, in the pancreas, adrenal medulla, placenta, stomach, testes and in the tissues of ACTH-producing tumors. It can be used as a diagnostic aid for hypophyseal function. Furthermore, CRH has been implicated in the onset of pregnancy, the 'fight or flight' response, and also many physiological disorders. CRH belongs to a family of peptides consisting of its mammalian analogue → urocortin, the amphibian peptides → sauvagine and → urotensin I, which bind with different affinities to two subtypes of the CRF receptor (CRFR), CRFR1 and CRFR2. CRFR is a G-protein-dependent receptor which acts mainly through G_S enhancing cAMP production. Even though the earliest actions of these peptides may have been connected with osmoregulation and diuresis, physiological effects associated with stress and anxiety, thermoregulation, vasoregulation, growth and metabolism, metamorphosis and reproduction have been established in various vertebrate species [W. Vale et al., *Science* **1981**, *213*, 1394; K. Eckart et al., *Curr. Med. Chem.* **1999**, *6*, 1035; A. J. Dunn, C. W. Berridge, *Brain Res. Rev.* **1990**, *15*, 71; D. A. Lovejoy, R. J. Balment, *Gen. Comp. Endocrinol.* **1999**, *115*, 1; P. A. Keller et al., *Bioorg. Med. Chem.* **2000**, *8*, 1213].

Corticostatin, GICACRRRFC^{10}PNSERFSGY C^{20}AVNGARYVRC^{30}CSRR (disulfide bonds: C^5–C^{20}/C^{10}–C^{30}/C^3–C^{31}), a 34-peptide with a high content of Arg and Cys first isolated from rabbit fetal and adult lung. It inhibits the → corticotropin-stimulated formation of corticosterone by adrenal cells *in vitro* [Q. Zhu et al., *Proc. Natl. Acad. Sci. USA* **1988**, *85*, 592; A. Singh et al., *Biochem. Biophys. Res. Commun.* **1988**, *155*, 524].

Corticotropin release-inhibiting factor, *CRIF*, FIDPELQRSW^{10}EEKEGEGVLM^{20}PE, a 22-peptide from the cryptic region of the prepro-thyreotropin releasing hormone (→ thyreoliberin) that fulfills the criterion of the endogenous corticostatin. It corresponds to the partial sequence prepro-TRH-(178–199). CRIF inhibits basal and corticoliberin-stimulated corticotropin synthesis and secretion in cultured primary anterior pituitary cells [E. Redei et al., *Endocrinology* **1995**, *136*, 3557].

Corticotropin, *adrenocorticotropin, adrenocorticotrophic hormone, **ACTH***, SYSMEHFRW G^{10}KPVGKKRRPV^{20}KVYPNGAEDE^{30}SAEA FPLEF, a 39-peptide hormone formed as a proteolytic cleavage product of pro-opiomelanocortin (POMC) in the adenohypophysis. ACTH is synthesized in response to stimulation by → corticoliberin, and stress. An opposite action in the sense of a corticostatin is exerted by the → corticotropin release-inhibiting factor. In the adrenal cortex, ACTH stimulates a Ca^{2+}-dependent process of the synthesis of glucocorticoids and mineralocorticoids. According to a proposal of Schwyzer, the partial sequence 11–18 is the receptor binding region, termed as "address" sequence 5–10 is the active site ("message"), the N-terminal tripeptide is the "amplifier", and the C-terminal fragment 25–39 ("envelope") is responsible for the antigenicity and transport. The species specificity is mainly located in the sequence 31–33. In porcine ACTH, Ser^{31} is substituted by Leu, whereas bovine and ovine ACTH is characterized by Gln^{33} instead of Glu. With regard to in-vitro corticosteroid-releasing activity, ACTH-(1–24) shows higher activity than native ACTH.

Native ACTH and active analogues are used as a diagnostic aid for adrenal cortex function and as therapeutic agents in the treatment of several disorders, such as insufficient function of the adrenal cortex, multiple sclerosis, collagen diseases, inflammatory rheumatic diseases, radicular pain syndrome, etc. The N-terminal sequence 1–13 is identical to that of α-MSH (→ melanotropin) [L. Proulx et al., *J. Endocrinol. Invest.* **1984**, *7*, 257].

Cortistatin, ***CST***, a neuropeptide related to → somatostatin released from prepro-cortistatin. The rat precursor protein is predicted to have 112 residues with a striking sequence homology to somatostatin at its distal C-terminal end. After removal of the 27-residue signal sequence, procortistatin could be further cleaved to rCST-29, rCST-14, PCKNFFWKTF^{10}SSCK (disulfide bond: C^2–C^{13}) and additional products by further cleavage. CST-29 and CST-14 show close similarities to the analogous → somatostatin (SST) peptides, such as SST-28 and SST-14. CST-14 shares 11 of the 14 residues of SST-14 despite being the products of separate genes. Furthermore, CST has also been cloned from humans, and mouse sources. CST is capable of binding to somatostatin receptors and shares many pharmacological and functional effects with SST, including the depression of neuronal activity. In addition, it shows many effects distinct from SST such as the induction of slow-wave sleep, reduction of locomotor activity and activation of cation-selective currents [A. D. Spier, L. de Lecea, *Brain Res. Rev.* **2000**, *33*, 228].

Crabolin, FLPLILRKIV^{10}TALa, a 13-peptide amide from the venom of European hornet. It lacks any lytic activity against cells and shows only minor mammalian toxicity [A. Argiolas, J. J. Pisano, *J. Biol. Chem.* **1984**, 10106; R. C. Hider et al., *Endeavour, New Series* **1988**, *12*, 60].

CREC family, a group of multiple EF-hand, low-affinity Ca^{2+}-binding proteins involved in the secretory pathway of mammalian cells. Until now, this family is known to comprise reticulocalbin, ERC-55/TCBP-49/E6BP, Cab45, calumenin, and crocalbin/CBP-50. In addition, similar proteins are found in quite diverse invertebrate organisms, e.g., DCB-45, and SCF in *Drosophila melanogaster*, SCF in *Bombyx mori*, CCB-39 in *Caenorhabditis elegans*, and Pfs40/PfERC in *Plasmodium falciparum*. The proteins of this family show rather low Ca^{2+} affinity, with K_d around 10^{-4}–10^{-3} M, and may participate in Ca^{2+}-regulated activities. There is evidence that some family members are involved in pathological activities, for example, in malignant cell transformation, putative participation in amyloid formation, and in mediation of the toxic effects of snake venom toxins [B. Honore, H. Vorum, *FEBS Lett.* **2000**, *466*, 11].

Cripto, *teratocarcinoma-derived growth factor*, ***TDGF***, a novel epidermal growth factor (EGF)-related peptide involved in mammary gland development and neoplasia. *CR-1* (*TDGF-1*) was isolated and sequenced from a human NTERA2/D1 embryonal carcinoma cDNA expression library, whereas the mouse homologue, *Cr-1* (*tdgf-1*) was isolated from a F-9 mouse embryonal carcinoma cDNA expression library. Cripto induces branching morphogenesis in mammary epithelial cells both *in vitro* and *in vivo* and inhibits the expression of various milk proteins [D. S. Salomon et al., *BioEssays* **1999**, *21*, 61].

Crustacean cardioactive peptide, ***CCAP***, PFCNAFTGCa, a 9-peptide amide first identified in the shore crab, *Carcinus maenas*. In the meantime, CCAP has been identified in many insect species. It stimulates the release of → adipokinetic hormone from the *corpus cardiacum* of locusts *in vitro*. Furthermore, it is thought to be a mediator in the neuronal pathway leading to the ecdysis behavior [J. Stangier et al., *Peptides* **1988**, *9*, 795; D. Veelaert et al., *Endocrinology* **1997**, *138*, 138].

C-type natriuretic peptide, ***CNP***, GLSKGCF GLK^{10}LDRIGSMSGL^{20}GC (p-CNP, disulfide bond: C^6–C^{22}), a 22-peptide that belongs to the → natriuretic peptides (ANP, BNP). The natriuretic activities of CNP are smaller compared with those of ANP and BNP. CNP does not occur practically in the blood circulation, but it has been detected in the vascular endothelium, kidney, intestinal tract and the cerebrospinal fluid where it fulfills endocrine functions. Porcine prepro-CNP contains 126 residues. After cleavage the Ala^{23}-Lys^{24} bond results pro-CNP (103 aa) that is converted proteolytically into the native 22-peptide CNP [T. Sudoh et al., *Nature* **1988**, *332*, 78; T. Sudoh et al., *Biochem. Biophys. Res. Commun.* **1990**, *168*, 863; G. McDowell et al., *Eur. J. Clin. Invest.* **1995**, *25*, 291; R.C. Fowkes, C.A. McArdle, *Trends Endocrinol. Metab.* **2000**, *11*, 333].

Cyclophilins, a family of → peptidyl prolyl *cis/trans* isomerases.

Cyclosporin A, ***CsA***, a homodetic cyclic undecapeptide (M_r 1,202 Da; for formula, see Fig. 3.23) produced by the fungus *Beauveria nivea* (previously named *Tolypocladium inflatum*). CsA exerts antifungal, immunosuppressive, antiparasitic, and anti-inflammatory activity. It is a highly effective agent for the treatment of autoimmune disorders and for preventing organ-transplant rejection. Until now, CsA has been the main component of 25 naturally occurring cyclosporins that differ in the basic primary structure of CsA by amino acid substitutions in positions 1, 2, 4, 5, 7, and 11. In addition, unmethylated peptide bonds have been found in positions 1, 4, 6, 9, 10, and 11. The first total chemical synthesis was described by Wenger in 1984, but since then several hundred CsA analogues have been chemically synthesized. The biosynthesis in *Beauveria nivea* is accomplished by the multienzyme cyclosporin synthetase (M_r ~ 1,400 Da), which is the largest integrated enzyme structure so far known, catalyzing about 40 reaction steps including the final assembly of the undecapeptide chain of CsA and its cyclization. By in-vitro synthesis, many new analogues of CsA became available which exerted remarkable immunosuppressive activity *in vitro*. CsA inhibits the activity of the → peptidyl prolyl *cis/trans* isomerase (PPIase) cyclophilin (M_r ~ 17.7 kDa), a cytosolic CsA binding protein. The immunosuppressive property of CsA is based on the capability of its complex with cyclophilin to prevent the expression of genes involved in the activation of T lymphocytes by interfering with the appropriate intracellular signaling pathways [B.D. Kahan (Ed.), *Cyclosporin: Biological Activity and Clinical Applications*, Grune & Stratton Inc., Orlando, FL, **1984**; R. Schindler (Ed.), *Cyclosporin in Autoimmune Diseases*, Springer, Berlin, **1985**; R.M. Wenger, *Helv. Chim. Acta* **1984**, *67*, 502; A. Lawen et al., *Biochimie* **1992**, *74*, 511].

Cyclotheonamides, cyclopeptides from the sponge *Theonella* with strong inhibitory activity against → thrombin and other serine proteases (cf. Section 7.2.2; **22**, **23**) [N. Fusetani et al., *J. Am. Chem. Soc.* **1990**, *112*, 7053; P. Wipf, *Chem. Rev.* **1995**, *95*, 2115].

Cystatins, a superfamily of cysteine peptidase inhibitors. They are tight-binding reversible inhibitors of many cysteine proteases and are not capable of inhibiting other proteases [A. J. Barret, *Trends Biol. Sci.* **1987**, *12*, 193].

Cytokines, local mediator proteins regulating the differentiation, proliferation, and activities of the various types of blood cells. This general term for these intercellular signaling agents was introduced by S. Cohen in 1974. Cytokines control both the communication between immune cells and the proliferation and differentiation of leukocytes. Furthermore, they support the induction of immune reactions, and can both amplify and stop inflammation reactions. They are produced only in very small concentrations by many different cells, but the main producers are the immune cells themselves. The message of the highly active cytokines is mediated through receptors on the surface of the target cells. Examples of cytokines are → interferons (IFN), → interleukins (IL), → colony-stimulating factors (CSF), → tumor necrosis factor (TNF), → transforming growth factor-β (TGF-β), and → fibroblast growth factors. The functions of the different factors overlap frequently. IL-1, IL-2, IL-4, IL-6, IFN-γ and TNF-α preferentially act in immune regulation. They promote the release of further cytokines, but also increase the effector function of mononuclear phagocytes, natural killer cells and cytotoxic T cells (killer T cells). IFN-α and IFN-β are antivirally acting proteins that inhibit virus reproduction. As with TGF-β, these interferons suppress the reproduction of cells and play an important role in the inhibition of blood formation (hemopoiesis). TGF-β and IL-10 are engaged in suppressing cytokine production. IL-3 and CSFs are hematopoietic growth factors causing formation of blood in spinal cord supported by many interleukins. The increasing importance of cytokines in the therapy of many diseases is emphasized. IL-20, which is structurally related to IL-10, is the most newly discovered member of the cytokines, and it appears to be an autocrine factor for keratinocytes that regulates their participation in inflammation processes [F. R. Balkwill, F. Burke, *Immunology Today* **1989**, 299; H. Kirchner et al., *Cytokine and Interferone: Botenstoffe des Immunsystems*, Spektrum Akademischer Verlag, Heidelberg, **1994**; H. Blumberg et al., *Cell* **2001**, *104*, 9].

D

Decorsin, a 39-residue polypeptide originally isolated from the leech *Macrobdella decora*. Decorsin is a potent integrin $\alpha_{IIb}\beta_3$ antagonist and inhibitor of platelet aggregation. It contains six Cys as well as six Pro, and a RGD adhesion site recognition sequence, which is found in adhesive proteins such as → fibrinogen, von-Willebrand factor, and vitronectin. Purified recombinant decorsin has been found to be indistinguishable from the native peptide [J. L. Seymour et al., *J. Biol. Chem.* **1990**, *265*, 10143; A. M. Krezel et al., *Prot. Sci.* **2000**, *9*, 1428].

Defensins, one of the best-studied families of → animal defense peptides. Defensins have been categorized into two groups, *α-defensins* ("classical D") and *β-defensins*. Both contain many Arg residues. Three disulfide bonds stabilize the β-sheet structures. The latter distinguish defensins from other antimicrobial peptides which form amphipathic helices. However, the location and connectivity of the Cys are different between the two groups. Furthermore, there are also differences in other conserved amino acids. The defensins are small cationic polypeptides with an average number of amino acid building blocks of between 29 to 34. They were first discovered in mammalian polymorphonuclear leukocytes and macrophages. This group includes human, rabbit, mouse, rat, and guinea pig peptides. *Human α-defensin, HNP-1 (human neutrophil peptide)*, CYCRIP ACI^{10}AGERRYGTCI^{20}YQGRLWAFCC^{30} (disulfide bonds: C^2–C^{30}/C^4–C^{19}/C^9–C^{29}) forms a triple-stranded β-sheet connected by a loop with a β-hairpin hydrophobic finger. Defensins bind electrostatically to membranes, causing the formation of multimeric pores and the leakage of ions and metabolites. A great variety of defensins from many different sources has been identified as a result of intensive screening experiments [R.I. Lehrer et al., *Annu. Rev. Immunol.* **1993**, *11*, 105; S.H. White et al., *Curr. Opin. Struct. Biol.* **1995**, *5*, 521; T. Ganz, R.I. Lehrer, *Pharmacol. Ther.* **1995**, *66*, 191; T. Ganz, R.I. Lehrer, *Curr. Opin. Hematol.* **1997**, *4*, 53].

Delta-sleep-inducing peptide, *DSIP*, H-Trp-Ala-Gly-Gly-Asp-Ala-Ser-Gly-Glu-OH, a 9-peptide first isolated from the blood of sleeping rabbits. It is widely distributed in the brain and other tissues. The concentration in the brain increases during hibernation. DSIP and appropriate analogues mediate sleep-like states (δ-slow-wave sleep) after intraventricular application into the rabbit brain. From various studies it can be concluded that DSIP passes the blood-brain barrier. DSIP causes a prolonged time of sleep in patients suffering from disturbed sleep. The phosphorylation of the hydroxyl group of Ser^7 results in a more potent and more proteolytically stable analogue [G.A. Schoenenberger, M. Monnier, *Proc. Natl. Acad. Sci. USA* **1977**, *74*, 1282; G.A. Schoenenberger, *Eur. Neurol.* **1984**, *23*, 321; M.V. Graf, A.J. Kastin, *Peptides* **1986**, 1165].

Deltorphins, 7-peptide amides isolated from the skin of the South American frog *Phyllomedusa bicolor.* Deltorphins contain D-Met or D-Ala, respectively, in position 2, and together with → dermorphin they are the first naturally occurring regulatory peptides that contain D-amino acids. *Deltorphin (DT, Dermenkephalin)* H-Tyr-D-Met-Phe-His-Leu-Met-Asp-NH_2, is released from a biosynthetic precursor additionally containing four copies of dermorphin. A propeptide from *Phyllomedusa bicolor* contains three copies of *[D-Ala^2]deltorphin I*, H-Tyr^1-D-Ala-Phe-Asp-Val^5-Val-Gly-NH_2, and one copy of *[D-Ala^2]deltorphin II*, H-Tyr^1-D-Ala-Phe-Glu-Val^5-Val-Gly-NH_2. However, in all cases the D-amino acids in position 2 are formed in a post-translational reaction. In contrast to dermorphin, DT binds to δ-receptors. The varying selectivity for the opioid receptors can be explained by charge effects and differences in the hydrophobicity of the C-terminal part of the peptides [A. Mor et al., *FEBS Lett.* **1989**, *255*, 2690; K. Richter et al., *Proc. Natl. Acad. Sci. USA* **1990**, *87*, 4836].

Depsipeptides, heterodetic peptides containing peptide bonds as well as ester bonds. Depsipeptides include homomeric *O-peptides* and *peptide lactones* (→ actinomycins, → etamycin) of hydroxyamino acids such as Ser, Thr, etc., and heteromeric peptides characterized by the replacement of some of the amino acids by hydroxy acids, also called *peptolides.* Most of the depsipeptides are heterodetic cyclic peptides containing alternating residues of α-amino and α-hydroxy acids, e.g., the fungal antibiotic → valinomycin. A highly symmetric cyclodepsipeptide containing α- and β-amino and hydroxy acids is → onchidin B. Another type of cyclic depsipeptide are the quinomycins known as DNA bisintercalators with potent antitumor activity. Most depsipeptides are metabolic products of microorganisms with usually strong antibiotic effects [B. Dietrich, P. Viout, J.-M. Lehn, *Macrocyclic Chemistry*, VCH, Weinheim, **1993**; O. Kuisle et al., *J. Org. Chem.* **1999**, *64*, 8063].

Dermatoxin, SLGSFLKGVG^{10}TTLASVGKV V^{20}SDQFGKLLQA^{30}GQ, a 32-peptide isolated from the skin of a single specimen of the tree frog *Phyllomedusa bicolor.* Dermatoxin proved to be bactericidal towards mollicutes (wall-less eubacteria) and Gram-positive eubacteria, and, additionally, to a lesser extent, towards Gram-negative eubacteria [M. Amiche et al., *Eur. J. Biochem.* **2000**, *267*, 4583].

Dermorphins, besides the → deltorphins the first examples of D-amino acid containing peptides found in animals. *Dermorphin*, H-Tyr^1-D-Ala-Phe-Gly-Tyr^5-Pro-Ser-NH$_2$, was first isolated in 1981 from the skin of the South American frog *Phyllomedusa sauvagei*. A similar peptide, *[Hyp^6]dermorphin*, was discovered in the skin of *Phyllomedusa rhodai* and *Phyllomedusa burmeisteri*. Dermorphin is released from a biosynthetic precursor containing several copies of dermorphin with the L-isomer only. The conversion of L-Ala into D-Ala occurs post-translationally. Dermorphin binds preferably to the μ-receptors. In the guinea pig ileum test, dermorphin is 57 times as active as Met-enkephalin, and when given intravenously it shows a significantly higher analgesic activity than morphine. With *[Lys^7]dermorphin*, *[Trp^4,Asn^7] dermorphin*, and *[Trp^4,Asn^5]dermorphin-(1–5)*, three additional dermorphin analogues were found in the skin of *Phyllomedusa bicolor.* The dermorphins are potent analgesics in rodents and primates, including humans [P.C. Montecuchi et al., *Int. J. Peptide Protein Res.* **1981**, *17*, 275; P. Melchiorri, L. Negri, *Gen. Pharmacol.* **1996**, *27*, 1099].

Diazepam-binding inhibitor peptide, *DBIP*, *anxiety peptide*, QATVGDVNTD^{10}RPGLLDLK, a 18-peptide isolated from human and rat brain extracts besides other neuroactive peptides. DBIP is formed by tryptic digestion of the diazepam-binding inhibitor (DBI). It interacts with the benzodiazepine receptor and appears to increase anxiety [P. Ferrero et al., *Neuropharmacology* **1984**, *23*, 1359; P.W. Gray et al., *Proc. Natl. Acad. Sci. USA* **1986**, *83*, 7547; P. De Stefanis et al., *Regul. Peptides* **1995**, *56*, 153].

Diptericin, a antibacterial 82-peptide carrying two Gal-GalNAc threonine side-chain substitutions isolated from the larvae of the dipteran *Phormia terranovae*. Synthetic unglycosylated diptericin was found to possess antibacterial activity against a wide range of Gram-negative bacteria. Unlike various other fast-acting antibacterial peptides, diptericin does not lyze normal mammalian cells [J.-L. Dimarcq et al., *Eur. J. Biochem.* **1988**, *171*, 17; P. Bulet et al., *Biochemistry* **1995**, *34*, 7394; M. Cudic et al., *Eur. J. Biochem.* **1999**, *266*, 549].

DnaK (*hsp70*, *BiP*), → molecular chaperones.

Dolichin, an antifungal protein (M_r ~ 28 kDa) isolated from the seeds of the field bean *Dolichos lablab*. It shows structural resemblance to chitinases which also exhibit antifungal activity. Dolichin shows antifungal activity against *Rhizoctonia solani, Fusarium oxysporum*, and *Coprinus comatus*. Furthermore, dolichin is able to inhibit HIV reverse transcriptase, and α- and β-glucosidases associated with HIV infection [X.Y. Ye et al., *Biochem. Biophys. Res. Commun.* **2000**, *269*, 155].

Drm-NPF peptide, SNSRPPRKND^{10}VNTM ADAYKF^{20}LQDLDTYYGD^{30}RARVRFa, a 36-peptide amide from the fruitfly related to the → neuropeptide Y family. It is expressed in larval and adult brain and gut [M. R. Brown et al., *Peptides* **1999**, *20*, 1035].

Dromyosuppressin, *DMS*, → FMRFamide-related peptides.

Drosomycin, a 44-peptide produced from *Drosophila melanogaster* that belongs to the insect → defensins. It shows significant homology with plant antifungal peptides from seeds of members of the family *Brassicaceae* [P. Fehlbaum et al., *J. Biol. Chem.* **1994**, *269*, 33159].

Drosulfakinins, *DSK*, → FMRFamide-related peptides.

Duramycins, analogues of B-type lantibiotics of the prototype → cinnamycin, exemplified by duramycin, duramycin B and duramycin C.

Dynorphins, *Dyn*, opioid peptides released from the precursor protein prepro-dynorphin, (prepro-enkephalin B). Besides various dynorphin sequences, prepro-D. (p-D) comprises also the sequences of the → neoendorphins and three copies of [Leu] enkephalin. Porcine *Big-D, p-D-(209–256)*, YGGFLRRIRP^{10}KLKWDNQKRY^{20}GGFLRR QFKV^{30}VTRSQEDPNA^{40}YYEELFDV, is an 48-peptide containing also the sequences of the shorter dynorphins, such as *dynorphin-A*, p-D-(209–225), YGGFLRRIRP^{10}KLK WDNQ, *dynorphin B, rimorphin*, p-D-(228–240), YGGFLRRQFK^{10}VVT, and *leumorphin*, p-D-(228–256), YGGFLRRQFK^{10}VVTR SQEDPN^{20}AYYEELFDV. The longer dynorphins were first detected in the hypothalamus, hypophysis, and duodenum of the pig. Later, they were isolated from adrenal medulla, guinea pig heart, and rat duodenum [A. Goldstein et al., *Proc. Natl. Acad. Sci. USA* **1979**, *76*, 6666; S. Tachibana et al., *Nature* **1982**, *295*, 339; D. L. Kilpatrick et al., *Proc. Natl. Acad. Sci. USA* **1982**, *79*, 6480].

E

Ecdysis-triggering hormones, *ETH*, peptide hormones involved in the ecdysis motor behavior that leads to the shedding of the old cuticle at the end of each molt. This pathway was first elucidated in the moth *Manduca sexta. Mas-ETH*, SNEAISPFDQ^{10} GMMGYVIKTN^{20}KNIPRMa, a 26-peptide hormone triggers ecdysis in developing insects. Besides the → crustacean cardioactive peptide (CCAP), Drm-ETH and the → eclosion hormone appear to be involved in this neuronal pathway. After gene encoding of the precursor for ETH-like peptides in *Drosophila*, both *Drm-ETH-1*, DDSSPGF FLK^{10}ITKNVPRLa, and *Drm-ETH-2*, GENF AIKNLK^{10}TIPRIa, have been found to induce premature eclosion after injection [D. Zitnan et al., *Science* **1996**, *271*, 88; Y. Park et al., *FEBS Lett.* **1999**, *463*, 133; J. V. Broeck, *Peptides* **2001**, *22*, 241].

Echistatin, ECESGPCCRN^{10}CKFLKEGTIC^{20} KRARGDDMDD^{30}YCNGKTCDCP^{40}RNPH KGPAT (disulfide bonds: C^2–C^{11}/C^7–C^{32}/ C^8–C^{37}/C^{20}–C^{39}), a RGD-containing 49-peptide originally isolated from the venom of the viper *Echis carinatus*. It has been shown that echistatin interacts with the osteoclast $\alpha_v\beta_3$ integrin receptor, and was capable of blocking osteoclast attachment to bone. Furthermore, echistatin inhibits bone reabsorption *in vitro*. Structure-activity relationship studies led to the conclusion that shorter and more potent antagonists of the β_3 integrin receptor may find clinical application in the blockade of bone reabsorption [M. Sato et al., *J. Bone Miner. Res.* **1994**, *9*, 1441].

Eclosion hormone, *EH*, a polypeptide produced in the nervous system and involved in eclosion behavior. The *Drosophila* gene has been characterized and the predicted neuropeptide sequence (80 aa) shows similarity to lepidopteran EH. EH is localized in two pairs of neurons designated VM cells in the cerebral ganglia. The neuronal pathway leading to ecdysis behavior is induced both by EH and the → ecdysis-triggering hormone [F.M. Horodyski et al., *Eur. J. Biochem.* **1993**, *215*, 221].

Edestin, a hexameric protein (M_r ~ 300 kDa) from hemp seeds (*Cannabis sativa*) belonging to the → globulins. Each of the six subunit consists of two nonidentical peptide chains linked by disulfide bridges.

EGF family, *epidermal growth factor family*, a member of the gastroenteropancreatic peptide families. This family comprises → epidermal growth factor (EGF), → transforming growth factor-α (TGF-α), amphiregulin (AR), heparin binding EGF-like growth factor (HB-EGF), betacellulin (BTC), neuroregulin 1, neuroregulin 2, neuroregulin 3, and → epiregulin (ER). The members of this family share six conserved Cys residues, have one or more EGF domains and a transmembrane domain. According to their receptor affinities, the members of the EGF family are divided into two groups. EGF, HB-EGF, TGF-α, AR, BTC and ER bind to ErbB1, whereas HB-EGF, ER, and BTC are bind to ErbB4 as well as to ErbB1. The remaining members are ligands for ErbB3 and ErbB4 [D. Zhang et al., *Proc. Natl. Acad. Sci. USA* **1997**, *94*, 9562].

Egg-laying hormone, ***ELH***, ISINQDLKAI^{10} TDMLLTEQIR^{20}ERQRYLADLR^{30}QRLLEKa, a 36-peptide amide from *Aplysia california*. ELH acts as a hormone facilitating the expulsion of the egg string, and as a neurotransmitter on cells of the abdominal ganglion [R.H. Scheller et al., *Cell* **1983**, *32*, 7].

Eglin c, a single-chain 70-peptide isolated from the leech *Hirudo medicinalis*. It belongs to the eglin family of protease inhibitors and inhibits effectively subtilisin, chymotrypsin, cathepsin G and leukocyte elastase. The potential therapeutic value of eglin c might lie in inflammatory processes [J. Dodt et al., *Biol. Chem. Hoppe Seyler* **1987**, *368*, 1447].

Eisenin, pGlu-Gln-Ala-OH, a tripeptide with immunological activity from the Japanese marine alga *Eisenia bicyclis Setchell*. It has importance in augmentation of natural cytotoxicity of peripheral blood lymphocytes in humans [T. Kojima et al., *J. Immunother.* **1993**, *13*, 36].

Elastin, a structural protein with rubber-like elastic properties. It is the main component of the elastic yellow connective tissue occurring, e.g., in the lungs and aorta. The amount of elastin is rather low in the inelastic white connective tissue of tendons. Elastin consists of 850–870 residues with a high content of Gly (27%), Ala (23%), Val (17%), and Pro (12%). It forms a three-dimensional network of fibers cross-linked by desmosine, lysinonorleucine, and isodesmosine [L. Robert, W. Hornebeck (Eds.), *Elastin and Elastases*, Volume 1, CRC Press, Boca Raton, FL, **1989**; D.R. Eyre et al., *Annu. Rev. Biochem.* **1984**, *53*, 717].

Elcatonin, an analogue of eel → calcitonin.

Eledoisin, <EPSKDAFIGL^{10}Ma, an 11-peptide amide from the salivary glands of the cephalopod (*Eledone moschata* and *E. aldrovandi*) belonging to the → tachykinin family. Its biological activities *in vitro* are similar to those of → substance P [V. Erspamer, A. Anastasi, *Experientia* **1962**, *18*, 58].

Enalapril, (*S*)-1-[*N*-(ethoxycarbonyl)-3-phenylpropyl]-L-alanyl-L-proline, an orally active inhibitor of the → angiotensin-converting enzyme (ACE). After oral administration, the prodrug enalapril is bioactivated by hydrolysis of the ethyl ester to yield enalaprilat, which is one of the most active → ACE inhibitors. Enalapril causes a drop in blood pressure, as does → captopril, in hypertensive patients.

Endomorphins, H-Tyr-Pro-Xaa-Phe-NH_2, 4-peptide amides with the highest known affinity and specificity for the μ-opiate receptor. *Endomorphin-1* (Xaa = Trp) shows 4000- and 15 000-fold preference over the δ- and κ-receptors, respectively, and may be a natural ligand for this receptor. *Endomorphin-2* (Xaa = Phe) shows potent μ-selective activity including analgesia [J. E. Zadina et al., *Nature* **1997**, *386*, 499; H. C. Champion et al., *Peptides* **1997**, *18*, 1393].

Endorphins, *opioid peptides, opiate-like peptides*, endogenous peptides with morphine-like activity. The name is related to *endo*genous mor*phine*. These morphinomimetic peptides are the natural ligands of the opiate receptor. The first discovered endogenous peptide with morphine-like activity was → enkephalin. Later partial fragments of → β-lipotropin, e.g. *α-endorphin*, YGGF MTSEKS^{10}QTPLVT, *β-endorphin*, TGGFMT SEKS^{10}QTPLVTLFKN^{20}AIIKNAYKKG^{30}E, *γ-endorphin*, YGGFMTSEKS^{10}QTPLVTL, and *δ-endorphin*, YGGFMTSEKS^{10}QTPLVT LFKN^{20}AIIKNAY, were isolated from pituitary material. Endorphins occur in the CNS, kidneys, cerebrospinal fluid, blood, ganglia of the intestinal tract, placenta and pituitary. All endorphins bear a common initial sequence, which corresponds to the primary structure of [Met]enkephalin. β-Endorphin is derived from → pro-opiomelanocortin (POMC), and is secreted in the hypophysis in response to stress and other stimuli. It has analgesic and lipolytic activity. The analgesic effect in the body is based on the secretion of β-endorphin. Acupuncture, for example, activates the central nervous endorphin system to cause an increase in the concentration of endorphins, and this leads to the elimination of sensitivity to pain. The hope that endorphins would allow analgesia to be separated from the development of addiction and dependence has not yet been realized. The term endorphin is used, in the narrower sense, not for all peptides with opiate-like effects, but preferentially for fragments of β-lipotropin. Therefore, → α-neoendorphin, → dynorphin, and enkephalin would not fit into this narrower definition [R. Guillemin et al., *C.R. Acad. Sci. (Paris)* **1976**, *282*, 783; L. Lazarus et al., *Proc. Natl. Acad. Sci. USA* **1976**, *73*, 2156; C. H. Li in: *Central and Peripheral Endorphins: Basis and Clinical Aspects*, E. Mueller, A. R. Genazzani (Eds.), p. 17, Raven Press, New York, **1984**].

Endothelin, ***ET***, human *hET-1*: CSCSSLMD KE^{10}CVYFCHLDII^{20}W (disulfide bonds: C^1–C^{15}/C^3–C^{11}), a strongly vasoactive peptide formed in the endothelial cells. Besides ET-1 first detected in the culture of endothelial cells, two other endothelins were found in humans after subsequent analysis of the endothelin encoding genes, named *ET-2* ([Trp^6,Leu^7]ET-1) and *ET-3* ([Thr^{2,5},Phe^4,Tyr^{6,14},Lys^7]ET-1), respectively. Only ET-1 is formed in endothelial cells, whereas ET-2 and ET-3 are released in other tissues, e.g., brain, kidney, suprarenal glands, and intestines. The biosynthetic precursor of hET-1, named *prepro-ET-1*, consists of 212 residues and provides after proteolytic processing the *big h-ET-1* which consists of 38 residues bearing the ET-1 sequence in the N-terminal part of the molecule. Big ET-1 shows 100 times lower in-vitro activity compared to ET-1. In human brain, big hET-1 is specifically converted by the endothelin-converting enzyme (ECE) into the fully active peptide by cleavage of the Trp^{21}-Val^{22} bond. All types of smooth muscle are contracted by ETs via activation of specific receptors. They can also relax the muscles, in some cases either by direct receptor activation or indirectly by releasing *endothelium-derived relaxing factors*, such as NO or eicosanoids. The receptors selective for ET-1 were designated ET_A with agonist potency ET-1 > ET-2 ≫ ET-3, whereas the nonselective ones (ET-1 = ET-2 = ET-3) were named ET_B. Generally, the ET_A receptors reside on smooth muscle cells and mediate vasoconstrictor responses, whereas the endothelial cells express the ET_B receptors. The latter mediate vasodilatory effects via the ET-induced release of endothelium-derived relaxing factors. Despite the low plasma concentration of ET, it behaves as a circulating mediator and can locally influence cells which are close to the ET-producing tissues. ET modulates chronotropy, ionotropy, bronchoconstriction, and neurotransmission. Furthermore, they act as regulators for other hormones and neurotransmitters. Intravenous application results in regulation of blood pressure (both pressor and depressor effect), local blood flow, and kidney function. The sequence of hET-1 is identical with p-ET-1, originally isolated from porcine aorta, and with dog, rat, and mouse ET-1. Furthermore, hET-2 is identical in sequence with dog ET-2, and hET-3 with rat ET-3, respectively. Structure similarities exist also between ET and the efficient vasoactive → sarafotoxins (SRTX) from the venom of the Israel snake *Atracta engadensis*, and other peptide neurotoxins, such as the bee venom → apamin or α-scorpion toxin. Furthermore, the vasoactive intestinal contractor, VIC, known as endothelin-β, belongs to the ET-SRTX group. ET shows high affinity to receptors and is the strongest vasoconstrictor presently known. From the properties of ET it can be concluded that it may also play a role in chronic diseases, such as hypertension. For this reason, the development of suitable ET antagonists for medical applications is of major importance. A cyclic pentapeptide isolated from the cultured broth of *Streptomyces misakiensis*, cyclo-(-D-Trp-D-Glu-Ala-D-Val-Leu-) has been found to be a potent endothelin antagonist with the selectivity for the ET_A receptor subtype. Starting from this natural lead compound, a synthetic analogue (BQ-123) has been obtained where the inhibitory potency is improved by more than two orders of magnitude. Human chymase specifically converts big ETs to novel smooth muscle-contracting 31-peptide ETs that may play a role in allergies and vascular diseases [M. Yanagisawa et al., *Nature* **1988**, *332*, 411; A. M. Doherty et al., *J. Med. Chem.* **1993**, *36*, 2587; K. Ishikawa et al., *J. Med. Chem.* **1992**, *35*, 2139; J. Hlavacek, R. Marcova, *Collect. Czech. Chem. Commun.* **1999**, *64*, 1211; H. Kido et al., *Biol. Chem.* **1998**, *379*, 885].

Endothelin-β, designation for the *vasoactive intestinal contractor* (*VIC*), a member belonging to the group of → endothelin, and → sarafotoxins.

Enkephalins, ***ENK***, naturally occurring 5-peptides acting as the endogenous ligands for the opiate receptor in the brain. *Methionine-enkephalin* ([Met]enkephalin), H-Tyr-Gly-Gly-Phe-Met-OH, and *leucine-enkephalin* ([Leu]enkephalin), H-Tyr-Gly-Gly-Phe-Leu-OH, are released by limited proteolysis from the precursor proteins prepro-enkephalin, that contains six copies of Met-enkephalin and one copy of Leu-enkephalin, and prepro-dynorphin (also named prepro-enkephalin B) which contains only three copies of Leu-enkephalin, but also → dynorphin and → neoendorphin. Enkephalin acts both as neurotransmitter and neuromodulator. Many analogues have been synthesized in an attempt to develop new anodynes that would be more potent and less addictive (→ endorphins) [J. Hughes et al., *Nature* **1975**, *258*, 577; M. Comb et al., *Nature* **1982**, *295*, 663].

Enniatins, cyclic depsipeptide antibiotics produced by strains of *Fusarium*. Important members are *enniatin A*, cyclo-(-D-Hyv-MeIle-)$_3$, *enniatin B*, cyclo-(-D-Hyv-Me-Val-)$_3$, and *enniatin C*, cyclo-(-D-Hyv-Me-Leu)$_3$ which all contain D-α-hydroxyisovaleric acid (D-Hyv) besides L-methylamino acids. Enniatins act as ionophores to form complexes with K^+ ions which allow their transport across membranes.

ENOD40 peptides, oligopeptides with 10 to 13 residues exhibiting mitogenic activity and acting as plant growth regulators. They are encoded by the *enod40* gene and characterized by a central Trp residue and the C-terminal sequence -Ile-His-Gly-Ser-OH. For example, the ENOD40 peptide of the pea (*Pisum sativum*) has the sequence: MKFLCWQKSI^{10}HGS. These peptides are believed to affect cell division by modulation of cellular responses to auxin [A. Schaller, *Plant Mol. Biol.* **1999**, *40*, 763].

Epidermal growth factor, ***h-EGF***, *urogastron*, ESYPGCPSSY^{10}DGYCLNGGVC^{20}MHIESL DSYT^{30}CNCVIGYSGD^{40}RCQTRDLRWW^{50} ELR (disulfide bonds: C^6–C^{20}/C^{14}–C^{31}/C^{33}–C^{42}), a hormone-like polypeptide growth factor stimulating cell proliferation. EGF is an effective mitogen for a wide variety of epidermal and epithelial cells *in vivo* and *in vitro*. A homologous polypeptide was isolated from human urine and named urogastron; this may play a role in the maintenance of gastrointestinal homeostasis *in vivo*. EGF is also used in wound healing. EGF from mouse submaxillary gland forms a complex that contains two molecules of EGF and two molecules of EGF binding protein (M_r ~ 29 kDa). Mouse EGF is first synthesized as a large precursor protein (1168 aa). Analogously to many other protein growth factors, EGF stimulates the proliferation and differentiation of their target cells by binding to their receptor tyrosine kinase consisting of a C-terminal cytoplasmic tyrosine kinase domain, a helical single-pass transmembrane segment and an extracellular cysteine-rich domain [S. Cohen, *J. Biol. Chem.* **1962**, *237*, 1555; H. Gregory, *Nature* **1975**, *257*, 325; J.A. Downie et al., *Annu. Rev. Biochem.* **1979**, *48*, 103; S. Tomic et al., *J. Biol. Chem.* **1995**, 21277].

Epidermin, a member of the Type-A family of → lantibiotics. Epidermin is a tetracyclic peptide (M_r ~ 2.2 kDa) isolated from *Staphylococcus epidermidis* with antimicrobial activity against *Propionibacterium acnes*, which is one of the causative factors of acne. Epidermin and its analogues, e.g. [Leu^6]epidermin (*gallidermin*), [Phe^1,Lys^2,Trp^4,Dha^5,Phe^6]epidermin (*mutacin B-Ny266*) are much shorter and differ at the C-terminus, but their 12 N-terminal residues are highly related to → nisin and → subtilin. Gallidermin produced by a *Staphylococcus gallinarum* strain has similar properties to epidermin, but shows slightly better antimicrobial activity against *Propionibacterium acnes*. Therefore, it is an interesting therapeutic agent for the treatment of acne [R.W. Jack, G. Bierbaum, H.-G. Stahl, *Lantibiotics and Related Peptides*, Springer, Berlin, Heidelberg, New York, **1998**].

Epiregulin, ***ER***, a single-chain 46-peptide belonging to the → EGF family. Epiregulin is one of the most recently described autocrine growth factors of this family found in normal human keratinocytes. It organizes epidermal structure by regulating keratinocyte proliferation and differentiation as well as the expression of → transforming growth factor-α (TGF-α), heparin-binding EGF-like factor (HB-EGF), and amphiregulin (AR) [Y. Shirakata et al., *J. Biol. Chem.* **2000**, *275*, 5748].

Erythropoietin, ***EPO***, a monomeric glycoprotein that stimulates the production and release of erythrocytes. It is synthesized in the kidney and liver of adults, but only in the liver of fetuses and neonates. A lack of oxygen or a shortage of red blood cells stimulates synthesis and secretion of EPO into the bloodstream. EPO completes the action of the classical → colony-stimulating factors. Human EPO consists of 165 residues (M_r ~ 30 kDa with carbohydrate), including four Cys and three potential glycosylation sites. The human gene is located on chromosome 7. EPO is quickly degraded and excreted by the kidneys. EPO is used in the treatment of anemia arising from kidney disease, and misused in doping. It is synthesized in large quantities by recombinant technology [K. Jacobs et al., *Nature* **1985**, *313*, 806; H. Pagel, W. Jelkmann, *Dtsch. Med. Wschr.* **1989**, *114*, 957].

Etamycin, *Viridogrisein*, Hypic-Thr-D-Leu-D-aHyp-Sar-Me_2Leu-Ala-PhSar-OH (lactone ring between Thr and PheSar), a heptapeptide lactone antibiotic (→ depsipeptides) produced by streptomycetes. The α-amino function of Thr is blocked by the hydroxypipecolic acid (Hypic), and further unusual building blocks are L-β-*N*-dimethylleucine (Me_2Leu), sarcosine (Sar), L-α-phenylsarcosine (PhSar), and D-*allo*-hydroxyproline (D-aHyp). Etamycin is active against Gram-positive bacteria and *Mycobacterium tuberculosis.*

Exendins, peptides isolated from the venom of the lizard (*Heloderma suspectum* and *H. horridum*) that are members of the secretin-glucagon-VIP family. *Exendin-4* (*H. suspectum*), HGEGTFTSDL^{10}SKQMEEE AVR^{20}LFIEWLKNGG^{30}PSSGAPPPSa, is a 39-peptide amide, and *exendin-3* (*H. horridum*) differs only in two positions (S^2 and D^3) from exendin-4. Exendin-4 and → glucagon-like peptide (GLP)-1 (7–36) amide bind with similar affinity to the GLP-1 receptor. N-terminal shortened sequences like exendin-(3–39), exendin-(5–39) and exendin-(9–39) are potent GLP-1 receptor antagonists [J. Eng et al., *J. Biol. Chem.* **1992**, *267*, 7402; H.C. Fehmann et al., Peptides **1994**, *15*, 453; C. Montrose-Rafizadeh et al., *J. Biol. Chem.* **1997**, *272*, 21201].

Exorphins, protein-derived opioid peptides of exogenous origin. They are mainly produced from food proteins during digestion. Especially, milk proteins, such as α-casein, β-casein, κ-casein, α-lactoglobulin, β-lactoglobulin and lactoferrin contain partial fragments that behave like opioid receptor ligands under in-vitro and in-vivo conditions (→ milk protein-derived opioid peptides). The most important members of this group are the → β-casomorphins. *Gluten-exorphins* are opioid peptides released from wheat gluten pepsin-elastase digest. The sequence of *gluten-exorphin A5*, H-Gly-Tyr-Tyr-Pro-Thr-OH, occurs 15 times in the primary structure of wheat gluten. The A5 peptide is selective for δ-receptors. After oral administration in rats, A5 stimulates postprandial insulin release, and this effect could be reversed by the opioid antagonist naloxone. Besides gluten exorphin A5, *A4* (H-Gly-Tyr-Tyr-Pro-OH), *B5* (H-Tyr-Gly-Gly-Trp-Leu-OH), and *B4* (H-Tyr-Gly-Gly-Trp-OH) are further gluten-exorphins. *Hemorphins* are peptides with opioid-like activity found after proteolytic treatment of hemoglobin. In the early 1990s it became clear that hemoglobin also serves *in vivo* as a source of bioactive peptides that might play a role in homeostasis. From this fact, a so-called tissue-specific peptide pool has been formulated which should be a novel system of peptidergic regulation besides hormonal and neuromodulatory systems [H. Teschemacher et al., *Inc. Biopoly.* **1997**, *43*, 99; S. Fukudome et al., *FEBS Lett.* **1997**, *412*, 475; V.T. Ivanov et al., *Inc. Biopoly.* **1997**, *43*, 171].

F

Farnesylated peptides, N- or S-substituted peptides with the farnesyl (Farn) moiety. Based on the hydrophobicity of this C_{15} unit isoprenoid farnesylated peptides are characterized by increased bioavailability. Farnesylated peptides are useful building blocks for the synthesis of special biological/pharmaceutical model targets, since the farnesyl group in proteins may participate in specific protein-protein interactions. In vivo, farnesylation is performed by the enzyme farnesyltransferase [G. Byk, D. Scherman, *Int. J. Peptide Protein Res.* **1996**, *47*, 333; D.M. Leonard, *J. Med. Chem.* **1997**, *40*, 2971; B. Bader et al., *Nature* **2000**, *403*, 223].

Ferredoxins, ***Fd***, iron-sulfur proteins involved in electron transfer processes. Highly homologous *2Fe-Fd* with 96 to 98 residues including four to six Cys occur in blue-green bacteria, higher plants and green algae. *4Fe-Fd* contain a single 4Fe-4S cluster as an active center. The iron atoms are bound to the only four Cys residues in the protein. These Fd have been found primarily in bacteria such as *Desulfovibrio gigas*. *8Fe-Fd* contain two identical 4Fe-4S clusters. Each of these clusters forms a cube and is covalently linked to four Cys residues in the peptide chain, which consists of about 55 residues including eight Cys. This type of Fd takes part in many electron transport processes in *Clostridia* and photosynthetic bacteria. A special type of Fd is the *high potential iron-sulfur protein* (*HiPIP*) isolated from some photosynthetic bacteria. It contains a single 4Fe-4S cluster and is characterized by a positive standard potential of about +350 mV; this is in contrast to the other Fd, which have standard potentials of about –420 mV.

Ferritin, an intracellular iron storage protein in mammals. Iron-free apoferritin consists of 24 partly carbohydrate-containing subunits (M_r ~ 450 kDa). The shell-like ferritin, with a diameter of 12 nm, forms a chamber with a diameter of 7 nm in which 4300 Fe(II) atoms can be deposited as iron-hydroxide-oxide micelles. Ferritin is the major storage depot for iron, and contains (together with the related → hemosiderin) 25% of the iron of the body.

α-Fetoprotein, ***AFP***, a glycoprotein (M_r ~ 70 kDa) belonging to the α-globulins that is formed in the serum of mammalian embryos (0.1 mg mL^{-1}) but occurs in serum of adults only in traces (< μg mL^{-1}). Synthesis is performed first by the yolk sac and later by the fetal liver. At birth, AFP is replaced by serum albumin. A higher concentration of AFP in adult serum detectable by radioimmunoassay is symptomatic of liver carcinoma [F. Jacob, *Ciba Foundation Symposium 96: Fetal Antigens and Cancer*, p. 4, Pitman, London, **1983**].

Fibrin, a globular protein belonging to the plasma proteins involved as main component in blood coagulation. Fibrin is formed from the soluble plasma protein → fibrinogen via a proteolytic reaction catalyzed by → thrombin. It aggregates spontaneously to form fibers with a banded structure that repeats every 225 Å. The rather fragile "soft clot" is converted to a "hard clot" by cross-linking of fibrin molecules catalyzed by the transamidase termed *fibrin-stabilizing factor* (*FSF* or XIII_a). This process is triggered by binding of Ca^{2+} and activated by thrombin. The blood clots consist of cross-linked fibrin forming an insoluble fibrous network which provides a stable wound closure [R. F. Doolittle, *Annu. Rev. Biochem.* **1984**, *53*, 195].

Fibrinogen, ***Fbg***, *factor I*, the only coagulable protein in the blood plasma consisting of three pairs of nonidentical but homologous peptide chains $[(A\alpha)_2(B\beta)_2\gamma_2]$ cross-linked by disulfide bonds and two pairs of N-linked oligosaccharides (***Formula***). A and B represent the → fibrinopeptides in the subunits Aα (610 aa) and Bβ (461 aa) whereas the γ-polypeptide chain contains 411 residues. The six peptide chains of fibrinogen are linked by 17 disulfide bonds, seven within each halve of the dimer and three joining these two protomers. The conversion of fibrinogen to → fibrin starts first by thrombin cleavage of the fibrinopeptide A from the Aα-chains, yielding fibrin I monomers. The second cleavage in the Bβ-chain is performed after the fibrin I monomers have polymerized end-to-end to form protofibrils. Fibrinogen occurs in the blood plasma of vertebrates and some arthropods. Human blood contains 2–3 gL^{-1} of fibrinogen [R. F. Doolittle, *Sci. Am.* **1981**, *245*, 126; R. F. A. Zwaal, H. C. Henker (Eds.), *Blood Coagulation*, Elsevier, Amsterdam, **1986**].

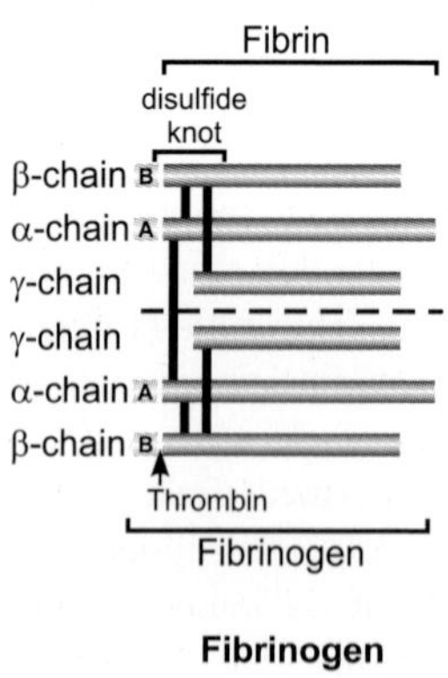

Fibrinogen

Fibrinopeptides, peptides proteolytically released from → fibrinogen during the conversion to → fibrin in the blood clotting process. With *fibrinopeptide A* (human), ADSGEGDFLA[10]EGGGVA, and *fibrinopeptide B* (human), <EEGVNDNEEG[10]FFSAR, two pairs of peptides exist that are cleaved by thrombin from the N-termini of the α and β chains of fibrinogen. Fibrinopeptide A ranges in the size from 14 to 21 residues depending on the species. The fibrinopeptides show vasoconstrictive activity. Fibrinopeptides gained importance by establishing a detailed phylogenetic tree for mammals.

Fibroblast growth factor, ***FGF***, *heparin binding growth factor*, ***HBGF***, a family of polypeptide growth factors belonging to the → cytokines. They are formed in various animal tissues such as the brain, hypophysis, placenta, retina, and thymus, and mediate as mitogens the growth of many mesodermal and neuroectodermal cells. Furthermore, FGF are involved in many differentiation processes. *Basic FGF* (M_r ~ 16 kDa; IP = 9.6) promotes for example the differentiation of PC-12 cells, increases the transcription of the c-*myc* and c-*fos* oncogenes, and promotes angiogenesis. *Acidic FGF* (M_r ~ 14.5 kDa; IP = 5–5.9) from brain and retina mediates via the same receptors almost the same biological actions as the basic FGF. Both FGF bind heparin, but only the complex with acidic FGF increases the biological activity. FGF show sequence similarity with IL-1 (→ interleukins). One member of the FGF family is the → keratinocyte growth factor.

Fibronectin, ***FN***, an extracellular adhesive glycoprotein mediating cell-matrix adhesion. FN is a dimer consisting of two similar (but not identical) subunits, each with almost 2500 residues. The subunits are joined together by a pair of disulfide bonds located near their C-termini. Each polypeptide chain is folded into a series of globular domains connected by flexible sequence segments. Special groups of these domains are responsible for binding of heparin, cells, and collagen. Analysis of the cell-binding domain led to the discovery of a specific tripeptide sequence (*Arg-Gly-Asp* or *RGD*) that is responsible for the cell-binding activity. The RGD sequence is a common motif in a variety of extracellular adhesive proteins, and is recognized by cell-surface receptors that bind the specific protein. Besides the RGD sequence, receptor binding may also depend on other parts of the sequence of the adhesive molecule. The RGD sequence is the most important site for recognition by a cell surface FN receptor, the $\alpha_5\beta_1$ integrin heterodimer. Further cell surface receptors of FN are heparan sulfate and chondroitin sulfate heterodimers, and glycolipids. The → integrins are the best characterized FN receptors in terms of binding specificity and functional roles. FN occurs in three different forms. The soluble dimeric form, termed *plasma FN*, circulates in the blood and other body fluids enhancing wound healing, blood clotting, and phagocytosis. *Cell-surface-FN* formed from oligomers of FN occurs transiently attached to the surface of cells. *Matrix FN* formed from highly insoluble FN fibrils is found in the extracellular matrix. In the last two FN forms, additional disulfide bonds stabilize these structures by cross-linking. In addition to its function for cell adhesion, FN is also important for cell migration [R.O. Hynes, *Sci. Am.* **1986**, *254*, 42; R.O. Hynes, K.M. Yamada, *J. Cell. Biol.* **1982**, *95*, 369; L.M. Schnapp et al., *J. Biol. Chem.* **1995**, *270*, 23196].

FK506 binding proteins, ***FKBP***, a family of the → peptidyl prolyl *cis/trans* isomerases (PPIases).

FMRFamide, H-Phe-Met-Arg-Phe-NH_2, FMRFa, the molluskan cardioexcitatory tetrapeptide amide. A variety of insect neuropeptides display C-terminal sequence similarities to FMRFa → FMRFamide-related peptides [D.A. Price, M.J. Greenberg, *Science* **1977**, *197*, 670].

FMRFamide-related peptides, ***FaRP***, insect neuropeptides with C-terminal sequence similarities to → FMRFamide. FaRP play important roles as neurohormones and as neuro- and myomodulators. It could be established that, for example, the majority of putative prepro-FMRFa gene-encoded peptides strongly enhanced twitch tension of larval body-wall muscles. In 1987, the first *Drosophila melanogaster* FaRP-gene was discovered and at the same time the 9-peptide amide, H-Asp-Pro-Lys-Gln-Asp-Phe-Met-Arg-Phe-NH_2, was isolated from adult fruitflies. The FMRFa precursor contains multiple, putative peptide sequences. For example, the 9-peptide sequence mentioned above occurs five times, and the sequence -Thr-Pro-Ala-Glu-Asp-Phe-Met-Arg-Phe- twice, in the precursor protein besides a couple of similar peptides. The *drosulfakinins* (*DSK*) show sequence similarity to other FaRPs and to gastrin- and CCK-like peptides. From the fruitfly sulfakinin peptides *DSK-I*, H-Phe-Asp-Asp-*Tyr*-Gly-His-Met-Arg-Phe-NH_2, and *DSK-II*, GGDDQFDD*Y*G^{10}HMRFa, contain a sulfated Tyr residue (*Tyr*/*Y*). *SchistoFLRFa*, also termed *myosuppressin*, H-Pro-Asp-Val-Asp-His-Val-Phe-Leu-Arg-Phe^{10}-NH_2, from the desert locust *Schistocerca gregaria* shows sequence similarity with the FaRPs, but it has a myoinhibitory activity on various visceral muscle preparations. The homologous neuropeptide from extracts of adult fruitflies

Dromyosuppressin (*DMS*), H-Thr-Asp-Val-Asp-His-Val-Phe-Leu-Arg-Phe^{10}-NH_2, contains also the C-terminal sequence -FLRFa, and only differs from cockroach and locust myosuppressins by the N-terminal residue. It has been localized in fruitfly CNS and gut tissues [L. E. Schneider, P. H. Taghert, *Proc. Natl. Acad. Sci. USA* **1988**, *85*, 1993; J. V. Broeck, *Peptides* **2001**, *22*, 241].

Follicle-stimulating hormone releasing hormone, → Gonadoliberin.

Folliliberin, → Gonadoliberin.

Follistatin, ***FS***, a 288-residue protein acting as an important mediator of cell secretion, development, and differentiation in a number of tissue and organ systems. It was first isolated from ovarian follicular fluid as a protein factor that is capable of suppressing the secretion of → follitropin by pituitary cells in culture, as does inhibin. FS regulates cellular differentiation and secretion through its potent capability to bind and bioneutralize activin, with which it is colocalized in many tissue systems [W. Vale et al., *Recent Prog. Horm. Res.* **1988**, *44*, 1; C. Peng, S. T. Mukai, *Biochem. Cell Biol.* **2000**, *78*, 261].

Follitropin, *follicle-stimulating hormone*, ***FSH***, a member of the → gonadotropins. FSH is a glycoprotein (M_r ~ 25 kDa) consisting of a α subunit (hFSH: 96 aa), which is identical to those of → thyrotropin, → lutropin (LH), and → chorionic gonadotropin, and a β subunit (hFSH: 111 aa). FSH causes spermatogenesis in male testes, and controls maturation of female follicle. Synthesis and release of FSH is regulated by → gonadoliberin [N. B. Schwartz, *Can. J. Physiol. Pharmacol.* **1995**, *73*, 675].

Formin-related protein, ***FRL***, a 160-kDa protein encoded by *frl* cDNA. The *frl* (formin-related gene in leukocytes) cDNA was isolated as a novel mammalian member of the formin gene family. FRL binds to Rac and regulates cell motility and survival of macrophages [S. Yayoshi-Yamamoto et al., *Mol. Cell. Biol.* **2000**, *20*, 6872].

Fractalkine, → chemokines.

FRL, acronym for → formin-related protein.

Fulicin, H-Phe-D-Asn-Glu-Phe-Val-NH_2, a neuroactive 5-peptide amide containing a D-amino acid in position 2 such as, e.g., → dermorphin, → deltorphins and → achatin. Fulicin was isolated from the ganglia of the African giant snail *Achatina fulica*. Structure activity studies concerning the possible role of the D-Asn residue have been performed in the *Achatina fulica* penis, resulting in increasing resistance to enzymatic digestion and stabilizing a conformation favorable for receptor interaction [N. Ohta et al., *Biochem. Biophys. Res. Commun.* **1991**, *178*, 486; M. Kobayashi et al. in: *Peptide Chemistry* **1992**, N. Yanaihara (Ed.), p. 353, ESCOM, Leiden, **1993**].

G

G proteins, *guanine nucleotide-binding proteins, GTP-binding regulatory proteins*, regulating proteins involved in cell signaling. G proteins are heterotrimers containing a α chain loosely bound to a βγ dimer, except for Ras proteins and the structurally unknown G_p. Generally, the subunits (M_r ~ 45 kDa) are homologous, and various G proteins share the same (or very similar) β (M_r ~ 37 kDa) and γ (M_r ~ 9 kDa) subunits. Several types of G proteins exist. The *stimulatory G protein* (G_s) is involved in enzyme activation mediating adenylate cyclase (AC) activation, whereas receptors inactivate AC via an *inhibitory G protein*

(G_i). On the other hand, G_q forms a link in the phosphoinositide cascade, *transducin* (G_t) activates cGMP phosphodiesterase in vertebrate rod cells, and G_{olf} is involved in olfactory sensory neurons in odorant signal transduction. The intracellular action of most polypeptide and catecholamine hormones is mediated by second messengers such as cAMP. For example, the binding of the hormone to certain seven-transmembrane receptors activates the $G_{s\alpha}$ subunit of the G protein to replace its bound GDP with GTP and activate AC to form cAMP [M. Rodbell, *Nature* **1980**, *284*, 17; A.G. Gilman, *Cell* **1984**, *36*, 577; Y. Kaziro et al., *Annu. Rev. Biochem.* **1991**, *60*, 349; D.E. Coleman et al., *Science* **1994**, *265*, 1405].

Galanin, *Gal*, GWTLNSAGYL^{10}LGPHAVG NHR^{20}SFSDKNGLTS^{30} (hGal), a 30-peptide originally isolated from colon and pituitary. In humans, Gal inhibits gastrointestinal motility and delays gastric emptying. Furthermore, it inhibits the release of → gastrin and → somatostatin from isolated perfused rat stomachs. The neuropeptide porcine Gal is a 29-peptide amide differing from hGal in six amino acid residues and the C-terminal moiety. The name is related to its N- and C-terminal amino acids (Gly and Ala) [K. Tatemoto et al., *FEBS Lett.* **1983**, *164*, 124; M. Bersani et al., *FEBS Lett.* **1991**, *283*, 189; L.W. Haynes, *Trends Biochem. Sci.* **1986**, *6*, 214].

Gallidermin, a naturally occurring analogue of → epidermin.

Galparan, GWTLNSAGYL^{10}LGPINLKALA^{20} ALAKKILa, a chimeric 27-peptide consisting of the N-terminal sequence 1–13 of → galanin linked to → mastoparan via a peptide bond. This synthetic peptide powerfully stimulates insulin secretion from isolated rat pancreatic islets in a reversible and dose-dependent manner. Since galparan incarnates an "address" to B cells, analogously to the galanin receptor recognizing the N-terminal sequence part 1–13 of galanin, and a "message" such as the mastoparan portion of galparan, this chimeric peptide may be useful in examining the molecular mechanism of insulin exocytosis in both normal and diabetic states [U. Langel et al., *Regul. Peptides* **1996**, *62*, 47; C.-G. Östenson et al., *Endocrinology* **1997**, *138*, 3308].

Gastric inhibitory polypeptide, *GIP*, YAEGT FISNY^{10}SIAMDKIHQQ^{20}DFVNWLLAQK^{30} GKKNDWKHNI^{40}TQ, a 42-peptide that inhibits gastric acid secretion and stimulates pancreatic insulin release in the presence of glucose (→ incretins). The human gene comprises 10 kb and is located on the long arm of chromosome 17. In humans, GIP derives from the 153-residue precursor prepro-GIP, whose proteolytic cleavage gives rise to the main product GIP besides other fragmentation products. Interestingly, the GIP-(7-42) fragment shows antibacterial activity. Processing of the 144-residue rat prepro-GIP yields GIP and N- and C-terminal flanking peptides of 22 and 59 residues, respectively. The rat GIP receptor (M_r ~ 52 kDa) is a 455-mer glycoprotein. The GIP receptor belongs to the family of G-protein-coupled receptors stimulated after activation of the adenylate cyclase/protein kinase A cascade. The human GIP receptor shows 41% homology with the GLP-1 receptors (→ glucagon-like peptides) [K. Yasuda, Y. Seino, *Jap. J. Clin. Med.* **1996**, *54*, 1078].

Gastrin family, a member of the gastroenteropancreatic peptide families. This family comprises the mammalian hormones → gastrin and → cholecystokinin, the protochordean neuropeptide → cionin, and the frog skin peptide → cerulein. The homology of this family is decisively concentrated in and around the well-defined active site, and the common C-terminal 4-peptide sequence (-Trp-Met-Asp-Phe-NH_2) [J. F. Rehfeld, *Physiol. Rev.* **1998**, *78*, 1087].

Gastrin, ***GT***, a gastrointestinal peptide hormone. GT is produced in the G cells of the antrum and duodenum as prepro-gastrin (101 aa) which yields after co- and post-translational processing mainly the bioactive peptides *G-71, G-34, G-17,* and *G-6.* The formation of prepro-gastrin is stimulated by partially digested protein, amino acids, and by the vagus nerve in response to stomach distension. Gastrin stimulates gastric acid and pepsinogen secretion, and it gives the name to the → gastrin family. The gene encodes only one prohormone with one active site that is processed to a number of bioactive molecular forms with the same C-terminal sequence. The cDNA-deduced amino acid sequence of human pro-gastrin comprises 80 residues. After co-translational cleavage of the 21-residue signal peptide from prepro-gastrin by signalase, pro-gastrin is transported to the Golgi apparatus where O-sulfation of Tyr^{66} by tyrosyl-protein sulfotransferase is performed. After trimming by the endoproteolytic enzyme proprotein convertase and carboxypeptidase E, peptidylglycine α-amidating monooxygenase forms the amidated bioactive peptide. Human *gastrin-71* has the following sequence: $SWKPRSQQPD^{10}$ $APLGTGANRD^{20}LELPWLEQQG^{30}PASHH$ *RR*↓$QLG^{40}PQGGPHLVAD^{50}PS$*KK*↓$QGPLE^{60}$ $EEEEAY^{S}GWMD^{70}Fa$ Y^S: Tyr(SO_3^-). The arrows ↓ indicate the dibasic processing sites providing *gastrin-17* and *gastrin-34,* respectively, with N-terminal pyroglutamic acid residues formed by subsequent glutaminyl cyclizations. In humans, more than 90% of the gastrins are gastrin-17, 5% are gastrin-34, and the remainder is a mixture of gastrin-71, gastrin-52, gastrin-14, and short C-terminal 7- and 6-peptide amide fragments. Nonamidated precursors are present in small percentages, mainly glycine-extended gastrin, and approximately one-half of the amidated gastrins are sulfated at Tyr. The synthetic *pentagastrin* (Gastrodiagnost®), Boc-β-Ala-Trp-Met-Asp-Phe-NH_2, is used as a diagnostic tool for maximum stimulation of the secretion of gastric acid in the analysis of gastric juice [J. S. Edkins, *J. Physiol. (Lond.)* **1906**, *34*, 133; H. Gregory et al., *Nature* **1964**, *204*, 931; J. F. Rehfeld, *Physiol. Rev.* **1998**, *78*, 1087].

Gastrin-releasing peptide, ***GRP***, hGRP: $VPLPAGGGTV^{10}LTKMYPRGNH^{20}WAVGH$ LMa, an intrapancreatic neuropeptide belonging to the → bombesin family. GRP has been shown to stimulate insulin secretion under in-vivo conditions, as well as in isolated islets of Langerhans and insulin-producing RINm5F cells. It could be established that GRP-stimulated insulin secretion is connected with raised cytoplasmic concentration of calcium from an intracellular Ca^{2+} pool through activation of phospholipase C. Furthermore, the GRP-induced mobilization of Ca^{2+} is potentiated by cAMP. GRP shows more potent effects than the other bombesin-like peptide → neuromedin B (NMB) in most of the assays mediated through the GRP receptor and the NMB receptor, respectively. Interestingly, both receptors can interact with both bombesin-like peptides. The distribution of GRP and NMB is overlapped in many brain regions and digestive tissues. In 1979, GRP was isolated as a mammalian counterpart of → bombesin from porcine gastric and intestinal tissues using bioassays for gastrin release [T. J. McDonald et al., *Biochem. Biophys. Res. Commun.* **1979**, *90*, 227; S. Karlsson, B. Ahren, *Peptides* **1996**, *17*, 909; H. Ohki-Hamazaki, *Prog. Neurobiol.* **2000**, *62*, 297].

Ghrelin, GSS^{Y}FLSPEHQ^{10}RVQQRKESKK^{20} PPAKLQPR, (human ghrelin), a 28-peptide with an essential n-octanoic moiety ($Y=CH_3$-$(CH_2)_6$-CO-) esterified with the hydroxy group of Ser^3. Ghrelin is found in the secretory granules of X/A-like cells in the submucosal layer of the stomach. It is the natural endogenous ligand for the → growth hormone secretagogue receptor (GHS-R). Ghrelin stimulates GH release from primary pituitary cells in a dose-dependent manner. Ghrelin is secreted from the stomach, circulates in the bloodstream, and acts directly and specifically on the pituitary. Furthermore, ghrelin induces adiposity and stimulates gastric acid secretion. In the hypothalamus, ghrelin is involved in the control of food intake due to its appetite-stimulating effect. Ghrelin is found to be the most powerful stimulator of appetite of all known peptides. The name 'ghrelin' is related to 'ghre', a word root in Proto-Indo-European languages for 'grow'. Ghrelin also stimulates the release of growth hormone (→ somatotropin). Therefore, the release of growth hormone might be regulated not only by the hypothalamic growth hormone-releasing hormone (→ somatoliberin), but also by ghrelin secreted from the stomach and the hypothalamus. The rat and human ghrelin precursors both contain 117 residues, and both active compounds are 28-peptides modified at Ser^3 by n-octanoic acid, differing only in two positions (rat ghrelin: [Lys^{11}, Ala^{12}] hghrelin). Ghrelin and → motilin share partial sequence homology, but motilin is not modified at Ser^3. In principle, motilin can activate GHS-R expressing cells, but its activity is very weak. It has been suggested that ghrelin and motilin might have evolved from a common ancestral peptide [M. Kojima et al., *Nature* **1999**, *402*, 656; C. Dieguez, F.F. Casanueva, *Eur. J. Endocrinol.* **2000**, *142*, 413; M. Kojima et al., *Trends Endocrinol. Metab.* **2001**, *12*, 118 (review)].

Gliadin, a simple protein occurring in cereals belonging to the → prolamines. Gliadin is rich in glutamine and forms, together with glutelins, the → glutens.

Glicentin, RSLQNTEEKS^{10}RSFPAPQTDP^{20} LDDPDQMTED^{30}KRHSQGTFTS^{40}DYSKYL DSRR^{50}AQDFVQWLMN^{60}TKRNKNNIA, a 69-peptide from the porcine small intestine. Glicentin corresponds to the N-terminal fragment 1–69 of proglucagon, and is secreted together with the two → glucagon-like peptides in the L-cells of the gut. The biological activity of glicentin is still a controversial issue. Glicentin-(33–39) corresponds to the sequence of *oxyntomodulin* [C. Orskov et al., *Endocrinology* **1986**, *119*, 1467; J.J. Holst, *Annu. Rev. Physiol.* **1997**, *59*, 257].

Glicentin-related pancreatic peptide, ***GRPP***, → glucagon.

Globin, the protein component of hemoglobin and myoglobin.

Globulins, a family of simple proteins insoluble in pure water but soluble in dilute salt solutions. The globulins occur in all animal and plant cells and body fluids, including serum and milk. They are precipitated by ammonium sulfate at various concentrations depending on the type of globulin. Many enzymes and glycoproteins belong to this group of proteins. The serum globulins are the most familiar members (→ plasma proteins).

Glucagon, HSQGTFTSDY^{10}SKYLDSRRAQ^{20} DFVQWLMNT, a 29-peptide hormone formed in the α-cells of the pancreas in response to a decrease in blood glucose concentration (see Section 3.3.1.5). It stimulates both glucose release through glycogenolysis and lipolysis. The most important target organ is the liver, where glucagon stimulates the formation of glucose from glycogen and ketone bodies. Glucagon inhibits fatty acid synthesis in hepatocytes, and it stimulates lipolysis in brown adipose tissue. Molecular biology studies have demonstrated that glucagon and its receptor are expressed in numerous other tissues. The human and rat glucagon gene consists of six exons and five introns and encodes *prepro-glucagon* (180 aa). Interestingly, each functional domain of the large precursor is encoded by a separate exon. Post-translational proteolytic cleavage of the precursor protein gives rise to the following main cleavage fragments: genuine pancreatic glucagon (sequence 33–61), → glucagon-like peptide-1 (GLP-1) (sequence 78–108) and GLP-2 (sequence 126–159). Prepro-glucagon cDNA from various tissues and organs shows identity in its sequence, but its post-translational processing can differ significantly. In the α-cells of the pancreatic islets the precursor protein is processed to release the pancreatic glucagon, whereas GLP processing remains incomplete. On the other hand, the L cells of the gut process the precursor in a different way to release GLPs, but glucagon remains as a part of glicentin. This prohormone fragment can be further processed to glucagon-related pancreatic peptide (GRPP) and oxyntomodulin. Glucagon and GLP-1 show an extensive sequence homology, but they exert their action through different specific receptors. Binding of glucagon to its receptor activates the adenylate cyclase/protein kinase A pathway. Glucagon belongs to the → secretin family. The total chemical synthesis of glucagon was performed by Wünsch et al. in 1968. Glucagon is used in the treatment of hypoglycemia. It can be applied parenterally, also using a portable pump, or in a depot form, nasally or as eye drops [E. Wünsch et al., *Chem. Ber.* **1968**, *101*, 3664; G.I. Bell et al., *Nature* **1983**, *304*, 368; J.J. Holst et al., *J. Biol. Chem.* **1994**, *269*, 18827; G.G. Nussdorfer et al., *Peptides* **2000**, *21*, 309].

Glucagon-like peptides, ***GLP***, peptide hormones belonging to the → secretin family. In mammals, the glucagon gene is not only expressed in the pancreas, but also in the intestinal L cells, where *proglucagon* (PG, 160 aa) provides the glucagon-containing peptide → glicentin (PG 1–69), the two glucagon-like peptides (GLP) and the so-called intervening peptide 2 (PG 111–123). The GLP are glucagon-like because of their ~ 50% sequence homology with glucagon. *GLP*-1, GLP-*1-(7–36)-amide,* (PG 78–107-NH_2), HAEGTFTSDV^{10}SSYL EGQAAK^{20}EFIAWLVKGR^{30}a was originally predicted to consist of 37 residues (PG 72–108; N-terminally extended sequence: H-His-Asp-Glu-Phe-Glu-Arg-), and, therefore, the secreted hormone is frequently alluded to as GLP-1-(7–36)-amide. The insulinotropic peptide amide is secreted from endocrine cells in the gut mucosa in response to meal ingestion. In addition, also small amounts (~ 20%) of the Gly-extended *GLP-1-(7–37)*, GLP-1-Gly, are formed, which has the same affinity to the GLP-1 receptor, but it is less resistant to enzymatic cleavage at the C-terminus compared with the amidated form. Each of the GLP interacts with a specific receptor, underlying that their actions diverge in spite of their close sequence homology. GLP-1 belongs to the → incretins. A prominent action of GLP-1 is to potentiate glucose-induced insulin secretion. The insulinotropic effect is preserved in patients with type 2 diabetes mellitus, in whom also the secretion of glucagon is inhibited. After GLP-1 infusion, blood glucose may be completely normalized. GLP-1 appears to regulate plasma glucose levels through various and independent mechanisms. It is an excellent candidate option for the treatment of patients with type 2 diabetes mellitus. Since GLP-1 is extremely rapidly metabolized *in vivo* initially by the action of dipeptidyl peptidase-IV (DPP-IV), for practical diabetes therapy either stable analogues or inhibitors for the degrading enzyme are promising solutions. The structure of *GLP-2* (PG 126–159), HADGRFSDEM^{10} NTILDNLAAR^{20}DFINWLIQTK^{30}ITDRa, was recently determined and confirmed by synthesis. GLP-2 is secreted from gut endocrine cells and promotes nutrient absorption. GLP-2 regulates gastric motility, gastric acid secretion, intestinal hexose transport, and increases the barrier function of the gut epithelium. It reduces mortality and decreases mucosal injury, cytokine expression, and bacterial septicemia in the setting of small and large bowel inflammation. It is assumed that GLP-2 is the intestinal growth factor of the intestinal L-cell. Based on its intestinotrophic effects, GLP-2 is considered for the treatment of a couple of gastrointestinal diseases that are connected with insufficient intestinal mucosal function. The finding that GLP-2 is less extensively degraded by DPP-IV than GLP-1 may provide access to an interesting spectrum of clinical applications [H.-C. Fehmann et al., *Endocr. Rev.* **1995**, *16*, 390; B. Hartmann et al., *Peptides* **2000**, *21*, 73; R. Perfetti, P. Merkel, *Eur. J. Endocrinol.* **2000**, *143*, 717; J. J. Holst, *Regul. Peptides* **2000**, *93*, 45; D. J. Drucker, *Endocrinology* **2001**, *142*, 521; A. Wettergren, *Dan. Med. Bull.* **2001**, *48*, 19].

Glutathione, ***GSH***, H-Glu(Cys-Gly-OH)-OH (reduced), a naturally occurring tripeptide in animals, most plants and bacteria. It acts as a biological redox agent, as coenzyme and cofactor, and as substrate in certain reactions catalyzed by the glutathione S-transferase. GSH scavenges free radicals and reduces peroxides, and is important in the lens and in certain parasites lacking catalase to remove H_2O_2.

Glutelins, a group of globular proteins from grain containing up to 45% Glu. Generally, the glutelins are insoluble in water, salt solutions and dilute ethanol, but at extreme pH values they are soluble in water. Main members of the glutelins are glutenin in wheat, hordenin in barley, and orycenin in rice.

Gluten, a mixture of approximately equal parts of → gliadin and glutelins. Gluten enables wheat and rye flours to form dough during bread making, whereas oat and rice grains are not suitable for baking due to the lack of gliadin.

Glycophorins, erythrocyte transmembrane proteins with a high carbohydrate content. *Glycophorin A* (131 aa; M_r ~ 31 kDa) is the major sialoglycoprotein of the erythrocyte membrane consisting of 60% carbohydrate by weight. Glycophorin A is a transmembrane glycoprotein bearing 15 O-linked oligosaccharides, and one that is N-linked. It consists of three domains. The N-terminal domain (72 aa) bears 16 carbohydrate chains (about 100 sugar residues), the transmembrane domain consists of 19 sequential predominantly hydrophobic residues, and the C-terminal domain (40 aa) in the cytosol has a high proportion of charged and polar residues. *Glycophorin B* (M_r ~ 23 kDa) and *glycophorin C* (M_r ~ 19 kDa) occur in the erythrocyte membrane, at low content [V. T. Marchesi et al., *Annu. Rev. Biochem.* **1976**, *45*, 667].

Glycyl-L-histidyl-L-lysine, *liver cell growth factor*, a growth-modulating plasma tripeptide. It produces a disparate set of responses ranging from stimulation of growth and differentiation to outright toxicity after addition at nanomolar concentrations to a wide group of cultured systems. It may act by facilitating copper uptake into cells. The Cu(II) complex of the tripeptide promotes wound healing [L. Pickart, S. Lovejoy, *Methods Enzymol.* **1987**, *147*, 314].

Gonadoliberin, → gonadotropin-releasing hormone.

Gonadotropin-releasing hormone, ***GnRH***, *gonadoliberin*, pGlu-His-Trp-Ser-Tyr-Gly-Leu-Arg-Pro-Gly^{10}NH_2, a 10-peptide amide formed in the hypothalamus which stimulates collectively synthesis and secretion of the gonadotropins, e.g., *luteinizing hormone* (LH), and *follicle-stimulating hormone* (FSH) in the adenohypophysis, that in turn regulates ovulation/spermatogenesis. Therefore, GnRH is identical with *luteinizing hormone-releasing hormone* (*LH-RH*) and *follicle-stimulating hormone-releasing hormone* (*FSH-RH*). The sequence of GnRH is synthesized as part of the precursor prepro-GnRH consisting of 92 residues from which GnRH is proteolytically released followed by amidation in a Cu^{2+}- and ascorbate-dependent reaction catalyzed by the peptidylglycine α-amidating monooxygenase. GnRH is found in the hypothalamus, but also, for example, in liver, heart, pancreas, kidneys, small intestine, adrenals, and gonads. Based on structure-activity studies, an enormous number of analogues has been synthesized and tested with the aim of developing GnRH analogues with potential use as nonsteroidal contraceptives or as fertility agents. A couple of GnRH agonists are currently used in the treatment of prostate cancer, precocious puberty, endometriosis, and other indications dependent on estrogen or testosterone. Analogues that are approximately 10-fold or more as potent as the native hormone are designated as "superagonists". The latter can actually reduce gonadotropin release, and therefore suppress ovulation or spermatogenesis. In a feedback inhibition the superagonists cause a decrease in content (down-regulation) of gonadotropin receptors. Antagonists are being tested as male and female contraceptive agents. GnRH was first isolated by Andrew Schally and co-workers in 1971 [A.V. Schally et al., *J. Biol. Chem.* **1971**, *246*, 7230; J.J. Nestor, Jr., B.H. Vickery in: *Annual Reports in Medicinal Chemistry*, R.C. Allen (Ed.), p. 211, Academic Press, San Diego, **1988**; F. Haviv et al., *J. Med. Chem.* **1993**, *36*, 363].

Gonadotropins, *gonadotropic hormones*, a family of protein hormones from the anterior pituitary gland and the placenta. Gonadotropins comprise → lutropin, → follitropin, → prolactin, → chorionic gonadotropin, and the human menopausal gonadotropin (urogonadotropin, hMG). Gonadotropins mediate their action via specific receptors in the theca and follicle cells of the ovary and corpus luteum, and the Leydig interstitial cells in the testes. The released cAMP stimulates as second messenger the production of steroid sex hormones [N.R. Mondgal (Ed.), *Gonadotropins and Gonadal Function*, Academic Press, New York, **1974**].

Goralatide, Ac-Ser-Asp-Lys-Pro-OH, a 4-peptide derivative isolated from fetal calf bone marrow acting as a physiological regulator of hematopoiesis. Goralatide inhibits the entry into the S-phase of murine and human hematopoietic stem cells. It may be derived from thymosin β4, which contains N-terminally the goralatide sequence [M. Lenfant et al., *Proc. Natl. Acad. Sci. USA* **1989**, *86*, 779; A. Masse et al., *Blood* **1998**, *91*, 441].

Gramicidins, a group of cyclic or linear peptide antibiotics produced by *Bacillus brevis* strains. *Gramicidin S*, cyclo-(-Val-Orn-

Leu-D-Phe-Pro-$)_2$, primarily acts on Gram-positive bacteria. Together with → tyrocidins, gramicidins are used as antibacterial drugs in a variety of formulations for the topical treatment of infections of the upper respiratory tract. Biosynthesis is performed by the gramicidin S synthetase, an enzyme complex of two subunits. Gramicidin S synthetase 1 (GS1, EC 5.1.1.11, M_r ~ 127 kDa, phenylalanine racemase) encoded by the *grs*A gene is responsible for activation and epimerization of the first amino acid L-Phe to D-Phe. The D-Phe enantiomer is transferred to gramicidin S synthetase 2 (GS2, M_r ~ 510 kDa), which is encoded by the *grs*B gene. GS2 is responsible for activating and coupling L-Pro, L-Val, L-Orn, and L-Leu to the appropriate pentapeptide according to the mechanism of the multiple carrier model at modular multienzymatic templates (cf. Section 3.2.3). Last but not least, the cyclization is carried out by two head-to-tail condensations between two pentapeptide sequences. *Gramicidins A–C* are linear 15-peptides with alternating D- and L-amino acid residues, a formyl group at the N-terminus, and a characteristic C-terminal ethanolamine moiety. They are ionophores forming channels through membranes transporting monovalent cations (cf. Fig. 3.2). Depending on the N-terminal amino acid (Val or Ile), a differentiation may be made between *[Val]Gramicidin A–C* and *[Ile]Gramicidin A–C*. *[Val]Gramicidin A*, HCO-Val^1-Gly-Ala-D-Leu-Ala^5-D-Val-Val-D-Val-Trp-D-Leu^{10}-Trp-D-Leu-Trp-D-Leu-Trp^{15}-NH-CH_2-CH_2-OH, is capable of transporting monovalent cations, e.g., alkali ions, NH_4^+, and H^+, through membranes. From electrophysiological studies, it can be concluded that the channel is formed by a dimer, and that the membrane conductivity fluctuates in such a way that the channel cycles between closed and open states. As shown by X-ray crystallography, the dimer forms a left-handed, antiparallel, double-stranded helix. Other models favor channel formation by left-handed, single-stranded helices with the two monomers associated with a head-to-head dimer via their N-termini. *Gramicidin B* has a Phe residue in position 11, whereas *Gramicidin C* has Tyr in this position [R. Sarges, B. Witkop, *J. Am. Chem. Soc.* **1964**, *86*, 1862; D.W. Urry et al., *Science* **1983**, *221*, 1064; A. Finkelstein, O.S. Andersen, *J. Membr. Biol.* **1981**, *59*, 155; D.A. Langs, *Science* **1988**, *241*, 188; T. Stein et al., *J. Biol. Chem.* **1996**, *271*, 15428].

Granuliberin R, FGFLPIYRRP^{10}ASa, a mast cell-degranulating 12-peptide amide first isolated from the skin of the frog *Rana rugosa* [T. Nakajima, T. Yasuhara, *Chem. Pharm. Bull.* **1977**, *25*, 2464].

GroEL *(hsp60, cpn60)*, → molecular chaperones.

Growth hormone release-inhibiting hormone, → somatostatin.

Growth hormone secretagogue receptor, *GHS receptor*, ***GHS-R***, a G protein-coupled seven-transmembrane receptor expressed in the pituitary, hypothalamus and hippocampus. GHS-R bind the endogenous ligands → ghrelin and → growth hormone secretagogues (GHSs) mediating the release of growth hormone by increasing the cellular Ca^{2+} concentration through inositol 1,4,5-trisphosphate signal transduction. GHS-R was identified by expression cloning in **1996** [A.D. Howard et al., *Science* **1996**, *273*, 974; K.K. McKee et al., *Mol. Endocrinol.* **1997**, *11*, 415; R.G. Smith et al., *Trends Endocrinol. Metab.* **1999**, *10*, 128].

Growth hormone secretagogues, ***GHS***, *growth hormone releasing peptides*, ***GHRP***, a group of synthetic peptides without structural similarities to growth hormone releasing hormone (GHRH, → somatoliberin) that stimulate the release of the growth hormone (→ somatotropin) when added to cultured pituitary cells or when injected intravenously into humans or rats. One of the first examples was the 6-peptide amide *SK&F 110679*, H-His-D-Trp-Ala-Trp-D-Phe-Lys-NH_2, showing about 10% of the somatoliberin activity. Further important members of this group are *GHRP-1* (H-Ala-His-D-βNal-Ala-Trp-D-Phe-Lys-NH_2), *GHRP-2*, *GHRP-6*, and *hexarelin*. Interestingly, non-peptide GHS such as *L-692,429* and *MK-0677* are also potent compounds in GH release. The GHS are orally active, as has been shown in both animals and humans. The actions of GHSs are mediated through the → growth hormone secretagogue receptor (GHS-R). The endogenous ligand for the GHS-R is → ghrelin. GHSs alone or in combination with → somatoliberin have been used for diagnosis and treatment of various forms of growth hormone deficiency [C.Y. Bowers et al., *Endocrinology* **1980**, *106*, 663; K. Cheng et al., *Endocrinology* **1989**, *124*, 2791; M.S. Akman et al., *Endocrinology* **1993**, *132*, 1286; R.G. Smith et al., *Science* **1993**, *260*, 1640; A.A. Patchett et al., *Proc. Natl. Acad. Sci. USA* **1995**, *92*, 7001; C.Y. Bowers, *Cell. Mol. Life Sci.* **1998**, *54*, 1316].

Growth hormone, ***GH***, *somatotropin*, a multifunctional hormone secreted by the anterior pituitary under the control of the hypothalamic growth hormone-releasing hormone, GHRH, (→ somatoliberin) and the appropriate release-inhibiting hormone → somatostatin. In addition, → ghrelin and → growth hormone secretagogues might be also involved in GH release. GH regulates overall body and cell growth, protein-carbohydrate-lipid metabolism and water-electrolyte balance. More information is given under the IUPAC-IUB name → somatotropin.

Growth hormone-releasing hormone, → somatoliberin.

Growth hormone-releasing peptides, → growth hormone secretagogues (GHS).

GRPP, acronym for glicentin-related pancreatic peptide.

GTP-binding proteins, → G proteins.

Guanine nucleotide-binding proteins, → G proteins.

Guanylin, PNTCEICAYA^{10}ACTGC (hG; disulfide bonds: C^4–C^{12}/C^7–C^{15}), a 15-peptide activating intestinal guanylate cyclase. Rat-guanylin differs only in position 2 (Asn) from hG. It has a function in regulating fluid and electrolyte absorption in the intestine [F.J. de Sauvage et al., *Proc. Natl. Acad. Sci. USA* **1992**, *89*, 9089; M.G. Currie et al., *Proc. Natl. Acad. Sci. USA* **1992**, *89*, 947].

Gurmarin, <EQCVKKDELC^{10}IPYYLDCCEP^{20}LECKKVNWWD^{30}HKCIG (disulfide bonds: C^3–C^{18}/C^{10}–C^{23}/C^{17}–C^{33}), a sweetness-suppressing 35-peptide from the leaves of the Asclepiad vine *Gymnema sylvestre*. Gurmarin suppresses the sweet taste responses of sucrose, glucose, glycine, and saccharin, without affecting the responses to salty, sour or bitter substances in laboratory animals, e.g., rats and mice. In humans, gurmarin has only a weak effect. Due to its capability to inhibit selectively the neural response to sweet tastants in rats, gurmarin has been used as a pharmaceutical tool in the investigation of sweet-taste transduction. *Gurmarin sylvestre* has found application in Ayurvedic medicine in the treatment of diabetes mellitus [J.I. Fletcher et al., *Eur. J. Biochem.* **1999**, *264*, 525; M. Ota et al., *Biopolymers* **1998**, *45*, 231].

H

Hamburger pentapeptide, *human IgE pentapeptide* (***HEPP***), H-Asp-Ser-Asp-Pro-Arg-OH, a 5-peptide related to the human IgE sequence 330–334 inhibiting cutaneous and systemic IgE-mediated allergic reaction in humans [R.H. Hamburger, *Science* **1975**, *189*, 389; G.S. Hahn, *Nature* **1986**, *324*, 283; B.M. Prenner, *Ann. Allergy* **1987**, *58*, 322].

Harzinins, a group of peptide antibiotics from *Trichoderma harzianum* belonging to the short-chain → peptaibols. They consist of 14 residues including three Aib-Pro motifs in the positions 4/5, 8/9 and 12/13. Harzinins show antagonistic activity against phytopathogenic fungi, increase the permeability of liposomes and form as the long-chain peptaibols voltage-dependent transmembrane channels [S. Rebuffat et al., *J. Chem. Soc., Perkin Trans.* **1995**, 1849].

Head activator, ***HA***, <EPPGGSKVIL^{10}F, an 11-peptide conserved from hydra to humans. HA is involved in the development of neuronal cells, and acts in hydra as an important factor in head regeneration. This morphogenic peptide was first isolated and sequenced from the coelenterates *Anthopleura elegantissima* and *Hydra vulgaris* in 1981. Later HA was also isolated from tissues of higher organisms, e.g., bovine or human hypothalamus and rat intestine. In *Hydra*, HA initiates head-specific growth and differentiation processes, and in mammals it acts as growth factor in neuronal development. The hydra low-affinity head activator receptor ($K_d \approx 1$ nM) from the multiheaded mutant of *Chlorohydra viridissima* has been characterized as a 200-kDa protein glycosylated mainly of the N-linked type [H.C. Schaller, H. Bodenmüller, *Proc. Natl. Acad. Sci. USA* **1981**, *78*, 7000; I. Franke et al., *Eur. J. Biochem.* **1997**, *244*, 940].

Heat-shock proteins, ***Hsp***, *stress-response proteins*, a group of highly conserved proteins in both prokaryotes and eukaryotes formed in response to hyperthermia or other noxious conditions. According to their molecular weight, the Hsp are divided into four categories: *Hsp90* (M_r ~ 80–90 kDa), *Hsp70* (M_r ~ 68–74 kDa), *Hsp60* (M_r ~ 58–60 kDa), and some relatively *small Hsp*, such as Hsp10. Mammalian cells in culture start to synthesize Hsp by raising the temperature to ~ 43 °C. In each group, closely related abundant proteins have been detected in normal cells. These help to prevent the accumulation of protein aggregates in bacterial cells, the cytoplasma of eukaryotic cells, and mitochondria by solubilizing and refolding denatured or misfolded proteins. Many Hsp, especially the best studied Hsp70, act as chaperonins (→ molecular chaperones) [O. Bensaude et al., *Nature* **1983**, *305*, 331; E.A. Craig, *CRC Crit. Rev. Biochem.* **1985**, *18*, 239; S. Lindquist, *Annu. Rev. Biochem.* **1986**, *55*, 1151].

Hemerythrin, an oxygen-transporting non-heme iron protein acting as a respiratory pigment in the blood cells of certain marine invertebrates. Hemerythrin of sipunculoid worms, for example, is an octamer (M_r ~ 108 kDa). Every polypeptide chain (M_r ~ 13.5 kDa; 2 Fe^{2+}) contains 113 residues and can bind one molecule of O_2.

Hemocyanins, oxygen-transporting porphyrin-free copper proteins in arthropod and molluscan hemolymph. Oxygen-loaded hemocyanins are blue, whereas the oxygen-free form is colorless. The hemocyanin of arthropods are hexamers (subunit: M_r ~ 75 kDa; 2 Cu^{+}) which are capable of aggregating to multi-hexamers. Molluscan hemocyanins are also oligomers from which each subunit (M_r ~ 400 kDa) contains 8 O_2-binding sites [W. P. J. Gaykema et al., *Nature* **1984**, *309*, 23].

Hemoglobin, ***Hb***, a vertebrate oxygen-transporting protein in the red blood cells (erythrocytes). The tetrameric Hb (M_r ~ 64.5 kDa) contains two pairs of polypeptide chains and four heme groups, and carries oxygen from the lungs to other tissues. The heme-free protein of Hb is termed *globin*. Human adult Hb consists of approximately 97.5% of HbA_1 ($\alpha_2\beta_2$) and 2.5% HbA_2 ($\alpha_2\delta_2$). The α-chains of humans contain 141 residues, and the β-chains 146 residues. Different forms of Hb occur during embryonic development. The tertiary structure of a Hb chain is folded in a manner that is very similar to that of → myoglobin (Mb). According to Mb, each of the four subunits of Hb noncovalently binds a single heme group. The latter is responsible for the red color of blood, and is the site at which each monomer of Hb binds one molecule of O_2. In Hb, the iron atom normally remains in the Fe(II) oxidation state, whether or not the heme is oxygenated (binds O_2). In deoxyHb the central Fe(II) is 5-coordinated by the four pyrrole N atoms of protoporphyrin IX and by the globin's proximal His. The sixth ligand of Fe(II) is ligated by O_2 upon oxygenation. Hb of all human races and even of chimpanzees is identical. Anomalies of Hb arise through point mutations, and the most frequent and best known is *sickle cell Hb* in which only Glu^6 of the normal β-chain is replaced by Val ($\alpha_2\beta_2{}^{6Glu \rightarrow Val}$). Medical symptoms are anemia and acute ischemia, tissue infraction and chronic failure of organ function [M. F. Perutz, *Sci. Am.* **1978**, *239*, 92; M. F. Perutz, *Q. Rev. Biophys.* **1989**, *22*, 139; W. A. Eaton, J. Hofrichter, *Adv. Prot. Chem.* **1990**, *40*, 63].

Hemopexin, a single-chain, heme-binding glycoprotein of the β-globulin fraction. Hemopexin (M_r ~ 57 kDa) contains about 23% carbohydrate. Hemopexin transports complex-bound heme into the hepatocytes, where heme is degraded.

Hemosiderin, a mammalian iron storage protein with a similar function to → ferritin. Hemosiderin is a nonheme iron protein deposited in the liver and spleen, especially in diseases such as pernicious anemia or hemochromatosis. The deposits are yellow to brown-red pigments. The iron content of hemosiderin is about 37%.

Hirudin, VVYTDCTESG^{10}QNLCLCEGSN^{20} VCGQGNKCIL^{30}GSDGEKNQCV^{40}TGEGT PKPQS^{50}HNDGDFEEIP^{60}EEY^{S}LQ [$Y^{S/63}$: Tyr(SO_3H); disulfide bridges: C^6–C^{14}/C^{16}–C^{28}/C^{22}–C^{39}], a single-stranded 65-peptide isolated from the salivary glands of the leech *Hirudo medicinalis*. Hirudin is the most potent natural inhibitor of the protease → thrombin. It is the most efficient member of thrombin inhibitors (K_i = 20 fM). Hirudin is a very effective anticoagulant, as thrombin acts mainly on → fibrinogen and stimulates the coagulation of blood. The binding of hirudin to thrombin involves a noncovalent, ionic interaction and is a two-step process. The rate-limiting step is the anchoring at the site distinct from the active site by the acidic domain of hirudin-(56–65). In the next step, the initial complex is rearranged to form a tighter complex in which the sequence around Lys^{47} covers the active center of thrombin. Today, hirudin is produced by large-scale recombinant (rec) technology in high purity, as for → Lepirudin®. Although rec-hirudin lacks the sulfate moiety at Tyr^{63}, CGP 39393 (Ciba-Geigy), for example, is a highly potent inhibitor (K_i = 230 fM). C-terminal fragments of 10–12 residues also act as antithrombotic agents [F. Markwardt, *Methods Enzymol.* **1970**, *19*, 924; J. B. Haycraft, *Proc. R. Soc. London*, **1984**, *B36*, 478; F. Markwardt, *Biomed. Biochim. Acta* **1985**, *44*, 1007; J. M. Maraganore et al., *Biochemistry* **1990**, *29*, 7065].

Hirullin, SDFEEFSLDD[10]IEQ, a 13-peptide originally isolated from the leech *Hirudinaria manillensis*. Hirullin acts as a hirudin-based fibrinogen recognition peptide inhibitor. It binds efficiently to thrombin [X. Qiu et al., *J. Biol. Chem.* **1993**, *268*, 20318].

Histones, relatively small basic globular proteins with a high content of Arg and Lys; these form preferentially the structure proteins of eukaryotic chromosomes. Histones are divided into five types based on the content of positively charged amino acids and the molecular weight. Additionally, these five types fall into two main groups, termed *H1 histones* and *nucleosomal histones*. *H1-histones* (215–220 aa; $M_r \sim$ 21.5–22 kDa) have been less conserved during evolution compared with the second main group. The *nucleosomal histones* are divided into *H2A* (129 aa; $M_r \sim 14$ kDa), *H2B* (125 aa; $M_r \sim 13.8$ kDa), *H3* (135 aa; $M_r \sim 15.3$ kDa) and *H4* (102 aa; $M_r \sim$ 11.3 kDa). They are responsible for coiling the DNA into nucleosomes. H3 and H4 are the most highly conserved of all known proteins, and form the inner core of the nucleosome [I. Isenberg, *Annu. Rev. Biochem.* **1979**, *48*, 159; R. S. Wu et al., *CRC Crit. Rev. Biochem.* **1986**, *20*, 201; J. Zlatnanova, K. van Holde, *Prog. Nucl. Acid Res. Mol. Biol.* **1996**, *52*, 217].

HIV protease, monomer: PQITLWORPL[10] VTIRIGGQLK[20]EALLDTGADD[30]TVLEEM NLPG[40]KWKPKMIGGI[50]GGFIKYRQYD[60] QIPVETCGHK[70]AIGTVLVGPT[80]PVNIIGR NLL[90]TQIGCTLNF, an aspartyl protease consisting of two identical monomers with 99 residues each that is essential for the maturation of human immunodeficiency virus (HIV). HIV is a retrovirus bearing a reverse transcriptase, which enables encoding viral RNA into DNA of the host's cells. HIV is the causative agent of AIDS. The spread of HIV has reached worldwide epidemic proportions, and has become of major concern in the search for effective drugs. HIV protease is responsible for the proteolytic maturation of the HIV structural proteins, reverse transcriptase, integrase, and RNase H, which are encoded by the viral *gag* and *pol* genes. The design of inhibitors with high potency for the HIV protease offers a logical approach to the formation of noninfectious viral particles. The HIV protease and an analogue of this enzyme, containing D-amino acids only, have been synthesized chemically [T. D. Meek et al., *Proc. Natl. Acad. Sci. USA* **1989**, *86*, 1841; R. L. C. Milton et al., *Science* **1992**, *256*, 1445].

HMG proteins, *high mobility group proteins*, a family of small chromosomal nonhistone proteins ($M_r < 30$ kDa) of eukaryotic organisms. The name is related to their high mobility in polyacrylamide gel electrophoresis. The highly conserved HMG contain ~ 25% basic and 30% acidic amino acids. The major members of this family constitute two pairs of homologous proteins: *HMG1* and *HMG2* ($M_r \sim 25$ kDa), which bind both to single- and double-stranded DNAs; and *HMG14* and *HMG17* ($M_r \sim 10$ kDa), which show a higher affinity to nucleosomes than to DNA. These proteins can be eluted from chromatin using 0.35 M NaCl solution, and are even soluble in 2% CF_3COOH [M. Bustin et al., *Biochim. Biophys. Acta* **1990**, *1049*, 231].

Human chorionic gonadotropin, → chorionic gonadotropin.

Human chorionic somatomammotropin, → chorionic mammotropin.

Hydrins, peptides isolated from the pituitary glands of *Xenopus laevis* and *Rana esculenta*, respectively. *Hydrin 1*: CYIQNCPRGG^{10}KR (disulfide bond: C^1–C^6), *hydrin 1′*, [deArg^{12}]*hydrin 1* and *hydrin 2*, [deArg^{12},deLys^{11}]*hydrin 1* show structural similarities to vasotocin, but lack any oxytocic and pressor activities. It is suggested that hydrins may be involved in the water-electrolyte regulation of amphibians [Y. Rouille et al., *Proc. Natl. Acad. Sci. USA* **1989**, *86*, 5272; S. Iwamuro et al., *Biochim. Biophys. Acta* **1993**, *1176*, 143].

Hylambatins, peptides isolated from the skin of the African frog *Hylambates maculates*. *Hylambatin*, DPPDPDRFYG^{10}MMa, increases plasma glucose and plasma insulin level. *Entero-hylambatin* differs from hylambatin 1 only in position 4 (Asn^4 instead of Asp^4) [H.-G. Guellner et al., *IRCS Med. Sci.* **1983**, *11*, 1072; L. Negri et al., *Regul. Peptides* **1988**, *22*, 13].

Hymenistatin I, ***HS-I***, cyclo-(-Pro-Pro-Tyr-Val-Pro-Leu-Ile-Ile-), causing an immunosuppressive effect in the humoral and cellular immune responses comparable to the action of → cyclosporin A (CsA). Based on studies of the comparative influence of HS-I and CsA on cytokine production, it could be concluded that the mechanisms of interaction with the immune system are different for the two compounds [M. Cebrat et al., *Peptides* **1996**, *17*, 191].

Hypocretins, ***HCRT***, alternative name for → orexins.

I

Iberiotoxin, <EFTDVDCSVS^{10}KECWSVCK DL^{20}FGVDRGKCMG^{30}KKCRCYQ (disulfide bonds: C^7–C^{28}/C^{13}–C^{33}/C^{17}–C^{35}), a 37-peptide isolated from the venom of the scorpion *Buthus tamulus*. Iberiotoxin shows 68% sequence homology with charybdotoxin. Both toxins inhibit the high conductance of a Ca^{2+}-activated K^+ channel, but they may bind at different sites and modulate the channel activity via different mechanisms [A. Galvez et al., *J. Biol. Chem.* **1990**, *265*, 11083; T. R. Jones et al., *J. Appl. Physiol.* **1993**, *74*, 1879].

IgE pentapeptide (human), → Hamburger pentapeptide.

IGF receptor, → insulin-like growth factors.

Immunoglobulin gene superfamily, a group of immune recognition molecules containing related structure elements which might be evolutionarily descended from a gene encoding an Ig-like domain. This superfamily includes, e.g., Ig molecules, T-cell receptors, MHC proteins, the CD2, CD4 and CD8 cell-cell adhesion proteins, some polypeptides of the CD3 complex associated with T-cell receptors, and the various Fc receptors containing one or more Ig-like domains. The Ig-like domain or Ig homology unit contains about 100 residues and is folded into the characteristic sandwich-like structure consisting of two antiparallel β sheets which are usually stabilized by a conserved disulfide bond [A. F. Williams, A. N. Barclay, *Annu. Rev. Immunol.* **1988**, *6*, 381].

Immunoglobulins, ***Ig***, *antibodies*, most abundant defense proteins occurring in blood plasma, lymph and many body fluids of all vertebrates. An Ig molecule is composed of two identical *light (L) chains* (each contains ~ 220 aa) and two identical *heavy (H) chains* (each contains ~ 440 aa). These polypeptide chains are held together by disulfide bonds and noncovalent interactions, such that they form a Y-shaped symmetric dimer. Ig molecules are glycoproteins. Each H chain has a N-linked carbohydrate. Papain and pepsin split Ig molecules into different fragments. Papain forms two separate and identical *Fab* (*fragment antigen binding*) *fragments*, each bearing one antigen-binding site, and one *Fc fragment* (so termed because it readily crystallizes). Pepsin produces one $F(ab')_2$ *fragment* consisting of two covalently linked *F(ab') fragments*, each of which is slightly larger than a *Fab fragment*. Humans have five classes of secreted Ig molecules termed: IgA, IgD, IgE, IgG, and IgM, each with its own class of H chain, designated α, δ, ε, γ and μ. There are only two types of L chain, κ and λ, from which Ig can have either κ or λ but not both. *IgA* [$(\alpha_2\kappa_2)_n$J; $(\alpha_2\lambda_2)_n$J; n=1, 2, or 3; $M_r \sim$ 360–720 kDa] occurs as monomers, dimers or trimers of its corresponding dimers. J is the *joining chain* ($M_r \sim$ 20 kDa) that participates in joining IgA chains to form dimers, and J is also involved in joining two of the pentamer's μ heavy chains in IgM. *IgD* ($\delta_2\kappa_2$; $\delta_2\lambda_2$; $M_r \sim$ 160 kDa), *IgE* ($\varepsilon_2\kappa_2$; $\varepsilon_2\lambda_2$; $M_r \sim$ 190 kDa), and *IgG* ($\gamma_2\kappa_2$; $\gamma_2\lambda_2$; $M_r \sim$ 150 kDa) exist only as $(L\text{–}H)_2$ dimers. IgG forms four subclasses (IgG1, IgG2, IgG3, and IgG4) differing in their γ chains. *IgM* [$((\mu_2\kappa_2)_5$J; $(\mu_2\lambda_2)_5$J; $M_r \sim$ 950 kDa consists of pentamers of the corresponding dimers. IgM is the first Ig to be secreted in response to an antigen, and is most effective against invading microorganisms. IgG is the most common Ig of all; it is equally distributed between the blood and the interstitial fluid, and its production starts 2–3 days after IgM first appears. IgG is the only Ig capable of crossing the placenta and providing the fetus with immunity. IgA is found in the intestinal tract, milk, colostrum, and in secretions such as sweat, salvia, and tears. IgE normally occurs in the blood in minute concentrations. It protects against parasites and is involved in allergic reactions. The function of IgD occurring in the blood also in minute concentrations is not well established [D.A. Davies, H. Metzger, *Annu. Rev. Immunol.* **1983**, *1*, 87; F.W. Alt et al., *Science* **1987**, *238*, 1079; D.R. Davies, S. Chacko, *Acc. Chem. Res.* **1993**, *26*, 421; A. Cattaneo, S. Biocca, *Trends Biotechnol.* **1999**, *17*, 115].

Incretins, *incretin hormones*, insulinotropic hormones secreted by the intestinal tract in response to food. The 'incretin effect' has been defined as the increased insulin response resulting from oral administration of glucose compared with the response elicited by an isoglycemic intravenous infusion. The most important incretins are → glucagon-like peptide 1 (GLP-1) and → gastric inhibitory polypeptide (GIP); these both account for more than 80% of the whole incretin effect. GLP-1 therapy is still an option for the treatment of type 2 diabetes. Incretins were first identified by Unger and Eisentraut, who studied the secretory response of islet β-cells and demonstrated that 50% of postprandial insulin response was triggered by the enteroinsular axis [N. McIntyre et al., *Lancet II* **1964**, 20; D. LeRoith et al. in: *Diabetes Mellitus*, Lippincott-Raven, Philadelphia, **1996**; American Diabetes Association Clinical Practice Recommendation, *Diabetes Care* **1996**, *19* (Suppl. 1) S, 1].

Indolicidin, ILPWKWPWWP^{10}WRRa, a linear cationic, tryptophan-rich antimicrobial 13-peptide amide (→antimicrobial animal peptides) from ox (*Bos taurus*). Indolicidin acts equally against Gram-negative and Gram-positive bacteria. It is also cytotoxic to T-cell lines [M.E. Selsted et al., *J. Biol. Chem.* **1992**, *267*, 4292; D. Hultmark, *Trends Genet.* **1993**, *9*, 178].

Inhibin, ***IHB***, a glycoprotein secreted by the gonads inhibiting FSH production and release. Other sites of synthesis are the brain and the placenta (in humans). *IHB-A* and *IHB-B* are heterodimers which are linked together by disulfide bonds. The α chains are identical and structural similarities exist between the α and β chains [N. Ling et al., *Endocr. Rev.* **1988**, *9*, 267].

Inhibiting factors, → statins.

Insect diuretic peptides, two structurally distinct families of diuretic peptides (DP) that stimulate primary urine production by insect Malpighian tubules and regulate water loss from the excretory system. CRH-related diuretic peptides and insect kinins have been found in a number of species, and they may be ubiquitous in insects. *CRH-related diuretic peptides* belong to the vertebrate → CRH/sauvagine/ urotensin I/urocortin superfamily. *Pea-DP*, TGSGPSLSIV^{10}NPLDVLRQRL^{20}LLEIARRR MR^{30}QSQDQIQANR^{40}EILQTIa, from *Periplaneta americana*, *Lom-DH*, MGMGPSLS IV^{10}NPMDVLRQRL^{20}LLEIARRRLR^{30}DAEE QIKANK^{40}DFLQQIa, *Acd-DP* (*Acheta domesticus*), *Mud-DP* (*Musca domestica*), *Mas-DH* (*Manduca sexta*), *Mas-DPII* (*Manduca sexta*) and *Tem-DH* (*Tenebrio molitor*) are members (not all sequences are shown) of this family. With the exception of Tem-DH, all peptides are amidated. These peptides increase urine production via a cAMP-dependent mechanism. Lom-DH has a hormonal function in the control of post-feeding diuresis in locusts. *Insect kinins*, which act via a Ca^{2+}-dependent mechanism to stimulate urine production, are characterized by the C-terminal sequence -Phe-Xaa-Yaa-Trp-Gly-NH_2 (where Xaa can be His, Asn, Tyr, Phe or Ser, and Yaa is Pro, Ser or Ala) [G.M. Coast, *Amer. Zool.* **1998**, *38*, 442].

Insulin family, a member of the gastroenteropancreatic peptide superfamilies. This family comprises → insulin, → the insulin-like growth factors, and → relaxin.

Insulin receptor, a transmembrane glycoprotein ($M_r \sim 300$ kDa) with tyrosine-specific protein kinase activity converting the extracellular → insulin signal into the cell. The receptor is a tetramer ($\alpha_2\beta_2$) and consists of two identical subunits (α: 719 aa; β: 620 aa) joined together by disulfide bonds. The catalytic domain of the tyrosine-specific protein kinase is exposed on the cytoplasmic side of the plasma membrane. When activated by insulin binding, it phosphorylates itself, thereby enhancing the activity of the kinase, and transfers the terminal phosphate group from ATP to the hydroxy group on a tyrosine residue of selected proteins in the target cell. Both polypeptide chains of the insulin receptor are encoded by a single gene, producing a precursor protein that is cleaved into the two disulfide-joined chains [A. Ullrich et al., *Nature* **1985**, *313*, 756].

Insulin, a 51-polypeptide hormone consisting of the A chain (21 aa) and the B chain (30 aa) linked by two disulfide bonds. It is biosynthetically derived from the single-chain, 84-residue precursor, named *proinsulin*. After three disulfide bridges have been formed within proinsulin, it is converted by proteolytic excision of the internal 33-residue C chain, or *C-peptide*, to the two-chained active hormone. The pancreatic hormone insulin is secreted from the β cells of the islets of Langerhans in response to high blood glucose levels. It stimulates glucose uptake by gluconeogenesis, protein synthesis, and lipogenesis. More details are given in Chapter 3.3.1.5 [F. Sanger et al., *Biochem. J.* **1955**, *59*, 509; J. Meienhofer et al., *Z. Naturforsch.* **1963**, *18b*, 1120; P. Sieber et al., *Helv. Chim. Acta* **1974**, *57*, 2617; G.I. Bell et al., *Nature* **1980**, *284*, 26; M.P. Czech (Ed.), *Molecular*

Basis of Insulin Action, Plenum Press, New York, **1985**; J.M. Conlon, *Peptides* **2001**, *22*, 1183].

Insulin-like growth factors, *IGF, somatomedins*, a group of polypeptides with structural and functional resemblance to → insulin and → relaxin belonging to the → insulin family. The IGF occurring in the blood of vertebrates show insulin-like activity as long-term effects, and stimulate cartilage growth. *IGF-1*, GPETLCGAEL^{10}VDALQFV CGD^{20}RGFYFNKPTG^{30}YGSSSRRAPQ^{40}TG IVDECCFR^{50}SCDLRRLEMY^{60}CAPLKPAK SA^{70} (disulfide bonds: C^{6}–C^{48}/C^{18}–C^{61}/C^{47}–C^{52}), is a single-chain 70-polypeptide (M_r ~ 7.6 kDa) containing three intrachain disulfide bridges. IGF-1 mediates, in interaction with other hormones, the effects of → somatotropin (growth hormone) on bone growth. In the same way as IGF-2, it possesses growth-promoting activity on chick embryo fibroblasts at a concentration of 10^{-9} M. *IGF-2*, AYRPSETLCG^{10}GELVDT LQFV^{20}CGDRGFYFSR^{30}PASRVSRRSR^{40}GI VEECCFRS^{50}CDLALLETYC^{60}ATPAKSE (disulfide bonds: C^{9}–C^{47}/C^{21}–C^{60}/C^{46}–C^{50}) is a single-chain 67-polypeptide (M_r ~ 7.5 kDa) containing three intrachain disulfide bridges. It participates in the fetal and embryonal development of the nerve system and the bones. The IGF-1 receptor (IGF receptor type I) resembles in structure the insulin receptor, whereas the IGF-2 receptor (IGF receptor type II) is more similar to the single-chain mannose-6-phosphate receptor [E.M. Spencer (Ed.), *Insulin-like Growth Factors, Somatomedins*, de Gruyter, Berlin, **1983**; W.S. Cohik, D.R. Clemmons, *Annu. Rev. Physiol.* **1993**, *55*, 131; R. Yamamoto-Honda, *J. Biol. Chem.* **1995**, *270*, 2729; D.E. Jensen et al., *J. Biol. Chem.* **1995**, *270*, 6555].

Integrins, a superfamily of cell surface glycoproteins that promote cellular adhesion. Integrins belong to the cell → adhesion molecules. Generally, cells interact with the extracellular matrix and other cells through cell adhesion receptors, including those of the integrin family. The interaction of integrins with extracellular matrices, e.g., → fibronectin and vitronectin is mediated via an Arg-Gly-Asp (RGD) sequence within adhesive proteins. In 1984, Pierschbacher and Ruoslahti demonstrated that cell adhesion mediated by fibronectin could be inhibited by the tripeptide Arg-Gly-Asp (RGD). The latter sequence, identified as the minimal binding site in fibronectin, is capable of mediating cell adhesion. In protein and DNA databases, several hundred RGD or homologous sequences could be identified. Most extracellular matrix proteins characterized to date contain RGD sequences. In 1987, the term 'integrin' was first suggested by Hynes to characterize a family of integral membrane receptors thought to link or "integrate" the intracellular cytoskeleton with extracellular matrix proteins. Integrins are type I heterodimeric proteins that consist of noncovalently associated α and β subunits. In mammals, 15 α chains and eight β chains can form over 20 distinct heterodimers. The α and β subunits are transmembrane, N-glycosylated glycoproteins characterized by large extracellular domains, a single hydrophobic transmembrane region, and a short cytoplasmic domain. The specificity of ligand binding is determined by the particular αβ combinations. The RGD sequence can be recognized by $\alpha_5\beta_1$, $\alpha_v\beta_1$, $\alpha_v\beta_3$, $\alpha_v\beta_5$, $\alpha_v\beta_6$, $\alpha_{IIb}\beta_3$, $\alpha_2\beta_1$, $\alpha_3\beta_1$ and $\alpha_4\beta_1$. Interestingly, some integrins are capable of binding a number of ligands; for example, $\alpha_v\beta_3$ can recognize fibrinogen, fibronectin, vitronectin, denatured collagen, and other proteins. Strategies have been developed for therapeutic modification of integrin function [M.D. Pierschbacher, E. Ruoslahti, *Nature*

1984, *309*, 30; R.O. Hynes, *Cell* **1992**, *69*, 11; E. Ruoslahti, *Annu. Rev. Cell. Dev. Biol.* **1996**, *12*, 697; M.A. Horton, *Exp. Nephrol.* **1999**, *7*, 178].

Interferons, ***IFN***, glycoproteins secreted by virus-infected vertebrate cells preventing viral proliferation largely by inhibition of protein synthesis in other infected cells. Three families of interferons are characterized: (a) *IFN-α* or *leukocyte IFN*, named according to the preferential formation by white blood cells, consists of a family of closely related proteins with sequence homology between 80 and 95%. In humans, all IFN-α genes are located on chromosome 9. Human IFN-α consists of 166–172 residues ($M_r \sim$ 16–27 kDa). This family is divided into two subfamilies (IFN-αI and IFN-αII or IFN-ω). (b) *IFN-β* or *fibroblast IFN* are preferentially produced by fibroblasts (connective tissue cells), but there are many other cell sources. Human IFN-β (166 aa; $M_r \sim$ 20 kDa) is encoded by a single gene on chromosome 9. (c) *IFN-γ* or *lymphocyte IFN* or *immune IFN* is a glycoprotein (143 aa; $M_r \sim$ 22 kDa) occurring *in vivo* as a dimer. The gene is located on chromosome 12, and has three introns. It is produced by T lymphocytes. IFN-γ enhances the cytotoxic activity of T lymphocytes, macrophages, and natural killer cells. The action of IFNs via inhibition of protein synthesis in infected cells is mediated by two independent pathways: (i) by inducing the production of a protein kinase (double-stranded RNA-activated inhibitor, DAI) which leads finally to the inhibition of ribosomal initiation; or (ii) by inducing the synthesis of (2′,5′)-oligoadenylate synthetase producing 2,5-A which activates the mRNA degrading RNase L. Generally, IFN belong to the most potent biological compounds, and represent a major defense against viral infection. These antiviral agents are active in concentrations as low as 3×10^{-14} M. As some cancers are virally induced, IFNs have also attracted attention as potential anti-cancer drugs. The clinical use of IFNs against viral infection and certain malignant tumors has become possible by the development of large-scale molecular cloning techniques. The IFNs were discovered by Isaacs and Lindenmann in 1957 [H. Kirchner, *Antiviral Res.* **1986**, *6*, 1; S. Pestka et al., *Annu. Rev. Biochem.* **1987**, *268*, 3017; G.C. Sen, P. Lengyel, *J. Biol. Chem.* **1992**, *267*, 5017; R.T. Dorr, *Drugs* **1993**, *45*, 177].

Interleukins, ***IL***, regulatory proteins with main functions in the immune system. Interleukins belong to the → cytokines. Those interleukins that are secreted by lymphocytes are also termed *lymphokines* (*T-cell cytokines*). According to recommendations of the *WHO-IUIS Nomenclature Subcommittee*, it is only permitted to designate a cytokine as an interleukin when: (a) it is purified and sequenced, and the gene is cloned; (b) it is a natural cell product; and (c) it plays a main function in the immune system. Some selected members of the interleukins should be described in brief. *Interleukin-1* (*IL-1*), *leukocyte activating factor* (*LAF*), is a small protein (M_r $\sim$ 17 kDa) that occurs in two forms (IL-1α and IL-1β) which differ significantly in their amino acid sequence (sequence homology: 22–26%). IL-1 acts via IL-1 receptors on a wide spectrum of target cells that are involved in immune and inflammation reactions. *Interleukin-2* (*IL-2*), *T-cell growth factor* (133 aa; $M_r \sim$ 15.5 kDa), is involved, e.g., in T-cell activation, improving IFN-γ production, and modulation of T-cell receptor expression. *Interleukin-3* (*IL-3*) belongs as a *multi-CSF* to the → colony stimulating factors. *Interleukin-4* (*IL-4*), *B-cell-stimulating factor-1* (*BSF-1*), is a highly glycosylated protein (130 aa; M_r ~ 25 kDa). It increases class II MHC molecules expression, and stimulates IgG_1- and IgE production by B cells. *Interleukin-5* (*IL-5*), *B-cell-stimulating factor-2* (*BSF-2*) is a homodimer (monomer: 115 aa; M_r ~ 25 kDa). It promotes proliferation and growth of B cells and eosinophils. *Interleukin-6* (*IL-6*), *B-cell-stimulating factor-2* (*BSF-2*), *hybridoma/plasmacytoma growth factor* (*HPGF*), *hepatocyte-stimulating factor* (*HSF*), *monocyte granulocyte inducer type 2* (*MGI-2*), *interferon-β_2* (*INF-β_2*), is a glycoprotein (M_r ~ 23–30 kDa) and plays an important role in immunological defense. The name IFN-β_2 is incorrect because of lacking antiviral activity and structural similarity with interferons. *Interleukin-10 (IL-10)* forms its own subfamily by comprising four other mammalian cytokines, as well as several viral products. *Interleukin-20 (IL-20)* with structural similarities to IL-10 was discovered in 2001. Its cDNA was found in a keratinocyte library. IL-20 belongs to the IL-10 subfamily and stimulates keratinocyte proliferation and differentiation [S. K. Durum, J. A. Schmidt, *Annu. Rev. Immunol.* **1985**, *3*, 263; K. A. Smith, *Science* **1988**, *240*, 1169; WHO-IUIS Nomenclature Subcommittee on Interleukin Designation, Nomenclature for Secreted Regulatory Proteins of the Immune System (Interleukins), *Immunology Today* **1992**, *13*, 118; H. Blumberg et al., *Cell* **2001**, *104*, 9].

Iron-sulfur proteins, *Fe-S-proteins*, a family of proteins occurring in all organisms. They contain iron-sulfur centers (iron sulfur clusters), and can be subdivided into *simple Fe-S-proteins*, like → ferredoxins, → rubredoxin, and *conjugated Fe-S-proteins*, e.g., Fe-S-molybdenum proteins, Fe-S-heme proteins. They are involved, for example, in nitrogen and carbon dioxide fixation, oxidative and photosynthetic phosphorylation. They are included in → nonheme iron proteins since the iron is not bound in a porphyrin ring system.

Isopeptide bond, a covalent cross-linking amide bond between the ω-amino group of diamino carboxylic acids such as lysine or ornithine and the ω-carboxyl function of amino dicarboxylic acids such as glutamate or aspartate. Isopeptide bonds occur, for example, in polymerized → fibrin and in native wool.

J

Jak proteins, proteins of the Janus kinase (Jak) family of nonreceptor protein kinases. Jak1 and Jak2, and Tyk2 are ~ 130-kDa proteins and contain an active tyrosine kinase-like domain, but lack SH2 and SH3 domains. Dimerized human growth hormone receptor (hGHR) (→ somatotropin) stimulates Jak2 to phosphorylate hydroxy groups of tyrosine residues on both the hGHR and itself.

Jaspamide, *Jaspakinolide*, a cyclodepsipeptide from extracts of *Jaspis* sp. It contains (*R*)-β-tyrosine and D-δ-bromotryptophan – two rare amino acids – in conjugation with a polyketide chain (***Formula***). Jaspamide shows potent cytotoxic, fungicidal and in-

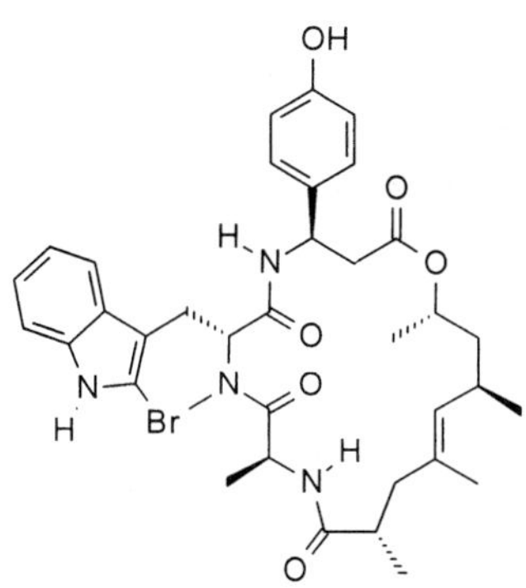

Jaspamide

secticidal actions [P. Wipf, *Chem. Rev.* **1995**, *95*, 2115].

Joining chain, J-chain, a protein chain (M_r ~ 20 kDa) involved in joining and stabilizing the dimeric IgA and pentameric IgM, respectively (→ immunoglobulins).

K

KALA amphipathic peptide, WEAKLA*KALA*^{10} *KALA*KHLA*KA*^{20}*LA*KALKACEA^{30}, a cationic amphipathic 30-peptide with a repeating KALA sequence. It binds to DNA, destabilizes membranes, and mediates transfection of plasmid DNA in various cell lines [T. B. Wyman et al., *Biochemistry* **1997**, *36*, 3008].

Kaliotoxin, GVEINVKCSG^{10}SPQCLKPCKD^{20} AGMRFGKCMN^{30}RKCHCTPK (disulfide bonds: C^8–C^{28}/C^{14}–C^{33}/C^{18}–C^{35}), a 38-peptide originally isolated from the venom of the scorpion *Androctonus mauretanicus mauretanicus*. Kaliotoxin shows 44% sequence homology with → iberiotoxin and charybdotoxin. It acts as a specific voltage-independent inhibitor of high-conductance Ca^{2+}-activated K^+ channels present in mollusk and rabbit nerve cells [M. Crest et al., *J. Biol. Chem.* **1992**, *267*, 1640; J. van Rietschoten et al., in: *Peptides: Chemistry, Structure and Biology*, R. S. Hodges, J. A. Smith (Eds.), p. 529, ESCOM, Leiden, **1994**].

Kallidin, *Kinin 10*, H-Lys-Arg-Pro-Pro-Gly-Phe-Ser-Pro-Phe-Arg^{10}-OH, a tissue peptide hormone belonging to the → plasma kinins. Kallidin is released from the precursor kallidinogen (M_r ~ 68 kDa) by kallikrein. It acts in similar manner to → bradykinin, and it is easily converted to bradykinin by proteolytic cleavage of the N-terminal lysine catalyzed by aminopeptidases.

Kallikreins, serine proteases that cleave kininogens (→ plasma kinins) to yield → bradykinin or bradykinin-like peptides. Kallikreins are proteolytically released from proenzymes, termed pro-kallikrein. An important inhibitor of the kallikreins is → aprotinin. The name kallikrein (Greek *kallikros*, pancreas) was first given for a protease found in the pancreas [K. D. Wuepper, C. G. Cochrane, *J. Exp. Med.* **1972**, *135*, 1].

Kassinin, DVPKSDQFVG^{10}LMa, a 12-peptide amide belonging to the → tachykinin family. Kassinin was isolated from methanol extracts of the skin of the African frog *Kassina senegalensis* in 1977. It causes contraction of smooth muscle preparations only in an activity of 0.5% compared with → substance P. [A. Anastasi et al., *Experientia* **1977**, *33*, 857].

Katacalcin, *PDN sequence*, ***PDN-21***, DMSSDLERDH^{10}RPHVSMPQNA^{20}N, the C-terminal flanking 21-peptide of the → calcitonin precursor. Besides mature calcitonin (CT), katacalcin is formed during processing of the precursor in C cells. Since medullary thyroid carcinoma (MTC) are the only known condition in which the plasma CT concentrations are consistently and significantly increased, the measurement of katacalcin or serum CT is generally used as a marker for this tumor. Assays based on katacalcin and CT not only help in diagnosis of MTC but also reflect the extent of the removal of the tumor and are useful in detecting early recurrence. Reports concerning the potent calcium-lowering activity of katacalcin have not yet been confirmed [I. MacIntyre et al., *Nature* **1984**, *308*, 84; S. J. Wimalawansa, *Crit. Rev. Neurobiol.* **1997**, *11*, 167].

Kemptide, H-Leu-Arg-Arg-Ala-Ser-Leu-Gly-OH, a synthetic 7-peptide substrate related to a partial sequence of the phosphorylation site in porcine liver pyruvate kinase. Kemptide is phosphorylated *in vitro* and *in vivo* by cAMP-dependent protein kinase, and *in vitro* by protein kinase C [B. E. Kemp et al., *Fed. Proc.* **1976**, *35*, 1384; J. Zhou, J. A. Adams, *Biochemistry* **1997**, *336*, 15733].

Kentsin, H-Thr-Pro-Arg-Lys-OH, a contraceptive 4-peptide first isolated from the oviductal lumen of progravid female hamsters [H. A. Kent, Jr., *Biol. Reprod.* **1973**, *8*, 38; H. A. Kent, Jr., *Fertil. Steril.* **1979**, *31*, 595].

Keramamide, a group of oxazole- and thiazole-containing weakly cytotoxic cyclopeptides from the sponge *Theonella* sp. Members of this group are *orbiculamide A* and the *keramamides B, C, D,* and *E*. They contain unusual building blocks such as vinlylogous and α-ketohomologous amino acids. The *keramamides F, G, H,* and *J* have similar structures, but the oxazole ring is replaced by a thiazole structure moiety. [P. Wipf, *Chem. Rev.* **1995**, *95*, 2115].

Keratinocyte growth factor, *KGF*, a → cytokine belonging to the → fibroblast growth factors. The natural hKGF (163 aa) shows mitogenic activity on keratinocytes and endothelial cells. KGF is involved as a specific paracrine factor in the regulation of the proliferation and differentiation of normal epithelial cells. The KGF receptor is a membrane-spanning tyrosine kinase [J. S. Rubin et al., *Proc. Natl. Acad. Sci. USA* **1989**, *86*, 802].

Keratins, a group of fibrous proteins occurring in wool, hair, hoofs, claws, horns and feather, but also in skin, connective tissues and intermediate filaments. *α-Keratins* occur in mammals, having around 30 various variants, whereas *β-keratins* are found in birds, reptiles, and in the silk of insects and arachnids. *α-Keratin* forms closely associated pairs of α-helices in which each pair is composed of a type I and a type II keratin chain twisted in parallel into a left-handed coil. The ~ 30-Å-wide protofilaments consist of two staggered and antiparallel rows of associated head-to-tail coiled coils. After the protofilaments dimerize to a ~ 50-Å-wide protofibril, four of these in turn form a microfibril. α-Keratins have a high content of Cys residues (2–16%) which form disulfide bridges cross-linking adjacent polypeptide chains. Insolubility and resistance to stretching are important properties of α-keratin. α-Keratins are subdivided into *"hard"* or *"soft"* keratins, depending on high or low Cys contents. Hard keratins occur, for example, in hair, horn and nail, whereas soft keratins are found in skin and callus. The elasticity of wool fibers and hair is based on the coiled coil's tendency to untwist after stretching and to recover its original conformation after relaxing the external force. *Silk fibroin* (M_r ~ 365 kDa) is an important *β-keratin*, in which the sequence -(Gly-Ser-Gly-Ala-Gly-Ala)$_n$- is repeated many times. The polypeptide chains form antiparallel β-pleated sheets in which the peptide chains extend parallel to the fiber axis. The relative instability of silk to stretch, plus its great flexibility, are caused by the strong covalent bonds between the extended peptide chain and the weak van der Waals' forces between the pleated sheets. Skin in higher animals contains an extensive network of *intermediate filaments* (*IF*) made of keratin forming protein fibers 100–150 Å in diameter. Keratin is mainly responsible for the toughness of this protective outer covering. IF are formed from four types of fibrous polypeptides. Type I IF, which occur primarily in epithelial cells, include two subfamilies of keratin, *acidic keratin* and *neutral* or *basic keratin* with M_r between 40 and 70 kDa [P. M. Steinert, D. A. D. Parry, *Annu. Rev. Cell. Biol.* **1985**, *1*, 41].

KGF receptor peptide, HSGINSSNAE^{10}VLA LFNVTEM^{20}DAGEY, a synthetic 25-peptide corresponding to the partial sequence 199–223 of the KGF receptor alternative exon. It blocks the mitogenic activity of KGF receptor (→ keratinocyte growth factor) and the interaction between KGF and its receptor [D.P. Bottaro et al., *J. Biol. Chem.* **1993**, *268*, 9180].

Kinetensin, H-Ile-Ala-Arg-Arg-His-Pro-Tyr-Phe-Leu-OH, a neurotensin-related 9-peptide obtained by pepsin cleavage of human, bovine and dog plasma albumins. It increases vascular permeability after intradermal injection into rats, and releases histamine from rat mast cells *in vitro*. Kinetensin shows sequence similarities to → neurotensin and → angiotensin I [M.H. Mogard et al., *Biochem. Biophys. Res. Commun.* **1986**, *136*, 983; R.E. Carraway et al., *J. Biol. Chem.* **1987**, *262*, 5968].

Kinins, → plasmakinins.

Kyotorphin, H-Tyr-Arg-OH, a neuroactive dipeptide first identified in the rat hypothalamus. Kyotorphin was named from the site of its discovery, Kyoto (Japan). When administered intracisternally to mice, it has an analgesic effect which is about 4.2 times more potent than that of [Met]enkephalin. Kyotorphin does not interact with an opioid receptor. It is thought to act by releasing enkephalin and to stabilize it against degradation [H. Takagi et al., *Nature* **1979**, *282*, 410; H. Takagi et al., *Eur. J. Pharmacol.* **1979**, *56*, 265].

L

Lactacystin, a microbial product from a *Streptomyces* strain which acts as a selective and potent irreversible inhibitor of the 20 S → proteasome. Lactacystin and the related β-lactone omuralide are able, as small molecules (***Formula***), to inhibit irreversibly the proteolytic activity of the 20 S → proteasome [S. Omura et al., *J. Antibiotics* **1991**, *44*, 113; E.J. Corey, W.D.Z. Li, *Chem. Pharm. Bull.* **1999**, *47*, 1].

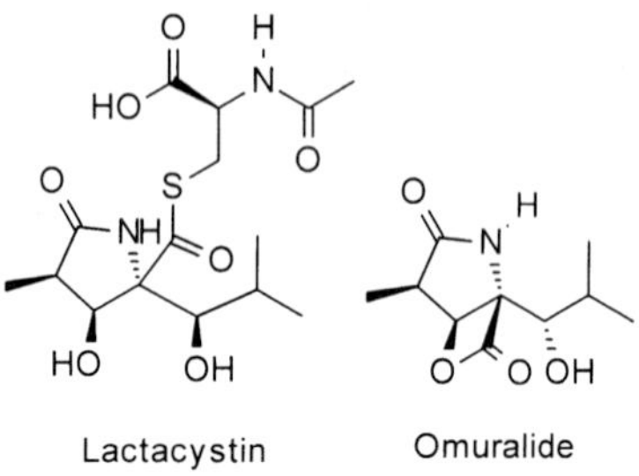

Lactacystins

α-Lactalbumin, one of the most important members of the → milk proteins. Human lactalbumin (123 aa; $M_r \sim 14$ kDa) occurs at a concentration of 0.14–0.6% in human milk. α-Lactalbumin acts as one subunit of the lactose synthase, but it is catalytically inactive by itself.

β-Lactam antibiotics, a family of antibiotics containing a β-lactam ring (2-azetidinone). *Monobactams* (name derived from *mono*cyclic *bact*erially produced β-lact*am* antibiotics) are β-lactam antibiotics, with only the lactam ring such as aztreonam and nocaricin A. The majority of β-lactam antibiotics have a second ring condensed to the β-lactam ring with basic types *penam, cepham, carbapenam,* and *clavan* (**Formula**). The most important groups of β-lactam antibiotics are the → penicillins and → cephalosporins. The β-lactam antibiotics inhibit the biosynthesis of bacterial cell walls.

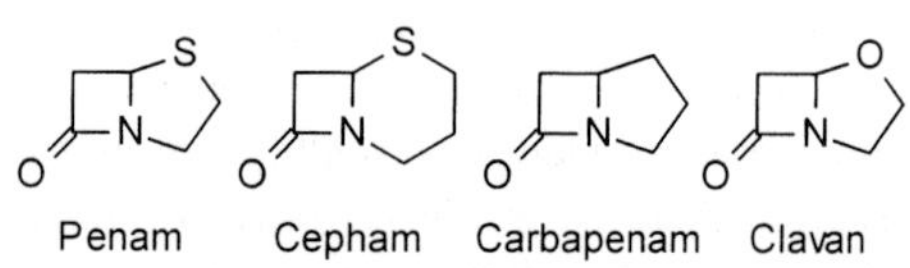

Basic types of β-lactam antibiotics

Further information is provided in Chapter 3.3.3.

Lactoferrin, an iron-binding protein, occurs in granulocytes and epithelial cells of glands. Lactoferrin ($M_r \sim$ 77–93 kDa) shows functional and structural similarities with → transferrin. Lactoferrin reversibly binds two atoms of iron. The antibacterial action of lactoferrin is based on its ability to withdraw iron which is essential for bacterial life.

Lactogenic hormone, → prolactin.

β-Lactoglobulin, in milk and colostrums occurring → globulin. β-Lactoglobulin (M_r $\sim$ 37 kDa) consists of two subunits, and shows structural similarity to the retinol-binding protein of blood plasma.

Lactorphins, → milk protein-derived opioid peptides.

Laminin, a large protein complex (1500 aa; M_r $\sim$ 850–1000 kDa) consisting of three polypeptide chains (A, B1, and B2). Laminin is an adhesive glycoprotein of the extracellular matrix. It occurs specifically in the basal lamina and enables epithelial cells to bind to underlying connective tissue. The three polypeptide chains are arranged in the shape of a cross, held together by disulfide bonds. The three chains form a long coiled-coil, α-helical domain that separates in different directions to form the three shorter arms of the cross. Laminin consists of various functional domains, among which one binds to type IV → collagen, one to the proteoglycan heparan sulfate, and others bind to the laminin receptor. It is thought that the glycoprotein *entactin* is tightly bound to each laminin molecule [J. Engel et al., *J. Mol. Biol.* **1981**, *150,* 97; G.R. Martin, R. Timpl, *Annu. Rev. Cell. Biol.* **1987**, *3,* 57; A. Utani et al., *J. Biol. Chem.* **1995**, *270,* 3292].

Lantibiotics, a class of peptides the name of which is derived from "***lan**thionine-containing **antibiotic** peptides*" belonging to the subgroup of → bacteriocins. They are characterized by the presence of the thioether amino acids lanthionine and 3-methyllanthionone that are generated via post-translational modification, and are produced by a wide variety of Gram-positive bacterial strains. These peptides consist of a minimum of 19 and a maximum of 38 residues. In the biosynthesis of lantibiotics, the precursors of these peptides are first modified and then proteolytically activated (see Section 3.3.3.2; Fig. 3.24). This may occur either before, concomitant with, or after the export of lantibiotics from the cell. The gene organization for the biosynthetic machinery is performed in clusters, including information for the antibiotic propeptide, the modification enzymes and accessory functions, e.g., special proteases and ABC transporter, immunity factors, and regulatory peptides. According to Jung, the lantibiotics are subdivided into two types: *Type-A* (→ nisin, → subtilin, → epidermin, → gallidermin, → pep5); and *type-B* (→ cinnamycin, → ancovenin, → duramycins). Type-A lantibiotics are elongated, helical amphiphiles in which action is directed to the formation of pores in the cytoplasmic membrane of susceptible bacteria, whereas type-B lantibiotics are more compact, forming globular structures, and generally interrupt a variety of membrane-derived enzyme functions [G. Jung, H.-G. Sahl (Eds.), *Nisin and Novel Lantibiotics*, ESCOM, Leiden, **1991**; A. Guder et al., *Biopolymers (Peptide Sci.)* **2000**, *55,* 62].

Lepirudin®, *[Leu^1,Thr^2]-63-desulfatohirudin*, a recombinant → hirudin (65 aa; M_r $\sim$ 6.9 kDa) first approved for the treatment of heparin-induced thrombocytopenia (HIT). Later, a new indication was found in the treatment of unstable angina. *Desirudin*® differs from Lepirudin only in the first two N-terminal residues (Val^1, Val^2), and it plays a definite role in thrombosis prophylaxis. Lepirudin directly inhibits the active site pocket and the fibrinogen binding site of free and clot-bound thrombin [A. Greinacher, N. Lubenow, *Circulation* **2001**, *103*, 1479].

Leptin, the protein product of the obese gene (*ob* gene) secreted from white adipose tissue cells [*leptos* (Greek): thin, small]. It is a blood-borne satiety factor that acts directly on the hypothalamus to regulate a large number of molecules implicated in energy homeostasis. The identification of single-gene mutant mice that were obese and hyperphagic first led to parabiosis experiments characterized by continuous transfusion of sera from wild-type (nonobese) mice into obese mutant mice. Obese mice homozygous for the recessive *ob* allele (*ob/ob* mice) responded to these experiments by decreasing body weight, whereas obese mice homozygous for the recessive *db* allele (*db/db* mice) were unresponsive. Leptin treatment attenuates the hyperphagia and obesity in *ob/ob* but not in *db/db* mice. The leptin receptor is localized in brain and other tissues in rodents and nonhuman primates. The *db/db* genotype seems to code for a deficient leptin receptor protein. Administration of mouse rec. leptin to *ob/ob* mice corrects both overeating and obesity. Furthermore, leptin reduces appetite and weight of mice with diet-induced obesity. The orexigenic peptide → ghrelin antagonizes leptin action through the activation of hypothalamic NY/Y1 receptor pathway [Y. Zhang et al., *Nature* **1994**, *372*, 425; R.V. Considine et al., *N. Engl. J. Med.* **1996**, *334*, 292; J.M. Friedman, J.L. Halaas, *Nature* **1998**, *395*, 763; J.E. Schneider, *Hormones and Behavior* **2000**, *37*, 258].

Leualacin, a cyclic pentadepsipeptide from *Hapsidospora irregularis*. Leualacin is a new Ca^{2+}-blocking agent (***Formula***) which is structurally completely different from clinically used compounds such as benzodiazepines, dihydropyridines, and verapamil derivatives. Leualacin competitively inhibits the specific binding of nitrendipines on microsomes of porcine heart [K. Hamano et al., *J. Antibiot.* **1992**, *45*, 899; U. Schmidt, J. Langner, *J. Chem. Soc. Commun.* **1994**, 2381].

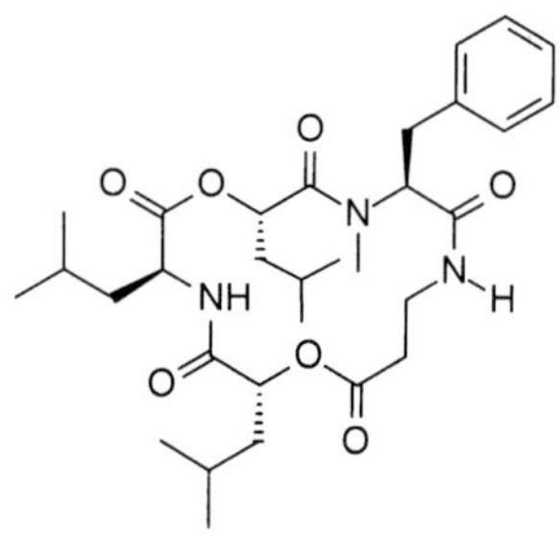

Leualacin

Leucokinins, insect neuropeptides originally isolated from head extracts of the Madeira cockroach *Leucophaea maderae*. *Leucokinin I*, H-Asp-Pro-Ala-Phe-Asn-Ser-Trp-Gly-NH_2, and three other members of this group (*Leucokinin II–Leucokinin IV*) are 8-peptide amides with both identical residues in position 1 and the same C-terminal tripeptide sequence. They stimulate contractions of the lower digestive tract of the cockroach. Further members of the leucokinins are *leucomyosuppressin* (*LMS*), pGlu-Asp-Val-Asp-His-Val-Phe-Leu-Arg-Phe^{10}-NH_2, and *leucopyrokinin* (*LPK*), pGlu-Thr-Ser-Phe-Thr-Pro-Arg-Leu-NH_2 [J. R. Nachmann et al., *Biochem. Biophys. Res. Commun.* **1986**, *137*, 936; J. van Brunt, *Biotechnology* **1987**, *5*, 31; G.M. Coast, *Peptides* **1996**, *17*, 327].

Leucomyosuppressin, → leucokinins.

Leucopyrokinin, → leucokinins.

Leucosulfakinins, tyrosine-sulfated neuropeptides isolated from the head of the cockroach *Leucophaea maderae*. *Leucosulfakinin I*, EEFEDY^{S}GHMR^{10}Fa, and *Leucosulfakinin II*, <ESDDY^{S}GHMRF^{10}a (Y^S: SO_3H) resemble each other and show some similarities to mammalian → cholecystokinin and → gastrin, but they do not belong to the CCK/gastrin family due to the discrepancies between the C-terminal tetrapeptide amide sequence [R. J. Nachman et al., *Science* **1986**, *234*, 71; R. J. Nachman et al., *Biochem. Biophys. Res. Commun.* **1986**, *140*, 357].

Leukemia inhibitory factor, ***LIF***, a variably glycosylated protein consisting of a polypeptide chain (hLIF: 180 aa) with pleiotropic action. Target cells of LIF are monocytes, macrophages, and their precursors. It promotes the entry of hematopoetic stem cells into the cell cycle.

Leukocyte interferon, IFN-α, → interferons.

Leukokinin, a specific leukophilic γ-globulin. Leukokinin binds to the leukocyte membrane and acts as precursor for → tuftsin.

Levitide, <EGMIGTLTSK^{10}RIKQa, a 14-peptide amide from the skin secretions of the South African frog *Xenopus laevis* acting as a neuropeptide [L. Poulter et al., *J. Biol. Chem.* **1988**, *263*, 3279].

LHRH, acronym of luteinizing hormone-releasing hormone, → gonadotropin-releasing hormone.

Liberins, IUPAC-IUB designation for releasing hormones and releasing factors. The suffix *-liberin* describes the corresponding releasing hormone of the hypothalamus, such as → corticoliberin.

Lipoproteins, noncovalent conjugates consisting of lipids and proteins occurring, for example, in cellular membranes, cell cytoplasm and blood plasma, and involved in transport and distribution of water-insoluble lipids in body fluids. The protein components of the lipoproteins are termed *apolipoproteins* or *apoproteins*. Apolipoproteins form amphipathic helices that coat lipoprotein surfaces. Human lipoproteins consist of at least nine apolipoproteins (A-I, A-II, B-48, B-100, C-I, C-II, C-III, D, E). The major classes of lipoproteins in human blood plasma are divided into *chylomicrons* (density <0.95 g cm^{-3}; major apolipoprotein: A-I, A-II, B-48, C-I, C-III, E), *very low-density lipoproteins, VLDL* (density: <1.006 g cm^{-3}; major apolipoproteins: B-100, C-I, C-II, C-III, E), *intermediate density lipoproteins, IDL* (density: 1.006–1.019 g cm^{-3}; major apolipoproteins: B-100, C-III, E), *low-density lipoproteins, LDL* (density: 1.019–1.063 g cm^{-3}; major apolipoprotein: B-100), and *high-density lipoproteins, HDL* (density: 1.063–1.210 g cm^{-3}; major apolipoproteins: A-I, A-II, C-I, C-II, C-III, D, E). Chylomicrons transport externally supplied triacylglycerols and cholesterol from the intestines to the tissues. VLDL, IDL and LDL are responsible for the transport of endogenous triacylglycerols and cholesterol from the liver to the tissues, whereas HDL transport endogenous cholesterol from the tissues to the liver. An excess of plasma LDL promotes atherosclerosis, which is also correlated with a low concentration of HDL [J. P. Kane, *Curr. Opin. Struct. Biol.* **1991**, *1*, 510; M. Rosseneu (Ed.), *Structure and Function of Apolipoproteins*, CRC Press, Boca Raton, FL, **1992**; D. Atkinson, *Curr. Opin. Struct. Biol.* **1992**, *2*, 482; R. M. Lawn, *Sci. Am.* **1992**, *266*, 54].

Lipotropic hormone, → lipotropin.

Lipotropin, *lipotropic hormone*, ***LPH***, a polypeptide hormone from the hypophysis stimulating the mobilization of lipids from lipid depots. *β-LPH* (91 aa; $M_r \sim 10$ kDa) is biosynthetically formed from the precursor → pro-opiomelanocortin (PMOC). β-LPH is released from PMOC in the anterior and intermediate lobes of the pituitary gland, whereas in the intermediate lobe only β-LPH is split to γ-LPH (corresponds to β-LPH 1–58) and → β-endorphin [J. Bogard et al., *J. Biol. Chem.* **1995**, *270*, 23038].

Litorin, pGlu-Gln-Trp-Ala-Val-Gly-His-Phe-Met-NH_2, a member of the → ranatensin family. Litorin was isolated from the skin of the Australian frog *Litoria aurea*. Litorin shows similar activities to → ranatensin [V. Erspamer, P. Melchiorri, *Trends Biochem. Sci.* **1980**, *1*, 391].

Liver cell growth factor, → glycyl-L-histidyl-L-lysine.

Locustakinin I, H-Ala-Phe-Ser-Ser-Trp-Gly-NH_2, a 6-peptide amide from *Locusta migratoria*. Locustakinin is a myotropic neuropeptide with sequence similarity to cockroach → leucokinins [L. Schoofs et al., *Regul. Peptides* **1992**, *37*, 49; G.M. Coast, *Peptides* **1996**, *17*, 327].

Luliberin, → gonadoliberin.

Luteinizing hormone, → lutropin.

Luteinizing hormone-releasing hormone, → gonadoliberin.

Lutropin, *luteinizing hormone*, ***LH***, a heterodimeric glycoprotein that belongs to the mammalian → gonadotropins. Human LH ($M_r \sim 23$ kDa) consists of a α subunit (96 aa), which is identical with those of → follitropin (FSH), → thyrotropin, → chorionic gonadotropin, and a β-chain (121 aa). The secretion of LH is regulated by → gonadoliberin. LH stimulates, together with FSH, growth and synthesis of sex hormones and spermatogenesis [N.B. Schwartz, *Can. J. Physiol. Pharmacol.* **1995**, *73*, 675].

Lymphokines, → interleukins.

Lymphotactin, → chemokines.

M

α_2-Macroglobulin, *α_2-antiplasmin*, a glycoprotein of the α_2-globulin fraction acting as an endogenous inhibitor of mechanistically different proteases. α_2-Macroglobulin consists of four identical subunits ($M_r \sim 720$ kDa). It binds tightly to a number of proteases such as → plasmin, → thrombin, chymotrypsin, but unlike other protease inhibitors it does not block the active sites of the enzymes. For example, α_2-macroglobulin forms an irreversible equimolar complex with plasmin that prevents it from binding to → fibrin. Generally, the interaction of α_2-macroglobulin with the appropriate proteases results in the formation of receptor recognition regions on α_2-macroglobulin which allow the binding of the α_2-macroglobulin-protease complex on various cells, especially on macrophages for elimination from the plasma. In humans, the average plasma concentration of α_2-macroglobulin is 220–380 mg per 100 mL.

Macrophage CSF, → colony-stimulating factors.

Maduropeptin, ***MDP***, an antitumor-acting chromoprotein isolated from the culture medium of *Actinomadura madurae*. Maduropeptin consists of an endiin-containing chromophore (***Formula***) and an acidic stabilizing protein ($M_r \sim 32$ kDa). The chromophore forms a labile 9-membered endiin ring without any homology to those of similar endiin-containing chromoproteins such as → neocarcinostatin, macromomycin, or kedarcidin. The maduropeptin chromophore shows antibacterial and cytotoxic activities itself. It cleaves dsDNA *in vitro* at specific biologically critically DNA sequences, whereas the protease-like apoprotein stabilizes and solubilizes the chromophore. Furthermore, maduropeptin catalyzes *in vitro* the degradation of histones to low-molecular weight peptides [N. Zein et al., *Biochemistry* **1995**, *34*, 11591].

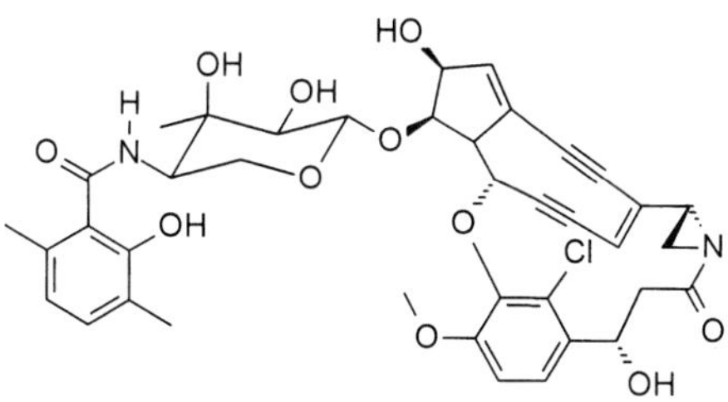

Maduropeptin

Magainins, amphibian-derived peptides with helical, amphiphilic structure resulting in high membrane affinity. They are synthesized in the skin glands of frogs and other amphibians. Magainins with 20 to 25 residues are shorter than → cecropins, but share the overall features of charge distribution, helicity and amphiphilic properties. The 23-peptides *magainin 1*, GIGKFLHSAG^{10}KFGKAFVGEI^{20}MKS, and *magainin 2*, [Lys^{10},Asn^{22}]magainin 1, from the South African clawed frog (*Xenopus laevis*) and other magainins are α-helical ionophores which dissipate ion gradients in membranes, causing lysis. *MSI-78*, GIGKFLKKAK^{10}KFGKAFVKIL^{20}KKa, a synthetic analogue based on the magainin consensus sequence, has been developed as a drug for topical treatment of Gram-positive infections. Another analogue, termed *pexiganan acetate*, has obtained approval for the treatment of diabetic foot ulcers [M. Zasloff, *Proc. Natl. Acad. Sci. USA* **1987**, *84*, 5449; D. Hirsh et al., *Biochemistry* **1996**, *35*, 12733].

Major proglucagon fragment, ***MPGF***, → glucagon.

Malformins, heterodetic cyclic pentapeptides with antibiotic and cytotoxic action. *Malformin A_1*, cyclo-(-D-Cys-D-Cys-Val-D-Leu-Xaa-) (disulfide bond: D-Cys^1-D-Cys^2; Xaa = Ile), is produced by *Aspergillus niger*. Further members of this group are *malformin B_1* (Xaa = aIle), *malformin B_2* (Xaa = Val), and *malformin C* (Xaa = Leu). Malformins cause malformation of the roots of cereals and other higher plants.

Maspin, a protein (375 aa; $M_r \sim 42$ kDa) with N-terminal methionine, C-terminal valine and eight cysteines. It is a member of the serpins. The cDNA of maspin was isolated from a normal human mammary epithelial cell library. In invasive breast carcinomas, the gene of maspin is down-regulated. Maspin is located in the cell membrane and extracellular matrix, and has been shown to have tumor suppressor activity. It does not behave as a classical inhibitory serpin against any known target protease [Z. Zou et al., *Science* **1994**, *263*, 526].

Mast cell degranulating peptide, ***MCDP***, IKCNCKRHVI^{10}KPHICRKICG^{20}KNa (disulfide bonds: C^3–C^{15}/C^5–C^{19}), a cationic 22-peptide amide. MCPD is a component of the bee venom which contains several biologically active nonpeptide substances as well as two more major peptides, the hemolytic peptide → melittin, the neurotoxic peptide → apamin, together with a number of minor peptides. Like apamin, MCDP blocks Ca^{2+}-dependent K^+ channels in neurons. MCDP causes mast cell degranulation and histamine release at low concentrations, and has anti-inflammatory activity at higher concentrations [A. Buku, *Peptides* **1999**, *20*, 415].

Mastoparan, INLKALAALA^{10}KKILa, a 14-peptide amide found in the venom of wasps. Mastoparan degranulates mast cells and induces the release of catecholamines and serotonin from adrenal chromaffin cells or platelets. *Mastoparan C*, [$Leu^{1,7}$, Val^{9}]mastoparan, and crabolin are constituents of the venom of European hornets. Like → mellitin, mastoparan shows hemolytic activity, and forms artificial ion channels in guest cell membranes [C. Hider, *Endeavour, New Series* **1988**, *12*, 60; A. Argiolas, J.J. Pisano, *J. Biol. Chem.* **1984**, *259*, 10106].

Matrix Gla protein, ***MGP***, a γ-carboxyglutamic acid(Gla)-containing protein that occurs associated with bone marrow. Besides → osteocalcin, also named *bone Gla protein* (BGP), MGP is the second Gla-containing protein in bone and contains 80% of Gla occurring in bone. hMGP is a vitamin K-dependent extracellular matrix protein that binds Ca^{2+} ions and is involved in the prevention of vascular calcification. Human MGP (84 aa; $M_r \sim 10.5$ kDa) contains five Gla residues and a disulfide bond (C^{54}–C^{60}). The total chemical synthesis of hMGP was described in 2001 [P.A. Price et al., *Biochem. Biophys. Res. Commun.* **1983**, *117*, 765; T.M. Hackeng et al., *Protein Sci.* **2001**, *10*, 864].

Mauritine A, a cyclopeptide alkaloid isolated from the root bark of *Ziziphus mauritiana* (***Formula***) The first total synthesis was performed using a novel cycloetherification methodology in 2000 [R. Tschesche et al., *Tetrahedron Lett.* **1972**, 2609; T. Laib et al., *Tetrahedron Lett.* **2000**, *41*, 7645].

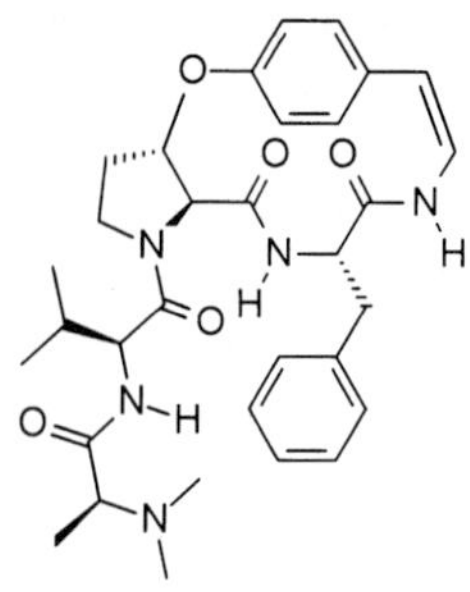

Mauritine A

Melanin-concentrating hormone, ***MCH***, DFDMLRCMLG[10]RVYRPCWQV (hMCH; disulfide bond: C^{7}–C^{16}), a 19-peptide from the mammalian hypothalamus. The sequence of hMCH was found to be the same as of the rat and mouse peptide. The physiological role of MCH in higher vertebrates has not been fully elucidated, but it may act as a neuromodulator and hypophysiotropic agent in the secretion of → corticotropin, growth hormone (→ somatotropin), and α-MSH (→ melanotropin). There is evidence that the hypothalamic MCH is involved in the regulation of body weight. Interestingly, the MCH receptor was reported to be the orphan receptor SLC-1. MCH was originally isolated from salmon pituitary glands (DTMRCMVGRV[10]YRPCWEV, disulfide bond: C^{5}–C^{14}). In lower vertebrates, salmon MCH alters pigmentation by inducing the aggregation of melanin granules in melanophores [H. Kawauchi et al., *Nature* **1983**, *305*, 321; J.M. Vaughan et al., *Endocrinology* **1989**, *125*, 1660; K.M. Knigge et al., *Peptides* **1996**, *17*, 1063; D.Qu et al., *Nature* **1996**, *380*, 243; J. Chambers et al., *Nature* **1999**, *400*, 261].

Melanocyte-stimulating hormone, → melanotropin.

Melanoliberin, *melanotropin-releasing hormone*, ***MRH***, *melanotropin-releasing factor*, ***MRF***, a hypothalamic peptide stimulating the release of → melanotropin. The hexapeptide fragment of → oxytocin, H-Cys-Tyr-Ile-Gln-Asn-Cys-OH, has been suggested for MRH.

Melanostatin, *melanotropin release-inhibiting hormone*, ***MRIH*** or ***MIH***, *melanotropin re-*

lease-inhibiting factor, ***MIF***, a hypothalamic peptide inhibiting the release of → melanotropin. The C-terminal tripeptide amide, H-Pro-Leu-Gly-NH_2, of → oxytocin has been suggested as a structural proposal for melanostatin. Melanostatin inhibits the release of α-MSH in the hypophysis. The analogue [*N*-Me-D-Leu^2]MRIH, named *pareptide*, increases the L-DOPA-induced behavior pattern. It shows antidepressant activity and has been used for the treatment of Parkinson's disease.

Melanotropin, *melanocyte-stimulating hormone*, ***MSH***, a peptide hormone produced in the pars intermedia of the hypophysis under the control of → melanoliberin and → melanostatin. Species lacking the pars intermedia (e.g., chicken, whale, porpoise) produce it in the neurohypophysis. Furthermore, MSH is formed by neurons of the central nervous system. Three different MSH sequences are parts of the biosynthetic precursor protein → pro-opiomelanocortin (POMC). Mammalian *α-MSH*, Ac-SYSMEHFRWG^{10}KPVa, exerts its activity at peripheral receptors that are mainly responsible for pigmentation. The acetyl group is important for melanotropic activity. Furthermore, α-MSH shows central effects on memory and thermal regulation. In frogs and certain lizards, the release of α-MSH from the pars intermedia of the pituitary causes a dark-colored background which results in the dispersion of melanin granula in the melanophores. Bovine *β-MSH*, DSGPYKMEHF^{10}RWGSPPKD, has slightly lower biological activity than α-MSH in the frog skin test. It probably occurs naturally, but is not supposed to have peripheral functions. Human β-MSH from the hypothalamus differs from bovine-β-MSH in two positions (Glu^2 and Arg^6). Various species contain two different β-MSH peptides. The same is true for a third MSH peptide found in the N-terminal part of POMC. *γ-MSH*, YVMGHFRWDR^{10}FG, occurs naturally and may have a function, like the N-terminal fragment of POMC, in promoting growth of the adrenal cortex and a hypertensive effect. Structure-activity relationship studies on α-MSH were started soon after the first isolation and structure determination in 1957. The linear analogue [Nle^4,D-Phe^7]α-MSH has greatly increased potency and prolonged duration of activity in various bioassays. For example, it causes increased pigmentation in the yellow mouse after reabsorption through the skin. Further research has been directed towards analogues with potential clinical use in the treatment of pigmentary disorders and melanoma [J. I. Harris, A. B. Lerner, *Nature* **1957**, *179*, 1346; A. Eberle, *Melanotropins: Chemistry, Physiology and Mechanisms*, Karger, Basel, **1988**; V. J. Hruby et al., *Ann. N. Y. Acad. Sci.* **1993**, *680*, 51].

Mellitin, GIGAVLKVLT^{10}TGLPALISWI^{20}KRKRQQ, a 26-peptide amide which comprises about 50% of the dried bee venom. It is synthesized as part of prepro-mellitin consisting of 70 residues. Removal of the signal sequence (21 aa) leaves pro-mellitin. The enzymatic release from the precursor occurs outside the venom gland cells. The hemolytic and surface tension-decreasing activity of mellitin is based on the distribution of hydrophobic residues in the N-terminal part and the hydrophilic residues in the C-terminal part. The resulting tenside character is probably a prerequisite for the pharmacological and biochemical action [R. C. Hider, *Endeavour, New Series* **1988**, *12*, 60].

Metalloproteins, proteins with bound metal ions. To this family of transport and storage proteins belong, e.g., → ferritin, →

transferrin, → iron-sulfur proteins, → metallothioneins.

Metallothioneins, highly conserved, metal ion-binding proteins. They are present in all vertebrates, invertebrates and fungi, and are implicated in heavy metal ion detoxification processes. The synthesis of metallothioneins is triggered by heavy metal ions (e.g., Cd^{2+}, Zn^{2+}, Cu^{2+}, Hg^{2+}, and other transition metals). The metallothioneins are small proteins (M_r ~ 6.5–7 kDa) consisting of about 60 residues with a high content of Cys (20–35%). Most of the mammals synthesize two isoforms [B. P. Monia et al., *J. Biol. Chem.* **1986**, *261*, 10957].

Metamorphosin A, ***MMA***, pGlu-Gln-Pro-Gly-Leu-Trp-NH_2, a 6-neuropeptide amide originally isolated from the anthozoan *Anthopleura elegantissima*. MMA triggers metamorphosis in *Hydractinia echinata*. The analogue [Lys^1,Pro^2]MMA, termed *He-LWamide II*, is a member of the neuropeptide family LWamides which is found in *Hydractinia echinata*, and is involved in the control of the morphogenetic process [T. Leitz et al., *Dev. Biol.* **1994**, *163*, 440; M. Gajewski et al., *Roux's Arch. Dev. Biol.* **1996**, *205*, 232].

Metorphamide, *adrenorphin*, H-Tyr-Gly-Gly-Phe-Met-Arg-Arg-Val-NH_2, an 8-peptide amide found in bovine brain and adrenal tissue. The sequence of metorphamide corresponds to bovine proenkephalin precursor-(206–213). Furthermore, metorphamide is similar in sequence to bovine adrenal medulla peptides. It shows high affinity to μ opioid receptors [H. Matsuo et al., *Nature* **1983**, *305*, 721; M. Sonders, E. Weber, *J. Neurochem.* **1987**, *49*, 671].

MHC molecules, → MHC proteins.

MHC proteins, *MHC molecules*, membrane-bound proteins encoded by the major histocompatibility complex (MHC; *histo* refers to tissue). With the *Class I MHC proteins* and the *Class II MHC proteins* two principal classes of MHC molecules were recognized long before the native function was understood. They act as the antigen-presenting markers by which the immune system is capable to distinguish body cells from invading antigens (Class I MHC proteins), and immune system cells from other cells (Class II MHC proteins). The MHC proteins are integral membrane glycoproteins with the biological function to bind peptides and present them as T-cell epitopes to T-cell receptors. In order to indicate their function in the discrimination between own and foreign, the products of the MHC gene cluster are also termed antigens. *Class I MHC molecules* (~ 350 aa; M_r ~ 44 kDa) are folded into five domains: a C-terminal cytoplasmic domain (~ 30 aa), a transmembrane domain (~ 40 aa), and three external domains (~ 90 aa each, designated α_3, α_2, α_1). Additionally, the Class I MHC proteins are invariably associated (1:1) with an extracellular, nonglycosylated small protein, termed *β_2-microglobulin* (*$\beta_2 m$*) (→microglobulins), which does not span the membrane, and is separately encoded by a gene on a different chromosome. The α_3 domain and β_2m, closest located to the membrane, are both homologous to an → immunoglobulin domain. *Class II MHC proteins* are heterodimeric transmembrane glycoproteins consisting of a α chain (M_r ~ 33 kDa) and a β chain (M_r ~ 28 kDa). Each (~ 230 aa) of the two chains forms two conserved immunoglobulin-like domains (α_1, α_2 and β_1, β_2, respectively). The formulae of both types are schematically shown in Fig. 3.4. Class I MHC proteins occur on most cells, whereas Class II MHC molecules are found on macrophages and B lymphocytes. The peptide fragments are presented to the T-cell receptor either via a Class I or Class II MHC molecule. As a rule, Class I MHC molecules present peptides to CD8 (predominantly cytotoxic) T cells, and Class II MHC molecules to CD4 (mainly helper) T cells. The genes encoding MHC proteins form a gene superfamily, since they have domains that are structurally similar to those in immunoglobulins and T-cell receptors [P. J. Bjorkman et al., *Nature* **1987**, *329*, 512; P. J. Bjorkman, P. Parham, *Annu. Rev. Biochem.* **1990**, *59*, 253; L. D. Barber, P. Parham, *Annu. Rev. Cell. Biol.* **1993**, *9*, 163; L. J. Stern, D. C. Wiley, *Structure* **1994**, *2*, 245; P. J. Lehner, P. Cresswell, *Curr. Opin. Immunol.* **1996**, *8*, 59].

Microcins, a highly diverse group of small antibiotic peptides from Gram-negative bacteria. The designation microcins was given in order to make a clear-cut distinction between the proteins → bacteriocins, such as colicins, cloacins and other enterobacterial bacteriocins, and low-molecular weight (<10 kDa) peptide antibiotics. Among the more than 20 peptides of this family *microcin B17* is, in many aspects, exceptional. Due to its unique post-translational modification pattern, it may be considered as a Gram-negative analogue of the → lantibiotics. *Microcin B17* is a 43-peptide (M_r 3093 Da) which contains both oxazole and thiazole rings formed from the modification of specific Ser, Cys, and Gly residues in the prepeptide sequence (***Formula***). Furthermore, more than two-thirds of the total amino acid residues consist of glycine. Microcin B17 is an inhibitor of DNA gyrases. *Microcin C_7* is a N-terminally acetylated 7-peptide (MRTGNAD) bearing another yet uncharacterized C-terminal modification and which inhibits protein biosynthesis. *Microcin E_{492}* (M_r ~ 5–7 kDa) is structurally and functionally related to the bacteriocins of the Gram-positive bacteria, and depolarizes cytoplasmic membranes. *Colicin V* (M_r 8741 Da), originally regarded as a colicin, consists of 88 unmodified amino acids and was long known as a pore-forming bacteriocin. It is processed from a 103-pre-peptide [R. James, C. Lazdunski, F. Pattus (Eds.), *Bacteriocins, Microcins and Lantibiotics*, Springer, Berlin, Heidelberg, New York, **1992**; A. Bayer et al., *Eur. J. Biochem.* **1995**, *234*, 414; D. Kaiser et al., *Pure Appl. Chem.* **1998**, *70*, 97].

Microglobulins, proteins with low molecular weight belonging to the → globulins. *α_1-Microglobulin* is a glycoprotein (183 aa) of the blood plasma, and has binding affinity to other proteins. It shows sequence homology to → β-lactoglobin and the retinol-binding protein. The nonglycosylated *β_2-microglobulin* (100 aa; M_r ~ 11.8 kDa) shows similarity to the → immunoglobulins. It occurs associated with the lymphocyte antigen CD1. Furthermore, it is a component of immunoglobulin Fc recep-

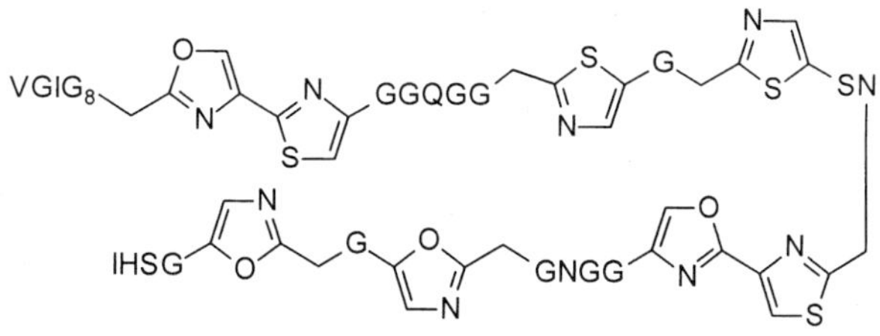

Microcin B17

tors of the gastric epithelium of newborn children. β_2-Microglobulin is noncovalently associated in a 1:1 ratio with the Class I → MHC proteins.

Micropeptin 90, a cyclic depsipeptide isolated from the blue-green alga *Microcystis aeruginosa* (*NIES-90*). Micropeptin 90 inhibits → plasmin and trypsin with IC_{50} of 0.1 and 2.0 µg mL^{-1}, respectively, but not papain, chymotrypsin, or elastase. Micropeptin 90 contains the unusual building blocks 3-amino-6-hydroxy-2-piperidone, and *N*-methyl-L-tyrosine. The amino group of Thr is acylated with glyceric acid 3-*O*-sulfate [K. Ishida et al., *Tetrahedron Lett.* **1995**, *36*, 3535].

Midkine, ***MK***, human MK, a neurotrophic 121-residue protein ($M_r \sim 13.2$ kDa) containing five intramolecular disulfide bonds. MK, a retinoic acid-inducible gene product, is a heparin-binding growth/differentiation factor. It induces neurite outgrowth of mammalian embryonic brain cells, and suppresses the growth of human HL60 leukemia cells. Furthermore, MK was found in senile plaques of Alzheimer's disease (→ amyloid-β), suggesting its involvement in the pathological process leading to senile plaque formation. The first three disulfide bonds in the highly basic protein comprise a N-terminal domain linked by a short peptide bridge to the C-terminal domain containing the remaining two disulfide bonds. MK was chemically synthesized applying the maximum protection strategy by Sakakibara and co-workers in 1996 [J.-I. Tsutsui et al., *Biochem. Biophys. Res. Commun.* **1991**, *176*, 792; H. Muramatsu et al., *Biochem. Biophys. Res. Commun.* **1994**, *203*, 1131; T. Inui et al., *J. Peptide Sci.* **1996**, *2*, 28].

Milk protein-derived opioid peptides, a group of opioid peptides derived from milk proteins which belong to the → exorphins. The → β-casomorphins were the first discovered peptides of this family. *α-Casein exorphins*, also termed *casoxins*, are derived from bovine α-casein. *α-Casein exorphin (1–7)*, H-Arg-Tyr-Leu-Gly-Tyr-Leu-Glu-OH, corresponds to the partial sequence 105–111 of bovine α_S-casein, and a fragment shortened by the C-terminal Glu show moderate typical opioid properties *in vitro*. *Casoxin D (1–7)*, H-Tyr-Val-Pro-Phe-Pro-Pro-Phe-OH, released upon peptic-chymotryptic digestion of a human casein, corresponds to human α_S-casein (158–164) and was reported to act as an opioid antagonist. From tryptic and peptic digests of bovine κ-casein preparations, two peptides have been isolated showing low opioid antagonist properties. *Bovine casoxin (1–6)*, H-Ser-Arg-Tyr-Pro-Ser-Tyr-OMe, a 6-peptide methyl ester, could be regarded as a low-affinity µ- and κ-selective opioid receptor antagonist. Furthermore, specific fragments of α-lactalbumin (*α-lactorphin*, H-Tyr-Gly-Leu-Phe-NH_2) and β-lactoglobulin (*β-lactorphin*, H-Tyr-Leu-Leu-Phe-NH_2), bearing N-terminal Tyr residues have been synthesized and act as µ-opioid receptor ligands with low potency. *Human lactoferrins A, B*, and *C*, peptides with six, five, and seven residues, isolated from the peptic digest of human lactoferrin, are µ-opioid receptor-selective antagonists with moderate potency [H. Teschemacher et al., *Inc. Biopoly.* **1997**, *43*, 99].

Mixanpril, *N*-[(2*S*,3*R*)-2-benzoylthiomethyl-3-phenylbutanoyl]-L-alanine (***Formula***), a dual inhibitor for the neutral endopeptidase-24.11 (NEP) and the → angiotensin-converting enzyme (ACE). The orally administerable mixanpril is important for the treatment of high blood pressure and cardiac defects [S. Turcaud et al., *Bioorg. Med. Chem. Lett.* **1995**, *5*, 1893].

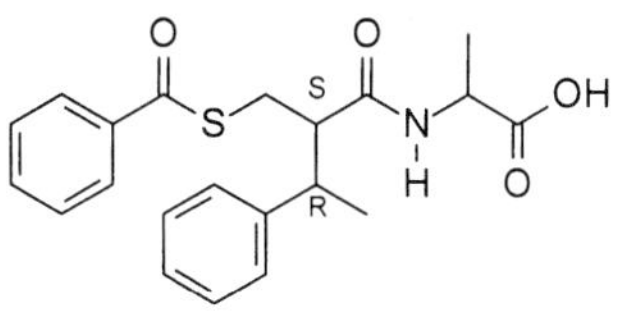

Mixanpril

Molecular chaperones, *protein-folding helper proteins*, a diverse set of abundant and ubiquitous protein families required for the correct folding, assembly, transport, and degradation of other proteins within the cell. The structures and modes of action of these families are significantly diverse. Some families are characterized by high specificity in their action, whereas others are very general. Furthermore, some, but not all, facilitate the binding and release of substrate proteins in an ATP-dependent manner. Many of the molecular chaperones are → heat-shock proteins (hsp). Common features of hsp are an interaction with non-native protein subunits, the stabilization of protein-folding intermediates, and the prevention of aggregation. The *small hsp* (*shsp*) have no ATPase activity, but they are characterized by a high capacity for substrate binding, the mechanism of which remains unknown. *DnaK* (*hsp70, BiP*) represents the *E. coli* homologue of an important class of hsp in the cytosol, endoplasmic reticulum, mitochondria, and chloroplasts. DnaK needs for its action the cooperation with the proteins *DnaJ* and *GrpE*. The binding and release of the substrate require ATP binding and hydrolysis. A large cage-like structure consisting of two rings of seven subunits, each surrounding a central cavity is characterized for *hsp60* (*GroEL, cpn60*), whereas the *hsp100* family forms a hexameric ring structure. The *chaperonin family* consists of two subfamilies. The *GroE chaperonins* occur in eubacteria, mitochondria, and chloroplasts (*Group I*), whereas the *TCP1 chaperonins* are located in archaebacteria and in the eukaryotic cytosol (*Group II*). The chaperonins are double-ring oligomeric proteins acting as containers for the folding of other protein subunits. They are acting in ATP-dependent folding and refolding of structurally unrelated proteins and are essential for cell viability at all temperatures. The *E. coli GroEL* (*hsp60, cpn60*) is a 14-mer of 58 kDa subunits in two rings. Its co-protein *GroES* (*hsp10, cpn10*) forms a heptameric ring of 10 kDa subunits [M. J. Gething, J. Sambrook, *Nature* **1992**, *355*, 33; R. I. Morimoto, A. Tissieres, C. Georgopoulos, *The Biology of Heat-Shock Proteins and Molecular Chaperones*, Cold Spring Harbor Laboratory Press, Cold Spring Harbor, **1994**; F. U. Hartl, *Nature* **1996**, *381*, 571; N. A. Ranson et al., *Biochem. J.* **1998**, *333*, 233].

Monellin, an intensively sweet protein from the West African berries *Dioscoreophyllum cumminsii*. On a weight basis, monellin is several thousand times more potent in sweetness than sucrose. It consists of two noncovalently associated polypeptide chains, A and B, with 44 and 50 residues, respectively. According to the X-ray crystal structure the natural protein consists of an anti-parallel β-sheet with five strands and an α-helix [T. Mizukoshi et al., *FEBS Lett.* **1997**, *413*, 409].

Morphine modulating neuropeptides, peptides found in relatively high concentrations in the periaqueductal gray and dorsal spinal cord of the mammalian nervous system. Known members are *A-18-Fa*, AGEGLSSPFW^{10}SLAAPQRFa, and → neuropeptide FF. Both peptides are capable of attenuating the analgesic effect of morphine [H. Y. T. Yang et al., *Proc. Natl. Acad. Sci. USA* **1985**, *82*, 7757; C. B. Goodman et al., *Peptides* **1996**, *17*, 389].

Motilin, ***Mot***, FVPIFTYGEL[10]QRMQEKER NK[20]GQ, a 22-peptide hormone stimulating the contraction of smooth muscle in the gastrointestinal tract. Motilin was originally isolated from porcine gut, and is mainly expressed and secreted by enterochromaffin cells of the small intestine. The sequence of motilin is highly conserved among species. Motilin is released from pro-motilin, besides motilin-associated peptides (MAP). Sequence comparison indicates that motilin and both → motilin-related peptide and → ghrelin share partial homology [J.C. Brown et al., *Can. J. Physiol. Pharmacol.* **1971**, *49*, 399; Z. Itoh, *Peptides* **1997**, *18*, 593].

Motilin-related peptide, ***MTLRP***, GSSFLSPE HQ[10]KAQQRKES, the amino-terminal 18-peptide sequence of the prepro-motilin-related peptide encoded by the mouse cDNA m46. MTLRP and → motilin share 22% identity and 39% similarity. Furthermore, there are also structural similarities to → ghrelin [C. Tomasetto et al., *Gastroenterology* **2000**, *119*, 395].

MPGF, acronym for major proglucagon fragment, → glucagon.

Myelopeptides, ***MP***, peptides with regulatory functions isolated from mammalian cells of the spinal cord. Two hexapeptides could be isolated from the supernatant of porcine spinal cord cell cultures. *MP-1*, H-Phe-Leu-Gly-Phe-Pro-Thr-OH, shows immunoregulatory activity, and *MP-2*, H-Leu-Val-Val-Tyr-Pro-Trp-OH, neutralizes the inhibitory effect of leukemia cells on the functional activity of T lymphocytes [R. V. Petrov et al., *Biosci. Rep.* **1995**, *15*, 1].

Myoglobin, an oxygen-binding protein preferentially occurring in skeletal muscle. Myoglobin consists of a single chain (153 aa; $M_r \sim 17$ kDa) and a heme that is tightly wedged in a hydrophobic pocket. The eight helices of myoglobin are linked by short peptide segments, forming a ellipsoidal molecule. The heme group binds O_2. The function of myoglobin is to store and transfer oxygen, i.e., from hemoglobin to respiratory enzymes [S.E.V. Phillips, *J. Mol. Biol.* **1980**, *142*, 531; M.F. Perutz, *Trends Biochem. Sci.* **1989**, *14*, 42].

Myosin, an essential protein component of contractile muscle together with → actin. Myosin forms almost entirely the vertebrate thick filaments and consists of six highly conserved polypeptide chains: two *heavy chains* (each $M_r \sim 230$ kDa), two pairs of *light chains*, termed *essential light chain* (ELC, each $M_r \sim 20$ kDa) and *regulatory light chain* (RLC, each $M_r \sim 20$ kDa). Myosin consists of a long fibrous α-helical tail formed by the C-terminal halves of the heavy chains, and two globular heads associated with one of each type of ELC and RLC. The fibrous part forms a left-handed parallel coiled coil ~ 1600 Å in length which ends in the two globular heads. The single form of myosin exists only at low ionic strengths, whereas under physiological conditions myosin aggregates to natural thick filaments consisting of several hundred myosin molecules packed end-to-end of their rod-like tails in a regular staggered fashion. The myosin heavy chain is acting as an ATPase, promoting muscle contraction by hydrolysis of ATP to ADP and P_i [W.F. Harrington, M.E. Rodgers, *Annu. Rev. Biochem.* **1984**, *53*, 35; H.M. Warrick, J.A. Spudich, *Annu. Rev. Cell. Biol.* **1987**, *3*, 379; I. Rayment et al., *Science* **1993**, *261*, 50].

Myosuppressin, → SchistoFLRFamide, → FMRFamide-related peptides.

N

Natriuretic peptides, a group of peptides isolated from the atrium of the mammalian heart exerting natriuretic-diuretic, vasorelaxant and other actions lowering blood pressure and controlling electrolyte homeostasis. Members of this group are the → atrial natriuretic peptide (ANP), the → brain natriuretic peptide (BNP), and the → C-type natriuretic peptide (CNP). The biological effects are mediated via specific natriuretic peptide receptors on the surface of the target cells. The receptor subtypes A/R_1(GC-A), B/R_1(GC-B), and C/R_2 have been characterized. After interaction with ANP, the type A receptor ($M_r \sim 120$ kDa) stimulates the formation of cGMP. Despite the fact that ANP is usually synthesized by the heart, this organ is simultaneously its target organ. The increase of cGMP concentration in the heart is the basis for its action. The type A receptor is also stimulated by BNP. CNP is the primary ligand for the structurally similar type B receptor, and also causes the formation of cGMP. The type C receptor ($M_r \sim 60$ kDa) is not a guanylate cyclase. It has been suggested that this receptor is coupled to the adenylate cyclase/cAMP system. Since all three peptides bind to this receptor with approximately similar affinity, it has been postulated that this receptor plays a role in the clearance of the peptides. The biological functions of ANP and BNP are quite similar. ANP occurs in high concentrations in the atrium, whereas BNP is preferentially located in the ventricles [G. McDowell et al., *Eur. J. Clin. Invest.* **1995**, *25*, 291; A.M. Richards et al., *Clin. Sci.* **1995**, *88*, 18; O. Lisy, J.C. Burnett, Jr., *Coronary Artery Disease* **1999**, *10*, 389].

Neocarcinostatin, a polypeptide antibioticum from *Streptomyces carzinostaticus* with strong anticancer activity. Neocarcinostatin consists of an endiin chromophore (***Formula***) and a peptide chain with 113 residues ($M_r \sim 10.7$ kDa). Neocarcinostatin has potential importance for the therapy of leukemia and both stomach and pancreatic cancer.

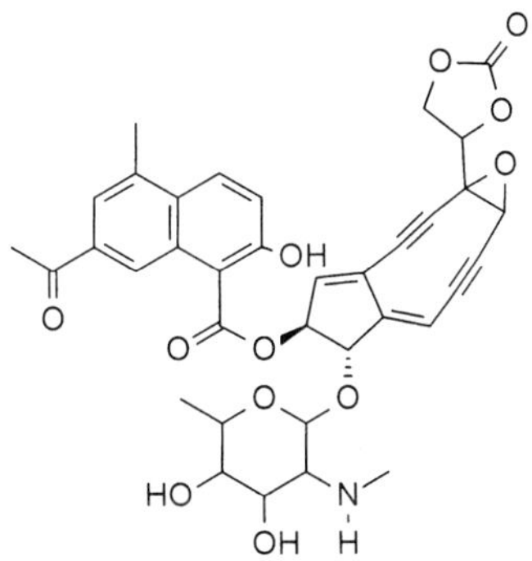

Neocarcinostatin

Neoendorphins, opioid peptides derived from the precursor protein pro-dynorphin (pro-enkephalin B). *α-Neoendorphin*, H-Tyr-Gly-Gly-Phe-Leu-Arg-Lys-Tyr-Pro-Lys^{10}-OH, corresponds to the partial sequence of human and porcine pro-dynorphin-(175–184), and contains the sequence of β-endorphin (de-Lys^{10}-α-endorphin). Both neoendorphins show potent activity in the guinea-pig ileum assay [K. Kangawa et al., *Biochem. Biophys. Res. Commun.* **1981**, *99*, 871; N. Minamino et al., *Biochem. Biophys. Res. Commun.* **1981**, *99*, 864].

Nerve growth factor, *NGF*, one of the most intensively studied protein growth factors, which was first discovered more than 50 years ago. It is the prototype of the larger family of neurotropins. NGF stimulates the proliferation and differentiation of cells of ectodermal and mesodermal origin. The activity had been identified both in tumor tissues by Levi-Montalcini in 1951, and eight years later by Cohen in various snake venoms one year prior to the identification in mouse submandibular glands by Cohen. In the submandibular tissue, mouse NGF consists of a 7S complex of three subunits (α, β, γ) and 1–2 g atoms of zinc, but this complex does not occur in other mouse tissues and in other species. The tightly associated β dimer acts as the active principle of NGF. The α and γ subunits actually inhibit the hormone action and must be dissociated for manifestation of biological activity. The β-protomer is formed from a larger precursor by limited proteolysis. NGF has a tertiary structure based on a cluster of three cystine disulfides and two very extended, but distorted β-hairpins. NGF binds as a dimer to the NGF receptor from which at least two cell-surface receptor types are expressed in a variety of neural and non-neural cells. It stimulates the proliferation and differentiation of its target cells by inducing autophosphorylation of the receptor tyrosine kinase, followed by activation of the tyrosine kinase to bind to and/or phosphorylate specific tyrosine residues on other cytoplasmic signaling proteins [R. Levi-Montalcini, V. Hamburger, *J. Exp. Zool.* **1951**, *116*, 233; S. Cohen, *Proc. Natl. Acad. Sci. USA* **1960**, *46*, 302; R.A. Bradshaw et al., *Protein Sci.* **1994**, *3*, 1901].

Neurokinins, ***NK***, mammalian members of the → tachykinin family. They are widely distributed in the central and peripheral nervous system and the two neurokinins (NKA and NKB) act as neurotransmitters or neuromodulators. The biological actions on many tissues are mediated via specific G-protein-coupled receptors. Among the three subtypes of NK receptors, NK1 is the preferred receptor for → substance P. *Neurokinin A, NKA* (also known as *substance K, neurokinin α*, and *neuromedin L*), H-His-Lys-Thr-Asp-Ser-Phe-Val-Gly-Leu-Met^{10}-NH_2, is the agonist for the NK2 receptor, whereas *neurokinin B, NKB* (also known as neurokinin β and *neuromedin K*), H-Asp-Met-His-Asp-Phe-Phe-Val-Gly-Leu-Met^{10}-NH_2, mediates its action through the NK3 receptor. Together with substance P, the NK play an important role in pain transmission, neurogenic inflammation, smooth muscle contraction, secretion, vasodilation, and activation of the immune system. The NK were isolated from porcine spinal cord extracts and synthesized by Munekata and co-workers in 1984 [E. Munekata et al., *Chem. Lett.* **1984**, 1013; K. Folkers et al., *Biochem. Biophys. Res. Commun.* **1984**, *118*, 405; J.E. Maggio, *Annu. Rev. Neurosci.* **1988**, *11*, 13; Z. Gao, N.P. Peet, *Curr. Medicinal Chem.* **1999**, *6*, 374; C.A. Maggi, *Trends Biochem. Sci.* **2000**, *21*, 173].

Neuromedin B, ***NMB***, H-Gly-Asn-Leu-Trp-Ala-Thr-Gly-His-Phe-Met^{10}-NH_2, a mammalian neuropeptide belonging to the → ranatensin family. Originally, NMB was purified from pig spinal cord and showed potent contractile activity against rat uterus. It has been shown to be present in the CNS, as well as in the gastrointestinal tract. The NMB receptor is a G-protein-coupled receptor with seven membrane-spanning regions. Agonist binding activates several intracellular signaling cascades, e.g., phospholipase activation, protein kinase C activation, and calcium mobilization leading to the expression of several genes, DNA synthesis, as well as cellular effects such as secretion. The pharmacological effects of NMB are smooth muscle contraction, exocrine and endocrine secretions of gastrointestinal tissues, pancreas and pituitary, various central effects and in-vitro effects. Interestingly, NMB also mediates its action through the receptor of the → gastrin-releasing peptide, which is another bombesin-like peptide in mammals. The N-terminally extended forms, NMB-30 and NMB-32, show similar activities to NMB [N. Minamino et al., *Biochem. Biophys. Res. Commun.* **1983**, *114*, 541; H.

Ohki-Hamazaki, *Prog. Neurobiol.* **2000**, *62,* 297].

Neuromedin C, ***NMC***, H-Gly-Asn-His-Trp-Ala-Val-Gly-His-Leu-Met^{10}-NH_2, a neuropeptide showing sequence identity with the C-terminal part of → gastrin-inhibitory polypeptide-(18–27) and sequence homology with the C-terminus of → bombesin.

Neuromedin N, ***NN***, H-Lys-Ile-Pro-Tyr-Ile-Leu-OH, a 6-peptide with a similar biological profile to → neurotensin (NT). NN is synthesized as part of a larger precursor which also contains neurotensin (NT) and neurotensin-like peptide. In the brain, processing of pro-NT/NN gives rise to NN and NT, whereas in the gut processing leads mainly to the formation of NT and *large NN*, a large peptide ending with the NN sequence at its C-terminus. NN together with NT are released upon depolarization of brain tissues. After food ingestion, intestinal NT and large NN are secreted into the circulation. There is evidence that both NN and NT bind with similar affinities to the three NT receptor subtypes. Large NN has been shown to have NT-like activity on intestinal preparations, although its pharmacological properties have not been elucidated in detail [N. Minamino et al., *Biochem. Biophys. Res. Commun.* **1984**, *120,* 542; P. R. Dobner et al., *Proc. Natl. Acad. Sci. USA* **1987**, *84,* 3516; J.-P. Vincent et al., *Trends Pharmacol. Sci.* **1999**, *20,* 302].

Neuromedin U, ***NMU***, (U derived from uterus), neuropeptides acting on the uterus smooth muscle. *NMU-8*, H-Tyr-Phe-Leu-Phe-Arg-Pro-Arg-Asn-NH_2, and *NMU-25*, FKVDEEFQGP^{10}IVSQNRRYFL^{20}FRPRNa, from porcine spinal cord stimulate the rat uterus smooth muscle. Furthermore, they also influence blood pressure in dogs and rats. *Rat NMU*, YKVNEYQGPV^{10}APSGGF FLFR^{20}PRNa, isolated from the small intestine exerts two-fold potent uterus stimulating activity in comparison to pig NMU-25 [N. Minamino et al., *Biochem. Biophys. Res. Commun.* **1985**, *130,* 1078].

Neuropeptide F, ***NPF***, PDKDFIVNPS^{10}DLV LDNKAAL^{20}RDYLRQINEY^{30}FAIIGRPRFa, a 39-peptide amide first isolated from the flatworm *Monieza expansa*. NPF shows sequence homology with members of the vertebrate → neuropeptide Y (NPY) family [A. G. Maule et al., *Parasitology* **1991**, *102,* 309].

Neuropeptide FF, ***NPFF***, H-Phe-Leu-Phe-Gln-Pro-Gln-Arg-Phe-NH_2, an 8-peptide amide initially detected in the bovine brain by using antisera directed against the molluscan peptide → FMRFamide. NPFF is involved in opiate-induced analgesia, morphine tolerance, and abstinence. Therefore, NPFF has been referred to be a member of the → morphine modulating neuropeptides [M. Allard et al., *Peptides* **1999**, *20,* 327; J. M. Zajac, *Trends Pharmacol. Sci.* **2001**, *22,* 63].

Neuropeptide Y family, ***NPY family***, a peptide family consisting of → neuropeptide Y (NPY), → pancreatic polypeptide (PP) and → peptide YY. The alternative designation → PP-fold family was proposed by Rehfeld in 1998. The → Drm-NPF peptide is also related to this family [J. F. Rehfeld, *Physiol. Rev.* **1998**, *78,* 1087].

Neuropeptide Y, *neuropeptide tyrosine*, ***NPY***, YPSKPDNPGE^{10}DAPAEDLARY^{20}YSALRH YINL^{30}ITRQRYa (porcine), a 36-peptide amide occurring as an abundant peptide in brain. The name is related to the two terminal tyrosine residues. NPY is widely distributed in the central and peripheral nervous systems. The highest concentrations in the brain are found in the hypothalamus. NPY is processed from a 97-prepropeptide. Four NPY receptor subtypes (Y_1,

Y_2, Y_4, Y_5) are known. An additional receptor, named y6 receptor, is a pseudo-gene in humans and rats. The Y_1 receptor (glycoprotein, $M_r \sim 70$ kDa) binds NPY, but not NPY-(13–36), whereas the Y_2 receptor (glycoprotein, $M_r \sim 50$ kDa) binds both NPY and the N-terminally shortened peptide. Porcine [Pro^{34}]NPY is an efficient agonist for the Y_1 receptor. NPY-(18–36) acts as an antagonist in membranes of the heart. Both receptors interact also with → peptide YY. NPY stimulates the intake of food via the Y_5 receptor. NPY inhibits the formation of cAMP which is stimulated by, e.g., forskolin or isoproterenol, and the calmodulin-stimulated phosphodiesterase. It increases intracellular Ca^{2+} concentration in vascular smooth muscle cells. However, many more biological and other effects of NPY have been described. According to a proposal of Rehfeld, NPY belongs to the → PP-fold family, but in the literature the designation → neuropeptide Y (NPY) family is also used for the same peptides [K. Tatemoto et al., *Nature* **1982**, *296*, 659; T. S. Gray, J. E. Morley, *Life Sci.* **1986**, *38*, 389; L. Grundemar, S. Bloon (Eds.), *Neuropeptide Y and Drug Developments*, Academic Press, New York, **1997**; D. R. Gehlert, *Neuropeptides* **1999**, *33*, 329].

Neuropeptide γ, ***NP*** *γ*, DAGHGQISHK^{10}R HKTDSFVGL^{20}Ma, a 21-neuropeptide discovered in an extract of the rabbit intestinal tract. NP_γ corresponds to γ-preprotachykinin-(72–92) [R. Kage et al., *J. Neurochem.* **1988**, *50*, 1412].

Neurophysins, ***NP***, precursor proteins and additionally transport molecules of the hormones → oxytocin (OT) and → vasopressin (VP). The NP (~ 95 aa; $M_r \sim 10$ kDa) contain the sequence of OT and VP, respectively, at the N-terminal end. OT is associated with NP I and VP with NP II. NP II contains additionally a C-terminal 39-glycopeptide with unknown function. The NP are synthesized in the hypothalamus and transported in neurosecretory vesicles down the axon to the posterior pituitary, which acts as a storage and release facility for VP and OT [H. Land et al. *Nature* **1983**, *302*, 342].

Neurotensin, ***NT***, <ELYENKPRRP^{10}YIL (hNT), a 13-peptide from brain and intestine, originally isolated from calf hypothalamus. In analogy to many other neuropeptides, NT shows a dual function as a neurotransmitter or neuromodulator in the nervous system and as a local hormone in the periphery. Hence, it acts as a neuromodulator of dopamine transmission, and on anterior pituitary hormone secretion. Furthermore, NT exerts potent hypothermic and analgesic effects in the brain. On the other hand, NT is a paracrine and endocrine modulator of the digestive tract and of the cardiovascular system of mammals. In addition, its action as a growth factor on a variety of normal and cancer cells is well documented. NT is synthesized as part of a larger precursor together with → neuromedin N and the *neurotensin-like peptide*, H-Lys-Leu-Pro-Leu-Val-Leu-OH. Neuromedin N exhibits a similar biological activity profile to NT mediated by the same NT receptor subtypes. Up to now, three subtypes of neurotensin receptors have been cloned. In 1990, the first NT receptor (NTS1, previously NTR-1) was cloned which belongs to the family of G-protein-coupled receptors. Six years later, two new NT receptors were cloned. One of these (NTS2) belongs to the G-protein-coupled receptor superfamily, whereas the other (NTS3) is structurally different. It is an entirely new type of neuropeptide receptor, which is identical to human gp95/sortilin [R. E. Carraway, S. E. Leeman, *Biochem. J.* **1973**, *248*, 6854; C. Pothoulakis et al., *Ann. N.Y. Acad. Sci.* **1998**, *840*, 635; J.-P. Vincent et al., *Trends Pharmacol.*

Sci. **1999**, *20*, 302; B.M. Tyler-McMahon, *Regul. Peptides* **2000**, *93*, 125].

Nisin, the prototype and best studied Type A of → lantibiotics. *Nisin A* is a heterodetic pentacylic peptide (M_r 3353 Da) formed by a single lanthionine (Ala-S-Ala) and four 3-methyllanthionine residues (Abu-S-Ala). In addition, nisin contains three unsaturated amino acid building blocks, 2,3-didehydroalanine (Dha) and two 2,3-didehydrobutyrines (Dhb). Starting from the N-terminal end, the rings are labeled A to E. The rings A, B, and C are all separated, whereas the rings D and C form a bicycle (***Formula***). *Nisin Z* is a naturally occurring variant differing from nisin A by a single amino acid exchange (His27Asn). Nisin is produced by strains of *Lactococcus lactis* and shows antimicrobial activity against a broad range of Gram-positive bacteria. Nisin is used as a food additive in over 50 countries. Although, nisin was discovered in the 1940s, the complete structure was elucidated in only 1970 by Gross [R.W. Jack, G. Bierbaum, H.-G. Sahl, *Lantibiotics and Related Peptides*, Springer, Berlin, Heidelberg, New York, **1998**; H.-G. Sahl, G. Bierbaum, *Annu. Rev. Microbiol.* **1998**, *52*, 41; G. Jung, *Angew. Chem. Int. Ed. Engl.* **1991**, *30*, 1051; E. Breukink, B. De Kruijff, *Biochim. Biophys. Acta* **1999**, *1462*, 223].

Nociceptin, ***NC***, *orphanin FQ*, ***OFQ***, further acronyms are ***NOC***, ***N/OFQ***, ***OFQ/N***, FGGFTGARKS^{10}ARKLANQ, a 17-peptide acting as an endogenous agonist of the nociceptin (ORL1) receptor. NC shows a clear resemblance to → dynorphin A. Both peptides have six residues in common, contain a very similar N-terminal tetrapeptide sequence, and possess a highly positively charged core. NC is synthesized as part of a larger precursor polypeptide together with → nocistatin which shows antinociceptive activity. NC is involved in the control of several biological activities not limited to nociception, but also including learning and memory, motivation, stress and anxiety, and the regulation of cardiovascular, hormonal, renal, and intestinal functions. Although NC acts at the molecular and cellular level in a similar way as opioids, it binds to the G-protein-coupled nociceptin (ORL1) receptor with 500- to 1000-fold higher affinity compared to the binding affinity to the κ-opioid receptor. Dynorphin A is inactive at the nociceptin (ORL1) receptor, but is a potent agonist of the κ-opioid receptor. Both the structural homology of the receptors and the chemical and physical similarities of the endogenous agonists require a high degree of functional selectivity for the recognition and activation of the two receptors. The nociceptin (ORL1) receptor was first named orphan opioid receptor-like receptor (ORL1) and was cloned from human, rat and mouse brain, and human lymphocytes. Besides ORL1, further acronyms are NCR, OFQR, NOR, and OP_4. From the cDNA-deduced amino acid sequences of the human nociceptin (ORL1) receptor it could be concluded that δ-, μ-, and κ-opioid receptors contain conserved regions,

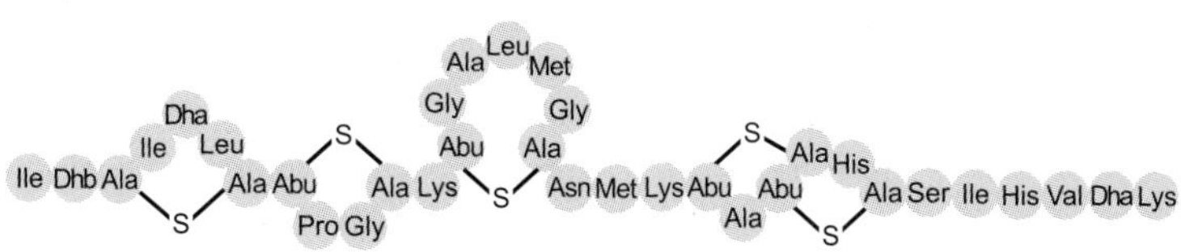

Nisin

especially, in the transmembrane helices and cytoplasmic loops. The broad pharmacological profile of NC has initiated the development of novel nonpeptidic nociceptin (ORL1) receptor ligands, antagonists and agonists which are protease-resistant and bioavailable [J.-C. Meunier et al., *Nature* **1995**, *377*, 532; R.K. Reinscheid et al., *Science* **1995**, *270*, 792; J.-C. Meunier et al., *Peptides* **2000**, *21*, 893; R.K. Reinscheid et al., *Peptides* **2000**, *21*, 901].

Nocistatin, TEPGLEEVGE^{10}IEQKQLQ, an endogenous 17-peptide isolated from bovine brains and encoded by the gene for the nociceptin/orphanin FQ precursor. The 176-residue precursor protein comprises two bioactive peptides, nocistatin and → nociceptin, which play opposing roles in the CNS. Simultaneous administration of nocistatin blocks allodynia and hyperalgesia induced by nociceptin. Nocistatin is widely present in the spinal cord and brain, and may play an important function in the CNS, including involvement in nociception, learning and memory. Human, rat, and mouse precursor produce larger counterparts containing 30, 35, and 41 residues that all show the antinociceptive activity [E. Okuda-Ashitaka et al., *Nature* **1998**, *392*, 286; E. Okuda-Ashitaka, S. Ito, *Peptides* **2000**, *21*, 1101].

Nucleoproteins, naturally occurring protein conjugates in which the hydroxy group of a Ser, Thr or Tyr residue is linked via a phosphodiester group to the 3′- or 5′-end of a nucleic acid. Nucleoproteins play decisive roles in important biological processes such as viral replication [B.A. Juodka, *Nucleosides Nucleotides* **1984**, *3*, 445; M. Salas, *Annu. Rev. Biochem.* **1991**, *160*, 39; L. Blanco et al., *Proc. Natl. Acad. Sci. USA* **1994**, *91*, 12198].

O

Octreotide, → somatostatin.

Omuralide, the related β-lactone of → lactacystin.

Onchidin B, a highly symmetric cyclodepsipeptide (→ depsipeptides) containing both α- and β-amino and hydroxy acids (***Formula***). It was first isolated from a South Pacific tunicate. Like → valinomycin, onchidin N forms a cyclic structure with the polar groups oriented toward the central cavity, whereas the remaining part of the molecule is relatively nonpolar. This structure enables them to complex ions acting as selective ion transporters through cellular membranes [R. Fernandez et al., *J. Am. Chem. Soc.* **1996**, *118*, 11635].

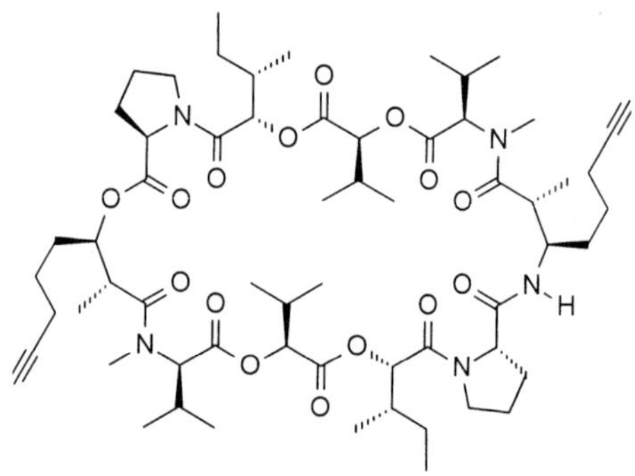

Onchidin B

Onconase, ***ONC***, a member of the RNase A superfamily with antitumor activity. ONC from oocytes of *Rana pipiens* consists of 104 residues and is an unusually stable protein. It shares only 30% identity with the bovine RNase A, but both proteins show a very similar tertiary structure [E. Notomista et al., *Biochemistry* **2000**, *39*, 8711].

Oncostatin M, ***OCM***, a single-chain glycoprotein produced by macrophages. Human OCM (127 aa; $M_r \sim 28$ kDa) shows significant homology to G-CSF (→ colony-stimulating factors). Oncostatin M inhibits cell proliferation in melanoma and other solid tumors. It also stimulates the production of IL-6 (→ interleukins) [T. M. Rose, A. G. Bruce, *Proc. Natl. Acad. Sci. USA* **1991**, *88*, 8641; T. J. Brown et al., *J. Immunol.* **1991**, *147*, 2175].

Ophthalmic acid, H-γ-Glu-Abu-Gly-OH, an acidic 3-peptide found in animal lens and in the brain of cattle and rabbits. Ophthalmic acid inhibits the γ-glutamylcysteine synthetase (→ glutathione) and the glyoxalase I. Ophthalmic acid occurs also together with norophthalmic acid (H-γ-Glu-Ala-Gly-OH) [S. G. Waley, *Biochem. J.* **1958**, *68*, 189; S. Tsuboi et al., *Anal. Biochem.* **1984**, *126*, 520].

Orcokinin, NFDEIDRSGF^{10}GFN, a myotropic 13-peptide from the abdominal nerve cords of the crayfish (*Orconectes limosus*). It acts as a highly potent stimulator of hindgut contractions [J. Stangier et al., *Peptides* **1992**, *13*, 859; D. Bungart et al., *Peptides* **1995**, *16*, 199].

Orexins, *hypocretins*, neuropeptides produced in the lateral hypothalamic area stimulating appetite and stereotypic behavior associated with feeding. They have been initially identified as endogenous ligands for an orphan G-protein-coupled receptor, now termed OX_1 receptor. Human (bovine, rat, mouse) *orexin-A* (also termed *hypocretin 1* because of its homology to secretin), <EPLPDCCRQK^{10}TCSCRLYELL^{20}HGAGNHAAGI^{30}LTLa (disulfide bonds: Cys^{6}– Cys^{12}/Cys^{7}–Cys^{14}), and human *orexin-B*, RSGPPGLQGR^{10}LQRLLQASGN^{20}HAAGILTMa, are derived from the 131-residue *prepro-orexin*, which shows sequence similarities with members of the secretin family. The orexins are distributed in the lateral hypothalamus region, which has been implicated in feeding behavior. Intracerebroventricular administration of orexin-A stimulates food consumption. The orexins play also roles in regulating drinking behavior, neuroendocrine function, and the sleep–wake cycle. The OX_1 receptor shows structural similarities to certain neuropeptide receptors, especially to the Y_2 neuropeptide Y receptor. The alternative name *hypocretin* reflects their hypothalamic origin and the similarity to secretin [T. Sakura et al., *Cell* **1998**, *92*, 573; L. de Lecea et al., *Proc. Natl. Acad. Sci. USA* **1998**, *95*, 322; T. Sakura, *Regul. Peptides* **1999**, *85*, 25; J. H. Lee et al., *Eur. J. Biochem.* **1999**, *266*, 831; W. K. Samson, Z. R. Resch, *Trends Endocrinol. Metab.* **2000**, *11*, 257; S. Taheri, S. Bloom, *Clin. Endocrinol.* **2001**, *54*, 421].

ORL1 receptor, acronym of orphan opioid receptor-like receptor (→ nociceptin).

Orphanin FQ → nociceptin.

Osteocalcin, also termed *bone Gla protein*, ***BGP***, a 49-residue Ca-binding protein (M_r ~ 5.8 kDa) comprising 10–20% of bone protein. It was found in bone organic matrix, dentin, and in other mineralized tissues. Beside the → matrix Gla protein, osteocalcin is the second Gla-containing protein in bone. Osteocalcin from various vertebrates has Gla residues at positions 17, 21, and 24, and contains a disulfide bond (C^{23}–C^{29}). Osteocalcin undergoes conformational change in the presence of Ca^{2+}, so that all Gla residues are located on the same side of the α-helix at a distance of 5.4 Å. This distance corresponds nearly to the spacing between adjacent calcium atoms in hydroxyapatite crystals. Osteocalcin is presumed to play a role in mineralization of bone, and appears to function as a highly specific osteoblastic marker formed during bone formation [J. W. Poser et al., *J. Biol. Chem.* **1980**, *255*, 8685; M. Nakao et al., *Peptide Res.* **1994**, *7*, 171; V.

Geoffroy et al., *J. Biol. Chem.* **1995**, *270*, 30973].

Osteogenic growth peptide, ***OGP***, ALKRQ GRTLY^{10}GFGG, a 14-peptide with regenerating bone marrow activity. OGP is identical to the C-terminal sequence of histone H4. It has been found in human as well as in animal serum. Synthetic OGP stimulates proliferation and alkaline phosphatase activity of osteoblastic cells *in vitro*. From the experimental data obtained it has been concluded that OGP is a key factor in the mechanism of the systemic osteogenic reaction to marrow injury [I. Bab, *Clin. Orthop.* **1995**, *313*, 64; I. Bab et al., *EMBO J.* **1992**, *11*, 1867; Z. Greenberg et al., *J. Cell Biochem.* **1997**, *65*, 359].

Osteonectin, a glycoprotein ($M_r \sim 30$ kDa) of the bone containing also phosphate groups. Osteonectin is located in the extracellular bone matrix. It binds to → collagen and hydroxyapatite, blocking the crystallization of mineralic bone components.

Ovalbumin, a member of the → albumins. Ovalbumin is a glycoprotein ($M_r \sim 44.5$ kDa) containing 3.2% carbohydrate (mannose and glucosamine) and phosphorylated serine residues. It is classified as ovalbumin A_1, ovalbumin A_2 and ovalbumin A_3, depending on the number of phosphorylated serine residues. Ovalbumin comprises 54% of the total protein amount of egg-white.

Ovokinin, H-Phe-Arg-Ala-Asp-His-Pro-Phe-Leu-OH, a vasorelaxing 8-peptide corresponding to → ovalbumin-(358–365). It shows relaxing activity for canine mesenteric artery, which is specifically blocked by the bradykinin B_1 antagonist [de-Arg^9, Leu^8]bradykinin [H. Fujita et al., *Peptides* **1995**, *16*, 785].

Oxyntomodulin, → glicentin.

Oxytocin, ***OT***, H-Cys-Tyr-Ile-Gln-Asn-Cys-Pro-Leu-Gln-NH_2 (disulfide bond: Cys^1–Cys^6), a member of the neurohypophyseal peptide hormones. The name has been derived from Greek: *oxys* (*oksys*=fast), and *tokos* (=give birth). In 1953, the heterodetic cyclic 9-peptide amide was structurally elucidated by the Nobel laureate Vincent du Vigneaud, and was chemically synthesized as the first peptide hormone one year later. OT mediates the contraction of the uterus, and stimulates milk ejection. Due to the structural similarities to → vasopressin, OT shows some vasopressin activity. OT is synthesized in the hypothalamus together with the precursor protein → neurophysin I, and transported via the tractus paraventriculo-hypophyseus to the posterior lobe. It is released proteolytically from the precursor in this storage site in response of an appropriate biological stimulation, and is secreted into the bloodstream. During the course of structure-activity studies, a large number of analogues with prolonged or dissociated biological effects, but also those with antagonistic properties have been synthesized [V. du Vigneaud et al., *J. Am Chem. Soc.* **1953**, *75*, 4879; V. J. Hruby, *Topics Mol. Pharmacol.* **1981**, *1*, 99].

P

Pancreastatin, ***PST***, GESRSEALAV^{10}DGAG KPGAEE^{20}AQDPEGKGEQ^{30}EHSQQKEE EE^{40}EMAVVPQGLF^{50}RGa (hPST), a 52-peptide derived from the sequence of the precursor human chromogranin A (→ chromogranins). Fragments of the precursor and hPST itself were isolated from human carcinoid liver metastases and from the liver metastases of a patient with insulinoma. Porcine PST isolated from the porcine pancreas consists of 49 residues. PST strongly inhibits glucose-induced and gastric inhibitory peptide-induced insulin secretion from rat pancreas. The inhibitory effect is restricted to the C-terminal part of PST [K. Tatemoto et al., *Nature* **1986**, *324*, 476; W. E. Schmidt et al., *Endocrinology* **1988**, *123*, 1395].

Pancreatic polypeptide, ***PP***, APLEPVYPGD^{10} DATPEQMAQY^{20}AAELRRYINM^{30}LTRPRYa (porcine PP), a 36-peptide amide belonging to the → neuropeptide Y family (→ PP-fold family). PP is formed in the PP cells of the pancreas, and the pulsatory release is coupled to an insulin pulse. In general, the plasma level of PP is increased by electric stimulation of the vagus, food intake and several other stimulating factors. PP inhibits exocrine secretion from the pancreas, which is stimulated, e.g., by secretin, caerulein, and cholecystokinin. The C-terminal tyrosine amide is essential for the biological activity. There is evidence that the active sequence of PP is located in the C-terminal region. The C-terminal hexapeptide is capable of stimulating the inhibition of PP on the exocrine pancreas secretion [J. R. Kimmel et al., *J. Biol. Chem.* **1975**, *250*, 9369].

Pancreozymin, → cholecystokinin.

Paramyosin, a muscle protein component of invertebrate thick filaments. It consists of two subunits ($M_r \sim$ 210–230 kDa).

Paraproteins, *'pathological proteins'*, abnormal → immunoglobulins or proteins occurring in increased quantities in various hematological disturbances (paraproteinemias). Bence-Jones proteins and amyloid proteins are well-known members. Depending on the origin of the cells from a common stem cell, immunoglobulins of the classes A, G, M, D, or E and one of the light chains of the Type κ and λ can occur as paraproteins, whereas the isolated occurrence of chains of the L- or H-type is very seldom.

Parathormone, → parathyroid hormone.

Parathyroid hormone, ***PTH***, *parathormone, parathyrin*, SVSEIQLMHN^{10}LGKHLNSME R^{20}VEWLRKKLQD^{30}VHNFVALGAP^{40}LAPR DAGSQR^{50}PRKKEDNVLV^{60}ESHEKSLGEA^{70} DKADVNVLTK^{80}AKSQ (hPTH), an endocrine 84-peptide hormone regulating the extracellular calcium concentration in animals. The regulation caused by PTH is effected through actions on kidney, intestine, and bone cell receptors. PTH is secreted in the parathyroid gland in response to subnormal serum calcium levels. It is expressed as a 115-peptide precursor (prepro-PTH) and secreted as an 84-peptide, but the major functions are associated with the N-terminal 34-residue sequence part. Up to now, three types of PTH receptors have been identified, characterized, and cloned. The PTH2 receptor binds exclusively PTH. The C-PTH receptor recognizes C-terminal PTH fragments, and the PTH/PTHrP receptor binds both PTH and → parathyroid hormone-related peptide (PTHrP). PTH receptor activation in bone, kidney, and intestine stimulates adenylate cyclase, phospholipase C and phospholipase D in target cells. PHT-based pharmaceuticals are used for treatment of bone loss due to osteoporosis and reversal of the hypercalcemic effect of malignancy [J. B. Collip, *J. Biol. Chem.* **1925**, *63*, 395; G. N. Hendy et al., *Proc. Natl. Acad. Sci. USA* **1981**, *78*, 7365; T. Kimura et al., *Biochem. Biophys. Res. Commun.* **1983**, *114*, 493; P. Morley et al., *Curr. Medicinal Chem.* **1999**, *6*, 1095].

Parathyroid hormone-related peptide, ***PTHrP***, originally discovered as a peptide thought to be responsible for tumor-associated hypercalcemia. Only the human gene of the autocrine/paracrine PTHrP is expressed to three PHTrH isoforms of 139, 141, and 173 residues. In analogy to the → parathyroid hormone (PTH) most of the known biological functions are exerted by the N-terminal PTHrP-(1–34), AVSEHQLL HD^{10}KGKSIQDKRR^{20}RFFLHHLIAE^{30}IHTA, with 60% sequence similarity to PTH-(1–34). The C-terminal part PTHrP-(107–139) was found to be a potent inhibitor of osteoclastic bone reabsorption. From the three identified receptors of PTH only the PTH/PTHrP receptor binds both PTHrP and PTH which activates by coupling to multiple G proteins both adenylate cyclase/protein kinase A- and phospholipase C/protein kinase C-dependent signaling cascades. PTHrP analogues which strongly inhibit PTHrP adenylate cyclase-stimulation seems to be useful for the treatment of malignancy-associated hypercalcemia in animal trials, but failed in human tests [L. J. Suva et al., *Science* **1987**, *237*, 893; N. Horiuchi et al., *Science* **1987**, *238*, 1566; B. E. Kemp et al., *Science* **1987**, *238*, 1586; S. Kitazawa et al., *Cancer* **1991**, *67*, 984; J. J. Orloff et al., *Endocr. Rev.* **1994**, *15*, 40; W. F. Schwidinger, *Endocrine* **1998**, *8*, 201; J. M. Mosely et al., *Proc. Natl. Acad. Sci. USA* **1987**, *84*, 5048; P. Morley et al., *Curr. Medicinal Chem.* **1999**, *6*, 1095].

Pareptide, → melanostatin.

Parvalbumins, water-soluble, monomeric proteins ($M_r \sim 12$ kDa) with high-affinity sites for Ca^2. The parvalbumins also bind Mg^{2+}competitively. They occur not only in skeletal muscle of fish and amphibia, but also in mammalian muscle. Parvalbumins contain six α-helical regions (A to F), and the binding sites for two Ca^{2+} are formed by helix-loop-helix motifs (EF hands) between helices C/D and E/F [C. H. Heizmann, *Experientia* **1984**, *40*, 910].

Parvulins, a group of → peptidyl prolyl *cis/trans* isomerases.

Pentagastrin, Boc-β-Ala-Trp-Met-Asp-Phe-NH_2, a synthetic 5-peptide derivative with the most potent sequence region of → gastrin. It shows the same qualitative effect as the parent hormone [O. Brawman-Mintzer et al., *Am. J. Psychiatry* **1997**, *154*, 700].

Pep5, a member of the Type-A family of → lantibiotics. Pep5 (34 aa; M_r 3488 Da) consists of three rings from which A is a single ring, whereas rings B and C form a bicycle. It is extremely basic, and the N-terminus is modified with a 2-oxobutyryl moiety. Pep5 is produced by strains of the bacterial species *Staphylococcus epidermidis*, and kills bacteria as do other Type-A lantibiotics by the formation of discrete, voltage-dependent channels (or pores) [R. W. Jack, G. Bierbaum, H.-G. Sahl, *Lantibiotics and Related Peptides*, Springer, Berlin, Heidelberg, New York, **1998**].

Pepstatin, isovaleryl-Val-Val-*Sta*-Ala-*Sta*-OH, a low-molecular weight inhibitor of acidic proteases (e.g., → renin, pepsin, and cathepsin D) isolated from the culture fluid of *Actinomycetes*. Pepstatin contains two building blocks of → statine (*Sta*). Pepstatin has proved a lead in the search for inhibitors of renin [H. Umezawa et al., *J. Antibiot.* **1970**, *23*, 259].

Peptaibolin, Ac-Leu-Aib-Leu-Aib-Phe-ol, an unusual short-sequence representative of the → peptaibols. It exhibits moderate antimicrobial activity against Gram-positive bacteria and yeasts. Total synthesis, 3D structure, and membrane-modifying properties of peptaibolin and selected analogues were described in 2001 [H. Hülsmann et al., *J. Antibiot.* **1998**, *51*, 1055; M. Crisma et al., *Tetrahedron* **2001**, *57*, 2813].

Peptaibols, a class of membrane-active peptide antibiotics produced by fungi of the genus *Trichoderma*. They contain a high proportion of α-aminoisobutyric acid (Aib, $C^{\alpha,\alpha}$-dimethylglycine) and other C^{α}-tetrasubstituted α-amino acids supporting α/3_{10}-helix. Another group of the Aib family contains Isovaline (Iva, C^{α}-methyl-C^{α}-ethylglycine). Furthermore, the peptaibols are N-terminally acylated and bear C-terminal amino-alcohol moieties like L-phenylalaninol (Phe-ol), valinol (Val-ol) or leucinol (Leu-ol). The longest members of the class (up to 20 building blocks), e.g., → alamethicin, are N-acetylated at the N-terminus, whereas the N-terminus of the shortest members (seven building blocks) is acylated with a fatty acid of eight or ten carbon atoms. An unusually short member of the peptaibols is → peptaibolin. The substitutions at the C^{α}-atoms, together with those of the terminal amino acid residues, cause a lack of local flexibility of the peptide chain and promote the formation of stable secondary structure domains (α- and 3_{10}-helices, respectively, and β turns). Peptaibols induce leakage of the cytoplasmic material which may lead to cell death. Long-sequence peptaibols form potential-dependent ion-conducting pores in liquid membranes. According to the carpet-like mechanism, short-sequence (lipo)peptaibols tend to float on the lipid bilayer. The peptaibols exert bacteriostatic, fungicidal, cytostatic, and hemolytic effects. Members of the peptaibols are, e.g., → alamethicin, → harzinins, → peptaibolin, and the lipopeptaibol, trichodecenin [E. Benedetti et al., *Proc. Natl. Acad. Sci. USA* **1982**, *79*, 7951; R. Nagaraj, P. Balaram, *Acc. Chem. Res.* **1981**, *14*, 356; R.F. Epand et al., *Eur. J. Biochem.* **1999**, *266*, 1021].

PeptiCLECs, cross-linked enzyme crystals, ***CLECs***, of proteases used as catalysts in peptide synthesis. CLECs are microcrystals grown from aqueous solution and cross-linked with a bifunctional agent. They are highly stable against autolysis and exogenous protease degradation, and are usable as catalysts in environments that are normally incompatible with enzyme activity such as aqueous-organic solvent mixtures, near-anhydrous organic solvents and increased temperature. CLECs offer the advantages of high product purity and the potential for enzyme recycling. Thermolysin-CLEC (T-CLEC), also termed PeptiCLEC-TR, and subtilisin-CLEC have been successfully used for peptide synthesis, especially for the preparation of dipeptides [R.A. Persichetti et al., *J. Am. Chem. Soc.* **1995**, *117*, 2732; Y.-F. Wang et al., *J. Org. Chem.* **1997**, *62*, 3488; S.J. Faulconbridge et al., *Organ. Process Res. Dev.* **2000**, *4*, 563].

Peptide 46, *human cellular tumor antigen p53 (361–382)*, GSRAHSSHLK^{10}SKKGQST SRH^{20}KK, a lysine-rich 22-peptide corresponding to the single-strand DNA end binding site of p53. It activates specific DNA binding of wild-type p53 *in vitro* and induces apoptosis in p53-expressing tumor cells [G. Selivanova et al., *Nature Med.* **1997**, *3*, 632].

Peptide 6A, H-Ala-Arg-Pro-Ala-Lys-OH, a 5-peptide with permeability-increasing properties originally derived from plasmin-degraded human → fibrinogen. Furthermore, it acts as a potent coronary vasodilator and effective pulmonary ACE inhibitor *in vitro* and *in vivo* [T. Saldeen et al., *Thromb. Res.* **1981**, *23*, 465; R. Moalli et al., *J. Pharmacol. Exp. Ther.* **1987**, *243*, 897].

Peptide 74, TMRKPRCGNP^{10}DVAN, a 14-peptide derived from a prosegment of the 72-kDa type IV collagenase. It inhibits tumor cell invasion of both A2058 fibrosarcoma and HT1080 melanoma cells. Peptide

74 could be a potential antimetastatic agent due to lacking cytotoxic effects [A. Melchiori et al., *Cancer Res.* **1992**, *52*, 2353].

Peptide 810, H-Gln-Asp-Leu-Thr-Met-Lys-Tyr-Gln-Ile-Phe^{10}-OH, a 10-peptide encoded by a gene fragment from human melanoma M14. It contains the antigenic epitope -Lys-Tyr-Gln-Ile- recognized by the human mAb L92. Peptide 810 is the first peptide recognized by cytotoxic T lymphocytes in melanoma patients, and should be interesting as a melanoma vaccine replacing whole cells [N. Morioka et al., *J. Immunol.* **1994**, *153*, 5650].

Peptide aptamers, a novel generation of molecules selected for their binding to a given target protein. Analogously to intracellular antibodies, they are capable of binding specifically to a given target protein both *in vitro* and *in vivo* with the potential to block selectively the function of their target protein. Peptide aptamers are selected from randomized expression libraries based on their in-vivo binding capacity to the appropriate target protein. They are powerful new tools for molecular medicine. Blocking the intracellular function of a target protein by peptide aptamers allows the investigation of distinct physiological and pathological processes within living cells. Furthermore, peptide aptamers meet the requirements for the development of novel diagnostic and therapeutic strategies with potential importance for a broad variety of various disease entities such as metabolic disorders, infections, and cancer [F. Hoppe-Seyler, K. Butz, *J. Mol. Med.* **2000**, *78*, 426 (review)].

Peptide E, *bovine adrenal medulla peptide 3200 BAM-3200*, YGGFMRRVGR^{10}PEWW MDYQKR^{20}YGGFL, a 25-peptide (M_r ~ 3.2 kDa) from the porcine adrenal medulla. It shows κ-opioid receptor agonistic activity [D.L. Kilpatrick et al., *Proc. Natl. Acad. Sci. USA* **1981**, *78*, 3265].

Peptide histidine isoleucine amide, *PHI*, *peptide with N-terminal histidine and C-terminal isoleucine amide*, H^{1}AGGVFTSDF^{10} SRLLGQLSAK^{20}KYLESLIa, a 27-peptide amide from porcine intestine belonging to the → secretin family. It shares the same peptide precursor with the → vasoactive intestinal peptide (VIP), and is found together with VIP, for example, in the neurons of the central nervous system, in the digestive tract, urogenital tract, lungs and the wall of the gallbladder. The biological activities are similar with those of VIP and → secretin [K. Tatemoto, V. Mutt, *Proc. Natl. Acad. Sci. USA* **1978**, *75*, 4115; **1981**, *78*, 6603; M. Bodner et al., *Proc. Natl. Acad. Sci. USA* **1985**, *82*, 3548].

Peptide leucine arginine, *PLR*, LVRGCWTK SY^{10}PPKPCFVR (disulfide bond: C^{5}–C^{15}), an immunomodulatory peptide isolated from the skin of the Northern Leopard frog (*Rana pipiens*). The name PLR reflects the N- and C-terminal residues. PLR elicits rapid, noncytolytic histamine release with a two-fold greater potency compared with → mellitin. PLR is capable of permeabilizing negatively charged unilamellar lipid vesicles, but not neutral vesicles – a result that is consistent with its nonhemolytic action. It inhibits the early development of granulocyte macrophage colonies from bone marrow stem cells [A.L. Salmon et al., *J. Biol. Chem.* **2001**, *276*, 10145].

Peptide M, DTNLASSTII^{10}KEGIDKTV, an 18-peptide corresponding to a sequence region of the retinal S-antigen. The latter is pathogenic for the induction of experimental autoimmune uveitis (EAU) [L.A. Donose et al., *Arch. Ophthalmol.* **1987**, *105*,

838; S. Nityanand et al., *J. Clin. Immunol.* **1993**, *13*, 352].

Peptide nucleic acids, ***PNA***, DNA mimicking peptides containing neutral amide backbone linkages. In PNA, the normal phosphodiester backbone of a DNA is replaced by *N*-(2-aminoethyl)glycine units. The standard nucleotide bases are connected by a methylene carbonyl linker to this backbone at the amino nitrogens (more details are given in Section 7.4.2) [T. Koch et al., *Nucleosides Nucleotides* **1997**, *16*, 1771; D.R. Corey, *Trends Biotechnol.* **1997**, *15*, 224; E. Uhlmann, *Biol. Chem.* **1998**, *379*, 1045; P.E. Nielsen, *Curr. Opin. Biotechnol.* **1999**, *10*, 71].

Peptide T, H-Ala-Ser-Thr-Thr-Thr-Asn-Tyr-Thr-OH, a partial sequence 185–192 of gp 10, the coat protein of HIV. The synthetic peptide, named peptide T according to its high Thr content, was shown to inhibit the binding of both isolated gp 120 and HIV-1 to the CD4 receptor. Later, it was reported that the invasion of healthy cells by HIV-1 requires the sequential interaction of the gp 120 envelope protein not only with the CD4 primary receptor, but also with a co-receptor which belongs to the chemokine receptor family of seven-helix transmembrane receptors. It has been suggested that CD4 binding triggers a conformational transition of gp 120, moving the regions V2, which contains the sequence of peptide T still closer to the co-receptor. Furthermore, peptide T shows potent chemotactic activity on human monocytes, which is correlated with the inhibition of CD4 binding. The minimum sequence of peptide T that retains most of the biological activity is the C-terminal 5-peptide (-Thr-Thr-Asn-Tyr-Thr-OH) that is probably representative for a widespread recognition motif [C.B. Pert et al., *Proc. Natl. Acad. Sci. USA* **1986**, *83*, 9254; D. Picone et al., *J. Peptide Sci.* **2001**, *7*, 197].

Peptide WE-14, WSKMDQLAKE^{10}LTAE, a 14-peptide derived from chromogranin A (→ chromogranins) and isolated from a human ileal carcinoid tumor [W.J. Curry et al., *FEBS Lett.* **1992**, *301*, 319].

Peptide YY, *peptide tyrosine tyrosine amide, peptide with N-terminal tyrosine and C-terminal tyrosine amide*, ***PYY***, YPAKPEAPGE^{10}D ASPEELSRY^{20}YASLRHYLNL^{30}VTRQRYa (porcine PYY), a 36-peptide amide which belongs to the → neuropeptide Y family/ → PP-fold family. In human cells PYY is formed, above all, in the duodenum, whereas in dog and in pig it is formed primarily in the mucous membranes of the ileum and colon. PYY inhibits pancreatic exocrine secretion [K. Tatemoto, *Proc. Natl. Acad. Sci. USA* **1982**, *79*, 5485].

β-Peptides, oligomers of β-amino acids, a class of artificial peptide analogues adopting predictable and reproducible folding patterns (further details are given in Section 7.4.3).

Peptidyl prolyl *cis/trans* isomerases, ***PPIases***, EC 5.2.1.8, ubiquitous and abundant enzymes ($M_r \sim$ 10–150 kDa) catalyzing prolyl peptide bond isomerization. Members of the new enzyme class are *cyclophilins (Cyp)*, *FK506 binding proteins (FKBPs)* including the subfamily of the ribose-bound *trigger factors*, and *parvulins (Par)*. The cyclophilins and FKBPs are the natural receptors for the immunosuppressants → cyclosporin A and FK506, respectively. The ribosome-bound trigger factors show weak sequence similarity to FKBPs but they are lacking any FK506-binding ability typically for other FKBPs. The parvulins are unrelated in their amino acid sequences to each other, show distinct substrate specificity, and also prove to be sensitive to different types of inhibitors. The human parvulin Pin 1 is a mitotic regulatory protein essential for G_2M transition of the eukaryotic cell cycle. Generally, the PPIases are composed of a catalytic core, which is also found as a single domain in the prototypic PPIases human Cyp 18, human FKBP12 and *E. coli* Par 10, and a different number of additional domains or short motifs covering other functional properties. In contrast to the → chaperones, PPIases are classical enzymes. Besides their putative function in accelerating slow folding steps, compartmentalized PPIases are likely to have pleiotropic effects within cells. It remains an open question whether Nature has evolved additional peptide bond isomerases. The first member of the PPIases was discovered by Gunter Fischer et al. in pig kidney in 1984 [G. Fischer et al., *Biomed. Biochim. Acta* **1984**, *43*, 1101; G. Fischer, *Angew. Chem. Int. Ed. Engl.* **1994**, *33*, 1415; A. Galat, S.M. Metcalfe, *Prog. Biophys. Mol. Biol.* **1995**, *63*, 67].

Peptidyl-glycyl-leucine-carboxamide, ***PGLa***, GMASKAGAIA^{10}GKIAKVALKA^{20}La, a cationic, highly potent antimicrobial peptide from the skin of the South African frog *Xenopus laevis* [D. Andreu et al., *Eur. J. Biochem.* **1985**, *149*, 531; A. Latal et al., *Eur. J. Biochem.* **1997**, *248*, 938].

β-Peptoids, oligomeric N-substituted β-aminopropionic acids (***Formula***).

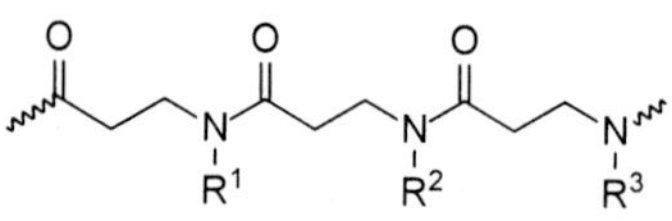

β-Peptoids

Peptoids, oligomers of N-substituted glycine building blocks (further information is given in Section 7.4.1) (***Formula***).

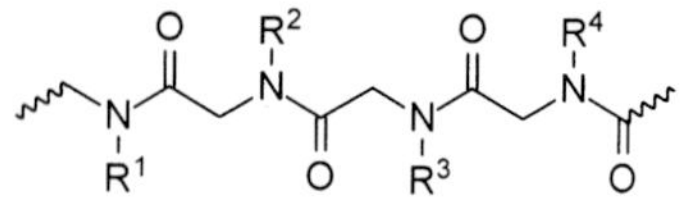

Peptoids

Perforin, *cytolysine*, pore-forming protein (M_r ~ 70 kDa) stored in secretory vesicles and released by cytotoxic T cells (killer T cells) after binding to antigen-bearing host cells. At the point of contact, the released perforin lyzes these target cells by aggregation and forming pores in the plasma membranes. The formation of lytic pores (5–16 nm) is performed by Ca^{2+}-dependent polymerization of perforin in the membrane analogously to the complement system. Perforin consists of two Cys-rich domains and an amphiphilic α-helix-forming region typically for cell-lyzing molecules [B. B. Herberman et al., *Annu. Rev. Immunol.* **1986**, *4*, 651].

Perisulfakinin, EQFDDY^{S}GHMR^{10}Fa, a sulfated 11-peptide amide [Y^S=Tyr(SO_3H)], originally isolated from the corpora cardiaca of the American cockroach (*Periplaneta Americana*). Perisulfakinin shows sequence similarity to → cholecystokinin and → gastrin. It stimulates hind gut contractions [J.A. Veenstra, *Neuropeptides* **1989**, *14*, 145].

Peroxinectin, a cell adhesion protein (M_r ~ 76 kDa) isolated from the blood cells of the fish *Pacifastacus leniusculus*. Peroxinectin seems to be the first cell adhesion protein in the blood of invertebrates. It shows also peroxidase activity because of a high degree of sequence similarity to myeloperoxidase [M.W. Johansson et al., *Biochem. Biophys. Res. Commun.* **1995**, *216*, 1079].

Petriellin A, a fungicidal 13-depsipeptide from the culture extracts of *Petriella sordida*. It contains 13 building blocks, most of which are noncoded amino acids (***Formula***). Petriellin A shows activity against *Ascobolus furfuraceus* and *Sordina fimicola*, but

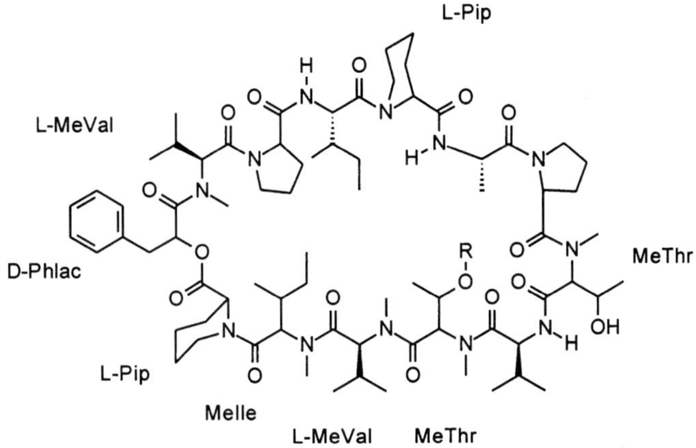

Petriellin A

not against *Candida albicans* [K. K. Lee et al., *J. Org. Chem.* **1995**, *60*, 5384].

PGLa, → peptidyl-glycyl-leucine-carboxamide.

Phallotoxins, heterodetic bicyclic 7-peptides from *Amanita phalloides* that form, together with the → amatoxins, the main toxic components of these mushrooms. The toxic effect of phallotoxins is most probably related to their capability to bind tightly to F-actin in the parenchymal cells of the liver, resulting in hemorrhagic shock and death within 2 to 5 h. The phallotoxins cause death in experimental animals after i.p. or i.v. injection, but not perorally. In general, the lethal doses are higher than those of the amatoxins, e.g., the LD_{50} values are about 2 mg kg^{-1} in white mice. All phallotoxins are derived from the same cyclic peptide backbone and consist of seven amino acids, cross-linked by tryptathionine between residues 3 and 6 (cf. Fig. 3.5 B and Section 3.1). The naturally occurring phallotoxins differ mostly in their number of hydroxyl groups in the side chain of L-leucine. Interestingly, → antamanide shows full protection of mice from death by phalloidin (0.5 mg kg^{-1}, injected not more than 1 h before or simultaneously with 5 mg of the toxic peptide per kg) [Th. Wieland, *Peptides of Poisonous Amanita Mushrooms*, Springer Series in Molecular Biology, Springer, Berlin, New York, **1986**; H. Faulstich et al., *Biochemistry* **1980**, *19*, 3334].

Pheromonotropin, KLSYDDKVFE[10]NVEFT PRLa, a pheromonotropic 18-neuropeptide amide originally isolated from head extracts of the armyworm larvae (*Pseudaletia separata*). It is also named *Pseudaletia* *m*elanization and *r*eddish *c*oloration *h*ormone (MRCH) because of its physiological function in lepidopteran insects [S. Matsumoto et al., *Biochem. Biophys. Res. Commun.* **1992**, *182*, 534].

PHONA, an analogue of → peptide nucleic acids (PNA) in which the peptide bond is replaced by a phosphonic acid ester bridge (***Formula***) [A. Peyman et al., *Angew. Chem. Int. Ed. Engl.* **1997**, *36*, 2809].

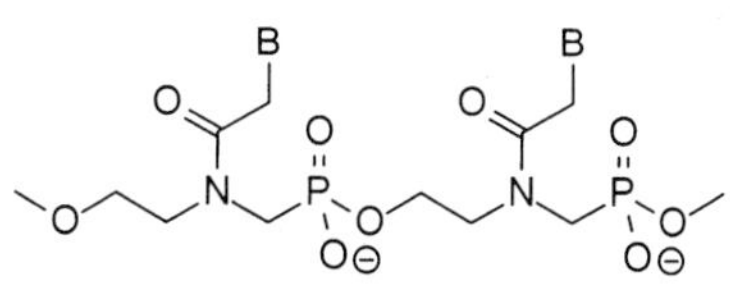

Phona

Phyllocerulein, pGlu-Glu-Tyr(SO_3H)-Thr-Gly-Trp-Met-Asp-Phe-NH_2, a 9-peptide amide originally isolated from the skin of the South American frog *Phyllomedusa sauvagi*. The pharmacological actions of Phyllocerulein are quite similar to those of → cerulein.

Phyllolitorin family, a subfamily of the → bombesin-like family. *Phyllolitorin*, pGlu-Leu-Trp-Ala-Val-Gly-Ser-Phe-Met-NH_2, and *[Leu^{8}]phyllolitorin* belong to this family. These peptides show similar activities as the members of the other subfamilies due to the related structure. In contrast to the other two subfamilies, mammalian homologues have not yet been found [V. Erspamer et al., *Peptides* **1985**, *6*, 7].

Phyllomedusin, pGlu-Asn-Pro-Asn-Arg-Phe-Ile-Gly-Leu-Met^{10}-NH_2, a 10-peptide amide originally isolated from the skin of the South American frog *Phyllomedusa bicolor*. Phyllomedusin belongs to the → tachykinins and shows similar pharmacological actions as → physalaemin, and → uperolein.

Physalaemin, <EADPNKFYGL^{10}Ma, an 11-peptide amide originally isolated from the skin extracts of the South American frog *Physalaemus fuscumaculatus*. It belongs to the → tachykinins. Physalaemin exerts a potent hypotensive action and stimulates extravascular smooth muscle [V. Erspamer et al., *Experientia* **1964**, *20*, 489; L. Bernardi et al., *Experientia* **1964**, *20*, 490].

Phytosulfokines, ***PSK***, mitogenic peptides produced and secreted by asparagus mesophyll cell cultures acting as plant growth regulators. *PKS-α*, H-Tyr(SO_3H)-Ile-Tyr(SO_3H)-Thr-Gln-OH, and *PKS-β*, the corresponding C-terminally truncated 4-peptide, from *Oryza sativa* and *Asparagus officinalis* are necessary for cell proliferation. The synthesis of the PSK requires both auxin and cytokinin in the culture medium. They are synthesized as prepro-phytosulfokines and, after release from the precursor and post-translational modification, the PKS interact with specific receptor proteins to mediate mitogenic activity [Y. Matsubayashi et al., *Planta* **1999**, *207*, 559; A. Schaller, *Plant Mol. Biol.* **1999**, *40*, 763].

Pigment dispersing hormone, ***PDH***, NSELINSLLS^{10}LPKNMNDAa, an 18-peptide amide from the CNS of the fruitfly. The name is related to crustacean PDH-like immunoreactivity. In fruitflies, PDH probably plays a role as a transmitter or modulator in the nervous system, especially in the visual system. Generally, it is involved as a mediator in neuronal pathways and rhythmic processes that are regulated on a circadian level [J.H. Park, J.C. Hall, *J. Biol. Rhythms* **1998**, *13*, 219; J.V. Broeck, *Peptides* **2001**, *22*, 241].

Pituitary adenylate cyclase activating polypeptide, ***PACAP***, hPACAP-38: HSDGIFTD SY^{10}SRYRKQMAVK^{20}KYLAAVLGKR^{30}YKQ RVKNKa, and its N-terminal fragment PACAP-(1–27) are neuropeptides originally isolated from ovine hypothalamus, but also occurring in humans and rats. PACAP-38 and PACAP-27 show considerable sequence homology to → vasoactive intestinal polypeptide (VIP) and stimulate adenylate cyclase more effectively than VIP. PACAP belong to the → secretin family. In mammals, PACAP has been located in a wide range of tissues and fulfils a variety of biological functions mediated via specific PACAP receptors. The distribution of PACAP-containing neuronal elements in the gut wall assumes their involvement in the regulation of both motor and secretory activities. PACAP-(6–38) acts as a selective antagonist. In lower vertebrates, especially fish, PACAP may function as a hypophysiotropic factor regulating pituitary hormone secretion [A. Miyata et al., *Biochem. Biophys. Res. Commun.* **1989**, *164*, 567; **1990**, *170*, 643; A. O. L. Wong et al., *Biochem. Cell Biol.* **2001**, *78*, 329].

Placenta lactogen, → chorionic mammotropin.

Plant defensins, antifungal peptides which are not related to mammalian or insect → defensins. They contain eight disulfide-linked Cys comprising a triple-stranded antiparallel β-sheet structure with only one α-helix [K. Thevissen et al., *J. Biol. Chem.* **1996**, *271*, 15018].

Plasma kinins, blood pressure-reducing peptide tissue hormones. They are formed from α-globulin fractions of the plasma (*kininogens*) by limited kallikrein proteolysis. In addition to plasma, kallikreins with various substrate specificities also occur in kidneys, pancreas, and other organs. All kallikreins are formed from inactive precursors. Members of this peptide hormone group are → bradykinin (kinin 9), → kallidin (kinin 10), and → methionyl-lysyl bradykinin. They are involved in the regulation of the blood flow through various vessel systems, as well as capillary permeability [J. E. Taylor et al., *Drug. Dev. Res.* **1989**, *16*, 1].

Plasmin, *fibrinolysin*, a serine protease catalyzing Lys-Xaa and Arg-Xaa bonds similar to that of → trypsin. Plasmin is the key protease in blood clot lysis, and its major natural substrates are → fibrinogen and → fibrin. Human plasmin is derived from → plasminogen, and is a two-chain protein consisting of the A or H chain (M_r ~ 65 kDa) and the B or L chain (M_r ~ 27.7 kDa). The active site is located in the B chain. The various molecular forms of plasmin are inactivated by protein inhibitors such as the Kunitz type, serpin, soybean, and limabean trypsin inhibitors. The most important, fast-acting inhibitor of plasmin in plasma is → α_2-antiplasmin, whereas → α_2-macroglobin forms stoichiometric complexes with plasmin, thereby inhibiting access to the active site of only large molecular mass substrates and inhibitors. Plasmin is readily prepared from plasminogen by activation with μ-plasminogen activator [F. J. Castellino in: *Handbook of Proteolytic Enzymes*, A. J. Barret, N. D. Rawlings, J. F. Woessner (Eds.), p. 190, Academic Press, San Diego, **1998**].

Plasminogen activators, → plasminogen, → streptokinase, → urokinase.

Plasminogen, the zymogen of → plasmin. The mature molecule of human plasminogen consists of 791 residues in a single peptide chain. One of the major glycoforms of plasminogen contains only O-linked sialylated trisaccharide on Thr^{346}, while the second glycoform possesses the same O-linked glycan and an Asn^{289}-attached bisialylated oligosaccharide on the only N-linked consensus sequence in plasminogen. The two glycoforms can be separated by gradient elution of plasminogen. Plasminogen is synthesized in the liver and secreted into the plasma. The most intensively studied in-vivo plasminogen activators are the tissue-type activator, called *t-plasminogen activator* (*tPA*) and the urokinase-type, *u-plasminogen activator* (*uPA*). tPA converts plasminogen to plasmin by cleavage of a single $Arg^{561}\downarrow Val^{562}$ bond, whereas uPA cleaves the sequence -Cys-Pro-Gly-$Arg^{560}\uparrow Val^{561}$-Val-Gly-Cys- that forms a small disulfide-bridged loop in plasminogen. Both plasminogen activators are used clinically for thrombolytic therapy. The *bacterial plasminogen activators* → streptokinase and staphylokinase acting in that capacity only via stoichiometric complexes with plasminogen or plasmin are also important due to their in-vivo thrombolytic potential. They activate plasminogen in solution bound to a variety of cell types, and when bound to → fibrin [F.J. Castellino, J.R. Powel, *Methods Enzymol.* **1981**, *80*, 365].

Platelet factor-4 related peptide, *hPF4-(58–70)*: PLYKKIIKKL[10]LES, a 13-peptide corresponding to the C-terminal sequence of platelet factor 4. Like the intact platelet factor, this peptide is capable of alleviating immunosuppression in mice. It acts also as an inhibitor of angiogenesis in the chicken chorioallantoic membrane. Human *[Gln[18]]-PF4-(15–22)*, termed *CT-112*, H-Thr-Thr-Ser-Gln-Val-Arg-Pro-Arg-OH, reduces the incidence of type II collagen-induced arthritis in mice and retards the disease progression [M.B. Zucker et al., *Proc. Natl. Acad. Sci. USA* **1989**, *86*, 7571; T.E. Maione et al., *Science* **1990**, *247*, 77].

Platelet-derived endothelial cell growth factor, ***PD-ECGF***, a monomeric polypeptide (482 aa; $M_r \sim 45$ kDa) with seven Cys residues and a potential N-glycosylation site. PD-ECGF is formed by thrombocytes. It is the most important mitogen for endothelial cells, and exerts chemotactic action on these cells.

Platelet-derived growth factor, ***PDGF***, a glycoprotein ($M_r \sim 30$ kDa) acting as multifunctional growth factor. It consists of the two subunits A (125 aa) and B (160 aa) that can be composed as AA, AB, or BB. PDGF stimulates proliferation of connective tissue cells and neuroglial cells via specific receptors that belong to the family of single-pass transmembrane glycoproteins with tyrosine-specific protein kinase activity. PDGF also plays a role in wound healing [C.D. Stiles, *Cell* **1983**, *33*, 653; R. Ross et al., *Cell* **1986**, *46*, 155; B. Westermark, *Acta Endocrinol.* **1990**, *123*, 131].

Plectin, a protein ($M_r \sim 300$ kDa) forming a plexus-like structure. It binds to intermediate filaments of various cell types.

Pleiotrophin, ***hPTN***, *heparin-binding growth-associated molecule* (*HBGAM*), a protein (136 aa; $M_r \sim 15.3$ kDa) with five intramolecular disulfide bonds. PTN belongs to the midkine family of developmentally regulated, secreted polypeptide growth factors expressed in a variety of established tumor cell lines, and exhibits 46% sequence homology to human → midkine. It promotes cell growth of NIH3T3 and NRK fibroblasts, epithelial cells, osteoblasts, and endothial cells. PTN induces neurite outgrowth from neuronal cells, and has been shown to exert angiogenic activity [Y.-S. Li et al., *Science* **1990**, *250*, 1690; P.G. Milner et al., *Biochemistry* **1992**, *31*, 12023; F. Czubayko et al., *Proc. Natl. Acad. Sci. USA* **1996**, *93*, 14753].

Pneumadin, human pneumadin, H-Ala-Gly-Glu-Pro-Lys-Leu-Asp-Ala-Gly-Val[10]-NH_2, and rat pneumadin, [Tyr[1]]hPneumadin, 10-peptide amides occurring in significant concentrations in both normal mammalian and human fetal lungs. Natural and synthetic pneumadin exert antidiuretic effects by releasing → vasopressin from the neurohypophysis [V.K. Batra et al., *Regul. Peptides* **1990**, *30*, 77].

Polymyxins, fatty acid-containing, branched cyclic peptides produced by *Bacillus polymyxa* possessing antibiotic activity against Gram-negative bacteria. Various peptides belong to this group. The basic sequence of polymyxins consists of ten amino acids: Dab[1]-Thr-Dab-Dab-Dab[5]-D-Phe-Leu-Dab-Dab-Thr[10], in which the cyclic heptapeptide part is achieved by coupling the C-terminal carboxy group of Thr to the γ-amino group of Dab (L-α,γ-diaminobutyric acid) in position 4. The N-terminal Dab residue is always acylated with methyloctanoic acid or isooctanoic acid, which is responsible for the strong amphiphilic, detergent-like properties of the peptide (cf. Section 3.3.3.1; **54**). Some naturally occurring analogues contain either Ser^5, Leu^6 or Ile^7. The activity of polymyxins is based on the high affinity of this rather basic peptide to the negatively charged core moiety of the lipopolysaccharide (LPS) in the outer leaflet of the Gram-negative bacterial outer membrane.

Porin, a pore-forming protein. It is a component of the mitochondrial outer membrane, which is reminiscent of the outer membrane of Gram-negative bacteria. The outer membrane contains large amounts of porin that form large aqueous channels through the lipid bilayer. Therefore, the outer membrane is freely permeable to inorganic ions, metabolites and proteins with $M_r < 10$ kDa.

Poststatin, H-Val-Val-Pos-D-Leu-Val-OH, a naturally occurring 5-peptide isolated from *Streptomyces viridochromogenes*. It acts as an inhibitor of prolyl endopeptidase. Pos is the abbreviation for (*S*)-3-amino-2-oxopentanoic acid, named L-postine. The α-ketoamide group seems to be necessary for the biological activity [M. Tsuda et al., *J. Antibiot.* **1996**, *49*, 909; H. H. Wassermann, A. K. Petersen, *Tetrahedron Lett.* **1997**, *38*, 953].

PP-fold family, a member of the gastroenteropancreatic peptide families according to the classification proposed by Rehfeld. This family comprises → pancreatic polypeptide (PP), → peptide YY, and → neuropeptide Y. The overall similarity in the primary structure varies between 45% and 70%. This similarity is connected to an almost identical and stable tertiary structure characterized by the PP-fold motif consisting of a polyproline-like helix (residues 1–8) and an amphiphilic α-helix (residues 15–30). Both helices are connected by a type I β-turn (residues 9–12), and the folded configuration is stabilized by hydrophobic interactions between the side chains of the α-helix residues and the N-terminal proline residues. Besides the term PP-fold family, the designation → *neuropeptide Y family* is also used in literature [J. F. Rehfeld, *Physiol. Rev.* **1998**, *78*, 1087].

PR-39, $RRRPRPPYLP^{10}RPRPPPFFPP^{20}RLPPRIPPGF^{30}PPRFPPRFPa$, an antimicrobial linear 39-peptide amide from porcine intestine with a high content of Pro and Arg residues. Like other → antimicrobial animal peptides and Pro- and Arg-rich peptides (→ Bac5, → Indolicidin), PR-39 is equally active against Gram-negative and Gram-positive bacteria. It has been assumed that they interact with membranes and disorganize them because of their basic nature [H. G. Boman et al., *Infect. Immun.* **1993**, *61*, 2978].

Prions, unprecedented infectious pathogens devoid of nucleic acids causing fatal neurodegenerative diseases in humans and animals. The majority (~ 80%) of prion diseases or transmissible spongiform encephalopathies (TSE) arise sporadically, about one-fifth is genetically induced, and around 1% has been transmitted between mammals both by inoculation with and dietary exposure to infected tissues. Human TSE include Creutzfeldt-Jacob disease (CJD), fatal familial insomnia, the Gerstmann-Sträussler-Scheinker syndrome, and a new variant Creutzfeldt-Jakob disease (nvCJD). The term *prion* is a short form derived from proteinaceous infectious particle, and was coined to describe the infectious agent that causes the neurodegenerative diseases. The *prion protein* (PrP) occurs in two physically and biologically distinct isoforms. The normal *cellular prion protein* (*PrP*^{C}) occurs as a constituent of the normal mammalian cells. It is a glycoprotein containing 209 residues, a disulfide bond (Cys^{179}–Cys^{214}), two N-linked glycosylation sites (Asn^{181}-Ile-Thr/Asn^{197}-Phe-Thr) and a glycosylphosphatidylinositol (GPI) anchor linked to the C-terminal residue Ser^{231}. PrP^C is normally attached to the cell membrane via the GPI anchor containing a heterogeneous sialylated glycan. Although the biological functions of PrP^C are not yet clear, it might be involved in synaptic function, circadian rhythm, copper transport, and signal transduction. According to the '*protein only*' hypothesis, a modified form of PrP^C triggers human TSE, as well as animal prion diseases such as bovine spongiform encephalopathy (BSE) and scrapie in sheep. The conversion of PrP^C into the abnormal, disease-causing scrapie-like ("Sc") or disease-causing isoform of the prion protein isoform, *PrP*^{Sc}, involves a significant conformational transition. The first fractions of PrP^{Sc} were isolated from the brain of Syrian hamsters with experimental scrapie. PrP^{Sc} is insoluble in detergents, contains an increased proportion of β-sheet and a protease-resistant core, whereas PrP^C is a soluble protein with a high content of α-helices and high susceptibility to proteolysis. Knowledge of the three-dimensional structure of PrP^C is essential for the interpretation of the transition to PrP^{Sc}. Various molecular models for the region of PrP^C, corresponding to the protease-resistant core of PrP^{Sc} (residues 90–231) have been proposed based on NMR structures. In 1996, the NMR structure of the mouse prion protein domain PrP(121–231) indicated that this domain contains two-stranded antiparallel β-sheet and three helices, leaving the last 14 amino acid residues without structure. Further NMR solution studies have been published for monomeric, cellular forms of PrP of the most widely used animals (mouse and Syrian hamster). Interestingly, in another report the C-terminal helix extends to residue 227, showing that only the last four residues are flexible. The NMR structure of the recombinant human prion protein hPrP(23–230), and the C-terminal fragments hPrP(90–230) and hPrP(121–230), indicates a globular domain (residues 125–228) containing three helices (144–154/173–194/200–228) and a short anti-parallel β-sheet (128–131/161–164). In 2001, the crystal structure of the human prion protein in dimeric form at 2 Å resolution was elucidated, and suggested a mechanism for oligomerization as an important step on the pathway of PrP^C → PrP^{Sc} conversion. The protease-resistant core of PrP^{Sc}, *PrP 27–30* ($M_r \sim$ 27–30 kDa) is the designation for a protein that results from limited digestion of PrP^{Sc} with proteinase K by truncation of the N-terminus. The N-terminal sequence of PrP 27–30 led to molecular cloning of PrP cDNA and characterization of the chromosomal gene. Translation of PrP^C-mRNA provides the *PrP*^{C} *precursor* consisting of 254 amino acids that yields, after post-translational modification and proteolytic processing, PrP^C. Furthermore, antiserum of PrP 27–30 has enabled the examination of brains from humans and animals with putative prion diseases for the presence of this protein. The studies of prions have elucidated a previously unknown mechanism of disease in humans and animals. A normal genetically encoded protein is converted to the disease-causing form PrP^{Sc} differing significantly in the tertiary structure from PrP^C. In contrast to pathogens carrying a nucleic acid genome, prions appear to encipher strain-specific properties in the three-dimensional structure of PrP^{Sc}. From transgenic investigations it has been assumed that PrP^{Sc} functions as a template upon which PrP^{C} is refolded into a nascent PrP^{Sc} molecule via a process promoted by another protein (protein X), which might function as a molecular chaperone. The PrP^{Sc} molecules are then believed to cause lesions that damage the brain and give rise to transmissible spongiform encephalopathies. It has been postulated that prion strains are different conformational forms of the same protein, since, for example, some TSE strains cause disease more quickly than others. A fluorescence immunoassay distinguishes different strains of prions based on different prion conformations. This test uses a labeled antibody binding only denatured PrP^{Sc}, takes only 8 h, and is very sensitive. Sodium phosphotungstate was used to isolate and concentrate PrP^{Sc} from the crude tissue. The protein-only hypothesis of TSEs was proposed by the 1997 Nobel laureate Stanley B. Prusiner in the 1980s [S. B. Prusiner et al., *Biochemistry* **1982**, *21*, 6942; S. B. Prusiner, *Science* **1997**, *278*, 245; S. B. Prusiner, *Proc. Natl. Acad. Sci. USA* **1998**, *95*, 13363; R. Zahn et al., *Proc. Natl. Acad. Sci. USA* **2000**, *97*, 145; K. J. Knaus et al., *Nature Struct. Biol.* **2001**, *9*, 770; P. M. Rudd et al., *Biochemistry* **2001**, *40*, 3759].

Proctolin, H-Arg-Tyr-Leu-Pro-Thr-OH, an excitatory neurotransmitter from the intestinal musculature of insects. Extremely low concentrations ($\sim 10^{-9}$ mol L^{-1}) cause violent contraction of the end of the gut. For the first isolation of proctolin, 125,000 cockroaches (*Periplaneta americana*) were required. Proctolin was the first insect neuropeptide to be discovered [A. N. Starratt, B. E. Brown, *Life Sci.* **1975**, *17*, 1253].

Proglucagon, the precursor of → glucagon, → GRPP, → MPGF, the → intervening peptides, → glicentin, and the → glucagon-like peptides.

Prolactin release-inhibiting hormone or factor, → prolactostatin.

Prolactin, ***PRL***, *lactogenic hormone, mammotropin, luteotropic hormone*, ***LTH***, *luteotropin*, a single chain proteohormone (198 aa, 3 disulfide bridges; $M_r \sim 22.5$ kDa) of the anterior pituitary that is responsible for lactation. It has been assumed that PRL is formed in the adenohypophysis under the control of → prolactoliberin and → prolactostatin. Nowadays, it has been reported that the → prolactin-releasing peptide (PrRP) fulfils the function to promote the release of PRL. PRL increases milk production in female mammals, and initiates maternal behavior. Analogously to → chorionic mammotropin, PRL has also luteotropic properties and stimulates the synthesis of progesterone, but inhibits the synthesis of estradiol and testosterone. Receptors of PRL are found, for example, in the mammary gland, brain, liver, kidneys, prostate, testes, ovaries, and lymphocytes. PRL can be displaced from the receptors on lymphocytes by → cyclosporin [W. S. Oetting et al., *J. Biol. Chem.* **1986**, *261*, 1649; P. C. Hiestand et al., *Proc. Natl. Acad. Sci. USA* **1986**, *83*, 2599].

Prolactin-releasing hormone or factor, → prolactoliberin.

Prolactin-releasing peptide, ***PrRP***, *hPrRP31*: SRTHRHSMEI^{10}RTPDIQPAWY^{20}ASRGIRP VGR^{30}Fa and the N-terminally truncated sequence (*PrRP20*) identified from the corresponding cDNA sequence have been found to promote the release of → prolactin (PRL). The bovine PrRP31 and PrRP20 were first isolated from hypothalamic extract. With the help of the bovine PrRP31 sequence the precursors of bovine, human and rat PrRP were cloned. PrRP is a new ligand for the orphan receptor hGR3 in the pituitary, and stimulates selectively the release of prolactin in a dose-dependent manner (10 pM to 100 nM). In contrast to other peptides with prolactin-releasing activity (→prolactoliberin), PrRP promotes only the release of PRL and cannot mimic the release of other pituitary hormones [S. Hinuma et al., *Nature* **1998**, *393*, 272; J.-C. Meunier et al., *Nature* **1998**, *393*, 211].

Prolactoliberin, *prolactin-releasing hormone*, ***PRH***, *prolactin-releasing factor*, ***PRF***, a hypothalamic substance (with for a long time an unknown structure) that stimulates the release of → prolactin in the adenohypophysis. Although the release of prolactin is stimulated by other peptides such as → thyroliberin (TRH), VIP, PACAP, oxytocin, and vasopressin, there is now evidence that the → prolactin-releasing peptide (PrRP) seems to be the real prolactoliberin.

Prolactostatin, *prolactin release-inhibiting hormone*, ***PRIH*** or ***PIH***, *prolactin release-inhibiting factor*, ***PRIF***, a hypothalamic substance of (until now) unknown structure that inhibits the release of → prolactin.

Prolamins, a group of reserve proteins occurring in cereals. They are soluble in 50–90% ethanol and can be so separated from the alcohol-insoluble → glutelins. Prolamins are globular proteins ($M_r \sim$ 10–100 kDa) with a high content of Glu (30–45%) and Pro (up to 15%). Members of the prolamins include → gliadin, → zein, and the Lys-lacking hordein (barley).

Proopiomelanocortin, ***POMC***, *proopiocortin*, a biosynthesis polyprotein (prepro-opiomelanocortin) that lacks the signal sequence and is formed in the distal and intermediary part of the hypophysis. POMC contains the sequences of → corticotropin (ACTH), → β-lipotropin, and γ-MSH (→melanotropin). The first two hormones are cleaved to smaller bioactive peptides in a tissue-specific manner (cf. Fig. 3.20). In the anterior lobe, POMC is proteolytically cleaved yielding the N-terminal fragment, ACTH, and β-lipotropin (β-LPH). However, in the intermediate lobe only, the latter polypeptide hormones are further cleaved to yield γ-MSH, α-MSH, corticotropin-like intermediate lobe peptide (CLIP), γ-LPH, and β-endorphin [R.E. Mains, B.A. Eipper, *J. Biol. Chem.* **1979**, *254*, 7885; H. Gainer et al., *Neuroendocrinology* **1985**, *40*, 171; S. Solomon, *Ann. N.Y. Acad. Sci.* **1999**, *885*, 22].

Protamines, a group of strongly basic globular proteins ($M_r \sim$ 1–5 kDa). They are characterized by a high content of Arg (80–85%). In particular, protamines occur in the sperm of fish, birds and molluscs, substituting there functionally the somatic histones. Members of the protamines include clupein (herring), salmin (salmon), sturin (sturgeon), and esocin (pike).

Proteasome, prosome, multimeric protease complex degrading preferentially by ubiquitination-targeted intracellular proteins. In both eukaryotes and archea, the proteasome is a multisubunit complex comprising four stacked rings each containing seven subunits. In eukaryotes, there exist 14 different but homologous subunits, and the proteasome shows different catalytic activities; in archea, the proteasome contains only two different types of subunits that possess only one catalytic activity. The proteasome of bacteria consists of two rings with six subunits. Of the two different bacterial subunits, only one is related to the eukaryote and archean subunits, whereas the other is an ATPase. Proteasomes are important in cellular regulation. They are, for example, responsible for the degradation of cyclins at key stages of the cell cycle, and for the removal of abnormal proteins during stress response. Furthermore, they degrade enzymes of the intermediary metabolism and are involved in the immune response [A.P. Arrigo et al., *Nature* **1988**, *331*, 192; J. Löwe et al., *Science* **1995**, *268*, 533; A. Lupas et al., *Mol. Biol. Rep.* **1997**, *24*, 125].

Protegrins, ***PG***, a family of Arg- and Cys-rich cationic peptides isolated from porcine leukocytes exhibiting a broad range of antimicrobial and antiviral activities. The protegrins are active against various Gram-positive and Gram-negative bacteria, and, furthermore, against HIV-1 and mycobacteria. The protegrins consist of 16–18 residues, including four invariant Cys which form two disulfide bonds and stabilize a rigid structure of two antiparallel β sheets and a β turn segment. The protegrins show similarities to the → tachyplesins. Five members of this family have been well characterized, and the sequence and numbering system of *PG-1* from pig (*Sus scrofa*) is as follows: RGGRLCYCRR^{10}RFCV CVGR (disulfide bonds: C^6–C^{15}/C^8–C^{13}) [V.N. Kokryakov et al., *FEBS Lett.* **1993**, *327*, 231; R.L. Fahrner et al., *Chem. Biol.* **1996**, *3*, 543].

Protein 7B2, a protein ($M_r \sim 21$ kDa) first isolated from porcine adenohypophysis, and characterized via cDNA sequence in humans (185 aa) and *Xenopus*. In analogy to → follitropin (FSH), protein 7B2 is released from the hypophysis by → gonadoliberin; the release is inhibited by inhibin and testosterone [G.J.M. Martens, *FEBS Lett.* **1988**, *234*, 160].

Protein C, vitamin K-dependent protein first purified from bovine plasma, and named protein C according to the third-eluted peak from an ion-exchange column. Human protein C precursor contains 461 residues. The mature protein C ($M_r \sim 62$ kDa) contains a light (L) chain (155 aa, $M_r \sim 21$ kDa) and a heavy (H) chain (262 aa; $M_r \sim 41$ kDa) linked together by a disulfide bond. The L chain consists of several separate domains, including a Gla-containing domain and two EGF-like domains, whereas the H chain forms the serine protease domain. The thrombin-thrombomodulin complex located on the surface of endothelial cells converts protein C to *activated protein C (APC)* by release of 12 residues from the N-terminus of the H chain. APC regulates the coagulation pathway by inhibiting the generation of thrombin via its selective proteolytic inactivation of coagulant cofactors Va and VIIIa [C.T. Esmon, K. Fukudome, *Semin. Cell Biol.* **1995**, *6*, 259; B. Dahlbäck, *Thromb. Res.* **1995**, *77*, 1].

Prothrombin, *thrombogen, factor II*, the zymogen of the serine protease → thrombin. Prothrombin occurs in all vertebrates, and is activated after initiation of the coagulation cascade at the site of vascular injury. Prothrombin consists of the thrombin domain, a Gla domain, and two kringle domains [J. Stenflo et al., *Proc. Natl. Acad. Sci. USA* **1974**, *71*, 2730].

Q

Quinqueginsin, a homodimeric protein ($M_r \sim 53$ kDa) with inhibitory activity against human immunodeficiency virus (HIV), fungi, ribonuclease and cell-free translation. Quinqueginsin has been isolated from the roots of American ginseng *Panax quinquefolium*. The inhibitory action towards HIV-1 reverse transcriptase was potentiated after chemical modification with succinic anhydride [H.X. Wang, T.B. Ng, *Biochem. Biophys. Res. Commun.* **2000**, *269*, 203].

R

RAMP, → receptor activity modifying protein.

Ranakinin, KPNPERFYGL^{10}Ma, an 11-peptide amide isolated from the brain of the frog *Rana ribibunda*. Ranakinin shows similarities to → physalaemin, and is a preferred agonist for the NK-1 tachykinin receptor [M.G. Kodjo et al., *Endocrinology* **1995**, *136*, 4535].

Ranalexin, FLGGLIKIVP^{10}AMICAVTKKC^{20} (disulfide bond: C^{14}–C^{20}), an antimicrobial 20-peptide isolated from the skin of the American bullfrog *Rana catesbeiana*. Ranalexin resembles structurally → polymyxin based on a similar 7-peptide ring system [P.D. Clark et al., *J. Biol. Chem.* **1994**, *269*, 10849; E. Vignal et al., *Eur. J. Biochem.* **1998**, *253*, 221].

Ranamargarin, DDASDRAKKF^{10}YGLMa, a 14-peptide amide isolated from the skin of the Chinese frog *Rana margaratae*. Its activity profile resembles that of → substance P and → physalaemin [L.Q. Tang et al., *Regul. Peptides* **1988**, *22*, 182].

Ranatachykinin A, KPSPDRFYGL^{10}Ma, a 11-peptide amide isolated from brain and intestine of the frog *Rana catesbeiana*. It causes strong effects in the guinea-pig ileum and rat duodenum contractility assays similar to → eledoisin [K. Kangawa et al., *Regul. Peptides* **1993**, *46*, 81].

Ranatensin family, a subfamily of the → bombesin-like family. To this family belong → ranatensin, → litorin, and → neuromedin B.

Ranatensin, <EVPQWAVGHF^{10}Ma, a 11-peptide amide belonging to the → ranatensin family. Ranatensin was first isolated from the skin of the American leopard frog, *Rana pipiens*. Ranatensin shows vasoactive and myotropic activities [T. Nakajima et al., *Fed. Proc.* **1970**, *29*, 282; T. Nakajima, *Trends Pharmacol. Sci.* **1981**, *2*, 202].

Ras inhibitory peptide, ***RIP***, H-Val-Pro-Pro-Pro-Val-Pro-Pro-Arg-Arg-Arg^{10}-OH, a 10-peptide corresponding to the sequence 1149–1158 of the guanine nucleotide-releasing factor hSos1. The latter is essential for the control of Ras activity. RIP specifically inhibits the binding of hSos1 to the protein human Grb2 which binds to activated receptor tyrosine kinase [N. Li et al., *Nature* **1993**, *363*, 85].

Ras proteins, a superfamily of plasma membrane-bound GTP-binding proteins that are involved in signal transduction pathways controlling cell growth, differentiation, apoptosis, and other events. *H-ras*, *N-ras* and *K-ras* are identified as human ras genes. *K-ras* is spliced to give two different variants, *K-ras4A* and *K-ras4B*. Each of the three functional ras genes encode highly related proteins known as *Ras p21* proteins that contain 188 or 189 residues. Whereas the first 85 residues of mammalian p21 Ras proteins of known sequences are identical, the following 80 residues differ slightly with at least 85% homology between any pair. The remaining part of the protein is highly variable except for the last four amino acids bearing the CAAX motif (C=Cys; A=any aliphatic amino acid; X=variable residue) which is present in all members of the Ras protein family. The synthesis of the Ras protein occurs in the cytosol and it is localized to the inner plasma membrane only after a series of post-translational modifications. The attachment of lipophilic groups to the C-terminus is a prerequisite for the biological function. Especially, farnesylation is necessary for membrane localization and its activity. Farnesylation of the cysteine residue of the CAAX motif occurs after the proteolytic cleavage of AAX followed by methyl esterification at the newly formed C-terminal Cys residue. The farnesylation of the appropriate SH group is catalyzed by the farnesyltransferase which transfers a C-15 farnesyl moiety from farnesyl pyrophosphate. After localization to the cell membrane, the Ras p21 proteins bind GTP and GDP and possess intrinsic GTPase activity. Ras acts as a molecular switch that cycles between the GDP-bound 'off' and the GTP-bound 'on' state. The lifetime of the two states is determined by *guanine nucleotide exchange factors* (*GEFs*) and *GTPase-activating proteins* (*GAP*). Oncogenic versions of Ras contain point mutations which cause blocking of the GTPase activity in the presence and absence of GAP. The resulting accumulation of Ras in the active form leads to uncontrolled cell growth, and this plays an important role in malignant transformations [M. Barbacid, *Annu. Rev. Biochem.* **1987**, *56*, 779; D.R. Lowy, B.M. Willumsen, *Annu. Rev. Biochem.* **1993**, *62*, 851; D.M. Leonard, *J. Med. Chem.* **1997**, *40*, 2971; A. Wittinghofer, H. Waldmann, *Angew. Chem. Int. Ed. Engl.* **2000**, *39*, 4192].

Receptor activity modifying protein, ***RAMP***, an accessory protein for seven-transmembrane domain receptors. *RAMP1* has 148 residues, a signal sequence and a single transmembrane protein close to the C-terminus. Co-expression of the *calcitonin receptor-like receptor* (*CRLR*) with RAMP1 was established to create novel CGRP receptors in all cell lines under examination, but expression of either protein alone was without success. Furthermore, it could be shown that *RAMP2* or *RAMP3* together with CRLR generate the receptor for → adrenomedullin. The RAMPs transport CRLR to the cell surface, define its pharmacology and determine its glycosylation state [L.M. McLatchie et al., *Nature* **1998**, *393*, 333; M. Aiyar et al., *J. Biol. Chem.* **1996**, *271*, 11325; S.M. Food, F.H. Marshall, *Trends Pharmacol. Sci.* **1999**, *20*, 184].

Relaxin, ***RLX***, a proteohormone with structural similarity to → insulin. Relaxin belongs to the → insulin family. Two genes (H1 and H2) have been found in humans, leading to *relaxin-1* and *relaxin-2*. Both human relaxins contain an A-chain with 24 residues bearing an intrachain disulfide bridge, and a B-chain with 32 residues connected by two interchain disulfide bridges. It is synthesized under the influence of progesterone primarily as a single-chain precursor (prorelaxin) and released in high concentrations in the corpus luteum of pregnant mammals. The plasma level increases significantly at the end of pregnancy, but falls rapidly after birth. Relaxin causes relaxation of collagenous tissue in the symphysis and ileosacral joints, resulting in enlargement of the pelvic girdle during pregnancy and, especially, at birth [B.E. Kemp, H.D. Niall, *Vitam. Horm.* **1984**, *41*, 79; P. Hudson et al., *EMBO J.* **1994**, *3*, 2333].

Release inhibiting hormones, *release inhibiting factors*, → statins.

Releasing hormones, *releasing factors*, → liberins.

Renin, *angiotensin-forming enzyme, angiotensinogenase*, an aspartyl protease with highly restricted substrate specificity. The only native substrate is angiotensinogen (→ angiotensins). It cleaves the peptide bond between two Leu residues of the minimally sized angiotensinogen substrate -Pro-Phe-His-*Leu*↓*Leu*-Val-Tyr-Ser-. A highly specific inhibitor for human renin is H-Pro-His-Pro-Phe-His-*Phe-Phe*-Val-Tyr-OH containing two Phe residues instead of the two Leu residues. Renins of all species are inhibited by → pepstatin. Active renin is produced from prorenin by cleavage of the N-terminal propeptide. It consists of two similar domains each containing the catalytical

Asp. The mature native renin ($M_r \sim$ 36–42 kDa) is a two-chain protein containing two disulfide bonds and two potential N-glycosylation sites formed proteolytically from the native single-chain enzyme. Renin is found in mammals, birds, amphibians, and teleosts. It is expressed mainly in the juxtaglomerular cells of the kidney. After secretion into the blood circulation, it reacts selectively with angiotensinogen to form angiotensin I [T. Ingami, *J. Hypertens.* **1989**, *7* (Suppl. 2), S3; A. Fukamizu, K. Murakami, *Trends Endocrinol. Metab.* **1995**, *6*, 279].

Resilin, a fibrillar protein from the exoskeleton of insects and other arthropods. The protein with a high Gly content, but lacking sulfur-containing amino acids and Trp, is located between the chitin lamellae and is responsible for the elasticity of the arthropod exoskeleton. An important component of resilin is trityrosine; this is formed by cross-linking of Tyr residues of one or more peptide chains and results in a three-dimensional lattice forming the rubber-like properties. Resilin shows functional similarities to → elastin of mammals.

Ricin, a phytotoxin from the seeds of *Ricinus communis.* It consists of an A chain ($M_r \sim$ 32 kDa) and a B chain (M_r $\sim$ 34 kDa) linked together by disulfide bonds. The B chain attaches the toxin to the cell surface, whereas the A chain carries the toxic activity. Like → abrin, ricin catalytically inactivates the eukaryotic large ribosome subunit. It has been reported that ricin is a complex consisting of two lectins (RCL I and RCL II) and two toxins (R. D and RCL IV).

Rigin, H-Gly-Gln-Pro-Arg-OH, a 4-peptide originally isolated from human IgG. Analogous to → tuftsin, rigin shows phagocytosis-stimulating activity [N. I. Veretennikova et al., *Int. J. Peptide Protein Res.* **1981**, *17*, 430].

Rubredoxin, a redoxin with functional similarity to → ferredoxin isolated from *Clostridium pasteurianum.* Rubredoxin is an → iron-sulfur protein ($M_r \sim$ 6 kDa; 52–54 aa) and contains, in contrast to ferredoxin, only one iron bound by coordination with four Cys residues.

S

Sanjoinine A, *frangufoline*, a cyclopeptide alkaloid (***Formula***). Sanjoinine A is the major bioactive component responsible for the sedative properties of 'Sanjoin', a plant seed of clinical importance in oriental medicine. Using a novel cycloetherification procedure, the asymmetric total synthesis of Sanjoinine A was described in 1999 [B. H. Han et al., *Pure Appl. Chem.* **1989**, *61*, 443; L. Laib, J. Zhu, *Tetrahedron Lett.* **1999**, *40*, 83].

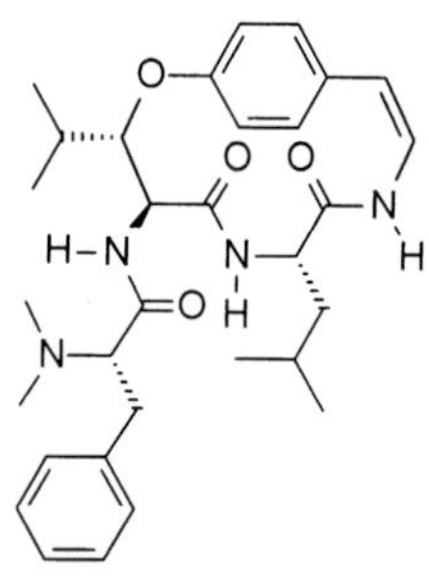

Sanjoinine A

Sarafotoxins, ***SRTX***, CSCKDMTDKE^{10}CLNF CHQDVI^{20}W (SRTX-A) (disulfide bonds: C^1–C^{15}/C^3–C^{11}), a group of four peptides (*SRTX-A, -B, -C, -E*) with a great structural similarity to → endothelin (ET). Each of the members of the ET-SRTX family contains four Cys, and about 60–70% of their 21 amino acid residues are identical. From the structural point of view, these strongly toxic snake venoms (*Atracta engadensis*) are ET analogues with similar biological effects. The SRTX cause strong vasoconstriction of coronary arteries. Death after intoxication with SRTX peptides is the result of cardiac ischemia or infarction [Y. Kloog et al., *Science* **1988**, *242*, 268; C. Takasaki et al., *Nature* **1988**, *335*, 302].

Saralasin, H-Sar-Arg-Val-Tyr-Val-His-Pro-Ala-OH, *[Sar^{1},Val^{5},Ala^{8}]angiotensin II*, a non-competitive antagonist of angiotensin II (→ angiotensins). Saralasin shows blood pressure-lowering activity in humans, and has found diagnostic application in renin-dependent hypertension. It inhibits the pressor effects of angiotensin II in rats [W. Wienen et al., *Mol. Pharmacol.* **1992**, *41*, 1081].

Sarmesin, H-Sar-Arg-Val-Tyr(Me)-Ile-His-Pro-Phe-OH, a synthetic 8-peptide acting as a competitive antagonist of the angiotensin II receptor (→ angiotensins) in rat smooth muscle preparations [M.N. Scanlon et al., *Life Sci.* **1984**, *34*, 317; W. Wienen et al., *Mol. Pharmacol.* **1992**, *41*, 1081].

Sauvagine, *SVG*, <EGPPISIDLS^{10}LELLRK MIEI^{20}EKQEKEKQQA^{30}ANNRLLLDTIa^{40}, a 40-peptide amide originally isolated from the leaf-frog, *Phyllomedusa sauvagei*. Sauvagine belongs to a family of related peptides including → corticoliberin, → urotensin-I, and → urocortin. It possesses only about 40% sequence similarity to the primary structure of rat urocortin, and about 50% with urotensin-I. Sauvagine causes a number of pharmacological activities on diuresis, the endocrine glands, and the cardiovascular system [P.C. Montecucchi et al., *Hoppe-Seyler's Z. Physiol. Chem.* **1979**, *360*, 1178; P.C. Montecucchi et al., *Int. J. Peptide Protein Res.* **1982**, *20*, 139; D.A. Lovejoy, R.J. Balment, *Gen. Comp. Endocrinol.* **1999**, *115*, 1].

SchistoFLRFamide, *myosuppressin*, → FMRFamide-related peptides.

Schistomyotropins, ***Scg-MT***, neuropeptides from the CNS of the desert locust (*Schistocerca gregaria*). The peptides were isolat-ed from the methanolic extract of about 7000 brains of the desert locust. The primary structures were elucidated as follows: *schistomyotropin-1* (*Scg-MT-1*) GAAPAAQFSP^{10}RLa, *schistomyotropin-2* (*Scg-MT-2*) TSSLFPHPRL^{10}a. Scg-MT-1 belongs to the FXPRLa family, whereas Scg-MT-2 is only related to this family and displays 10 times less activity compared to Scg-MT-1 in stimulating cockroach hindgut motility [D. Veelaert et al., *Biochem. Biophys. Res. Commun.* **1997**, *241*, 530].

Scyliorhinin I, H-Ala-Lys-Phe-Asp-Lys-Phe-Tyr-Gly-Leu-Met^{10}-NH_2, a 10-peptide amide isolated from the intestine of the dogfish *Scyliorhinus caniculus* with tachykinin-like activity [J.M. Conlon et al., *FEBS Lett.* **1986**, *200*, 111; R. Patacchini et al., *Eur. J. Pharmacol.* **1993**, *250*, 311].

Secretin family, a member of the gastroenteropancreatic peptide families. This family comprises → secretin, → glucagon and → glucagon-like peptides (both encoded by one gene), → gastric inhibitory polypeptide (GIP), → vasoactive intestinal polypeptide (VIP) and → peptide histidine isoleucine amide (both encoded by one gene), → growth hormone-releasing hormone, → pituitary adenylate cyclase-activating polypeptides (PACAP). Helospectine I and II and helodermin from the toxins of *Heloderma horridum* and *Heloderma suspectum*, respectively, and the → exendins are not present in mammals [J.F. Rehfeld, *Physiol. Rev.* **1998**, *78*, 1087].

Secretin, HSDGTFTSEL^{10}SRLREGARLQ^{20} RLLQGLVa (human secretin), a 27-peptide amide released by gastric acid from S cells in the duodenum. It was originally discovered in the duodenal mucosa as a hormone enhancing the secretion of bicarbonate, enzymes, and K^+ from the exocrine pancreas. Together with other peptides, secretin forms the → secretin family. For a long time, secretin was believed to exist only as the 27-peptide amide. In the mid-1980s the immediate precursor of amidated S-27, the glycine-extended *S-28*, and *S-30*, extended by Lys-Arg, were identified in porcine gut extracts as additional secretins with full biological activity. Although the existence of S-28 and S-30 is not surprising, the discovery of *S-71* was unexpected. The secretin gene encodes *prepro-secretin*, consisting of 132 to 134 residues, depending on the species. The precursor protein is normally processed to the three bioactive secretin peptides of almost similar size (S-27, -28, and -30) by endoproteolytic cleavages and variable C-terminal trimming. In contrast, the bioactive S-71 is produced after RNA splicing, whereas the middle sequence of prepro-secretin is removed on the gene level. S-71 contains the sequence of nonamidated S-27, followed by a Gly-Lys-Arg extension and a further C-terminal extension of 41 residues. Human and rat secretin receptors show 81% identity, and are coupled with G proteins. The activation of the receptor stimulates the adenylate cyclase/protein kinase A cascade and enhances Ca^{2+} concentration in various cellular systems [V. Mutt et al., *Eur. J. Biochem.* **1970**, *15*, 513; G. Gafvelin et al., *Proc. Natl. Acad. Sci. USA* **1990**, *87*, 6781; G.G. Nussdorfer et al., *Peptides* **2000**, *21*, 309].

Secretoneurin, ***SN***, a 33-peptide derived from secretogranin II (chromogranin C, → chromogranins). SN is widely distributed throughout the central and peripheral nervous systems. Known functions of SN include chemotaxis of monocytes and endothelial cells, and inhibition of endothelial cell proliferation. In addition, SN is a potent chemoattractant for human eosinophils. Since it can be co-released with → substance P and → calcitonin gene-related peptide from sensory afferent c-fibers by capsaicin, it might represent another member of the group of inflammatory neuropeptides [R. Kirchmair et al., *Neuroscience* **1993**, *53*, 359; S. Dunzendorfer et al., *Blood* **1998**, *91*, 1527; C.J. Wiedermann, *Peptides* **2000**, *21*, 1289].

Selenocysteine, ***Sec, U***, $HSeCH_2$-$CH(NH_2)$-COOH, the 21st proteinogenic amino acid which has been discovered in the past 15 years. Selenocysteine is an amino acid building block in several dozen proteins. The Sec residues are ribosomally incorporated into the proteins by a unique tRNA, $tRNA^{Sec}$, bearing an UCA anticodon. The latter is specified by a particular (in the mRNA) UGA codon which is normally the *opal* stop codon.

Shaker peptides, → conotoxins.

Sodefrin, H-Ser-Ile-Pro-Ser-Lys-Asp-Ala-Leu-Leu-Lys^{10}-OH, a 10-peptide first isolated from the abdominal gland of the cloaca of the male red-bellied newt *Cynops pyrrhogaster.* Sodefrin is the first discovered amphibian pheromone that exerts a specific female-attracting activity [S. Kikuyama et al., *Science* **1995**, *267*, 1643].

Somatoliberin, *somatotropin-releasing hormone*, ***SRH***, *growth hormone-releasing hormone*, ***GRH***, $YADAIFTNSY^{10}$ RKVLGQL $SAR^{20}KLLQDIMSRQ^{30}GESNQERGA^{40}RAR$ La (hSRH), a 44-peptide amide from the hypothalamus that stimulates the release of somatotropin (growth hormone) in the hypophysis. Human SRH is synthesized as part of the precursor prepro-hSRH. A cryptic 30-peptide, named *anorectin*, prepro-h-SRH-(78–107) <$EVDSMWAEQK^{10}QMELE$ $SILVA^{20}LLQKHSRNSQ^{30}a$, results from post-translational processing of the precursor, in addition to the sequence of SRH. It reduces food intake after injection into the third ventricle. The application of SRH for the treatment of children with somatotropin deficiency is very important. Structure-activity studies led to the conclusion that SRH can be shortened at the C-terminus. In children, hSRH-(1–29)-amide still has 50% of the activity of the native hormone [L.A. Frohamn, J.-O. Jansson, *Endocrinol. Rev.* **1986**, *7*, 223; K. Arase et al., *Endocrinology* **1987**, *121*, 1960; N. Ling et al., *Annu. Rev. Biochem.* **1985**, *54*, 403].

Somatomedins, ***SM***, previous term for → insulin-like growth factors (IGF). *SM C* is identical with IGF-1, *SM A* appears to be the deamidated form of IGF-1, but according to other sources it should be also a mixture consisting largely of ILF-1 and ILF-2. SM A was also named *sulfation factor* since it should cause an increase in sulfur incorporation into cartilage. Further elder terms are *thymidine factor, non-suppressible insulin-like activity in serum (NSILA-S)* or *multiplication stimulating activity (MSA)*. *SM B* is structurally completely different from IGF.

Somatostatin family, a member of the gastroenteropancreatic peptide families. Somatostatin, and → corticostatin belong to this family.

Somatostatin, ***SST***, somatotropin release-inhibiting hormone, ***SIH***, AGCKNFFWKT^{10} FTSC (disulfide bond: C^3–C^{14}), a heterodetic cyclic 14-peptide released by the hypothalamus. SST is a potent inhibitor of → somatotropin (growth hormone) release, but also inhibits the secretion of several other hormones, such as glucagon, thyreotropin, insulin, cholecystokinin, gastrin. In addition, it is a regulator of many other biological processes. SST is synthesized as prepro-SST which is processed at the C-terminus to form SST-14. Besides in the hypothalamus, SST-14 and SST-like peptides including SST-28 have been found in the stomach, pancreas, and also, for instance, in the central and peripheral nervous systems, thymus, gastrointestinal tract, ovaries, and even in plants (tobacco, spinach). SST infusions have been applied to the treatment of bleeding peptic ulcers and gastrointestinal lesions, for preventive treatment of stress ulcers, and in the healing of fistulae of the small intestine and gallbladder. Many analogues have been synthesized. For example, *octreotide*, H-D-Phe-Cys-Phe-D-Trp-Lys-Thr-Cys-Thr-ol (disulfide bond: Cys^2–Cys^7) is 70 times more potent than SST in inhibiting somatotropin release *in vivo* 15 min after administration, and it is characterized by a long duration of action after intramuscular administration. This analogue is used in the treatment of somatotropin- and thyrotropin-secreting pituitary tumors, carcinoid tumors and in further indications. More recently, two additional analogues, namely *lanreotide*, H-D-βNal-Cys-Tyr-D-Trp-Lys-Val-Cys-Thr-ol (disulfide bond: C^2–C^7), and *vapreotide*, H-D-Phe-Cys-Tyr-D-Trp-Lys-Val-Cys-Trp-OH (disulfide bond: C^2–C^7) have become available for clinical use [P. Brazeau et al., *Science* **1973**, *179*, 77; C. Johansson et al., *Digestion* **1981**, *22*, 126; C.H.S. McIntosh, *Life Sci.* **1985**, *17*, 2043; C. Scarpignato, I. Pelosini, *Chemotherapy* **2001**, *47* (Suppl. 2), 1].

Somatotropin release inhibiting peptide, → somatostatin.

Somatotropin, *growth hormone*, ***GH***, *somatotropic hormone*, ***STH***, a single-chain proteohormone ($M_r \sim 21.5$ kDa) formed as pro-STH in the adenohypophysis. Human STH consists mainly of a four-helix bundle (191 aa, two disulfide bridges). The biosynthesis is regulated by → somatoliberin and → somatostatin. Additionally, → ghrelin derived from stomach and hypothalamus might be involved in the release of STH. STH controls mainly, but together with other hormones (→ insulin, thyroxin, etc.), growth, differentiation and permanent renewal of body substances. Human STH causes its receptor to dimerize. This ligand-induced dimerization has important implications for the mechanism of signal transduction. STH is highly species-specific. Therefore, hSTH only and no other mammalian STHs are active in humans. High levels of STH result in excessive growth (*gigantism*), whereas GH deficiency causes insufficient growth (*dwarfism*). Recombinant hSTH is used in the therapy of hypophyseal dwarfism, in the treatment of Turner's syndrome, in the healing of wounds, and for increasing mass and density of bones [C. H. Li, J. D. Yamashiro, *Am. Chem. Soc.* **1970**, *92*, 7608; C. Carter-Su, *Annu. Rev. Physiol.* **1996**, *58*, 187; E. E. Muller et al., *Physiol. Rev.* **1999**, *79*, 511].

Somatotropin-releasing hormone, → somatoliberin.

Somatotropin-releasing peptides, → growth hormone-releasing peptides.

Spantide I, [D-Arg^{1},D-Trp^{7,9},Leu^{11}] substance P, an antagonist of → substance P [K. Folkers et al., *Br. J. Pharmacol.* **1984**, *83*, 449].

Speract, ***SAP-1***, H-Gly-Phe-Asp-Leu-Asn-Gly-Gly-Gly-Val-Gly^{10}-OH, a 10-peptide isolated from sea urchin eggs. Speract stimulates the respiration, motility, and cyclic nucleotide metabolism of sea urchin spermatozoa [D. L. Garbers et al., *J. Biol. Chem.* **1982**, *257*, 2734].

Splenopentin, ***SP-5***, H-Arg-Lys-Glu-Val-Tyr-OH, a 5-peptide with sequence similarity to thymopentin (→ thymopoietin) [T. Audhya et al., *Proc. Natl. Acad. Sci. USA* **1984**, *81*, 2847].

Statherin, *human salivary statherin*, an acidic Tyr-rich phosphoprotein secreted mainly by salivary glands. Statherin is a 43-peptide ($M_r \sim 5.4$ kDa). It shares important function together with other salivary phosphoproteins in providing an environment for recalcification and tooth enamel stabilization. Furthermore, statherin possesses high affinity for calcium phosphate minerals such as hydroxyapatite, and is a potent inhibitor of calcium phosphate precipitation. Circular dichroism studies of both the synthetic and native statherin revealed that it adopts an amphiphilic helical conformation of the N-terminus connected to a long poly-L-proline type II segment, that, in turn, is linked to an extended β-strand [T. L. Gururaja, M. J. Levine, *Peptide Res.* **1996**, *9*, 283].

Statine, ***Sta***, (3*S*,4*S*)-4-amino-3-hydroxy-6-methylheptanoic acid, an unusual amino acid occurring as a building block, e.g., in → pepstatin [D. H. Rich et al., *J. Org. Chem.* **1978**, *43*, 3624].

Statins, *release inhibiting factors, release inhibiting hormones*, neurohormones synthesized in the small-cell region of the hypothalamus and transported via the bloodstream to the anterior pituitary. Members of the statins include → melanostatin, → somatostatin, and → prolactostatin. They inhibit the secretion of melanotropin, somatotropin, and prolactin, resp. Together with the corresponding → liberins, the statins regulate the levels of the three pituitary hormones.

Stichodactyla toxin, ***ShK***, RSCIDTIPKS^{10}RCTAFQCKHS^{20}MKYRLSFCRK^{30}TCGTC (disulfide bonds: C^{3}–C^{35}/C^{12}–C^{28}/C^{17}–C^{32}), a 35-peptide originally isolated from the sea anemone *Stichodactyla helianthus*. The ShK-neurotoxin inhibits the specific binding of dendrotoxin I to rat brain membranes. ShK acts as a voltage-dependent K^+ channel (A channel) blocker, and may become a useful molecular probe for studying potassium channels [J. Pohl et al., *Lett. Peptide Sci.* **1994**, *1*, 29; O. Castaneda et al., *Toxicon* **1995**, *33*, 603].

Streptavidin, a tetrameric protein (M_r ~ 60 kDa) from *Streptomyces avidinii* with a high affinity to biotin ($K \sim 10^{-15}$M). Streptavidin is used in various laboratory techniques for detecting biotinylated molecules, and is preferred to → avidin due to its more favorable isoelectric point and lack of glycosylation.

Streptokinase, a protein (416 aa, M_r ~ 47 kDa) produced by β-hemolytic streptococci. Streptokinase is not generally considered to be a proteolytic enzyme, but is one of the most potent exogenous activators of human → plasminogen. → Plasmin alone is not able to catalyze activation of plasminogen, but the streptokinase-plasmin complex serves as a very potent plasminogen activator [L.A. Schick, F.J. Castellino, *Biochem. Biophys. Res. Commun.* **1974**, *57*, 47].

Substance K, → neurokinins.

Substance P, *SP*, RPKPQQFFGL^{10}Ma, a 11-peptide amide exerting a wide spectrum of biological effects. SP is a mammalian member of the → tachykinin family. It belongs to a large number of peptides occurring both in the gastrointestinal tract and in the brain. SP is released from the precursor protein β-prepro-tachykinin. Stimulation of the smooth muscle, salivary secretion, and lowering of the blood pressure due to vasodilation are important biological actions besides other effects mediated preferentially via the NK1 receptor. SP functions as a neurotransmitter in various brain regions, suppresses the action of morphine and endorphins, and plays a protective role against stress-determined disturbances. From structure-activity relationships it can be concluded that the C-terminal 5-peptide represents the active center. Acylated SP-(7–11) compounds are the shortest-acting SP analogues. <Glu-Phe-Phe-Gly-Leu-Met-NH_2 is one of the most active analogues in the vasodepressor-response assay. SP plays important roles in pathological processes, such as nociception, inflammation, cancer, and psychiatric disturbances. The development of antagonists including nonpeptide antagonists has opened new avenues for pharmaceutical research with high therapeutic importance. *Spantide I*, [D-Arg^1,D-Trp^{7,9}, Leu^{11}]SP, is the standard antagonist that binds preferentially to the NK1 and NK2 receptors. It shows a weak spasmogenic and a poor histamine-releasing effect [U.S. von Euler, B. Pernow (Eds.), *Substance P*, Raven Press, New York, **1977**; B. Pernow, *Pharmacol. Rev.* **1983**, *35*, 85; Z. Gao, N.P. Peet, *Curr. Medicinal Chem.* **1999**, *6*, 374; S.E. Leeman, S.L. Ferguson, *Neuropeptides* **2000**, *34*, 249].

Subtilin, a member of the Type A → lantibiotics with a pentacyclic structure similar to → Nisin. Subtilin is a 32-peptide (M_r 3317 Da) produced from *Bacillus subtilis*, and has antimicrobial activity against both the vegetative cells of a broad range of Gram-positive bacteria as well as against the outgrowth of *Bacillus* spp. and *Clostridium* spp. endospores [R.W. Jack, G. Bierbaum, H.-G. Sahl, *Lantibiotics and Related Peptides*, Springer-Verlag, Berlin, Heidelberg, New York, **1998**].

Surfactin, a branched cyclic lipopeptide from *Bacillus subtilis*. The 3-hydroxy-12-methyltridecanoic acid, $(CH_3)_2CH\text{-}(CH_2)_8CH(OH)$-

CH_2-COOH, is linked via an amide bond to the N-terminal Glu of the peptide part Glu-Leu-D-Leu-Val-Asp-D-Leu-Leu, and the cycle is formed by an ester bridge between the carboxy group of the terminal Leu residue to the β-hydroxy group of the tridecanoic acid. Surfactin is the most powerful bio-detergent so far detected, based on its pronounced amphiphilicity. The biosynthesis is catalyzed by a synthetase complex consisting of two subunits with M_r of 300 and 600 kDa, respectively.

Systemin, AVQSKPPSKR^{10}DPPKMQTD, the first peptide signaling molecule in plants that is involved in the attenuation of defense responses as a consequence of wounding by herbivorous insects. Systemin was first isolated in 1991 from leaves of tomato plants, based on its capability to induce the accumulation of protease inhibitors that interfere with protein digestion of attacking pests. Interestingly, the inducing activity of systemin is not only restricted to protease inhibitors, but also extends to the expression of about 20 proteins involved in plant defense. In the meantime, similar peptides have been found in other species of the *Solanaceae* that differ only in some positions of the sequence from the peptide of *Lycopersicon esculentum* shown above. Systemin is first formed as a larger precursor and, after release from prosystemin, is secreted into the apoplast in response to the wound stimulus [C.A. Ryan, G. Pearce, *Annu. Rev. Cell Dev. Biol.* **1998**, *14*, 1; A. Schaller, *Plant Mol. Biol.* **1999**, *40*, 763].

T

Tachykinin family, a family of peptides sharing a common C-terminal sequence (-Phe-Xaa-Gly-Leu-Met-NH_2) that comprises the mammalian tachykinins → substance P, and the → neurokinins A and B as well as nonmammalian tachykinins, such as → bufokinin, → eledoisin, → physalaemin, → kassinin, → uperolein. Multiple receptors, termed NK_1, NK_2 and NK_3, of the mammalian peptides belong to the superfamily of G-protein-coupled receptors that have distinct pharmacological features [C. Ohnmacht, Jr., et al., *Annu. Rep. Med. Chem.* **1997**, *32*, 71; C.J. Swain in: *Progress in Medicinal Chemistry*, G.P. Ellis, D.K. Luscomber, A.W. Oxford (Eds.), p. 57, Elsevier, Amsterdam, **1998**; Z. Gao, N.P. Peet, *Curr. Medicinal Chem.* **1999**, *6*, 374].

Tachyplesins, bicyclic peptides containing 16 to 18 residues from the horseshoe crab. They are processed from a 77-residue precursor protein. *Tachyplesin I*, KWCFRVCYR G^{10}ICYRRCRa (disulfide bonds: C^3–C^{16}/ C^7–C^{12}), and the other tachyplesins contain two disulfide bridges which stabilize a rigid structure of two antiparallel β-sheets and a β-turn segment. Tachyplesins are stored in high concentrations in the crab hemocytes. They are active against Gram-negative and Gram-positive bacteria. Tachyplesins show similarity in size, charge and overall structure to the → protegrins [S. Iwanaga et al. in: *Antimicrobial Peptides*, H.G. Boman, J. Marsh, J.A. Goode (Eds.), p. 160, Ciba Foundation Symposium 186, Chichester, John Wiley & Sons, **1994**].

Tentoxin, cyclo-(-MeAla-Leu-MeΔ^{Z}Phe-Gly-), a phytotoxic cyclic peptide isolated from the phytopathogenic fungus *Alternaria alternata*. Tentoxin is a potential natural herbicide that induces chlorosis in many plants, and binds specifically to the soluble part CF_1 of the chloroplastic proton ATPase. At low concentration, tentoxin is a powerful inhibitor, whereas above 10 μM it stimulates the enzyme both *in vitro* and *in vivo* [M.G. Klotz, *Physiol. Plant.* **1988**, *74*, 575; C. Sigalat et al., *FEBS Lett.* **1995**, *368*, 253].

Tertiapin, ALCNCNRIII^{10}PHMCWKKCGK^{20} Ka (disulfide bonds: C^3–C^{14}/C^5–C^{18}), a 21-peptide amide from the honey bee (*Apis mellifera*). It acts as a blocker of special K^+ channels. Tertiapin blocks the G-protein-gated channel (GIRK1/4) and the ROMK1 channel with nanomolar affinity [W. Jin, Z. Lu, *Biochemistry* **1998**, *37*, 13291; X. Xu, J.W. Nelson, *Proteins Struct. Funct. Genet.* **1993**, *17*, 124].

Thrombin receptor activator peptide 6, ***TRAP-6***, H-Ser-Phe-Leu-Leu-Arg-Asn-OH, a 6-peptide corresponding to the partial sequence 42–47 of the → thrombin receptor with receptor activating potential. TRAP-6 causes platelet aggregation, mediates tyrosine phosphorylation, inhibits cAMP formation, and increases cytosolic Ca^{2+} concentration [R.R. Vasallo et al., *J. Biol. Chem.* **1992**, *267*, 6081; L.F. Brass et al., *J. Biol. Chem.* **1994**, *269*, 2943].

Thrombin, *factor IIa, fibrinogenase*, a serine protease capable of producing a thrombus or blood clot. It consists of two chains (M_r ~ 36.5 kDa). The A chain (human: 36 aa) of α-thrombin, without known function, is linked by a single disulfide bond to the catalytic B chain (259 aa) which is homologous to the catalytic domains of trypsin and chymotrypsin. Six additional Cys residues within the B chain form three intrachain disulfide bridges. Furthermore, the B chain is glycosylated on Asn^{60}. Thrombin is derived from its zymogen prothrombin through cleavage by factor Xa in the presence of cofactor Va, Ca^{2+} and a phospholipid surface. Thrombin is the last protease in the coagulation cascade, and cleaves → fibrinogen to yield fibrin monomers that polymerize to form the basis of the blood clot. The specificity of thrombin is trypsin-like, but much more restricted with respect to the P_1 residue Arg in substrates and inhibitors. Thrombin also causes platelets to aggregate, modulates neurite growth, and is mitogenic for various cell types. Many of these activities are mediated via a G-protein-coupled receptor [M.T. Stubbs, W. Bode, *Thromb. Res.* **1993**, *69*, 1; R.J. Grand et al., *Biochem. J.* **1996**, *313*, 353].

Thrombomodulin, ***TM***, a glycoprotein (M_r ~ 74 kDa) acting *in vivo* as a clot-limiting factor. TM specifically binds → thrombin, leading to a conformation with decreased capability to catalyze clot formation, but with a more than 1000-fold increased capacity to activate → protein C [B.F. Le Bonniec, C.T. Esmon, *Proc. Natl. Acad. Sci. USA* **1991**, *88*, 7371].

Thrombospondin, ***TS***, a glycoprotein consisting of three identical, disulfide-linked chains (M_r ~ 450 kDa) released from thrombocytes by treatment with → thrombin. TS is the major constituent of platelet α-granules. It is synthesized and secreted by a variety of cells, and functions in the regulation of cell growth and migration connected with processes such as wound healing, hemostasis, angiogenesis, and morphogenesis. Besides promoting attachment and spreading in the sense of an adhesive protein, TS often also displays anti-adhesive properties. The 6-peptide H-Gly-Gly-Trp-Ser-His-Trp-OH acts as a TS analogue and binds specifically to → fibronectin. It inhibits fibronectin-promoted adhesion of breast carcinoma and melanoma cells to gelatin or type I collagen substrates [M.D. Kosfeld et al., *J. Biol. Chem.* **1991**, *266*, 24257; J.M. Sipes et al., *J. Cell. Biol.* **1993**, *121*, 469].

Thymic hormones, ***TH***, *thymus factors*, in all likelihood in the thymus produced hormones that have been shown to induce specific T-cell-marker precursor cells and to promote T-lymphocyte functions. Members of the TH are → thymopoietin, → thymosins, → thymulin, → TP-5 and the thymic humoral factor, a polypeptide with M_r ~ 3.2 kDa [J.-F. Bach in: *Thymic Factor Therapy*, N. A. Byrom, J. R. Hobbes (Eds.), Volume 16, p. 21; Raven Press, New York, **1984**].

Thymopentin, ***TP-5***, H-Arg-Lys-Asp-Val-Tyr-OH, a synthetic 5-peptide corresponding to the sequence 32–36 of → thymopoietin II. TP-5 exerts the same biological activities as thymopoietin. It is administered to strengthen the unspecific immune system.

Thymopoietin, ***TP***, PEFLEDPSVL^{10}TKEKLK SELV^{20}ANNVTLPAGE^{30}QRKDVYVELY^{40}LQ SLTALKR (TP II), a group of peptide hormones from the thymus. All members consist of 49 residues with varying residues in position 1, 2, 34, and 43. For example, *TP I* is characterized as [Gly^1,Gln^2,Ser^{43}]TP II. *TP III* (spleen) corresponds to [His^{43}]TP II, whereas *splenin* (spleen) corresponds to [Glu^{34},His^{43}]TP II. The TP induce the differentiation of T lymphocytes. Interestingly, TP II 29–41 shows the same activity as TP II. The active sequence of TP represents → thymopentin (TP-5) corresponding to TP-II-(32–36) [D. H. Schlesinger et al., *Cell* **1975**, *5*, 367; G. Goldstein et al., *Science* **1979**, *204*, 1309; T. Abiko, H. Sekino, *Chem. Pharm. Bull.* **1987**, *35*, 2016].

Thymosin α_1, → thymosins.

Thymosins, polypeptides originally isolated from tissues of vertebrates and postulated to possess hormonal or immunomodulating functions. The designation 'thymosins' is related to the first isolation from thymus tissue preparations. α- and β-thymosins are the main members of this peptide family. *α-Thymosin* includes *prothymosin α* (109 aa), the structurally related *parathymosin α* (101 aa), and the 28-peptide *thymosin α_1*, Ac-SDAAVDTSSE^{10}ITTKDLKEKK^{20}EVVEE AEQ. The term *β-thymosin* is restricted to a group of homologous peptides with an average M_r ~ 5 kDa. In mammals, the most commonly occurring member of this group is *thymosin β_4* (*Tβ_4*) containing 43 residues, which is accompanied by *Tβ_9* (calf), *Tβ_9^{Met}* (pig), *Tβ_{10}* (human, horse, rat, cat and mouse) or *Tβ_{13}* (whale). Tβ_4 has been shown to be active in various biological assays, e.g., the terminal deoxynucleotidyl transferase activity *in vivo* and *in vitro* is induced or the hypothalamic secretion of → luliberin is stimulated. Tβ_4 forms a 1 : 1 complex with G-actin, thereby sequestering its polymerization and playing an important role in the regulation of the actin polymerization in many cells. Recently, thymosin α_1, in combination with low doses of → interferon or → interleukin (IL-2), has found application in the treatment of cancer [T. K. L. Low, A. L. Goldstein, *Thymus* **1984**, *6*, 27; W. Voelter in: *Peptides 1992*, C. H. Schneider, A. N. Eberle (Eds.), p. 103, ESCOM, Leiden, **1993**; E. Garaci et al., *Int. J. Immunopharmacol.* **2000**, *22*, 1067; T. Huff et al., *Int. J. Biochem. Cell Biol.* **2001**, *33*, 205].

Thymulin, *thymic factor, serum thymic factor,* ***STF***, pGlu-Ala-Lys-Ser-Gln-Gly-Gly-Ser-Asn-OH, a Zn^{2+}-binding 9-peptide from the epithelium of the thymus. It plays a role in the differentiation of thymocytes. The synthesis of thymulin is stimulated by → somatotropin [J. F. Bach et al., *Bull. Inst. Pasteur* **1978**, 325; T. L. K. Low, A. L. Goldstein, *Springer Semin. Immunpathol.* **1979**, *2*, 169].

Thyroglobulin, a globular glycoprotein (M_r ~ 660 kDa) synthesized in the thyroid gland. Thyroglobulin is the precursor protein of the thyroid hormones triiodothyronine (T_3) and thyroxine (T_4). During post-translational modification, approximately 20% of the 140 Tyr residues are iodinated using I^- and H_2O_2 catalyzed by iodoperoxidase yielding 2,5-diiodo-Tyr residues. Coupling of two modified residues, rearrangement, followed by lysosomal proteolysis, results in the formation of about five or six molecules of T_3 and T_4. Synthesis of thyroglobulin and release of the thyroid hormone is stimulated by → thyrotropin.

Thyroliberin, *thyrotropin-releasing hormone*, ***TRH***, pGlu-His-Pro-NH_2, produced in the paraventricular nucleus of the hypothalamus, stimulates biosynthesis and secretion of → thyrotropin (TSH) from the anterior pituitary. It is central in regulating the hypothalamic-pituitary-thyroid (HPT) axis. Furthermore, TRH influences the release of other hormones, e.g., → prolactin, → growth hormone, → vasopressin, → insulin and also the classic neurotransmitters norepinephrine (noradrenaline) and epinephrine (adrenaline). In addition, it is found in many brain loci outside the hypothalamus, and also outside the CNS in the gastrointestinal tract, pancreas, ovary, placenta, testis, seminal vesicles, retina, and prostate. This widespread distribution of TRH underlines a diverse range of functions also outside the HPT axis. The discovery and characterization of TRH as the first identified hypothalamic-releasing hormone by Guillemin and Schally in 1969 provided ultimate confirmation of the basic principles of neuroendocrinology, and was followed by the discovery of other releasing hormones. In 1986, the rat prepro-TRH sequence (255 aa; $M_r \sim 29$ kDa) was identified. The precursor contains an N-terminal 25-residue leader sequence, five copies of the TRH progenitor -Gln-His-Pro-Gly- flanked by paired basic amino acids (Lys-Arg or Arg-Arg), four non-TRH peptides between the TRH progenitors, an N-terminal flanking peptide, and a C-terminal flanking peptide. In the course of the processing of pro-TRH, Gln-His-Pro-Gly is then amidated, catalyzed by peptidylglycine α-amidating monooxygenase (PAM) using the C-terminal Gly as the amide donor, and the Gln residue undergoes cyclization to pyroglutamate to yield TRH. Besides TRH, other pro-TRH-derived peptides are formed with potential biological functions. In the cryptic region of the prepro-TRH the sequence of the 22-peptide → corticotropin release-inhibiting hormone has been elucidated [R. Guillemin, *Science* **1978**, *202*, 390; A.V. Schally, *Science* **1978**, *202*, 18; E.A. Nillini, K.A. Sevarino, *Endocr. Rev.* **1999**, *20*, 599].

Thyrotropin, *thyroid-stimulating hormone*, ***TSH***, a α/β glycoprotein (human TSH: α chain: 96 aa; β chain: 112 aa) formed in the adenohypophysis with structural similarities to the → gonadotropins. The α chain is identical with the appropriate α chains of both → follitropin and → lutropin. TSH stimulates synthesis and secretion of the thyroid hormones T_3 and T_4 in the thyroid gland and the formation of → thyroglobulin. The human TSH-receptor consists of 744 residues. The release of TSH is regulated by → thyroliberin and inhibited via a negative feedback mechanism by high doses of T_3 and T_4 [J.A. Magner, *Endocr. Rev.* **1990**, *11*, 354].

Thyrotropin-releasing hormone, → thyroliberin.

Transducin, G_T, a trimeric (α,β,γ) → G protein ($M_r \sim 82$ kDa) in the vertebrate eye transducing visual stimuli by coupling the light-induced conformational change of rhodopsin to the activation of specific phosphodiesterase. $G_{T\alpha}$ activates the cGMP phosphodiesterase that hydrolyzes cGMP to GMP. The resulting drop of the cGMP level causes an electric change in the photoreceptor cell [L. Stryer, *Sci. Am.* **1987**, *257*, 42].

Transferrin, a non-heme protein carrying iron in the blood. Human transferrin (M_r ~ 80 kDa) comprises about 5% of the plasma proteins. Cell-surface transferrin receptors deliver transferrin with bound iron to peripheral endosomes by receptor-promoted endocytosis. The release of bound iron is induced by the low pH in the endosomes. The iron-free transferrin (*apoferritin*) remains linked to the receptor and is recycled back to the plasma membrane as receptor–apoferritin complex. The neutral pH induces the dissociation of apoferritin from the receptor, thus picking up more iron and starting the cycle again [A. Dautry-Varsat et al., *Proc. Natl. Acad. Sci. USA* **1983**, *80*, 2258].

Transforming growth factor-β, *TGF-β*, a protein (M_r ~ 25 kDa) consisting of two identical peptide chains (each 112 aa) linked together by disulfide bridges. TGF-β is regarded as the prototype of a group of related protein factors that regulate cell growth and differentiation via the interaction with different membrane receptors. It potentiates or inhibits the response of most cells to other growth factors, depending on the cell type. Furthermore, TGF-β regulates differentiation of some cell types, e.g., cartilage differentiation *in vitro*. It was originally identified as a → cytokine in transformed fibroblasts [J. Massague, *Cell* **1987**, *49*, 437; T.F. Deuel, *Annu. Rev. Cell Biol.* **1987**, *3*, 443].

Transforming growth factor-α, *TGF-α*, a polypeptide controlling, together with → transforming growth factor-β (TGF-β), the transformation of various cell lines. Rat TGF-α-(1–50), VVSHFNKCPD^{10}SHTQYCFHGT^{20}CRFLVQEEKP^{30}ACVCHSGYVG^{40}VRCEHADLLA^{50} (disulfide bonds: $C^8–C^{21}/C^{16}–C^{32}/C^{34}–C^{43}$) and human TGF-α are single polypeptides with homology to → epidermal growth factor. TGF-α is formed in excessive amounts in the skin of individuals with psoriasis, a skin disease characterized by epidermal hyperproliferation [J.A. Feild et al., *Biochem. J.* **1992**, *283*, 91; J.R. Goldenring et al., *Regul. Peptides* **1993**, *43*, 37].

Tremerogens, peptide hormones from *Tremella mesenterica* containing a sulfur-linked farnesyl moiety involved in the zytotene stages. For example, *Tremerogen A-13*, EGGGNRGDPS^{10}GVC^{Farn}, is produced by the a-type cell, whereas *Tremerogen A-10* is formed by the A-type cell [Y. Sakagami et al., *Science* **1981**, *212*, 1525].

Triflavin, a cysteine-rich polypeptide (M_r ~ 7.5 kDa) isolated from *Trimeresurus flavoviridis* snake venom that belongs to the family of RGD-containing peptides, termed *disintegrins*. Triflavin inhibits the adhesion of tumor cells to matrix proteins through binding to multiple → integrin receptors expressed on human hepatoma cells.

Trigger factor, → peptidyl prolyl *cis/trans* isomerases.

Tritrpticin, VRRFPWWWPF^{10}LRR, a 13-peptide, designated to its three consecutive Trp residues, with a broad antimicrobial activity spectrum. It has potential importance as a therapeutic agent for a variety of infectious diseases [C. Lawyer et al., *FEBS Lett.* **1996**, *390*, 95].

Tropomyosin, a protein of the striated muscle thin filament. Tropomyosin is associated with → actin both in muscle and the cytoskeleton of other cell types. Tropomyosin is a homodimer consisting of two 284-residue α helical subunits that wrap around each other forming a parallel coiled coil. Tropomyosin wraps in the groove of the F-actin helix in a head-to-tail manner so that one tropomyosin molecule contacts seven consecutive actin monomer units. Furthermore, each tropomyosin molecule binds to a single → troponin TnT molecule at its head-to-tail joint. The tropomyosin-troponin complex is involved in the regulation of muscle contraction [F. G. Whitby et al., *J. Biol. Chem.* **1992**, *227*, 441; A. S. Zot, J. D. Potter, *Annu. Rev. Biophys. Chem.* **1987**, *16*, 535].

Troponin, ***Tn***, the central regulatory protein of skeletal muscle contraction in higher vertebrates. Troponin consists of three subunits: *TnC* ($M_r \sim 18$ kDa), a Ca^{2+}-binding protein with about 70% homology to → calmodulin, *TnI* ($M_r \sim 24$ kDa), with binding capacity to → actin, and *TnT* ($M_r \sim 37$ kDa) containing the binding site for → tropomyosin. Muscle contraction is initiated by an increase in Ca^{2+} concentration. Ca^{2+} binds to TnC, resulting in its conformational change that causes tropomyosin to move deeper into the thin filament groove exposing actin's myosin head-binding sites leading to switch on muscle contraction. TnT is important for early diagnosis of acute cardiac infarction [K. A. Satyshur et al., *Acta Crystallogr.* **1994**, *D50*, 40].

Tuftsin, H-Thr-Lys-Pro-Arg-OH, a natural phagocytosis stimulating 4-peptide which was discovered by Victor Najjar's group at Tufts University in 1970. Tuftsin corresponds to the 289–292 sequence in the CH2 domain of the Fc fragment of leukokinin, a leukophilic fraction of a α-globulin. It is released under the catalysis of a specific enzyme (leukokininase) located in the outer membrane of the neutrophils. It is assumed by circumstantial evidence only that the processing should be performed by the successive action of two enzymes: the splenic tuftsin endocarboxypeptidase cleaves the Arg^{292}-Glu^{293} bond, whereas the leucokininase cleaves the Lys^{288}-Thr^{289} bond. The primary biological effect of tuftsin, after binding to specific cell-surface receptors, is the stimulation of the functions of macrophages and polymorphonuclear (PMN) cells. Interestingly, the tetrapeptide H-Thr-Glu-Pro-Arg-OH was isolated from leukokinin of a patient with tuftsin deficiency. Much attention was paid to the synthesis of various analogues of this short peptide since antimicrobial, antiviral, and antitumor activities promise useful therapeutic effects [V. A. Najjar, K. Nishioka, *Nature* **1970**, *228*, 672; I. Z. Siemion, A. Kluczyk, *Peptides* **1999**, *20*, 645].

Tumor necrosis factor-α, ***TNF-α***, *cachectin*, a single chain protein (hTNF-: 157 aa; $M_r \sim 17$ kDa) containing an intrachain disulfide bridge. TNF-α is a monocyte-macrophage-derived → cytokine first detected by its in-vivo antitumor activities and in-vitro cytotoxicity to certain transformed cell lines. Furthermore, it causes a growth-enhancing effect on various cells and thus functions as a bifunctional growth regulator like, for example, IFN-α (→ interferons). TNF-α shares some biological activities with IL-1 (→ interleukins) like a mediator of inflammatory processes [B. Bharat et al., *Nature* **1985**, *318*, 665; K. J. Tracy, A. Cerami, *Annu. Rev. Cell Biol.* **1993**, *9*, 317].

Tumor necrosis factor-β, ***TNF-β***, ***lymphotoxin***, a glycoprotein (171 aa; unglycosylated: $M_r \sim 19$ kDa) produced by mitogen-stimulated lymphocytes. TNF-β causes cytostasis of some tumor cell lines and cytolysis of other transformed cells. TNF-β and → tumor necrosis factor-α (TNF-α) show 36% sequence homology. Both TNF bind to the same receptors, TNF receptor 1 ($M_r \sim 55$ kDa) and TNF receptor 2 ($M_r \sim 75$ kDa), respectively [K. J. Tracey, A. Cerami, *Annu. Rev. Cell Biol.* **1993**, *9*, 317].

Tyrocidins, a group of cyclic peptide antibiotics produced from *Bacillus brevis* strains. Derived from *tyrocidin A*, cyclo-(-Val^1-Orn-Leu-D-Phe-Pro^5-Phe-D-Phe-Asn-

Gln-Tyr^{10}-), *tyrocidin B* corresponds to [Trp^6]tyrocidin A, *tyrocidin C* to [Trp^{6}D-Trp^7]tyrodin A, and *tyrocidin D* to [Trp^6,D-Trp^7,Trp^{10}]tyrocidin A. The biosynthetic thiotemplate mechanism is similar to that of → gramicidin S, except that three synthesizing enzyme components TycABC are involved. Tyrocidin and gramicidin S primarily act on Gram-positive bacteria, and are used as antibacterial drugs in different formulations (*tyrothricin*) for the topical treatment of infections, e.g., sore throats. Open-chain analogues of the tyrocidins are without any antibacterial activity.

U

Uperolein, <EPDPNAFYGL^{10}Ma (*U.I*), an 11-peptide amide from the skin of the Australian frog *Uperoleia rugusa* belonging to the → tachykinin family. Uperolein shows activities similar to those of → phyllomedusin. *Uperolein II*, [Ala^2,Lys^5,Thr^6]uperolein, could be isolated from the same source [A. Anastasi et al., *Experientia* **1975**, *31*, 394].

Urocortin, *UCN*, DNPSLSIDLT^{10}FHLLRTL LEL^{20}ARTQSQRERA^{30}EQNRIIFDSV^{40}a (hUCN), a mammalian 40-polypeptide amide representing a tetrapodal form of → urotensin-I. UCN is the second endogenous mammalian ligand of the → corticoliberin receptors. It produces behavioral and physiological effects that are quantitatively similar to corticoliberin. Human UCN possesses only about 44–59% sequence identity with urotensins-I found in various fishes. Furthermore, it has only 33% identity with the presumptive closest known nonmammalian homologue → sauvagine. Recently, a third mammalian corticoliberin-related 38-peptide, named UCN II, has been identified [J. Vaughan et al., *Nature* **1995**, *378*, 287; Y. Takahashi et al., *Peptides* **1998**, *19*, 643; T.M. Reyes et al., *Proc. Natl. Acad. Sci. USA* **2001**, *98*, 2843; K.H. Skelton et al., *Regul. Peptides* **2000**, *93*, 85].

Urokinase-type plasminogen activator, ***uPA***, *u-plasminogen activator, urokinase*, a serine protease with extremely limited substrate specificity. It cleaves the Arg^{560}-Val^{561} bond within the partial sequence -Cys-Pro-Gly-Arg-Val-Val-Gly-Gly-Cys- comprising the small disulfide-bridged loop in → plasminogen, and synthetic substrates based on this sequence. The active enzyme consists of an epidermal growth factor (EGF)-like module and a serine protease domain separated by an unusually long (16 aa) linker region. It is released from a single-chain proenzyme (411 aa) after removal of a 20-residue signal peptide, followed by → plasmin or → kallikrein cleavage the Lys^{158}-Ile^{159} bond giving rise to two polypeptides linked together by Cys^{148}–Cys^{279}. Post-translational modification is performed by glycosylation at Asn^{302}, and fucosylation at Thr^{18}. The EGF-like module of uPA is the binding determinant to its specific glycolipid-anchored cell surface uPA-receptor [K. Dano et al., *Adv. Cancer Res.* **1985**, *44*, 139].

Urotensin I, *UT I*, NDDPPISIDL^{10} TFHLL RNMIE^{20}MARIENEREQ^{30}AGLNRKYLDE^{40} Va, a 41-polypeptide from the urophysis of the white sucker, *Catostomus commersoni*. Similar peptides have been found, for example, in shark, carp, trout, trout sole, maggy sole and European flounder. Urotensin I is part of a family of related peptides including → corticoliberin, → sauvagine and → urocortin in vertebrates, and the diuretic peptides in insects (→ insect diuretic peptides) [D.A. Lovejoy, R.J. Balment, *Gen. Comp. Endocrinol.* **1999**, *115*, 1].

Urotensin II, *UT II*, ETPDCFWKYC^{10}V (disulfide bond: C^5–C^{10}), a human 11-peptide acting as a potent vasoconstrictor and agonist for the orphan receptor GRP14. The mRNA of UT II was found in the spinal cord. In-situ hybridization studies led to the conclusion that the urotensin II precursor gene is actively expressed in motor neurons. There is evidence that UT II may exert important physiological functions in humans. Before the discovery of the human peptide, UT II was reported as a hederodetic cyclic 12-peptide in fish (AGTADCFWKY^{10}CV, disulfide bond: C^6–C^{11}) which contracts smooth muscles and exerts hypertensive and osmoregulatory effects in fish and birds [Y. Coulouarn et al., *Proc. Natl. Acad. Sci. USA* **1998**, *95*, 15803; R. S. Ames et al., *Nature* **1999**, *401*, 282].

V

Valinomycin, cyclo-(-D-Val-Lac-Val-D-Hyv-)$_3$, a symmetric cyclodepsipeptide isolated from *Streptomyces fulvissimus.* This fungal antibiotic is formed by three repeating units of a sequence of D-Val, L-lactic acid (Lac), Val and D-hydroxyisovaleric acid (D-Hyv) (cf. Fig. 3.23). Valinomycin acts as an ion carrier which selectively transports potassium ions across membranes. K^+ is coordinated by six Val carbonyl oxygen atoms. The amino acid side chains provide the overall hydrophobic exterior, enabling valinomycin to cross the hydrophobic core of the membrane. Within the membrane, valinomycin alternates between loaded and unloaded states through which ions are transported across the membrane. Valinomycin shows the best K^+/Na^+ selectivity of all K^+-ionophores known till date. It is especially active against *Mycobacterium tuberculosis* [R. M. Izatt et al., *Chem. Rev.* **1985**, *85*, 271; B. Dietrich et al., *Macrocyclic Chemistry*, VCH, Weinheim, **1993**].

Valorphin, H-Val-Val-Tyr-Pro-Trp-Thr-Gln-OH, a 6-peptide isolated from bovine hypothalamus tissue possessing opiate activity. Valorphin corresponds to the partial sequence 32–38 of the bovine hemoglobin β-chain, and may belong to the hemorphins (→ exorphins) [V. Brantl et al., *Eur. J. Pharmacol.* **1986**, *125*, 309; J. Erchegyi et al., *J. Peptide Protein Res.* **1992**, *39*, 477].

Vancomycin, a glycopeptide antibiotic from a fermentation broth of the actinomycete *Streptomyces orientalis,* later renamed *Nocardia orientalis* and finally reclassified as *Amycolatopsis orientalis.* It exhibits lethal properties against all tested strains of *Staphylococcus* and other Gram-positive bacteria. Despite recent incidences of bacterial

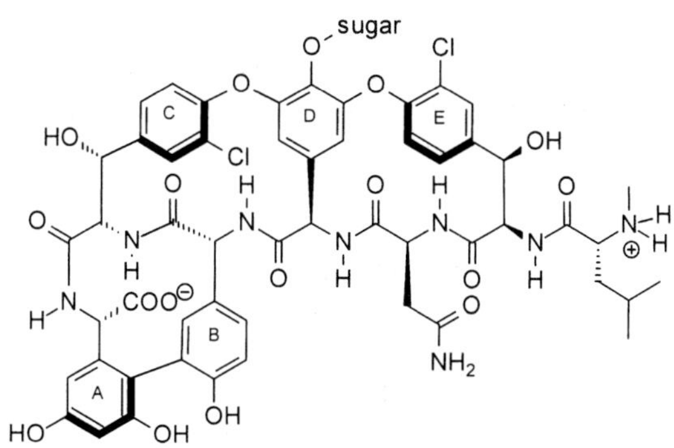

Vancomycin

resistance to vancomycin, it became almost legendary because of its performance against methicillin-resistant *S. aureus* (MRSA). Vancomycin and the closely related *teicoplanin* are typically administered parenterally to treat severe staphylococcal infections, including MRSA. Worldwide sales of vancomycin in 1997 were $ 417 million. Vancomycin was discovered by Eli Lilly in 1956. The complete structure (***Formula***) was established only in 1982, followed by the elucidation of the crystal structure in 1996, and the total synthesis of vancomycin aglycone and vancomycin itself in 1999. Vancomycin consists of seven amino acids containing in total five aromatic rings. The sugar components are D-glucose and L-vancosamine. Vancomycin binds reversibly to the -Lys-D-Ala-D-Ala- sequence of the peptidoglycan monomer and prevents the transglycosidase from polymerizing the peptidoglycan monomers; this results in cell death. The biosynthesis proceeds through construction of the amino acid building blocks, formation of the linear heptapeptide according to multienzyme thiotemplate mechanism, oxidative coupling processes, and finally glycosidations [C.M. Harris, T.M. Harris, *J. Am. Chem. Soc.* **1982**, *104*, 4293; M. Schäfer et al., *Structure* **1996**, *4*, 1509; K.C. Nicolaou et al., *Angew. Chem. Int. Ed.* **1999**, *38*, 2096].

Vasoactive intestinal contractor, ***VIC***, known as → endothelin-β, a member of the ET-SRTX group (→ endothelin, → sarafotoxins) [K. Saido et al., *J. Biol. Chem.* **1989**, *264*, 14613].

Vasoactive intestinal peptide, ***VIP***, HSDAVF TDNY^{10}TRLRKQMAVK^{20}KYLNSILNa, an 28-peptide amide belonging to the → glucagon-secretin-VIP family. VIP was first isolated from porcine intestine. It is a potent vasodilator, a major growth stimulator, a bronchodilator and a neuronal survival-promoting agent. VIP is widely distributed in the peripheral and central nervous systems, and acts both as neurotransmitter and as a hormone. Thus, it is not only an intestinal peptide but, in accordance with its early given abbreviation, it is a **v**ery **i**mportant **p**eptide. VIP is co-synthesized with → peptide histidine isoleucine amide (PHI) on the same peptide precursor. Multiple receptors for VIP have been molecularly cloned, and this has broadened the understanding of their mechanisms of action. VIP-based drug design and noninvasive innovative delivery modes are interesting, with the emphasis on tumor diagnosis and treatment, neuroprotection and impotence treatment [S.I. Said, V. Mutt, *Eur. J. Biochem.* **1972**, *28*, 199; I. Gozes et al., *Curr. Med. Chem.* **1999**, *6*, 1019].

Vasopressin, ***VP***, *antidiuretic hormone*, ***ADH***, H-Cys-Tyr-Phe-Gln-Asn-Cys-Xaa-Gly-NH_2 (disulfide bond: Cys^1–Cys^6), a member of the neurohypophyseal peptide hormones. VP has antidiuretic activity, and in higher doses it increases blood pressure. In bovine and other mammalian hormones Xaa=Arg (AVP), but in the porcine peptide Xaa=Lys (LVP). VP shows strong structural similarities to → oxytocin, and has a slight oxytocin effect. VP deficiency causes *diabetes insipidus*. Because of inadequate resorption by the kidneys, in this disorder up to 20 L of urine is passed per day. VP is synthesized by the magnocellular cells, mainly in the nucleus supraopticus, and also in the paraventricular nuclei of the hypothalamus as neurophysin II precursor protein bearing the VP sequence following the N-terminal signal sequence. In this form it is transported through the supraoptico-hypophyseal tract to the posterior pituitary, where it is stored and then released proteolytically into the blood after an adequate stimulus. The biological actions of AVP are mediated by three receptor subtypes: the V_{1a} and V_{1b} receptors that activate phospholipases, and the V_2 receptor that activates adenylate cyclase. The three receptors are heptahelical membrane proteins activating intracellular effectors via G proteins. Very active analogues with agonistic, dissociated, and antagonistic properties have been synthesized which also allow for nasal application. The commercial drug *[1-deamino-D-Arg^8]VP* (*DDAVP*) has 400 times the effect of native AVP on the kidneys with practically negligible effect on blood pressure. *[1-deamino-penicillinamine, 2-O-methyltyrosine]AVP* belongs to the widely used antagonists. Many useful analogues have been synthesized, e.g., those with increased V_1-antagonistic potency and reduced V_2 agonism (antidiuretic effects), but also analogues with CNS activity, such as *[de-Gly^9-NH_2]VP*. Sequence determination and the first chemical synthesis were performed by du Vigneaud in 1953/54, together with those of oxytocin [V. J. Hruby, D. Patel in: *Peptides: Synthesis, Structures, and Applications*, E. Gutte (Ed.), p. 261, Academic Press, San Diego, **1995**; M. Birnbaumer, *Trends Endocrinol. Metab.* **2000**, *10*, 406].

Viomycin, *viocin, vinactin A, florimycin, tuberactinomycin B*, a cyclic 5-peptide (***Formula***, see p. 542) antibiotic from the culture broth of various *Streptomyces* sp. such as *S. puniceus*, *S. floridae* and *S. vinaceus*. Viomycin is active against Gram-negative bacteria and has found therapeutic application against *Mycobacterium tuberculosis* [S. Noda et al., *J. Antibiot.* **1972**, *25*, 427].

Viridogrisein, → etamycin.

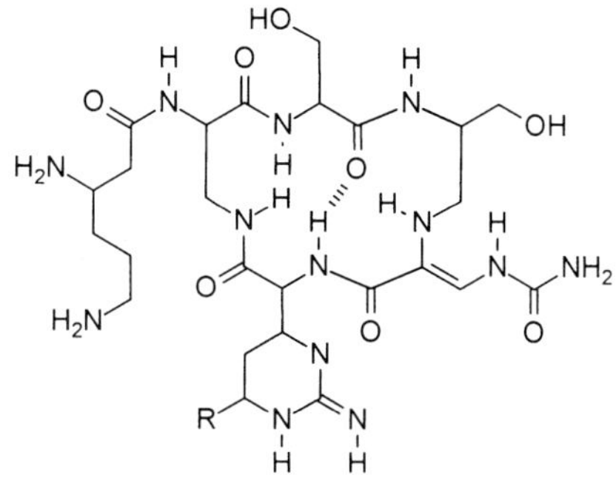

R = OH Viomycin
R = H Tuberactinomycin B

Viomycins

W

WWamides, biologically active 7-peptide amides originally isolated from ganglia of the African snail *Achatina fulica*. The name is related to the both terminal Trp(W) residues. *WWamide-1*, H-Trp-Lys-Glu-Met-Ser-Val-Trp-NH_2, shows inhibitory activity on a central neuron of the snail. Furthermore, it exhibits peripheral modulatory effects on muscular contractions in various tissues of the snail and other molluscs. *WWamide-2* corresponds to [Arg^2]WWamide-1 and *WWamide-3* to [Gln^3]WWamide-1, respectively [H. Minakata et al., *FEBS Lett.* **1993**, *323*, 104].

X

Xenin 25, MLTKFQTKSA^{10}RVKGLSFHPK^{20}RPWIL, a 25-peptide from the human gastric mucosa. Xenin 25 stimulates the exocrine pancreatic secretion. The C-terminal sequence 19–25 is in common with amphibian → xenopsin [G. F. Feurle et al., *J. Biol. Chem.* **1992**, *267*, 22305; G. F. Feurle et al., *J. Peptide Res.* **1997**, *49*, 324].

Xenopsin, pGlu-Gly-Lys-Arg-Pro-Trp-Ile-Leu-OH, an 8-peptide from the skin extract of the South African frog *Xenopus laevis*. It shows sequence homology with both the C-terminal parts of → xenin 25 and → neurotensin, respectively. Xenopsin shares a number of its biological activities with neurotensin [I. Sures, M. Crippa, *Proc. Natl. Acad. Sci. USA* **1984**, *81*, 380; C. L. Bevins, M. Zasloff, *Annu. Rev. Biochem.* **1990**, *59*, 395].

Xenopsin-related peptides, ***XRP***, two peptides from the avian gastric extracts with a C-terminal hexapeptide sequence in common with → xenopsin. *XRP 1*, [His^1,Pro^2] xenopsin, and *XRP 2*, H-Phe-[His^1,Pro^2] xenopsin show xenopsin-like activities on gastric regulation [R. E. Carraway et al., *Regul. Peptides* **1988**, *22*, 303].

Xenopus Xfin protein, a Cys_2-His_2 zinc finger protein in which the Zn^{2+} ions are tetrahedrally coordinated by the His and Cys residues. According to the first NMR structure of α zinc finger by P. E. Wright, it forms a compact globule containing a two-stranded antiparallel β sheet and one α helix that are stabilized by the Zn^{2+}-ion [A. Klug, D. Rhodes, *Trends Biochem. Sci.* **1987**, *12*, 464].

Y

Yeast mating factor α, *a-factor, pheromone a-factor*, WHWLQLKPGQ[10]PMY, a 13-peptide acting as a pheromone of mating type α cells of *S. cerevisiae.* It inhibits DNA synthesis, thereby preventing cell division of mating type a cells [D. Stoetzler et al., *Eur. J. Biochem.* **1976**, *69*, 397; F. Naider, J. M. Becker, *CRC Crit. Rev. Biochem.* **1986**, *21*, 225].

Z

Zeamatin, a 27-peptide produced by *Zea mays* seeds with antifungal activity. Zeamatin causes the release of cytoplasmic material from *C. albicans* and *Neurospora crassa*, which results in hyphal rupture [W. K. Roberts, C. P. Selitrennikoff, *J. Gen. Microbiol.* **1990**, *136*, 1771].

Zein, a protein mixture from the glue of maize (M_r ~ 10–22 kDa). It has a high content of Glu (23%) and Leu (19%), but contains no Trp and Lys.

Zinc finger, a DNA binding motif first discovered in *Xenopus* transcription factor IIIA (TFIIIA) by A. Klug. This 344-residue protein contains nine repeating domains, each containing 30 residues folded into a single structural unit around a Zn^{2+} ion that links two invariant Cys and His residues (***Formula***). The first elucidated NMR structure of a single zinc finger was the → *Xenopus* Xfin protein. In some zinc fingers, the two Zn^{2+}-coordinating His residues are replaced by two additional Cys residues. Despite the structural diversity in zinc fingers, in all cases Zn^{2+} appears to stabilize relatively small globular domains without the need for much larger hydrophobic cores [A. Klug, D. Rhodes, *Trends Biochem. Sci.* **1987**, *12*, 464; J. M. Berg, *J. Biol. Chem.* **1990**, *265*, 6513].

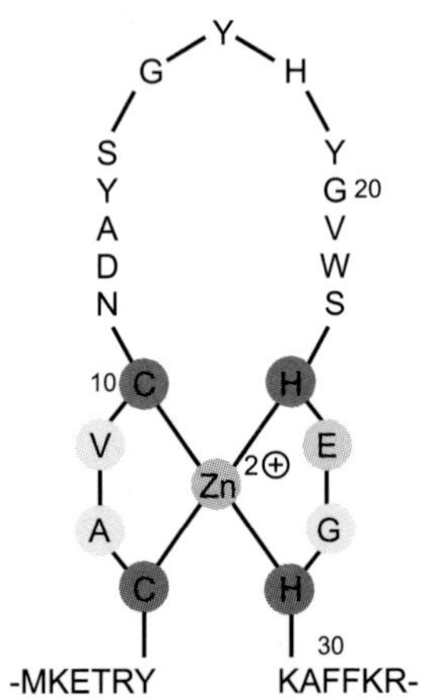

Zinc-finger motif

Zymogens, inactive precursors of enzymes, usually proteases. Zymogens are converted in active proteases by limited proteolysis. Examples are the zymogens of the blood coagulation enzymes → plasminogen and → prothrombin.

Index

b

c